Large-Scale Molecular Systems

Quantum and Stochastic Aspects—
Beyond the Simple Molecular Picture

NATO ASI Series

Advanced Science Institutes Series

A series presenting the results of activities sponsored by the NATO Science Committee, which aims at the dissemination of advanced scientific and technological knowledge, with a view to strengthening links between scientific communities.

The series is published by an international board of publishers in conjunction with the NATO Scientific Affairs Division

A	**Life Sciences**	Plenum Publishing Corporation
B	**Physics**	New York and London
C	**Mathematical and Physical Sciences**	Kluwer Academic Publishers
D	**Behavioral and Social Sciences**	Dordrecht, Boston, and London
E	**Applied Sciences**	
F	**Computer and Systems Sciences**	Springer-Verlag
G	**Ecological Sciences**	Berlin, Heidelberg, New York, London,
H	**Cell Biology**	Paris, Tokyo, Hong Kong, and Barcelona
I	**Global Environmental Change**	

Recent Volumes in this Series

Volume 252—Laser Systems for Photobiology and Photomedicine
edited by A. N. Chester, S. Martellucci, and A. M. Scheggi

Volume 253—Condensed Systems of Low Dimensionality
edited by J. L. Beeby, P. K. Bhattacharya,
P. Ch. Gravelle, F. Koch, and D. J. Lockwood

Volume 254—Quantum Coherence in Mesoscopic Systems
edited by B. Kramer

Volume 255—Vacuum Structure in Intense Fields
edited by H. M. Fried and Berndt Müller

Volume 256—Information Dynamics
edited by Harald Atmanspacher and Herbert Scheingraber

Volume 257—Excitations in Two-Dimensional and Three-Dimensional Quantum Fluids
edited by A. F. G. Wyatt and H. J. Lauter

Volume 258—Large-Scale Molecular Systems: Quantum and Stochastic Aspects—
Beyond the Simple Molecular Picture
edited by Werner Gans, Alexander Blumen, and Anton Amann

Series B: Physics

Large-Scale Molecular Systems

Quantum and Stochastic Aspects—Beyond the Simple Molecular Picture

Edited by

Werner Gans

Free University of Berlin
Berlin, Germany

Alexander Blumen

University of Bayreuth
Bayreuth, Germany

and

Anton Amann

Laboratory of Physical Chemistry
ETH–Zentrum
Zürich, Switzerland

Plenum Press
New York and London
Published in cooperation with NATO Scientific Affairs Division

Proceedings of a NATO Advanced Study Institute on
Large-Scale Molecular Systems: Quantum and Stochastic Aspects—
Beyond the Simple Molecular Picture,
held March 25–April 7, 1990,
in Acquafredda di Maratea, Italy

Library of Congress Cataloging in Publication Data

NATO Advanced Study Institute on Large-Scale Molecular Systems: Quantum
and Stochastic Aspects—Beyond the Simple Molecular Picture (1990: Acquafredda di Maratea, Italy)
    Large-scale molecular systems: quantum and stochastic aspects—beyond the
simple molecular picture / edited by Werner Gans, Alexander Blumen, and Anton Amann.
        p.   cm.—(NATO ASI series. Series B, Physics; vol. 258)
    "Proceedings of a NATO Advanced Study Institute on Large-Scale Molecular
Systems: Quantum and Stochastic Aspects—Beyond the Simple Molecular Picture, held March 25–April 7, 1990, in Acquafredda di Maratea, Italy"—T.p. verso.
    "Published in cooperation with NATO Scientific Affairs Division."
    Includes bibliographical references and index.
    ISBN-13: 978-1-4684-5942-5   e-ISBN-13: 978-1-4684-5940-1
    DOI: 10.1007/978-1-4684-5940-1
    1. Quantum theory—Congresses. 2. Statistical physics—Congresses. 3.
Molecular theory—Congresses. 4. Stochastic processes—Congresses. 5.
Chemistry, Physical and theoretical—Congresses. I. Gans, W., date. II. Blumen,
Alexander. III. Amann, A., 1956–   . IV. North Atlantic Treaty Organization.
Scientific Affairs Division. V. Series: NATO ASI series. Series B, Physics; v. 258.
QC173.96.N384   1990   91-15447
530.1′2—dc20   CIP

© 1991 Plenum Press, New York
Softcover reprint of the hardcover 1st edition 1991
A Division of Plenum Publishing Corporation
233 Spring Street, New York, N.Y. 10013

Preface

This NATO Advanced Study Institute centered on large–scale molecular systems: Quantum mechanics, although providing a general framework for the description of matter, is not easily applicable to many concrete systems of interest; classical statistical methods, on the other hand, allow only a partial picture of the behaviour of large systems. The aim of the ASI was to present both aspects of the subject matter and to foster interaction between the scientists working in these important areas of theoretical physics and theoretical chemistry.

The *quantum–mechanical* part was mostly based on the operator–algebraic formulation of quantum mechanics and comprised quantum statistics of infinite systems with special emphasis on macroscopic observables, equilibrium conditions, irreversibility on the one hand, symmetry breaking for molecules in the radiation field and macroscopic quantum phenomena in the theory of superconductivity (BCS–theory) on the other hand. In addition, phase–space methods for many–body systems were also presented. *Statistical physics* was the main topic in the other lectures of the School; much emphasis was put on the statistical features of macroscopic ("large") systems, the lectures dealt with mass and energy transport im polymers, in gels and in microemulsions, with aggregation and growth phenomena, with relaxation in complex, correlated systems, with conduction and optical properties of polymers, and with the means of describing disordered systems, above all fractals and related hierarchical models. These two main areas of the Institute were bridged by a series of lectures concerned with quantum Brownian motion and with quantum mechanical models of population and phase relaxation.

In all domains, the advances attained in the research of the last years are remarkable, and the Institute helped to disseminate this knowledge. Algebraic quantum mechanics, when applied to systems with infinitely many degrees of freedom, gives rise to *classical* properties or observables in a natural way; this was put in evidence in such diverse fields as the theory of superconductivity, of irreversible dynamics and of molecules coupled to their environment. These classical observables also establish the link with well–known phenomena in classical physics. Furthermore, the appearance of scaling laws and hierarchical models such as fractals provides a unified framework for the study of complex, irregular systems. For these, applications such as porous media, micellar structures, Langmuir–Blodgett films and gels abound. The strengthening of the already existing connections between the quantum–mechanical and the statistical approaches will be one of the important tasks of the future.

The generous financial support provided by the NATO Scientific Affairs Division has made our NATO ASI possible, for which we are very grateful. We also thank the Deutsche Bunsengesellschaft für Physikalische Chemie for their financial support.

The local organization benefitted very much from the organisational skills of Mr. Guzzardi and his staff of the Hotel Villa del Mare in Acquafredda di Maratea. Mr. Guzzardi also took the "official photograph" of the meeting, and it is a pleasure to thank him for allowing us to reproduce it here.

Berlin/Bayreuth/Zürich   W. Gans, A. Blumen, A. Amann

November 1990

Contents

A. The Course

B. The Seminars

PART A

The Course

MOLECULES COUPLED TO THEIR ENVIRONMENT

Anton Amann

Laboratory of Physical Chemistry
ETH-Zentrum
CH-8092 Zürich

1. INTRODUCTION

Physical systems are never closed in a strict sense: Even a small coupling to another system (its "environment") may change its behavior. An instructive example has been proposed in ([1]: p. 98). There the influence of a Sirian beetle (8.3 10^{16} m away) on a gas at normal conditions in a cube of 10 cm length is estimated. The beetle's walk of just 1 cm changes the (classical mechanical) computation such that the position of an individual particle of the gas is changed by approx. 10 cm after 10^{-6} seconds. The cause seems almost negligible and nevertheless the effect is enormous. For quantum systems the situation is particularly intricate. Any two quantum systems - even entirely separated ones without any interacting force - are Einstein-Podolsky-Rosen correlated and pure states of the joint system do not necessarily restrict to pure states of the constituents (cf. [2-6]).

Any "small" system (e.g., a molecule) has an environment which has to be described in quantum theory by infinitely many degrees of freedom. The environment encorporates the radiation and gravitation field. The radiation field in a cavity of finite extent consists already of infinitely many harmonic oscillators. Hence the coupling to the environment embeds the small system into an infinite one which may *qualitatively* change its outward shape. It is therefore inadmissible to think of *isolated* small systems without checking the environmental influence.

The traditional Hilbert space formalism as codified in [7] refers to *strictly closed systems* with only finitely many degrees of freedom. It cannot deal with infinite systems in a rational manner. Nevertheless only modest formal modifications (with dramatic physical consequences) are necessary to make it applicable to open systems and systems with infinitely many degrees of freedom. The irreducibility postulate must be omitted and the Fock space description must be considered as one description among others. The quantum mechanical structure is not changed but one gets rid of superfluous restrictions.

Boson systems (e.g., a system consisting of harmonic oscillators) are described by the **Canonical Commutation Relations (CCR)**. Any degree of freedom of the CCR corresponds to a (Heisenberg) pair of a position and a momentum operator. For finitely many degrees of freedom only physically equivalent representations arise. It is therefore legitimate to consider a particular irreducible or reducible representation. The situation changes *drastically* for infinitely many degrees of freedom. There it is always possible to construct physically *inequivalent* (= disjoint) representations and representations which admit classical observables or superselection rules. A classical observable commutes with all "observables" of the system in question (e.g., all position and momentum operators) but is itself an appropriate limit of such operators. Classical observables can be constructed by pasting together disjoint representations.

The *model Hamiltonian* for the joint system {molecule and environment} used in the present paper is a *spin-boson Hamiltonian*. The "molecule" is restricted to two important states (i.e., may be viewed as a spin) whereas the environment is described by an infinity of harmonic oscillators (bosons) with a kind of dipole-coupling between the spin and the bosons. The parameters entering into this Hamiltonian are the level splitting of the isolated molecule and the coupling constants between the molecule and the respective modes.

This spin-boson Hamiltonian is a prototype for the coupling of a small system to a boson field. It can be derived in various situations and is, for example, the relevant Hamiltonian for the system {molecule and radiation field}, where the molecule has been cut down to a 2-level system. An important characterizing constant of a particular set of parameters is the strength of the so-called *infrared-singularity* (in short: IR-singularity). In the terminology of [8], three situations are distinguished: "Subohmic", "ohmic" and "superohmic". Superohmic systems show no IR-singularity at all. A naive estimate for the coupling of a molecule to the radiation field leads to ohmic parameters (cf. the discussion in section 3). The problem studied here can be summarized as follows: *"Do physically inequivalent representations (and hence classical observables) of the spin-boson system exist and how does inequivalence depend on the molecular and environmental parameters entering in the spin-boson Hamiltonian?"*

2. SUPERSELECTION RULES AND CLASSICAL OBSERVABLES

The "superposition principle" of quantum mechanics says that the superposition

$$\psi = c_1 \psi_1 + c_2 \psi_2 \tag{1}$$

of two state vectors ψ_1 and ψ_2 with coefficients c_1, $c_2 \in \mathbb{C}$ (properly normalized) is again a legitimate state vector of the system in question. The superposition principle leads immediately to strange situations ("paradoxes") if it is applied to macroscopic systems. Consider, for example, a billiard ball in a box [9], [10]. Such a ball can perfectly be described in quantum mechanics by coherent states ("wave packets"). The open question there is: *"Why do superpositions of such coherent states (e.g., a standing wave) not exist?"* or *"Are superpositions of coherent states particularly unstable?"*.

Common sense excludes such superpositions. In the formalism of quantum mechanics they may be excluded in an ad hoc manner by classifying state vectors into sectors. If $\mathscr{H}$ is the basic Hilbert space of the system in question, sectors are (orthogonal) subspaces of $\mathscr{H}$ and superpositions of state vectors in different sectors are "forbidden".

Interestingly enough, also microscopic systems seem to possess superselection rules. This is the more interesting, since one avoids there the difficult problem of describing a macroscopic system quantum mechanically. Typical microscopic superselection rules are the charge of an elementary particle or the handedness and the knot type of a molecule: Two macromolecules with different knot type (cf., e.g., the respective articles in [11]) seem to be separated by a superselection rule. Similarly, one has the impression that molecules with different chirality cannot be superposed. If this impression is "true" experimentally or not remains undecided (see [12-14]).

Superselection rules are intimately connected with *classical observables*, i.e., observables which in a pure state have a dispersion-free value. Classical observables are described in the formalism of (algebraic) quantum mechanics by a self-adjoint operator, say Ξ, which commutes with all other operators of the system in question. The eigenspaces of Ξ (the set of eigenvectors with respect to a given eigenvalue) are the sectors of the corresponding superselection rule: Vectors in different eigenspaces cannot be superposed.

Chirality is a particularly interesting example for a classical observable (cf. [15-17]), with two possible expectation values only, namely ± 1, called D/L, R/S depending on nomenclature. It may be used as a *heuristic* example here and is critically discussed in the next article [14] of this same proceedings volume.

If a molecule is chiral or not seems to depend on the energy difference (level splitting) between the ground state and the first excited state, respectively[1]. Small level splitting (such as around 10^{-70} atomic units for alanine [15]) seems to induce chirality whereas with a large level splitting (such as around 10^{-7} atomic units in NHDT, a derivate of ammonia) chiral states do not arise. One may of course prepare states of NHDT (cf. [18], where ammonia is used which is better accessible experimentally) which are not invariant under space-inversion symmetry, but those "pseudochiral" states may be arbitrarily superposed (no superselection rule is present). Certain pseudochiral states correspond to the chemist's pyramid. These states are *non*-stationary. The proper eigenstates (ground state and first excited state) of ammonia or NHDT are stationary and *do not possess a nuclear frame*. Thus it seems as if a kind of phase-transition occurred when the level splitting is changed (by considering different molecules, if necessary).

The existence or non-existence of superpositions of chiral states has nothing to do with the correct argument of [19] which says that chiral molecules - once prepared - are stable for a very long time. The question of chirality is seen here from the other end: *"Why have superpositions of chiral states not (yet) been prepared ?"* and *"Are superpositions of chiral states particularly unstable ?"*. The situation is completely analogous to that of billiard balls discussed above.

Consider now a classical observable, say Ξ, with the eigenvalues $\lambda_\pm = \pm 1$ and the spectral decomposition

$$\Xi = P_+ - P_- \tag{2}$$

Assume ξ_+ and ξ_- to be eigenstates of Ξ

$$\Xi \xi_+ = +1 \, \xi_+, \tag{3a}$$

$$\Xi \xi_- = -1 \, \xi_-. \tag{3b}$$

Then, using selfadointness of Ξ, a simple calculation shows that for every other observable T,

$$\langle \xi_+ | T \xi_- \rangle = \langle \Xi \xi_+ | T \xi_- \rangle = \langle \xi_+ | (\Xi T) \xi_- \rangle = \tag{4a}$$

$$= \langle \xi_+ | (T \Xi) \xi_- \rangle = \langle \xi_+ | T (\Xi \xi_-) \rangle = - \langle \xi_+ | T \xi_- \rangle \tag{4b}$$

holds. That is, $\langle \xi_+ | T \xi_- \rangle$ vanishes, where $\langle \cdot | \cdot \rangle$ denotes the scalar product of the underlying Hilbert space. Considering now a (formal) superposition

$$\xi = c_+ \xi_+ + c_- \xi_- \tag{5a}$$

---

1 In the Born-Oppenheimer "approximation" this is related to the height of the barrier in an associated double-minimum potential.

$$|c_+|^2 + |c_-|^2 = 1, \tag{5b}$$

it is a consequence of the above that for an arbitrary operator T one has

$$\langle \xi | T \xi \rangle = \tag{6a}$$

$$= |c_+|^2 \langle \xi_+ | T \xi_+ \rangle + |c_-|^2 \langle \xi_- | T \xi_- \rangle + (c_+)^* c_- \langle \xi_+ | T \xi_- \rangle + c_+ (c_-)^* \langle \xi_- | T \xi_+ \rangle = \tag{6b}$$

$$= |c_+|^2 \langle \xi_+ | T \xi_+ \rangle + |c_-|^2 \langle \xi_- | T \xi_- \rangle. \tag{6c}$$

Hence, the expectation value of the state $\langle \xi | \cdot \xi \rangle$ is a (convex) sum of the expectation values $\langle \xi_+ | T \xi_+ \rangle$ and $\langle \xi_- | T \xi_- \rangle$. Thus the state (= expectation value functional) corresponding to the state vector ξ is *mixed*, and not pure. The formal superposition of ξ_+ and ξ_- is incoherent, i.e., it gives a mixture which in case of chirality does not represent the superposition of left and right handed states but a racemate. Therefore the eigenspaces of Ξ are sectors. Superpositions between state vectors in different sectors are not *coherent* superpositions. This fact is usually expressed by saying that *"superpositions are forbidden"*.

Here the notions "superselection rule" and (the slightly stronger and more precise) "classical observable" will be used interchangeably, i.e., a superselection rule without a corresponding classical observable will not be considered as superselection rule.

3. THE SPIN-BOSON MODEL

The spin-boson model consists of a two-level system and (infinitely) many harmonic oscillators. The two-level system is described by Pauli matrices

$$\sigma_1 = \begin{pmatrix} 0 & 1 \\ 1 & 0 \end{pmatrix} , \quad \sigma_2 = \begin{pmatrix} 0 & -i \\ i & 0 \end{pmatrix} , \quad \sigma_3 = \begin{pmatrix} 1 & 0 \\ 0 & -1 \end{pmatrix} \tag{7}$$

whereas the N harmonic oscillators are described by boson operators a_n and a_n^* (its adjoint), indexed by $n = 1, 2, ..., N$ and fulfilling the commutation relations

$$[a_n, a_{n'}] = 0 \quad , \quad [a_n, a_{n'}^*] = \delta_{nn'}. \tag{8}$$

They may be replaced by position and momentum operators Q_n and P_n, $n = 1, 2, ..., N$, using the defining relations

$$a_n =: (2\hbar)^{-\frac{1}{2}} \{ (m(n)\,\omega(n))^{\frac{1}{2}} \, Q_n + i \, (m(n)\,\omega(n))^{-\frac{1}{2}} \, P_n \} , \quad n = 1, 2, ..., N. \tag{9}$$

Here $m(n)$ and $\omega(n)$ are the masses and frequencies of the respective harmonic oscillators. The number of modes N may be finite or (countably) infinite.

The *discrete spin-boson model* is characterized by the Hamiltonian

$$H = \hbar \varepsilon \sigma_1 \otimes \mathbf{1} \tag{10a}$$

$$+ \, \mathbf{1} \otimes \sum_{n=1}^{N} \hbar \, \omega(n) \, a_n^* a_n \tag{10b}$$

$$+ \hbar \, \sigma_3 \otimes \sum_{n=1}^{N} \lambda(n) \, \{ a_n + a_n^* \} \tag{10c}$$

$$+ \mathbb{1} \otimes \sum_{n=1}^{N} \sum_{n'=1}^{N} \eta_{nn'} \{a_n + a_n^*\} \{a_{n'} + a_{n'}^*\}. \tag{10d}$$

This Hamiltonian consists of a part (10a) for the isolated molecule with level splitting $2\hbar\varepsilon$, a part (10b) for the free field, linear coupling terms (10c) and quadratic coupling terms (10d). The corresponding *continuous spin-boson model* has continuously many modes (the sum in (10) is replaced by an integral). Physically the discrete and continuous versions should be equivalent. The mode frequencies $\omega(n)$ are assumed to be strictly positive.

The quadratic terms (10d) with scalar constants $\eta_{nn'}$ are extremely important. They can be incorporated into linear ones by a Bogoliubov transformation (introducing new boson operators and new coupling constants $\lambda(n)$ [20]). In realistic situations, there arise quadratic terms with nontrivial ("outerdiagonal") spin part. Such terms are difficult to discuss, since they cannot be incorporated into the linear term (10c) by a Bogoliubov transformation as before. In case of the coupling {molecule ↔ radiation field}, the simpler form (10a-c) arises if the dipole approximation ("long wavelength approximation" [21]) is made or if the A^2-term is omitted altogether (where A is the transverse vector-potential of the field, cf. [15]).

Here "the spin-boson Hamiltonian" is defined to be of the form (10a-c) without any quadratic terms, which are thought to be incorporated as far as this is possible. *It is one of the most important **model** Hamiltonians for the interaction of a small system (a "molecule") with an environment.* It is a semirealistic **model** Hamiltonian because the assumptions (two-level system and simple quadratic terms (10d)) cannot easily be estimated.

The coupling constants $\lambda(n)$ are assumed to fulfill

$$\sum_{n=1}^{N} |\lambda(n)|^2 < \infty, \tag{11a}$$

$$\sum_{n=1}^{N} \frac{|\lambda(n)|^2}{|\omega(n)|} < \infty. \tag{11b}$$

The latter condition guarantees that the spin-boson Hamiltonian is bounded below.

The invariance of the spin-boson Hamiltonian (10) under the "space-inversion" symmetry ι

$$\iota(\sigma_1) := \sigma_1 \tag{12a}$$

$$\iota(\sigma_2) := -\sigma_2 \tag{12b}$$

$$\iota(\sigma_3) := -\sigma_3 \tag{12c}$$

$$\iota(a_n) := -a_n, \quad n \in \mathbb{N}. \tag{12d}$$

is readily verified.

Note that the Hamiltonian (10) is *formal* insofar, as there exist different realizations of the boson commutation relations (8) on "different" Hilbert spaces. In a particular realization it has to be checked if the spin-boson Hamiltonian can be defined. For finitely many degrees of freedom N there exist only physically equivalent realizations of the boson commutation relations (see the discussion below) and the spin-boson Hamiltonian is defined. For infinitely many degrees of freedom there exist myriads of *physically inequivalent* realizations, some of them admitting a spin-boson Hamiltonian and some of them not. The problem then is how to choose the "physically relevant" realizations. *For the realizations studied here it is **assumed** that the spin-boson Hamiltonian exists and that it has a ground state vector.* The latter assumption, in particular, is too restrictive. Nevertheless, it allows to get a first insight into the problem of understanding the partially classical behavior of molecules.

Historically, the *Fock* representation of the boson commutation relations (8) played an important role. Tensoring the Fock Hilbert space $\mathscr{H}_F$, which refers to the boson field alone, with a two-dimensional Hilbert space gives the Hilbert space

$$\mathscr{H} := \mathbb{C}^2 \otimes \mathscr{H}_F \tag{13}$$

for the system consisting of the two-level system *and* the boson field. The Hamiltonian (10b) for the harmonic oscillators of the field alone as well as the spin-boson Hamiltonian can be properly defined on this Hilbert space. It is characteristic for this Fock realization, that the field Hamiltonian (10b) has a ground state, the vacuum. The spin-boson Hamiltonian, on the other hand, may or may not have a ground state in the Fock realization, depending on the level splitting ε, the frequencies $\omega(n)$ and the coupling constants $\lambda(n)$. This ground state, if it exists, is *unique* and *invariant under space-inversion*.

Consider again the *heuristic* discussion on chirality above: The Fock realization corresponds to the situation of ammonia (a unique ground state invariant under space-inversion). For chiral molecules one expects two ground states, which are transformed one into another by space-inversion, and hence are *not* invariant under space inversion.

In order that different ground states arise, which are separated by a superselection rule, it is necessary that the *infrared singularity*

$$\sum_{n=1}^{N} \frac{|\lambda(n)|^2}{\omega(n)^2} = \infty \tag{14}$$

holds. Its "strength" is measured by the coefficient ζ in the relation

$$\sum_{n=1}^{N} |\lambda(n)|^2 \, \exp\{-\omega(n)|t|\} \ \propto \ |t|^{-\zeta} \qquad \text{(for large } |t|\text{).} \tag{15}$$

Here "$\propto$" means "proportional to". For ζ strictly larger than 2, no IR-divergence arises. In the terminology of Leggett [8] this situation is called "superohmic". The case $\zeta = 2$ characterizes an ohmic IR-singularity. For $1 < \zeta < 2$ the IR-singularity gets even stronger ("subohmic").

Incorporating the quadratic terms (10d) into linear ones by a Bogoliubov transformation may have an influence on the strength of the IR-singularity and possibly make it disappear altogether. For the joint system {molecule and radiation field} one arrives at an ohmic IR-singularity if the A^2-term is omitted (see [15] and [22]). The IR-singularity disappears if the dipole approximation is imposed. The dipole approximation eliminates all higher multipoles of the respective multipolar Hamiltonian [23], [21]. This whole matter is very delicate and not yet fully discussed.

4. A PRECISE FORMULATION OF THE SPIN-BOSON MODEL

WEYL OPERATORS

The boson operators a_n and a_n^* are unbounded. Therefore they are not precisely characterized by the commutation relations (8) (see [24]: Section VIII.5). It is useful to replace them by unitary "Weyl operators" $W(f), f \in \mathbb{C}^N$, defined by

$$W(f) := \exp\left\{ \sum_{n=1}^{N} a_n^* f(n) - \sum_{n=1}^{N} f(n) a_n \right\} \tag{16}$$

with the Canonical Commutation Relations (CCR)

$$W(f) W(f') = W(f+f') \exp\left\{ - i \operatorname{Im}\langle f|f'\rangle \right\} \tag{17}$$

Here $\operatorname{Im}\langle f|f'\rangle$ denotes the imaginary part of the complex scalar product

$$\langle f|f'\rangle := \sum_{n=1}^{N} f(n)^* f'(n). \tag{18}$$

A representation of the canonical commutation relations (17) on a Hilbert space $\mathcal{H}$ is called *regular* if the mappings

$$\mathbb{R} \ni \kappa \mapsto W(\kappa f), \quad f \in \mathbb{C}^N, \tag{19}$$

are strongly continuous. By Stone's theorem ([24]: Theorem VIII.8) this is equivalent to saying that there exist generators for the unitary groups (19). Regular representations of the CCR can therefore be written as in (16) which then defines the boson operators a_n. Regular representations of the CCR (17) are nothing but a precise formulation of the boson commutation relations (8).

For infinitely many degrees of freedom, $N = \infty$, "$\mathbb{C}^\infty$" should at least contain the vectors f of the form

$$f = (f(1), f(2), f(3), \dots , f(M), 0, 0, 0, \dots) \in \text{“}\mathbb{C}^\infty\text{”}, \tag{20}$$

where M is arbitrary. For the formulation of the spin-boson (Heisenberg) dynamics (see below) it is useful to consider a slightly larger class $\mathcal{H}$ of vectors ("test functions"), namely consisting of complex-valued functions on the natural numbers

$$f: \mathbb{N} \to \mathbb{C} \tag{21}$$

fulfilling

$$\left(\sum_{n=1}^{\infty} |f(n)|^2 \, (1 + \omega(n)^{-1}) \right) < \infty . \tag{22}$$

DISJOINT REPRESENTATIONS AND THE GENERATION OF CLASSICAL OBSERVABLES

The *-algebraic view* invites to consider as observables not just spin operators and Weyl operators, but also *polynomials and adjoints* of them. This gives a certain "reservoir" of observables, a *-algebra*. The "*" refers to the adjoint and algebra means that multiplication and addition of operators as well as multiplication of operators with complex scalars is possible. Note that multiplication of Weyl operators gives a Weyl operator again (apart from a scalar factor, see (17)), and note that every

2×2-matrix is a linear combination of the Pauli matrices and the unit matrix σ_0. Therefore the respective *-algebra of the spin-boson model consists of linear combinations of operators

$$\sigma_j \otimes W(f), \quad j = 0,1,2,3, \tag{23}$$

where f is an arbitrary element of $\mathscr{K}$ (or $\mathbb{C}^N$ for finitely many degrees of freedom). This *-algebra can be formulated *abstractly*, i.e., without reference to Hilbert space representations (see [25]: 5.2.8). It generates a uniquely defined C*-algebra $\mathscr{A}$,

$$\mathscr{A} := \mathscr{M}_2 \otimes \Delta(\mathscr{K}) \tag{24}$$

where $\Delta(\mathscr{K})$ is the C*-algebra generated by the canonical commutation relations (17). Taking the C*-algebra $\mathscr{A}$ instead of the *-algebra discussed is but a kind of completion (like proceeding from the rational numbers to the reals).

A C*-algebra $\mathscr{C}$ is a *-algebra of operators equipped with a norm $\| \cdot \|$ (a "length" of operators) fulfilling

$$\| A B \| \leq \| A \| \| B \|, \quad A, B \in \mathscr{C}, \tag{25}$$

and the crucial C*-condition

$$\| A^* A \| = \| A \|^2, \quad A \in \mathscr{C}. \tag{26}$$

Such a norm is extremely useful from a *mathematical* point of view. Nevertheless, the respective norm topology has no particular *physical* significance. The physically relevant topology is the σ-weak topology to be mentioned below.

Def. 1: A *representation* of a C*-algebra $\mathscr{C}$ on a Hilbert space $\mathscr{H}$ associates to every element A of $\mathscr{C}$ a Hilbert space operator $\pi(A)$ and is characterized by the properties

$$\pi(c A + B) = c \, \pi(A) + \pi(B) \tag{27a}$$

$$\pi(A B) = \pi(A) \, \pi(B) \tag{27b}$$

$$\pi(A^*) = \pi(A)^*, \quad A, B \in \mathscr{C}, \quad c \in \mathbb{C}. \tag{27c}$$

Representations of the boson C*-algebra $\Delta(\mathscr{K})$ are nothing else than representations of the Weyl operators on a Hilbert space fulfilling the CCR commutation rules. Hence the CCR C*-algebra $\Delta(\mathscr{K})$ is but a concise description of the boson commutation rules (8). An analogous statement holds for the spin-boson C*-algebra $\mathscr{A}$.

"Physics" comes in with representations. Different representations refer to different physical situations or contexts. Physical inequivalence of representations is intimately connected with the generation of classical observables.

Def. 2: A linear expectation value functional

$$T \rightarrow \phi(T)$$

on a set of observables $\{T\}$ *represented on a Hilbert space* $\mathscr{H}$ is said to be a *physical state* (pure or mixed) if there exists a *positive trace-class operator D with trace one* acting on the Hilbert space $\mathscr{H}$, such that the expectation value $\phi(T)$ for any "observable" (operator)[2] T of the system is given by the trace $\text{Tr}(DT)$ of DT.

---

[2] Here "observables" and "operators describing the system" are sometimes almost used in the same sense. Not every (self-adjoint) operator ascribed to a certain system can be considered as a physically relevant "measurable" quantity or some quantity which - albeit not measurable - gives an interesting physical structure to the system [26].

10

*Note that two different positive trace-class operators D_1 and D_2 may implement the same expectation values (i.e., the same state) on the observables of the system! This is the case if the observables do not generate linearly all operators on the Hilbert space $\mathcal{H}$. If the positive trace-class operator with trace 1 is **uniquely** determined by the state ϕ, one could call it a **density operator**.*

A sequence[3] $(T_j)_{j \in \mathbf{N}}$ of (represented) observables is said to *converge*[4] to an observable T, if the expectation values $\mathrm{Tr}(T_j D)$ converges to $\mathrm{Tr}(TD)$ for every positive trace-class operator D. *That is, the convergence is defined with respect to the physical states of a certain representation. In mathematical terms, this kind of convergence is called σ-weak convergence or convergence with respect to the σ-weak topology.*

As an illustration for this limit process consider a lattice spin system (a model for a magnet) with spin matrices in z-direction σ_{zj} indexed by $j = 1, 2, \ldots$. Then a *global magnetization operator* is defined by

$$\lim_{N \to \infty} \frac{1}{N} \sum_{j=1}^{N} \sigma_{zj} \tag{28}$$

where the limit is taken precisely in the σ-weak sense, i.e., with respect to expectation values.

The spin-boson C*-algebra $\mathcal{A}$ refers (essentially) to finitely (but arbitrary) many modes of the field. All observables referring to infinitely many degrees of freedom and hence all new and interesting observables arise in certain representations: As possible candidates apart from the global magnetization above consider temperature, chemical potential, chirality, or a global momentum [27].

Def. 3: Let π be a representation of a C*-algebra $\mathcal{C}$ on a Hilbert space $\mathcal{H}$. The *von Neumann algebra* $\{\pi(\mathcal{C})\}''$ consists of all operators acting on $\mathcal{H}$ which are σ-weak limits of sequences of operators in $\pi(\mathcal{C})$. The *center* $\mathcal{X}(\mathcal{M})$ of a von Neumann algebra $\mathcal{M}$ consists of those of its elements which commute with *all* other operators in $\mathcal{M}$.

The von Neumann algebra with respect to a given representation is a "reservoir" for all physically relevant observables. It contains (the spectral projections of) all the physically interesting observables with respect to a certain representation, i.e., the observables referring to a certain physical context. Depending on the specific representation, one may have classical observables or not. Classical observables are contained in the center of the respective von Neumann algebra.

The "replacement" of the original C*-algebra by a von Neumann algebra is the physically relevant point in algebraic quantum mechanics. In ordinary Hilbert-space quantum mechanics only one particular representation is considered, namely the Fock representation. The essential element of algebraic quantum mechanics is that various physically relevant representations exist apart from the Fock representation. None of the above global observables (magnetization, temperature, ...) is contained in a Fock description ! Therefore the Fock representation leads to empirically wrong results.

Def. 4: Two representations π_1 and π_2 of a C*-algebra (e.g., the spin-boson C*-algebra) acting on Hilbert spaces $\mathcal{H}_1$ and $\mathcal{H}_2$ are called *physically equivalent* (*quasiequivalent*) if for every given positive trace-class operator D_1 on $\mathcal{H}_1$ there exist a positive trace-class operator D_2 acting on $\mathcal{H}_2$ such that

---

[3] If the Hilbert space is not separable or if the operators T_j are not uniformly bounded, one has to use nets instead of sequences. Here all Hilbert spaces are assumed to be separable.

[4] This convergence is thought to be physically sensible.

$$\text{Tr}(D_1 \pi_1(A)) = \text{Tr}(D_2 \pi_2(A)), \tag{29}$$

which means that *the physical states with respect to the representations π_1 and π_2 can be identified.*

If the above condition is fulfilled, one can show (see [28]: Section 10.3) that there exists a (σ–weakly continous) *-isomorphism (i.e., a bijective mapping respecting the *-algebraic operations addition, multiplication, adjoint) such that

$$\tau_{12}: \{\pi_1(\mathscr{C})\}'' \rightarrow \{\pi_2(\mathscr{C})\}'' \tag{30a}$$

$$\tau_{12}(\pi_1\{A\}) = \pi_2\{A\}, \quad A \in \mathscr{C}, \tag{30b}$$

hold. Therefore one has a correspondence of the observables defined with respect to the two representations π_1 and π_2. Conversely, the existence of a *-isomorphism (30) implies that the physical states with respect to π_1 and π_2 correspond. In mathematical nomenclature, such representations are called *quasiequivalent* ([28]: Section 10.3).

> *Theorem 1 (Stone - von Neumann): For finitely many degrees of freedom any two regular representations of the Weyl canonical commutation relations (or equivalently: representations of the respective C*-algebra $\Delta(\mathbb{C}^N)$) are quasiequivalent.*

In other words: For finitely many degrees of freedom there exists essentially only one representation of the Weyl CCR. The same holds for the spin-boson C*-algebra $\mathscr{A}$ if the respective boson field has only finitely many modes (the two-level system is irrelevant in this respect). It is therefore legitimate (but by no means necessary and not always useful) to restrict attention to the (irreducible) Schrödinger representation or the (irreducible) Fock representation of the Weyl relations. As a corollary of the Stone-von Neumann theorem one infers that for finitely many degrees of freedom there do *not* exist classical observables.

The Stone-von Neumann theorem is true only for systems with finitely many degrees of freedom. For $N = \infty$ there are myriads of (pairwise) *disjoint* representations of the CCR and all kinds of classical observables can arise. In particular, the *Fock representation* (cf. [24], [25] and [29]) is by far not the only physically interesting representation of the Weyl relations.

Def. 5: Representations π_1 and π_2 of a C*-algebra are called *disjoint* ([28]: Section 10.3) if there exists a representation π on a Hilbert space $\mathscr{H}$ and a projection Z in the center $\mathscr{X}\{\pi(\mathscr{C})\}''$ of $\{\pi(\mathscr{C})\}''$ such that

$$\pi_1 \cong Z \pi(\cdot) Z \tag{31a}$$

$$\pi_2 \cong (1 - Z)\pi(\cdot)(1 - Z) \tag{31b}$$

holds. Here it is understood, that $Z \pi(\cdot) Z$ acts on the Hilbert space $Z \mathscr{H}$ (analogously for $(1 - Z)$)and the point «·» in $\pi(\cdot)$ stands for an arbitrary element of $\mathscr{C}$.

Disjointness of representations means in particular, that one may *construct* a classical observable – namely based on the central projections Z and $(1 - Z)$ – by pasting together the representations π_1 and π_2 into a single one – namely π. The sectors $Z\mathscr{H}$ and $(1 - Z)\mathscr{H}$ support the representations π_1 and π_2, respectively. Hence two disjoint representations allow to construct a two-valued classical observable and in a similar way M disjoint representations allow to construct an M-valued classical observable. Analogously one may construct a classical observable with continuous spectrum by pasting together (in a process of integration instead of a sum) uncountably many (pairwise) disjoint representations indexed, e.g., by elements of $\mathbb{R}^3$ when constructing a classical position or momentum operator [27], [30]. The chirality observable mentioned corresponds to the simplest case of a two-valued classical observable.

THE DYNAMICS OF THE SPIN-BOSON MODEL WITH LEVEL SPLITTING ZERO

Consider the spin-boson system (infinitely many degrees of freedom) with fixed frequencies and coupling constants but variable level splitting $(2\hbar)\varepsilon$. Denote the Hamiltonian (10a-c) by

$$H^\varepsilon = \hbar\,\varepsilon\,\sigma_1 \otimes 1 + H^0 \tag{32}$$

where H^0 is the Hamiltonian (10b-c) of the spin-boson system with zero level splitting. An additional index N will be used if the first N modes are taken into account only.

For finitely many degrees of freedom it is possible to calculate the Heisenberg time evolution

$$\alpha_t^{0,N}(T) := \exp\left\{iH_N^0\,\frac{t}{\hbar}\right\}\,T\,\exp\left\{-iH_N^0\,\frac{t}{\hbar}\right\}, \quad t \in \mathbb{R}, \tag{33}$$

for arbitrary observables T. To state the result (see [31], [32], [33], [34] and [35]), an obvious matrix notation is used instead of the tensor product notation such that the following correspondence

$$\begin{pmatrix} a & b \\ c & d \end{pmatrix} \otimes S \equiv \begin{pmatrix} aS & bS \\ cS & dS \end{pmatrix} \tag{34}$$

holds for arbitrary scalars a, b, c, d and arbitrary observables S of the field alone. It is sufficient to compute or define the dynamics $\{\alpha_t^{0,N} \mid t \in \mathbb{R}\}$ on generating elements of the type

$$\begin{pmatrix} a\,W(f_1) & b\,W(f_2) \\ c\,W(f_3) & d\,W(f_4) \end{pmatrix}. \tag{35}$$

One gets

$$\alpha_t^{0,N}\begin{pmatrix} a\,W(f_1) & b\,W(f_2) \\ c\,W(f_3) & d\,W(f_4) \end{pmatrix} = \tag{36a}$$

$$= \begin{pmatrix} a\,\exp\{-i\,\mathrm{Im}\,\langle\zeta_N^{-t}|f_1\rangle\}\,W(f_1^t) & b\,W(f_2^t+\zeta_N^t) \\ c\,W(f_3^t-\zeta_N^t) & d\,\exp\{+i\,\mathrm{Im}\,\langle\zeta_N^{-t}|f_4\rangle\}\,W(f_4^t) \end{pmatrix}, \tag{36b}$$

$$f_j \in \mathbb{C}^N, \quad j = 1, 2, 3, 4, \qquad a, b, c, d \in \mathbb{C}, \tag{36c}$$

where the definitions

$$\zeta_N^t := (\zeta^t(1), \zeta^t(2), ..., \zeta^t(N)) := \tag{37a}$$

$$:= \left(\frac{\lambda(1)}{\omega(1)}\{1 - e^{it\omega(1)}\}, \frac{\lambda(2)}{\omega(2)}\{1 - e^{it\omega(2)}\}, ..., \frac{\lambda(N)}{\omega(N)}\{1 - e^{it\omega(N)}\}\right), \tag{37b}$$

$$\zeta_N^t \in \mathbb{C}^N, \quad t \in \mathbb{R}, \tag{37c}$$

$$\langle\zeta_N^t|f\rangle := \sum_{n=1}^N \zeta^t(n)^*\,f(n), \tag{37d}$$

$$f_j^t := e^{it\omega_N} f_j := \tag{37e}$$

$$:= (e^{it\omega(1)} f(1), e^{it\omega(2)} f(2), ..., e^{it\omega(N)} f(N)), \quad j = 1, 2, 3, 4, \tag{37f}$$

have been used. An additional index N for the vectors/functions f_j^t has been omitted.

The Heisenberg time evolution allows to make the limit

$$\alpha_t^0 \begin{pmatrix} a\, W(f_1) & b\, W(f_2) \\ c\, W(f_3) & d\, W(f_4) \end{pmatrix} := \lim_{N \to \infty} \alpha_t^{0,N} \begin{pmatrix} a\, W(f_1) & b\, W(f_2) \\ c\, W(f_3) & d\, W(f_4) \end{pmatrix} \tag{38a}$$

$$f_j \in \mathscr{K}, j = 1, 2, 3, 4, \qquad a, b, c, d \in \mathbb{C}. \tag{38b}$$

This defines the dynamics of the spin-boson model with infinitely many degrees of freedom (where some minor continuity assumptions have to be made). The result is completely analogous to (36), the only difference being that the test functions in (37) are now elements of $\mathscr{K}$ instead of the finite-dimensional vector space $\mathbb{C}^N$. Note that this sort of limit, though almost trivial with respect to the Heisenberg dynamics, is very delicate on the level of Hamiltonians.

THE DYNAMICS OF THE SPIN-BOSON MODEL WITH NONZERO LEVEL SPLITTING

Hamiltonians may be introduced with respect to certain representations. A Hamiltonian H_π corresponding to the dynamics α^0 of the spin-boson model with respect to a representation π of the spin-boson algebra $\mathscr{A}$ has to fulfill the condition

$$\pi(\alpha_t^0(A)) := \exp\left\{ iH_\pi \frac{t}{\hbar} \right\} \pi(A) \exp\left\{ -iH_\pi \frac{t}{\hbar} \right\}, \quad t \in \mathbb{R}, \quad A \in \mathscr{A}. \tag{39}$$

Note that not every representation admits a Hamiltonian. The existence of a Hamiltonian may be considered as a criterion for physically meaningful representations. The full dynamics of the spin-boson model (inclusive the level splitting term (10a)) can then be defined by

$$T \to \exp\left\{ i(H_\pi + \hbar \varepsilon \sigma_1 \otimes 1) \frac{t}{\hbar} \right\} T \exp\left\{ -i(H_\pi + \hbar \varepsilon \sigma_1 \otimes 1) \frac{t}{\hbar} \right\}, \quad t \in \mathbb{R}, \tag{40}$$

where T is an arbitrary observable contained in the von Neumann algebra $\{\pi(\mathscr{A})\}''$. This dynamics on $\{\pi(\mathscr{A})\}''$ does *not* depend on the specific form of H_π (which is not unique).

The triple $(\mathscr{A}, \mathbb{R}, \alpha^0)$ consisting of the C*-algebra $\mathscr{A}$ and the dynamics α^0 (38) is called a C*-system. Since arbitrary distinguished Weyl operators $W(f_1)$ and $W(f_2)$ fulfill ([25]: Theorem 5.2.8)

$$\| W(f_1) - W(f_2) \| = 2 \tag{41}$$

the mappings

$$t \to \alpha_t^0(A), \quad t \in \mathbb{R}, \quad A \in \mathscr{A}, \tag{42}$$

are *not* norm-continuous. In mathematics and mathematical physics, on the other hand, this sort of norm-continuity is often assumed to hold for dynamical groups of C*-algebras. This condition, which is referred to as *pointwise norm-continuity*, is sometimes fulfilled in physical systems (e.g., for quantum spin systems as in section 6.2 of [25]). In general, it is too strong. Boson systems and spin-boson systems are *not* pointwise norm-continuous.

For pointwise norm-continuous C*-systems, the perturbation of the dynamics by an operator of the C*-algebra (such as $\hbar \varepsilon \sigma_1 \otimes 1$) can be done abstractly, i.e., without

14

reference to a Hilbert space representation (see [25]). For C*-systems without this property, such as the spin-boson C*-system, one must refer to specific representations to get a dynamics. Note that even then one cannot expect that the perturbed dynamics (40) leaves globally invariant the (represented) C*-algebra $\pi(\mathscr{A})$, i.e.,

$$\exp\left\{i(H_\pi + \hbar\varepsilon\sigma_1\otimes 1)\,\frac{t}{\hbar}\right\}\,\pi(\mathscr{A})\,\exp\left\{-i(H_\pi + \hbar\varepsilon\sigma_1\otimes 1)\,\frac{t}{\hbar}\right\} \not\subseteq \pi(\mathscr{A}) \quad (43)$$

may happen [36]. Hence the perturbed dynamics "α^ε" for the spin-boson model with nonzero level splitting $2\hbar\varepsilon$ is not defined on $\mathscr{A}$ but only in specific representations. On the C*-level it might be called a *pseudodynamics*.

5. GROUND STATES OF THE SPIN-BOSON MODEL

The notion of a ground state can be "naturally" defined in algebraic quantum mechanics.

Def. 6: Let $(\mathscr{C}, \mathbb{R}, \theta)$ be a C*-system. Then a state ϕ on $\mathscr{C}$ is called a *ground state* (cf. [25]: 5.3.19) if there exist

- a representation $\pi: \mathscr{C} \twoheadrightarrow \mathscr{B}(\mathscr{H})$ on a Hilbert space $\mathscr{H}$,

- a self-adjoint operator (the "Hamiltonian") H acting on $\mathscr{H}$ such that

$$\pi(\theta_t(A)) = \exp\left\{iH_\pi\,\frac{t}{\hbar}\right\}\,\pi(A)\,\exp\left\{-iH_\pi\,\frac{t}{\hbar}\right\}, \quad t\in\mathbb{R}, \quad (44a)$$

- a (normalized) eigenvector ξ of H_π with eigenvalue E_0 such that

$$\phi(A) = \langle\xi\,|\,\pi(A)\,\xi\rangle, \quad A\in\mathscr{C}, \quad (44b)$$

$$H_\pi \geq E_0\,\mathbf{1}. \quad (44c)$$

The definition of an *"algebraic ground state"* is perfectly adapted to the corresponding notion in ordinary Hilbert space quantum mechanics. The new aspect here is that the representation π of the observables and the Hamiltonian H_π are no more uniquely determined. The Hamiltonian H_π and its spectral structure depend on π. *Different ground states may be elements of different sectors, i.e., be separated by a superselection rule.*

Since the spin-boson system $(\mathscr{A}, \mathbb{R}, \alpha^0)$ is not pointwise norm-continuous and since the full dynamics (40) is not defined on $\mathscr{A}$, it is necessary to adapt the ground state definition a little bit:

Def. 7: Let $(\mathscr{A}, \mathbb{R}, \alpha^0)$ be the spin-boson C*-system. Then a state ϕ on $\mathscr{A}$ is called a *ground state* with respect to the (pseudo)dynamics "α^ε" if there exist

- a representation $\pi: \mathscr{A} \twoheadrightarrow \mathscr{B}(\mathscr{H})$ on a Hilbert space $\mathscr{H}$,

- a self-adjoint operator (the "Hamiltonian") H_π acting on $\mathscr{H}$ such that

$$\pi(\alpha_t^0(A)) := \exp\left\{iH_\pi\,\frac{t}{\hbar}\right\}\,\pi(A)\,\exp\left\{-iH_\pi\,\frac{t}{\hbar}\right\}, \quad t\in\mathbb{R}, \quad A\in\mathscr{A}, \quad (45a)$$

- a (normalized) eigenvector ξ of $(H_\pi + \hbar\varepsilon\sigma_1\otimes 1)$ with eigenvalue E_0 such that

$$\phi(A) = \langle\xi\,|\,\pi(A)\,\xi\rangle, \quad A\in\mathscr{C}, \quad (45b)$$

$$(H_\pi + \hbar\varepsilon\sigma_1\otimes 1) \geq E_0\,.\mathbf{1}. \quad (45c)$$

For the spin-boson model with zero level splitting it is relatively simple to determine the respective ground states (cf. [15], [16] and [37]):

If the infrared-singularity (14) holds, one gets precisely two disjoint *pure* ground states. Disjoint is to say that there exists a classical "chirality observable" with eigenvalues ±1 and such that the corresponding eigenspaces (sectors) contain each a unique ground state vector. Formal superposition of these ground state vectors leads to mixed states.

Suppose now that the parameters of the spin-boson model are chosen in a way such that the infrared-singularity (14) does *not* hold. Then the (corresponding) pure states of before do still exist. They are no more disjoint and each superposition of them results in a *pure* ground state again. Furthermore every pure ground state is of this form.

For nonzero level splitting, the situation is much more complicated. Even for *one* mode of the field the ground states of the spin-boson model are not explicitly known.

6. *GROUND STATES OF THE SPIN-BOSON MODEL AS LIMITS OF THERMIC STATES*

The notion of a *Gibbs thermic equilibrium state* can be generalized to the algebraic scheme and is called β-KMS-state there, β being the inverse temperature. Every Gibbs state is a KMS-state. Conversely, KMS-states can frequently be constructed as appropriate (infinite volume, ...) limits of Gibbs states.

The KMS-states of the spin-boson model with zero level splitting have been studied in [31-35]. Under certain continuity assumptions (which exclude boson condensation) there exists precisely one KMS-state for every inverse temperature β.

The *perturbation theory of KMS-states* is relatively simple: There is an explicit (see [25]: 5.4.5) bijective correspondence between β-KMS states with respect to "α^{ε}" and β-KMS states with respect to α^0 on $\mathscr{A}$. *Corresponding β-KMS states are never separated by a superselection rule, i.e., lie in the same sector.* Therefore there exists precisely one β-KMS state with respect to the perturbed dynamics "α^{ε}".

For ground states the situation is totally different. No simple perturbation theory is at hand. For $\varepsilon = 0$ it was mentioned that two ground states exist in different sectors if the IR-singularity holds. Nevertheless the slightest change of ε may effect that H^{ε} admits only one ground state.

For the coupling of a *single* molecule to a field, superselection rules and classical observables have to be discussed with respect to pure states (e.g., ground states) but never with respect to thermic states [16]. Here the (admittedly too small) class of ground states is studied.

At least on a heuristic basis, ground states can be introduced as temperature to zero limits of thermic equilibrium states (KMS-states). The latter are easier to discuss than algebraic ground states themselves. Introducing an additional perturbation

$$h\sigma_3 \otimes \mathbf{1} \tag{46}$$

(with $h \to \pm 0$ later on) offers the possibility to get different "ground states". This program has been carried out in a very interesting paper [38](cf. also [39 – 41]).

The set-up of Spohn's paper then is as follows:

Keeping fixed the coupling parameters, the coupling strength and the frequencies of the spin-boson model and varying only ε and h, Spohn introduces β-KMS states $\omega_\beta^{\varepsilon,h}$ with respect to the dynamics "$\alpha^{\varepsilon,h}$" and shows that the limits

$$\phi^{\varepsilon,h}(A) := \lim_{\beta \to \infty} \omega_\beta^{\varepsilon,h}(A), \qquad A \in \mathscr{A}, \tag{47a}$$

$$\phi_\pm^\varepsilon(A) := \lim_{h \to \pm 0} \phi^{\varepsilon,h}(A), \qquad A \in \mathscr{A}, \tag{47b}$$

exist such that the symmetry ι ("space inversion") fulfills

$$\phi_+^\varepsilon(A) = \phi_-^\varepsilon(\iota(A)), \qquad A \in \mathscr{A}, \tag{48a}$$

$$\phi_-^\varepsilon(A) = \phi_+^\varepsilon(\iota(A)), \qquad A \in \mathscr{A}. \tag{48b}$$

The states $\phi_\pm^\varepsilon$ are per definitionem called ground states. For

$$\phi_+^\varepsilon(\sigma_3 \otimes 1) = 0 \tag{49}$$

the states ϕ_+^ε and ϕ_-^ε coincide and can be described by state vectors in the Fock representation with the Hilbert space (13). For

$$\phi_+^\varepsilon(\sigma_3 \otimes 1) \neq 0 \tag{50}$$

one has

$$\phi_-^\varepsilon(\sigma_3 \otimes 1) = -\phi_+^\varepsilon(\sigma_3 \otimes 1) \tag{51}$$

which implies that at least 2 "ground states" exist. Then the inversion symmetry ι of the spin-boson problem is said to be **broken**. *Note that the underlying dynamics and its Hamiltonian are still invariant under inversion, since h is finally set to zero.*

The difficult problem is of course to check wether $\phi_+^\varepsilon(\sigma_3 \otimes 1)$ vanishes or not! The technique used translates the spin-boson model into a 1-dimensional continuous Ising model (see [38-41]). For the presentation of Spohn's results an additional parameter, the *"coupling strength"* will be introduced[5], replacing the coupling constants $\lambda(n)$ in the Hamiltonian (10) by $\sqrt{\rho}\,\lambda(n)$. This parameter ρ just allows to scale simultaneously all coupling constants. It is normalized in a suitable way. Typical "phase diagrams" for ohmic and subohmic IR-singularity are given in figures 1 and 2. The symmetry is broken for parameters in the shaded region. For the superohmic situation (no IR-singularity) the "space-inversion" symmetry is never broken. The dotted line in these figures refer to the phase diagram proposed in [15], which is the same for ohmic and subohmic IR-singularity. This phase diagram is based on a (variational principle) calculation with Hartree product states.

The phase diagrams of Spohn and Pfeifer coincide qualitatively in the subohmic situation: The symmetry is broken for small (but nonzero) level splitting independent of the coupling strength ρ. In the ohmic case, Spohn predicts that the space-inversion symmetry is only broken if the coupling strength exceeds a certain value.

---

[5] The nomenclature of Spohn is adapted to the present paper: There the letter α is used for the coupling strength.

17

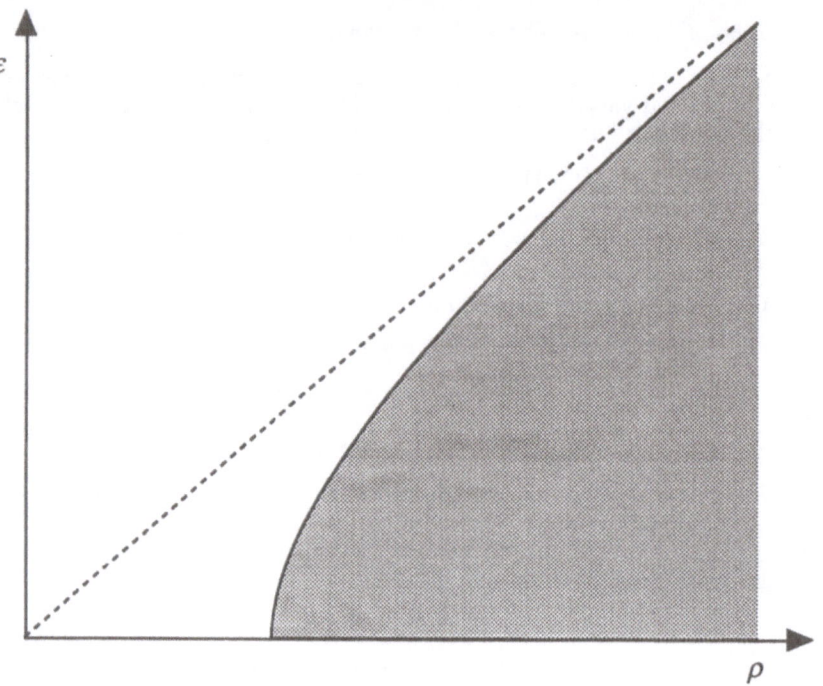

Figure 1. Spohn's phase diagram for the spin-boson model with *ohmic* infrared singularity.

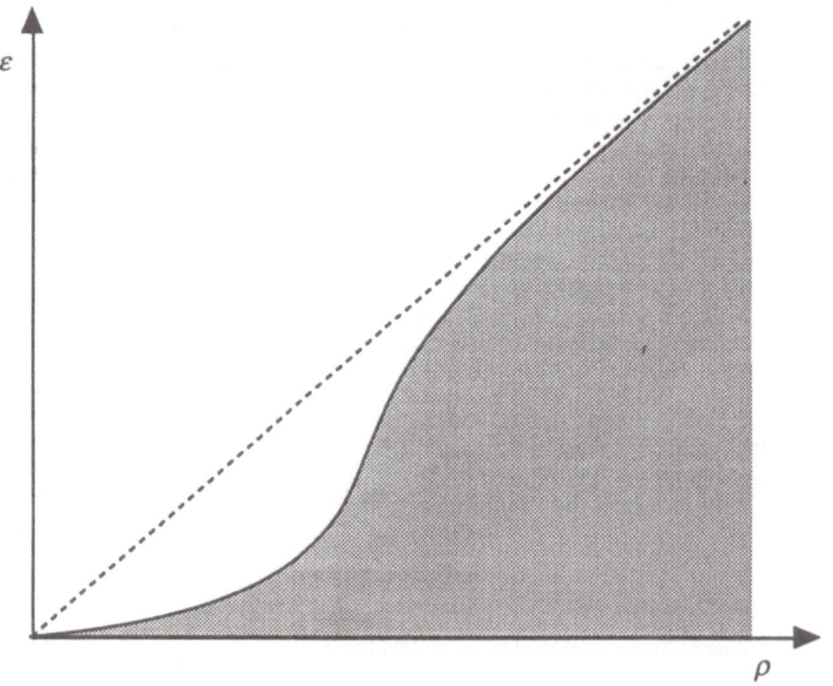

Figure 2. Spohn's phase diagram for the spin-boson model with *subohmic* infrared singularity

Consider again the *heuristic* discussion on chirality in section 2: For small level splitting (as with chiral molecules) the space-inversion symmetry may be broken and two "chiral" ground states arise. For large level splitting (as with ammonia) the space-inversion symmetry is not broken and the ground state is unique (at least in Spohn's setting).

7. ALGEBRAIC GROUND STATES OF THE SPIN-BOSON MODEL

For pointwise norm-continuous C*-systems it is no problem at all to show that the temperature to zero limit of thermic equilibrium states is a ground state in the algebraic sense, i.e., given by an eigenvector of the respective Hamiltonian with an eigenvalue at the lower end of the Hamiltonian's spectrum (see [25]: Proposition 5.3.23). Similarly, the additional perturbation (45) could be dealt with. The main advantage of pointwise norm-continuous C*-systems is the existence of a generator δ (see [42]: chapter 3), related to the Hamiltonian H_π of the system *in a suitable representation* π by

$$\pi(\delta(A)) = i\,[H_\pi, \pi(A)]\,, \quad A \in D(\delta), \tag{52}$$

where $D(\delta)$ is the domain of definition of δ. Recall that expression (52) arises in the von Neumann–Liouville equation. The perturbed dynamical groups related to perturbed Hamiltonians $H + P$ (with P some selfadjoint element of $\mathscr{A}$) can then be defined (see [25]: section 5.4).

For C*-systems *without norm-continuity properties*, the situation is not clear a priori. Recall that the spin-boson dynamics "α^ε" is not even defined on the underlying C*-algebra $\mathscr{A}$. Hence KMS-states, ground states and other concepts must be introduced with the help of suitable representations and a generator (52) does not exist. Nevertheless an *auxiliary pointwise norm-continuous C*-system* $(\mathscr{D}^\varepsilon, \mathbb{R}, \delta^\varepsilon)$ can be introduced such that there exists a continuous 1-1 correspondence between KMS-states and ground states on $\mathscr{A}$ with respect to the (pseudo)dynamics "α^ε" and KMS-states and ground states on the C*-algebra $\mathscr{D}^\varepsilon$ with respect to the dynamics δ^ε (see [43], [44]). Refining this construction a little bit to encorporate the additional perturbation (45) then leads to the following

Theorem 2: Spohn's ground states of the spin-boson model are ground states in the algebraic sense, i.e., are eigenstates of the spin-boson Hamiltonian with an eigenvalue at the lower end of the Hamiltonian's spectrum.

Recall that the spin-boson Hamiltonian, referring to a system with infinitely many degrees of freedom, is only a formal operator which has to be defined in every sector. For different ground states the respective spin-boson Hamiltonians correspond to the same formal Hamiltonian (10a-c) but nevertheless look pretty different.

Spohn's construction of ground states of the spin-boson model is very special and may therefore not exhaust all algebraic ground states in the sense of definition (7). It might be imaginable that an algebraic ground state ϕ exists with

$$\phi(\sigma_3 \otimes 1) \neq 0 \tag{53}$$

for parameters ρ and ε where Spohn predicts that the symmetry remains unbroken (i.e., in the unshaded region of figures 1 and 2). Then

$$\iota(\sigma_3 \otimes 1) = -(\sigma_3 \otimes 1) \tag{54}$$

implies that the ground state ϕ and its "mirror-image"

$$A \mapsto \phi(\iota(A))\,, \quad A \in \mathscr{A}, \tag{55}$$

do not coincide. This would mean that space-inversion symmetry ι is broken with respect to the ground state ϕ, but not with respect to ground states of Spohn. The following theorem tells us that this is not the case.

> *Theorem 3 ([45]): Consider the discrete spin-boson model and assume that its parameters are chosen such that $\phi_+^\varepsilon(\sigma_3 \otimes \mathbf{1}) = 0$, i.e., no symmmetry-breaking in the sense of Spohn occurs. Then any algebraic ground state ϕ of the spin-boson model with respect to these chosen parameters fulfills $\phi(\sigma_3 \otimes \mathbf{1}) = 0$.*

It has not (yet) been proven that the ground states (46) of Spohn are disjoint, i.e., that they lie in different sectors, separated by a superselection rule. Nevertheless, a generalized positivity argument in the sense of ([46]: Section XIII.12) shows that for a large class of representations of $\mathscr{A}$ containing ground state vectors this is indeed the case (see [45]). This supports the following

> *Conjecture: Let ϕ_1 and ϕ_2 be **pure** ground states of spin-boson system with parameters $\varepsilon, h \in \mathbb{R}$, $\varepsilon \neq 0$. Then ϕ_1 and ϕ_2 are disjoint or equal.*

8. CONCLUDING REMARKS

The spin-boson model with Hamiltonian (10a-c) is a semirealistic model for the coupling of a small system to a bosonic environment. It is *semirealistic* since only two levels of the small system are considered and since certain terms quadratic in the boson operators are omitted. Hence any results of this model should be estimated with caution.

Nevertheless, the spin-boson *model* shows that molecular superselection rules can at least in principle be *derived* without any ad hoc assumptions. It is noteable that this derivation starts with a genuine quantum system ! Classical observables arise when representations of the original C*-algebra $\mathscr{A}$ are considered. Recall that different representations (corresponding to different physical contexts) give rise to different classical observables (sometimes only trivial ones, i.e., multiples of the identity operator).

Here it was supposed that the representations considered admit a Hamiltonian and a ground state. The latter condition is very restrictive. It might be, that the phase diagram of the spin-boson model looks differently if other conditions are imposed on the representations investigated. Classical observables and superselection rules abound even if the IR-singularity does not hold. The only problem is to choose physically relevant classical observables.

The Fock representation of ordinary Hilbert space quantum mechanics is definitely not sufficient for these purposes. It does *not* admit any classical observables. Many physically relevant representations, particulary representations associated to thermic equilibrium states, are physically inequivalent to the Fock representation. Furthermore, already scaling transformations (scaling position and momentum operators (9), but leaving invariant the Heisenberg commutation relations) lead out of Fock space.

ACKNOWLEDGMENT

I am indebted to Prof. H. Primas and Prof. H. Spohn for various discussions.

REFERENCES

[1] Borel, E.: *Introduction géométrique à quelques théories physiques*, Paris. Gauthier-Villars (1914).

[2] Schrödinger, E.: *Discussion of probability relations between separated systems*, Proc. Cambr. Phil. Soc. **31**, 555 - 563 (1935).

[3] Schrödinger, E.: *Die gegenwärtige Situation in der Quantenmechanik*, Naturwissenschaften **23**, 807 - 812, 823 - 828, 844 - 849 (1935).

[4] Schrödinger, E.: *Probability relations between separated systems*, Proc. Cambr. Phil. Soc. **32**, 446 - 452 (1936).

[5] Clauser, J. F. and A. Shimony: *Bell's theorem: experimental tests and implications*, Rep. Prog. Phys. **41**, 1881 - 1927 (1978).

[6] Aspect, A., G. Grangier and G. Roger: *Experimental test of Bell's inequalities using time-varying analyzers*, Phys. Rev. Lett. **49**, 1804 - 1807 (1982).

[7] Neumann, J. v.: *Mathematische Grundlagen der Quantenmechanik*, Berlin. Springer (1932).

[8] Leggett, A. J., S. Chakravarty, A. T. Dorsey, M. P. A. Fisher, A. Garg and W. Zwerger: *Dynamics of the dissipative two-state system*, Rev. Mod. Phys. **59**, 1 - 85 (1987).

[9] Einstein, A.: *Elementare Überlegungen zur Interpretation der Quantenmechanik*, In: *Scientific Papers, presented to Max Born*. Edinburgh. Oliver Boyd (1953),

[10] Born, M.: *Albert Einstein, Hedwig und Max Born, Briefwechsel 1916 - 1955*, München. Nymphenburger Verlagshandlung (1969).

[11] Amann, A., L. Cederbaum and W. Gans: *Fractals, Quasicrystals, Chaos, Knots and Algebraic Quantum Mechanics*, Dordrecht. Kluwer (1988).

[12] Quack, M.: *On the measurement of the parity violating energy difference between enantiomers*, Chem. Phys. Lett. **132**, 147 - 153 (1986).

[13] Quack, M.: *Structure and dynamics of chiral molecules*, Angew. Chem. Int. Ed. Engl. **28**, 571 - 586 (1989).

[14] Amann, A.: *Theories of molecular chirality: A short review*, in this same volume.

[15] Pfeifer, P.: *Chiral Molecules - a Superselection Rule Induced by the Radiation Field*, Zürich. Thesis ETH-Zürich No. 6551, ok Gotthard S+D AG (1980).

[16] Amann, A.: *Chirality as a classical observable in algebraic quantum mechanics*, In: *Fractals, Quasicrystals, Chaos, Knots and Algebraic Quantum Mechanics*. Ed. by A. Amann, L. Cederbaum and W. Gans. Dordrecht. Kluwer (1988), Pp. 305 - 325.

[17] Wightman, A. S. and N. Glance: *Superselection rules in molecules*, Nucl. Phys. B (Proc. Suppl.) **6**, 202 - 206 (1989).

[18] Kukolich, S. G., J. H. S. Wang and D. E. Oates: *Molecular beam maser measurements of relaxation cross sections in NH_3*, Chem. Phys. Lett. **20**, 519 - 524 (1973).

[19] Hund, F.: *Zur Deutung der Molekelspektren III*, Z. Phys. **43**, 805 - 826 (1927).

[20] van Hemmen, J. L.: *A note on the diagonalization of quadratic Boson and Fermion Hamiltonians*, Z. Phys. B **38**, 271 - 277 (1980).

[21] Cohen-Tannoudji, C., J. Dupont-Roc and G. Grynberg: *Photons and Atoms*, New York. John Wiley & Sons (1989).

[22] Pfeifer, P.: *Estimates of the A^2-term in the interaction of matter with radiation*, J. Phys. A **14**, L129 - L132 (1981).

[23] Craig, D. P. and T. Thirunamachandran: *Molecular Quantum Electrodynamics. An Introduction to Radiation-Molecule Interactions*, London. Academic Press (1984).

[24] Reed, M. and B. Simon: *Methods of Modern Mathematical Physics. Volume I: Functional Analysis*, New York. Academic Press (1972).

[25] Bratteli, O. and D. W. Robinson: *Operator Algebras and Quantum Statistical Mechanics Vol. 2*, New York. Springer (1981).

[26] Amann, A.: *Observables in W*-algebraic quantum mechanics*, Fortschr. Phys. **34**, 167 - 215 (1986).

[27] Amann, A. and U. Müller-Herold: *Momentum operators for large systems*, Helv. Phys. Acta **59**, 1311- 1320 (1986).

[28] Kadison, R. V. and J. R. Ringrose: *Fundamentals of the Theory of Operator Algebras. Volume II: Advanced Theory*, New York. Academic Press (1986).

[29] Guichardet, A.: *Symmetric Hilbert Spaces and Related Topics. Lecture Notes in Mathematics Vol. 261*, Heidelberg. Springer (1970).

[30] Amann, A.: *Broken symmetries and the generation of classical observables in large systems*, Helv. Phys. Acta **60**, 384 - 393 (1987).

[31] Nachtergaele, B.: *Exakte resultaten voor het Spin-Boson model*, Leuven. Thesis, Katholieke Universiteit Leuven (1987).

[32] Fannes, M., B. Nachtergaele and A. Verbeure: *Quantum tunneling in the spin-boson model*, Europhys. Lett. **4**, 963 - 965 (1987).

[33] Fannes, M., B. Nachtergaele and A. Verbeure: *Tunneling in the equilibrium state of a spin-boson model*, J. Phys. A **21**, 1759 - 1768 (1988).

[34] Fannes, M., B. Nachtergaele and A. Verbeure: *The equilibrium states of the spin-boson model*, Commun. Math. Phys. **114**, 537 - 548 (1988).

[35] Fannes, M.: *Temperature states of spin-boson models*, In: *Quantum Probability and Applications IV. Lecture Notes in Mathematics Volume 1396*. Ed. by L. Accardi and W. von Waldenfels. Berlin. Springer (1989),

[36] Fannes, M. and A. Verbeure: *On the time evolution automorphisms of the CCR-algebra for quantum mechanics*, Commun. Math. Phys. **35**, 257 - 264 (1974).

[37] Primas, H.: *An introduction into algebraic quantum mechanics*, in preparation. (1990).

[38] Spohn, H.: *Ground state(s) of the spin-boson Hamiltonian*, Commun. Math. Phys. **123**, 277 - 304 (1989).

[39] Spohn, H. and R. Dümcke: *Quantum tunneling with dissipation and the Ising model over R*, J. Stat. Phys. **41**, 389 - 423 (1985).

[40] Spohn, H.: *Models of statistical mechanics in one dimension originating from quantum ground states*, In: *Statistical Mechanics and Field Theory: Mathematical Aspects*. Ed. by T. C. Dorlas Hugenholtz, N.M. and Winnink, M. Berlin. Springer (1986),

[41] Fannes, M. and B. Nachtergaele: *Translating the spin-boson model into a classical system*, J. Math. Phys. **29**, 2288 - 2293 (1988).

[42] Bratteli, O. and D. W. Robinson: *Operator Algebras and Quantum Statistical Mechanics Vol. 1*, New York. Springer, 2nd revised edition (1987).

[43] Amann, A.: *Perturbation theory of boson dynamical systems*, to appear in: J. Phys. A (1990).

[44] Amann, A.: *Perturbation theory of C*-systems without norm-continuity properties*, preprint (1990).

[45] Amann, A.: *Ground states of the spin-boson model*, preprint (1990).

[46] Reed, M. and B. Simon: *Methods of Modern Mathematical Physics. Volume IV: Analysis of Operators*, New York. Academic Press (1978).

THEORIES OF MOLECULAR CHIRALITY: A SHORT REVIEW

Anton Amann

Laboratory of Physical Chemistry
ETH-Zentrum
CH-8092 Zürich

1. PRELIMINARY REMARKS

The discussion of chirality from a *quantum* mechanical point of view has a long history. It was first observed by [1] that "chiral" states - once prepared - are stable for a long time, depending of course on the height of the respective double-minimum potential in the Born-Oppenheimer description. There "chiral" states are simply meant to be states which are localized in one of the potential's wells.

Hund's result refers to an isolated molecule and is generally accepted. Note that it does not answer the following questions: *"Why have superpositions of chiral states not (yet) been prepared ?"* and *"Are superpositions of chiral states particularly unstable ?"* Problems of this sort are summarized here as the *"problem of chirality"*.

Chirality in this respect need not but can be discussed *without any reference to the Born-Oppenheimer approximation*. Chiral states are not necessarily localized states. Furthermore, it should be realized that *molecular states without a nuclear frame* can be prepared in specific situations (cf. [2]).

Completely analogous "problems" arise for billiard balls [3-4], cats [5-7], planets and other objects of everyday life [8]:

"Why is it impossible to prepare a billiard ball in a box as a standing wave ?"

"Why is it impossible to prepare a superposition of two states of a cat, one referring to a dead and one referring to a living animal ?"

"A quantum description of the solar system by coherent states (wave packets) is perfectly possible. Why do superpositions of such coherent states not exist ?"

Hence the above "problems" may be reformulated:

"Why do classical properties exist in a quantum world ?"

"Can one derive classical properties out of quantum stuff ?"

From a pragmatic point of view there is no problem at all: For a conversation about quantum mechanics one needs a Boolean language, i.e., a "metalanguage" with a two-valued logic. The recipe than runs as follows: "Simply encorporate all classical phenomena into this Boolean region of discourse and restrict quantum mechanics to proper quantum phenomena."

Large-Scale Molecular Systems, Edited by W. Gans *et al.*
Plenum Press, New York, 1991

Here the discussion of chirality is considered as a *by-product* of the discussion of systems with infinitely many degrees of freedom. Adapted to chirality, the *general line of thinking is as follows:*

The superposition of handed states is destabilized by the "influence" of the environment; destabilization depends on the molecular and environmental parameters (level splitting or barrier height of the molecule, ...).

There is no strict separation between the classical and the quantum world: A change of the conditions (molecular and environmental parameters) may lead to the generation or disappearance of a classical property [9].

All outer influences (radiation field, gravitation field, collisions with other molecules, etc.) should in principle be discussed jointly.

An interesting experiment has been proposed by M. Quack (cf. [10], [11]). It refers to an *isolated chiral molecule seen in the Born-Oppenheimer approximation.* It is assumed that the electronic ground state potential is of double-well form whereas the electronic excited state potential is thought to have a unique minimum. Hence chiral states of the whole molecule {electrons and nuclei} may exist for the electronic ground but not for the excited state. The non-chiral electronic excited state is expected to decay to the superposition of the handed states with nonzero probability. Handed states and their superpositions can be distinguished since the respective selection rules are different.

The mentioned experiment has not (yet) been performed. Even as a "Gedanken-experiment" it is interesting in its own right. Several pseudo-objections referring to the low energy difference between the ground and the first excited state (of the electronic ground state potential) turn out to be unsound. The experiment's essential message says that the *"superposition of handed states may exist but nobody has yet tried to prepare it"*.

The theoretical status of the discussion is not yet fully convincing. Hence the mentioned experiment would perhaps be difficult to interpret *if* it is meant to distinguish between different views concerning chirality. It seems difficult, for example, to discuss it for a molecule coupled to a field:

"What is an excited state of an infinite system?"

"Is it possible to break up a classical observable by considering excited states?"

Superselection rules and classical observables (notions which are identified here) may be introduced in different ways. Two main views are exemplified. Consider a potentially classical observable and arbitrary eigenvectors Ψ_a and Ψ_b (with different eigenvalues). Then "classical" could mean that

❏ the superposition

$$2^{-\frac{1}{2}} (\Psi_a + \Psi_b) \tag{1}$$

decays into its components Ψ_a and Ψ_b with certain probabilities, or that

❏ the superposition

$$2^{-\frac{1}{2}} (\Psi_a + \Psi_b) \tag{2}$$

describes a **classical** mixture of the states described by the state vectors Ψ_a and Ψ_b.

The latter view has been used and explained in [9] without further discussion. The superposition of states in different sectors is then simply forbidden whereas the former view advocates that this superposition is existent but unstable. These two lines of thinking may perhaps be reconciled, i.e., they are not as far from each other as one might believe at first sight.

2. MOLECULAR ENVIRONMENTS

The coupling of a molecule to its environment is from the very beginning an almost ill-defined problem: "The" environment, consisting of the rest of the universe, can never be given a precise description. It must therefore be replaced by a model environment which mimics certain aspects of the real situation. One such aspect is the large - eventually infinitely large - number of the environment's degrees of freedom. But large systems and in particular the joint system {molecule and environment} may be extremally vulnerable to small perturbations. This suggests that any additional environmental effect - not yet contained in the model environment - will again considerably disturb the discussion and lead to completely different results. Hence it is doubtable if the coupling of a molecule to its environment may ever be treated in a sensible way.

On the other hand, there is no large variety of model environments at hand and environments like

- the radiation field

- the gravitation field

- phonons (if the molecule is embedded into a semirigid matrix of other molecules)

- or a "heat bath"

are treated (or may at least be treated) as systems consisting of infinitely many *harmonic oscillators* (see e.g. [12-14]). Sometimes it is suitable or even necessary to replace the *quantum* environment by a corresponding classical one, i.e., a stochastic process (see e.g. [15], [16], [17]) such as a Brownian motion or a Wiener noise; from there it is no more a long way to environments made up of "collisions with surrounding particles" (cf. e.g. [18-22], [11], [23]) or to Onsager-type reaction field theories (cf. [16], [20]).

Seen from a theoretical viewpoint, it is indispensible to take into consultation the environmental effects when discussing a molecule: The latter may be Einstein-Podolsky-Rosen correlated (in short: EPR-correlated) to the environment and ceases to be an individual object [24-27]. Such EPR-correlations arise if neither the molecule nor its environment can be described as classical systems, where "classical" means that any expectation value with respect to a pure state is dispersion-free [27]. In other words: The molecule can no more be described by a pure state (i.e., a state vector in the ordinary irreducible Hilbert-space description) but has to be described by a mixed state (a density matrix) which depends on the properties of the environment in a way that makes it a euphemism to speak of environmental *effects*: The influence of the environment cannot be coped with by introduction of just an additional potential, some interaction or similar procedures. Nevertheless, the chance remains that some significant part of the joint system {molecule and environment} behaves again as an object, i.e., it is not EPR-correlated to the rest of the system. Such a significant part will be called a *quasimolecule*. If one is lucky, the quasi-molecule behaves similarly as the isolated molecule does apart from some restrictions on the states and observables: Superpositions of certain state vectors may no more be "allowed" and certain observables turn out to be classical, i.e. their expectation value is dispersion-free with respect to all "allowed" states.

As an illustration consider an ion in the gas phase which is brought into an aqueous solution. The solvated ion carries some layers of water molecules; the complex consisting of the ion, some water molecules and (part of) the hydrogen bond interaction with the aqueous environment corresponds to the quasimolecule above.

The isolated quantum-mechanical molecule is the molecule of the theorists whereas the (partially classical) quasimolecule is the molecule of the experimentalists. The spin-boson model discussed in [9] shows that the derivation of a partially classical quasimolecule is possible starting with a genuine quantum mechanical system {molecule and bosonic environment}: Originally the molecule is treated as a two-level system. If the "space-inversion" symmetry of this joint system is broken one has to replace the ground state and the excited state of the isolated two-level system by two ground states separated by a superselection rule. This is to say that the respective ground state vectors are eigenstates (with different eigenvalue) of the corresponding superselection rule operator Ξ [9]. The *quasi-two-level system* has a trivial Hamiltonian (since the new ground states are degenerate) and its observables consist only of 2×2-matrices commuting with Ξ.

In the following, some specific molecular environments will be critically discussed.

3. CAN THE RADIATION FIELD GENERATE CLASSICAL MOLECULAR OBSERVABLES?

The spin-boson Hamiltonian models the coupling of a small system to its environment. It is in particular applicable to the special situation where the coupling of a molecule to the radiation field is investigated.

The generation of chiral disjoint molecular states by the radiation field in terms of the spin-boson model was first proposed and discussed in the pioneering work of P. Pfeifer [28]. There a variational principle with respect to Hartree product states was used. As a result, Pfeifer arrived at the phase diagram of figure 1. The parameter ε is the level splitting of the isolated two-level system. The parameter ρ describes the coupling strength between the two-level system and the field [9]. The inversion symmetry is *expected* to be broken *with respect to ground states* for parameters in the shaded region. Hence for such parameters one expects two ground states separated by a superselection rule. Otherwise the ground state of the spin-boson system is unique.

This phase diagram supports the heuristic idea that an almost degeneracy of the ground state of the isolated molecule leads to a classical observable "chirality" in the joint system {molecule and environment}.

Pfeifer's phase diagram is only asymptotically correct [9] (for large coupling strength). The correct phase diagrams are those proposed by Spohn (see figures 1 and 2 in [9]). The main new aspect there concerns the case of an *ohmic* Infrared Singularity (in the terminology of [13]). In this situation a superselection rule separating ground states can but arise if the coupling strength exceeds a certain critical value. For a *subohmic* IR-singularity there is at least a qualitative agreement between the phase diagrams of Pfeifer and Spohn [9]. *Recall that all these phase diagrams refer to ground states.*

The question if there arises an IR-singularity at all for a neutral molecule coupled to the radiation field is very delicate. If the A^2-term is omitted (A being the transverse vector potential of the radiation field), one arrives at an ohmic IR-singularity [28]. Keeping the A^2-term but introducing the dipole approximation ("long wavelength approximation" [29]) eliminates the IR-singularity altogether. A discussion of the coupling {molecule $\leftrightarrow$ radiation field} without any approximation is difficult. But even in the ohmic situation, the strength of the coupling to the radiation field seems to be much too low to get beyond the critical value of the coupling strength [13], [30].

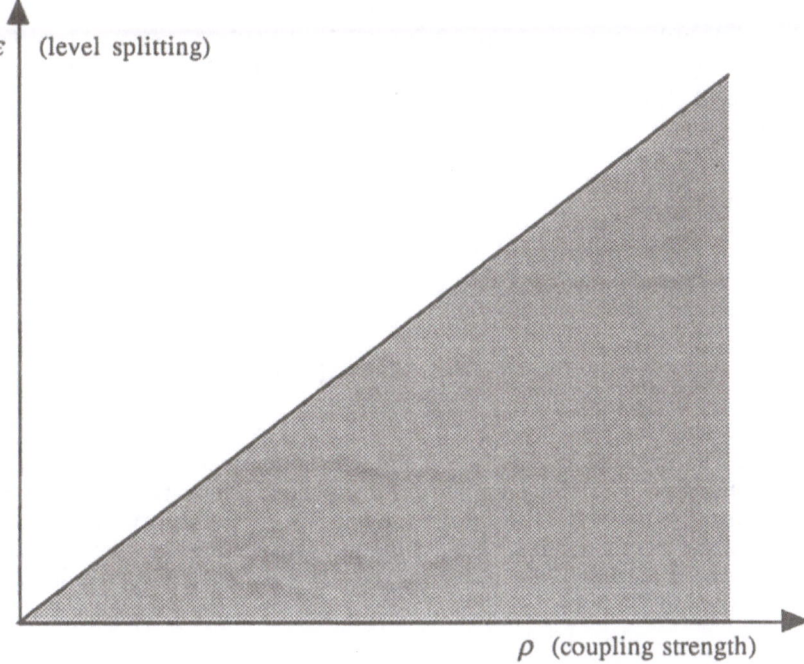

ε (level splitting)

ρ (coupling strength)

Figure 1. Pfeifer's phase diagram for the spin-boson model with an (ohmic
or subohmic) infrared-singularity. The space-inversion symmetry is broken
for parameters in the shaded region.

The main *critique* of Pfeifer's approach (and of [9]) is the use of *ground states*.[1] It
can be made plausible that one should use *pure* states for the discussion of symmetry-
breaking in *single* molecules [31]. Ground states are essentially pure states, i.e.,
classical mixtures of pure states[2], but not every pure state is a ground state or even a
stationary state. *There is no reason to believe that eigenstates of a Hamiltonian are of
particular importance and should be preferred to nonstationary states.* Therefore the
search for sectors containing a ground state is nothing but a first approach to the
problem of finding classical observables in an infinite system. Any system with
infinitely many degrees of freedom admits myriads of sectors (cf. [9]). *The problem is
not to find classical observables but much more to choose physically sensible ones.*
Even if there is no IR-divergence at all (e.g., if the cavity of the radiation field has finite
lenght), there might arise different sectors, superselection rules and classical
observables.

---

[1] But note that the superselection rule does not only refer to a particular ground state and its "mirror-
image", but *to all states (stationary or nonstationary) in the respective sector and their "mirror-
images"*. Hence the condition critcized demands only that there is a ground state in the sector studied.

[2] This should at least be so in physically relevant models.

Another point is the two-level approximation. It is based on the view that higher levels do not disturb too much the reasonings and calculations proposed here in context of the spin-boson model. This might or might not be true. An interesting conjecture is presented in [32]. There the following spin J-boson Hamiltonian

$$H = \hbar \frac{\varepsilon}{J} \, \sigma_1 \otimes 1 \tag{3a}$$

$$+ \, 1 \otimes \int_{\mathbb{R}^3} \hbar \omega(k) \, a_k^* a_k \, dk \tag{3b}$$

$$+ \hbar \frac{1}{J} \, \sigma_3 \otimes \int_{\mathbb{R}^3} \lambda(k) \{ a_k^* + a_k \} \, dk \tag{3c}$$

is discussed. The matrices $\sigma_j, j = 1, 2, 3$, are spin J matrices with $[\sigma_x, \sigma_y] = i \, \sigma_z$ plus cyclic permutations. The Hamiltonian (3) corresponds to a $(2J+1)$-level molecule where the energy difference is the same (namely $\hbar\varepsilon$) for arbitrary neighbored levels. The promised conjecture for the ohmic situation (proven in the subohmic case) now states that with increasing J the phase diagram of this system approaches the phase diagram proposed by Pfeifer.

4. IS MOLECULAR CHIRALITY INDUCED BY COLLISIONS WITH NEIGHBOR MOLECULES?

Collisions of a given chiral molecule with other particles may be responsible for the destabilization of the superposition of handed states (cf. [18], [11] and [23]). The dynamics proposed ordinarily consists of two parts, namely the free dynamics of the isolated molecule and some particular dynamics describing collisions. It is then a standard assumption that a collision of the considered chiral molecule produces a *"dephasing"*, i.e., the molecule is *localized* in one of the wells of its double-well potential.

The dynamics used is a linear density operator dynamics, possibly a semigroup, e.g., of Bloch type. Taking the "localized" states, say Ψ_L and Ψ_R, as basis vectors, the "dephasing" is described by the shrinking of the outer-diagonal elements of the respective density matrix.

The dephasing-assumption is perhaps supported by experimental results (cf. [33-35] and [2]). The first three cited papers refer to the pressure-dependence of the ammonia inversion spectrum, the last one to molecular beam experiments. Ammonia under high pressure shows a "collision"-broadening of the inversion line and the approximate resonant frequency tends to zero at a pressure of about 2 atmospheres. One interpretation of these facts might be that the totally symmetric form of ammonia is destabilized by collisions. Unfortunately, its *theoretical* support is relatively weak. This is by no way a shortcoming of the authors advocating that approach. The deeper reason for this is of course that a molecule's localization is a *measurement-type process* which is notoriously difficult to derive.

It would be desirable to replace the density-matrix dynamics by a dynamics of *pure* states in the sense of (1) (cf., e.g., the dynamics commented upon in the next section). Recall that the decomposition of a mixed state into pure states is highly nonunique. Hence the shrinking of the outer-diagonal elements of the density matrix *with respect to a certain basis* need not necessarily be interpreted as a dephasing with respect to this basis.

5. *STOCHASTIC PERTURBATION OF MOLECULAR PARAMETERS AND REACTION FIELDS*

Collisions with neighbor particles may be described by the perturbation of the molecule's parameters. One might, for example, think of a double-minimum electronic ground state potential which is perturbed in a stochastic way. This gives rise to some stochastic differential equation which sends pure states (of the isolated molecule) into pure states. The respective dynamics need not be linear, i.e., it might be that it is *impossible* to extend the solution of the stochastic differential equation in a linear way from atomic projections (corresponding to pure states) to arbitrary density matrices.

Depending on the particular stochastic process, its coupling to the molecule and the molecular parameters, one might expect that localization takes place in the sense of (1). The results of [36-39] might be interpreted as a hint in that direction. In these papers it is shown that *slightly perturbed* double- or multiple-well potentials have localized eigenfunctions. Of course, these results cannot be directly applied to the problem of chirality *as defined above*, since they do in no way forbid the superposition of localized (eigen-) states. A similar remark could be made concerning solutions of the chirality-problem based on *weak interactions / parity violation / weak neutral current* (cf. [40], [10] and [41]). If the Hamiltonian looses its space-inversion symmetry[3] by weak interaction terms, then the eigenfunctions have a tendency to be "localized" but superposition is *not* forbidden !

In [20] a mechanism for the stabilization of localized states is presented: It is argued that localization leads to a nonzero dipole moment $\mu = <\hat{\mu}>$; this dipole moment polarizes the surroundings which, in turn, creates a *reaction field* $\mathscr{E}_R$ which is collinear with $\hat{\mu}$ so that the interaction $-\hat{\mu}\,\mathscr{E}_R$ is negative and tends to stabilize the nonsymmetric state under consideration. It would be interesting to discuss this approach with a semirealistic model.

The derivation of a stochastic differential equation for a two-level system is given in [16] and [42]. There the starting point is a spin-boson Hamiltonian. The respective (von Neumann-Liouville) differential equation for spin- and boson operators can be transformed into an ordinary stochastic differential equation for expectation values *if the quantum environment is replaced by a classical one, i.e., if it is assumed that the algebra of observables describing the environment is a commutative algebra.* Equivalently [43], one had to assume that all pure states of the joint system {molecule and environment} factorize, i.e., are product states with respect to molecular and environmental observables. This is indispensible if one wants the molecule (the spin) to be an *object*, i.e., if one requests that there are *no* EPR-correlations between the molecule and its environment. As a result one gets a Schrödinger-Langevin stochastic differential equation with a feedback mechanism (a *reaction field*) which may possibly describe measurement-type processes required above.

---

[3] One should not speak of a *broken* symmetry here. "Broken symmetry" means that the Hamiltonian is invariant under the discussed symmetry but not its eigenstates (e.g., ground states or thermic states).

6. CONCLUDING REMARKS

The "problem of chirality" refers to the more general problem of reconciling quantum mechanics and the commonplace experience which suggests that classical observables (with dispersion-free expectation values) exist. Here it is regarded as a matter of (in)stability: Superpositions of handed states are thought to be unstable under *outer* perturbations, i.e., under the influence of the molecular environment. Though plausible, this view has still the status of a conjecture. None of the mentioned environments have been shown to generate classical observables in the sense of (1) or (2).

The experimental situation is not convincing, either. All presently available experimental evidence is compatible with the idea that chirality – and other above mentioned candidates – are classical observables. Nevertheless it might be that just nobody really tried to prepare a superposition of handed states. Eventually these unstable states could be stabilized by screening all outer influences (radiation field, collisions, etc.).

ACKNOWLEDGMENT

I am indebted to Prof. H. Primas and Prof. M. Quack for various discussions.

REFERENCES

[1] Hund, F.: *Zur Deutung der Molekelspektren III*, Z. Phys. **43**, 805 - 826 (1927).

[2] Kukolich, S. G., J. H. S. Wang and D. E. Oates: *Molecular beam maser measurements of relaxation cross sections in NH$_3$*, Chem. Phys. Lett. **20**, 519 - 524 (1973).

[3] Einstein, A.: *Elementare Überlegungen zur Interpretation der Quantenmechanik*, In: *Scientific Papers, presented to Max Born*. Edinburgh. Oliver Boyd (1953),

[4] Born, M.: *Albert Einstein, Hedwig und Max Born, Briefwechsel 1916 - 1955*, München. Nymphenburger Verlagshandlung (1969).

[5] Schrödinger, E.: *Discussion of probability relations between separated systems*, Proc. Cambr. Phil. Soc. **31**, 555 - 563 (1935).

[6] Schrödinger, E.: *Die gegenwärtige Situation in der Quantenmechanik*, Naturwissenschaften **23**, 807 - 812, 823 - 828, 844 - 849 (1935).

[7] Schrödinger, E.: *Probability relations between separated systems*, Proc. Cambr. Phil. Soc. **32**, 446 - 452 (1936).

[8] Primas, H.: *Zur Quantenmechanik makroskopischer Systeme*, In: *Wieviele Leben hat Schrödingers Katze ?* Ed. by J. Audretsch and K. Mainzer. Mannheim. B.I.-Wissenschaftsverlag (1990),

[9] Amann, A.: *Molecules coupled to their environment*, in this same volume.

[10] Quack, M.: *On the measurement of the parity violating energy difference between enantiomers*, Chem. Phys. Lett. **132**, 147 - 153 (1986).

[11] Quack, M.: *Structure and dynamics of chiral molecules*, Angew. Chem. Int. Ed. Engl. **28**, 571 - 586 (1989).

[12] Ford, G. W., J. T. Lewis and R. F. O'Connell: *Quantum Langevin equation*, Phys. Rev. A **37**, 4419 - 4428 (1988).

[13] Leggett, A. J., S. Chakravarty, A. T. Dorsey, M. P. A. Fisher, A. Garg and W. Zwerger: *Dynamics of the dissipative two-state system*, Rev. Mod. Phys. **59**, 1 - 85 (1987).

[14] Smedt, P. d., D. Dürr, J. L. Lebowitz and C. Liverani: *Quantum system in contact with a thermal environment: Rigorous treatment of a simple model*, Commun. Math. Phys. **120**, 195 - 231 (1988).

[15] Ford, G. W., M. Kac and P. Mazur: *Statistical mechanics of assemblies of coupled oscillators*, J. Math. Phys. **6**, 504 - 515 (1965).

[16] Primas, H.: *Induced nonlinear time evolution of open quantum objects*, To be published in: *Proceedings of the NATO ASI "Sixty-two Years of Uncertainty: Historical, Philosophical, Physics Inquiries into the Foundations of Quantum Physics", Erice (Italy), 5. - 15. August 1989*. Ed. by A. I. Miller. New York. Plenum Press (1990),

[17] Funck, P.: *Die Landau-Lifschitz-Gleichung als nichtlineare Schrödingergleichung. Diplomarbeit ETH-Zürich*. Unpublished (1989).

[18] Harris, R. A. and L. Stodolsky: *On the time-dependence of optical activity*, J. Chem. Phys. **74**, 2145 - 2155 (1981).

[19] Harris, R. A. and R. Silbey: *On the stabilization of optical isomers through tunneling friction*, J. Chem. Phys. **78**, 7330 - 7333 (1983).

[20] Claverie, P. and G. Jona-Lasinio: *Instability of tunneling and the concept of molecular structure in quantum mechanics: The case of pyramidal molecules and the enantiomer problem*, Phys. Rev. A **33**, 2245 - 2253 (1986).

[21] Meyer, R. and R. R. Ernst: *Hydrogen transfer in double minimum potential: Kinetic properties derived from quantum dynamics*, J. Chem. Phys. **86**, 784 - 801 (1987).

[22] Nielaba, P., J. L. Lebowitz, H. Spohn and J. L. Vallés: *Behavior of a quantum particle in contact with a classical heat bath*, J. Stat. Phys. **55**, 745 - 767 (1989).

[23] Silbey, R. and R. A. Harris: *Tunneling of molecules in low-temperature media: an elementary description*, J. Phys. Chem. **93**, 7062 - 7071 (1989).

[24] Einstein, A., B. Podolsky and N. Rosen: *Can quantum-mechanical description of physical reality be considered complete ?*, Phys. Rev. **47**, 777 - 780 (1935).

[25] Aspect, A., J. Dalibard and G. Roger: *Experimental realization of Einstein-Podolsky-Rosen-Bohm Gedankenexperiment: A new violation of Bell's inequalities*, Phys. Rev. Lett. **49**, 91 - 94 (1982).

[26] Primas, H.: *Chemistry, Quantum Mechanics, and Reductionism. Perspectives in Theoretical Chemistry*, 2nd revised ed. Berlin. Springer (1983).

[27] Primas, H.: *Mathematical and philosophical questions in the theory of open and macroscopic quantum systems*, To be published in: *Proceedings of the NATO ASI "Sixty-two Years of Uncertainty: Historical, Philosophical, Physics Inquiries into the Foundations of Quantum Physics", Erice (Italy), 5.- 15. August 1989*. Ed. by A. I. Miller. New York. Plenum Press (1990),

[28] Pfeifer, P.: *Chiral Molecules - a Superselection Rule Induced by the Radiation Field*, Zürich. Thesis ETH-Zürich No. 6551, ok Gotthard S+D AG (1980).

[29] Cohen-Tannoudji, C., J. Dupont-Roc and G. Grynberg: *Photons and Atoms*, New York. John Wiley & Sons (1989).

[30] Amann, A.: *Molecules Coupled to Their Environment. Habilitationsarbeit, ETH-Zürich*. Unpublished (1990).

[31] Amann, A.: *Chirality as a classical observable in algebraic quantum mechanics*, In: *Fractals, Quasicrystals, Chaos, Knots and Algebraic Quantum Mechanics*. Ed. by A. Amann, L. Cederbaum and W. Gans. Dordrecht. Kluwer (1988), Pp. 305 - 325.

[32] Spohn, H., R. Stückl and W. Wreszinski: *Localisation for the spin J-boson Hamiltonian*, preprint (1990).

[33] Bleaney, B. and J. H. N. Loubser: *Collision broadening of the ammonia inversion spectrum at high pressures*, Nature **161**, 522 - 523 (1948).

[34] Anderson, P. W.: *On the anomalous line-shapes in the ammonia inversion spectrum at high pressures*, Phys. Rev. **75**, 1450 (1949).

[35] Margenau, H.: *Inversion frequency of ammonia and molecular interaction*, Phys. Rev. **76**, 1423 - 1429 (1949).

[36] Jona-Lasinio, G., F. Martinelli and E. Scoppola: *New approach to the semiclassical limit of quantum mechanics*, Commun. Math. Phys. **80**, 223 - 254 (1981).

[37] Graffi, S., V. Brecchi and G. Jona-Lasinio: *Tunneling instability via perturbation theory*, J. Phys. A **17**, 2935 - 2944 (1984).

[38] Simon, B.: *Semiclassical analysis of low lying eigenvalues. IV. The flea on the elephant*, J. Functional Analysis **63**, 123 - 136 (1985).

[39] Jona-Lasinio, G. and P. Claverie: *Symmetry breaking and classical behaviour*, Progr. Theor. Phys. Suppl. **86**, 54 - 59 (1986).

[40] Harris, R. A. and L. Stodolsky: *Quantum beats in optical activity and weak interactions*, Phys. Lett. B **78**, 313 - 317 (1978).

[41] Barron, L. D.: *Symmetry and molecular chirality*, Chem. Soc. Rev. **15**, 189 - 223 (1986).

[42] Primas, H.: *The measurement process in the individual interpretation of quantum mechanics*, In: *Quantum Theory without Reduction*. Ed. by M. Cini and J.-M. Lévy-Leblond. Bristol. IOP Publishing Ltd. (1990),

[43] Raggio, G.: *States and Composite Systems in W*-Algebraic Quantum Mechanics*, Zürich. Thesis ETH-Zürich No. 6824, ADAG AG (1981).

CONDENSED COOPER PAIRS
AND MACROSCOPIC QUANTUM PHENOMENA

Alfred Rieckers

Institut für Theoretische Physik
D-7400 Tübingen

Abstract

We give in the first section an introduction into the basic notions of operator algebraic quantum theory with the emphasis on the state space of the quasi-local algebra. Representation theory and decomposition theory are related with the macroscopic distinguishability (disjointness) of states. In section 2 we use this formalism to discuss superconductivity in terms of a gauge covariant BCS-model in which the operators of the macroscopic phase and of the condensed Cooper pairs play a decisive role. Two weakly coupled BCS-superconductors constitute a model for the Josephson junction, in which the condensed Cooper pairs tunnel across the isolation barrier of the junction and provide one component of the two-fluid model. The Josephson relations are microscopically derived as operator equations.

In the last section the problem of a coherent superposition of two junction states which are macroscopically different (by having different phase differences), leads to an unconventional extension of the observable algebra to include the number operators which have been shown to be incompatible with the phase operators. In this extended framework the semi-phenomenological energy terms (connected with the capacitance and self-inductance) can be added to the previous junction Hamiltonian. We then have a rigorous formalism by means of which the actual discussions of macroscopic quantum phenomena can be reconsidered, clarified, and confirmed to a surprisingly large extent.

Large-Scale Molecular Systems, Edited by W. Gans *et al.*
Plenum Press, New York, 1991

1. Operatoralgebraic Quantum Theory

Traditional Quantum Mechanics

The statistical set up of traditional quantum mechanics is formulated in terms of a separable Hilbert space $\mathcal{H}$ [1,2]. If one considers all bounded observables, then they usually make up all of $B_{sa}(\mathcal{H})$, the self-adjoint part of all bounded operators $B(\mathcal{H})$ in $\mathcal{H}$. This is so, if the system consists of a fixed number of various microscopic particles. The set of all considered states is assumed to be given by all density matrices, which are mathematically trace class operators $\rho \in \mathcal{T}(\mathcal{H})$ with $\mathrm{tr}[\rho] = 1$ (*normalization*) and $\mathrm{tr}[\rho A^* A] \geq 0$ for all $A \in B(\mathcal{H})$ (*positivity*), and constitute the set $\mathcal{T}^1_+(\mathcal{H})$. Since the expectation of a (generalized) observable $A \in B(\mathcal{H})$ in the state ρ is $\mathrm{tr}[\rho A]$, normalization means that the unit operator obtains the expectation value unity and positivity guarantees a positive expectation for all positive observables $B = A^* A$, the latter form being equivalent for B to have a positive spectrum.

The statistical *description* for *traditional quantum mechanics* is, therefore, the tripel

$$(\mathcal{T}^1_+(\mathcal{H}), \ B(\mathcal{H}), \ tr[\cdot\cdot]) \ . \tag{1.1}$$

Every $\rho \in \mathcal{T}^1_+(\mathcal{H})$ has a representation

$$\rho = \sum_{n=1}^{\infty} \lambda_n |\Phi_n)(\Phi_n| \tag{1.2}$$

where $\lambda_n \in [0,1]$, $\sum_{n=1}^{\infty} \lambda_n = 1$ and $\Phi_n \in \mathcal{H}$ with $\|\Phi_n\| = 1$. A representation of the type (1.2) is only unique, if $(\Phi_n|\Phi_m) = 0$ for $n \neq m$, and if the λ_n are not degenerate, and is then the spectral representation. If the latter condition is dropped, then there are infinitely many representations of the form (1.2) for a given ρ, if ρ is not simply a one-dimensional projection $|\Phi)(\Phi|$, $\Phi \in \mathcal{H}$, $\|\Phi\| = 1$ [3]. In the latter case ρ is called *pure*, since it cannot be further decomposed. If $\rho_{1/2} \in \mathcal{T}^1_+(\mathcal{H})$ and $\lambda \in [0,1]$, then $\rho = \lambda \rho_1 + (1-\lambda)\rho_2$ is again in $\mathcal{T}^1_+(\mathcal{H})$ and is called an incoherent superposition (or *mixture*) of ρ_1 and ρ_2. Since the set of mixtures of ρ_1 and ρ_2 is geometrically the line segment between ρ_1 and ρ_2, we remark that $\mathcal{T}^1_+(\mathcal{H})$ contains all line segments between its points and thus is a *convex set*. The pure states are by definition just these points of the convex set which are never interior points of line segments in $\mathcal{T}^1_+(\mathcal{H})$ and constitute the so-called *extreme boundary* $\partial_e \mathcal{T}^1_+(\mathcal{H})$ of $\mathcal{T}^1_+(\mathcal{H})$. (We use here the notation of [4]. In [5] the extreme boundary is denoted by an E.)

If $\rho_{1/2} \in \partial_e T_+^1(\mathcal{H})$ then there are $\Phi_{1/2} \in \mathcal{H}$ (unique up to phases) such that $\rho_{1/2} = |\Phi_{1/2})(\Phi_{1/2}|$. A *coherent superposition* of ρ_1 and ρ_2 is a *pure* state of the form $\rho = |\Phi)(\Phi|$, where $\Phi = c_1 \Phi_1 + c_2 \Phi_2$ for some $c_{1/2} \in \mathbb{C}$. If $\rho_1 \neq \rho_2$, then Φ_1 and Φ_2 are not linearly dependent and the vectors Φ of the coherent superpositions are all of the unit vectors in the (complex) plane which is spanned by Φ_1 and Φ_2. The same set is obtained if Φ_2 is replaced by a vector in this plane which is orthogonal to Φ_1 (again denoted by Φ_2) from which $\|\Phi\|^2 = |c_1|^2 + |c_2|^2 = 1$ results. Cancelling a common phase of c_1 and c_2 it remains a three-parametric variety of coefficients which is isomorphic to the surface of a unit ball with real dimension 3. Forming all mixtures of the corresponding pure states we arrive at a convex norm-closed subset $E(\rho_1, \rho_2)$ of $T_+^1(\mathcal{H})$ which is - in a sense specified below - isomorphic to the whole of the three-dimensional unit ball $B(3)$.

By construction, $E(\rho_1, \rho_2)$ consists of those density matrices which have a decomposition (1.2) into pure states of the $\Phi_1 - \Phi_2$ plane. Since the coefficients λ_n are positive, *any* (1.2)-decomposition of such a $\rho \in E(\rho_1, \rho_2)$ contains pure states only of the $\Phi_1 - \Phi_2$-plane. This tells us that $E(\rho_1, \rho_2)$ is a so-called *face* of $T_+^1(\mathcal{H})$: If $E(\rho_1, \rho_2) \ni \rho = \lambda \rho_1' + (1 - \lambda)\rho_2'$ for $\lambda \in [0, 1]$, $\rho_{1/2}' \in T_+^1(\mathcal{H})$, then $\rho_{1/2}' \in E(\rho_1, \rho_2)$. Faces are closed under decompositions and make up the state spaces of statistical sub-theories. Since according to Dirac's unrestricted superposition principle[6] any (traditional) quantum theory should have a state space which is closed under the coherent-superpositions of its pure states, $E(\rho_1, \rho_2)$ is the smallest state space, which contains ρ_1 and ρ_2, of a sub-theory.

Two-Level Systems

Let us study the three-ball state space in more analytical terms. The smallest non-trivial Hilbert space is $\mathbb{C}^2$ and $\mathcal{B}(\mathbb{C}^2) = T(\mathbb{C}^2) = \mathcal{M}^{(2)}$, the set of all 2×2-matrices with basis $\{\sigma^i; 0 \leq i \leq 3\}$, where $\sigma^0 = 1$ and σ^i, $1 \leq i \leq 3$, are the usual Pauli spin matrices. If $\rho = \sum_{i=0}^3 x_i \sigma^i$ shall be in $T_+^1(\mathbb{C}^2)$, then normalization implies $x_0 = 1/2$. Positivity is equivalent to positive eigenvalues and for 2×2-matrices with a positive trace equivalent to a positive determinant. Since $\det(\rho) = \det \begin{pmatrix} \frac{1}{2} + x_3 & x_1 - ix_2 \\ x_1 + ix_2 & \frac{1}{2} - x_3 \end{pmatrix} = \frac{1}{4} - \sum_{i=1}^3 x_i^2$, positivity requires that (x_1, x_2, x_3) be within the real 3-ball with radius $1/2$. If $\lambda \rho + (1 - \lambda)\rho'$ is a mixture, the corresponding 3-vector is $\lambda(x_1, x_2, x_3) + (1 - \lambda)(x_1', x_2', x_3')$, which characterizes an *affine* mapping.

It is now clear that the mapping

$$\mathcal{T}^1_+(\mathbb{C}^2) \ni \rho = \sum_{i=0}^{3} x_i \sigma^i \to 2(x_1, x_2, x_3) \in B(3) \tag{1.3}$$

is an affine bijection which maps the extreme boundary $\partial_e \mathcal{T}^1_+(\mathbb{C}^2)$ onto $\partial_e B(3)$, the latter coinciding with the surface of $B(3)$.

$B(3)$ is not only a convex set but also a topological space (with, e.g., the metric topology). The smallest σ-algebra, which contains all open sets, is made up by the so-called Borel sets. (A measure theory on state spaces is described in [5], Ch. 4.1.) If μ is any Borel probability measure on $B(3)$, we can form the *bary-center* for this "mass"-distribution, which again is in $B(3)$ in virtue of convexity. This integration may be transferred to $\mathcal{T}^1_+(\mathbb{C}^2)$ by the inverse mapping of (1.3) to give

$$\bar{\rho} := \int_{\mathcal{T}^1_+} \rho d\mu(\rho) \in \mathcal{T}^1_+(\mathbb{C}^2) . \tag{1.4}$$

Thus, mixing of states is possible by means of *any* probability measure on the state space, discrete or continuous. Let us mix rotated states!

If $\mathbb{C}^2$ signifies the state space spanned by one lower energy eigenstate $\chi^1 = \begin{pmatrix} 1 \\ 0 \end{pmatrix}$ and one higher energy eigenstate $\chi^2 = \begin{pmatrix} 0 \\ 1 \end{pmatrix}$ (as, e.g., for a 2-level atom), the raising operator $\sigma^+ = \begin{pmatrix} 0 & 0 \\ 1 & 0 \end{pmatrix}$ maps χ^1 onto χ^2, and the lowering operator $\sigma^- = (\sigma^+)^* = \begin{pmatrix} 0 & 1 \\ 0 & 0 \end{pmatrix}$ gives the reverse transformation. The basis elements of $\mathcal{M}^{(2)}$ may be written in terms of these operators:

$$\sigma^0 = 1 = \sigma^- \sigma^+ + \sigma^+ \sigma^- \tag{1.5}$$

$$\sigma^1 = (\sigma^- + \sigma^+) \tag{1.6}$$

$$\sigma^2 = (\sigma^- - \sigma^+)/i \tag{1.7}$$

$$\sigma^3 = \sigma^- \sigma^+ - \sigma^+ \sigma^- . \tag{1.8}$$

The *symmetries* we consider in this lecture correspond only to unitary operators U (and not to anti-unitary operators which are also possible according to Wigner's theorem[7,8,9,10,11,12]). In the Heisenberg picture they give the transformation

$$\alpha : \mathcal{B}(\mathcal{H}) \to \mathcal{B}(\mathcal{H}) \tag{1.9}$$

$$\alpha(A) := U \, A \, U^*, \quad A \in \mathcal{B}(\mathcal{H}) .$$

Obviously α is linear, adjoint preserving ($\alpha(A^*) = \alpha(A)^*$) and multiplicative ($\alpha(AB) = \alpha(A)\alpha(B)$). We say α is a *-*automorphism* of $\mathcal{B}(\mathcal{H})$, and write $\alpha \in Aut(\mathcal{B}(\mathcal{H}))$. A gauge transformation of the first kind changes the phase of σ^+:

$$\alpha_\vartheta(\sigma^+) := e^{i\vartheta}\sigma^+ \ , \quad \vartheta \in [0, 2\pi) \ . \tag{1.10}$$

By the stipulation that α_ϑ be in $Aut(\mathcal{M}^{(2)})$ we find

$$\begin{aligned}
\alpha_\vartheta(\sigma^-) &= e^{-i\vartheta}\sigma^- \\
\alpha_\vartheta(\sigma^0) &= \sigma^0 \\
\alpha_\vartheta(\sigma^1) &= \cos\vartheta \ \sigma^1 + \sin\vartheta \ \sigma^2 \\
\alpha_\vartheta(\sigma^2) &= -\sin\vartheta \ \sigma^1 + \cos\vartheta \ \sigma^2 \\
\alpha_\vartheta(\sigma^3) &= \sigma^3 \ .
\end{aligned} \tag{1.11}$$

Thus, for calculations, there is no much need for the corresponding (*implementing*) unitary $U_\vartheta = \begin{pmatrix} 1 & 0 \\ 0 & e^{i\vartheta} \end{pmatrix}$. The gauge transformations in the Schrödinger picture ν_ϑ are chosen to cancel the Heisenberg action:

$$\mathrm{tr}[\nu_\vartheta(\rho)\alpha_\vartheta(A)] = \mathrm{tr}[\rho \ A] \qquad \begin{array}{l} A \in \mathcal{B}(\mathcal{H}) \\ \rho \in \mathcal{T}^1_+(\mathcal{H}) \ . \end{array} \tag{1.12}$$

Thus,

$$\nu_\vartheta(\rho) = U^*_{-\vartheta}\rho \ U_{-\vartheta} = U_\vartheta\rho \ U^*_\vartheta \ , \tag{1.13}$$

and concerning the dependence of ϑ, both α_ϑ and ν_ϑ represent homomorphically the one-dimensional *torus group* T. The formulas (1.11) can be used to calculate $\nu_\vartheta(\rho)$ explicitly and show that ρ is rotated around the 3-axis.

The gauge average is now (with (1.3))

$$\bar{\rho} = \int_0^{2\pi} \nu_\vartheta(\rho)\frac{d\vartheta}{2\pi} = \frac{1}{2}\sigma^0 + x_3\sigma^3 \ \in \mathcal{T}^1_+(\mathbb{C}^2) \ . \tag{1.14}$$

If ρ has non-vanishing x_1- or x_2-coordinates, (1.14) is a decomposition of the gauge-invariant state $\bar{\rho}$ into the non-invariant states $\{\nu_\vartheta(\rho) \ ; \ \vartheta \in [0, 2\pi)\}$ by means of the continuous measure $d\vartheta/2\pi$ (the Haar measure of T).

Composition of Two-Level Systems

Let us assume here that we have a family of two-level systems which are indexed by $k \in \mathcal{R}$, where $\mathcal{R}$ is denumerable. For every finite subset $\Lambda \subset \mathcal{R}$ we consider the composite system which is formed by the two level systems with indices in Λ. The corresponding Hilbert space is

$$\mathcal{H}_\Lambda := \bigotimes_{k \in \Lambda} \mathbb{C}_k^2 \ . \tag{1.15}$$

The algebra of observables has the form

$$\mathcal{A}_\Lambda := \mathcal{B}(\mathcal{H}_\Lambda) = \bigotimes_{k \in \Lambda} \mathcal{M}_k^{(2)} \ . \tag{1.16}$$

Assuming a total ordering in $\mathcal{R}$ (and thus in Λ) we may define the creation operators

$$a_k^* := \bigotimes_{l \leq k-1} \sigma_l^3 \sigma_k^+ \ , \tag{1.17}$$

where l is here restricted to Λ. Then - using (1.5) - we have the *canonical anti-commutation* relations (CAR)

$$a_k a_l^* + a_l^* a_k = \delta_{kl} \, 1 \ , \ \forall \ k, l \in \Lambda \ . \tag{1.18}$$

The traditional description

$$(\mathcal{T}_+^1(\mathcal{H}_\Lambda), \mathcal{A}_\Lambda, \mathrm{tr}[\cdot \cdot]) \tag{1.19}$$

is thus also appropriate for Fermions which have a one-particle Hilbert space of the dimension

$$|\Lambda| := \text{ cardinality of } \Lambda \ . \tag{1.20}$$

Being a traditional quantum theory the unrestricted superposition principle is valid as large as $|\Lambda|$ may be.

Superselection Rules in Hilbert Space Quantum Theories

Assume that the basic Hilbert space has a direct sum decomposition

$$\mathcal{H} = \bigoplus_{n \in M} \mathcal{H}_n \ , \ M \subset \mathbb{N} \ , \tag{1.21}$$

into certain Hilbert spaces $\mathcal{H}_n$, which are viewed as subspaces of $\mathcal{H}$ by a natural embedding. In the case of a Fock space, n is, e.g., the particle number. If one takes now as algebra of observables

$$\mathcal{A} = \bigoplus_{n \in M} \mathcal{B}(\mathcal{H}_n) \ , \tag{1.22}$$

then this is strictly smaller than $\mathcal{B}(\mathcal{H})$, if $|M| > 1$. The description

$$(\mathcal{T}_+^1(\mathcal{H}), \mathcal{A}, tr[\cdot\cdot]) \tag{1.23}$$

looks very much like a traditional one, but has essentially new features. If C_n is the projection of $\mathcal{H}$ onto $\mathcal{H}_n$, then any operator of the form

$$Z = \bigoplus_{n \in M} c_n C_n \ , \ c_n \in \mathbb{C} \ , \tag{1.24}$$

is in $\mathcal{A}$, iff $(c_n) \in l^\infty(M)$ (i.e., is a bounded sequence). Denote by

$$\mathcal{A}' := \{A \in \mathcal{B}(\mathcal{H}); [A, B]_- = 0 \ \forall \, B \in \mathcal{A}\} \tag{1.25}$$

the *commutant* of $\mathcal{A}$ (in the Hilbert space $\mathcal{H}$). It is easily seen that

$$\mathcal{A}' = \{Z \in \mathcal{B}(\mathcal{H}); Z \text{ has the form } (1.24)\} \tag{1.26}$$

$$= \mathcal{A}' \cap \mathcal{A} =: \mathcal{Z} \tag{1.27}$$

where (1.27) defines the *center* of $\mathcal{A}$. For every $\rho \in \mathcal{T}_+^1(\mathcal{H})$ and every $A \in \mathcal{A}$ it holds

$$tr[\rho A] = tr[\rho \bigoplus_n C_n A \bigoplus_m C_m] = tr\left[\left(\bigoplus_n C_n \rho C_n\right) A\right] = \sum_n \lambda_n tr[\rho_{C_n} A] \tag{1.28}$$

with $\lambda_n = tr[\rho C_n]$ and the general definition of a *perturbed* density operator (with $B \in \mathcal{A}$)

$$\rho_B := \left\{ \begin{matrix} B\rho B^* / tr[\rho B^* B] & \text{, if defined} \\ \rho & \text{, otherwise} \end{matrix} \right\} . \tag{1.29}$$

Observe that ρ_{C_n} is the unique density operator in $\mathcal{T}_+^1(\mathcal{H}_n)$, which gives the restriction of ρ to $\mathcal{B}(\mathcal{H}_n)$. Thus (1.28) expresses the unique decomposition of the state given by ρ into states given by elements in $\mathcal{T}_+^1(\mathcal{H}_n)$. In general, however, it does not hold $\rho = \sum_n \lambda_n \rho_{C_n}$. Let, e.g., $\Psi = \bigoplus_n a_n \Psi_n$, $a_n \in \mathbb{C}$, $\Psi_n \in \mathcal{H}_n$ and all vectors be normalized. For $\rho = |\Psi)(\Psi|$ it holds $\rho_{C_n} = |\Psi_n)(\Psi_n|$ and $\lambda_n = |a_n|^2$. Thus the state given by $|\Psi)(\Psi|$ has also the density operator $\sum_n |a_n|^2 |\Psi_n)(\Psi_n|$ and is a mixture (if more than one a_n is non-zero). This is usually considered the most spectacular aspect of the description (1.23) since a linear combination of state vectors in different $\mathcal{H}_n$'s does not lead to a pure state and, therefore, does not satisfy the requirements of a coherent superposition. The observables in $\mathcal{Z}$ are called *super-selection rules*[13,14,15] and are considered to divide $\mathcal{H}$ into the *sector spaces* $\mathcal{H}_n$. Since the observables in $\mathcal{A}$ leave the sectors invariant, the relative phases between different Ψ_n's are not detectable via $\mathcal{A}$-expectations.

To get in touch with the operator algebraic notions let us call here the $\mathcal{T}_+^1(\mathcal{H}_n)$ *sectors* and consider them as subsets of the total state space. A density operator $\rho \in \mathcal{T}_+^1(\mathcal{H})$ is in $\mathcal{T}_+^1(\mathcal{H}_n)$, iff $\rho_{C_n} = \rho$. If for such a ρ there is a decomposition $\rho = \lambda\rho' + (1 - \lambda)\rho''$, then also $\rho' = \rho'_{C_n}$ and $\rho'' = \rho''_{C_n}$, so that $\mathcal{T}_+^1(\mathcal{H}_n)$ is a face of the state space. If $\rho \in \mathcal{T}_+^1(\mathcal{H}_n)$ and $B \in \mathcal{A}$, then the perturbed state ρ_B (1.29) has also the property

$$\rho_{BC_n} = \rho_{C_n B} = \rho_B$$

and thus is again in $\mathcal{T}_+^1(\mathcal{H}_n)$. Such a face is called *invariant* or a *folium*. The decomposition (1.28) of a state into states of the folia of the sectors is a special case of the so-called *central decomposition* (cf. (1.60)ff.).

In this description the superselection rules (or *classical observables*) together with the resulting decomposition of the state space into folia arise by ansatz. One has simply combined different traditional descriptions which refer to essentially different systems, e.g., for systems with different kinds of particles. This appears so artificial that it has been systematically studied rather lately[15].

One main point of this lecture is to demonstrate that merely by the transition to infinitely many degrees of freedom also for a single species of particles superselection sectors show up and that it is a typical collective phenomenon if an equilibrium state has components in different sectors.

Quasi-Local Algebra

Let us take up the composition of two-level systems over the finite index set $\Lambda \subset \mathcal{R}$. The decisive new step which leads from traditional to algebraic quantum theory is to consider not only one region, but the whole family $\mathcal{L} := \{\Lambda \subset \mathcal{R} \; ; \; |\Lambda| < +\infty\}$. In order that the corresponding traditional descriptions in the Λ's combine to a description of a macroscopic system associated with the infinite set $\mathcal{R}$, they have to satisfy compatibility requirements. If $A \in \mathcal{A}_\Lambda$ and $\Lambda \subset \Lambda'$, then A should also be in some sense an observable in $\mathcal{A}_{\Lambda'}$, which is to be expressed by an embedding mapping

$$\eta_{\Lambda',\Lambda} : \mathcal{A}_\Lambda \to \mathcal{A}'_\Lambda$$

$$\eta_{\Lambda',\Lambda}(A) := A \otimes 1_{\Lambda'\setminus\Lambda}, \quad \forall A \in \mathcal{A}_{\Lambda'} .$$

(1.30)

Here 1_Λ is the unit operator in $\mathcal{A}_\Lambda$ and $\Lambda' \setminus \Lambda$ is the complement of Λ in Λ'. The mapping $\eta_{\Lambda',\Lambda}$ is easily seen to preserve the algebraic and $*$-operations and to act injectively, what are by definition the properties of a *-isomorphism*. It holds also for $\Lambda \subset \Lambda' \subset \Lambda''$

$$\eta_{\Lambda'',\Lambda'} \circ \eta_{\Lambda',\Lambda} = \eta_{\Lambda'',\Lambda} \qquad (1.31)$$

and

$$\eta_{\Lambda',\Lambda}(1_\Lambda) = 1_{\Lambda'} \ . \qquad (1.32)$$

The basic elements of the global algebra (connected with $\mathcal{R}$) are now introduced as the local observables of some $\mathcal{A}_\Lambda$, which are iteratively embedded into the algebras of larger and larger regions. In other words, we consider the families

$$(A_\Lambda) := \{A_\Lambda \in \mathcal{A}_\Lambda; \ \Lambda \in \mathcal{L}, \ ex. \ \Lambda_0, \ \text{such that} \ A_\Lambda = \eta_{\Lambda,\Lambda_0}(A_{\Lambda_0}) \ \forall \ \Lambda \supset \Lambda_0\} \ . \quad (1.33)$$

Observe that the elements in (A_Λ) are specified only for $\Lambda \supset \Lambda_0$. The algebraic operations are defined pointwise:

$$c(A_\Lambda) + (B_\Lambda) := (cA_\Lambda + B_\Lambda), \ c \in \mathbb{C}, \qquad (1.34)$$

$$(A_\Lambda)(B_\Lambda) := (A_\Lambda B_\Lambda) \qquad (1.35)$$

$$(A_\Lambda)^* := (A_\Lambda^*) \qquad (1.36)$$

$$\| (A_\Lambda) \| := \lim_\Lambda \| A_\Lambda \|_{\mathcal{A}_\Lambda} \ . \qquad (1.37)$$

Concerning the limit in (1.37), observe that $\mathcal{L}$ is a directed index set by the inclusion relation and thus may index a net, and that $\| \eta_{\Lambda',\Lambda}(A_\Lambda) \|_{\mathcal{A}_{\Lambda'}} = \| A_\Lambda \|_{\mathcal{A}_\Lambda}$ for $\Lambda \subset \Lambda'$ as a consequence of the *-isomorphism property of $\eta_{\Lambda',\Lambda}$ (cf. [5], p. 44). Altogether we have constructed a *-algebra

$$\hat{\mathcal{A}}_0 := \{(A_\Lambda) \ \text{of the form} \ (1.33)\} \ , \qquad (1.38)$$

which is abstract in the sense that its elements are not operators in a Hilbert space. The unit in $\hat{\mathcal{A}}_0$ is (1_Λ) and the null is (0_Λ). $\| (A_\Lambda) \| = 0$ does not imply $A_\Lambda = 0_\Lambda$ for all $\Lambda \in \mathcal{L}$. Thus $\| \cdot \|$ is only a semi-norm. It satisfies, however,

$$\| (A_\Lambda)^*(A_\Lambda) \| = \| (A_\Lambda) \|^2 \ , \qquad (1.39)$$

since $\| A_\Lambda^* A_\Lambda \|_{\mathcal{A}_\Lambda} = \| A_\Lambda \|_{\mathcal{A}_\Lambda}^2$ for all $\Lambda \in \mathcal{L}$. Equation (1.39) is called the C^*-property of a semi-norm. If $\mathfrak{I}_0$ is the two-sided ideal in $\hat{\mathcal{A}}_0$ of elements with vanishing semi-norm, then the *local algebra*

$$\mathcal{A}_0 := \hat{\mathcal{A}}_0 / \mathfrak{I}_0 \qquad (1.40)$$

is a normed *-algebra and its norm completion

$$\mathcal{A} := \bar{\mathcal{A}}_0^{\|\cdot\|} \tag{1.41}$$

is the *quasi-local algebra*. It is a complete *-algebra, the norm of which fulfills the C^*-property, in other words: $\mathcal{A}$ is a C^*-*algebra*[5,16,17,18]. The mathematical construction we have outlined is called the C^*-*inductive limit*[16]. Given any family of C^*-algebras $\mathcal{A}_\Lambda; \Lambda \in \mathcal{L}$, which possesses embedding *-isomorphisms with the properties (1.31) and (1.32), the indicated C^*-inductive limit construction can be performed. In general the $\Lambda \in \mathcal{L}$ are subsets of an overcountable set. (The pioneering article [19] uses $\Lambda \subset \mathbb{R}^4$ for applications in relativistic quantum field theory, whereas $\Lambda \subset \mathbb{R}^3$ in [20] for non-relativistic many-body physics.) This construction provides also the direct embedding *-isomorphisms

$$\begin{aligned} \eta_\Lambda &: \mathcal{A}_\Lambda \to \mathcal{A} \\ \eta_\Lambda(A) &= (A_{\Lambda'}) + \mathfrak{I}_0 \ , \quad A \in \mathcal{A}_\Lambda \ , \end{aligned} \tag{1.42}$$

where

$$A_{\Lambda'} := \left\{ \begin{array}{cc} \eta_{\Lambda',\Lambda}(A) & , \ \text{for } \Lambda' \supset \Lambda \\ 0 & , \ \text{otherwise} \end{array} \right\}$$

It holds for all $\Lambda' \supset \Lambda$

$$\eta_{\Lambda'} \circ \eta_{\Lambda',\Lambda} = \eta_\Lambda \ . \tag{1.43}$$

Since $\mathcal{A}_0$ is viewed as a part of $\mathcal{A}$, we can write for this norm-dense sub-algebra

$$\mathcal{A}_0 = \bigcup_{\Lambda \in \mathcal{L}} \eta_\Lambda(\mathcal{A}_\Lambda) \ . \tag{1.44}$$

In the special case of our two-level systems one writes for the quasi-local algebra

$$\mathcal{A} = \bigotimes_{k \in \mathcal{R}} \mathcal{M}_k^{(2)} \ . \tag{1.45}$$

The direct embedding may here be visualized as

$$\eta_\Lambda(A) = A \otimes 1_{\mathcal{R}\backslash\Lambda} \ , \ A \in \mathcal{A}_\Lambda \ , \tag{1.46}$$

suggesting

$$\eta_\Lambda(1_\Lambda) = 1 \ , \quad \forall \Lambda \in \mathcal{L} \ , \tag{1.47}$$

which in fact also holds for the general quasi-local algebras.

A product element

$$A = \otimes A_k \ , \quad A_k \in \mathcal{M}_k^{(2)} \ ,$$

is in $\mathcal{A}$ only if $\| A_k - 1_k \| \to 0$ for $k \to \infty$.

One can show that $\mathcal{A}$ is *simple* and that it has, therefore, a *trivial center*. Thus, it appears at first sight, that by means of the quasi-local algebra no classical features are describable. These may be discovered, however, as a structure in the state space.

State Space of the Quasi-Local Algebra

Together with the Hilbert space we have lost the density operators as state concept for the quasi-local algebra, but have retained the families $\{ \rho_\Lambda \in \mathcal{T}_+^1(\mathcal{H}_\Lambda) \ ; \ \Lambda \in \mathcal{L} \}$ of local density operators. For consistency it should hold, if $\Lambda \subset \Lambda'$,

$$\mathrm{tr}[\rho_\Lambda A_\Lambda] = \mathrm{tr}[\rho_{\Lambda'} \eta_{\Lambda',\Lambda}(A_\Lambda)] \ \forall \ A_\Lambda \in \mathcal{A}_\Lambda \ , \tag{1.48}$$

which we write

$$\rho_\Lambda = \eta_{\Lambda,\Lambda'}^*(\rho_{\Lambda'}) \ . \tag{1.49}$$

By duality this mapping is continuous in the topologies given by the convergence of the expectation values. If $A \in \mathcal{A}_0$, then by (1.44) there is a Λ and an $A_\Lambda \in \mathcal{A}_\Lambda$, such that $A = \eta_\Lambda(A_\Lambda)$. Since $\eta_\Lambda(A_\Lambda) \subset \eta_{\Lambda'}(A_{\Lambda'})$ for $\Lambda \subset \Lambda'$, there is also an $A_{\Lambda'} \in \mathcal{A}_{\Lambda'}$ with $A = \eta_{\Lambda'}(A_{\Lambda'})$. By the injectivity of η_Λ and $\eta_{\Lambda'}$, the elements A_Λ and $A_{\Lambda'}$ are unique in the respective algebras $\mathcal{A}_\Lambda$ and $\mathcal{A}_{\Lambda'}$. Thus it must hold $A_{\Lambda'} = \eta_{\Lambda',\Lambda}(A_\Lambda)$ (use (1.43)). Thus (1.48) gives a well-defined expectation value

$$< \varphi; A >:= \mathrm{tr}[\rho_\Lambda A_\Lambda] \ , \text{ for } A \in \mathcal{A}_0 \ . \tag{1.50}$$

As a functional on $\mathcal{A}_0$ φ is linear, normalized and positive. One can show that it is then continuous in the norm topology of $\mathcal{A}_0$ and can uniquely be extended to the whole of $\mathcal{A}$ (cf. [5], Prop. 2.3.11).

The set $\mathcal{S} = \mathcal{S}(\mathcal{A})$ of all *states* on $\mathcal{A}$ is given by all linear, normalized, positive functionals. It is a convex, w^*-compact set (where the w^*-topology is given by the convergence of the expectation values). The subset $\mathcal{S}_0$ of $\mathcal{S}$, with elements defined by a compatible family of local density matrices - satisfying (1.49) - consists of the so-called *locally normal states*. If the $\mathcal{A}_\Lambda$ are finite-dimensional, it holds $\mathcal{S}_0 = \mathcal{S}$.

The most important method to construct a compatible family of local density matrices goes via the thermodynamic limit. One starts with a family $\{\sigma_\Lambda \in \mathcal{T}_+^1(\mathcal{H}_\Lambda) \; ; \; \Lambda \in \mathcal{L}\}$ which satisfies

$$\lim_{\substack{K \in \mathcal{L}}} \eta_{\Lambda',K}^*(\sigma_K) =: \rho_{\Lambda'} \; , \; \forall \, \Lambda' \in \mathcal{L} \; , \qquad (1.51)$$

where $\rho_{\Lambda'} \in \mathcal{T}_+^1(\mathcal{H}_{\Lambda'})$, if the limit exists. Then one obtains for $\Lambda \subset \Lambda'$

$$\begin{aligned}
\eta_{\Lambda,\Lambda'}^*(\rho_{\Lambda'}) &= \eta_{\Lambda,\Lambda'}^*(\lim_K \eta_{\Lambda',K}^*(\sigma_K)) \\
&= \lim_K \eta_{\Lambda,\Lambda'}^* \circ \eta_{\Lambda',K}^*(\sigma_K) \\
&= \lim_K \eta_{\Lambda,K}^*(\sigma_K) = \rho_\Lambda \; ,
\end{aligned}$$

where the continuity of the $\eta_{\Lambda,\Lambda'}^*$ and the dual of (1.31) have been employed.

The calculation of (1.51) is a central task of quantum statistical mechanics, especially for equilibrium states. If

$$\sigma_\Lambda = \exp[-\zeta_\Lambda - \beta H_\Lambda] \; , \; \Lambda \in \mathcal{L} \; , \qquad (1.52)$$

(where $\zeta_\Lambda \in \mathbb{R}$ is determined by the normalization) and (1.51) holds for an *absorbing sub-family* $\mathcal{N} \subset \mathcal{L}$ (which eventually dominates every $\Lambda \in \mathcal{L}$), then the state determined by the ρ_Λ's is called *limiting Gibbs state*. Its calculation is so difficult that many methods in quantum field theory and many body physics have been invented to circumvent the direct proof of (1.51).

Simple but useful examples for (1.51) are provided by product states. Let be specified an *arbitrary* sequence $\{\rho_k \in \mathcal{T}_+^1(\mathbb{C}^2) \; ; \; k \in \mathcal{R}\}$ and set

$$\rho_\Lambda := \bigotimes_{k \in \Lambda} \rho_k \; , \; \forall \, \Lambda \in \mathcal{L} \; . \qquad (1.53)$$

Then (1.51) is satisfied with $\sigma_K \equiv \rho_K$. Especially if $\rho_k = \rho \in \mathcal{T}_+^1(\mathbb{C}^2)$, for all $k \in \mathcal{R}$, then call the resulting state $\varphi(\rho) \in \mathcal{S}$.

For further analysis of the convex set $\mathcal{S}$ let us consider certain of its *faces* (i.e. convex subsets $E \subset \mathcal{S}$, so that $E \ni \varphi = \lambda\varphi_1 + (1 - \lambda)\varphi_2 \; , \; \varphi_{1/2} \in \mathcal{S} \; , \; \lambda \in [0,1]$ implies $\varphi_{1/2} \in E$). Since the elements of $\mathcal{S}$ are norm-continuous functionals, their linear combinations are bounded and have a finite norm. We only consider faces which are closed in this norm topology. If $\mathcal{T} \subset \mathcal{S}$, then set

$$E(\mathcal{T}) := \{smallest \; (norm - closed \; face) \; containing \; \mathcal{T} \; \} \; . \qquad (1.54)$$

A state $\varphi \in \mathcal{S}$ is in the extreme boundary $\partial_e \mathcal{S}$, iff $E(\varphi) = \{\varphi\}$ (the one-point set), and then is called *pure*. For $\varphi \in \mathcal{S}$ and $B \in \mathcal{A}$ define the perturbed state φ_B (by its action on $A \in \mathcal{A}$)

$$< \varphi_B;\ A >:= \left\{ \begin{matrix} < \varphi;\ B^*AB > / < \varphi;\ B^*B > & ,\ \text{if defined} \\ < \varphi;\ A > & ,\ \text{otherwise} \end{matrix} \right\}. \qquad (1.55)$$

A face which is invariant under all perturbations with $B \in \mathcal{A}$ is called *folium*[21] and denoted by F. It generalizes the notion of a superselection sector. For $\varphi \in \mathcal{S}$ denote

$$F(\varphi) := \{ smallest\ folium\ containing\ \varphi \}. \qquad (1.56)$$

Verify that

$$F(\varphi) = E(\{\varphi_B;\ B \in \mathcal{A}\})\ ! \qquad (1.57)$$

If $\rho, \rho' \in \mathcal{T}_+^1(\mathbb{C}^2)$, $\rho \neq \rho'$, then you cannot get by a perturbation from one product state $\varphi(\rho)$ to the other $\varphi(\rho')$, so that it follows

$$F(\varphi(\rho)) \cap F(\varphi(\rho')) = \varnothing . \qquad (1.58)$$

Thus we have overcountably-many non-intersecting folia in $\mathcal{S}$, if $\mathcal{A}$ is $\otimes \mathcal{M}_k^{(2)}$.

The analogous assertion for a general quasi-local algebra is obtained by identifying in it a C^*-sub-algebra which is isomorphic to $\otimes \mathcal{M}_k^{(2)}$ [17]. Many of these non-intersecting folia are w^*-dense in $\mathcal{S}$, an interwoven structure which is beyond imagination.

Why is the appearance of so many disjoint folia in the state space of the quasi-local algebra of utmost conceptual importance? The answer is that we have, by this form of the thermodynamic limit, now clear-cut notions to compare the states from the macroscopic-classical point of view and find here a rich structure merely by the largeness of the system. From the very definition of the folia it follows that two states $\varphi, \psi \in \mathcal{S}$ in different folia are macroscopically different. More precisely we define: φ and ψ are *disjoint*, if $F(\varphi) \cap F(\psi) = \varnothing$, are *quasi-equivalent*, if $F(\varphi) = F(\psi)$, and φ is *quasi-contained* in ψ, if $F(\varphi) \subset F(\psi)$. If we denote

$$\mathcal{F} := \{ F \subset \mathcal{S};\ F\ \text{is a folium} \} \cup \varnothing , \qquad (1.59)$$

then the inclusion relation induces a partial order in $\mathcal{F}$, which makes it to a complete lattice, with largest element $\mathcal{S}$ and least element $\varnothing$. This lattice is distributive and has a complement relation (negation) in a classical fashion: for $F \in \mathcal{F}$, $F^\perp$ is the unique element in $\mathcal{F}$ with $F \wedge F^\perp = \emptyset$ and $F \cup F^\perp = \mathcal{S}$. Altogether $\mathcal{F}$ is a Boolean algebra, and this demonstrates undoubtedly that the folia express classical properties.

(Here *property* may even be taken as a formalized notion, namely as an element of the (non-Boolean) lattice $\mathcal{E}$ of all (norm-closed) faces in $\mathcal{S}$. This lattice is of the same structure as the projection lattice of quantum logics, has a typical non-classical orthocomplement and is orthomodular. $\mathcal{F}$ is a Boolean sub-lattice of $\mathcal{E}$, that is $\mathcal{F} \subset \mathcal{E}$ and all lattice operations of $\mathcal{F}$ are those of $\mathcal{E}$ restricted to $\mathcal{F}$. A characteristic feature of $\mathcal{E}$ is that for two pure states $\varphi, \psi \in \partial_e \mathcal{S}$ the faces $E(\varphi, \psi) = E(\varphi) \vee E(\psi)$ are affinely isomorphic to $B(3)$, iff φ and ψ are quasi-equivalent, just as in the case of Hilbert space quantum theory.)

We have thus in the case of a quasi-local algebra $\mathcal{A}$ a description $(\mathcal{S}, \mathcal{A}, < \cdot \; ; \; \cdot >)$ which combines in a natural manner quantum mechanical and classical features. We have intentionally discussed these features at first in terms of relations in the state space, because superposition and decomposition of states are in our opinion the most fundamental and operational notions of a statistical theory. Let us proceed to further notions of this kind!

A state $\varphi \in \mathcal{S}$ is called *factorial*, if $F(\varphi)$ is atomic in $\mathcal{F}$ (i.e. dominates only the smallest element $\emptyset \in \mathcal{F}$). In our interpretation a factorial state is classically pure, it cannot be decomposed further by means of classical properties. Let be μ a Borel probability measure on $\mathcal{S}$, which decomposes a state $\omega \in \mathcal{S}$, i.e.,

$$\omega = \int_\mathcal{S} \varphi \; d\mu(\varphi) \; . \tag{1.60}$$

The *support* supp μ of μ is the smallest w^*-closed subset of $\mathcal{S}$ with measure unity. If for a Borel set $E \subset \mathcal{S}$ it holds $\mu(E) > 0$, one may define the state

$$\omega_E := \int_E \varphi \; d\mu(\varphi)/\mu(E) \; . \tag{1.61}$$

The measure μ in (1.60) is called *sub-central*, if for all Borel sets E, with $\mu(E) > 0$ and $\mu(\mathcal{S} \setminus E) > 0$, ω_E is disjoint to $\omega_{\mathcal{S} \setminus E}$. A sub-central measure decomposes ω into classically different components. The finest of all sub-central measures which decompose ω is uniquely given and called the *central measure* μ_ω of ω. Under a certain separability assumption which is satisfied for locally normal states, the support of μ_ω consists of factor states which necessarily are pair-wise disjoint. If ω^β is a limiting Gibbs state, given by the local canonical density operators (1.52), then it is locally normal, and the support of its central measure defines the classically pure equilibrium states, that is the *pure phases*.

Representations

A representation π of a C^*-algebra $\mathcal{A}$ consists of a $*$-homomorphism

$$\pi : \mathcal{A} \to \mathcal{B}(\mathcal{H}) \,, \tag{1.62}$$

which associates with every $A \in \mathcal{A}$ a bounded operator $\pi(A) \in \mathcal{B}(\mathcal{H})$ in a Hilbert space $\mathcal{H}$, where the abstract algebraic operations are transformed into the corresponding Hilbert space operations.

For any $\rho \in \mathcal{T}_+^1(\mathcal{H})$ the definition

$$< \varphi; A >:= \operatorname{tr}[\rho \pi(A)] \,, \quad A \in \mathcal{A} \,, \tag{1.63}$$

defines a state φ on $\mathcal{A}$. The set of all states of this form is a folium $F_\pi \subset \mathcal{S}$. Two representations π and π' are *disjoint*, if $F_\pi \cap F_{\pi'} = \varnothing$, are *quasi-equivalent*, if $F_\pi = F_{\pi'}$, and π is *quasi-contained* in π', if $F_\pi \subset F_{\pi'}$.

For a representation π let be

$$\mathcal{M}_\pi := \overline{\pi(\mathcal{A})}^w \,, \tag{1.64}$$

where the closure of the represented C^*-algebra is taken in the weak operator topology in the representation Hilbert space. A C^*-algebra (in a Hilbert space), which is weakly closed, is called a *von Neumann algebra*. For the von Neumann algebra $\mathcal{M}_\pi$ in $\mathcal{H}$ one obtains special states by the use of density matrices, the so-called *normal states* $\mathcal{S}_n(\mathcal{M}_\pi)$. F_π is affinely isomorphic to $\mathcal{S}_n(\mathcal{M}_\pi)$. If π and π' are quasi-equivalent one has, therefore, an affine isomorphism between $\mathcal{S}_n(\mathcal{M}_{\pi'})$ and $\mathcal{S}_n(\mathcal{M}_\pi)$, which gives (by duality) a $*$-isomorphism

$$\eta : \mathcal{M}_\pi \overset{\text{onto}}{\longrightarrow} \mathcal{M}_{\pi'} \tag{1.65a}$$

with

$$\eta(\pi(A)) = \pi'(A) \,, \quad \forall A \in \mathcal{A} \,. \tag{1.65b}$$

By the algebraic isomorphy the center $\mathcal{Z}_\pi$ of $\mathcal{M}_\pi$ is mapped by η onto the center $\mathcal{Z}_{\pi'}$ of $\mathcal{M}_{\pi'}$. Now with every sub-folium $F \subset F_\pi$, $F \in \mathcal{F}$, there is associated the smallest central projection $P_F \in \mathcal{Z}_\pi$ which satisfies[21]

$$< \varphi; P_F >= 1 \,, \quad \forall \varphi \in F_\pi \,. \tag{1.66}$$

(Observe that we have by the correspondence $F_\pi \leftrightarrow S_n(\mathcal{M}_\pi)$ the possibility to extend a state $\varphi \in F_\pi$ to a normal state on $\mathcal{M}_\pi$!). The classical structure of F_π is the Boolean sublattice

$$\mathcal{F}_\pi := \{F \in \mathcal{F} ; \ F \subset F_\pi\} . \tag{1.67}$$

One can show that the mapping

$$\mathcal{F}_\pi \ni F \to P_F \in \mathcal{Z}_\pi \tag{1.68}$$

is a lattice ortho-isomorphism between $\mathcal{F}_\pi$ and all projections in $\mathcal{Z}_\pi$. (In an analogous manner one can map all faces in F_π onto the projections of $\mathcal{M}_\pi$.)

Thus, the weak closure of $\pi(\mathcal{A})$ is not an unmotivated extension of the represented quasi-local algebra $\mathcal{A}$, but makes explicit the already existing structure in the distinguished folium F_π which is associated with the representation π. The notion of quasi-equivalence puts just those representations into one class, which have affinely isomorphic sets of density operator states.

The famous *GNS-construction* is a prescription how to build a representation Hilbert space $\mathcal{H}_\varphi$ over an arbitrarily given state $\varphi \in S$, be it pure or mixed. That is, one introduces for every $C \in \mathcal{A}$ classes

$$\Psi_C := \{B \in \mathcal{A} ; \ < \varphi ; \ (B - C)^*(B - C) >= 0\} \tag{1.69}$$

of algebra elements B, whose difference to C is not detectable in the state φ (and neither its perturbations). Now the scalarproduct

$$(\Psi_C | \Psi_{C'}) :=< \varphi ; \ C^* C' > \tag{1.70}$$

for these classes is a well-defined positive-definite sesquilinear form on $\{\Psi_C; \ C \in \mathcal{A}\}$, the norm-closure of which is $\mathcal{H}_\varphi$. The representation of $\mathcal{A}$ in $\mathcal{H}_\varphi$ is introduced by setting

$$\pi_\varphi(A)\Psi_C := \Psi_{AC} , \ \ \forall A, C \in \mathcal{A} , \tag{1.71}$$

and extending this to $\mathcal{H}_\varphi$. Since then

$$\Psi_C = \pi_\varphi(C)\Psi_1 , \ \ C \in \mathcal{A} , \tag{1.72}$$

a dense set of vectors in $\mathcal{H}_\varphi$ is gained by applying the represented algebra elements onto

$$\Omega_\varphi := \Psi_1 . \tag{1.73}$$

Thus Ω_φ is a so-called cyclic vector of this representation. The *GNS-triple* $(\pi_\varphi, \mathcal{H}_\varphi, \Omega_\varphi)$ is characterized uniquely by the facts that Ω_φ is cyclic for $\pi_\varphi(\mathcal{A})$ and that

$$(\Omega_\varphi | \pi_\varphi(A)\Omega_\varphi) = <\varphi\,;\,A> , \quad \forall\, A \in \mathcal{A} . \tag{1.74}$$

If there is another representation π in a Hilbert space $\mathcal{H}$ with cyclic vector Ω, such that

$$(\Omega | \pi(A)\Omega) = <\varphi\,;\,A> , \quad \forall\, A \in \mathcal{A}$$

then there is a unitary operator

$$U : \mathcal{H}_\varphi \to \mathcal{H} \tag{1.75}$$

with

$$U\,\pi_\varphi(A)\Omega_\varphi = \pi(A)\Omega , \quad \forall\, A \in \mathcal{A} , \tag{1.76}$$

and

$$\pi(A) = U\,\pi_\varphi(A)U^* , \quad \forall\, A \in \mathcal{A} . \tag{1.77}$$

Thus, the unitary equivalence, expressed by (1.77), extends in virtue of (1.76) (with $A = 1$) also to the cyclic vectors, and both representations are to be identified.

One can show that the folium of π_φ is equal to $F(\varphi)$ (the smallest one containing φ). Thus the disjointness etc. of states takes over to their GNS-representations.

Assume that there is given a distinguished state ω, e.g., a limiting Gibbs state, the classical properties of which should be analyzed. This can be done most effectively if its central decomposition

$$\omega = \int_{\mathcal{T}_\omega} \varphi \; d\mu_\omega(\varphi) \tag{1.78}$$

is known, where we assume that the central measure μ_ω has the closed set $\mathcal{T}_\omega$ as its support, and that the GNS-Hilbert spaces $\mathcal{H}_\varphi$ are separable for all $\varphi \in \mathcal{T}_\omega$ (which is known to hold for locally normal states φ). Then one can show that the GNS-Hilbert space $\mathcal{H}_\omega$ is (unitarily equivalent to) the direct integral

$$\mathcal{H}_\omega = \int_{\mathcal{T}_\omega}^{\oplus} \mathcal{H}_\varphi d\,\mu(\varphi) , \tag{1.79}$$

with vectors

$$\mathcal{H}_\omega \ni \Psi = \int_{\mathcal{T}_\omega}^{\oplus} \Psi_\varphi d\,\mu_\omega(\varphi) , \tag{1.80}$$

where $\varphi \to \Psi_\varphi \in \mathcal{H}_\varphi$ is a μ_ω-measurable family of vectors with

$$\| \Psi \|^2 := \int_{\mathcal{T}_\omega} \| \Psi_\varphi \|^2 \, d\mu_\omega(\varphi) < +\infty . \tag{1.81}$$

(More precisely, Ψ is the class of vectors modulo vectors with zero norm.) For $\Psi, \Psi' \in \mathcal{H}_\omega$ the scalarproduct is

$$(\Psi|\Psi') = \int_{\mathcal{T}_\omega} (\Psi|\Psi') d\mu_\omega(\varphi) . \tag{1.82}$$

The cyclic vector has the decomposition

$$\Omega_\omega = \int_{\mathcal{T}_\omega}^{\oplus} \Omega_\varphi d\mu_\omega(\varphi) , \tag{1.83}$$

where the Ω_φ are the cyclic vectors for the states φ. The representation operators act as

$$\pi_\omega(A)\Psi = \int_{\mathcal{T}_\omega}^{\oplus} \pi_\varphi(A)\Psi_\varphi d\mu_\omega(\varphi) , \quad A \in \mathcal{A} . \tag{1.84}$$

Thus all $\pi_\omega(A)$, $A \in \mathcal{A}$, leave the component spaces $\mathcal{H}_\varphi$ invariant and multiply with each other component-wise. Since in every component $\pi_\varphi(A)$ the same $A \in \mathcal{A}$ is represented, the components are interrelated with each other. By going over to the weak closure

$$\mathcal{M}_\omega := \mathcal{M}_{\pi_\omega} = \overline{\pi_\omega(\mathcal{A})}^w , \tag{1.85}$$

this interrelation vanishes and the general element $M \in \mathcal{M}_\varphi$ has the form

$$M = \int_{\mathcal{T}_\omega}^{\oplus} M_\varphi d\mu_\omega(\varphi) , \quad M_\varphi \in \mathcal{M}_\varphi , \tag{1.86}$$

where $\varphi \to M_\varphi$ is an arbitrary, measurable, (essentially) norm-bounded operator function. Most importantly, every central element

$$Z \in \mathcal{Z}_\omega = \mathcal{M}_\omega \cap \mathcal{M}'_\omega \tag{1.87}$$

has the form

$$Z = \int_{\mathcal{T}_\omega}^{\oplus} c_\varphi 1_\varphi d\mu_\omega(\varphi) , \tag{1.88}$$

where $\varphi \to c_\varphi \in \mathbb{C}$ is in $\mathcal{L}^\infty(\mathcal{T}_\omega, \mu_\omega)$, i.e. is an (essentially) bounded function. Comparing this with (1.24) we remark the close analogy to the former superselection rules. $Z \in \mathcal{Z}_\omega$ is a projection, iff $\varphi \to c_\varphi$ in (1.88) is a characteristic function of a measurable set (unique only up to μ_ω-null sets) in $\mathcal{T}_\omega$. The family of equivalence classes of sets is the canonical form of a Boolean lattice[22]. By means of (1.68) this lattice is ortho-isomorphic to $\mathcal{F}_\varphi$, the set of all sub-folia of $F(\varphi)$. To identify these sub-folia experimentally requires in principle only the elementary statistical operations of mixing and decomposition. In this sense it is just the way we have learned to look at quantum states, which discloses the classical structure of macroscopic systems.

Coherent superpositions

In order to destillate the essential features from the formulation of coherent state superpositions in traditional quantum mechanics we need the notion of a support of a state. For this we assume there is given an arbitrary state $\varphi \in \mathcal{S}$ on the quasi-local C^*- algebra $\mathcal{A}$ and a representation π of $\mathcal{A}$ such that φ is given by a density operator ρ in the representation space. If $\mathcal{M}_\pi = \overline{\pi(\mathcal{A})}^w$ is the representation von Neumann algebra, then

$$S_\varphi(\pi) := \inf\left\{P = P^2 = P^* \in \mathcal{M}_\pi \;;\; \mathrm{tr}[\rho P] = 1\right\} \in \mathcal{M}_\pi \qquad (1.89)$$

is the π-support of φ. If $\rho_i = |\Psi_i)(\Psi_i| \in \mathcal{T}_1^+(\mathcal{H})$, $1 \le i \le 3$, describe pair-wise different pure states in a traditional quantum theory over the Hilbert space $\mathcal{H}$, then coherence is — as we have recalled at the beginning of this lecture — expressible as the linear dependence

$$c_1\Psi_1 + c_2\Psi_2 + c_3\Psi_3 = 0 \;,\; 0 \ne c_i \in \mathbb{C} \qquad (1.90)$$

for the state vectors $\Psi_i \in \mathcal{H}$. Every pair from the three vectors spans the same plane in $\mathcal{H}$, the projection onto which we denote by P. This situation is equivalently expressed in terms of the supports (in the self-representation) $S_i = |\Psi_i)(\Psi_i|$

$$S_i \wedge S_j = 0 \;,\quad 1 \le i \ne j \le 3 \qquad (1.91a)$$

$$S_i \vee S_j = P \;,\quad 1 \le i \ne j \le 3 \;. \qquad (1.91b)$$

It is easy to check that (1.91) is not valid, if Ψ_1 and Ψ_2 are taken from different sectors in a theory with super-selection rules. (Then $S_3 = S_1 \oplus S_2$ and (1.91a) is not valid.)

We now define the general coherence relation K for three states $\varphi_i \in \mathcal{S}$, $1 \le i \le 3$, in close analogy to (1.91) by means of the π-supports, where π is an arbitrary representation with $\varphi_i \in F_\pi$:

$$K(\varphi_1, \varphi_2, \varphi_3) \;,\; \text{if}\;\; S_i(\pi) \wedge S_j(\pi) = 0 \;,\quad 1 \le i \ne j \le 3 \qquad (1.92a)$$

$$S_i(\pi) \vee S_j(\pi) = P \;,\quad 1 \le i \ne j \le 3 \qquad (1.92b)$$

where P is some projection in $\mathcal{M}_\pi$ [23].

Observe that the φ_i need not be pure states, as is appropriate for macroscopic systems! First one checks that K is independent from π. Then one states that not all of the S_i are allowed to commute with each other if (1.92) should be valid. This shows that K is a genuine quantum mechanical relation. Further one finds that (1.92) implies $F(\varphi_1) = F(\varphi_2) = F(\varphi_3)$, so that the states have to be classically equivalent[23].

If two of the φ_i are pure, then so is also the third state, and the supports are pairwise unitarily equivalent (by means of unitary elements in $\mathcal{M}_\pi$). If $\varphi_{1/2}$ are fixed pure states, then $\{\varphi \in \mathcal{S} ; K(\varphi_1, \varphi_2, \varphi)\}$ is equal to the extreme boundary $\partial_e E(\varphi_1, \varphi_2)$ of the smallest norm-closed face $E(\varphi_1, \varphi_2)$ containing $\varphi_{1/2}$, if $F(\varphi_1) = F(\varphi_2)$ [24]. As we have mentioned after (1.59), also in the general C^*-algebraic description $\partial_e E(\varphi_1, \varphi_3)$ is isomorphic to the surface of the 3-dimensional unit ball $B(3)$, if $\varphi_{1/2}$ are pure and quasi-equivalent. (If $\varphi_{1/2}$ are pure and disjoint, $E(\varphi_1, \varphi_2)$ is isomorphic to a line segment.) Altogether, we find that the K-relation has all features which one would expect from physical intuition.

Let us mention that (1.92) may be expressed also by means of the faces $E(\varphi_i)$ without any representation π at all. This form can then be used to investigate coherence in even more general statistical theories. For JB-algebras this is carried through in [24], where the set $\partial_e E(\varphi_1, \varphi_2)$ of the coherent superpositions of two pure states $\varphi_{1/2}$ is isomorphic to the surface of an n-ball, n ranging from 3 to infinity.

2. Superconductivity

We treat here superconductivity in terms of a simple BCS-model for the electrons [25,26,27,28]. The one-electron states are indexed by the tuples $(k, \sigma) \in \mathcal{K} \times \{\uparrow, \downarrow\} =: \mathcal{R}$, where $\mathcal{K}$ is a denumerable set of momenta (taken from a shell around the Fermi surface), and σ denotes the spin-index. The directed sets of subsets $\Lambda \subset \mathcal{K}$, with cardinality $|\Lambda| < +\infty$, is again denoted by $\mathcal{L}$. For every $\Lambda \in \mathcal{L}$ we set

$$\mathcal{A}_\Lambda := \bigotimes_{k \in \Lambda} \left(\mathcal{A}_{k\uparrow} \otimes \mathcal{A}_{-k\downarrow} \right), \quad \mathcal{A}_{k\sigma} \equiv \mathcal{M}^{(2)} \qquad (2.1)$$

which induces in the C^*-inductive limit the (quasi-local) electron field algebra $\mathcal{A}$ (as in (1.41)). Introducing the electron creation operators $c_{k\sigma}^*$ as in (1.17), we write the local pairing Hamiltonian in the form

$$H_\Lambda := \sum_{\sigma,k\in\Lambda} \epsilon c^*_{k\sigma} c_{k\sigma} - (g/|\Lambda|) \sum_{k,k'\in\Lambda} c^*_{k\uparrow} c^*_{-k\downarrow} c_{-k'\downarrow} c_{k'\uparrow} \ , \tag{2.2}$$

where ϵ is the averaged kinetic energy and g the averaged coupling energy with $|2\epsilon/g| < 1$ [27]. The local particle number operator is

$$N_\Lambda := \sum_{\sigma,k\in\Lambda} c^*_{k\sigma} c_{k\sigma} \ , \tag{2.3}$$

and the local reduced Hamiltonian

$$H^r_\Lambda := H_\Lambda - \mu \, N_\Lambda \tag{2.4}$$

has the form (2.2) with ϵ replaced by $\epsilon^r := \epsilon - \mu$, where from now on we fix μ to the zero point value ϵ_F. The local grand canonical density operator

$$\sigma^\beta_\Lambda = \exp\left[-\zeta_\Lambda - \beta H^r_\Lambda\right], \quad \zeta_\Lambda \in \mathbb{R} \ , \tag{2.5}$$

is invariant under gauge transformations

$$\alpha_\vartheta(c^*_{k\sigma}) = e^{i\vartheta} c^*_{k\sigma} \ , \quad \vartheta \in [0, 2\pi) \ , \tag{2.6}$$

(cf. also (1.10)). The proof of the convergence of $\{\sigma^\beta_\Lambda \ ; \ \Lambda \in \mathcal{L}\}$ has been worked out in terms of generating functions in [29], where a technique similar to the large deviation principle[30] was anticipated. This method has been modified in [31] in order to show (1.51) directly. To be precise one should state that both of the mentioned convergence proofs refer to a restricted electron algebra, the so-called pair algebra[32,33]. The convergence on the full electron algebra $\mathcal{A}$ (where the σ^β_Λ are no more diagonalizable) was carried through in the unpublished diploma thesis [34], where the free energy minimum principle plays a decisive role. From this we have the result

$$\lim_{\Lambda'\in\mathcal{L}} \eta^*_{\Lambda,\Lambda'}(\sigma^\beta_{\Lambda'}) =: \rho^\beta_\Lambda = \int_0^{2\pi} \rho^{\beta,\vartheta}_\Lambda \, d\vartheta/2\pi \tag{2.7}$$

with

$$\rho^{\beta,\vartheta}_\Lambda = \exp\left[-\xi_\Lambda - \beta\left(\epsilon^r N_\Lambda - g\sum_{k\in\Lambda}(we^{-i\vartheta}c^*_{k\uparrow}c^*_{-k\downarrow} + we^{i\vartheta}c_{-k\downarrow}c_{k\uparrow})\right)\right], \quad \xi_\Lambda \in \mathbb{R} \ . \tag{2.8}$$

Here $w = w(\beta) = 0$, for $\beta \leq \beta_c = 2\text{arctanh}(2\epsilon/g)/\epsilon$ and increases monotonically to the maximal value $w(\infty) = \left(1 - (2\epsilon/g)^2\right)^{\frac{1}{2}}$. We henceforth assume $\beta_c < \beta < +\infty$.

The merit of the technique in [31] has been to make the integral in (2.7) directly interpretable as a conservation law for Cooper pairs outside Λ. Since Λ is arbitrary this
means a conservation law for Cooper pairs "at infinity", which are to be interpreted
as *condensed* Cooper pairs. Eq. (2.7) defines a compatible family of density matrices
in virtue of (1.51) ff., and (2.8) does so by the product state property. Thus, the
decomposition (2.7) being valid for all $\Lambda \in \mathcal{L}$, gives rise to the desintegration

$$\omega^\beta = \int_0^{2\pi} \omega^{\beta\vartheta} d\vartheta / 2\pi \tag{2.9}$$

for the associated states on $\mathcal{A}$. In fact, (2.9) can be identified as the central decomposition of the limiting Gibbs state ω^β and brings into play the macroscopic phase
angle ϑ.

It is evident that ω^β is still invariant under gauge transformations, whereas

$$\omega^{\beta,\gamma} \circ \alpha_\vartheta = \omega^{\beta,\gamma-2\vartheta} \tag{2.10}$$

signifies spontaneous symmetry breaking for the pure phase states.

In virtue of (1.79) - (1.84) it holds

$$\left(\pi_\beta, \mathcal{H}_\beta, \Omega_\beta\right) = \int_0^{2\pi\oplus} \left(\pi_{\beta\vartheta}, \mathcal{H}_{\beta\vartheta}, \Omega_{\beta\vartheta}\right) d\vartheta / 2\pi \tag{2.11}$$

for the GNS-triples of ω^β, respectively $\omega^{\beta,\vartheta}$. In the weak closure $\overline{\pi_\beta(\mathcal{A})}^w =: \mathcal{M}_\beta$ is
especially contained the phase operator

$$\Theta^\beta := \int_0^{2\pi\oplus} (\vartheta/2)\pi_{\beta\vartheta}(1)d\vartheta / 2\pi \tag{2.12}$$

which — more specifically — is an element of the center $\mathcal{Z}_\beta$ of $\mathcal{M}_\beta$.

Beside the l.h.s. of (2.11), also $(\pi_\beta \circ \alpha_\vartheta, \mathcal{H}_\beta, \Omega_\beta)$, for all $\vartheta \in [0, 2\pi)$, is a GNS-
triple of ω^β, and we have by (1.77) a (strongly continuous) family of unitary operators
U_ϑ^β with

$$U_\vartheta^\beta \Omega_\beta = \Omega_\beta \tag{2.13}$$

and

$$\pi_\beta(\alpha_\vartheta(A)) = U_\vartheta^\beta \pi_\beta(A) U_\vartheta^{\beta*} \tag{2.14}$$

for all $\vartheta \in [0, 2\pi)$ and all $A \in \mathcal{A}$. The self-adjoint generator N^β of this family is
the renormalized particle number operator corresponding to this finite temperature
representation. Let us choose an element in $\mathcal{M}_\beta$ of the form

$$M = \int_0^{2\pi\oplus} M_\vartheta \, d\vartheta/2\pi \ , \quad M_\vartheta \in \pi_{\beta\vartheta}(\mathcal{A}_\Lambda) \tag{2.15}$$

where $\Lambda \in \mathcal{L}$ is arbitrary but fixed. The family of represented matrices $\vartheta \to M_\vartheta$ is assumed to satisfy the (periodicity) condition $M_{\vartheta=0} = \lim M_\vartheta$, $\vartheta \to 2\pi$ and to be $d\vartheta$-a.e. differentiable, so that the family of derivatives defines the element $(\partial/\partial\vartheta)M \in \mathcal{M}_\beta$. With this it holds[35]

$$N^\beta M \, \Omega_\beta = [\pi_\beta(N_\Lambda), \ M]_- \Omega_\beta + (2\partial/i\partial\vartheta) \, M \, \Omega_\beta \ . \tag{2.16}$$

If therein $M_\vartheta = \pi_{\beta\vartheta}(c_{k\sigma}^{(*)})$, the derivation term vanishes and the commutator produces the eigenvalues $\underset{(+)}{-} 1$. If on the other hand [35]

$$M = w - \lim_{\Lambda \in \mathcal{L}} \pi_\beta \left(\sum_{k \in \Lambda} c_{k\uparrow}^* c_{-k\downarrow}^* \right) /|\Lambda| = \int_0^{2\pi\oplus} w e^{i\vartheta} \pi_{\beta\vartheta}(1) \, d\vartheta/2\pi =: s^*(\beta) \ , \tag{2.17}$$

then the commutator in (2.16) vanishes and the derivative leads to the eigenvalue $+2$. Thus the averaged Cooper pair creation operator is counted by the macroscopic part of N^β. It is also this part which leads to the non-vanishing commutator

$$[\Theta^\beta, N^\beta]_- \subset i\pi_\beta(1) \ . \tag{2.18}$$

As we shall show later on, (2.18) is the basic relation for macroscopic quantum phenomena. Since there is a long lasting discussion on the meaning of (2.18), let us give some mathematical details. Since Θ^β is bounded, it is defined on all vectors $\psi = \int_0^{2\pi\oplus} \psi_\vartheta \, d\vartheta/2\pi \in \mathcal{H}_\beta$. The problem is to find a dense domain for the unbounded N^β which is invariant under Θ^β. A natural choice is to select all absolutely continuous functions $\vartheta \to \psi_\vartheta \in \mathcal{H}_{\beta\vartheta}$ which vanish at the boundaries of $[0, 2\pi)$. There is no problem with the uncertainty relation [37], since the eigenvectors of N^β are not in this domain. (This domain is no core for N^β but allows for a family of self-adjoint extensions, from which the condition $N^\beta\Omega_\beta = 0$ selects just one. From (2.25) below it follows that the spectrum of N^β is equal to $\mathbb{Z}$, as it should be [38].)

For more information on the particle structure in the temperature representation we need the limiting dynamics. In the Heisenberg picture one finds[36,39,40,41] for $A \in \mathcal{A}_\Lambda$

$$s - \lim_{\Lambda' \in \mathcal{L}} \pi_\beta\left(e^{itH_{\Lambda'}} A e^{-itH_{\Lambda'}}\right) = e^{itH_\Lambda^\beta} \pi_\beta(A) \, e^{-itH_\Lambda^\beta} =: \tau_t^\beta\left(\pi_\beta(A)\right) \ , \tag{2.19}$$

with

$$H_\Lambda^\beta = \epsilon\, \pi_\beta(N_\Lambda) - g \sum_{k\in\Lambda} \Big(s(\beta)\pi_\beta\big(c_{k\uparrow}^* c_{-k\downarrow}^*\big) + s^*(\beta)\pi_\beta\big(c_{-k\downarrow} c_{k\uparrow}\big)\Big) . \tag{2.20}$$

In spite of corresponding to the well-known model Hamiltonian in the BCS-theory[28] (with the mean-field there being replaced by a non-trivial central observable $s^*(\beta)$ here), H_Λ^β is not the correct Hamiltonian of the representation. The latter is obtained by a unitary implementation of the limiting dynamics which leaves (as in the case of the gauge transformation) Ω_β stable. The associated self-adjoint generator K^β satisfies for the M of (2.15)

$$K^\beta M\, \Omega_\beta = \big[H_\Lambda^\beta,\ M\big]_- \Omega_\beta + (2\mu\partial/i\partial\vartheta)\, M\,\Omega_\beta . \tag{2.21}$$

In the case (2.17) we obtain from the derivative the eigenvalue 2μ, typical for a "ground pair". The other eigen-excitations are constructed by means of the Bogoliubov transformation

$$u_k\pi_\beta(c_{k\uparrow}) - r_k s(\beta)\pi_\beta(c_{-k\downarrow}^*) =: \gamma_{k0} \quad \in \mathcal{M}_\beta$$

$$u_k\pi_\beta(c_{-k\downarrow}) - r_k s(\beta)\pi_\beta(c_{k\uparrow}^*) =: \gamma_{k1} \quad \in \mathcal{M}_\beta \tag{2.22}$$

with

$$u_k = \big(1 + \epsilon^r/E\big)^{1/2}/2^{1/2} \ , \quad v_k = \big(1 - \epsilon^r/E\big)^{1/2}/2^{1/2}$$

$$r_k = v_k/w \ , \quad E = \big(\epsilon^{r2} + \Delta^2\big)^{1/2} \ , \quad \Delta = wg \ .$$

Insertion into (2.20) leads to

$$H_\Lambda^\beta = \sum_{k\in\Lambda}(E + \mu)\big(\gamma_{k0}^* \gamma_{k0} + \gamma_{k1}^* \gamma_{k1}\big) + \text{const.} \ . \tag{2.23}$$

Defining for $n \in \mathbb{Z}$ and $\Lambda_1,\dots,\Lambda_4 \in \mathcal{L}$

$$\phi(n,\Lambda_1,\dots,\Lambda_4) := s^{*n}(\beta) \prod_{k\in\Lambda_1}\gamma_{k0}^* \prod_{k'\in\Lambda_2}\gamma_{k'0} \prod_{l\in\Lambda_3}\gamma_{l1}^* \prod_{l'\in\Lambda_4}\gamma_{l'1}\Omega_\beta \tag{2.24}$$

one arrives at

$$\left.\begin{matrix} N^\beta \\ K^\beta \end{matrix}\right\}\phi(n,\Lambda_1,\dots,\Lambda_4) = \left\{\begin{matrix} 2n + |\Lambda_1| + |\Lambda_3| - |\Lambda_2| - |\Lambda_4| \\ 2n\mu + (E+\mu)(|\Lambda_1| + |\Lambda_3| - |\Lambda_2| - |\Lambda_4|) \end{matrix}\right\}\phi(n,\Lambda_1,\dots,\Lambda_4) .$$
$$\tag{2.25}$$

The fact that the energy eigenvectors (2.25) are also particle number states comes again from N^β counting the condensed Cooper pairs and the quasi-particles (2.22) (together with their holes) correctly. Without the $s^{*n}(\beta)$ (2.24) would not constitute a basis of $\mathcal{H}_\beta$.

There are many hints in the literature for the necessity of the $s^*(\beta)$-operators[42-46], but no consistent treatments of these quantities (cf., however, [47]). In (2.25) all energy eigenvalues are infinitely degenerated and thus belong to the essential spectrum of K^β. Traditional mathematical stability theorems imply only that this spectrum is invariant under relative compact perturbations, which is much too restricted for the physical applications. By means of the Connes theory[17,48] we obtain the stability of this spectrum under all perturbations which are given by operators in $\mathcal{M}_\beta$. As a hint we remark that $\mathcal{M}_\beta$ is isomorphic to $\mathcal{Z}_\beta \otimes \mathcal{N}_\beta$, where $\mathcal{N}_\beta$ is a Connes factor of type III_λ , $\lambda = \exp(-4\beta\, E(\beta))$. The total dynamics splits accordingly into $\tau_t^\beta = \tau_t^{\beta s} \circ \tau_t^{\beta q}$, where the quasi-particle dynamics $\tau_t^{\beta q}$ is connected with the Connes spectrum, and the condensate dynamics $\tau_t^{\beta s}$ is equally robust. Explicitly we have

$$\tau_t^\beta(\Theta^\beta) = \tau_t^{\beta s}(\Theta^\beta) = \Theta^\beta + \mu\, t\, \pi_\beta(1) \tag{2.26}$$

$$\tau_t^\beta(s(\beta)) = \tau_t^{\beta s}(s(\beta)) = e^{-i2\mu t} s(\beta) \tag{2.27}$$

$$\tau_t^\beta(\gamma_{k0}) = \tau_t^{\beta q}(\gamma_{k0}) = e^{-i(E+\mu)t}\gamma_{k0} \quad . \tag{2.28}$$

Altogether we have derived here a two-fluid model consisting of the quasi-particle and condensed Cooper pair components.

Josephson Junction

Reflecting on the physical significance of broken gauge symmetry in superconductors Josephson arrived in 1962 [45] at the conclusion that two *weakly* coupled specimen are needed. In fact, first one must have a phase *difference*, and second the different phase values should persist over a sufficiently long period to enable experiments, but nevertheless should be correlated with each other. Thus one is led to look for tunneling condensed Cooper pairs which provide as a weak super-current the phase-correlation without making the total system homogeneous.

In order to give a formalization of this idea[50,51] we start from the combined system of two BCS-models (denoted by a and b) without any interaction. The composite local regions are $\Lambda = (\Lambda_a, \Lambda_b)$ (and constitute a directed family again denoted by $\mathcal{L}$)

with associated local electron field algebras

$$\mathcal{A}_\Lambda := \mathcal{A}_{\Lambda a} \otimes \mathcal{A}_{\Lambda b} \ . \tag{2.29}$$

The C^*-inductive limit algebra

$$\mathcal{A} = \mathcal{A}_a \otimes \mathcal{A}_b \tag{2.30}$$

is the tensor product of the single quasi-local algebras (where there is only one C^*-cross norm since the $\mathcal{A}_{a/b}$ are nuclear[52]). The central decomposition of the uncoupled equilibrium state

$$\omega = \omega_a^\beta \otimes \omega_b^\beta = \int\int_0^{2\pi} \omega^{\beta\vartheta_a} \otimes \omega^{\beta\vartheta_b} \ d\vartheta \tag{2.31}$$

involves the Haar measure $d\vartheta = d\vartheta_a d\vartheta_b/(2\pi)^2$ on the two-dimensional torus $T \times T$. The GNS-triple

$$(\pi_\beta, \mathcal{H}_\beta, \Omega_\beta) = \int\int_0^{2\pi}{}^\oplus (\pi_{\beta\vartheta}, \mathcal{H}_{\beta\vartheta}, \Omega_{\beta\vartheta}) \ d\vartheta \tag{2.32}$$

decomposes into the direct integral over the triples belonging to the pure phase states $\omega^{\beta\vartheta} = \omega^{\beta\vartheta_a} \otimes \omega^{\beta\vartheta_b}$, $\vartheta := (\vartheta_a, \vartheta_b)$. A central element of the von Neumann algebra

$$\mathcal{M}_\beta := \overline{\pi_\beta(\mathcal{A})}^\omega \tag{2.33}$$

has then the general form

$$Z = \int\int^\oplus c_\vartheta \pi_{\beta\vartheta}(1) \ d\vartheta \in \mathcal{Z}_\beta = \mathcal{M}_\beta \cap \mathcal{M}'_\beta \ , \tag{2.34}$$

with $\vartheta \to c_\vartheta \in L^\infty([0, 2\pi)^2, d\vartheta)$.

The question of which interaction describes the arrangement of a Josephson junction in appropriate terms is not a matter of the microscopic theory alone: The single superconductors are in special states with a specific (quasi-) particle structure, and the concept of a "weak coupling" refers to collective properties of the constituent particles, especially to mean tunneling frequencies (which for themselves are connected with a mean separation distance between the particles at both sides of the junction). The Hamiltonian is then by definition an effective one in the temperature representation (with β here always chosen larger than β_c of a and b), and should have the form

$$K'^\beta = K^\beta + W^\beta \ , \quad K^\beta = K_a^\beta + K_b^\beta \ , \tag{2.35}$$

where the $K^\beta_{a/b}$ are given by (2.21), and in terms of (2.17) and (2.22) we have

$$
\begin{aligned}
W^\beta &= g_s(s_a^* s_b + s_b^* s_a) \\
&\quad + \sum_\sigma \sum_{k,l \in X} (g_{kl}\gamma_{k\sigma a}^* \gamma_{l\sigma b} + \bar{g}_{kl}\gamma_{l\sigma b}^* \gamma_{k\sigma a}) \\
&=: W_s^\beta + W_q^\beta \in \mathcal{M}_\beta ,
\end{aligned}
\tag{2.36}
$$

with $X \in \mathcal{L}$ fixed. The perturbation W^β contains the dominating terms according to the fundamental principles of quantum mechanics (in close analogy to the radiating atom). Cross terms of the form $\gamma^* s$ would violate particle conservation of the total system, and those of the form $\gamma^* \gamma^* s$ would be of higher order and contradict the fact that quasi-particles cannot condense. We consider this as a strong evidence for the persistence of the two-fluid model in the presence of a weak coupling. As for every element in $\mathcal{M}_\beta$ we have also for W^β a decomposition in its sector components

$$
W_s^\beta + W_q^\beta = \int \int^\oplus [W_s^\vartheta + \pi_{\beta\vartheta}(W_q^\vartheta)] \; d\vartheta
\tag{2.37}
$$

with $W_q^\vartheta \in \mathcal{A}_X$ and

$$
W_s^\vartheta = 2g_s w_a w_b \cos(\vartheta_a - \vartheta_b) .
\tag{2.38}
$$

The perturbed equilibrium state ω'^β may be obtained from ω^β by a convergent Dyson expansion in powers of W^β. It may also be calculated as the limit of the local composite equilibrium states perturbed by the local approximations of W^β [53]. Its central decomposition is

$$
\omega'^\beta = \int \int \omega'^{\beta\vartheta} \exp[-\zeta - 2\beta g_s w_a w_b \cos(\vartheta_a - \vartheta_b)] \; d\vartheta , \quad \zeta \in \mathbb{R} ,
\tag{2.39}
$$

where the perturbed pure phase states $\omega'^{\beta\vartheta}$ have the compatible family of local density operators

$$
\rho_\Lambda'^{\beta\vartheta} = \exp[-\zeta_\Lambda - \beta(H_{\Lambda_a}^{r\beta\vartheta_a} + H_{\Lambda_b}^{r\beta\vartheta_b} + W_q^\vartheta)] , \quad \zeta_\Lambda \in \mathbb{R} ,
\tag{2.40}
$$

if $\Lambda \supset X$. One concludes that ω'^β is quasi-equivalent to ω^β (and here is realizable by a vector in $\mathcal{H}_\beta$). Thus ω'^β has the same classical "properties" as ω^β. The cosine-potential shows up in ω'^β only in terms of a statistical weight for the phase-difference distribution.

The coupled dynamics

$$
\tau_t'^\beta(M) := e^{itK'^\beta} M \, e^{-itK'^\beta} , \quad M \in \mathcal{M}_\beta , \; t \in \mathbb{R} ,
\tag{2.41}
$$

gives rise to an automorphism group in $\mathcal{M}_\beta$ and may also be obtained as the limiting dynamics of the local perturbed BCS-systems[50]. Since W_s^β commutes as a central observable with all $M \in \mathcal{M}_\beta$, it has no influence on the dynamics in $\mathcal{M}_\beta$, the (weakly closed) electron field algebra. After the diagonalization of K^β (by means of the Bogoliubov transformations in a and b), the total dynamics factorizes into the two-fluid components

$$\tau_t'^\beta = \tau_t^{\beta s} \circ \tau_t^{\beta q} \;, \tag{2.42}$$

where $\tau_t^{\beta s} \in \mathrm{Aut}(\mathcal{Z}_\beta)$ contains no interaction at all (since W^β commutes with all central elements). Thus we derive from (2.26)

$$\tau_t'^\beta \left(\Theta_a^\beta - \Theta_b^\beta \right) = \Theta_a^\beta - \Theta_b^\beta + (\mu_a - \mu_b)\, t\pi_\beta(1) \;, \tag{2.43}$$

the first Josephson relation. Correspondingly there is no dressing for the condensed Cooper pair operators, whereas for the quasi-particle operators

$$\tau_t'^\beta(\gamma_{k\sigma a}) = \sum_{n=0}^\infty i^n \int_0^t dt_n \dots \int_0^{t_2} dt_1 \left[W_q^\beta(t_1), \dots \left[W_q^\beta(t_n)\, , \gamma_{k\sigma a}(t) \right] \dots \right] =: \gamma_{k\sigma a}'(t) \;, \tag{2.44}$$

with the time dependence in the series being given by the uncoupled dynamics. The norm-convergent perturbation series thus provides a dressing with infinitely many γ_a- and γ_b-particles and may lead to a complicated spectrum. But one knows from the afore-mentioned stability considerations that the uncoupled quasi-particle eigenvalues are still part of the perturbed spectrum. (The Connes spectrum of the factor $\mathcal{N}_{\beta a} \otimes \mathcal{N}_{\beta b}$ — cf. the remark after (2.25) — is $\{\lambda^n \;;\; n \in \mathbb{N} \;,\; \lambda = \exp(-4\beta E)\}$, if there are integers $n_{a,b}$ with $E_{a,b} = n_{a,b}E$, and is $\mathbb{R}_+$ otherwise.)

In order to get the total current between the superconductors one needs time-dependent particle numbers. In spite of the coupled equilibrium state ω'^β being no more invariant under the separate gauge transformations, we have the separate particle number operators $N_{a/b}^\beta$, since we are still in the uncoupled GNS-representation. These unbounded operators cannot be in $\mathcal{M}_\beta$, but their spectral projections could be in this von Neumann algebra. This is, however, not the case, and $N_{a,b}^\beta$ are not *affiliated* with $\mathcal{M}_\beta$. There is no topology in which (the bounded functions of) the particle number operators could be approximated by elements from $\mathcal{M}_\beta$ and we have thus no canonical prescription for calculating their dynamical behaviour. Since both particle number and energy operator are conceptually connected with the temperature representation and have the cyclic vector Ω_β as zero-value eigenvector, a natural ansatz is

$$N_{a,b}^{\beta}(t) := e^{itK'^{\beta}} N_{a,b}^{\beta} e^{-itK'^{\beta}} , \qquad (2.45)$$

which is defined on a common dense domain for all $t \in \mathbb{R}$. With this it follows for the particle current

$$
\begin{aligned}
I^{\beta}(t) :&= d(N_a^{\beta} - N_b^{\beta})/2dt \\
&= d\, N_a^{\beta}(t)/dt = i\left[W^{\beta}, N_a^{\beta}(t)\right] \\
&= w - \lim_{\Lambda \in \mathcal{L}} \pi_{\beta}\Big([iW_{\Lambda}, N_{\Lambda a}(t)]\Big) \\
&= I_s^{\beta}(t) + I_q^{\beta}(t) ,
\end{aligned}
\qquad (2.46)
$$

where $W_{\Lambda} \in \mathcal{A}_{\Lambda}$, $N_{\Lambda a}(t) \in \mathcal{A}_{\Lambda a}$, and the local perturbed dynamics is employed in the approximation formula. The quasi-particle current therein is

$$I_q^{\beta}(t) = \sum_{\sigma} \sum_{k,l \in X} i\big(g_{kl}\gamma_{k\sigma a}^{l*}(t)\gamma_{l\sigma b}^{l}(t) - \bar{g}_{kl}\gamma_{l\sigma b}^{l*}(t)\gamma_{k\sigma a}^{l}(t)\big) \qquad (2.47)$$

and the super current has the form (second Josephson relation)

$$I_s^{\beta}(t) = 4g_s w_a w_b \sin\left[2(\Theta_a^{\beta}(t) - \Theta_b^{\beta}(t))\right] \in \mathcal{Z}_{\beta} . \qquad (2.48)$$

Observe that this part stems from the central cosine-potential not commuting with the macroscopic part of N_a^{β}, the only dynamical effect of this potential up to now. In other words, we see that "phase coherence" across the junction is in fact mediated by our weak tunneling interaction and is, also formally, the reason for the super current, *if* one performs a totally unconventional extension of the field formalism. Since (2.48) is an operator relation its content is much stronger than the usual second Josephson relation in terms of a special expectation value: we get the sine-current in all states on $\mathcal{M}_{\beta}$ in which the macroscopic phase difference is (almost) sharp, irrespectively of the quasi-particle distribution. This is a theoretical assertion which is completely different from the usual microscopic treatment by lowest order perturbation theory. In a certain sense it is satisfying that this strange kind of electrodynamic laws — with a directed current for $\mu_a = \mu_b$ and an alternating current for $\mu_a - \mu_b$ being a constant different from zero (insert (2.43) into (2.48)!) — requires an uncommon theoretical extension formalism. It is claimed that for tunnel junctions the Josephson relations are experimentally well established[54].

Altogether we have obtained a neat derivation of the Josephson relations from the weak coupling interaction. The usual coupling Hamiltonians in the literature are, however, treated on the same footing as the internal pairing interactions, which would suggest to take an extensive interaction in performing the thermodynamic limit. (The essential difference between a weak and a strong coupling is not expressible by the numerical values of the coupling constants, but given by the $|\Lambda|^n$-asymtotics: if the norm of the interaction is $O(|\Lambda|^0)$ we call it weak, if it is $O(|\Lambda|)$, we call it strong.) In our group we have also analyzed the strong coupling version of (2.36). In [55] the strong coupling between the two superconductors is combined with the weak coupling to particle reservoirs in terms of grand canonical equilibrium states. The current then always exhibits a periodic behaviour with a frequency which is only proportional to the electrostatic potential difference if the coupling energy is larger than $3 \cdot 10^{-6}$ eV (for a Nb_3Ge junction). Below this value there is a strongly non-linear behaviour with a bifurcation point. The supercurrent for itself shows a doubly periodic structure with non-harmonic side-extrema in this very low coupling region, which goes over to a sine-current at the bifurcation point by something like a period-doubling. Here one has only a (modified) ac-Josephson current.

A current oscillation would be a special effect if the model has a strong coupling to the reservoirs. Most of the microscopic treatments are indeed of this type. A rigorous discussion in terms of expectation values in the thermodynamic limit was given by Hepp and Lieb. They found a Hopf-bifurcation for increasing coupling constant of the pair-tunneling term, where the condensate densities start to oscillate. These density oscillations have, however, a constant phase difference between the a and b superconductors. Moreover, there is no internal pairing interaction, so that the whole effect would not be specific for superconductivity. The situation has been thoroughly analyzed by Unnerstall who tested several strong reservoir couplings (always performing the so-called singular coupling limit in order to get a global influence of the reservoirs onto the junction): he could only derive the dc-Josephson effect (as a stationary sine-current at zero voltage), and this only if pairs were fed from the reservoirs into the junction, and not single electrons.

As a conclusion we are left with the described weak coupling case as a rigorous model which provides us with both the dc- and ac-Josephson effect and gives an affirmative answer to Josephson's question "whether there can be any behaviour of two superconductors that is intermediate between those characteristic of complete separation and complete union?"[49].

3. Macroscopic Quantum Phenomena

In his theoretical investigation on superfluidity and superconductivity F. London[58] used the notion of "macroscopic quantum phenomena" to express the fact that a macroscopic number of quantum mechanically described microscopic degrees of freedom cooperate in producing the spectacular collective phenomena. These ideas were taken up by Mercereau[59], as he studied current superpositions at the SQUID (superconducting quantum interference device) which consists of two Josephson junctions 1,2 connected in parallel by means of superconducting wires. In this way very subtle interference phenomena arose and led to extremely accurate measuring devices for the magnetic field. The basic theoretical considerations are rather simple if one works in terms of current expectation values. Assuming well prepared current states, the phase differences $\delta_{1/2} = \vartheta_{a1/2} - \vartheta_{b1/2}$ at the two junctions should almost be sharp, so that the expectations of (2.48) in the two junction states have the form

$$I_{1/2} = I_c \sin \delta_{1/2} , \tag{3.1}$$

where the time-dependence is not made explicit. The total current has the value

$$I = I_1 + I_2 = I_c 2 \sin \left(\frac{\delta_1 + \delta_2}{2} \right) \cos \left(\frac{\delta_1 - \delta_2}{2} \right) \tag{3.2}$$

and may be biased to give the sine-term the value unity.

If there is a magnetic field threading through the junctions and the loop which is formed by the two arms of the parallel array, model independent considerations lead to a proportionality between the difference $\delta_1 - \delta_2$ and the total magnetic flux Φ_T through the loop:

$$I = I_c 2 \cos(\pi \Phi_T / \Phi_0) , \tag{3.3}$$

where $\Phi_0 = hc/2e$ is the flux quantum. The interference oscillations in (3.3) with varying Φ_T are modulated by the depencence of I_c on Φ_J, the flux through one junction, and which behaves as $\sin(\pi \Phi_J/\Phi_0)/(\pi \Phi_J/\Phi_0)$. Since the formula (3.2) has been experimentally well established, Mercereau considers this as a direct confirmation of the phase properties of the macroscopic quantum states. The superposition of the two supercurrents in (3.2) is compared with the interference of two macroscopic de Broglie waves. The formulations in [59], that the SQUID effects "provide the experimenter the opportunity of grappling with quantum mechanics first hand" strongly suggest that Mercereau interprets the situation as a coherent superposition of two macroscopic quantum states.

Just this point of view was denied by Leggett[60] who declared the macroscopic variables which are relevant for the SQUID experiment to be purely classical. He starts for himself a discussion of macroscopic quantum coherence (MQC) which had such a great influence on the literature that we cannot avoid analyzing his reasonings in some detail. We do this, however, in terms of our operator algebraic set-up using a family of local algebras $\mathcal{A}_\Lambda$, $\Lambda \in \mathcal{L}$, as introduced in Section 1.

If the local but large system ($|\Lambda'|$ of order 10^{23}) is in a coherent superposition of (say) two states, the state vector has the form

$$\Psi_{\Lambda'} = c_1 \Psi^1_{\Lambda'} + c_2 \Psi^2_{\Lambda'} , \quad c_{1/2} \in \mathbb{C} , \tag{3.4}$$

where the vectors are from the Hilbert space $\mathcal{H}_{\Lambda'}$ which is canonically associated with this local, traditional quantum description. If $\Lambda' = \Lambda \cup K$, $\Lambda \cap K = \varnothing$, and if $\Psi^i_{\Lambda'} = \Psi^i_\Lambda \otimes \Psi^i_K$, $i = 1, 2$, we obtain by means of the restriction maps (1.49) for $A \in \mathcal{A}_\Lambda$

$$\mathrm{tr}_\Lambda \left[\eta^*_{\Lambda,\Lambda'}(|\Psi_{\Lambda'})(\Psi_{\Lambda'}|)A \right] = (\Psi_{\Lambda'}|\eta_{\Lambda',\Lambda}(A)\Psi_{\Lambda'})$$

$$= \mathrm{tr}_\Lambda \left[(|c_1|^2 |\Psi^1_\Lambda)(\Psi^1_\Lambda| + |c_2|^2 |\Psi^2_\Lambda)(\Psi^2_\Lambda|)A \right] + 2\mathrm{Re} \left[\bar{c}_1 c_2 (\Psi^1_\Lambda|A\Psi^2_\Lambda)(\Psi^1_K|\Psi^2_K) \right] \tag{3.5}$$

which is equal to a mixed state iff $(\Psi^1_K|\Psi^2_K) = 0$. This observation is the background for Leggett's attempt to formulate a criterion for (3.4) being a *macroscopic* coherent superposition, in which the component state vectors $\Psi^{1/2}_{\Lambda'}$ are essentially different from each other in some sense. To express the mixedness of the restriction $\rho_\Lambda = \eta^*_{\Lambda,\Lambda'}(|\Psi_{\Lambda'})(\Psi_{\Lambda'}|)$, Leggett uses the local entropy

$$S(\rho_\Lambda) = -\mathrm{tr}_\Lambda \left[\rho_\Lambda \ln \rho_\Lambda \right] . \tag{3.6}$$

For every Λ' he defines[60]

$$\delta_{\Lambda'} := S(\rho_{\Lambda'}) / \left[\min_{\Lambda \subset \Lambda'} \left(S(\rho_\Lambda) + S(\rho_K) \right) \right] \tag{3.7}$$

and calls "the *disconnectivity* D of a many particle state the largest $|\Lambda'|$ for which $\delta_{\Lambda'}$ is smaller than some small fraction a." The formal properties of D are not worked out and the D-values which are communicated for special states are not formally derived in [60]. In order to get an idea of the meaning of D, let us assume that $\rho_{\Lambda'}$ is almost pure, so that $S(\rho_{\Lambda'})$ is small. If the denominator in (3.7) is to be different from zero, then for every partition $\Lambda \cup K = \Lambda'$, ρ_Λ or ρ_K should be mixed, which is the case if, e.g., the $\Psi_{\Lambda'}^{1/2}$ are product states with pair-wise orthogonal components. If this is so for macroscopic values of $|\Lambda'|$, then one would certainly agree about $\Psi_{\Lambda'}^{1/2}$ calling *macroscopically different*, which here is in accordance with a *large disconnectivity*.

Up to now we can only state that the concept of disconnectivity for a superposition of two (product) states is much in the spirit of our formalism: one considers the whole family of restrictions of states and deals with asymptotic relations for large $|\Lambda'|$-values. But then it would be a matter of self-consistence to study the limit $|\Lambda'| \to \infty$ explicitly. Let us test the disconnectivity, as far as we can understand it, in the thermodynamic limit.

First the weaker condition

$$\lim_{K \in \mathcal{L}} (\Psi_K^1 | \Psi_K^2) = 0 \tag{3.8}$$

now is sufficient to get a mixed state ρ_Λ in (3.5), and this holds then for all $\Lambda \in \mathcal{L}$. Under this assumption the denominator in (3.7) is greater than zero for all $\Lambda \subset \Lambda'$ and all $\Lambda' \in \mathcal{L}$. But then there is no principal difference between $S(\rho_\Lambda)$ and $S(\rho_{\Lambda'})$ anymore, and the nominator is of the same order: $\delta_{\Lambda'}$ has no proper tendency to get small. Only the requirement that the denominator stays greater than zero, respectively that ρ_Λ is always mixed, is connected with

$$\lim_{\Lambda' \in \mathcal{L}} \left(\Psi_{\Lambda'}^1 | \eta_{\Lambda,\Lambda'}(A) \Psi_{\Lambda'}^2 \right) = 0 \tag{3.9}$$

for all $\Lambda \in \mathcal{L}$ and all $A \in \mathcal{A}_\Lambda$. Assuming that the $|\Psi_K^{1/2})(\Psi_K^{1/2}|$, $K \in \mathcal{L}$, satisfy (1.51) and define two pure states $\psi_{1/2}$ on $\mathcal{A}$, relation (3.9) enforces the disjointness of $\psi_{1/2}$ (as can be shown by the reverse reasoning of [61], Lemma 3). We thus have the feeling that Leggett is struggling with the concept of disjointness. But if the $\psi_{1/2}$ are really disjoint, coherence is broken, as we have explained at the end of Section 1.

It is of course a pragmatic decision in which experimental situation one is allowed to use theoretically the thermodynamic limit. If one insists in (3.5), that $|\Lambda'|$ be macroscopically large but finite, and the $\Psi_{\Lambda'}^{1/2}$ be macroscopically different, then the interference terms in (3.5) are unmeasurably small, as is well known in the theory of measurement. In the operator algebraic approach one neglects these interference terms in order to get a clear view on the consequences of the fact that one has macroscopically different states. These consequences are non-trivial theoretical insights, as we have tried to describe in Section 1.

In [60] Leggett applies the notion of disconnectivity to superconducting states. In [60], eq. (3.6), he forms a superpositions of states with just one Cooper pair on the right resp. left side of the Josephson junction. This cannot refer to the Mercereau experiment where one has a superposition of two total states of the two junctions in the SQUID. In order to comment on another ominous remark in [60] let us perform the low temperature limit $\beta \to +\infty$ in (2.8) to obtain (recalling the vacuum $\otimes_{k \in \Lambda} \binom{1}{0}_k$)

$$\rho_\Lambda^{\infty,\vartheta} = \bigotimes_{k \in \Lambda} \left| \left(u_k + v_k e^{-i\vartheta} c_{k\uparrow}^* c_{-k\downarrow}^* \right) \binom{1}{0}_k \right) \left(\left(u_k + v_k e^{-i\vartheta} c_{k\uparrow}^* c_{-k\downarrow}^* \right) \binom{1}{0}_k \right| , \quad (3.10)$$

i.e., the BCS-ground state. In [60] the disconnectivity of such a state is communicated to be 2, with the consequence not to be usable for MQC. Here, the original use of disconnectivity for the superposition of two states is apparently shifted to one component alone and looses the partial meaning we could discern in it.

Let us analyze the superposition of states which stands behind the current superposition at the SQUID (3.2) in terms of our reconstructed quantum theory in the thermodynamic limit. Here we need a two-fluid picture over the interacting reference vector Ω'_β which results from the decomposition

$$\mathcal{M}_\beta = \mathcal{Z}_\beta \otimes \mathcal{N}_\beta , \quad (3.11)$$

where $\mathcal{N}_\beta$ is the factorial von Neumann algebra generated by the quasi-particle operators $\{\gamma_{k0a}, \gamma_{l1a}, \gamma_{k0b}, \gamma_{l1b}\}$. From (2.39) and (2.40) we can deduce in the same way as in [36] that

$$\mathcal{H}_\beta = \mathcal{H}_{\beta s} \otimes \mathcal{H}_{\beta q} , \quad (3.12)$$

where

$$\mathcal{H}_{\beta s} := \overline{\mathcal{Z}_\beta \Omega'_\beta} , \quad (3.13)$$

and

$$\mathcal{H}_{\beta q} := \overline{\mathcal{N}_\beta \Omega'_\beta} . \quad (3.14)$$

We assume now that any of the two junctions in the SQUID is in a vector state ψ_i belonging to the vector

$$\Psi_i := F_i Q_i \Omega'_\beta , \quad i = 1, 2 , \quad (3.15)$$

with $Q_i \in \mathcal{N}_\beta$ and $F_i \in \mathcal{Z}_\beta$. Interpreting the SQUID as a realization of a superposition of the junction states we arrive at the state ψ_3 with the vector

$$\Psi_3 = c_1 \Psi_1 + c_2 \Psi_2 , \quad c_{1/2} \in \mathbb{C} . \quad (3.16)$$

66

In order to get almost sharp phase differences for the junction states we choose the F_i to be smoothed characteristic functions over disjoint intervals around δ_1 and δ_2 so that the $\Psi_{1/2}$ are orthogonal. We then have for the super current operator (2.48)

$$\left(\Psi_i \mid I_s^\beta \Psi_j\right) \approx \delta_{ij} I_0 \sin \delta_i \ , \tag{3.17}$$

with $I_0 = 4g_s w_a w_b$, and thus

$$\left(\Psi_3 \mid I_s^\beta \Psi_3\right) \approx \sum_{i=1}^{2} |c_i|^2 I_0 \sin \delta_i \ . \tag{3.18}$$

If $|c_i|^2 = 1/2$, $i = 1, 2$, we regain (3.2) with $I_c = I_0/2$. The expectation of the quasi-particle current is

$$\begin{aligned}
\left(\Psi_3 \mid I_q^\beta \Psi_3\right) &= \sum_{i=1}^{2} |c_i|^2 \parallel F_i \Omega_\beta' \parallel^2 \left(Q_i \Omega_\beta' \mid I_q^\beta Q_i \Omega_\beta'\right) \\
&= \sum_{i=1}^{2} |c_i|^2 \left(Q_i \Omega_\beta' \mid I_q^\beta Q_i \Omega_\beta'\right) / \parallel Q_i \Omega_\beta' \parallel^2 \ ,
\end{aligned} \tag{3.19}$$

where we have used the product structure of $\mathcal{H}_\beta$ and the normalization of the Ψ_i , $i = 1, 2$. We find only for the super current a δ_i-dependence (in contrast to the usual treatments which have it also for I_q^β [54,62]).

If $P_i \in \mathcal{Z}_\beta$ and $R_i \in \mathcal{N}_\beta$ are the support projections of the selfadjoint operators F_i and Q_i, then

$$S_i = P_i \otimes R_i \in \mathcal{M}_\beta$$

are the supports of the ψ_i , $i = 1, 2$, (since Ω_β' is a separating vector for $\mathcal{M}_\beta$). The R_i are unitarily equivalent if $0 < R_i < 1$ (because $\mathcal{N}_\beta$ is a factor of type III), but the central P_i are orthogonal, which prevents equivalence (for central operators). Thus, $K(\psi_1, \psi_2, \psi_3)$ is not valid and the interference terms $(\Psi_1 \mid B\Psi_2)$ vanish for all $B \in \mathcal{M}_\beta$.

In $\mathcal{M}_\beta$ are, however, not included all observables of physical interest, and one should study quantum coherence on a larger algebra which contains the bounded functions of the particle number operators and of other symmetry generators. From the usual Schrödinger representation one infers that the addition of the phase-derivatives leads from $\mathcal{Z}_\beta$ to $\mathcal{B}(\mathcal{H}_{\beta s}) =: \mathcal{B}_{\beta s}$. The required extension of $\mathcal{N}_\beta$ is seen from the commutator in (2.16) being written as

$$[\pi_\beta(N_{\Lambda a}), M] \Omega_{\beta a} = [\pi_\beta(N_{\Lambda a}) - j(\pi_\beta(N_{\Lambda a}))] M \Omega_{\beta a} \ , \tag{3.20}$$

where $j(A) \in \mathcal{N}'_\beta$ for all $A \in \mathcal{N}_\beta$ (and is introduced in the so-called Tomita-Takesaki theory). Since other generators which implement internal symmetries have a similar structure, it seems natural to include all of $\mathcal{N}'_\beta$ into the extension. This leads to the algebra

$$\left(\mathcal{N}_\beta \cup \mathcal{N}'_\beta\right)'' = \left(\mathbb{C}\pi_\beta(1)\right)' = \mathcal{B}(\mathcal{H}_{\beta q}) =: \mathcal{B}_{\beta q} \,, \tag{3.21}$$

because $\mathcal{N}_\beta$ is a factor with center $\mathcal{N}_\beta \cap \mathcal{N}'_\beta = \mathbb{C}\pi_\beta(1)$. Altogether we have the extended observable algebra

$$\mathcal{M}^e_\beta = \mathcal{B}_{\beta s} \otimes \mathcal{B}_{\beta q} = \mathcal{B}(\mathcal{H}_\beta) \,. \tag{3.22}$$

The next step is to extend the ψ_i to $\mathcal{M}^e_\beta$, which may be done by means of the vectors Ψ_i, $1 \leq i \leq 3$. Observe, however, that there are many other vectors for the same ψ_i which would lead to other extensions. In other words: the exact form (3.15) of the appropriate state vectors is fixed only by the use of the extended set of observables. This aspect occurred already in dealing with the energy eigenstates (2.24) (where the exterior generator K^β was necessary to determine them, cf. also[64]). If we have chosen the vectors (3.15), (3.16) for the extended states ψ^e_i, then we have the supports

$$S^e_i = \mid \Psi_i)(\Psi_i \mid \in \mathcal{M}^e_\beta \,, \ 1 \leq i \leq 3 \,. \tag{3.23}$$

These satisfy the coherence relations (1.92) in the projection lattice of $\mathcal{M}^e_\beta$ and establish the validity of $K(\psi^e_1, \psi^e_2, \psi^e_3)$. Formally this situation is the same as in traditional quantum mechanics (cf. the beginning of section 1), but here it takes place on a high level of integration involving many cooperating microscopic degrees of freedom. This may be seen, e.g., from those elements of $\mathcal{M}^e_\beta$ which give non-vanishing interference terms, the most important ones of which are related with the particle number difference

$$N^\beta := \left(N^\beta_a - N^\beta_b\right)/2 \,, \tag{3.24}$$

which is the canonically conjugate to $\Theta^\beta_a - \Theta^\beta_b =: \Delta^\beta$ with (cf. (2.18))

$$[\Delta^\beta, N^\beta]_- \subset i\, \pi_\beta(1) \,. \tag{3.25}$$

Then $\exp(is N^\beta) \in \mathcal{M}^e_\beta$ for $s \in \mathbb{R}$, and $(\Psi_1 \mid \exp(is N^\beta)\Psi_2)$ is different from zero for appropriately chosen s, where the decisive parts are the differential operators in N^β. Let us characterize this special form of quantum coherence as follows:

Definition: A coherence relation $K(\psi_1^e, \psi_2^e, \psi_3^e)$ of density matrix states on the extended observable algebra $\mathcal{M}_\beta^e$ is called *macroscopic*, if two states in it have mutually disjoint restrictions to the quasi-local algebra $\mathcal{A}$.

We have renounced to require the ψ_i^e , $1 \leq i \leq 3$, to be pure since at finite temperatures it is impossible to make the quasi-particle components pure, whereas the superfluid components are usually assumed to be well preparable.

According to the above definition our junction states which describe the Mercereau experiment, realize MQC. The superposed SQUID state with state vector Ψ_3 exhibits a version of "macroscopic disconnectivity D" by having component states which are disjoint on $\mathcal{A}$ and by being for itself a pure state on $\mathcal{M}_\beta^e$. Both features, however, can be combined only by the use of different algebras of observables.

For the experimental verification of MQC at the SQUID the current operators (in $\mathcal{M}_\beta$) alone are not sufficient. From the theoretical treatment, the current arises, however, from the variation of N^β in time. The relation

$$\left(\Psi_3 \mid I_s^\beta \Psi_3 \right) = \left(\Psi_3 \mid \Delta N_s^\beta \Psi_3 \right) / \Delta t \sim \cos(\delta_1 - \delta_2) \tag{3.26}$$

with

$$N_s^\beta = (\partial/\partial\vartheta_a - \partial/\partial\vartheta_b)/i \tag{3.27}$$

is then conceptually related with the commutator in (3.25). If $\delta_1 - \delta_2$ is small, ΔN_s^β gets a large expectation. The fact that a particle number variation depends reciprocally on the variation of one other dynamical variable — which is at the origin of the Mercereau interference patterns — is in any case a very peculiar property of the system. In our opinion the preceding discussion of the SQUID states gives some justification for the verbal remarks both of Mercereau and Leggett.

It is certainly desirable to get a more manifest experimental consequence of the incompatibility of N_s^β and Δ^β. If one includes a capacitance C into the junction model, the super charge imbalance operator $Q^\beta = 2e \, N_s^\beta$ enters the Hamiltonian directly via the term[65,66,67]

$$K_C^\beta := Q^{\beta 2}/2C \ . \tag{3.28}$$

Then one has also an additional voltage term

$$U^\beta := Q^\beta/C \ , \tag{3.29}$$

which does not commute with I_s^β. In [68] we have described the relationship to the external voltage $(\mu_a - \mu_b)/2e$ which belongs to the charge of electrons in the Fermi sea, which have moments "below" the shell we have taken into account for the BCS-model. There is a long standing discrepancy in the literature (cf. references in [68]) between models which use the one or the other type of voltage, whereas for a complete description both types are needed.

Let us take into account also another energetic term which arises from the externally applied super-current I_x. I_x is used to bias the junction into a prescribed current state and is a c-number, because it feels no back reaction of the system. Since Δ^β is in our units ($\hbar \equiv 1$) an energy (cf. (2.42)), the Hamiltonian connected with I_x has the form[65,67,69]

$$K_x^\beta := -I_x \Delta^\beta / 2e \ . \tag{3.30}$$

The new junction model is then given by the total Hamiltonian

$$\hat{K} := K'^\beta + K_C^\beta + K_x^\beta \ , \tag{3.31}$$

where we still use $\mathcal{H}_\beta$ of (2.11) and (3.12), and K'^β of (2.35). The Heisenberg dynamics

$$\hat{\tau}_t(A) := \exp(it\hat{K}) A \exp(-it\hat{K}) \tag{3.32}$$

is in any case a (strongly continuous) automorphism group of $\mathcal{M}_\beta^e = \mathcal{B}(\mathcal{H}_\beta)$, if $\hat{K}$ can be defined as a self-adjoint operator, what we assume in the following discussion. The remarkable feature is, that the new junction model is defined in the Hilbert space of the old one in spite of being a strong perturbation (via the unbounded operator K_C^β) of the old one. The new equilibrium state is here a stationary state of $\hat{K}$. But let us first check the phase and current dynamics in the Heisenberg picture. Merely by using the commutation relations we find

$$\dot{\Delta}^\beta(t) = i[\hat{K}, \Delta^\beta(t)] = (\mu_a - \mu_b)\pi_\beta(1) + (2e)U^\beta(t) \ . \tag{3.33}$$

Since U^β is incompatible with Δ^β, we have here a quantum dynamics for the phase with a dynamical dressing by operators in $\mathcal{B}_{\beta s}$. Only the operator relation (and not an equation for the expectation values) reveals the fundamental difference to (2.43) which is purely classical (taking place in $\mathcal{Z}_\beta \subset \mathcal{B}_{\beta s}$). For the current we get

$$\hat{I}(t) = i[\hat{K}, N^\beta(t)] = I_q^\beta(t) + I_s^\beta(t) + I_x \pi_\beta(1) \ , \tag{3.34}$$

where still

$$I_s^\beta(t) = 4g_s w_a w_b \sin(2\Delta^\beta(t)) \ . \tag{3.35}$$

But now the super current operator is also not in $\mathcal{Z}_\beta$ and its expectations may show a complicated non-harmonic behaviour.

The investigation of this quantum dynamics is usually performed in terms of a wave equation which we have also in our framework by splitting

$$\hat{K} = \hat{K}_s + \hat{K}_q \qquad (3.36)$$

with $(\gamma := \vartheta_a + \vartheta_b)$

$$\hat{K}_s = -i(\mu_a - \mu_b)\frac{\partial}{\partial \delta} - i(\mu_a + \mu_b)\frac{\partial}{\partial \gamma} + 2g_s w_a w_b \cos \Delta^\beta + K_C^\beta + K_x^\beta \qquad (3.37)$$

and $\hat{K}_q$ containing the commutators in (2.21) with the BCS-"model Hamiltonians" (2.23) for a and b and W_q^β of (2.36). Neither $\hat{K}_s$ nor $\hat{K}_q$ is positive definite (the latter by the finite temperature), and we have no true ground states but only (metastable) stationary states. A sufficient condition for the time invariance of $\Psi = \Psi_s \otimes \Psi_q \in \mathcal{H}_{\beta s} \otimes \mathcal{H}_{\beta q}$ is

$$\hat{K}_s \Psi_s = 0 = \hat{K}_q \Psi_q . \qquad (3.38)$$

In order to get in touch with the usual discussions we first drop the unphysical γ-dependence and then set $\mu_a = \mu_b$. Then there remains only a second order δ-derivative in $\hat{K}_C^\beta$ and the washboard potential $2g_s w_a w_b \cos \delta - I_x \delta/2e$. If δ varied in the whole of $\mathbb{R}$, we would have a one-dimensional Schrödinger equation for $\Psi_s = \Psi_s(\delta)$. In how far the range of δ can be extended is discussed in various articles (cf. [69] and references therein) and we will not comment on this. The Ψ_s-component of a stationary state should be localized around a local minimum of the washboard potential. For the tunneling probability out of this state to a region with lower potential values one can in fact use the usual (approximate) formulas which are β-independent, since the whole formalism in $\mathcal{H}_{\beta s} \equiv L^2([0, 2\pi), d\delta)$ is so. This indicates in a drastic manner, how the macroscopic quantum formalism concerning certain collective variables is separated from the normal degrees of freedom of the many-body system. In the course of this tunneling the zero expectation of the phase derivative $\dot{\Delta}^\beta$ in the stationary state changes into a non-vanishing value, and voltage shows up by the second term in (3.33). In reality one expects a coupling of the collective variables to the temperature dependent surroundings which we have not in our model. (A coupling of the quasi-particle coordinates would not do.) The macroscopic quantum tunneling (MQT in the literature) is then superimposed by thermal activation and should dominate only by low temperatures.

In order to observe MQT the Josephson junction is cooled to millikelvin temperatures and current-biased in the zero voltage state. The elapsed time between the application of the current bias and the appearance of a voltage across the junction is measured in dependence of the temperature. At low temperatures this time is in fact β- independent and confirms MQT.[70,71]

Another experimental confirmation of the macroscopic quantum theory consists in revealing the discrete (approximate) energy eigenstates of the macroscopic wave function by the application of a microwave radiation to the junction. This increases the rate for MQT, but only for discrete frequencies.[72]

In order to find MQC one has investigated a superconducting loop interrupted by a Josephson tunnel junction, a so-called ring SQUID. We can describe this by substituting the phase difference δ by the total flux Φ and replacing[69] the external current term by $(\Phi - \Phi_x)^2/2L$, where L is the selfinductance of the ring and Φ_x the externally applied flux. For $\Phi_x = \Phi_0/2$ the resulting potential has two absolute minima at the same level. This enables the most popular form of MQC concerning the two one-well ground state wave functions $\Psi_{1/2}$ which are localized around the two minima. The possibility of a coherent superposition by the tunneling through the potential barrier lifts this degeneracy and leads to two lowest eigenstates, the symmetric superposition being the ground state and the anti-symmetric one being the first excited state. If the system is prepared in the left well this is not the true ground state and a coherent oscillation sets in. Since Φ is a classical macroscopic observable (for the original field algebra) $\Psi_{1/2}$ are always macroscopically different, i.e. disjoint, irrespectively of how far in the Φ scale the two-well minima are separated from each other (this is not so clearly expressed in the usual discussion). The spectacular effect which has not been observed up to now would be the oscillation between the disjoint states. This would give a decisive hint that the condensed charge Q^β is a relevant observable, so that the (weakly closed) field algebra has, in fact, to be extended. The flux is then no longer an actualized observable in the coherent state superpositions[73]. (In a *microscopic* two-well system the strange phenomenon is the realization of the localized chiral states [74].) From this problem a large field of research has originated which investigates the influence of noise on the localization of states.[75,76] This, however, is just not of fundamental *conceptual* importance for macroscopic quantum phenomena, since their collective structure must be a robust one anyway, if one writes down a macroscopic wave function.

Altogether we have seen how the operator algebraic formalism provides a book-keeping for the relevant observables at any stage of the description: the local observables are combined to the quasi-local algebra. This gives the possibility to introduce macroscopically different states and to break Dirac's superposition principle. The macroscopic frame condition (like the temperature) select a special folium and a special representation. The classical collective observables are then in the center of the weakly closed represented observable algebra. Up to this stage this is a standard machinery of the operator algebraic quantum theory. For the Josephson junction we have performed a further extension of the von Neumann algebra $\mathcal{M}_\beta$ in the temperature representation to $\mathcal{M}_\beta^e$ which covers the charge and the dynamical voltage operators. Especially the capacitance energy gave rise to a Heisenberg dynamics which induced a quantum dressing of the previously classical collective observables. The somewhat surprising result at the end of our deduction has been that the described macroscopic Heisenberg dynamics is equivalent to a macroscopic wave equation exactly in the form as it is widely discussed in the more heuristic approach and partially confirmed by the experiments. Nevertheless, there are still many mathematical[77] and physical aspects to be clarified before we could claim to understand *macroscopic quantum phenomena*.

References

1. J. von Neumann, "Mathematische Grundlagen der Quantenmechanik", Springer, Berlin (1932)

2. G. Ludwig, "Die Grundlagen der Quantenmechanik", Springer, Berlin (1954)

3. G. Ludwig, "Einführung in die Grundlagen der Theoretischen Physik III", Vieweg, Braunschweig (1976)

4. E. Alfsen, "Compact Convex sets and Boundary Integrals", Springer, Berlin (1971)

5. O. Bratteli and D.W. Robinson, "Operator Algebras and Quantum Statistical Mechanics I, II", Springer, New York (1979, 1981)

6. P.A.M. Dirac, "The Principles of Quantum Mechanics", Clarendon Press, Oxford (1930)

7. E.P. Wigner, "Gruppentheorie und ihre Anwendungen auf die Quantenmechanik der Massenspektren", Vieweg, Braunschweig (1931)

8. G.G. Emch, "Algebraic Methods in Statistical Mechanics and Quantum Field Theory", Wiley-Interscience, New York (1972)

9. A. Rieckers, "Fundamentals of Algebraic Quantum Theory", in: "Groups, Systems and Many-Body Physics", P. Kramer and M. Dal Cin, eds., Vieweg, Braunschweig (1980)

10. A. Rieckers, in: "Group Theoretical Methods in Physics VI", P. Kramer and A. Rieckers, eds., Springer, Berlin (1978)

11. A. Rieckers and H. Roos, Ann. Inst. Henri Poincaré 50:95 (1989)

12. A. Rieckers, in: "Semesterbericht Funktionalanalysis", SS 1988: 169, Mathematisches Institut der Universität Tübingen (1988)

13. G.C. Wick, A.S. Wightmann, and E.P. Wigner, Phys. Rev. 88: 101 (1952)

14. J.M. Jauch, Helv. Phys. Acta 33:711 (1960)

15. H. Primas, "Chemistry, Quantum Mechanics, and Reductionism", Springer, New York (1981)

16. S. Sakai, "C*-Algebras and W*-Algebras", Springer, Berlin (1971)

17. G.K. Pedersen, "C*-Algebras and their Automorphism Groups", Academic Press, London (1979)

18. M. Takesaki, "Theory of Operator Algebras I", Springer, New York (1979)

19. R. Haag and D. Kastler, J. Math. Phys. 5:848 (1964)

20. D. Ruelle, "Statistical Mechanics", Benjamin, New York (1969)

21. R. Haag, R.V. Kadison, and D. Kastler, Commun. Math. Phys. 16:81 (1970)

22. R. Sikorski, "Boolean Algebras", Springer, Berlin (1960)

23. G. Raggio and A. Rieckers, Int. J. Theor. Phys. 22:267 (1983)

24. S. Zanzinger, this Proceedings

25. J. Bardeen, L.N. Cooper, and J.R. Schrieffer, Phys. Rev. 108: 1175 (1957)

26. G. Rickayzen, "Theory of Superconductivity", J. Wiley, New York (1965)

27. D.J. Thouless, Phys. Rev. 117:1256 (1960)

28. M. Tinkham, "Introduction to Superconductivity", McGraw-Hill, Tokyo (1975)

29. W. Thirring, Commun. Math. Phys. 7:181 (1968)

30. J.T. Lewis, V.A. Zagrebnoc, and J.V. Pulé, Helv. Phys. Acta 61:1063 (1988)

31. W. Fleig, Acta Phys. Austr. 55:135 (1983)

32. P.W. Anderson, Phys. Rev. 112:1900 (1958)

33. H. Koppe and B. Mühlschlegel, Z. Phys. 151:613 (1958)

34. W. Hauser, Dipl. Thesis, Tübingen (1987)

35. A. Rieckers and M. Ullrich, Acta Phys. Austr. 56:131 (1985)

36. A. Rieckers and M. Ullrich, Acta Phys. Austr. 56:259 (1985)

37. P. Carruthers and M.M. Nieto, Rev. Mod. Phys. 40:411 (1968)

38. F. Rocca and M. Sirugue, Commun. Math. Phys. 34:111 (1973)

39. A. Rieckers, J. Math. Phys. 25:2593 (1984)

40. E. Duffner and A. Rieckers, Z. Naturforsch. 43a:521 (1988)

41. P. Bona, J. Math. Phys. 29:2223 (1988)

42. N.N. Bogoliubov, Lectures on Quantum Statistics I, II", Gordon and Breach, New York (1967, 1970)

43. J. Bardeen, Phys. Rev. Lett. 9:147 (1962)

44. J.R. Schrieffer, "Theory of Superconductivity", Benjamin, New York (1964)

45. B.D. Josephson, Phys. Lett. 1:251 (1962)

46. V. Ambegaokar, in: "Superconductivity I, II", R.D. Parks, ed., Dekker, New York (1969)

47. G.G. Emch and M. Guenin, J. Math. Phys. 7:915 (1966)

48. A. Connes, Ann. Sci. Ecole Norm. Sup. Paris (4)6:133 (1973)

49. B.D. Josephson, in: R.D. Parks[46]

50. A. Rieckers and M. Ullrich, J. Math. Phys. 27:1082 (1986)

51. A. Rieckers, in: "Frontiers of Nonequilibrium Statistical Physics", G.T. Moore and M.O. Scully, eds., Plenum Press, New York (1986)

52. R.V. Kadison and J.R. Ringrose, "Fundamentals of the Theory of Operator Algebras I, II", Academic Press, New York (1983, 1986)

53. M. Ullrich, Rep. Math. Phys. 23:67 (1986)

54. A. Barone and G. Paterno, "Physics and Application of the Josephson Effect", Wiley and Sons, New York (1982)

55. E. Duffner, Z. Phys. B 63:37 (1986)

56. T. Unnerstall, J. Stat. Phys. 54:379 (1989)

57. B.D. Josephson, Rev. Mod. Phys. 36:216 (1964)

58. F. London, "Superfluids I, II", Wiley, New York (1950)

59. J.E. Mercereau, in: R.D. Parks[46]

60. A.J. Leggett, Prog. Theor. Phys. 69:80 (1980)

61. K. Hepp, Helv. Phys. Acta 45:237 (1972)

62. B.D. Josephson, Rev. Mod. Phys. 46:251 (1974)

63. M. Takesaki, "Tomita's Theory of Modular Hilbert Algebras and its Applications", Springer, Berlin (1970)

64. R. Haag, Nuovo Cim. 25:287 (1962)

65. P.W. Anderson, in: E.R. Caianiello, "Lectures on the Many Body Problem", Academic Press, London (1963)

66. U. Eckern, G. Schön, and V. Ambegaokar, Phys. Rev. B 30: 6419 (1984)

67. J. Clarke and G. Schön, Europhys. News 17:94 (1986)

68. T. Unnerstall and A. Rieckers, Phys. Rev. B 39:2173 (1989)

69. A.O. Caldeira and A.J. Leggett, Ann. Phys. 149:374 (1983)

70. R.F. Voss and R.A. Webb, Phys. Rev. Lett. 47:647 (1981)

71. M.H. Devoret, J.M. Martinis, and J. Clarke, Phys. Rev. Lett. 55:1908 (1985)

72. J.M. Martinis, M.H. Devoret, and J. Clarke, Phys. Rev. Lett. 55:1543 (1985)

73. A.J. Leggett and A. Garg, Phys. Rev. Lett. 54:857 (1985)

74. A. Amann, this Proceedings

75. A.J. Leggett, S. Chakravarty, A.T. Dorsey, M.P.A. Fisher, A. Garg, and W. Zwerger, Rev. Mod. Phys. 49:1 (1987)

76. R. Silbey and R.A. Harris, J. Phys. Chem. 93:7072 (1989), and R. Silbey, this Proceedings

77. T. Unnerstall, Ph.D. Thesis, Tübingen (1990). (There the mathematical frame in the case of extensive junction couplings and reservoir couplings is analyzed. The somewhat negative result seems to provide further arguments for our weak coupling model.)

NON-EQUILIBRIUM STATISTICAL MECHANICS:

DYNAMICS OF MACROSCOPIC OBSERVABLES

Geoffrey L. Sewell

Department of Physics
Queen Mary and Westfield College
Mile End Road, London E1, 4NS

Abstract

We present an approach to non-equilibrium statistical thermodynamics within the framework of the quantum theory of infinite systems. This framework is natural for a mathematically precise formulation of irreversible processes, since it is only by idealising many-particle systems as infinite that Poincare' cycles can be eliminated. We start by illustrating how this idealisation can indeed lead to both irreversibility and macroscopic causality by a treatment of a simple, exactly solvable model, corresponding to a 'heavy' particle, that interacts with a 'heat bath'. Specifically, we show that the motion of the heavy particle conforms to a classical, deterministic, irreversible law, in the limit where the ratio of its mass to that of an atom of the bath tends to infinity: this limit serves to characterise the macroscopicity of the large particle. We next explain why, for a *general* treatment of infinite systems, it is neccessary to extend the standard quantum-mechanical framework, designed for finite assemblies of particles, to a form based on the algebraic structure of its observables (cf. [1-4]). We provide a brief, self-contained account of this generalised quantum mechanical framework and demonstrate how, by contrast with the traditional one for finite systems, it accommodates a precise distinction between microscopic and macroscopic quantities and even between different levels of macroscopicality. We employ it to obtain a general quantum statistical thermodynamical formulation both of equilibrium states and phases and of irreversible deterministic processes. In this way, we obtain a non-linear generalisation of the Onsager reciprocity relations for a class of irreversible processes in continuum mechanics.

Large-Scale Molecular Systems, Edited by W. Gans *et al.*
Plenum Press, New York, 1991

1. Introduction

Non-equilibrium statistical mechanics is concerned with the connection between the microscopic and macroscopic laws governing the dynamical behaviour of many-particle systems. The problem of establishing the nature of this connection is, of course, a very old one, whose study has been fraught with paradoxes, such as the fact that the macroscopic evolution is generally irreversible, by the second law of thermodynamics, whereas the underlying microscopic dynamics is reversible. However, although there is an enormous literature on the subject, dating back to Boltzmann's time, it is only in relatively recent years that a framework has been constructed, within which the theory can be systemmatically developed in a form that reveals the old paradoxes to be pseudo-problems. This is the so-called algebraic framework [1-4], which has been set up for the purpose of treating the properties of N-particle systems in the thermodynamic limit, in which N and the volume become infinite, while the density remains finite. The advantage gained from this idealisation is that it serves to reveal, in mathematically precise terms, intrinsic properties of these systems, that would otherwise be masked by finite size-effects. For example, it permits the characterisation of phase transitions by thermodynamic singularities and symmetry changes, which could not emerge, in a strict mathematical sense, from a theory of finite systems. As regards non-equilibrium phenomena, this framework accommodates a theory of irreversibility by removing the Poincare' cycles, that obstruct such a theory for finite systems, and thus leading to time-correlation functions with 'good' decay properties, corresponding to dissipative processes.

Let us now specify the problem involved in passing from the microscopic to the macroscopic level of description. We take the view that, at the former level, the dynamics of a many-particle system is governed by quantum, rather than classical, laws, since the very stability of matter [5], as well as phenomena such as superconductivity and laser radiation, depend in an essential way on quantum mechanics. Thus, we assume that, at the microscopic level, the dynamics of the system is governed by the Schrödinger equation* for its state ψ, which in a standard notation, takes the form

$$i\hbar\frac{d\psi}{dt} = H\psi \qquad (1.1)$$

Since this equation is invariant under time-reversals $t \to -t$, $\psi \to \psi^*$, the microdynamics is reversible. Furthermore, according to the standard interpretation of quantum theory, this dynamics is probabilistic. On the other hand, it is an empirical fact that, at the macroscopic level, we generally have particular subsets of variables $a = (a_1, \ldots, a_k)$, which evolve according to *self-contained, deterministic* laws, of the form

$$\frac{da}{dt} = \phi(a), \qquad (1.2)$$

that conform to the second principle of thermodynamics and thus imply an *irreversible* approach to equilibrium. Here, a may be either a discrete set of variables, such as the velocity components of a macroscopic particle, or a set of space dependent fields, of

---

* Here, for pedagogical simplicity, we are describing the situation within the framework of the standard quantum theory of finite systems. The problem can be formulated analogously (cf. §5) within the framework of the generalised quantum theory, presented in §3, which is applicable to infinite systems.

the hydrodynamical type. The basic problems of non-equilibrium statistical thermo-dynamics are

(I) to obtain a quantum mechanical characterisation of the macroscopic variables a that conform to the above-described phenomenological laws;

(II) to determine the origin of the irreversible and deterministic character of these laws; and

(III) to infer general properties of the form of these laws from those of the underlying quantum mechanical ones.

We remark that the macroscopic description is a highly contracted one, in that the number of variables on which it is based is extremely small by comparison with the number of microscopic degrees of freedom of the system. This implies that the values of the macro-variables a do not suffice to determine the microstate of the system. Thus, given the initial value of a, which can presumably be obtained by measurement, some statistical assumptions are needed to determine the initial microstate. The problem then is to extract the the phenomenological law governing a, in the subsequent evolution of the system, from the microscopic dynamics.

To be more specific, we may regard the problem as consisting of the following steps.

(A) To characterise a set of macroscopic quantum mechanical observables $\hat{a} = (\hat{a}_1, \ldots, \hat{a}_k)$, that we take to be candidates for a self-contained phenomenological law. We note here that the macroscopicality of these observables should be represented by some dimensionless size parameter $\Gamma \gg 1$, which enters into their description [6]-[8]. This could be, for example, the number of particles in a subvolume of a fluid, in a hydrodynamical situation, or the ratio of the mass of a Brownian particle to that of a molecule of the host fluid. Further, the distinction between the macroscopic observables and the microscopic ones becomes sharp only in the limit $\Gamma \to \infty$.

(B) To obtain conditions under which the time-dependent expectation value $a(t)$ of $\hat{a}$, for evolution of the system from a suitable class of initial microstates, does indeed conform to an irreversible phenomenological law of the form (1.2), at least in the limit, dicussed in (A), where $\Gamma \to \infty$.

(C) To obtain conditions under which the time-dependent dispersion of $\hat{a}$ vanishes in this limit.

Thus, the objective is to find conditions under which the observables $\hat{a}$ evolve according to a closed dynamical law, which is both irreversible and deterministic, i. e. dispersion-free in the macroscopic limit $\Gamma \to \infty$.

We shall now substantiate the contention made at the beginning of this article that a finite system cannot exhibit the required irreversibility properties, and that therefore one must pass to the idealisation of an infinite system in order to obtain these properties. To this end, we represent the observables and states of a finite system, in a standard way, by the self adjoint operators, A, and density matrices, ρ, respectively, in a Hilbert space, $\mathcal{H}$, the expectation value of the observable A for the state ρ being $Tr(\rho A)$. Here, ρ is a pure state if it is a one-dimensional projector. The dynamics of the system is governed by its Hamiltonian operator H, i.e. its energy observable, according to the principle that the evolution of the state over time t is given, in the Schrödinger picture, by the unitary transformation $\rho \to \exp(-iHt/\hbar)\rho\exp(iHt/\hbar)$.

Thus, the time-dependent expectation value of an observable A, for evolution of the system from an initial state ρ, is

$$\langle A \rangle_t = Tr[\exp(-iHt/\hbar)\rho \exp(iHt/\hbar)A] \qquad (1.3)$$

Furthermore, the Helmholtz free energy of the system at inverse temperature $(k\beta)^{-1}$, with k Boltzmann's constant, is

$$\Phi = -\beta^{-1}\ln(Tr(\exp(-\beta H))) \qquad (1.4)$$

Assuming that this is finite, the operator $\exp(-\beta H)$, and hence H, must have a discrete spectrum [9]. Thus, the Hamiltonian takes the form

$$H = \sum E_n P_n \qquad (1.5)$$

where P_n is the eigenprojector for the energy level E_n. It follows now from (1.3) and (1.5) that

$$\langle A \rangle_t = \sum_{m,n} Tr(P_m \rho P_n A) exp(i\omega_{m,n} t), \ \ with \ \ \omega_{m,n} = \frac{E_n - E_m}{\hbar} \qquad (1.6)$$

Since the r.h.s. of this equation is a convergent sum of simply periodic functions of t, the time-dependent expectation value of *any* observable, whether macroscopic or not, must be quasi-periodic and hence cannot evolve irreversibly to an equilibrium value. We conclude from this that any hope of a theory of irreversibility has to rest on the properties of the infinite system model.

Our objective, then, will be to formulate an approach to non- equilibrium statistical thermodynamics within the framework of the quantum theory of infinite systems. We start, in §2, with a treatment of a simple, solvable model, corresponding to a 'heavy' particle, B, harmonically coupled to an infinite chain of atoms, which plays the role of a 'heat bath'. We are able to obtain the principal properties of this model by elementary methods, which do not require the full apparatus of the quantum theory of infinite systems. We investigate the dynamics of B, in the limit where the ratio of its mass that of an atom tends to infinity and under initial conditions where the chain is in a thermal equilibrium state, uncorrelated to B. The result we obtain is that, under these conditions, the motion of B, on an appropriate time-scale, corresponds to a classical, deterministic, dissipative law, governed by a frictional force proportional to its speed. Thus, the model does indeed exhibit irreversibility and macroscopic determinism in the motion of B. This tells us that the quantum theory of infinite systems provides a framework within which a theory of irreversible statistical thermodynamics is feasible.

In order to pursue the theory in a general context, we present, in §3, a simple account of the algebraic formulation of quantum theory, which is applicable both to finite and to infinite systems. The essential reason why we need this formulation , in the case of an infinite system, is that there are (infinitely) many inequivalent Hilbert space representations of its observables: by contrast, a finite systems admits only one irreducible representation. Thus, the model of the infinite system has to be based on the algebraic structure of its observables, rather than on any particular Hilbert space representation of them. This model is constructed so as to provide a precise formulation of the observables, states and dynamics of the system, in a form corresponding to a 'natural' generalisation of the standard quantum theory of finite systems.

We pass on, in §4, to a general formulation of equilibrium statistical thermodynamics, within the framework of the infinite system model. This is centred on global intensive observables, that are densities of conserved quantities. The resultant theory leads to classical thermodynamics *with phase structure* (cf.[4,Ch.4]), a structure that could not be accommodated within the traditional framework of finite systems.

Our treatment of non-equilibrium statistical thermodynamics is then presented in §5. This is concerned with the the problem of obtaining the restrictions imposed by quantum theory on the form of the phenomenological laws of continuum mechanics, e. g. hydrodynamics or heat conduction, which are in general non-linear. Thus, our essential aims are similar to those of the Onsager theory [10], though now we have the advantage of a formalism that admits a systemmatic and much more general treatment of the problem than was available at the time of that theory. Our treatment is based on the assumptions that the macroscopic observables are local densities, on a certain large distance (hydrodynamic) scale, of the conserved quantities that form the basis of the equilibrium thermodynamics; and that these observables evolve according to a phenomenological law with suitable scaling properties. Furthermore, we assume a generalised version of Onsager' s celebrated regression hypothesis [10]. Specifically, we assume that the linearised dynamics, corresponding to weak perturbations of the phenomenological law, takes the same form as the regressions of spontaneous fluctuations of the macro- observables about that law. This enables us to define local equilibrium, in a hydrodynamical sense, in terms of the correlations of these fluctuations. On the basis of the scheme we have just described, we obtain generalisations both of the Onsager reciprocity relations [10] and of the Onsager-Maschlup fluctuation theory [11] to situations where the phenomenological law is non-linear.

We shall conclude, in §6, with a brief discussion of other, related developments in the subject.

There will be two Appendices, which establish points that are often taken for granted, though there does not appear to be any explicit proofs of them in the literature. In the first Appendix, we shall sketch a proof of the differentiability of the specific entropy function of the densities of the extensive conserved quantities, constituting the thermodynamical variables of an infinite system: this signifies, for example, that one can define the inverse temperature as the derivative of the entropy w.r.t. the energy. In the second Appendix, we shall derive a much used formula relating the second derivatives of the specific entropy to correlations of fluctuations of macroscopic observables.

We shall aim throughout to emphasise the conceptual structure of the theory, rather than the mathematical technicalities, which we shall try to keep to a minimum, referring to other works for rigorous proofs. In particular, a rigorous treatment of the model of §2, within the frameworks of [14], [15], would lead to the same results that we obtain here.

2. Solvable Model Exhibiting Irreversibility and Macroscopic Causality

2.1. The Model: We take as our model, Σ a semi-infinite harmonic chain, transversely vibrating in a plane, such that the first particle, which we term B, is of mass M, and the rest of the particles are of mass m. This is a type of model that has been extensively studied for various purposes [12-16].

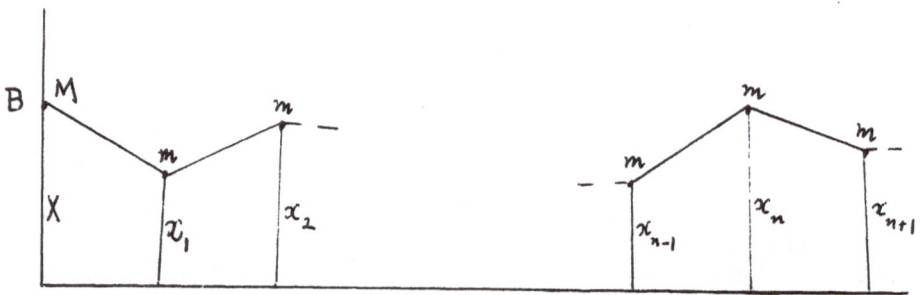

We assume that $M >> m$, with the view that B is considered as macroscopic, in a sense that will presently be made precise, and that the rest of the particles be regarded as 'atoms'. We denote the displacement and momentum of B by X and P, and that of the n'th atom by x_n and p_n, respectively. We assume that the formal Hamiltonian, governing the dynamics of the system, is

$$H = \frac{P^2}{2M} + \frac{1}{2}a(X - x_1)^2 + \sum_1^\infty [\frac{p_n^2}{2m} + \frac{1}{2}a(x_{n+1} - x_n)^2]$$

with $a > 0$. We denote the mass ratio m/M by α and, for simplicity, we employ units for which for which $m = a = 1$. Thus, the above Hamiltonian reduces to the form

$$H = \frac{1}{2}\alpha P^2 + \frac{1}{2}(X - x_1)^2 + \frac{1}{2}\sum_1^\infty [p_n^2 + (x_{n+1} - x_n)^2] \qquad (2.1)$$

and the velocity observable of the particle B is

$$V = \alpha P \qquad (2.2)$$

Our objective will be to extract the dynamics of B from that of the entire system Σ, subject to specified initial conditions, on a certain 'macroscopic' time-scale and in a limit where the ratio, α, of the atomic to the B-mass tends to zero. We shall generally denote the expectation value of an observable Q of Σ for the state under consideration, whether pure or mixed, by $\langle Q \rangle$ or, sometimes, $\overline{Q}$; and we shall denote the Heisenberg operator, representing the evolute of Q at time t by $Q(t)$. Thus, the time-dependent expectation value of Q, for evolution from an initial state, whose means are represented by $\langle . \rangle$, is

$$\langle Q \rangle_t = \overline{Q}(t) = \langle Q(t) \rangle \qquad (2.3)$$

We introduce also an auxiliary model, $\Sigma^{(0)}$, consisting of the atomic chain alone, with the first particle coupled to the fixed origin, O, instead of B, with potential energy $\frac{1}{2}x_1^2$. Thus, the formal Hamiltonian for $\Sigma^{(0)}$ is

$$H^{(\prime} = \frac{1}{2}\sum_0^\infty [p_n^2 + (x_{n+1} - x_n)^2], \; with \; x_0, p_0 \equiv 0 \qquad (2.4)$$

82

We shall denote the Heisenberg operator for the evolute of an observables Q of $\Sigma^{(0)}$ by $Q^{(0)}(t)$.

As stated above, our objective will be to extract the dynamics of B from that of the entire system Σ. The initial conditions we assume are the following.

(I.1) The states of the particle B and the chain $\Sigma^{(0)}$ are uncorrelated.

(I.2) The chain $\Sigma^{(0)}$ is in thermal equilibrium at temperature $(k\beta)^{-1}$, k being Boltzmann's constant. As we shall see in §4, the equilibrium condition for an infinite system is given by the fluctuation-dissipation relation of Kubo, Martin and Schwinger (KMS), which we may express as follows here. Let A, B be observables of $\Sigma^{(0)}$, let $F_{\pm}(t)$ be the functions defined by

$$F_{\pm}(t) = \langle [A^{(0)}(t), B]_{\pm} \rangle^{(0)}, \tag{2.5}$$

where $[\ ,\ ]_{\pm}$ denote commutator and anticommutator, respectively, and let $\hat{F}_{\pm}$ be their Fourier transforms, i. e.

$$\hat{F}_{\pm}(u) = \int F_{\pm}(t) \exp(-iut) dt, \tag{2.6}$$

Then the KMS equilibrium condition, for inverse temperature β is that

$$\hat{F}_{+}(u) = \hat{F}_{-}(u) \coth(\frac{1}{2}\beta\hbar u) \tag{2.7}$$

for arbitrary observables A, B of $\Sigma^{(0)}$.

(I.3) The state of B is given by a Gaussian wave-packet, for which the means, dispersions and correlation of X, V are given by

$$\langle X \rangle = X_0; \ \langle V \rangle = V_0 \tag{2.8}$$

$$\Delta X \equiv \langle (X - X_0)^2 \rangle^{1/2} = \alpha^{1/2} l; \ \Delta V \equiv \langle (V - V_0)^2 \rangle^{1/2} = \alpha^{1/2} \hbar / l \tag{2.9}$$

where l is some fixed length, independent of α; and

$$\langle [X - X_0, V - V_0]_{+} \rangle = 0 \tag{2.10}$$

Note that these conditions are compatible with the Heisenberg principle, by virtue of (2.2).

The results we shall obtain, subject to these conditions are the following.

(R.1) Let τ be the time on the scale for which the unit interval corresponds to α^{-1} on the t-scale, and let $v(\tau)$ be the Heisenberg operator corresponding to the time-dependent velocity observable of B on this scale, i. e.

$$v(\tau) = V(\tau/\alpha) \tag{2.11}$$

Then, in the limit $\alpha \to 0$, which represents the macroscopicality of B, v evolves according to the irreversible, deterministic, i. e. dispersion-free, law

$$\frac{dv(\tau)}{d\tau} = -v(\tau) \tag{2.12}$$

(R.2) Let $w(\tau)$ be the fluctuation of the velocity of B about its mean, as rescaled by the factor $\alpha^{-1/2}$, i. e.

$$w(\tau) = \alpha^{-1/2}(v(\tau) - \langle v(\tau) \rangle) \tag{2.13}$$

Then, in the limit $\alpha \to 0$, w describes a classical Brownian motion, according to the formula given by the Langevin equation

$$\frac{dw(\tau)}{d\tau} = -w(\tau) + f(\tau), \tag{2.14}$$

where f is a Gaussian, stochastic force, with zero mean and autocorrelation function

$$\langle f(\tau)f(\tau') \rangle = \beta^{-1}\delta(\tau - \tau'), \tag{2.15}$$

subject to the initial condition that $w(0) = 0$.

Comments.(1) Since the frictional forces in (R.1,2) are of precisely the same form, namely *minus the velocity*, these results conform to Onsager's hypothesis [10] that the dynamical laws governing the dterministic evolution of the macroscopic variables and the regression of their fluctuations are identical. Further, the fluctuation process (R.2) is of the form assumed by Onsager and Maschlup [11].

(2) The dissipation occuring in (R.1) arises as a result of B radiating its energy away down the atomic chain.

(3) The scaling factor $\alpha^{-\frac{1}{2}}$ in the definition of w reflects the fact that the dispersion in the thermal fluctuation of P is $O(\alpha^{-\frac{1}{2}})$ and therefore that of V is $O(\alpha^{\frac{1}{2}})$.

(4) The results (R1,2) could be obtained under conditions more general than (I.1-3) (cf.[17]). For example, it is not essential that, at $t = 0$, the subsystems B, $\Sigma^{(0)}$ be completely decorrelated or that $\Sigma^{(0)}$ be in thermal equilibrium. It is neccessary, however, that there is sufficient initial randomness to exclude, for example, coherent incoming waves that could lead to a boost, rather than a retardation, of the velocity of B.

2.2. Equations of Motion. Since Σ is a harmonic system, its Heisenberg equations of motion take the same form as the Hamilton equations of a classical version of the model, and thus, by (2.1), are given by the following formulae.

$$\frac{dX(t)}{dt} = V(t); \quad \frac{dV(t)}{dt} = \alpha(x_1(t) - X(t)) \tag{2.16}$$

and

$$\frac{dx_n(t)}{dt} = p_n(t); \quad \frac{dp_n(t)}{dt} = x_{n+1}(t) - 2x_n(t) + x_{n-1}(t) + X(t)\delta_{n,1} \tag{2.17}$$

We note also that, by (2.4), the Heisenberg equations for $\Sigma^{(0)}$ are

$$\frac{dx_n^{(0)}(t)}{dt} = p_n^{(0)}(t); \quad \frac{dp_n^{(0)}(t)}{dt} = x_{n+1}^{(0)}(t) - 2x_n^{(0)}(t) + x_{n-1}^{(0)}(t) \tag{2.18}$$

We see from (2.16) that, in order to obtain the dynamics of the particle B, we need to solve for $x_1(t)$ in terms of X, V and the initial conditions. For this purpose, we pass to a normal mode description for $\Sigma^{(0)}$, introducing

$$x(k) = \sum_{n=0}^{\infty} x_n \sin kn; \quad p(k) = \sum_{n=0}^{\infty} p_n \sin kn \tag{2.19}$$

with x_0, $p_0 = 0$; or conversely

$$x_n = \frac{2}{\pi} \int_0^\pi dk x(k) \sin kn; \; p_n = \frac{2}{\pi} \int_0^\pi dk p(k) \sin kn \qquad (2.20)$$

It follows now from (2.19) that the equations (2.17) transform to

$$\frac{dx(k,t)}{dt} = p(k,t); \; \frac{dp(k,t)}{dt} + \omega(k)^2 x(k,t) = X(t)\sin k, \qquad (2.21)$$

where

$$\omega(k) = 2\sin(k/2) \qquad (2.22)$$

By (2.18) and (2.19), the equations of motion for $\Sigma^{(0)}$ are obtained by replacing $x(k,t)$, $p(k,t)$ by $x^{(0)}(k,t)$, $p^{(0)}(k,t)$ in (2.21) and removing the last term there. Hence, it follows from (2.21) and (2.22) that

$$x(k,t) = x^{(0)}(k,t) + \cos(k/2) \int_0^t \sin(\omega(k)s) X(t-s) ds \qquad (2.23)$$

with

$$x^{(0)}(k,t) = x(k)\cos(\omega(k)t) + \frac{p(k)}{\omega(k)}\sin(\omega(k)t) \qquad (2.24)$$

We now use (2.21) and (2.21) to obtain the form of $x_1(t)$, namely

$$x_1(t) = x_1^{(0)}(t) - \int_0^t \dot{\phi}(s) X(t-s) ds \qquad (2.25)$$

where

$$\phi(s) = \frac{2}{\pi} \int_0^\pi \cos^2(k/2)\cos(\omega(k)s) dk$$

i. e., putting $u = \omega(k)$ and using (2.22) and the evenness of cos,

$$\phi(s) = \frac{1}{\pi} \int_{-2}^2 (1 - u^2/4)^{1/2}\cos(us) du \qquad (2.26)$$

We see from this formula that $\phi(0) = 1$ and thus that the integral in (2.25) is equal to

$$\phi(t)X(0) - X(t) + \int_0^t \phi(s)\dot{X}(t-s) ds,$$

and that consequently, by (2.16) and (2.25), the equation of motion for the velocity of B is

$$\frac{dV(t)}{dt} = -\alpha \int_0^t \phi(s)V(t-s) ds + \alpha x_1^{(0)}(t) - \alpha\phi(t)X(0) \qquad (2.27)$$

This is a generalised Langevin equation, the three terms on the r.h.s. representing a retarded frictional force, a stochastic force and a 'memory' of the initial position of B. In order to obtain the stochastic properties of the second term, we first note that, by the stationarity of the equilibrium state of $\Sigma^{(0)}$, $\langle x^{(0)}(t) \rangle$ is a constant; and assuming that the equilibrium state preserves the invariance of the dynamics under the reversals $x, p \rightarrow - x, -p$, (cf. (2.4)), this constant must vanish, i. e.

$$\langle x^{(0)}(t) \rangle = 0 \qquad (2.28)$$

Further, we can obtain the time-correlation functions for $x_1(t)$ from the dynamics of $\Sigma^{(0)}$ and the KMS condition (2.7). For it follows from (2.19), (2.20), (2.24) and (2.26) that $x_1^{(0)}(t)$ is a time-dependent linear combination of the x_n' s and the p_n' s, and that the coefficient of p_1 is $-\dot{\phi}(t)$. Hence, by (2.5), (2.7), (2.26) and the canonical commutation relations $[x_j, p_k]_{\pm} = i\hbar\delta_{jk}$, we obtain the formulae

$$\langle [x_1^{(0)}(t), x_1]_- \rangle = i\hbar\dot{\phi}(t) \qquad (2.29)$$

and

$$\langle [x_1^{(0)}(t), x_1]_+ \rangle = \psi(t) \equiv \int_{-2}^{2} (1 - u^2/4)^{1/2} \hbar u \coth(\frac{1}{2}\beta\hbar u)\cos(us) \qquad (2.30)$$

Since, by (2.4), the Hamiltonian H^0 of $\Sigma^{(0)}$ is quadratic in the coordinates and momenta of its particles, the fluctuations of $x_1^{(0)}(t)$ correspond to a Gaussian process, and therefore its properties are completely determined by the formulae (2.28)-(2.30).

We note now that it follows from (2.27)-(2.30) that the dynamics of B is governed by the forms of the functions ϕ and ψ. Most importantly, it follows from (2.26) and (2.30) that, by the Riemann-Lebesgue lemma,[*] $\phi(t)$, $\psi(t) \rightarrow 0$ as $t \rightarrow \pm\infty$. We shall see that these decay properties of ϕ, ψ are the key to the irreversibility in the motion of B. Moreover, this decay could not possibly arise in a finite chain, since there the displacement of each atoms would be a finite sum of periodic contributions, due to normal modes, and would thus be quasi-periodic. (cf. the discussion of §1.)

2.3. The Macroscopic Description of B. Since α is the ratio of the atomic mass to that of B, we characterise the macroscopicality of the latter particle by the condition that $\alpha << 1$ and idealise it by passing to the limit $\alpha \rightarrow 0$, in a sense that we have to specify precisely.

Thus, we start by noting that, by (2.27), the rate of change of $V(t)$ is proportional to α. This suggests that the time scale appropriate to a macroscopic description should have $t = \alpha^{-1}$ as its unit. Accordingly, we explore the dynamics of B on this scale. Thus, to pass to the macroscopic description, we define the time to be $\tau = \alpha t$ and the velocity of B to be

$$v_\alpha(\tau) = V(\tau/\alpha) \qquad (2.31)$$

We also define $\xi_\alpha(\tau)$ to be the 'stochastic force' $x_1^{(0)}(t)$, as represented on the macroscopic scale, i. e.

$$\xi_\alpha(\tau) = x_1(\tau/\alpha) \qquad (2.32)$$

it follows from (2.26)-(2.32) that the generalised Langevin equation for B takes the following form in the macroscopic description.

$$\frac{dv_\alpha(\tau)}{d\tau} + \int_0^{\tau/\alpha} ds\phi(s)v_\alpha(\tau - \alpha s) = \xi_\alpha(\tau) - \phi(\tau/\alpha)X(0) \qquad (2.33)$$

---

[*] This states that, if $\int_a^b |f(u)|du$ is finite, then $\int_a^b f(u)\exp(iut) \rightarrow 0$ as $t \rightarrow \pm\infty$.

and that, in view of the stationarity of the equilibrium state of Σ_0, the stochastic properties of the Gaussian fluctuating force $\xi_\alpha(\tau)$ are given by

$$\langle \xi_\alpha(\tau) \rangle = 0 \tag{2.34}$$

$$\langle [\xi_\alpha(\tau), \xi_\alpha(\tau')]_- \rangle = i\hbar \dot{\phi}(\alpha^{-1}(\tau - \tau')) \tag{2.35}$$

and

$$\langle [\xi_\alpha(\tau), \xi_\alpha(\tau')]_+ \rangle = \psi(\alpha^{-1}(\tau - \tau')) \tag{2.36}$$

Our aim now is to obtain the properties of $v_\alpha(\tau)$ for each fixed τ in the limit that α tends to zero. For this purpose, we note the following key properties of the functions ϕ, ψ, which from the formulae (2.24), (2.27), together with the above-mentioned Riemann-Lebesgue lemma and standard Fourier analysis.

(P.1) The functions ϕ, ψ and their derivatives are bounded, and $\phi(t)$ and $\psi(t)$ are tend to zero as $t \to \pm\infty$

(P.2) $\int_0^T \phi(s)ds$ and $\int_0^T \psi(s)ds$ are bounded functions of T and tend to 1 and β^{-1}, respectively, as $T \to \infty$.

These are the properties that will enable us to obtain the phenomenological (2.12) and the fluctuation dynamics (2.14). To indicate how they lead to (2.12), for example, let us first pass naively to the limit $\alpha \to 0$ of (2.33), by putting $\alpha = 0$ in the second terms on the l.h.s. and r.h.s. of that formula, and also in equations (2.35), (2.36) for the autocorrelation functions for the $\xi_\alpha(\tau)$. Then one sees that, for $\tau > 0$, it follows from (P.1,2) that the second terms on the l.h.s. and r.h.s. reduce to $v(\tau)$, 0, respectively, and that the autocorrelation functions for $\xi_\alpha(\tau)$ vanish. In other words, the generalised Langevin equation (2.33) reduces to the deterministic law (2.12). In the next two subsections, we shall provide a systemmatic derivation of (2.12) and (2.14).

2.4. The Phenomenological Law. We commence our treatment of the generalised Langevin equation (2.33) by integrating it from 0 to t, thereby converting it into the integral equation, which is amenable to the passage to the limit $\alpha \to 0$ (cf.[15]).

$$v_\alpha(\tau) + \int_0^\tau d\sigma k_\alpha(\tau - \sigma)v_\alpha(\sigma) = v_\alpha(0) + \int_0^\tau d\sigma(\xi_\alpha(\sigma) - \phi(\sigma/\alpha)X(0)) \tag{2.37}$$

where

$$k_\alpha(\sigma) = \int_0^{\sigma/\alpha} ds\phi(s) \tag{2.38}$$

It follows from the property (P2) of ϕ that the function k_α is uniformly bounded w.r.t. α, and that

$$k_\alpha(\sigma) \to 1 \; as \; \alpha \to 0, \; \forall \sigma > 0 \tag{2.39}$$

We shall formulate the solution of (2.37) in terms of Green functions. For this, we note that it follows from standard arguments in the theory of Volterra integral equations that, since k_α is uniformly bounded, the equation

$$g_\alpha(\tau, \sigma) + \int_\sigma^\tau k_\alpha(\tau, \sigma')g_\alpha(\sigma', \sigma) = 1 \tag{2.40}$$

has a unique solution for g_α, given that $\tau \geq \sigma \geq 0$ and that this is bounded, uniformly w.r.t. α, for $0 \leq \sigma \leq \tau \leq \tau_0$, for any fixed and finite τ_0. Further, the solution of (2.37) may be expressed in terms of g_α by the formula

$$v_\alpha(\tau) = g_\alpha(\tau, 0)v_\alpha(0) + \int_0^\tau d\sigma g_\alpha(\tau, \sigma)(\xi_\alpha(\sigma) - \phi(\sigma/\alpha)X(0)) \qquad (2.41)$$

It also follows from (2.39),(2.40) and the uniform boundedness of k_α that [15]

$$g_\alpha(\tau, \sigma) \to g_0(\tau, \sigma) \ as \ \alpha \to 0 \qquad (2.42)$$

where g_0 is the solution of the formal limit of equn. (2.40), as $\alpha \to 0$, i. e.

$$g_0(\tau, \sigma) + \int_\sigma^\tau d\sigma' g_0(\sigma', \sigma) = 1$$

and therefore is given by the formula

$$g_0(\tau, \sigma) = \exp(-(\tau - \sigma)) \qquad (2.43)$$

We now employ equation (2.41), together with the initial conditions (I.1-3) of §2.1, to obtain the mean and dispersion of $v_\alpha(\tau)$, and then pass to the limit $\alpha \to 0$ for these quantities. Thus, we note that, by equns. (2.41), (2.31), (2.34) -(2.36) and conditions (I.1), (I.3),

$$\bar v_\alpha(\tau) = g_\alpha(\tau, 0)V_0 - \int_0^\tau d\sigma g_\alpha(\tau, \sigma)\phi(\sigma/\alpha)X_0 \qquad (2.44)$$

and, as the equilibrium state of $\Sigma^{(0)}$ is stationary and thus $\langle \xi_\alpha(\sigma)\xi_\alpha(\sigma')\rangle$ depends only on the difference $\sigma - \sigma'$

$$(\Delta v_\alpha(\tau))^2 = (g_\alpha(\tau, 0))^2 \alpha \hbar^2/l^2$$

$$+ \int_0^\tau \int_0^\tau d\sigma d\sigma' g_\alpha(\tau, \sigma)g_\alpha(\tau, \sigma')[\frac{1}{2}\psi(\alpha^{-1}(\sigma - \sigma')) + \phi(\sigma/\alpha)\phi(\sigma'/\alpha)\alpha l^2] \qquad (2.45)$$

It follows from these equations, together with (2.42), (2.43) and the properties (P1,2) of ϕ, ψ, specified in §2.3, that

$$\bar v_\alpha(\tau) \to v(\tau) \equiv V_0 \exp(-\tau) \ as \ \alpha \to 0 \qquad (2.46)$$

and

$$\Delta v_\alpha(\tau) \to 0 \ as \ \alpha \to 0 \qquad (2.47)$$

These two equations signify that, in the limit where α tends to zero, the velocity of B becomes dispersion-free and satisfies the phenomenological law (2.12). Thus the model exhibits irreversibility and macroscopic causality, as represented by that law.

2.5. The Fluctuation Dynamics. We define the velocity fluctuation observable to be (cf. (2.13))

$$w_\alpha(\tau) = \alpha^{-1/2}(v_\alpha(\tau) - \bar v_\alpha(\tau)), \qquad (2.48)$$

the normalisation factor $\alpha^{-1/2}$ representing the fact that the equilibrium dispersion of v is of the order of $\alpha^{1/2}$, i.e. $M^{-1/2}$. It follows from equns. (2.31), (2.41) and (2.48) that w_α is given explicitly by the equation

$$w_\alpha(\tau) = \alpha^{-1/2} g_\alpha(\tau, 0)(V(0) - \overline{V}(0))$$

$$+ \alpha^{-1/2} \int_0^\tau d\sigma g_\alpha(\tau, \sigma)[\xi_\alpha(\sigma) - \phi(\sigma/\alpha)(X(0) - \overline{X}(0))] \qquad (2.49)$$

We note that the fluctuation process w_α is completely specified by the autocorrelation function $\langle \xi_\alpha(\tau)\xi_\alpha(\tau')\rangle$ since, by condition (I.3), the initial state of Σ is Gaussian and, by (2.48), $\langle w_\alpha(\tau)\rangle = 0$. By condition (I.1) and equations (2.9), (2.10), (2.32), (2.34)-(2.36) and (2.49), this autocorrelation function is given by

$$\langle w_\alpha(\tau)w_\alpha(\tau')\rangle = \frac{\hbar^2}{l^2} g_\alpha(\tau)g_\alpha(\tau') + \int_0^\tau d\sigma \int_0^{\tau'} d\sigma' \times$$

$$(g_\alpha(\tau, \sigma)g_\alpha(\tau', \sigma')[\frac{1}{2\alpha}\psi(\alpha^{-1}(\sigma - \sigma')) + \frac{i\hbar}{2\alpha}\dot{\phi}(\alpha^{-1}(\sigma - \sigma')) + l^2\phi(\sigma/\alpha)\phi(\sigma'/\alpha)] \quad (2.50)$$

In order to pass to the limiting form of this equation, as $\alpha \to 0$, we observe that, by (2.26) and (2.30), $\alpha^{-1}\phi(\tau/\alpha)$, $\alpha^{-1}\psi(\tau/\alpha)$ tend to $1, 2\beta^{-1}$, respectively as $\alpha \to \infty$. Hence, in view of equns. (2.42), (2.43), together with the properties (P.1,2) of ϕ, ψ and the uniform boundedness of g_α, noted after (2.40), it follows from (2.50) that

$$\lim_{\alpha \to \infty} \langle w_\alpha(\tau)w_\alpha(\tau')\rangle = (\Delta w(0))^2 \exp(-(\tau + \tau')) + \beta^{-1}\int_0^{min(\tau,\tau')} d\sigma \exp(-(\tau + \tau' - 2\sigma))$$

$$(2.51)$$

We can now compare this result with that obtained for the classical stochastic process specified in (R.2). For the latter process is also Gaussian [18], and it is a simple matter to infer from (2.14),(2.15) that its autocorrelation function $\langle w(\tau)w(\tau')\rangle$ is identical to the r.h.s. of (2.51). Hence, the fluctuation fluctuation process w_α reduces to the classical one, w, of (R.2) in the limit where α tends to zero.

3. The Generalised Quantum Mechanical Framework

3.1. The Need To Go Beyond The Traditional Framework. We have seen in §'s 1,2 that a quantum theory of irreversible processes requires the idealisation of many-particle systems as infinite. We shall now explain why this requires a drastic extension of the standard formulation of quantum mechanics [19,20], that was designed for finite systems.

Thus, we first recall that, in the traditional formulation, the pure states and observables of a system are represented by the vectors and self-adjoint operators, respectively. For a system consisting of a finite number of particles, a classic theorem of Von Neumann [21] establishes that, given the algebraic structure imposed by the canonical commutation relations (CCR), there is only one way in which this representation can be achieved. To be specific, let q, p be self-adjoint operators in a Hilbert space $\mathcal{H}$, such that

$$qp - pq = i\hbar I \qquad (3.1);$$

and assume that this representation of the CCR satisfies the irreducibility condition

---

* If this latter condition were not fulfilled, the restrictions of q, p to $\mathcal{H}_0$ would provide a subrepresentation, in that space, of the original one. Thus, in general, the reducible representations may be decomposed into the ireducible ones, and therefore the latter may be regarded as basic.

that $\mathcal{H}$ contains no subspace $\mathcal{H}_0$ such that q, p map $\mathcal{H}_0$ into $\mathcal{H}_0$.* Similary, let q', p' be self-adjoint operators in another Hilbert space, $\mathcal{H}'$, that also provide an irreducible representation of the CCR,

$$q'p' - p'q' = i\hbar I' \qquad (3.2)$$

Then Von Neumann's theorem tells us that the two representations, and thus all irreducible ones, of the CCR are equivalent, i. e. that there is a unitary transformation $U : \mathcal{H} \rightarrow \mathcal{H}'$, such that

$$q = U^{-1}q'U; \ p = U^{-1}p'U$$

This implies that it suffices to use the Schrödinger representation, where $\mathcal{H}$ is the Hilbert space of square-integrable functions $f(x)$ of the real variable x and

$$qf(x) = xf(x); \ pf(x) = -i\hbar\frac{df(x)}{dx}, \ \forall f \in \mathcal{H}, \qquad (3.3)$$

since any other representation would yield identical predictions.

The essential uniqueness of the irreducible representation of the observables extends to finite systems of particles and also to finite spin systems, such as the Heisenberg ferromagnet. Thus, for such a system, the properties ensuing from the algebraic structure of the observables are completely captured by any of their irreducible representations. It is this that justifies the standard procedure of employing the Schrödinger formulation (3.3).

The situation changes completely when one passes to systems with infinite numbers of degrees of freedom [1-4]. For then, the conditions underlying Von Neumann's theorem becomes inapplicable and the algebraic structure of the observables can support inequivalent Hilbert space representations: in fact, it generally supports an infinity of them! Indeed, the physical picture obtained from a mathematical treatment of infinite systems of particles is that each representation corresponds to a family of locally different but globally equivalent states; while different representations correspond to globally different families of states. Thus, for example, in the case of a Heisenberg ferromagnet, there is a ground state corresponding to each polarisation direction and a Hilbert space representation whose vectors correspond to local modifications of that state (cf.[4, §2.3]). The representations so obtained are globally different from one another, since they correspond to different polarisation directions.

In view of this situation, the properties of an infinite system are governed by the algebraic structure of its obervables, rather than any particular representations of them.

3.2. The General Scheme. Let us now specify, in very general, convention-free terms, the essential ingredients of a model of a physical system. We assume that such a model, Σ, consists of three components $(\mathcal{O}, \mathcal{S}, \mathcal{D})$, corresponding to its observables, states and dynamics, respectively. Here, $\mathcal{O}$ is a set of quantities which can, in principle, be measured; and we assume that, in each specific case, $\mathcal{O}$ is equipped with a certain algebraic structure, e. g. that of the CCR.

The set of states, $\mathcal{S}$, serve to provide expectation values of the observables. Thus, they are functionals on $\mathcal{O}$ such that, if $A \in \mathcal{O}$ and $\rho \in \mathcal{S}$, then $\rho(A)$ is a real number, representing the expectation value of the observable A when the system is in the state ρ, i. e.

$$\rho(A) = \langle A \rangle_\rho$$

Finally, $\mathcal{D}$ is assumed to be a set of transformations $\{\tau(t_1, t_2)\}$ of the states $\mathcal{S}$, where $\tau(t_1, t_2)$ corresponds to the evolution over the interval from t_1 to t_2. Thus, if ρ is the state at time t_1, then $\tau(t_1, t_2)\rho$ is the state at time t_2.

It follows from this description of Σ that the model is determined by the structures assumed for $\mathcal{O}, \mathcal{S}$ and $\mathcal{D}$. We shall now specify these structures first for finite and then for infinite quantum systems.

3.3. Finite Systems. According to the standard formulation of quantum mechanics [20], the observables of a finite system correspond to the self-adjoint operators in a (separable) Hilbert space $\mathcal{H}$. In fact, one loses nothing by formulating the model in terms of the bounded observables, since the unbounded ones can always be built up from linear combinations of these *; and further, by so doing, one gains the mathematical advantage that the bounded operators, and only these, are defined on all the vectors in $\mathcal{H}$. Thus, defining $\mathcal{A}$ to be the set, $\mathcal{L}(\mathcal{H})$, of bounded operators in $\mathcal{H}$, the (bounded) observables of Σ are represented by the self-adjoint elements of $\mathcal{A}$.

The states $\mathcal{S}$ of Σ are the linear functionals on $\mathcal{A}$, represented by the density matrices, ρ, in $\mathcal{H}$ according to the formula

$$\rho(A) \equiv Tr(\rho A) \tag{3.4}$$

In particular, the pure states correspond to the one-dimensional projectors, so that, if P_ψ is the projector for the normalised vector ψ, then, by (3.4),

$$P_\psi(A) = (\psi, A\psi)$$

The dynamics of Σ is governed by the Hamiltonian H, a self-adjoint operator in $\mathcal{H}$, whose precise form depends on the constitution of the system. The evolution of the state over time t corresponds to the unitary transformation $\tau(t)$ of $\mathcal{S}$ given by

$$\rho_t \equiv \tau(t)\rho = \exp(-iHt/\hbar)\rho\exp(iHt/\hbar) \tag{3.5}$$

Thus, the time-dependent expectation value of an observable A, for the evolution of the system from an initial state ρ, is $\rho_t(A)$.

The Algebraic Description. We shall now express the structure of the model in terms of the general scheme of §3.2. Thus, the (bounded) observables, $\mathcal{O}$, are the self-adjoint elements of the $\mathcal{A} \equiv \mathcal{L}(\mathcal{H})$; the states $\mathcal{S}$ are the functionals on $\mathcal{A}$, represented by the density matrices according to the formula (3.4); and the dynamics is given by the unitary transformations $\tau(t)$ of $\mathcal{S}$, defined by (3.5). The model is therefore specified by $(\mathcal{A}, \mathcal{S}, \tau)$. We now note following simple algebraic properties, which will subsequently recur in the model of an infinite system, of these three components of Σ.

(A.1) $\mathcal{A}$ is closed w.r.t. binary addition and multiplication, i. e. if A, B belong to $\mathcal{A}$, then so too do $A + B$ and AB. Here, the operation of addition is associative, invertible and commutative, while that of multiplication is associative and also distributive w.r.t. addition.

(A.2) $\mathcal{A}$ is closed w.r.t. multiplication by complex numbers, i.e. if λ is a complex number and A belongs to $\mathcal{A}$, then so too does λA. Further, this multiplication is distributive w.r.t. the binary operations of (A.1) and satisfies the condition that

---

* This follows from the spectral theorem [20], which permits us to resolve any self-adjoint operator into a linear combination of its spectral projectors, which are, of course, bounded.

$$\lambda_1(\lambda_2 A) = (\lambda_1 \lambda_2) A$$

(A.3) $\mathcal{A}$ is closed w.r.t. the transformation $A \rightarrow A^*$, which is termed an *involution* by virtue of its properties

$$(A^*)^* = A; \ (A+B)^* = A^* + B^*; \ (AB)^* = B^* A^*; \ (\lambda A)^* = \overline{\lambda} A^*$$

where $\overline{\lambda}$ is the complex conjugate of λ.

In view of (A.1-3), $\mathcal{A}$ is termed a *-algebra* (or star or involutive algebra): it is an algebra by virtue of (A.1,2) and a *-algebra by virtue of the further property (A.3).

It follows easily from our definition of S that the states possess the following simple properties.

(S.1) Each state, ρ, satisfies the conditions that

$$\rho(A^* A) \geq 0 \ (positivity);$$

$$\rho(\lambda_1 A_1 + \lambda_2 A_2) \equiv \lambda_1 \rho(A_1) + \lambda_2 \rho(A_2) \ (linearity);$$

and

$$\rho(I) = 1 \ (normalisation)$$

(S.2) The set of states, S, is *convex*, i.e. if w_1, w_2 are positive numbers, whose sum is unity, and ρ_1, ρ_2 are states, then $w_1 \rho_1 + w_2 \rho_2$ is also a state. In other words, mixtures of states are states.

Finally, the dynamical law (3.5) possesses the following properties.

(τ.1) The transformations τ possess the group property that

$$\tau(t_1)\tau(t_2) = \tau(t_1 + t_2)$$

(τ.2) They also satisfy the condition

$$\tau(t)(w_1 \rho_1 + w_2 \rho_2) = w_1 \tau(t)(\rho_1) + w_2 \tau(t)(\rho_2),$$

which signifies that if ρ_1, ρ_2, evolve to $\rho_1(t)$, $\rho_2(t)$, respectively, in time t, then the mixture $w_1 \rho_1 + w_2 \rho_2$ evolves correspondingly to $w_1 \rho_1(t) + w_2 \rho_2(t)$.

Comments. (1) The above formulation of the dynamics is that of the Schrödinger picture, in which the evolution is carried by the states. One could equivalently use the Heisenberg picture, in which the dynamical evolution is carried by the observables. For, defining $\alpha(t)$ to be the transformation of $\mathcal{A}$ defined by

$$\alpha(t)A = A(t) \equiv \exp(iHt/\hbar) A \exp(-iHt/\hbar), \qquad (3.6)$$

it follows from (3.4)-(3.6) that

$$(\tau(t)\rho)(A) \equiv \rho(\alpha(t)A),$$

which signifies that the time-dependent expectation values of the observables are the same in the two pictures. It follows easily from (3.6) that $\alpha(t)$ preserves the *-algebraic structure of $\mathcal{A}$, and for that reason it is termed an *automorphism* of $\mathcal{A}$. It also follows from (3.6) that α possesses the group property that

$$\alpha(t_1)\alpha(t_2) = \alpha(t_1 + t_2) \tag{3.7}$$

(2) The model can be extended to unbounded observables, as represented by the unbounded self-adjoint operators, Q, in $\mathcal{H}$, with the expectation value of Q for the state ρ being defined by the obvious generalisation of (3.4), i. e.

$$\rho(Q) = Tr(\rho Q),$$

provided that the r.h.s. of this formula is well-defined, i. e. that $Tr(\rho|Q|)$ is finite: for given ρ, it will not be so for all unbounded observables Q.

Note on Microscopic Reversibility. Assuming that $\mathcal{H}$ is the Hilbert space of square-integrable functions on some space X, e.g. that of the configurations of an N-particle system, we introduce the Wigner time-reversal operator, i.e. the transformation, T, of $\mathcal{H}$ defined by

$$Tf(x) = f^*(x) \tag{3.8}$$

Correspondingly, we define the transformation R of the observables by

$$RA = TAT \tag{3.9}$$

Extending this definition to unbounded observables, in particular to a position, q, and a momentum, p, we see from (3.3),(3.8) and (3.9) that

$$Rq = q: \quad and \quad Rp = -p$$

Thus, as R reverses velocities and leaves positions unchanged, it corresonds to a time-reversal. Assuming now that the Hamiltonian is of the standard form

$$H = -\frac{\hbar^2}{2m}\sum_{j=1}^{N}\Delta_j + V(x_1,...,x_N),$$

it follows from (3.6), (3.8) and (3,9) that

$$R\alpha(t)R\alpha(t) = I \tag{3.10}$$

or equivalently

$$R\alpha(t)R = \alpha(-t) \tag{3.10}'$$

This is the *principle of microscopic reversibility.*

3.4. Infinite Systems. Let Σ be a infinitely extended system of particles of a given species in a space X, which may be the d-dimensional Euclidean space R^d or a lattice, such as Z^d. We assume that, in general, we are dealing with states where the mean particle density is finite and non-zero, and thus where the number of particles is infinite. We formulate the model of Σ so that, for each bounded region, its states and observables reduce to those of finite system of the given species, confined to that region; and so that its dynamics is given by a certain infinite volume limit of that of such a finite system.

We start by defining L to be the set of all bounded, open regions of X. For each such region, Λ, we formulate the model of a finite system, $\Sigma(\Lambda)$, of the given species, confined to Λ, according to the prescription of §3.3. Thus, denoting by $\mathcal{H}(\Lambda)$

the Hilbert space of the pure states of this finite system, we construct its algebra of bounded observables to be a *-algebra in $\mathcal{H}(\Lambda)$ and its Hamiltonian as a self-adjoint operator, $H(\Lambda)$, in that space. The construction, which we shall specify explicitly in §3.5, is made in such a way that the following conditions (L.1-4), concerning the local structure of the observables, are fulfilled.

(L.1) If $\Lambda \subset \Lambda'$, then $\mathcal{H}(\Lambda) \subset \mathcal{H}(\Lambda')$ and $\mathcal{A}(\Lambda) \subset \mathcal{A}(\Lambda')$, i. e. $\Sigma(\Lambda)$ is a subsystem of $\Sigma(\Lambda')$. This is termed the *isotony* condition.

(L.2) If Λ, Λ' are mutually disjoint, then the algebras $\mathcal{A}(\Lambda), \mathcal{A}(\Lambda')$, which, by (L.1), are subalgebras of $\mathcal{A}(\Lambda \cup \Lambda')$, intercommute. This is termed the *local commutativity* condition and signifies that observables in disjoint regions are simultaneously measurable.

We define the algebra of observables of Σ to be the union of the local algebras, $\mathcal{A}(\Lambda)$, i. e. $\mathcal{A} = \cup_{\Lambda \in L} \mathcal{A}(\Lambda)$. The bounded observables are then the self-adjoint elements of $\mathcal{A}$. We note that the algebra $\mathcal{A}$ satisfies the conditions (A.1-3) of §3.3.

(L.3) The algebra $\mathcal{A}$ is equipped with a group of automorphisms $\{\sigma(x)|x \in X\}$, which represent space translations, i. e. $\sigma(x)\sigma(x') = \sigma(x + x')$ and which satisfy the covariance condition that $\sigma(x)\mathcal{A}(\Lambda) = \mathcal{A}(\Lambda + x)$.

We define the states of Σ to be functionals on $\mathcal{A}$, that satisfy the conditions (S.1,2) of §3.3 and reduce to states of the finite systems $\Sigma(\Lambda)$ in the respective regions Λ. They must also satisfy certain dynamical conditions, that we shall specify presently. Thus, a state, ρ, of Σ is given by a family of density matrices ρ_Λ in the respective Hilbert spaces $\mathcal{H}(\Lambda)$, according to the prescription

$$\rho(A) = Tr_\Lambda(\rho_\Lambda A) \ \forall A \in \mathcal{A}(\Lambda) \tag{3.11}$$

where Tr_Λ is the Trace over $\mathcal{H}(\Lambda)$. In view of the (L.1), it follows that this specification of ρ depends on the fulfillment of the following consistency condition.

(L.4) If $A \in \mathcal{A}(\Lambda)$ and $\Lambda \subset \Lambda'$, then

$$Tr_\Lambda(\rho_\Lambda A) = Tr_{\Lambda'}(\rho_{\Lambda'} A)$$

We denote by S_0 the family of all the ρ, defined according to the above prescription. One sees easily that, if $B \in \mathcal{A}$ and ρ belongs to S_0, then so too does the functional ρ_B, defined by

$$\rho_B(A) = \rho(B^\star A B)/\rho(B^\star B) \tag{3.12}$$

Evidently, ρ_B corresponds to a localised modification of ρ.

At this stage, we assert that, in order to be a candidate for a *physical* state a functional ρ should not only possess the S_0 properties, but must also support a dynamics, given by the infinite volume limit of that of $\Sigma(\Lambda)$. Here, it is important to emphasise that, in general, not all elements of S_0 can support such a dynamics: for catastrophies can arise in which an S_0-class functional ρ can evolve, in a finite time, to one carrying an infinity of particles in some finite region, in which case the equations of motion become indeterminate [22].Thus, we must formulate the physical states and dynamics of Σ together.

To this end, we proceed as follows [23]. We let ρ belong to S_0 and A be an arbitrary element of $\mathcal{A}$, and thus of $\mathcal{A}(\Lambda)$ for Λ sufficiently large. The time translate of A, for the system $\Sigma(\Lambda)$, is therefore

$$A_\Lambda(t) = \exp(iH(\Lambda)t/\hbar)A\exp(-iH(\Lambda)t/\hbar), \tag{3.13}$$

and thus, by (3.11), the time-dependent expectation value of A for the evolution of $\Sigma(\Lambda)$ from the initial state ρ_Λ, induced by ρ on the region Λ, is $\rho(A_\Lambda(t))$. We demand that, for a physical state, ρ, of Σ, the time-dependent expectation value of A, for evolution from that state, is the limit of the corresponding quantity for $\Sigma(\Lambda)$, as Λ increases to cover the whole space X.

Accordingly, we define the set $\mathcal{S}$ of physical states by the following conditions.

(PH.1) $\mathcal{S} \subseteq \mathcal{S}_0$

(PH.2) For $\rho \in \mathcal{S}$, and $A \in \mathcal{A}$,

$$\lim_{\Lambda \uparrow X} \rho(A_\Lambda(t)) = \rho_t(A), \tag{3.14}$$

where $\rho_t \in \mathcal{S}$. Thus, physical states evolve into physical states, and the resultant time-dependent expectation values of the observables are the infinite volume limits of the corresponding ones for $\Sigma(\Lambda)$. We define the transformation of $\mathcal{S}$ that takes ρ to ρ_t to be $\tau(t)$, i. e.

$$\tau(t)\rho = \rho(t) \tag{3.15}$$

(PH.3) $\tau(t)$ possesses the group property

$$\tau(t_1)\tau(t_2) = \tau(t_1 + t_2), \tag{3.16}$$

signifying that an evolution over time t_1, followed by one over time t_2, is equivalent to an evolution over time $t_1 + t_2$.

(PH.4) If ρ belongs to $\mathcal{S}$, then so too does ρ_B, defined by (3.12), for all B in $\mathcal{A}$. Thus, $\mathcal{S}$ is stable under local perturbations of state.

(PH.5) $\mathcal{S}$ is the largest set satisfying the conditions (PH.1-4), i.e. it is the union of all such sets.

Thus (PH.1-5) specify both the states and dynamics of the model in terms of the interactions, that are buried in the form of $H(\Lambda)$, which enters into these conditions via (3.13) and (3.14). We see that it follows from these specifications that $\mathcal{S}$ and τ satisfy the conditions (S.1,2) and (τ.1) of §3.3. Thus, the algebraic structure of the infinite system, like that of a finite one, possesses the properties (A.1-3), (S.1,2) and (τ.1).

The formulation we have just presented corresponds to a Schrödinger picture, with the dynamics carried by the states. We can pass from it to a Heisenberg picture, where the dynamics is carried by the observables, according to the prescription that the evolute, $A(t)$, of A is defined by

$$\rho(A(t)) = \rho_t(A) \tag{3.17}$$

Here, $A(t)$ is a functional on the states*, rather that an observable in the sense of

---

* In fact [23], $A(t)$ belongs to a larger algebra, namely the Von Neuman algebra dual to the state space, and $\alpha(t)$, defined by (3.18), is an automorphism of that algebra.

being an element of $\mathcal{A}$. Further, defining $\alpha(t)$ as the mapping that takes A to $A(t)$, i.e.

$$\alpha(t)A = A(t), \tag{3.18}$$

we see from (3.16)-(3.18) that α possesses the group property

$$\alpha(t_1)\alpha(t_2) = \alpha(t_1 + t_2) \tag{3.19}$$

Microscopic Reversibility. It follows from our definitions that Σ inherits from its finite counterparts $\Sigma(\Lambda)$ the microscopic reversibility property (3.10), i.e. that $\mathcal{A}$ is equipped with a time reversal operator, R, that satisfies the condition

$$R\alpha(t)R\alpha(t) = I \tag{3.20}$$

Unbounded Observables. As in the case of a finite system, the model extends to unbounded observables. Thus [24], if Q is an unbounded observable for the region Λ, as represented by a self-adjoint operator in $\mathcal{H}(\Lambda)$, then its expectation value for the state ρ is

$$\rho(Q) = Tr_\Lambda(\rho_\Lambda Q), \tag{3.21}$$

provided that the r.h.s. is well-defined, i.e. that $Tr_\Lambda(\rho_\Lambda(|Q|)$ is finite.

Decay of Correlations. The time-correlation function, for a state ρ, of two observables A, B is defined as

$$\rho(A(t)B) - \rho(A(t))\rho(B)$$

For the infinite system model, unlike the finite one, this function *can* decay with time, as we saw in the model of § 2.

Symmetries. An automorphism γ of $\mathcal{A}$ that commmutes with the time translations $\alpha(t)$ is termed a *symmetry* of the system. Similarly, if G is a group and $\{\gamma(g)|g{\in}G\}$ is a representation* of G in the automorphisms of $\mathcal{A}$, such that $\gamma(g)$ and $\alpha(t)$ intercommute, then $\gamma(G)$ is termed a *symmetry group* of Σ. In this case, a state ρ is termed G-invariant if

$$\rho(\gamma(g)A) = \rho(A) \; \forall A{\in}\mathcal{A}, \; g{\in}G$$

Thus, for example, the space translation group $\sigma(X)$, as specified in (L.3), is a symmetry group if the interactions are translationally invariant; and, in this case, a state ρ is translationally invariant if $\rho(\sigma(x)A){\equiv}\rho(A)$.

3.5. The Construction. We shall now construct the above algebraic structure explicitly for the case where Σ is a system of particles of one species, which may be bosons or fermions, and X is a Euclidean continuum. We describe the system in terms of a quantised field ψ_s, where the index s refers to the spin. This field satisfies the canonical anticommutation or commutation relations

$$[\psi_s(x), \psi_i^*(x')]_\pm = \delta(x - x')\delta_{ss'}; \; [\psi_s(x), \psi_{s'}(x')]_\pm = 0 \tag{3.22}$$

---

* This means that the automorphisms $\gamma(g)$ retain the algebraic properties of G, i.e. that $\gamma(g_1)\gamma(g_2){\equiv}\gamma(g_1g_2)$.

according to whether the system consists of fermions or bosons. In view of the δ singularity, $\psi(x)$ cannot be an operator in a Hilbert space. To remedy the situation, we have to 'smear out' this quantised field against square-integrable test functions, defining

$$\psi_s(f) = \int \psi_s(x)f(x)dx, \ \forall f \in L^2(X) \tag{3.23}$$

Thus, by (3.22) and (3.23), the smeared field satisfies the relations

$$[\psi_s(f), \psi_{s'}(g)^*]_\pm = \langle g, f \rangle \delta_{ss'}; \ [\psi_s(f), \psi_{s'}(g)]_\pm = 0 \tag{3.24}$$

where the angular brackets denote the $L^2(X)$ inner product. In fact, it is the smeared field $\psi(f)$ and the relations (3.24), rather than $\psi(x)$ and (3.22), that is the basis of the formulation; and (3.23) should be regarded as no more than a heuristic expression of how one arrives at $\psi(f)$.

In order to provide a Hilbert space formulation of the quantum field ψ, we proceed as follows. We define a Hilbert space, $\mathcal{H}$, such that

(1) $\{\psi(f) | f \in L^2(X)\}$ are operators in $\mathcal{H}$ and satisfy the anticommutation or commutation relations (3.24).

(2) $\mathcal{H}$ contains a vector Φ, such that, on the one hand,

$$\psi_s(f)\Phi \equiv 0 \tag{3.25}$$

and, on the other hand, $\mathcal{H}$ is generated by application to Φ of the polynomials in the $\psi_s(f)^{*}$' s.

$\mathcal{H}$, so defined is termed the Fock space. One sees immediately from (1) and (2) that Φ corresponds to a vacuum state, the $\psi(f)$'s to destruction operators, which annihilates that state vector, and the $\psi(f)^{*}$' s to creation operators, whose action on the vacuum produces particle-carrying states.

For each bounded open region Λ, we define $\mathcal{H}(\Lambda)$ to be the subspace of $\mathcal{H}$ generated by application to Φ of the polynomials in the $\psi_s(f)^{*}$' s for which f has support in Λ, i. e. vanishes outside Λ.

We define the unitary operators, $V(\theta), W(x)$, representing gauge transformations and space translations, respectively, in $\mathcal{H}$ by the formulae

$$V(\theta)\psi_{s_1}(f_1)\cdots\cdot\psi_{s_k}(f_k)\Phi = \psi_{s_1}(f_1)\cdots\cdot\psi_{s_k}(f_k)\Phi\exp(ik\theta) \tag{3.26}$$

and

$$W(x)\psi_{s_1}(f_1)\cdots\cdot\psi_{s_k}(f_k)\Phi = \psi_{s_1}(f_{1,x})\cdots\cdot\psi_{s_k}(f_{k,x})\Phi \tag{3.27}$$

where θ, x run through the real numbers and the space X, respectively, and

$$f_x(x') \equiv f(x' - x) \tag{3.28}$$

Thus, formally, by (3.23), V, W implement gauge transformations and space translations, respectively according to the rules

$$V(\theta)\psi(x)V(\theta)^{-1} = \psi(x)\exp(i\theta)$$

and

$$W(x)\psi(x')W(x)^{-1} = \psi(x + x')$$

We now formulate the algebraic description of the model as follows. For each bounded open region Λ, we define $\mathcal{A}_0(\Lambda)$ to be the algebra of bounded observables in $\mathcal{H}(\Lambda)$ and $\mathcal{A}(\Lambda)$ to be the subalgebra of $\mathcal{A}_0(\Lambda)$ consisting of those operators that commute with the the gauge transformations $V(\theta)$. We take $\mathcal{A}(\Lambda)$, the gauge invariant subalgebra of $\mathcal{A}_0(\Lambda)$ to be the algebra of observables for the region Λ, in accordance with the principle of second quantisation. It now follows from this definition that the algebras $\mathcal{A}(\Lambda)$ and the Hilbert spaces $\mathcal{H}(\Lambda)$ satisfy the conditions (L1,2,4)* of §3.4. Thus, we may define $\mathcal{A}$ as we did there, namely as the union of the local algebras $\mathcal{A}(\Lambda)$. Further, defining $\sigma(x)$ as the automorphism of $\mathcal{A}$ implemented by the unitary $W(x)$, i. e.

$$\sigma(x)A = W(x)AW(x)^{\star} \tag{3.29}$$

it follows from this formula and (3.28) that $\{\sigma(x)|x\in X\}$ is group of automorphisms, that represents space translations in accordance with the requirements of (L.3). The construction therefore satisfies (L.1-4).

Note. The states ρ formulated here do *not* correspond to density matrices in the Fock space $\mathcal{H}$ except in the special case where the total number of particles is finite [25]. Hence, in general, it is only in a local sense that the states are described in the Fock representation.

Finally, in order to implement the dynamical scheme (PH.1-5), we formulate the local Hamiltonians $H(\Lambda)$ in a standard way. Thus, assuming that the particles of the system are of mass m and that the potential energy of interaction between a pair of them at points x, x' is $V(x - x')$,

$$H(\Lambda) = \frac{\hbar^2}{2m} \int_{\Lambda} \nabla\psi^{\star}(x).\nabla\psi(x)dx + \int_{\Lambda}\int_{\Lambda} \psi^{\star}(x)\psi^{\star}(x')V(x-x')\psi(x')\psi(x)dx'dx \tag{3.30}$$

with Dirichlet boundary conditions on the surface of Λ. The states $\mathcal{S}$ and the dynamical group τ or α can now be formulated from this formula according to the prescription (PH.1-5).

---

* Note that the restriction to the gauge invariant subalgebra is essential for the fulfillment of (L.2) in the case of fermions. For this latter algebra is generated by product fields $\psi^{\star}(x)\psi(x')$, and these satisfy the relevant commutation rules, even when ψ is governed by an anticommutation law.

4. Equilibrium Statistical Thermodynamics with Phase Structure

Our aim now is to formulate the statistical thermodynamics of infinite systems, within the framework of §3. For this purpose, we start by reviewing the standard theory of equilibrium states of finite systems, and extracting from this the features that are generaliseable to infinite ones. The picture obtained in this way has been rigorously established for lattice systems. For continuous ones, there are still some technical problems to be resolved*, though we shall generally ignore these here. We employ units in which Boltzmann's constant is unity.

4.1. Equilbrium States of Finite Systems. We shall employ the model of a finite system, Σ, as formulated in §3.3. Thus, the algebra, $\mathcal{A}$, of bounded observables of Σ is that of the bounded operators in a Hilbert space, $\mathcal{H}$, the states correspond to the density matrices in $\mathcal{H}$ and the dynamics is governed by the Hamiltonian operator, H, in the manner described there.

The canonical equilibrium state at inverse temperature β is then given by the density matrix

$$\rho_c = \exp(\beta(\Phi - H)) \tag{4.1}$$

where

$$\Phi = -\beta^{-1}\ln Tr(\exp(-\beta H)) \tag{4.2}$$

is the Helmholtz free energy. We remark that the formula (4.1) cannot possibly be generalised to infinite systems, because Φ would be infinite for these. For this reason, we shall characterise ρ_c, as defined above, by properties that can also be similarly formulated for infinite systems.

The Thermodynamical Stability Condition. We define a state ρ of Σ to be thermodynamically stable if it minimises the free energy $F(\rho)$, as defined by the formula

$$F(\rho) = Tr(\rho H + \beta^{-1}\rho\ln\rho) \tag{4.3}$$

with $Tr(\rho H)$ and $-Tr(\rho\ln\rho)$ representing the energy and entropy, respectively. Now $F(\rho)$ attains its minimum value, uniquely, for $\rho = \rho_c$, (cf.[4,Ch.3,Appendix A]). Hence, we conclude that the canonical equilibrium state is completely characterised by the thermodynamical stability condition.

The KMS Fluctuation-Dissipation Condition. It was shown by Kubo [26] and by Martin and Schwinger [27] that the canonical state satisfies the condition

$$\rho(A(t)B) = \rho(BA(t + i\hbar\beta)) \; \forall A, B \in \mathcal{A} \tag{4.4}$$

or, equivalently,

$$\int \rho[A(t), B]_-\exp(-i\omega t)dt = \tanh(\tfrac{1}{2}\beta\hbar\omega) \int \rho[A(t), B]_+\exp(-i\omega t)dt \; \forall A, B \in \mathcal{A} \tag{4.4}'$$

This is the KMS condition. We note that the integral on the l.h.s. of this last equation represents the component of frequency ω of the linear response of the system to an external perturbation, while that on the r.h.s. corresponds to spontaneous fluctuations at that frequency: hence the term fluctuation-dissipation condition. Its significance as

---

* cf. the discussion in [4,§3.8]

a general characterisation of equilibrium was first proposed by Haag, Hugenholtz and Winnink [28].

For a finite system, the KMS condition, like that of thermodynamical stability, is a *characterisation* of the canonical state ρ_c, i.e. it is satisfied, uniquely, by that state (cf.[4,Ch.3,Appendix A]).

We conclude, therefore, that *the canonical equilibrium state of a finite system is equivalently characterised by the thermodynamical stability and by the KMS conditions.*

4.2. Equilibrium States of Infinite Systems. In order to formulate thermodynamical stability for an infinite system, we first note that the energy, entropy and free energy of are extensive variables and so are generally infinite for such a system. On the other hand, we can define functionals of the states, representing the global densities of these variables and thence define the thermodynamically stable states.

The Thermodynamic Functionals. We denote by S_X the set of translationally invariant states of the infinite system, Σ, as defined at the end of §3.4, and assume that the interactions in the system are translationally invariant. In this case, we may define the functionals $\hat{e}, \hat{s}, \hat{f}$ on S_X, representing the global densities of energy, entropy and free energy, respectively, by the formulae [1,4]

$$\hat{e}(\rho) = \lim_{\Lambda \uparrow X} \frac{\rho(H(\Lambda))}{|\Lambda|} \tag{4.5}$$

$$\hat{s}(\rho) = \lim_{\Lambda \uparrow X} \frac{-Tr(\rho_\Lambda \ln \rho_\Lambda)}{|\Lambda|} \tag{4.6}$$

and

$$\hat{f} = \hat{e} - T\hat{s} \tag{4.7}$$

where and $|\Lambda|$ is the volume of Λ, on a scale where the mean interparticle spacing is unity and $T = \beta^{-1}$ is the temperature.

Global Thermodynamical Stability (GTS).* We term a translationally invariant state of Σ globally thermodynamically stable (GTS) if it minimises the free energy density $\hat{f}$.

The KMS Condition. We formulate this condition exactly as for a finite system. Thus, ρ is a KMS state if (and only if)

$$\rho(A(t)B) = \rho(BA(t + i\hbar\beta)) \ \forall A, B \in \mathcal{A} \tag{4.8}$$

The physical significance of the KMS property is that it characterises those states for which the (infinite) system Σ behaves as a *thermal reservoir*, in that it drives finite systems to which it is coupled into thermal equilibrium at inverse temperature β [29].

The GTS and KMS States. It follows from this last observation that the KMS states are those that satisfy the equilibrium conditions demanded by the Zeroth Law of Thermodynamics, whereas the GTS states are defined as those that conform to the equilibrium requirements of the Second Law. One might guess from these thermodynamical characterisations that the GTS and KMS conditions are closely related, and

---

* There is also another kind of thermodynamical stability, namely the local one, which turns out to be equivalent to the KMS property [30,31].

possibly equivalent. In fact, the properties that have been established for the states satisfying these conditions are the following (cf.[4,Ch.3] and articles referred to there).

(1) The sets, $\mathcal{G}_T$ and $\mathcal{K}_T$, of states satisfying the GTS and KMS condition, repectively, are not empty. Further, by contrast with the situation for finite systems, each of these sets may consist of more than one state.

(2) The GTS and KMS states are all stationary, i.e. they satisfy the condition that $\rho(A(t))$ is constant w.r.t. t.

(3) For systems with short range interactions, $\mathcal{G}_T$ consists precisely of the translationally invariant KMS states. For systems with certain long range interactions, there may be translationally invariant KMS states that are not GTS: these are metastable.

In view of (1)-(2) and the above thermodynamic characterisations of $\mathcal{G}_T, \mathcal{K}_T$, we take $\mathcal{G}_T$ to be the equilibrium states of Σ.

(4) The sets $\mathcal{G}_T, \mathcal{K}_T$ are convex*, i.e. mixtures of GTS (resp. KMS) states are GTS (resp.KMS) (cf.(S.2),§3.4). We denote by $\mathcal{E}(\mathcal{G}_T)$ the set of extremal elements of $\mathcal{G}_T$, i.e. those that are not mixtures, $w_1\rho_1 + w_2\rho_2$, of different GTS states ρ_1, ρ_2.

(5) Of all the GTS states, the extremals are the only ones that provide sharp definitions of global intensive observables, in the following sense. Let $\tilde{A}(\Lambda)$ be the space average of an arbitrary local observable A over the region Λ, i.e.

$$\tilde{A}(\Lambda) = \frac{1}{|\Lambda|} \int_\Lambda A(x)dx \qquad (4.9)$$

where $A(x) = \sigma(x)A$ is the space translate of A. Then $\rho(\in \mathcal{G}_T)$ satisfies the condition

$$\lim_{\Lambda \uparrow X}(\rho(\tilde{A}(\Lambda)^2) - \rho(\tilde{A}(\Lambda))^2) = 0 \ \forall A = A^\star \in \mathcal{A} \qquad (4.10)$$

if and only if $\rho \in \mathcal{E}(\mathcal{G}_T)$.

This characterisation of the extremal GTS states by macroscopic sharpness corresponds to the phenomenological one of a pure phase [1]. Accordingly, we assume that the pure phases are given by the extremal GTS states [1].

(6) Suppose that the system has a symmetry, represented by an automorphism group $\gamma(G)$, as described at the end of §3.4. Then $\mathcal{G}_T$ is invariant, as a set, under $\gamma(G)$, i.e. if $\rho \in \mathcal{G}_T$ and ρ_g is its transform, defined by $\rho_g(A) \equiv \rho(\gamma(g)A)$, then $\rho_g \in \mathcal{G}_T$. Nonetheless, it is possible that some GTS states ρ may lack this symmetry (cf.[1,4]), i.e. $\rho(\gamma(g)A) \neq \rho(A)$. In this case, we have a *symmetry breakdown*. A typical example is a ferromagnetic state of a system with isotropic interactions: for there the pure phases are polarised and thus lack the rotational symmetry of the interactions. Note that, by contrast, the canonical state of a finite system, i.e. its unique thermodynamically stable state at a given temperature, possesses all the symmetry of the interactions (cf.(4.1)).

Comment on Phase Transitions. It is well known that it is essential to pass to the thermodynamic limit of infinite systems in order to characterise phase transitions by singularities in thermodynamic potentials (cf.[4,P.94]). Our above remarks concerning symmetry breakdown show that the infinite system model is also needed for a characterisation of these transitions by symmetry changes.

---

* See (S.2),§3.3 for the definition of 'convex'

4.3. Connection between Statistical and Classical Thermodynamics

The Thermodynamic Variables. In the above formulation, the equilibrium states are those that minimise the free energy density functional $\hat{f} = \hat{e} - T\hat{s}$. Equivalently [4,Ch.3], they are the states that maximise the entropy density $\hat{s}$ for fixed value e of the energy density $\hat{e}$, the temperature entering as a Lagrange multiplier. Thus since, as we have already noted, the GTS condition, for given T or e, may yield more than one state, it is clear that the energy variable e does not suffice to label the equilibrium states. This is in accordance with classical thermodynamics, where the equilibrium states of a system are specified, at the macroscopic level, by the density of not only the energy but also of a certain set of extensive conserved quantities, e.g. the polarisation, in the case of a ferromagnet.

We have, then, the problem of characterising the macroscopic variables that serve to label the GTS states, i.e. that provide a *complete* specification of the equilibrium states. Our basic assumption is that these are just the variables that provide the thermodynamic description and thus are densities of extensive, conserved quantities [32,§3]. On this basis, we provide a statistical mechanical characterisation of a complete set of thermodynamical variables in the following way.

Extensive Conserved Quantities. Let $\{Q(\Lambda)|\Lambda\in L\}$ be a family of observables for the bounded regions of the system. We term Q an extensive conserved quantity if

$$Q(\Lambda) + Q(\Lambda') = Q(\Lambda\cup\Lambda') \ for \ \Lambda,\Lambda' \ disjoint \ (extensivity)$$

and

$$[Q(\Lambda), H(\Lambda)]_- = 0 \ (conservation)$$

both of these formulae being subject to 'surface corrections', precisely defined in [4,Ch.4]. Correspondingly, we define $\hat{q}$, the global density of Q, to be the functional on the translationally invariant states, $\mathcal{S}_X$, given by

$$\hat{q}(\rho) = \lim_{\Lambda\uparrow X} \frac{\rho(Q(\Lambda))}{|\Lambda|} \tag{4.11}$$

Complete Set of Thermodynamic Variables. Now let Q be a set of extensive conserved quantities $(Q_1,..\,,Q_n)$, with Q_1 the energy, i.e. $Q_1(\Lambda) = H(\Lambda)$ and let $\hat{q}$ be the set of functionals $(\hat{q}_1,.\,,\hat{q}_n)$ on $\mathcal{S}_X$, corresponding to their global densities according to the formula (4.11). We term $\hat{q}$ a *complete set of thermodynamic variables* if

(C.1) for each given value q of $\hat{q}$, there is precisely one state $\rho\in\mathcal{S}_X$ that maximises the entropy density $\hat{s}$; and

(C.2) there is no (proper) subset of $(\hat{q}_1,..\,,\hat{q}_n)$ that satisfies (C.1).

Thus, (C.1) signifies that the values q of $\hat{q}$ completely label the thermodynamically stable states, as given by the maximum entropy principle; while (C.2) ensures that there is no redundancy in this labelling.

We shall also make the following simplifying assumption, which will be used only for our treatment of the fluctuations of the macro-observables.

(C.3) The $Q_i(\Lambda)$'s intercommute, for fixed Λ, i.e.

$$[Q_i(\Lambda), Q_j(\Lambda)]_- = 0$$

up to surface effects. This assumption will not always be fulfilled, even though the $Q_i(\Lambda)$ commute with $H(\Lambda \equiv Q_1(\Lambda)$. For example, it fails if $Q_1(\Lambda), Q_2(\Lambda), Q_3(\Lambda)$ are the components of the magnetic moment for the region Λ, since,in this case,

$$[Q_1(\Lambda), Q_2(\Lambda)]_- = i\hbar Q_3(\Lambda)$$

Thermodynamic Potentials: Entropy and Pressure. We define the entropy density, as a function of q by the equation

$$s(q) = max\{\hat{s}(\rho)|\hat{q}(\rho) = q\} \tag{4.12}$$

Since, as we shall prove in Appendix A, s is differentiable, we may define the thermodynamical variables $\theta = (\theta_1, .. , \theta_n)$, conjugate to $(q_1, .. , q_n)$, by the standard formula

$$\theta_j = \theta_j(q) = \frac{\partial s(q)}{\partial q_j} \tag{4.13}$$

In particular, the temperature is given by

$$T^{-1} = \frac{\partial s}{\partial e} \equiv \frac{\partial s}{\partial q_1} \tag{4.14}$$

The thermodynamic pressure, $p(\theta)$, is given, as a function of θ by the formula [1]

$$p(\theta) = max(\hat{s}(\rho) - \theta.\hat{q}(\rho) \equiv \lim_{\Lambda \uparrow X} \frac{\ln Tr(\exp(-\theta.Q(\Lambda)))}{|\Lambda|} \tag{4.15}$$

where the dot represents the n-dimensional scalar product. Note that the maximisation principle involved here is just the GTS condition w.r.t. the effective energy density functional $\hat{e} - T\sum_2^n \theta_j \hat{q}_j$. Further, by (4.12), this definition of p signifies that it is the Legendre transform of the entropy, i.e.

$$p(\theta) = max_q(s(q) - \theta.q)) \tag{4.15}'$$

Phase Structure. Although the entropy is differentiable, the pressure $p(\theta)$ can have discontinuities in its first differential coefficients [1,4]. Furthermore, we have proved that these discontinuities occur at precisely those values of θ for which there is more than one state satisfying the GTS condition that $(\hat{s} - \theta.\hat{q})$ be maximised [4,Ch.4]. Thus, the phase structure of the system is determined by the discontinuities in the first derivatives of the pressure, since the values of θ that admit phase coexistence are those where these discontinuities occur. This, of course, accords with the assumptions of classical thermodynamics. However, it is noteworthy that this result could be derived only from quantum statistics for the infinite system model, since, as remarked in §4.2, a finite system could not admit either thermodynanamical singularities or coexistence of different equilibrium states.

We note here that, in the single phase regions, where p is differentiable, it follows from (4.15)' that

$$q_j = q_j(\theta) = -\frac{\partial p(\theta)}{\partial \theta_j}, \tag{4.16}$$

which is the inverse relation to (4.13). It follows from these two equations that

$$\sum_{k=1}^{n} \frac{\partial^2 s}{\partial q_i \partial q_k} \frac{\partial^2 p}{\partial \theta_k \partial \theta_j} = -\delta_{ij}$$

Thus, defining

$$B_{ij} = \frac{\partial^2 s}{\partial q_i \partial q_j} \tag{4.17}$$

and

$$C_{ij} = \frac{\partial^2 p}{\partial \theta_i \partial \theta_j} \tag{4.18}$$

and denoting the matrices $[B_{ij}]$ and $[C_{ij}]$ by B and C, respectively,

$$C = -B^{-1} \tag{4.19}$$

Fluctuations of Macroscopic Observables. We represent the equilibrium fluctuations of $Q(\Lambda)$ by the observables

$$\xi(\Lambda) = |\Lambda|^{-1/2}(Q(\Lambda) - \rho(Q(\Lambda))), \tag{4.20}$$

the normalisation factor $|\Lambda|^{1/2}$ representing the fact that the fluctuations in extensive observables for a volume V are of the order of $V^{1/2}$ (cf.[32,§111). As we shall show in Appendix B, the correlations of the fluctuation observables $\xi(\Lambda)$ in a pure equilibrium phase are related to the above matrices B and C by the formula

$$\rho(\xi_i(\Lambda)\xi_j(\Lambda)) = C_{ij} = -(B^{-1})_{ij} \tag{4.21},$$

subject to certain analyticity conditions. This formula is, of course, standard for finite systems (cf.[32,§111).

Moreover, assuming that the equilibrium state has suitably short range spatial correlations, as is natural for a pure phase, it has been proved [33] that the statistical properties of the fluctuations $\xi(\Lambda)$ become Gaussian in the hydrodynamic limit. Thus, by (4.21),

$$\lim_{\Lambda \uparrow X} \rho(\xi_{j_1}(\Lambda) .. .\xi_{j_m}(\Lambda) = \Sigma \Pi_{rs}(C_{j_r j_s}) \tag{4.22}$$

if m is even and zero if it is odd.

5. Macrostatistics and Irreversible Thermodynamics

We now address ourselves to the fundamental problem of determining the relationship between the dynamical laws exhibited by many-particle systems at the microscopic and macroscopic levels. Thus, since it is an empirical fact that certain sets of macroscopic variables $a = (a_1, \ldots, a_n)$ obey self-contained dynamical laws of the form

$$\dot{a} = \phi(a), \tag{5.1}$$

the problem is to obtain restrictions on the form of ϕ imposed by the demands of microphysics.

5.1. The Onsager Theory. Onsager [10] provided a classic treatment of this problem, based on the traditional statistical mechanics of large, but finite systems, for situations close to equilibrium. For such cases, it was assumed that $\dot{a}$ is linear in the thermodynamic conjugates of the a_j's, given by the first derivatives of the entropy density, $s(a)$, so that (5.1) reduces to the form

$$\dot{a}_i = \sum_{j=1}^{n} \Lambda_{ij} \frac{\partial s}{\partial a_j} \tag{5.2}$$

where the Λ's are constants. Onsager obtained the important result that the matrix Λ must be symmetric, i.e. that

$$\Lambda_{ij} = \Lambda_{ji}, \tag{5.3}$$

on the basis of the following general assumptions.

(a) The underlying microscopic laws are reversible.

(b) The characteristic time-scale for the macroscopic law (5.2) is enormously long by comparison with that for microscopic processes, e.g. collisions.

(c) The regressions of spontaneous fluctuations of the macroscopic variables a about equilibrium follow precisely the same law, (5.2), as that governing the dynamics of the system when slightly displaced from equilibrium by external means. This assumption permits one to reduce the problem to fluctuation theory.

(d) The static fluctuations of the macrovariables are determined by the Einstein formula, $P = const.\exp(S)$, relating the probability distribution P, for a, to the entropy S.

Note. The assumption (c) is satisfied by the model of §2, as follows from the results (R1,2) stated there. In general, however, this assumption, though plausible, is far from innocuous. For the fluctuations in intensive variables a of an N-particle system are generally of order $N^{-1/2}$ (cf.[32,§111]), whereas the values of a in the phenomenological law (5.2) are of order $N^{(0)}$.

Critique of the Onsager Theory. Apart from the issue raised in the last Note, Onsager's theory has the following deficiencies.

(1) Although the variables a are said to be macroscopic, the theory provides no mathematical characterisation of macroscopic quantities.

(2) The theory is restricted to the linear regime, close to thermal equilibrium.

(3) The theory is classical at all levels, whereas the very stability of matter depends, in an essential way, on quantum mechanics at the microscopic level [5].

105

5.2. The Present Strategy. Our aim now is to formulate an approach ([34,35]) to non-equilibrium statistical thermodynamics that is designed to incorporate the essential physical ideas of Onsager's theory in a way that is free from the above limitations (1)-(3). For this purpose, we formulate the connection between the phenomenological laws and the underlying quantum mechanical ones within the general framework of §'s 3 and 4.

We first consider the questions of the specification of the variables, a, that enter into the phenomenological law (5.1), and the characterisation of their macroscopicality. Our basic assumption is that these are just versions of the thermodynamic variables, $\hat{q}$, of §4, restricted to local subvolumes large enough to contain enormous numbers of particles. This leads us to a hydrodynamical picture, where the dynamical variables are the space averages over these regions of the local densities of the extensive conserved quantities, Q, constituting the thermodynamic variables. In order to sharpen the picture, and characterise the macroscopicality of these regions, we pass to the so-called *hydrodynamic limit* where the numbers of particles in them become infinite. We note that the macroscopic variables here are 'almost conserved quantities', whose dynamical evolution is governed by 'surface effects'. Consequently, the ratio of the macroscopic time-scale to the microscopic becomes infinite in the hydrodynamic limit. Thus, one sees from this qualitative discussion how our scheme may overcome the above limitations (1) and (3) of the Onsager theory. In order to also surpass (2), we need rather more detailed consideration of the phenomenological laws.

In the treatment that follows, we shall formulate the scheme we have just described for the case where the phenomenological laws are continuum mechanical ones, such as those of hydrodynamics or heat conduction. Again we shall assume that the thermodynamical observables of the system, comprising a complete set of densities of extensive conserved quantities, Q, are $\hat{q} = (\hat{q}_1, .. , \hat{q}_n)$. We also assume the following local properties at the microscopic and macroscopic levels.

(A) Each $\hat{q}_j$ has a local density $\hat{q}_j(x)$, i.e.

$$Q_j(\Lambda) = \int_\Lambda \hat{q}(x) \tag{5.4}$$

and, further, the time evolute, $\hat{q}_j(x,t)$, of this density satisfies a local conservation law

$$\frac{\partial \hat{q}_i(x,t)}{\partial t} + div\hat{c}_i(x,t) = 0, \tag{5.5}$$

$\hat{c}_i$ being the current density corresponding to $\hat{c}_i$.

(B) In a suitable space-time scaling and hydrodynamic limit, the evolution of $\hat{q}$ reduces to a classical deterministic law of the form

$$\dot{q}_t = \mathcal{F}(q_t), \tag{5.6}$$

where $q_t(x)$ is the expectation value of $\hat{q}_t(x) \equiv \hat{q}(x,t)$ in the relevant scaling and $\mathcal{F}$ is a differential operator with respect to the spatial coordinates.

(C) The state of the system always satisfies a condition of *local equilibrium*, which we shall specify precisely in §5.4. The local equilibrium hypothesis provides a crucial link between the linear theory, which pertains to situations close to (global) thermal equilibrium (cf.§5.3), and the general non-linear theory.

Our aim, then, will be do determine the restrictions, imposed by the underlying quantum mechanics, on the transport coefficients occurring in $\mathcal{F}$. The principal further hypotheses that we shall bring to bear on this objective are essentially the Onsager assumptions (a) and (c) of §5.1, transferred to the present setting. Thus, our strategy will be to first provide a treatment of the phenomenological law (5.6) and then to relate properties of the ensuing dynamics to those of the fluctuations of the macroscopic quantum field $\hat{q}_t$.

To be specific, we shall concentrate here on the case where the phenomenological law (5.6) reduces to non-linear diffusions of the following form.

$$\dot{q}_{i,t}(x) = \sum_{j=1}^{n} \nabla.(\Lambda_{ij}(\theta(x,t))\nabla\theta_j(x,t)) \; for \; i = 1,2,..,n \tag{5.7}$$

where $q_{i,t}(x) \equiv q_i(x,t)$ and $\theta(x,t)$ is *defined* as the equilibrium value of the thermodynamic variables θ corresponding to the local value $q_t(x)$ of q, i.e. by (4.13),

$$\theta_i(x,t) \equiv \theta_i(q_t(x)) \equiv \frac{\partial s(q(x,t))}{\partial q_i} \tag{5.8}$$

A special feature of this phenomenological dynamics, which we shall exploit, is that it is invariant under scale transformations $x \rightarrow \lambda x, t \rightarrow \lambda^2 t$. The principal results we shall obtain are that

(1) Λ is symmetric, i.e.

$$\Lambda_{ij}(\theta(x,t)) = \Lambda_{ji}(\theta(x,t)) \tag{5.9}$$

which amounts to a non-linear generalisation of the Onsager relations, by virtue of the θ-dependence of L; and

(2) the fluctuations about that deterministic law constitute a Markov process, of a generalised Onsager-Maschlup form.

Note on Distributions. In general, the densities $\hat{q}_{i,t}(x)$ are highly singular, in that their commutators contain δ-functions. Consequently, we must assume that they, and likewise their expectation values $q_{i,t}(x)$, are *generalised functions, or distributions*, in the sense of L.Schwartz [36]. This means essentially that, like the quantum field ψ of §3, they are properly defined only when 'smeared out' against suitably well-behaved functions $f(x)$ according to the formula

$$\hat{q}_{i,t}(f) = \int \hat{q}_{i,t}(x)f(x)dx; \; q_{i,t}(f) = \int q_{i,t}(x)f(x)dx \tag{5.10}$$

To be precise, we assume this formula for the so-called smooth, fast- decreasing functions f. These constitute the set, $\mathcal{S}(X)$ of functions that are differentiable to all orders and fall off faster than any inverse powers of $|x|$ as $|x| \rightarrow \infty$. Thus, $\hat{q}_{i,t}, q_{i,t}$ are linear functionals on $\mathcal{S}(X)$. We assume that they are continuous* and thus belong to the space of functionals,$\mathcal{S}'(X)$, termed the *tempered distributions* on X. This is a basic assumption of quantum field theory [37].

---

* For a concise and relatively simple treatment of the Schwartz spaces $\mathcal{S}, \mathcal{S}'$, and their continuity properties, see [37,Ch.2]

Thus, $\hat{q}_t$ belongs to the space of n-dimensional vectors, whose components belong to $S'(X)$. We denote this space by $S_n'(X)$. Correspondingly, we define $S_n(X)$ to be the space of n-dimensional vectors, whose components belong to $S(X)$. Thus, q_t is a (continuous) linear functional on $S_n(X)$, according to the prescription that, if

$$f = (f_1, .. , f_n) \in S_n(X),$$

then

$$q_t(f) = \sum_{i=1}^{n} q_{i,t}(f_i) \tag{5.11}$$

5.3. The Linear Case. We turn first to the simplest case, namely that of the linear version of the diffusion process (5.7), which we assume to be applicable in situations close to thermal equilibrium. In this case, (5.7) reduces to the form

$$\dot{q}_i(x,t) = \sum_{j=1}^{n} \Lambda_{ij} \Delta(\theta_j(x,t)) \; for \; i = 1, 2, .. , n \tag{5.12}$$

where Λ is the equilibrium value of that matrix and Δ is the Laplacian. Further, on linearising the r.h.s. with respect to q, this equation takes the form

$$\dot{q}_i(x,t) = \sum_{j,k=1}^{n} \Lambda_{ik} B_{kj} \Delta q_j(x,t) \tag{5.13}$$

where B_{ij} is defined by (4.17). Thus, the linearised phenomenological equations may be expressed concisely as

$$\dot{q}_t = K \Delta q_t, \tag{5.14}$$

where

$$K = \Lambda B, \tag{5.15}$$

this product being the usual one for n-dimensional matrices.

Solution of the Phenomenological Equations. As a first step in our treatment of the statistical thermodynamics underlying the linear diffusion equation (5.14), we formulate its general solution. By our observation at the end of §5.2, (5.14) should be considered as an equation of motion in $S_n'(X)$. Its solution is simply

$$q_t = T(t)q \; \forall t \geq 0 \tag{5.16}$$

where

$$T(t) = \exp(K \Delta t) \tag{5.17}$$

Hence, by (5.10),(5.11), (5.16) and (5.17), the time-dependent smeared field $q_t(f)$ is given by

$$q_t(f) = q(T^\star(t)f) \; \forall t \geq 0, \tag{5.18}$$

where

$$T^\star(t) = \exp(K^\star \Delta t) \tag{5.19}$$

Thus, (5.18) serves to express the time evolution of q_t in terms of the action of $T^\star(t)$ on the test functions f. We shall employ a parallel formalism in our treatment of the dynamics of the quantum field $\hat{q}_t$.

The Large Scale Quantum Fluctuations. Our aim now is to relate the above phenomenological dynamics to that of the *large scale* fluctuations of the quantum field

$\hat{q}_t$, in keeping with the principle that hydrodynamical-type properties arise only on such a scale. Thus, we formulate the dynamics of $\hat{q}$ on a length scale, L, that is enormously large by comparison with the interparticle spacing: in the hydrodynamic limit, L will tend to infinity. Since, as noted after (5.10), the phenomenological dynamics is invariant under scale transformations $x \to \lambda x, t \to \lambda^2 t$, we must employ a time-scale L^2 to go with our length scale L. Accordingly, we define the smeared fluctuation field on these space and time scales to be

$$\xi_L(f,t) = \hat{q}(f_L, L^2 t) - \rho(\hat{q}(f_L, L^2 t)) \tag{5.20}$$

where

$$f_L(x) = L^{-d/2} f(x/L), \tag{5.21}$$

ρ is the equilibrium state, d is the dimensionality of the space X and the normalisation factor $L^{d/2}$ corresponds to the $|\Lambda|^{1/2}$ of (4.20). We formulate the quantum dynamical properties of ξ_L in terms of the correlation functions

$$W_L^{(k)}(f_1, \ldots, f_k; t_1, \ldots, t_k) = \rho(\xi_L(f_1, t_1), \ldots \xi_L(f_k, t_k)) \tag{5.22}$$

where now $f_1, \ldots f_k$ are arbitrary elements of $S_n(X)$, and not, as previously, $S(X)$-class components of vectors in that that space. Our treatment of the fluctuation dynamics will be based on the following assumptions.

(I) $W_L^{(k)}$ is continuous in all its arguments. This is a standard hypothesis of quantum field theory [37].

(II) $W_L^{(k)}$ converges to a limit $W^{(k)}$, as $L \to \infty$, i.e.

$$\lim_{L \to \infty} W_L^{(k)}(f_1, \ldots, f_k; t_1, \ldots, t_k) = W^{(k)}(f_1, \ldots, f_k; t_1, \ldots, t_k) \tag{5.23}$$

This is just the assumption that there is a well-defined fluctuation dynamics in the hydrodynamic limit $L \to \infty$. Note that it follows from distribution theory [36] that (I) and (II) imply the continuity of W in the f's, but not neccessarily in the t's. The next assumption is needed for dynamical continuity in the hydrodynamic limit.

(III) $W^{(k)}$ is continuous in all its arguments.

(IV) Defining the static correlation function

$$W_s^{(2)}(f,g) = W^{(2)}(f,g;0,0), \tag{5.24}$$

we assume that $W_s^{(2)}$ vanishes if the supports of f, g, i.e. the regions where these functions are non-zero, are disjoint. Thus,

$$W_s^{(2)}(f,g) = 0 \; if \; (supp \; f) \cap (supp \; g) = \emptyset \tag{5.25}$$

Since, by (5.21), the disjointness of the supports of f, g implies the infinite separation of those of f_L, g_L in the limit $L \to \infty$, we see from (5.20)-(5.23), this assumption signifies that $\hat{q}(x)$, $\hat{q}(x')$ become uncorrelated as $|x - x'| \to \infty$, and so represents a characteristic property of a pure phase (cf.[2,Pp.293-5]).

(V) For $k > 2$, $W^{(k)}$ decomposes into products of two point functions $W^{(2)}$, i.e.

$$W^{(k)}(f_1, \ldots, f_k; t_1, \ldots, t_k) = \sum \Pi_{i<j} W^{(2)}(f_i, f_j; t_i, t_j) \tag{5.26}$$

109

This is an assumed extension of the result (4.22) to the dynamics of fluctuations. It represents a condition on the decay of microscopic correlations, now both spatial and temporal. This is also a natural assumption in the case of a pure phase, for there space-time correlations are generally of short range (cf.[2,Pp.293-5] and [8] for a concrete example).

Consequences of (I)-(V). The following consequences of our assumptions (I)-(V) have been established in Ref.[34]. We shall sketch the proofs below.

(A) Assuming (I), (II) and (IV), $W_s^{(2)}$ takes the form

$$W_s^{(2)}(f,g) = -\langle f, B^{-1}g \rangle \equiv - \sum_{i,j=1}^{n} (B^{-1})_{ij} \int f_i(x) g_j(x) dx, \qquad (5.27)$$

where $f = (f_1, \dots, f_n)$, $g = (g_1, \dots, g_n)$ and B is given by (5.13).

(B) (1) Assuming (I)-(IV),

$$W^{(2)}(f,s;g,t) = W^{(2)}(g,t;f,s) \qquad (5.28)$$

(2) Under the further assumption of (V), $W^{(k)}(f_1, \dots, f_k; t_1, \dots, t_k)$ is invariant under permutations $f_i, t_i \rightleftharpoons f_j, t_j$.

Comment. The significance of (B2) is that, under the given assumptions, the correlation functions $W^{(k)}$ correspond to a *classical* stochastic process for the random variable $\xi_{cl}(f,t)$, with

$$W^{(k)}(f_1, \dots, f_k; t_1, \dots, t_k) = \mathbf{E}(\xi_{cl}(f_1,t_1) \dots \xi_{cl}(f_k,t_k)), \qquad (5.29)$$

where $\mathbf{E}$ denotes expectation value. By (5.21)-(5.23) and (5.29), the r.h.s. of this last equation, and hence the ξ_{cl} process, is invariant under scale transformations $x \rightarrow \epsilon x, t \rightarrow \epsilon^2 t$. Further, by the space-time translational invariance of ρ, this process is also invariant under $x \rightarrow x + x_0, t \rightarrow t + t_0$. Hence, it is invariant under $x \rightarrow x_0 + \epsilon x, t \rightarrow t_0 + \epsilon^2 t$ for arbitrary x_0, t_0 and $\epsilon(>0)$.

Sketch of Proof of (B). The key here is the KMS condition (4.4). By (5.22), this implies that

$$W_L^{(2)}(f,g;t,0) = W_L^2(g,f;0,t+i\hbar\beta L^{-2})$$

The assumptions (I)-(IV) permit us to pass to the limiting form of this formula as $L \rightarrow \infty$, with the result that [34]

$$W^{(2)}(f,g;t,0) = W^{(2)}(g,f;0,t)$$

By time-translational invariance of ρ, this implies part (1) of (B). Part (2) follows immediately from this result and assumption (V). Note that the derivation of (1), hence also of (2), depends crucially on the condition that the ratio, L^{-2}, of the microscopic time scale to the macroscopic one vanishes in the hydrodynamic limit.

Sketch of Proof of (A). The translational invariance of ρ implies that $W_s^{(2)}$ takes the form

$$W_s^{(2)}(f,g) = \int f(x) T(x-x') g(x') dx dx'$$

110

with $T \in \mathcal{S}_n(X)$, the class of distributions defined above. In view of (IV), it follows from this formula and standard distribution theory that $T(x - x')$ is a finite linear combination of $\delta(x - x')$ and its derivatives. The invariance, ensuing from (II), of $W_s^{(2)}(f,g)$ under scale transformations $x \to \lambda x$, implies that T cannot contain derivatives of δ and thus that $W_s^{(2)}$ is of the form

$$W_s^2(f,g) = \langle f, Dg \rangle$$

for some constant matrix $D = [D_{ij}]$, which, by part (1) of (B), must be Hermitian. Since, as we shall show in Appendix B, (4.21) implies that

$$D = -B^{-1},$$

this result implies (5.27).

The Regression Hypothesis. We formulate our version of the Onsager hypothesis relating the phenomenological law to the fluctuation dynamics as the following assumption.

(VI) The regression of fluctuations of the classical process ξ_{cl} follows the same law as that governing the phenomenological evolution of q. Thus, by (5.18) and (5.29),

$$\mathbf{E}(\xi_{cl}(f,t)\xi_{cl}(g,0)) = \mathbf{E}(\xi_{cl}(T(t)^* f, 0)\xi_{cl}(g,0)), \tag{5.30}$$

and hence, in view of (5.27) and (5.29),

$$\mathbf{E}(\xi_{cl}(f,t)\xi_{cl}(g,0)) = -\langle T^*(t)f, B^{-1}g \rangle \tag{5.31}$$

Microscopic Reversibility. We assume henceforth that the system satisfies the principle of microscopic reversibility and that both the $\hat{q}_i$'s and the state ρ are invariant under the time reversal operator R of §'s 2,3.* Hence, in view of the stationarity of ρ, it follows from (5.22), (5.23) and (5.27) that

$$W^{(2)}(f,g;t,0) = W^{(2)}(g,f;0,t) = W^{(2)}(g,f;-t,0) = W^{(2)}(g,f;t,0)$$

i.e., by (5.29),

$$\mathbf{E}(\xi_{cl}(f,t)\xi_{cl}(g,0)) = \mathbf{E}(\xi_{cl}(g,t)\xi_{cl}(f,0)) \tag{5.32}$$

The Onsager Relations. We shall now prove that, under the above assumptions, the transport coefficients Λ_{ij} satisfy the Onsager relations

$$\Lambda_{ij} = \Lambda_{ji}, \tag{5.33}$$

The essential ingredients of the proof are the formulae (5.31) and (5.32), which represent macroscopic equilibrium and microscopic reversibility, respectively. It follows from those formulae that

$$\langle T^*(t)f, B^{-1}g \rangle = \langle T^*(t)g, B^{-1}f \rangle$$

---

* Generalisation to the case where some of the $\hat{q}_i$ are odd w.r.t. time reversal is straightforward and leads to Casimir's [38] extension of the Onsager theory.

On differentiating this equation w.r.t. t, for $t = 0$, and using (5.19), we obtain

$$\langle K^\star \Delta f, B^{-1} g \rangle = \langle K^\star \Delta g, B^{-1} f \rangle$$

Since B is Hermitian, it follows from this formula, together with (5.27) and the identity $\int f \Delta g dx \equiv \int g \Delta f dx$, that $K B^{-1} = B^{-1} K^\star$, and, by (5.15), this is equivalent to the required result, namely $\Lambda = \Lambda^\star$.

The Markovian Fluctuations. By (5.27),(5.29) and (5.30), the stochastic process $(\xi)_{cl}$ is governed by the forms of B and $T^\star$. Since the latter possesses the semigroup property $T^\star(t)T^\star(s) = T^\star(t + s)$, it is rather a simple matter [33] to infer from (5.29) and (5.30) that the process is *Markovian*. In other words, it carries no 'memory', in the sense that, if $\mathbf{E}(.|\xi_{cl}(s))$, $\mathbf{E}(.|\xi_{cl}(s' \le s))$ respectively denote conditional expectation values, given ξ_{cl} at time s and at all times up to and including s, then

$$\mathbf{E}(\xi_{cl}(f_1, t_1) .. \ \xi_{cl}(f_k, t_k) | \xi_{cl}(s)) = \mathbf{E}(\xi_{cl}(f_1, t_1) .. \ \xi_{cl}(f_k, t_k) | \xi_{cl}(s' \le s)), \quad for \ t_1, .. , t_k \ge s$$

Thus, in view of (5.26) and (5.29), we have a classical, Gaussian, Markov process. This result corresponds to the basic assumptions underlying the Onsager-Maschlup fluctuation theory [11].

Scale Invariance and Local Equilibrium As noted in the discussion following equn. (5.28), the ξ_{cl} process is invariant under scale transformations cum space-time translations $x \to x_0 + \epsilon x$, $t \to t_0 + \epsilon^2 t$, for arbitrary x_0, t_0 and positive ϵ. Hence, defining

$$\xi_{cl}^{(\epsilon)}(f, t) = \xi_{cl}(f_\epsilon, t_0 + \epsilon^2 t) \tag{5.34}$$

with

$$f_\epsilon(x) = \epsilon^{-d/2} f(x/\epsilon), \tag{5.35}$$

it follows that $\mathbf{E}(\xi_{cl}^\epsilon(f_1, t_1) ... \ \xi_{cl}^{(\epsilon)}(f_k, t_k))$ is independent of ϵ. Thus, by (5.30),

$$\mathbf{E}(\xi_{cl}^{(\epsilon)}(f, t) \xi_{cl}^{(\epsilon)}(g, 0)) = \langle T^\star(t) f, bg \rangle \tag{5.36}$$

for all positive ϵ. For 'small' ϵ, this represents the equilibrium condition, at the hydrodynamical level, in the neighbourhood of the point x_0 and time t_0. We shall therefore term it a local, or even *hydrolocal* equilibrium condition (cf.[39])in the limit $\epsilon \to 0$.

5.4. The Nonlinear Case. The nonlinear macroscopic dynamics, as represented by (5.6) or (5.7), stems from the evolution of the system in states that are, in general, far from equilibrium. Therefore the theory of §(5.3), as it stands, is inapplicable here, since it is heavily dependent on the proximity of the state to thermal equilibrium. However, the condition of hydrolocal equilibrium, which we have just described for the linear theory, can be extended to states far from global equilibrium; and we shall show that the assumption of this condition enables us to extend the theory of §5.3 to the non-linear case. It will, in fact, be formulated in terms of the macroscopic observables only and thus demands much less than a 'mixing' hypothesis that would require all the observables in each macroscopic subregion to adjust almost instantaneously, on the macroscopic time scale, to the prevailing local value of q.

We assume, then, that the system evolves from an initial state ρ_L, parametrised by the length L and having the property that $\rho(\hat{q}(x))$ is of the form $F(x/L)$ for some smooth function F, so that L is indeed the characteristic length scale for spatial variations of the macroscopic field $\langle \hat{q}(x) \rangle$. In order to smear out the field $\hat{q}(x)$ on this scale, we define

$$f^{(L)}(x) = L^{-d} f(x/L) \tag{5.37}$$

for $f \in \mathcal{S}_n(X)$ and

$$\hat{q}_t^{(L)}(f) = \hat{q}_t(f^{(L)}) \equiv \sum_{i=1}^{n} \hat{q}_{i,t}(f_i^{(L)}), \tag{5.38}$$

the r.h.s. being as in (5.10). We assume that

$$\lim_{L \to \infty} \rho_L(\hat{q}^L(f, t)) = q_t(f) = \sum_{i=1}^{n} \int q_{i,t}(x) f_i(x) dx, \tag{5.39}$$

where $q_t(x)$ evolves according to (5.7). We also define the fluctuation observables

$$\xi_{q,L}(f, t) = \hat{q}(f_L, L^2 t) - \rho_L(\hat{q}(f_L, L^2 t)), \tag{5.40}$$

with f_L as in (5.21). The properties of the fluctuation process are then represented by the functions

$$W_{q,L}^{(k)}(f_1, \cdots, f_k; t_1, \cdots, t_k) = \rho_L(\xi_{q,L}(f_1, t_1) \cdots \xi_{q,L}(f_k, t_k)) \tag{5.41}$$

Our extension of the theory of §5.3 to the non-linear case will be based on an Onsager-type hypothesis relating the dynamics of the small perturbations of the macroscopic field $q(x, t) \equiv q_t(x)$ to the regressions of fluctuations about the law (5.7).

The Perturbed Macroscopic Dynamics. Let q_t be the time-dependent macroscopic field, satisfying the law (5.7), which, in view of (5.8), is equivalent to

$$\dot{q}_{i,t}(x) = \sum_{j=1}^{n} \nabla.(K_{ij}(q(x, t)) \nabla q_{j,t}(x)), \tag{5.42}$$

where

$$K(q) = \Lambda(q) B(q) \tag{5.43}$$

and (cf.(5.14))

$$B_{ij}(q) = \frac{\partial^2 s(q)}{\partial q_i \partial q_j} \tag{5.44}$$

Let y_t be a perturbation to q_t, so that (5.42) is also satisfied with q_t replaced by $q_t + y_t$. Then the equation of motion for y_t is obtained by taking the differences between the forms of (5.42) for the variables q_t and $q_t + y_t$. Thus, in the case where y_t is a sufficiently weak perturbation for the *linearised* theory in this variable to be applicable, we obtain the equation of motion

$$\dot{y}_t(x) = \nabla.(K(q(x, t)) \nabla y_t(x) + F(x, t) y_t(x)) \equiv \Phi_q y_t(x) \tag{5.45}$$

where F is the matrix, whose elements are vectors in X, given by

$$F_{ij}(x, t) = \sum_{l=1}^{n} \frac{\partial K_{il}(q(x, t))}{\partial q_l} \nabla q_l(x, t) \tag{5.46}$$

The solution of (5.45) may be expressed in the form

$$y_t = T_q(t) y \tag{5.47}$$

where $y \equiv y_0$ and $\{T_q(t)|t \geq 0\}$ is the family of linear transformations generated by Φ_q, i.e.

$$\frac{dT_q(t)}{dt} = \Phi_q T_q(t) \tag{5.48}$$

Defining now the smeared perturbation, $y_t(f)$, by (5.11), with q replaced by y, it follows from (5.45), (5.47) and (5.48) that

$$y_t(f) = y(T_q^\star(t)f) \tag{5.49}$$

where $T_q^\star(t)$ is the dual of $\Phi_q^\star(t)$, and is defined by the formulae

$$\frac{dT_q^\star(t)}{dt} = T_q^\star(t)\Phi_q^\star; \quad T_q^\star(0) = I \tag{5.50}$$

and

$$\Phi_q^\star f(x) = \nabla.(K^\star(q(x,t)\nabla f(x)) - F^\star(x,t).\nabla f(x) \; \forall f \in \mathcal{S}_n(X) \tag{5.51}$$

Hydrolocal Limit of Perturbed Dynamics. In order to formulate the perturbed dynamics in the neighbourhood of a space-time point (x_0, t_0), we define

$$f^{(\epsilon)}(x) = \epsilon^{-d}f(x/\epsilon) \tag{5.52}$$

and

$$y_t^{(\epsilon)}(f) = y_{t_0+\epsilon^2 t}(f^{(\epsilon)}) \tag{5.53}$$

for arbitrary $f \in \mathcal{S}_n$. Thus, for small ϵ, $y_t^{(\epsilon)}(f)$ represents the perturbative field, y, in the neighbourhood of x_0, t_0. It follows from (5.49)-(5.53) that

$$y_t^{(\epsilon)}(f) = y_0^{(\epsilon)}(T_q^{(\epsilon)\star}(t)f) \tag{5.54}$$

where $T_q^{(\epsilon)\star}$ is defined by the formulae

$$\frac{dT_q^{(\epsilon)\star}(t)}{dt} = T_q^{(\epsilon)\star}(t)\Phi_q^{(\epsilon)\star}; \quad T_q^{(\epsilon)\star}(0) = I \tag{5.55}$$

and

$$\Phi_q^{(\epsilon)\star}g(x) = \nabla.(K^\star(q(x_0 + \epsilon x, t_0 + \epsilon^2 t)\nabla g(x)) - \epsilon F^\star(x_0 + \epsilon x, t_0 + \epsilon^2 t).\nabla g(x) \; \forall g \in \mathcal{S}_n(X) \tag{5.56}$$

It now follows from this last equation that, in the *formal* limit where ϵ tends to zero, $\Phi_q^{(\epsilon)\star}$ reduces to the form

$$\Phi_q^{(0)\star} = K^\star(q(x_0, t_0))\Delta \tag{5.57}$$

By (5.19), this is precisely the generator of the $T^\star$ arising in the linear theory, when the q value for the relevant equilibrium state is $q(x_0, t_0)$. We now assume that

$$\lim_{\epsilon \to 0} T_q^{(\epsilon)\star}(t)f = T_q^{(0)\star}f$$

and that equation (5.55) remains valid at $\epsilon = 0$. Thus we obtain the formula

$$T_q^{(0)\star}(t) = \exp(K^\star(q(x_0, t_0)\Delta t) \tag{5.58}$$

On comparing this with (5.18), we see that the perturbed macroscopic evolution reduces, in the hydrolocal limit, to the linear one around the equilibrium state for which q takes the prevailing local value, $q(x_0, t_0)$. This dynamics is represented by the form of $y_t^{(0)}$, as defined by the formula

$$y_t^{(0)}(f) = y_0(T_q^{(0)\star}(t)f) \tag{5.59}$$

The Fluctuation Dynamics. The fluctuations about the macroscopic law (5.42) are represented by the correlation functions $W_{q,L}$, defined by (5.41). We base our treatment of the fluctuations on the assumption that the conditions (I)-(V), formulated in §5.3 for the linear case, are still valid here, with W_L replaced by $W_{q,L}$. Thus, in the hydrodynamic limit, the fluctuation dynamics is represented by the functions

$$W_q^{(k)}(f_1, \cdots, f_k; t_1, \cdots, t_k) = \lim_{L \to \infty} W_{q,L}^{(k)}(f_1, \cdots, f_k; t_1, \cdots, t_k) \tag{5.60}$$

In order to formulate local properties of the fluctuations, in the neighbourhood of (x_0, t_0), we define

$$W_q^{(\epsilon, k)}(f_1, \cdots, f_k; t_1, \cdots, t_k) = W_q^{(k)}(f_{\epsilon,1}, \cdots, f_{\epsilon,k}; t_0 + \epsilon^2 t_1, \cdots, t_0 + \epsilon^2 t_k) \tag{5.61}$$

where f_ϵ is as specified by (5.35). We assume that, as in the linear case (cf. discussion at the end of §5.3), $W_q^{(\epsilon)}$ converges, in the hydrolocal limit $\epsilon \to 0$, to the corresponding correlation function for fluctuations around the equilibrium state at the prevailing local value $q(x_0, t_0)$ of q, i.e.

$$\lim_{\epsilon \to 0} W_q^{(\epsilon, k)}(f_1, \cdots, f_k; t_1, \cdots, t_k) = E_0(\xi_{cl}(f_1, t_1) \cdots \xi_{cl}(f_k, t_k)) \tag{5.62}$$

where the r.h.s. is precisely as specified in §5.3, with the subscript 0 indicating that the q value of the equilibrium state is $q(x_0, t_0)$. This equation represents our hydrolocal equilibrium hypothesis.

The Generalised Onsager Relations. We also introduce the local version of the regression hypothesis (5.30). Specifically, we assume that the regressions of the local fluctuations are governed by the same law as the perturbed local macroscopic dynamics. In view of (5.58) and (5.59), this assumption takes the form

$$E_0(\xi_{cl}(f, t)\xi_{cl}(g, 0)) = E_0(\xi_{cl}(\exp(K^\star(q(x_0, t_0))\Delta t)f, 0)\xi_{cl}(g, 0)) \quad \forall t \geq 0 \tag{5.63}$$

The non-linear problem is therefore reduced to the linear one, based on the equilibrium value $q(x_0, t_0)$ of q. It therefore follows immediately from the theory of §5.3 that the transport coefficients Λ satisfy the reciprocity relations

$$\Lambda_{ij}(x_0, t_0) = \Lambda_{ji}(x_0, t_0) \tag{5.64}$$

Since (x_0, t_0) is arbitrary, this is equivalent to the result stated in (5.9).

The Gaussian Markov Process. In §5.3, we showed that the fluctuation process around an equilibrium state reduced to a classical, Gaussian, Markovian form in the hydrodynamic limit. Our prove of that result was based, in part, on the KMS conditions, which were used in order to establish the formulae (5.27) and (5.28). In general, however, when we have fluctuations about a nonequilibrium state, we cannot

general, however, when we have fluctuations about a nonequilibrium state, we cannot use the KMS condition, except at the hydrolocal level. Instead, we invoke the assumption that that the observables $\hat{q}_L(f, t)$ all intercommute in the limit $L \rightarrow \infty$, in the sense that the functions $W_{q,L}^{(k)}(f_1, \ldots, f_k; t_1, \ldots, t_k)$ become invariant under permutations $(f_i, t_i) \rightleftharpoons (f_j, t_j)$ in this limit and thus, by (5.60), that $W_q^{(k)}$ is invariant under these permutations. This is based on the intercommutativity of the $Q_i(\Lambda)$'s, up to surface effects, together with the standard hypothesis [2,P.176] that time-translates $A(t), B(t')$ of observables A, B of an infinite system intercommute, due to dispersive effects, when $|t - t'|$ becomes infinite. Thus, under this assumption on the invariance of $W_q^{(k)}$ under the above permutations, this function takes the form

$$W_q^{(k)}(f_1, \ldots, f_k; t_1, \ldots, t_k) = \mathbf{E}_q(\xi_{cl}(f_1, t_1) \ldots \xi_{cl}(x_k, t_k)), \qquad (5.65)$$

where again ξ_{cl} is a classical stochastic process, now attached to the profile $q(x, t)$. Moreover, it follows from our various assumptions that this stochastic process is Gaussian and Markovian, though not, in general, stationary. Thus, it constitutes a nonstationary generalisation of the Onsager-Maschlup process.

6. Concluding Remarks

The scheme presented here has been designed as an approach to nonequilibrium statistical thermodynamics, based on the quantum theory of infinite systems. The algebraic framework, within which this scheme is cast, contains the structures that admit precise distinctions between different levels of macroscopicality and permit the decay of time correlation functions needed for a theory of irreversibility.

The processes treated in §5 are, of course, rather special in the following respects.

(a) The assumed phenomenological law (5.7) has simple scaling properties, which we exploited. By contrast, certain other macroscopic laws, including that given by the Navier-Stokes hydrodynamical equation, do not have any such scaling properties.

(b) We assumed, at various stages, that the macroscopic evolution was 'smooth', thereby excluding the possibility of shock waves. We also assumed that the macroscopic observables $\hat{q}_t^{(L)}(f)$, defined in (5.38), become dispersion-free* in the hydrodynamic limit, and thus excluded the possibility of turbulence.

(c) The theory of §5 was restricted to the continuum mechanics of closed systems, though, of course, the nonequilibrium statistical mechanics of open systems also covers a vast wealth of fascinating phenomena (cf.[7]).

Other approaches to the subject, within the framework of statistical mechanics of infinite systems, have progressed in somewhat different directions. Thus, Goderis, Verbeure and Vets [40] have obtained Onsager relations for equilibrium fluctuations, in certain models, of macroscopic observables that are in general neither intercommuting nor conserved. They have also obtained these relations [41] for the macroscopic fluctuations of *open* lattice systems, where the dissipation comes from the interaction between these systems and their reservoirs.

---

* This assumption is implied by that given by (5.60) for the fluctuation process.

A different approach, involving many people, has been centred on classical stochastic models (cf. [42] for a review). Although this does not touch on the problem of the connection between the quantum theory of matter and classical, irreversible thermodynamics, it has led to some insights into the hydrodynamic limit and even into the microscopic origin of exotica such as shock waves [43].

Thus, we do have at least some of the essential structures required for a systemmatic and mathematically precise treatment of the enormously difficult subject of nonequilibrium statistical thermodynamics.

Appendix A

We base the proof of the differentiability of the entropy function on the following two established results.

(1) The entropy function s is concave [4,Ch.4, Appendix A], i.e.

$$s(\lambda q + (1 - \lambda)q') \geq \lambda s(q) + (1 - \lambda)s(q') \; for \; 0 < \lambda < 1 \qquad (A.1)$$

(2) The pressure function p is *strictly* convex [44], i.e.

$$p(\lambda \theta + (1 - \lambda)\theta') < \lambda p(\theta) + (1 - \lambda)p(\theta') \; for \; 0 < \lambda < 1, \; \theta \neq \theta' \qquad (A.2)$$

Both of these properties of these thermodynamic potentials stem from thermodynamical stability.

In order to utilise these properties, we first note that, by standard convex analysis [45], a tangent plane to the graph of a concave function $s(q)$ at the point q corresponds to a vector, ϕ, such that, for any other point q',

$$s(q') - s(q) \leq \phi.(q' - q) \qquad (A.3)$$

From this, it follows easily that the set, Δ, of tangent vectors at q is convex, i.e. if $0 < \lambda < 1$ and ϕ_1, ϕ_2 belong to Δ, then so too does $\lambda \phi_1 + (1 - \lambda)\phi_2$. Further [45], s is differentiable at q if and only if this set consists of a *single* element θ, in which case, θ is the gradient of s at q. Thus, to prove the differentiability of s, it suffices to show that Δ consists of a single element.

Now (A.1) signifies that, for arbitrary q',

$$s(q') - \phi.q' \leq s(q) - \phi.q \; \forall \phi \in \Delta$$

In view of (4.15)', this implies that

$$p(\phi) = s(q) - \phi.q \; \forall \phi \in \Delta \qquad (A.4)$$

Thus, if Δ contained different elements ϕ, ϕ', then it would follow from (A.4) and the convexity of Δ that

$$p(\lambda \phi + (1 - \lambda)\phi') = \lambda p(\phi) + (1 - \lambda)p(\phi')$$

in contradiction to the strict convexity property (A.2). This rules out the possibility that Δ might consist of more than one element and thus establishes that s is differentiable.

Appendix B

Our aim here is to derive the formula (4.21) and to justify the assertion in the last line of our proof of (A) in §5.2. The additional assumptions we make for this purpose are the following.

(1) We assume that the local densities, $\hat{q}_i(x)$ of the extensive conserved quantities satisfy commutation rules of the form

$$[\hat{q}_i(x), \hat{q}_j(x')]_- = i\hbar \, div \hat{c}_{ij}(x)\delta(x - x') \qquad (B.1)$$

where $\hat{c}_{ij}$ may be regarded as a local current density. This assumption provides a generalisation of the continuity equation (5.5) and specifies the sense in which the components of $Q(\Lambda) = \int_\Lambda \hat{q}(x)dx$ intercommute, up to surface effects. Thus, defining the effective local Hamiltonian, occurring in (4.15),

$$K(\Lambda) = \theta.Q(\Lambda),$$

and now denoting by $A(t)$ the evolute of an observable A for the dynamics governed by K, rather than H, we have a continuity equation of the same form as (5.5), i.e.

$$\frac{\partial \hat{q}_i(x, t)}{\partial t} + div\hat{c}_i(x, t) = 0,$$

though the current $\hat{c}_i$ is not the same as in §5. On integrating the last equation against the test function f, we obtain

$$\frac{\partial \hat{q}_i(f, t)}{dt} = \hat{c}_i(\nabla f, t), \qquad (B.2)$$

where

$$\hat{q}_i(f, t) = \int \hat{q}_i(x, t)f(x)dx; \quad \hat{c}_i(\nabla f, t) = \int \hat{c}_i(x, t).\nabla f(x)dx$$

(2) We assume that the system is in a single phase region, where the pressure is analytic in θ. Thus, for ϕ in a certain neighbourhood Δ of the origin, $p(\theta + \phi)$ is analytic in ϕ. In order to relate this assumption to the form of the equilibrium state ρ_θ, corresponding to θ, we note that, by (4.16),

$$\hat{q}_i(\rho_\theta) = -\frac{\partial p(\theta)}{\partial \theta_i}$$

and further, by the translational invariance of ρ_θ, $\rho_\theta(\hat{q}_i(f))$ is equal to the l.h.s. of this equation times $\int f(x)dx$. Thus, choosing f so that this integral is unity, and replacing θ by $\theta + \phi$, with $\phi \in \Delta$,

$$\rho_{(\theta+\phi)}(\hat{q}_i(f)) = -\frac{\partial p(\theta + \phi)}{\partial \phi_i} \qquad (B.3)$$

It follows from this formula and our analyticity assumption that $\rho_{(\theta+\phi)}(\hat{q}_i(f))$ is analytic in ϕ in the region Δ.

(3) Let $\phi_\theta^{(h)}$ be the equilibrium, i.e. KMS, state obtained from ϕ_θ by a local hamiltonian perturbation, h, of the dynamics as given by $K = \theta.Q$ (cf.[46] for a precise formulation of this). We assume that, for perturbations ϕ of the thermodynamic variables θ that do not lead out of the single phase region,

$$\rho_{(\theta+\phi)}(A) = \lim_{\Lambda \uparrow X} \rho_\theta^{(\phi.Q(\Lambda))}(A) \qquad (B.4)$$

In other words, we assume that, from a local point of view, there is no difference between a global change ϕ in the thermodynamic variables θ and a local change by the same amount over the region Λ, in the limit where Λ covers the whole space X. To be precise, we employ a modification of condition (B.4), confined to the macroscopic observables, which we formulate this as follows. We introduce an $S(X)$-class function v_R that is unity for $|x| \leq R$, zero for $|x| \geq R + b$, for some fixed b, and which decreases smoothly from unity to zero as $|x|$ increases from R to $R + b$. We then define

$$\rho_\theta^{(\phi,R)} = \rho_\theta^{(h)}, \; for \; h = \phi.\hat{q}(v_R) \qquad (B.5)$$

$\rho_\theta^{(\phi,R)}$ thus represents a thermal state of an *open* system, located in the sphere $|x| \leq R$, for which the value of the thermodynamic variables is $(\theta + \phi)$, whereas it is θ for the 'heat bath' formed by the rest of Σ. We assume the following modification of (B.4).

$$\rho_{(\theta+\phi)}(\hat{q}_i(f)) = \lim_{R \to \infty} \rho_\theta^{(\phi,R)}(\hat{q}(f)) \qquad (B.6)$$

for arbitrary $f \in S(X)$.

(4) We assume that the analytic properties of $\rho_{\theta+\phi}$, specified in (2), are shared by $\rho_\theta^{(\phi,R)}$ in the single phase region. To be precise, we assume that there is a complex open environment of Δ in which both $\rho_{(\theta+\phi)}(\hat{q}_i(f))$ and $\rho_{\theta+\phi}(\hat{q}_i(f))$ are analytic and uniformly bounded, for R sufficiently large. This implies, by Vitali's theorem, that we may differentiate (B.6) w.r.t. ϕ_j, and interchange the limit there with the differential operation. Thus, invoking (B.3),

$$\frac{\partial^2 p(\theta)}{\partial \theta_i \partial \theta_j} = -\lim_{R \to \infty} \frac{\partial \rho_\theta^{\phi,R}(\hat{q}(f))}{\partial \theta_j}, \; at \; \phi = 0 \qquad (B.7)$$

Further, it follows from our definition of $\rho_\theta^{\phi,R}$ that the derivative on the r.h.s. of this equation is equal to [46]

$$-\int_0^1 [\rho_\theta(\hat{q}_i(f)\hat{q}_j(v_R, it\hbar)) - \rho_\theta(\hat{q}_i(f))\rho_\theta(\hat{q}_j(v_R))]dt$$

Hence, we may rewrite (B.7) in the form

$$\frac{\partial^2 p(\theta)}{\partial \theta_i \partial \theta_j} = -\lim_{R \to \infty} \int_0^1 [\rho_\theta(\hat{q}_i(f)\hat{q}_j(v_R, it\hbar)) - \rho_\theta(\hat{q}_i(f))\rho_\theta(\hat{q}_j(v_R))]dt \qquad (B.8)$$

We shall subsequently drop the suffix θ from the ρ.

Sketched Derivation of (4.21). By (B.2),

$$\hat{q}_j(v_R, it\hbar) = \hat{q}_j(v_R) + i\hbar \int_0^t \hat{c}_j(\nabla v_R, is\hbar)ds$$

Hence, defining

$$\xi(f, t) = \hat{q}(f, t) - \rho(\hat{q}(f)), \tag{B.9}$$

we may express the integral in (B.8) as

$$\rho(\xi_i(f)\xi_j(v_R)) + i\hbar \int_0^1 dt \int_0^t ds \rho(\xi_i(f)\hat{c}_j(\nabla v_R, is\hbar))$$

Furthermore, the integrand here vanishes, since ρ and $\xi(f)$ are even, whereas $c(v_R, is)$ is odd under time reversals: the oddness of $c(v_R, 0)$ follows from its definition and the evenness of $\hat{q}$, while the argument is does not change under time reversals, since the formal generator of 'imaginary time translations' of observables is $\hbar^{-1}[\theta.Q, .]_-$ and Q is even. Consequently, (B.8) reduces to the form

$$\frac{\partial^2 p(\theta)}{\partial \theta_i \partial \theta_j} = \lim_{R \to \infty} \rho(\xi_i(f)\xi_j(v_R)) \tag{B.10}$$

By the translational invariance of ρ and our specifications of f, v_R, this equation reduces formally to

$$\frac{\partial^2 p(\theta)}{\partial \theta_i \partial \theta_j} = \int \rho(\xi_i(x)\xi_j(0))dx \tag{B.11}$$

where $\xi(x) = \hat{q}(x) - \rho(\hat{q}(x))$. Moreover, it also follows from the translational invariance of ρ that the r.h.s.'s of (B.11) and (4.21) are equal. This completes our derivation of this latter equation.

Finally, we note that (B.11) may be equivalently expressed as

$$\frac{\partial^2 p(\theta)}{\partial \theta_i \partial \theta_j} = L^{-d} \int \rho(\xi_i(x/L)\xi_j(0))dx \tag{B.12}$$

for arbitrary positive L. On combining this formula with (4.21) and (5.21)-(5.24), we see that the matrix D occurring in the derivation of (A), in §4, must be equal to B^{-1}.

References

1. D. Ruelle: 'Statistical Mechanics: Rigorous Results', W. A. Benjamin, New York, 1969

2. G. G. Emch: 'Algebraic Methods in Statistical Mechanics and Quantum Field Theory', Wiley, New York, London, 1972

3. W. Thirring: 'Quantum Mechanics of Large Systems', Springer, New York, Vienna, 1980

4. G. L. Sewell: 'Quantum Theory of Collective Phenomena', Clarendon Press, Oxford, 1989

5. E. H. Lieb and W. Thirring: Phys. Rev. Lett. **35**, 687, 1975

6. N. G. Van Kampen: Can. J. Phys. **39**, 551, 1961

7. K. Hepp and E. H. Lieb: Helv. Phys. Acta **45**, 237, 1973

8. G. L. Sewell: J. Math. Phys. **26**, 2324, 1985

9. J. Dixmier: Ann. Math. **51**, 387, 1950

10. L. Onsager, Phys. Rev. **37**, 405, 1931; and **38**, 2265, 1931

11. L. Onsager and S. Maschlup: Phys. Rev. **18**, 1505, 1953; and **91**, 1512, 1953

12. G. W. Ford, M. Kac and P. Mazur: J. Math. Phys. **6**, 504, 1965

13. J. T. Lewis and L. C. Thomas: 'How to Make a Heat Bath', Pp. 97-123 of 'Functional Analysis and its Applications', Ed. A. M. Arthur, Oxford, Clarendon, 1975

14. G. L. Sewell: Ann. Phys. **85**, 336, 1974

15. E. B. Davies: Commun. Math. Phys. **39**, 91, 1974

16. P. L. Torres: J. Math. Phys. **18**, 301, 1977

17. G. L. Sewell: 'Statistical Mechanical Considerations of Local Equilibrium and Hydrodynamics', Pp. 1-14 of 'Local Equilibrium in Strong Interaction Physics', Ed. D. K. Scott and R. M. Weiner, World Scientific Publ. Co., 1985

18. G. E. Uhlenbeck and L. S. Ornstein: Phys. Rev. **36**, 823, 1930

19. P. A. M. Dirac: 'Principles of Quantum Mechanics', Clarendon Press, Oxford, 1958

20. J. Von Neumann: 'Mathematical Foundations of Quantum Mechanics',

Princeton University Press, 1955

21. J. Von Neumann: Math. Annalen **104**, 570, 1931

22. C. Radin: Commun. Math. Phys. **54**, 69, 1977

23. G. L. Sewell: Lett. Math. Phys. **6**, 209, 1982

24. G. L. Sewell: J. Math. Phys. **11**, 1868, 1970

25. G. F. Dell' Antonio, S. Doplicher and D. Ruelle: Commun. Math. Phys. **2**, 223, 1966

26. R. Kubo: J. Phys. Soc. Japan **12**, 570, 1957

27. P.C. Martin and J. Schwinger: Phys. Rev. **115**, 1342, 1977

28 R. Haag, N. M. Hugenholtz and M. Winnink: Commun. Math. Phys **5**, 215, 1967

29. A. Kossakowski, A. Frigerio, V. Gorini and M. Verri: Commun. Math. Phys. **57**, 97, 1977

30. H. Araki and G. L. Sewell: Commun Math. Phys. **52**, 103, 1977

31. G. L. Sewell: Commun. Math. Phys. **55**, 53, 1977

32. L. D. Landau and E. M. Lifschitz: 'Statistical Physics', Pergamon, London, New York, Paris, 1959

33. D. Goderis and P.Vets: Commun. Math. Phys. **122**, 249, 1989

34. G. L. Sewell: 'Quantum Macrostatistics and Irreversible Thermodynamics', to be published in the proceedings of 'Quantum Probability and Applications, V', held at Heidelberg in 1988

35. G. L. Sewell: 'Macrostatistics and Nonequilibrium Thermodynamics', to be published in the proceedings of the symposium on 'Stochastic Processes, Physics and Geometry', held at Ascona, 1988

36. L. Schwartz: 'Theorie des Distrbutions', Tome I, Hermann, Paris, 1950; Tome II, Hermann, Paris, 1951

37 R. F. Streater and A. S. Wightman: 'PCT, Spin and Statistics and All That', W. A. Benjamin, New York, Amsterdam, 1964

38. H. B. G. Casimir:Rev. Mod. Phys. **17**, 343, 1945

39. H. Narnhofer and G. L. Sewell; Commun. Math. Phys. **71**, 1, 1980

40. D. Goderis, A. Verbeure and P. Vets: J. Stat. Phys. **56**, 721, 1989

41. D. Goderis, A. Verbeure and P. Vets: 'Glauber Dynamics of Fluctuations and the Onsager Theory', Preprint, 1989

42.A. De Masi, N. Janiro, A. Pellegrinotti and E. Presutti: 'A Survey of the Hydrodynamical Properties of Many-Particle Systems', Pp. 123-294 of 'Nonequilibrium Phenomena II: From Stochastics to Hydrodynamics', Ed. J. L. Lebowitz and E. W. Montroll, North Holland, Amsterdam, 1984

43. W. D. Wick: J. Stat. Phys. **38**, 1015, 1985

44. R. B. Griffiths and D. Ruelle: Commun. Math. Phys. **23**, 169, 1971

45. R. T. Rockafeller: 'Convex Analysis', Princeton University Press, Princeton, 1970

46. H. Araki: Publ. R.I.M.S. **9**, 165, 1973

QUANTUM BROWNIAN MOTION

Philip Pechukas

Department of Chemistry
Columbia University
New York, NY 10027 U.S.A.

INTRODUCTION

Brownian motion is perhaps the simplest dissipative process, and of course the classical theory of it is well understood. One starts from the Langevin equation

$$M\ddot{Q} = - \eta\dot{Q} + F(t); \tag{1}$$

here η is the friction constant of the (one-dimensional) Brownian particle and $F(t)$ is the memoryless Gaussian random force on it, sufficiently strong to drive the particle to equilibrium at temperature T, $\langle F(t)F(t')\rangle = 2\eta kT\,\delta(t-t')$. Equivalently, one starts from the Fokker-Planck equation

$$\partial\rho/\partial t + \partial/\partial Q\ (\rho\dot{Q}) + \partial/\partial P\ (\rho\dot{P}) = \eta kT\ \partial^2\rho/\partial P^2, \tag{2a}$$

$$\dot{Q} = P/M, \quad \dot{P} = - \eta P/M, \tag{2b}$$

which describes the relaxation to equilibrium of ρ, the phase space density of the Brownian particle, by a continuity equation for the deterministic part of the motion supplemented by a momentum diffusion term for the effect of the random force.

It is natural to ask, where does one start if the Brownian particle is quantum-mechanical? What is the correct Langevin (i.e., Heisenberg) or Fokker-Planck (i.e., Schrödinger) equation for quantum Brownian motion? There have been many answers to this question, none of which seems to me entirely satisfactory. Here I briefly discuss some of this work and then suggest another model for quantum Brownian motion, a model which attempts to mimic the classical motion of a heavy particle traveling through a dense bath of light thermal particles and suffering frequent brief, uncorrelated binary collisions with them.

PREVIOUS WORK

Look at the Langevin equation (1) and ignore the random force for the moment; is there any way to get simple friction, $M\ddot{Q} = - \eta\dot{Q}$, out of a Hamiltonian? Yes, if the Hamiltonian is time-dependent:

Large-Scale Molecular Systems, Edited by W. Gans *et al.*
Plenum Press, New York, 1991

$$H(t) = e^{-\eta t/M}(P^2/2M) \qquad (3)$$

will do. In 1941 Caldirola proposed the obvious quantum analogue, based on the Hamiltonian $\hat{H}(t) = e^{-\eta t/M}(-\hbar^2/2M \; \partial^2/\partial Q^2)$.[1] The trouble with this proposal has to do with translation in time. There is no trouble in classical mechanics: if $Q(t)$ is a possible classical trajectory under Hamiltonian (3), so is $Q(t+\tau)$. But in the quantum version, $\psi(t)$ and $\psi(t+\tau)$ do not both satisfy the Schrödinger equation: what happens to a Brownian particle, starting in a given quantum state, depends on when it starts, which is unacceptable.

Here is another proposal, unpublished for good reason, which avoids the problem with translation in time by starting from a time-independent Lagrangian:

$$L = (M/2)[\dot{Q}^2 + (\eta/M) \; Q\dot{Q} \; \ln(\dot{Q}^2) - (\eta/M)^2 \; Q^2]. \qquad (4)$$

Believe it or not, this gives the correct damped classical trajectories $Q(t)$. Use this Lagrangian in Feynman's path integral version of quantum mechanics to propagate the Brownian particle. Now the trouble has to do with translation in space: $\psi(Q,t)$ and $\psi(Q+a,t)$ do not both propagate correctly under Lagrangian (4). What happens to a Brownian particle depends nontrivially on where in space the particle happens to be, which is unacceptable.

Consider now the nonlinear Schrödinger equation proposed by Kostin in 1972:[2]

$$i\hbar \; \partial\psi/\partial t = (-\hbar^2/2M) \; \partial^2\psi/\partial Q^2 - (i\hbar\eta/2M) \; [\ln(\psi/\psi^*) - \langle \ln(\psi/\psi^*) \rangle] \; \psi \qquad (5)$$

where $*$ means complex conjugate and $\langle \; \rangle$ denotes the usual expectation value over ψ. Then if $\hat{Q}$ and $\hat{P}$ are the usual coordinate and momentum operators, it follows that $d\langle\hat{Q}\rangle/dt = \langle\hat{P}\rangle/M$ and $d\langle\hat{P}\rangle/dt = -\eta\langle\hat{P}\rangle/M$. A potential term can be added in the normal way to this "Schrödinger" equation, so we can discuss damped motion of a quantum particle in a field of force. There's the trouble: if $\phi_n(Q)$ happens to be an eigenfunction of the normal Schrödinger $\hat{H}$, without the nonlinear friction term, then $\phi_n \exp(-iE_n t/\hbar)$ satisfies both the normal time-dependent Schrödinger equation and also the Kostin equation. In this proposal, friction doesn't affect stationary states, no matter how highly excited they are. That's not right.

The three attempts above are all attempted shortcuts to a theory: more or less inspired guesswork. The proper, if painful, way to do it is from a full model of Brownian particle plus bath. One approach, mainly by chemists,[3] starts from the full quantum kinetic equation of a particle in a fluid, expands appropriately in the small mass ratio m/M where m is the mass of the fluid particles, and derives a reduced description of the Brownian particle motion from the leading term in the expansion. The conclusion is quite simple: it is that the Wigner function associated with the Brownian particle density operator satisfies the classical Fokker-Planck equation. In operator form,

$$\dot{\hat{\rho}} = - (i/\hbar) \; [\hat{H}_0,\hat{\rho}] - (i\eta/2\hbar M) \; [\hat{Q},\{\hat{P},\hat{\rho}\}] - (\eta kT/\hbar^2) \; [\hat{Q},[\hat{Q},\hat{\rho}]] \qquad (6)$$

where $\hat{\rho}$ is the Brownian particle density operator, $\hat{H}_0$ the free-particle Hamiltonian $\hat{P}^2/2M$, and $\{\;,\;\}$ means anticommutator. The first term on the right side is of course free motion, the second term is the friction, and the last term is momentum diffusion.

Brownian motion theory built on Eq. (6) has many nice features. First, if $\hat{\rho}$ is Hermitian and Tr $\hat{\rho} = 1$ at some instant in time, these properties are preserved by the motion, as they should be. Second, the averages behave correctly: if $Q(t) = \text{Tr } \hat{Q}\hat{\rho}(t)$ and $P(t) = \text{Tr } \hat{P}\hat{\rho}(t)$, then $\dot{Q} = P/M$ and $\dot{P} = -\eta P/M$. Third, the motion is Markovian, as in the classical case. Fourth, the motion behaves correctly under translation in space or in time. And fifth, the equilibrium density operator $\hat{\rho} = \exp(-\beta\hat{H}_o)$ is stationary.

It is a shame, then, that Eq. (6) must be rejected on grounds of mathematical inconsistency: it produces negative probabilities. If $\hat{\rho}(t)$ is to be a density operator, we must have $(\psi, \hat{\rho}(t)\psi) \geqslant 0$ for all time and any ψ. In particular, if $\hat{\rho}\psi = 0$ at some instant and for some ψ, we must have $(\psi, \dot{\hat{\rho}}\psi) \geqslant 0$. But if $\hat{\rho}\psi = 0$,

$$(\psi, [\hat{Q}, \{\hat{P}, \hat{\rho}\}]\psi) = (\hat{Q}\psi, \hat{\rho}\hat{P}\psi) - (\hat{P}\psi, \hat{\rho}\hat{Q}\psi), \tag{7}$$

and this may be negative imaginary. Then the friction term in Eq. (6) makes a negative contribution to $(\psi, \dot{\hat{\rho}}\psi)$. The momentum diffusion term, proportional in this case to the positive quantity $(\hat{Q}\psi, \hat{\rho}\hat{Q}\psi)$, helps, but in general not enough. Unacceptable "density" operators cannot be avoided; the evolution equation (6) is not of the proper form required by the theory of quantum dynamical semigroups.[4]

Another approach, mainly by physicists and supported by a voluminous recent literature,[5] models the Brownian particle as a point mass with an infinite collection of harmonic oscillators, of various frequencies, hanging off it. Because the minimum of each harmonic well is at the instantaneous position of the Brownian particle, this model has linear coupling between the oscillator coordinates and the coordinate of the Brownian particle. Linear coupling to a background of oscillators is a useful and realistic approximation in many physical situations; that is not in dispute. But is it a good model for Brownian motion of a particle in a fluid? Formally, yes; the oscillator motion can be integrated out, and with an appropriate distribution of oscillator frequencies one gets, classically, precisely the Langevin equation (1) for the Brownian particle, provided one assumes the oscillators are initially in thermal equilibrium with respect to the instantaneous position of the Brownian particle. In other words, this is an adiabatic model, which is as it should be: ponderous Brownian particle interacts with a swarm of speedy little oscillators. Then the low frequency oscillators should not be important to the theory; but leave them out and there is no friction, in the sense that the time integral of the random force autocorrelation function then vanishes,

$$\int_0^\infty dt \ \langle F(0)F(t)\rangle = 0. \tag{8}$$

That's because the "random force" on the Brownian particle from each harmonic oscillator is perfectly periodic in time, with average zero--a strange model for the collision of a Brownian particle with individual fluid particles in which, if the interaction is purely repulsive, the product $F(t)F(t')$ is positive during the entire brief duration of each collision.

Never mind that the physics is suspect; the model can be quantized and solved and one can calculate properties of the quantum Brownian particle. Calculate for instance its average kinetic energy when it reaches thermal equilibrium with the oscillators hanging off it. Get infinity. This infinity is not trivial; it arises from infinitely sharp localization of the Brownian particle by its interaction with the harmonic bath: the reduced density matrix of the Brownian particle at equilibrium is diagonal in coordinate space. Can one then trust the predictions of the oscillator model for problems where localization is the important issue, such as tunneling in condensed media?

BINARY COLLISION MODEL

Classical Brownian motion can be derived from the following model, with appropriate scaling: a heavy mass M interacts by finite-range repulsive forces with a dense thermal gas of otherwise free, light masses m; the range of interaction is short compared to the mean interparticle separation, so only binary collisions need be considered; because the background gas is ideal, successive binary collisions are uncorrelated. The Brownian limit is $m \to 0$, M fixed. The momentum carried by a thermal gas particle, and therefore the momentum transferred to M in each collision, is $O(m^{\frac{1}{2}})$, sometimes positive, sometimes negative. The number of collisions M suffers per unit time must then be $O(m^{-1})$ or M will not diffuse. But the gas particle speed is only $O(m^{-\frac{1}{2}})$; the gas density must therefore scale as $m^{-\frac{1}{2}}$ to get this high collision frequency. The range of interaction must then be scaled as well, as $m^{\frac{1}{2}}$, to preserve the binary collision approximation: $V_m(|q-Q|) = V(|q-Q|/m^{\frac{1}{2}})$.

The straightforward quantum version of this model is not quite what's needed. The problem is, no matter how short the range of interaction, the quantum Brownian particle is simultaneously "in collision" with all gas particles lying within a thermal wavelength. A two-state absorption/emission model perhaps better mimics the classical behavior of frequent uncorrelated binary collisions: the quantum Brownian particle "eats" gas particles, with conservation of linear momentum, and cannot eat a second before spitting out the first; the Brownian limit $m \to 0$ is a "feeding frenzy" in which the absorption/emission rate scales as m^{-1}.

I am ashamed that I am not yet able to work out the Brownian limit of this model properly. Still, to see that it may be worth doing, consider a sloppy Fermi Golden Rule version. One Brownian particle and N bath particles move in one dimension, around a finite ring so we may use momentum eigenstates. $\hat{H} = \hat{H}_o + \hat{V}$ where $\hat{H}_o$ is the free particle Hamiltonian $\hat{P}^2/2M + \Sigma \, \hat{p}_j^2/2m$ and $\hat{V}$ is the absorption/emission operator,

$$\hat{V} \, |P,p_1p_2\cdots p_j\cdots p_N> = \underset{j}{\Sigma} \, g(p_j - mP/M) \, |P+p_j,p_1p_2\cdots p_{j-1}p_{j+1}\cdots p_N> \quad (9)$$

The amplitude g for absorption of particle j, with momentum p_j, by the Brownian particle, with momentum P, can only depend on the relative velocity of the two particles; thus, $g(p_j - mP/M)$. We'll make g real.

Ignore the free motion for the moment and calculate the change in the density operator $\hat{\rho}$, for Brownian particle plus bath, to second order in the elapsed time dt:

$$\hat{\rho}(dt) \cong \hat{\rho} - (i/\hbar)[\hat{V},\hat{\rho}]dt - (1/2\hbar^2)[\hat{V},[\hat{V},\hat{\rho}]](dt)^2 + \cdots \quad (10)$$

The term proportional to dt will disappear after tracing over the bath variables; for the second order term, imagine that $dt \to 0$ and the strength of $\hat{V} \to \infty$ with $\hat{V}^2 dt$ fixed in magnitude. Or just cross out one of the dt's. Anyway, what we'll study is the rate equation

$$\dot{\hat{\rho}} \, \alpha - [\hat{V},[\hat{V},\hat{\rho}]] \, \alpha \, \hat{V}\hat{\rho}\hat{V} - (\hat{V}^2\hat{\rho} + \hat{\rho}\hat{V}^2)/2. \quad (11)$$

Assume the gas is in thermal equilibrium, so $\hat{\rho}$ has the following form:

$$\hat{\rho} = \hat{\rho}_o \, \underset{j}{\Pi} \, [e^{-\beta\hat{p}_j^2/2m}/Tr \, e^{-\beta\hat{p}_j^2/2m}] + (\hat{\rho}_+/N) \, \underset{k}{\Sigma} \, \underset{j \neq k}{\Pi} \, [e^{-\beta\hat{p}_j^2/2m}/Tr \, e^{-\beta\hat{p}_j^2/2m}]$$

$$\quad (12)$$

Here $\hat{\rho}_o$ and $\hat{\rho}_+$ are reduced density operators for the Brownian particle:
$\hat{\rho}_o$ characterizes the "empty" particle, $\hat{\rho}_+$ the particle when "full". Each
of the thermal bath operators in the sum over k (second term in Eq. (12))
of course acts in its own space of (N-1)-particle states, that with the
kth particle missing. Tr $\hat{\rho}$ = 1 implies that Tr $(\hat{\rho}_o + \hat{\rho}_+)$ = 1, where the
second trace is over one-particle states of the heavy mass.

Plug this form into rate equation (11), trace over bath states, go to
the continuum limit, and ignore inessential numerical factors; you get
absorption/emission kinetic equations for the matrix elements of $\hat{\rho}_o$ and $\hat{\rho}_+$
of the following form:

$$d<P'|\hat{\rho}_o|P>/dt = \int dp \; [g(p-mP'/M)g(p-mP/M)<P'+p|\hat{\rho}_+|P+p>$$
$$- n \, e^{-\beta p^2/2m} \{g^2(p-mP'/M) + g^2(p-mP/M)\} \; <P'|\hat{\rho}_o|P>/2] \tag{13a}$$

$$d<P'|\hat{\rho}_+|P>/dt = \int dp [n e^{-\beta p^2/2m} g(p-m(P'-p)/M)g(p-m(P-p)/M)<P'-p|\hat{\rho}_o|P-p>$$
$$- \{g^2(p-m(P'-p)/M)+g^2(p-m(P-p)/M)\}<P'|\hat{\rho}_+|P>/2] \tag{13b}$$

Here n (aside from numerical factors) is the density of bath particles in
number per thermal wavelength (thermal wavelength of the bath particles,
that is). The interpretation of these equations, as gain/loss equations
from absorption and emission, is transparent.

Now, scaling. The small parameter is ε^2 = m/M. The gas particle
momentum is $O(\varepsilon)$, so write p = εu. The width of the absorption/emission
amplitude function g(p) must also be $O(\varepsilon)$; otherwise, the momentum of
particles emitted will differ in order of magnitude from that of particles
absorbed, which won't do. So g is a function of u − εP whose width is
$O(1)$. Leave n alone; since the thermal wavelength scales as $\varepsilon^{-\frac{1}{2}}$, this
means that the actual gas density (particles/unit length) decreases, as
$\varepsilon^{\frac{1}{2}}$. I believe that this is the wrong scaling, if one were doing this
model right, but of course we are not. Finally, scale the magnitude of
g so the absorption/emission rate goes as ε^{-2}. With the notation

$$\hat{a}(u,\varepsilon) = e^{i\varepsilon u \hat{Q}/\hbar} \; \tilde{g}(u - \varepsilon \hat{P}), \tag{14}$$

where the amplitude function $\tilde{g}$ is now independent of ε, the kinetic
equations, in operator form, read as follows:

$$\dot{\hat{\rho}}_o = \varepsilon^{-2} \int du \; [\hat{a}^\dagger \hat{\rho}_+ \hat{a} - n \, e^{-\beta u^2/2M} \{\hat{a}^\dagger \hat{a}\hat{\rho}_o + \hat{\rho}_o \hat{a}^\dagger \hat{a}\}/2] \tag{15a}$$

$$\dot{\hat{\rho}}_+ = \varepsilon^{-2} \int du \; [n \, e^{-\beta u^2/2M} \hat{a}\hat{\rho}_o \hat{a}^\dagger - \{\hat{a}\hat{a}^\dagger \hat{\rho}_+ + \hat{\rho}_+ \hat{a}\hat{a}^\dagger\}/2] \tag{15b}$$

Note first that we have fast relaxation, on a time scale $O(\varepsilon^{-2})$,
obtained by setting ε = 0 in the definition (14) of $\hat{a}$:

$$\dot{\hat{\rho}}_o \cong \varepsilon^{-2}[(\int du \; \tilde{g}^2(u))\hat{\rho}_+ - (\int du \; n \, e^{-\beta u^2/2M} \tilde{g}^2(u))\hat{\rho}_o] \cong - \dot{\hat{\rho}}_+ \tag{16}$$

After this relaxation, $\hat{\rho}_o$ and $\hat{\rho}_+$ are simply proportional to each other,

$$(\int du \; \tilde{g}^2(u))\hat{\rho}_+ = (\int du \; n \, e^{-\beta u^2/2M} \tilde{g}^2(u))\hat{\rho}_o, \tag{17}$$

and each may be written as the appropriate fraction of the total Brownian
density operator $\hat{\rho} = \hat{\rho}_o + \hat{\rho}_+$: $\hat{\rho}_o = f_o \hat{\rho}$, $\hat{\rho}_+ = f_+ \hat{\rho}$.

The total density operator $\hat{\rho}$ relaxes on a time scale independent of ε:
$\dot{\hat{\rho}} = O(1)$. Expanding the operator $\hat{a}(u,\varepsilon)$ to second order in ε (and adding
in the free motion under $\hat{P}^2/2M$ which we have so far ignored), we get finally
this kinetic equation for quantum Brownian motion:

$$\dot{\rho} = - (1/\hbar) \; [\hat{H}_0, \hat{\rho}] - i\Gamma \; [\hat{Q}, \{\hat{P}, \hat{\rho}\}] - A \; [\hat{Q}, [\hat{Q}, \hat{\rho}]] - B \; [\hat{P}, [\hat{P}, \hat{\rho}]] \quad (18)$$

This looks very much like the Fokker-Planck equation (6), with an additional term: the first term is free motion, the second term the friction, the third term momentum diffusion, and the fourth term--the additional term-- is coordinate diffusion. The constants Γ, A, and B are all positive, as follows:

$$\Gamma = - (1/\hbar) \int du \; \tilde{g}(u) \tilde{g}'(u) \; u \; (f_+ - f_o \; n \; e^{-\beta u^2/2M})$$
$$= (\beta/2\hbar M) \int du \; \tilde{g}^2(u) \; u^2 \; f_o \; n \; e^{-\beta u^2/2M} \quad (19a)$$

$$A = (1/2\hbar^2) \int du \; \tilde{g}^2(u) \; u^2 \; (f_+ + f_o \; n \; e^{-\beta u^2/2M}) \quad (19b)$$

$$B = (1/2) \int du \; \tilde{g}'^2(u) \; (f_+ + f_o \; n \; e^{-\beta u^2/2M}) \quad (19c)$$

Neither A nor B can vanish separately: momentum and coordinate diffusion go hand-in-hand in this absorption/emission model of Brownian motion. One can verify that the problem with positivity that led us to reject the Fokker-Planck equation (6) does not arise with Eq. (18): it is of the approved form.[4] Everything is O.K.

DISCUSSION

Except of course it isn't. We have done a shoddy analysis of an ill-specified model and there is no point in going further with it. The previous section ought to be regarded simply as evidence that a mathematically consistent, properly Markovian, and not wildly unphysical description of quantum Brownian motion is possible.

How to do it right? The rigorous derivation of Markovian kinetic equations from Hamiltonian dynamics has been thoroughly reviewed by Spohn.[6] Scaling is always required: there must be a large parameter in the Hamiltonian to produce fast relaxation. Here the large parameter ε^{-1} must multiply the interaction V between Brownian particle and bath, to give an absorption/emission rate proportional to ε^{-2}. Put one Brownian particle and $N(\varepsilon)$ bath particles on a ring whose length scales as the thermal wavelength of the bath particles, $\varepsilon^{-\frac{1}{2}}$; scale the bath momenta, $p = \varepsilon u$; momentum eigenstates are $|P, u_1 \ldots u_N\rangle$, where P can be regarded as continuous, the u's as discrete and independent of ε. Then the Hamiltonian is

$$\hat{H} = \hat{H}_B + \hat{H}_b + \varepsilon^{-1}\hat{V}(\varepsilon), \quad \hat{H}_B = \hat{P}^2/2M, \quad \hat{H}_b = \sum_j \hat{u}_j^2/2M, \quad (20a)$$

$$\hat{V}(\varepsilon) = \sum_j (\hat{v}_j(\varepsilon) + \hat{v}_j^\dagger(\varepsilon)), \quad (20b)$$

$$\hat{v}_j(\varepsilon)|P, u_1 \ldots u_N\rangle = g(u_j - \varepsilon P)|P + \varepsilon u_j, u_1 \ldots u_{j-1} u_{j+1} \ldots u_N\rangle, \quad (20c)$$

and the ε-dependence of the problem is only through the interaction and in $N(\varepsilon)$. $\hat{V}(\varepsilon) = \hat{V}(0) +$ first order terms linear in the Brownian operators $\hat{Q}$ and $\hat{P}$. Fast relaxation is induced by $\hat{V}(0)$, and I believe the appropriate form of the density operator is the steady state so generated from a thermal bath:

$$\hat{\rho} = \hat{\rho}_B \; [\; \delta \int_{-\infty}^0 dt \; e^{\delta t} \; e^{i\hat{V}(0)t/\hbar\varepsilon} \; e^{-\beta \hat{H}_b} \; e^{-i\hat{V}(0)t/\hbar\varepsilon}] \quad (21)$$

$\hat{V}(0)$ is not completely trivial: by the rules of the game, $[\hat{v}_j, \hat{v}_k^\dagger] \neq 0$. Nevertheless, progress can be made: the spectrum of $\hat{V}(0)$ is quasi-continuous, running from $-O(N^{\frac{1}{2}})$ to $+O(N^{\frac{1}{2}})$. To get relaxation on a time scale of order ε^{-2}, and bath correlation functions that decay on this same time scale, it's evident from Eq. (21) that $N(\varepsilon)$ must scale as ε^{-2},

128

and therefore that the spatial density of bath particles must scale as ϵ^{-1}, as in classical Brownian motion.

So that's the problem; what's the solution?

REFERENCES

1. P. Caldirola, Nuovo Cimento 18:393 (1941).
2. M. D. Kostin, J. Chem. Phys. 57:3589 (1972).
3. I. Oppenheim and V. Romero-Rochin, Physica 147A:184 (1987).
4. R. Alicki and K. Lendi, "Quantum Dynamical Semigroups and Applications", Lecture Notes in Physics 286, Springer-Verlag, New York (1987).
5. Reviewed by H. Grabert, P. Schramm, and G.-L. Ingold, Physics Reports 168:115 (1988).
6. H. Spohn, Rev. Mod. Phys. 53:569 (1980).

LOCALIZATION CRITICAL EXPONENTS

J. L. Skinner,* T.-M. Chang and J.D. Bauer+

Department of Chemistry
Columbia University
New York, NY 10027

*Author to whom correspondence should be addressed; Permanent address: Department of Chemistry, University of Wisconsin, Madison, WI 53706
+Present address: Lawrence Livermore Laboratory, Livermore, CA 94550

The problem of Anderson localization has generated intense interest for over three decades. It can serve as a simple model for understanding the dynamics of vibrational or electronic excitons in molecular and inorganic crystals, as well as the transport of electrons in doped semiconductors. So far, most studies have been focused on determining the critical disorder and the correlation (localization) length exponent, ν. Although initially there has been substantial disagreement about the value of the critical disorder for different model problems, more recently a consensus seems to have been reached, particularly for the original diagonally disordered Anderson model with a rectangular probability distribution of site energies,[1,2,3] and perhaps for the quantum site percolation model.[2,4] On the other hand, the case of the localization length exponent is much less clear. Using an ϵ-expansion technique ($\epsilon = d - 2$, where d is the dimension of space), Wegner[5] found that $\nu = 1/\epsilon + O(\epsilon^3)$, which gives $\nu = 1$ in three dimensions. More recently, Wegner[6] discovered that this result is in error, and a correct calculation yields $\nu = 1/\epsilon - (9/4)\zeta(3)\epsilon^2 + O(\epsilon^3)$, where $\zeta(x)$ is the Riemann zeta function ($\zeta(3) = 1.202...$). This result when evaluated for $\epsilon = 1$ gives an (unphysical) value of $\nu \approx -1.7$. Numerical work has also led to ambiguous results. Another interesting set of critical exponents, π_k ($k = 2, 3, 4...$), which will be described below, and which have received somewhat less attention, involve the inverse participation ratio and its generalizations.[5,7] In this paper, we summarize numerical calculations of ν, π_2, π_3, and π_4, and comment on the relationship among the different π_k, which in this case implies a multifractal structure to the critical wavefunctions.

Large-Scale Molecular Systems, Edited by W. Gans *et al.*
Plenum Press, New York, 1991

We consider a modified Anderson model described by a tight-binding Hamiltonian defined on a 3-D simple cubic lattice:

$$H = \sum_i \epsilon_i |i\rangle\langle i| + J \sum_{<i,j>} |i\rangle\langle j|, \tag{1}$$

where $|i\rangle$ are orthogonal site states, J is the hopping matrix element, and the sum is over nearest neighbors only. The disorder is introduced by assuming that the site energies, ϵ_i, are uncorrelated random variables described by a Gaussian distribution with mean zero and variance Σ^2. A dimensionless disorder parameter, σ, is defined as $\sigma = \Sigma/J$. For a particular realization of the disorder one can diagonalize the Hamiltonian and express the eigenstates, $|\mu\rangle$, as $|\mu\rangle = \sum_i c_{i\mu}|i\rangle$. For the above Hamiltonian, the density of states is symmetric around the band center ($E = 0$). For $\sigma > \sigma_c$ (σ_c is the critical disorder), all states are localized, and as σ is decreased below σ_c, extended states appear at the band center. Therefore the localization length at the band center, $\xi(\sigma)$, diverges as $\sigma \to \sigma_c^+$ with a critical exponent defined by $\xi(\sigma) \sim (\sigma - \sigma_c)^{-\nu}$.

Focusing only at the band center, the inverse participation ratio, $P^{(2)}(\sigma) \equiv P(\sigma)$, and its generalizations are defined as[7]

$$P^{(k)}(\sigma) = \frac{\langle \sum_{i,\mu} |c_{i\mu}|^{2k} \delta(E_\mu) \rangle_\sigma}{\langle \sum_\mu \delta(E_\mu) \rangle_\sigma}, \tag{2}$$

where E_μ are the eigenvalues, and the brackets indicate configurational averaging for a particular value of the disorder, σ. The inverse participation ratio is a measure of the inverse of the number of sites that "participate" in the eigenstates.[7] Let us consider a finite system with $N = b^3$ sites, and the corresponding $P_b^{(k)}(\sigma)$. It is clear that in the infinite cell limit $P_\infty^{(k)}(\sigma)$ are zero for extended states and finite for localized states; the critical exponents π_k are defined by[7] $P_\infty^{(k)}(\sigma) \sim (\sigma - \sigma_c)^{\pi_k}$, for $\sigma > \sigma_c$. Conversely, for a finite system, $P_b^{(k)}(\sigma)$ are finite for all σ. From the finite-size scaling argument,[8] one expects the finite and infinite system values of $P^{(k)}(\sigma)$ to be related to each other by

$$P_b^{(k)}(\sigma) = P_\infty^{(k)}(\sigma) Y_k(b/\xi(\sigma)), \tag{3}$$

where $\xi(\sigma)$ is the (infinite system) localization length mentioned above. Generally, the scaling functions Y_k are unknown, but the limiting behaviors are readily determined. Since $P_b^{(k)}(\sigma)$ for the finite system is always finite, the singularity of $P_\infty^{(k)}(\sigma)$ at $\sigma = \sigma_c$ must be cancelled by $Y_k(b/\xi(\sigma))$, which implies that Y_k must behave like a power law for small argument. In particular, this implies that $P_b^{(k)}(\sigma_c) \sim b^{-\pi_k/\nu}$, which leads to a convenient method for determining π_k/ν, as will be discussed below.

For a fractal object such as a percolation cluster, the number of sites on the cluster inside a volume b^d goes like $N \sim b^D$, where D is the fractal dimension of the cluster. Since the inverse participation ratio is the *inverse* of the number of sites that participate in the eigenstates, from the above scaling of $P_b^{(2)}(\sigma_c)$ it is natural to associate a fractal dimension $D = \pi_2/\nu$ with the critical eigenstates.[9] There have been several suggestions as to the value of D for three-dimensional localization problems. It is believed[10] that for two dimensions or fewer, even with very small amounts of disorder, all states are localized, except under unusual circumstances.[11] This result leads to the conjecture that $D = 2$.[12] In fact, numerical work by Soukoulis and Economou shows that $D = 1.7 \pm 0.3$,[13] in agreement with this conjecture.

Schreiber[9] also used the value $D = 2$ as a criterion for localization in determining the critical disorder. Finally, the ϵ-expansion gives[6] $D = 2 - \epsilon + 3\zeta(3)\epsilon^4 + O(\epsilon^5)$, which in three dimensions leads to $D \approx 4.6$.

Once π_2/ν is known, from the finite-size scaling hypothesis one can define a generalized phenomenological renormalization transformation by[14]

$$\frac{P_b(\sigma)}{b^{-\pi_2/\nu}} = \frac{P_{b'}(\sigma')}{b'^{-\pi_2/\nu}}. \tag{4}$$

The fixed point of the above transformation gives an estimate of the critical disorder, and the correlation length exponent ν is obtained from

$$\nu = \frac{ln(b/b')}{ln(\lambda_b/\lambda_{b'})}, \tag{5}$$

where

$$\lambda_b = \frac{\partial}{\partial \sigma}\left(\frac{P_b(\sigma)}{b^{-\pi_2/\nu}}\right)\bigg|_{\sigma=\sigma_c}. \tag{6}$$

For each pair of finite-sized systems with cell sizes b and b', this equation can be used to find an estimate for ν.

Our calculations are performed on five different cell sizes for a variety of disorders. For each value of σ, we generate several random configurations for the ϵ_i. For each configuration we diagonalize the Hamiltonian (assuming periodic boundary conditions) and then calculate the $P_b^{(k)}(\sigma)$, for $k = 2, 3$, and 4 as defined above. From other numerical work, it is known that for this model $\sigma_c \approx 6$ ($\sigma_c = 6.03 \pm 0.14$[1]). We therefore calculated $P_b^{(k)}(6)$ for the various cell sizes. For each value of k the data points lie on a straight line whose slope gives π_k/ν.[15] In addition to the intrinsic uncertainty in these slopes from the linear fits, we can include a contribution to the error estimates from the uncertainty in σ_c. This analysis yields[15] $D = \pi_2/\nu = 1.43 \pm 0.10$, $\pi_3/\nu = 2.16 \pm 0.19$, and $\pi_4/\nu = 2.62 \pm 0.27$. This value for D is in agreement with the numerical result of $D = 1.7 \pm 0.3$,[13] but in strong disagreement with the conjecture that $D = 2$,[12] and in stronger disagreement with the ϵ-expansion result of $D = 4.6$.[6]

In fact there exists other numerical evidence supporting the conclusion that $D < 2$. Schreiber[9] calculated the inverse participation ratio for the same model, and plotted lnP_b vs. lnb for different values of the disorder, finding straight lines in each case. However, in his analysis of the data he assumed that $D = 2$ was the fractal dimensionality at criticality, from which he calculated that the critical disorder at the band center is $\sigma_c = 4.79 \pm 0.10$. More recent work shows that this value of σ_c is substantially lower than the correct value.[1,16] Indeed, Schreiber found that D is a monotonically decreasing function of σ, so that at $\sigma \approx 6$, D would be substantially less than 2. The disagreement of our value with the ϵ-expansion result[6] of $D = 4.6$ is not altogether surprising, since in this instance the expansion does not appear to be converging nicely. Expanding the series to first order in ϵ, one obtains (for three dimensions) $D = 1 + O(\epsilon^2)$, while retaining the quartic term, one obtains $D = 4.6 + O(\epsilon^5)$. However, a Pade/Borel resummation of the ϵ-expansion as suggested by Paladin and Vulpiani,[17] but with the corrected fourth-order term,[6] yields a value close to our numerical result.

One can define generalized fractal dimensions by[17,18] $D_k = \pi_k/\nu(k - 1)$, with $D_2 = D$, as defined previously. For a homogeneous fractal $D_k = D$ for all k.[17,18]

From the values obtained in this study one sees that the D_k are not the same for $k = 2, 3$ and 4, implying a multifractal structure to the critical eigenstates.[17,18] One should also note that the ϵ-expansion results of Wegner[6] also predict a multifractal structure, although as in the case of D, the numerical estimates ($\pi_3/\nu = 18.0; \pi_4/\nu = 54.7$) are not accurate. We also note that in the multifractal language D_2 is known not as the fractal dimension but the correlation dimension.[17]

An estimate for the exponent ν was calculated by employing the generalized phenomenological renormalization transformation defined in Eq. 6. For each cell size data for $P_b(\sigma)$ was fit to a second order polynomial, which for each pair of cells crossed at $\sigma \approx 6$, as expected. The slopes of the fitted curves at the crossing were then used to determine ν for that pair from Eq. 7.[15] A weighted average of these values yields $\nu = 0.97 \pm 0.05$. The error estimate of 0.05 must be considered a lower bound to the true error since it is conceivable that underlying the apparent random scatter in our values for ν, lies a trend that would become manifest for larger cell sizes, and an extrapolation rather than an average would be appropriate. We also note that a recent numerical analysis by Schreiber has yielded a value of $\nu = 1.0 \pm 0.1$.[19] Experiments on compensated or highly disordered doped semiconductors also suggest that $\nu \approx 1$.[20]

Acknowledgements

We thank S. Evangelou for interesting discussion, and for bringing our attention to two references.[6,17] A preliminary account of this report, which is adapted from an article submitted for publication in Physical Review B,[21] has been published in a conference proceeding.[22] This research is supported by the National Science Foundation under Grant Nos. CHE 89-10749, DMR 86-03394 and CHE 83-51207.

References

[1] B. Bulka, M. Schreiber, and B. Kramer. *Z. Phys. B* **66**, 21 (1987).

[2] L.J. Root and J.L. Skinner. *J. Chem. Phys.* **89**, 3279 (1988).

[3] A.D. Zdetsis, C.M. Soukoulis, E.N. Economou, and G.S. Grest. *Phys. Rev. B* **32**, 7811 (1985).

[4] C.M. Soukoulis, E.N. Economou, and G.S. Grest. *Phys. Rev. B* **36**, 8649 (1987).

[5] F. Wegner. *Nucl. Phys.* **B280[FS18]**, 210 (1987).

[6] F. Wegner. *Nucl. Phys.* **B316**, 663 (1989).

[7] F. Wegner. *Z. Phys. B* **36**, 209 (1980).

[8] B. Derrida and L. De Seze. *J. Physique* **43**, 475 (1982).

[9] M. Schreiber. *Phys. Rev. B* **31**, 6146 (1985).

[10] E. Abrahams, P.W. Anderson, D.C. Licciardello, and T.V. Ramakrishnan. *Phys. Rev. Lett.* **42**, 673 (1979).

[11] D.H. Dunlap, K. Kundu, and P. Phillips. (preprint).

[12] M.H. Cohen and E.N. Economou amd C.M. Soukoulis. *Phys. Rev. Lett.* **51**, 1202 (1983).

[13] C.M. Soukoulis and E.N. Economou. *Phys. Rev. Lett.* **52**, 565 (1984).

[14] M.N. Barber and W. Selke. *J. Phys. A* **15**, L617 (1982).

[15] T.-M. Chang, J. Bauer, and J.L. Skinner. (unpublished).

[16] C.M. Soukoulis, A.D. Zdetsis, and E.N. Economou. *Phys. Rev. B* **34**, 2253 (1986).

[17] G. Paladin and A. Vulpiani. *Phys. Rep.* **156**, 147 (1987).

[18] C. Castellani and L. Peliti. *J. Phys. A* **19**, L429 (1986).

[19] M. Schreiber. *J. Non-Cryst. Sol.* **97/98**, 221 (1987).

[20] R.F. Milligan and G.A. Thomas. *Ann. Rev. Phys. Chem.* **36**, 139 (1985).

[21] J.D. Bauer, T.-M. Chang, and J.L. Skinner. (unpublished).

[22] J.L. Skinner, J.D. Bauer, and T.-M. Chang. *J. Luminescence* **45**, 333 (1990).

STOCHASTIC MODELS OF POPULATION AND PHASE RELAXATION

J. L. Skinner,[*][+] H. M. Sevian,[+] M. Aihara[**] and B. B. Laird[++]

Department of Chemistry
Columbia University
New York, NY 10027

[*]Author to whom correspondence should be addressed
[+]Permanent address: Department of Chemistry, University of
Wisconsin, Madison, WI 53706
[**]Permanent address: Department of Physics, Faculty of Liberal
Arts, Yamaguchi University, Yamaguchi 753 Japan
[++]Present address: IFF Theorie III, Kernforschungsanlage,
Postfach 1913, 5170 Juelich, West Germany

The time constants T_1 and T_2 were introduced many years ago to describe relaxation in nuclear magnetic resonance.[1,2] To be explicit, let us consider a collection of spin 1/2 particles in a static magnetic field in the z direction. Each of the spins can of course be found in either of two quantum states, up or down. In thermal equilibrium, this produces a nonzero z-component of the magnetization. If the system is prepared in a nonequilibrium state, the longitudinal (z) and transverse (x or y) magnetizations relax in time to the appropriate equilibrium values. In the simple phenomenological model of Bloch, these longitudinal and transverse components decay exponentially with time constants T_1 and T_2 respectively. Focusing instead on the 2 x 2 density matrix for the two spin states, T_1 and T_2 also describe the exponential decay to equilibrium of the diagonal and off-diagonal elements respectively. For that reason T_1 and T_2 are called the population and phase relaxation times, respectively.

The notation and concepts of NMR have been adopted by other fields, including vibrational and optical spectroscopy.[3,4,5,6,7,8,9] In this case it is often useful to model a molecular system simply by two quantum levels representing a particular pair of vibronic states. Here again the decay to equilibrium of the elements of the 2 x 2 density matrix are described phenomenologically by T_1 and T_2. As with NMR, both of these times can be measured experimentally. For example, in optical spectroscopy, where the equilibrium population of the upper state is approximately zero, T_1 is simply the lifetime of this state. T_2 can be measured in absorption, holeburning or coherent transient experiments.

In all of the above fields, relaxation of the two quantum states, or two-level-system (TLS), is caused by interactions with the condensed phase environment, and there are at least two possible theoretical avenues for calculating the relaxation rate constants. One approach, appropriate, for example, for optical or vibrational spectroscopy of molecules in crystals, describes the environment with a quantum mechanical heat bath.[8,9] The TLS-bath interaction in general will have terms that are both diagonal and off-diagonal in the TLS states, and to lowest order, the off-diagonal terms will cause transitions between the levels (population relaxation), while diagonal terms lead only to phase relaxation. The quantum mechanical models have the appealing feature that detailed balance is described properly in that the ratio of the up and down rate constants is equal to the ratio of the equilibrium populations.

A second theoretical approach has mainly been applied to NMR relaxation in liquids.[1,2] In this case, one models the effects of the environment by stochastic perturbations of the TLS. That is, one assumes that the classical motion of a tremendous number of solvent molecules produces additive contributions to the TLS perturbations, which can be represented by random variables with certain statistical properties. As in the case of the completely quantum mechanical model discussed above, there are random fluctuations that are both diagonal and off-diagonal in the TLS states. This stochastic approach should really be considered an infinite temperature model, since it leads to equal populations of the ground and excited states in equilibrium.

In both the quantum mechanical and stochastic models, if one calculates $1/T_1$ and $1/T_2$ to second order in the fluctuations, one finds that[1,2,10,11]

$$\frac{1}{T_2} = \frac{1}{2T_1} + \frac{1}{T_2'}, \tag{1}$$

where, as discussed above, the T_1 contribution to T_2 is due to off-diagonal interactions, and $1/T_2'$, which arises only from the diagonal interaction terms, is called the "pure dephasing" rate, since it produces a contribution to dephasing over and above that due to population relaxation. The interpretation of Eq. 1 is that whatever causes populations to relax must also cause the phase coherence to disappear, but that there can be additional mechanisms producing dephasing that do not affect the populations. The above relation between T_1 and T_2 seems to be accepted in all fields of spectroscopy. One interesting consequence is that since $1/T_2' \geq 0$, this implies that $T_2 \leq 2T_1$.

In a recent paper, Budimir and Skinner[12] considered a stochastic model with both off-diagonal and diagonal fluctuations. They calculated $1/T_1$ and $1/T_2$ to fourth order in the fluctuations. Defining $1/T_2'$ to be the contribution to $1/T_2$ due solely to diagonal fluctuations, they found that in general Eq. 1 is not correct. A particularly transparent illustration of this is obtained for the special case of off-diagonal fluctuations only, when, by definition, $1/T_2' = 0$. In this case they found that in general $2T_1 \neq T_2$, which is in disagreement with Eq. 1, and hence in disagreement with the conventional wisdom that everything that produces population relaxation must also produce phase relaxation, leading to the equality of T_1 and T_2 (except for the factor of 2, which is supposedly well understood). In fact, our results show that off-diagonal fluctuations can lead (through higher-order interactions) to what might be misconstrued as pure dephasing. Even more surprising is the fact that for one model of off-diagonal fluctuations, with certain parameter values we find that $T_2 > 2T_1$! This is a very strange result indeed, which if correct, will certainly have important theoretical, if not experimental implications.

However, our fourth-order theoretical results for T_1 and T_2 must be viewed with caution for two reasons. The first is simply that they are series expansions in the strengths of the fluctuations. While it seems reasonable that there should be a range of parameters when the fourth-order term makes a modest correction to the second-order result, but the sixth-order term is unimportant, there is no guarantee that this is correct. That is, maybe the coefficient of the sixth-order term is huge, so that whenever the fourth-order term is important, the sixth-order term is more important. Or maybe the expansion is not analytic at all! Secondly, our results for T_1 and T_2 only describe the asymptotic long time decay of the density matrix elements to equilibrium. At short times, since the fluctuations have a finite correlation time, the true decay will be nonexponential. If in fact the decay is nonexponential until the density matrix elements have nearly fully relaxed, then the whole concept of relaxation times will not have experimental significance.

In order to ascertain to what extent our strange results represent a possibly important physical effect, rather than simply an amusing mathematical curiosity, we have performed[13] numerical simulations of the same stochastic model that we previously studied analytically. After averaging over different random realizations of the stochastic process, these numerical results provide the exact time evolution of the density matrix elements for all times. We find[13] that as long as the strength of the fluctuations is not too large, our analytic results are in excellent agreement with the numerical simulations, even for the parameters that lead to $T_2 > 2T_1$. Furthermore, one sees that for these same parameters, the time dependence of the density matrix elements is already exponential after decaying only a few percent of the original value. Therefore, for the models considered, our analytic results are both correct and meaningful.

We have also generalized our analytical results for the stochastic model in two ways:[14] we have extended the calculation to sixth order in perturbation theory, and for each cumulant we have calculated the exact rather than asymptotic time dependence. We found that these analytical results are again in excellent agreement with the numerical simulations, even for short times. We have also shown that T_2 can be more than 20% larger than $2T_1$ and still be meaningful.

The breakdown of Eq. 1 has so far been shown only for the stochastic model, which is strictly only applicable at infinite temperatures, since it leads to equal equilibrium populations of the two states. While it may be a reasonable model for NMR, it is not suitable for vibrational and optical spectroscopy, since in these cases the equilibrium populations of the two states are very different. In order to have a proper theory of T_1 and T_2 for this case, one must consider a model where the bath is treated quantum mechanically. We have, in fact, generalized the theory in this direction,[11] and we find that again T_2 can be greater than $2T_1$, showing that the breakdown of the usual inequality is not simply an artifact of the infinite temperature assumption inherent in the stochastic model.

What makes our result of $T_2 > 2T_1$ so surprising is that it shows that phase coherence persists even after populations have relaxed. One might wonder whether such a situation produces a physically acceptable density matrix—one that is positive (non-negative) for all times. One can show[11] that the positivity of the density matrix does produce a bound on $T_2/2T_1$ that depends on temperature in such a way that at $T = \infty$, T_2 can be infinitely large, while at $T = 0$, $T_2 \leq 2T_1$. Our analytic results at finite temperature are in agreement with this bound.

In fact, however, there are more stringent requirements on the density matrix. Since in the problems discussed in this paper, the density matrix represents a reduced description of the system, and is obtained from averging over a stochastic process, one can argue that the density matrix must be completely positive, which is mathematically a much stronger restriction than simple positivity.[15,16,17,18] If the Bloch equations are valid from $t = 0$, then the evolution of the density matrix forms a quantum dynamical semigroup.[15,16,17,18] From this, coupled with the requirement of complete positivity, one can show that in fact $T_2 \leq 2T_1$![15,18] How can one reconcile this proof that $T_2 \leq 2T_1$ with our result that $T_2 > 2T_1$? The Bloch equations are only asymptotically valid (for times long compared to the correlation time of the fluctuations) and so it is only in the white noise limit that they are valid for all times. Fox has shown that in the white noise limit, second-order perturbation theory is exact,[19] and therefore, in this limit there is no disagreement with the requirement of complete positivity. Nonetheless, for colored noise the Bloch equations are only asymptotically valid, and therefore the evolution of the density matrix does not form a semigroup, which invalidates one of the assumptions in the proof that $T_2 \leq 2T_1$. Thus for realistic models (those with colored rather than white noise), the requirements of complete positivity do not appear to impose restrictions on T_1 and T_2, and indeed, it is only for colored noise that we find that $T_2 > 2T_1$.

There is also an argument based on the Heisenberg uncertainty principle that leads to $T_2 \leq 2T_1$. Suppose that the lifetimes of both the upper and lower levels are $2T_1$. According to many textbook discussions, this leads to an uncertainty in the energy of each level of $\Delta E = \hbar/2T_1$, and therefore an uncertainty in the transition energy of $\Delta E = \hbar/T_1$. This gives an uncertainty in the frequency of the transition of $\Delta \nu = 1/2\pi T_1$, which is the accepted relation between the FWHM linewidth of a transition and T_1. One can argue that $\Delta E = \hbar/T_1$ really represents a minimum uncertainty, and that other processes could produce $\Delta \nu \geq 1/2\pi T_1$. Since the absorption spectrum is the Fourier transform of the dipole-dipole correlation function, which decays approximately exponentially with a time constant T_2, this leads to the well-known relation that $\Delta \nu = 1/\pi T_2$. Combining these last two statements gives $T_2 \leq 2T_1$! However, unlike the position-momentum uncertainty relation, which is a rigorous statement about expectation values, we believe that, at least in this context, the time-energy uncertainty principle is simply a qualitative statement about Fourier transforms. Thus if a function decays with a characteristic time constant, Δt, then its Fourier transform will have a width, $\Delta \omega$, with $\Delta t \Delta \omega \approx 1$. Since as discussed above, the relevant Fourier transform for the lineshape is of the dipole correlation function, the statement of the time-energy uncertainty principle should really be $\Delta E T_2 \approx \hbar$, the quantitative statement of which is just $\Delta \nu = 1/\pi T_2$. So we do not believe that the "uncertainty principle" imposes a restriction on the relative values of T_1 and T_2.

Acknowledgements

This research is supported by the National Science Foundation under Grant Nos. CHE 89-10749, DMR 86-03394 and CHE 83-51207. This paper was adapted to a large extent from the article by Sevian and Skinner.[13]

References

[1] A. Abragam. *The Principles of Nuclear Magnetism*. Oxford, London (1961).

[2] C.P. Slichter. *Principles of Magnetic Resonance, 2nd ed.* Springer-Verlag, Berlin (1978).

[3] D. Oxtoby. *Adv. Chem. Phys.* **40**, 1 (1979).

[4] D. Oxtoby. *Adv. Chem. Phys.* **47 (part 2)**, 487 (1981).

[5] L. Allen and J.H. Eberly. *Optical Resonance and Two-Level Atoms*. Wiley, New York (1975).

[6] M.D. Fayer. *In: Spectroscopy and Excitation Dynamics of Condensed Molecular Systems*. Ed. V.M. Agranovich and R.M. Hochstrasser. North-Holand, Amsterdam (1983).

[7] J.L. Skinner and D. Hsu. *Adv. Chem. Phys.* **65**, 1 (1986).

[8] J.L. Skinner and D. Hsu. *J. Phys. Chem.* **90**, 4931 (1986).

[9] J.L. Skinner. *Ann. Rev. Phys. Chem.* **39**, 463 (1988).

[10] A.G. Redfield. *Adv. Mag. Reson.* **1**, 1 (1965).

[11] B.B. Laird and J.L. Skinner. (unpublished).

[12] J. Budimir and J.L. Skinner. *J. Stat. Phys.* **49**, 1029 (1987).

[13] H.M. Sevian and J.L. Skinner. *J. Chem. Phys.* **91**, 1775 (1989).

[14] M. Aihara, H.M. Sevian, and J.L. Skinner. *Phys. Rev. A* (in press).

[15] V. Gorini, A. Kossakowski, and E.C.G. Sudarshan. *J. Math. Phys.* **17**, 821 (1976).

[16] G. Lindblad. *Commun. Math. Phys.* **48**, 119 (1976).

[17] H. Spohn and J.L. Lebowitz. *Adv. Chem. Phys.* **38**, 109 (1978).

[18] R. Alicki and K. Lendi. *Quantum Dynamical Semigroups and Applications*. Springer-Verlag, Berlin (1987).

[19] R.F. Fox. *Phys. Rep.* **48**, 179 (1978).

CLASSICAL AND QUANTUM, LATTICE AND CONTINUUM PERCOLATION

J. L. Skinner,* J. G. Saven, J. R. Wright and L.J. Root[+]

Department of Chemistry
Columbia University
New York, NY 10027

*Author to whom correspondence should be addressed; Permanent
address: Department of Chemistry, University of Wisconsin
Madison, WI 53706
[+]Permanent address: Department of Physics, Barnard College
New York, NY 10027

The concepts of percolation theory have been useful in describing the transport of matter or energy in condensed matter.[1] Both lattice and continuum models have received considerable attention. For lattice problems, Monte Carlo simulation has been quite effective in determining critical thresholds and exponents. Alternatively, real-space renormalization group methods have also been successful.[2] We have devised a renormalization group method, which is based on the finite-size scaling hypothesis,[3] and which can be generalized easily to study quantum percolation (to be described below).

The method[4] is based on calculating the connectedness correlation length, ξ, by Monte Carlo methods for a series of different finite sized systems, and then using phenomenological renormalization[5] to estimate critical thresholds and exponents. In particular, we have studied site percolation on a simple cubic lattice in three dimensions. Random configurations consistent with a given occupation probability p are generated. Two occupied sites that are nearest neighbors on the lattice are said to be directly connected. For each configuration, the connectivity Δ_{ij} is defined to be 1 if both sites are occupied and on the same cluster of connected sites, and 0 otherwise. The connectedness correlation function, $g_{ij}(p)$ is essentially an average over the disorder of the connectivity, and the connectedness correlation length is obtained from the second moment of this correlation function: $\xi(p)^2 = \Sigma_{ij}g_{ij}r_{ij}^2/\Sigma_{ij}g_{ij}$. $\xi(p)$ is calculated for four different cell sizes. Phenomenological renormalization gives an estimate of the critical threshold for each pair of cells, and then these estimates are extrapolated using an ansatz due to Nightingale.[5] This procedure results in $p_c = 0.308 \pm 0.004$ for the percolation threshold,[4] which is in good agreement with the accepted result of $p_c = 0.3117 \pm 0.0003$.[6]

We have also studied the (site) quantum percolation model for a simple cubic lattice. As in the usual (classical) problem, sites are occupied at random with probability p. With each site is associated a quantum state, the collection of which form an orthonormal set. A tight-binding Hamiltonian is then defined such that there is a constant non-zero Hamiltonian matrix element between two sites only if they are both occupied and nearest neighbors on the lattice. The quantum percolation threshold occurs when the first eigenstate of the Hamiltonian becomes extended over the entire (infinite) system. With an appropriate generalization of the connectivity, the ideas explained above can be used to solve this quantum percolation problem. That is, in the classical problem two sites are connected as long as there is at least one path connecting the two sites. In quantum mechanics, the amplitudes for different paths can interefere. This led us to define a quantum connectivity for two sites i and j by[4] $\Delta_{ij} = P_{ij}/\sqrt{P_{ii}P_{jj}}$, where $P_{ij} = \Sigma_{ij}c_{i\mu}^2 c_{j\mu}^2$, and where the $c_{i\mu}$ are the expansion coefficients of the eigenstates, $|\mu\rangle$, in the site states. This definition ensures that the quantum connectivity is bounded above by the classical connectivity of 1. With this definition in hand, for a given finite size one can then generate random configurations, diagonalize the Hamiltonian exactly, calculate the quantum connectivity, average over the disorder to obtain the quantum connectedness correlation function, take the second moment to define the quantum connectedness correlation length, perform the phenomenological renormalization transformation to obtain estimates for p_q, and then extrapolate. This procedure yields $p_q = 0.48 \pm 0.01$,[4,7] which is close to the result of the "strip" method, $p_q = 0.44 \pm 0.01$.[8] We see that the quantum percolation threshold is well above the classical threshold, which can be attributed to interference effects. This also shows that quantum percolation is really not percolation at all, since the threshold is significantly higher than the classical threshold of geometrical connectivity–it is simply Anderson localization.

The above techniques can also be extended to continuum percolation problems. In the classical version one is interested in the density at which randomly centered overlapping spheres percolate. Our method for classical site percolation can be easily applied to this problem.[9] If the sphere diameter is d, we find that the critical (reduced) density is $\rho_c = 0.646 \pm 0.007$, which is in agreement with several other studies (see for example[10,11]) and which we believe is the most accurate determination of this quantity to date.

An interesting variation on this problem occurs when the spherical particles have a hard core diameter σ, with $\sigma \leq d$. In this case the centers of the spheres may not be closer than σ, but if they are closer than d they are said to be directly connected. One is interested in the percolation threshold as a function of the dimensionless ratio σ/d. This problem has been studied by Bug et al.[12] by Monte Carlo simulation. They find an interesting nonmonotonic trend in ρ_c–for small σ/d it is a decreasing function, and then rises rapidly as σ/d approaches 1. This behavior can be understood as follows: For a given density, when σ is increased from $\sigma = 0$ this "pushes" the spheres apart, causing more pairs to become connected, and hence lowering the percolation threshold. However as σ/d approaches 1, the particles must be very densely packed in order for their thin "skins" to overlap. Indeed at $\sigma/d = 1$ one expects percolation to occur at random dense packing.

We have studied this same problem with our finite-size scaling approach. For each finite size, we generate random configurations with a hard sphere Monte Carlo algorithm, and then identify connected clusters. Averaging over the disorder then leads to the connectedness correlation length, from which we obtain the critical thresholds. As a function of σ/d we find[9] similar results to those of Bug et al.,[12] although for some values of σ/d their data points (no error bars are reported) fall outside our error bars, which we believe to be quantitatively accurate. (The results of Bug et al. appear to suffer from a somewhat arbitrary criterion for percolation.) Our results appear to extrapolate smoothly to the dense random packing density at $\sigma/d = 1$.

From our point of view, more interesting is the continuum quantum percolation problem. First considering the case of overlapping spheres, we associate a quantum state $|i\rangle$ with the ith sphere, and define a tight-binding Hamiltonian to have constant non-zero matrix elements between two sites that are directly connected. From this point on we can use the concept of quantum connectvity, introduced by us for the lattice[4,7] (and other[13]) problems, to calculate the quantum connectedness correlation length. We find the quantum percolation threshold is $\rho_q = 1.08 \pm 0.03$. As in the lattice problem, we see that this is well above the classical threshold.

Turning finally to the problem of continuum quantum percolation with hard core interactions, we have calculated the quantum percolation threshold as a function of σ/d, finding that for all σ/d $\rho_q > \rho_c$. ρ_q as a function of σ/d shows the same nonmonotonic dependence as the classical threshold, although in this case the initial fall is substantially more pronouced, presumably because the quantum threshold occurs at higher density, and this enhanced connectivity due to pushing apart spheres is increased. As σ/d approaches 1, it appears that the quantum threshold again approaches that of dense random packing. A full description of this problem and of our results will be published elsewhere.[9]

Acknowledgements

This research is supported by the National Science Foundation under Grant Nos. CHE 89-10749, DMR 86-03394, CHE 83-51207, and CHE87-00522.

References

[1] D. Stauffer. *Introduction to Percolation Theory.* Taylor and Francis London, (1985).

[2] P.J. Reynolds, H.E. Stanley, and W. Klein. *J. Phys. A* **11**, L199 (1978).

[3] M.N. Barber. *in: Phase Transitions and Critical Phenomena, vol. 8, ed. C. Domb and J. Lebowitz.* Academic, London, (1983).

[4] L.J. Root, J.D. Bauer, and J.L. Skinner. *Phys. Rev. B* **37**, 5518 (1988).

[5] M.P. Nightingale. *Physica* **83 A**, 561 (1976).

[6] D.W. Herrmann and D. Stauffer. *Z. Phys. B* **44**, 339 (1981).

[7] L.J. Root and J.L. Skinner. *J. Chem. Phys.* **89**, 3279 (1988).

[8] C.M. Soukoulis, E.N. Economou, and G.S. Grest. *Phys. Rev. B* **36**, 8649 (1987).

[9] J.G. Saven, J.R. Wright, J.L. Skinner, and B.J. Berne. (unpublished).

[10] S.W. Haan and R. Zwanzig. *J. Phys. A* **10**, 1547 (1977).

[11] Y.C. Chiew and G. Stell. *J. Chem. Phys.* **90**, 4956 (1989).

[12] A.L.R. Bug, S.A. Safran, G.S. Grest, and I. Webman. *Phys. Rev. Lett.* **55**, 1896 (1985).

[13] J.D. Bauer, V. Logovinsky, and J.L. Skinner. *J. Phys. C* **21**, L993 (1988).

TUNNELING AND RELAXATION IN LOW TEMPERATURE

SYSTEMS

Robert Silbey

Department of Chemistry and
Center for Materials Science and Engineering
Massachusetts Institute of Technology
Cambridge, MA 02139 USA

I. **Introduction**: The tunneling of heavy atoms in molecular systems, both in
the gas phase and in condensed phases, is an archetypical quantum
phenomenon, and as such, has been of great interest to physical scientists for
the last 60 years.   The standard textbook treatment [1] is to consider a
symmetric double well potential, and calculate the effect of tunneling on the
splitting of the energy levels which would be degenerate in the absence of
tunneling.   This one dimensional system can be solved using the WKB
approximation, or by simple numerical procedures.   Such a one dimensional
model does not contain all the physics of tunneling in real systems, since in
most cases, the tunneling system is coupled to other coordinates in the
molecule or condensed phase.   This coupling can lead to relaxation
(broadening of the energy levels), dephasing, and in some cases to symmetry
breaking.   By the latter, I mean that the energy level splitting due to
tunneling is zero when coupling to the other modes is considered, even
though it is non zero in the absence of that coupling.   This kind of symmetry
breaking has been considered in detail in the last few years by a number of
workers [2,3,4,5]; in the present lecture, I will follow the work of Harris and
Silbey [4], and proceed in a didactic fashion, making contact with other work
as it is necessary.   The lecture will begin with a description and simplification
of the Hamiltonian.   I will then describe a simple perturbation treatment, and
why it breaks down for strong coupling.   I will then introduce a different
perturbation theory, applicable for the strong coupling regime, and show how
this breaks down. Finally I will introduce a variational-perturbation
technique which circumvents the problems associated with both previous
attempts, and which interpolates from weak to strong coupling without too
many problems.   This procedure gives good agreement with the results of
other calculations using path integral techniques [2,3], and describes the
essential physics in a simple manner.

II. **Hamiltonian**:   Consider a tunneling system represented by an effective
one dimensional symmetric double well potential, V(Q), where Q is the
tunneling coordinate.   The Hamiltonian for this system in the absence of
coupling to other modes is

$$H_T = (1/2) P^2 + V(Q) \qquad\qquad 1.$$

where P is the momentum (and the coordinate and momentum are mass
weighted so that the mass does not explicitly appear).   The eigenvalues for this
Hamiltonian will be either symmetric (+) or antisymmetric(-) with respect to

changing Q to -Q, or to exchanging the left hand (L) side of the potential with the right hand side (R). In particular, for deep potentials, the lowest pair of states will be denoted |+> and |->, and their splitting given by 2K. The next higher doublet in the double well will be at an energy Ω above the ground doublet (Ω is effectively a vibrational quantum in the left or right well).

The other coordinates in the system (either the other vibrational coordinates in a large molecule or the other coordinates in a condensed phase) will be denoted by q_n. These will be taken to be harmonic modes and coupled to the tunneling coordinate by a term linear in the q_n. The total Hamiltonian is then given by:

$$H = H_T + (1/2) \sum_n \{p_n^2 + \omega_n^2 q_n^2\} + \sum_n q_n F_n(Q) \qquad 2.$$

where F is an operator in Q space which may be different for each harmonic mode. In particular, we expect that the lowest frequency acoustic modes will be very weakly coupled to the tunneling coordinate, which is after all quite localized on the scale of acoustic phonon wavelength.

We will be interested in the case of a deep double well potential so that $\Omega \gg K$, and in the case of low temperatures so that $kT \ll \Omega$. In this case, we can restrict our attention to the lowest doublet in the double well, so that H_T can be replaced by

$$H_T = K\{ |-><-| - |+><+| \} = K\{|L><R| + |R><L|\} \qquad 3.$$

where we have introduced the localized states |L> and |R> which are linear combinations of the eigenstates |+> and |->. The coupling term F(Q) in H can now be replaced by its matrix elements in the |+> basis (or in the L,R basis), so that the Hamiltonian has now been simplified to

$$H = K\{ |-><-| - |+><+| \} + (1/2) \sum_n \{p_n^2 + \omega_n^2 q_n^2\} +$$

$$\sum_n q_n \{\lambda_n[|+><-| + |-><+|] + \mu_n [|-><-| - |+><+|]\} \qquad 4.$$

In the theory of relaxation dynamics [6] the term proportional to |+><-| + |-><+| is a population relaxation term while the term proportional to |-><-| - |+><+| is a "pure" dephasing term. At very low temperatures, the latter term is usually unimportant relative to the former term, so we will neglect it in the following. The Hamiltonian is now

$$H = K\{ |-><-| - |+><+| \} + (1/2) \sum_n \{p_n^2 + \omega_n^2 q_n^2\} +$$

$$\sum_n q_n \lambda_n [|+><-| + |-><+|] \qquad 5.$$

and is identical to the spin-Boson Hamiltonian treated extensively in the literature [2,3,4,5]. Note that we can also represent H in the L,R basis:

$$H = K\{ |L><R| + |R><L| \} + (1/2) \sum_n \{p_n^2 + \omega_n^2 q_n^2\} +$$

$$\sum_n q_n \lambda_n [|L><L| - |R><R|] \qquad 6.$$

From this we see that the coupling is a localization term, i.e. fluctuations in the q_n lead to instantaneous energy differences between the left and right sides of the double well, which tend to localize the particle. This term is sometimes said to cause a friction which slows the tunneling rate. The term in K, on the other hand, tends to delocalize the particle, i.e. allow it to tunnel from one side to the other.

We will be concerned with the following question: suppose the particle is in state |L> at the initial time, what is the probability it is in |L> at a time t later? This gives us a measure of the tunneling rate. A more convenient function is the difference in probabilities for being on the left and being on the right which is the average value at time t of the operator

$$\hat{P} = |L><L| - |R><R| \qquad\qquad 7.$$

Note that in the absence of coupling, the average of this operator is given by

$$<P(t)> \; = <L| \{e^{+iHt}\, \hat{P} e^{-iHt}\} \; |L> = \cos(2Kt) \qquad\qquad 8.$$

where I have taken the initial time to be t=0. This is simple coherent oscillation between the left and right states since the particle was initially put in a coherent superposition of eigenstates, |+> and |->.

III. **Weak Coupling Perturbation Theory**: Using standard second order time dependent perturbation theory with the coupling term as the perturbation (Fermi Golden Rule), we find that P(t) obeys the following equation of motion [7]:

$$<\ddot{P}(t)> \; + (2K)^2 < P(t) > + (1 + A)\, \Gamma < \dot{P}> \; = 0 \qquad\qquad 9.$$

where A = exp(-2K/kT), and Γ is the population relaxation rate from the upper state (|->) to the lower (|+>) calculated in the usual way:

$$\Gamma = \int_{-\infty}^{\infty} dt < V(t)V(0)> \exp(-i2Kt) \;, \qquad [\hbar = 0] \qquad\qquad 10.$$

with $V = \Sigma_n q_n \lambda_n$, the heat bath assumed to be in a canonical distribution at temperature T, and the time dependence given by the zeroth order part of the Hamiltonian. Note that the equation for <P(t)> is that of the damped harmonic oscillator. The coupling to the other modes has introduced damping into the dynamics of the tunneling particle.

The solution for < P(t) > given that the particle is on the left at t =0, is

$$<P(t)> \; = \exp\{(1+A)\Gamma t/2\} \cos [\{(2K)^2 - (1+A)^2\Gamma^2\}^{1/2}t] \qquad\qquad 11.$$

which reduces to equ(8) when Γ = 0.

For 2K > (1+A)Γ, this represents underdamped harmonic motion, while for 2K < (1+A)Γ, it represents overdamped motion. Note that for very large Γ, there are two decay rates, one given approximately by (1+A)Γ and the other by $4K^2/[(1+A)\Gamma]$, the latter being very small compared to the former. Thus in the case of large Γ, there is a slow decay of <P(t)> to zero. We can look at this in the frequency domain and see that for 2K> (1+A)Γ, there are two peaks at non zero frequency, each with width (1+A)Γ. For 2K < (1+A)Γ, the peaks have merged, and are both centered at zero frequency, one being very broad and one very narrow. As the coupling increases, the narrow peak becomes still narrower (motional narrowing). All of this looks correct, except that for large Γ we have done second order perturbation theory (i.e. weak coupling) using a perturbation which produces a result larger than the initial splitting (2K). In other words, in this case we have predicted a width of the levels larger than their separation $((1+A)\Gamma > 2K)$. This suggests that the perturbation theory cannot be trusted any longer.

IV. Strong Coupling Perturbation Theory

In the case that the bath-tunneling particle coupling is strong, we should treat it exactly and treat the tunneling term (K) by perturbation theory. If we neglect the term in K in H, the Hamiltonian is that of two uncoupled sets of states, those built on |L> and those built on |R>. The bath modes, q_n, are coupled linearly but differently in the two subspaces. In other words, the bath modes are displaced harmonic oscillators, displaced to the right in the R subspace and displaced to the left in L. The eigenstates can easily be found for each of these:

$$\Phi_L^{(i)} = |L> \prod_n \phi^{(i_n)}(q_n + \lambda_n/\omega_n^2) \qquad \qquad \text{12a.}$$

$$\Phi_R^{(i)} = |R> \prod_n \phi^{(i_n)}(q_n - \lambda_n/\omega_n^2) \qquad \qquad \text{12b.}$$

where $\phi^{(i)}$ is the ith harmonic oscillator eigenfunction. The eigenvalues are (in the absence of K) given as

$$E^{(i)} = - \sum_n \lambda_n^2/(2\omega_n^2) + \sum_n (i_n + \frac{1}{2})\omega_n \qquad \qquad \text{13.}$$

Since the energy levels are doubly degenerate, with one state in the R subspace and one in the L subspace, we must use degenerate perturbation theory.

The perturbation is now given by the term containing K in H. If we now calculate the first order correction to the energies at zero temperature, we find that the lowest state in the |+> subspace and the lowest state in the |-> subspaces (these are the exact symmetries of H) are split by 2K' where

$$K' = K \exp - (\sum_n \lambda_n^2/\omega_n^3) = K \exp - \int_0^\infty d\omega J(\omega)/\omega^2 \qquad \qquad \text{14.}$$

and we have introduced the spectral strength function

$$J(\omega) = \sum_n \lambda_n^2/\omega_n \, \delta(\omega - \omega_n) \qquad \qquad \text{15.}$$

The term multiplying K in equ(14) can be considered to be a Franck Condon factor if one is a molecular spectroscopist or a Debye Waller factor if one is a solid state scientist.

For a heat bath of d dimensional phonons interacting with the tunneling system (using deformation potential coupling), $J(\omega) \sim \omega^d$ as $\omega \to 0$. This gives rise to a decrease in the effective tunneling matrix element, K', but not to symmetry breaking as long as d >1. However, if the coupling is Ohmic [2-5] so that $J \sim \omega$ as $\omega \to 0$, then K' is zero for any coupling to the bath because of the logarithmic singularity at $\omega=0$ in the integral of equ(14). This is the so-called orthogonality catastrophe or infra red singularity which is known from the study of X-Ray absorption line shapes [7,8]. Ohmic coupling is predicted for the coupling of certain localized systems to electron hole pairs in metals, for example, and this is the reason that this Hamiltonian has been of interest to those people studying tunneling in Josephson junctions.

Note however that our result is that any non-zero strength coupling will give rise to an effective tunneling matrix element that is zero for an

Ohmic bath. But, this result comes about from coupling to the low frequency modes ($\omega \to 0$), and we have assumed that the correct zeroth order wavefunctions are the product states given in equ(12). That is, we have assumed that the bath modes are fast compared to the tunneling motion, so that the bath always knows in which well the tunneling particle sits. But, it is the slowest modes that are giving us the symmetry breaking. We must therefore be very careful to pick the correct approximate zeroth order states in the low frequency regime in order to obtain sensible results. In order to examine this, we will turn to the variation-perturbation method.

V. Variation-Perturbation Method

From the above brief discussion, we expect that if the vibrational frequency is larger than the effective tunneling splitting ($\omega > 2K_{effective}$), the mode will be able to follow the tunneling particle. If on the other hand, the mode frequency is smaller than the tunneling splitting, we expect the mode not to be able to follow the tunneling particle. Therefore we expect the slow modes to be much less displaced than the fast modes. In order to be self consistent , we introduce variational states:

$$\Psi_L^{(i)} = |L> \prod_n \phi^{(i_n)}(q_n + f_n/\omega_n^2) \qquad \text{16a.}$$

$$\Psi_R^{(i)} = |R> \prod_n \phi^{(i_n)}(q_n - f_n/\omega_n^2) \qquad \text{16b.}$$

with i_n an integer and calculate the ground state energy of H. We then minimize that energy with respect to variations in the f_n. We find that

$$f_n = \lambda_n/[1 + 2\tilde{K}/\omega_n] \qquad \text{17.}$$

where

$$\tilde{K} = K \exp(- (\sum_n f_n^2/\omega_n^3) \qquad \text{18.}$$

Notice that the best value of f_n is a function of the effective tunneling splitting which is in turn dependent on the $\{f_n\}$, so that the values of these parameters must be determined self consistently. Our expectation of the behaviour of the displacements with respect to frequency is borne out by equ(17): if $2\tilde{K}/\omega_n \ll 1$, the mode is fully displaced, while if $2\tilde{K}/\omega_n \gg 1$ the mode is essentially undisplaced.

For the case of three dimensional phonons, this method gives only a quantitative change in the dependence of the tunneling splitting on coupling strength. However, in the case of Ohmic coupling, we find a qualitative change from the results of the last section. We now find that the effective tunneling matrix element is given by

$$\tilde{K} = K (2K/\omega_c)^\gamma \qquad \text{19.}$$

where ω_c is a cutoff frequency in the bath mode spectrum, and γ is related to the Ohmic density of states strength function $J(\omega) = \eta\omega$ by $\gamma = \frac{(2\eta/\pi)}{[1-(2\eta/\pi)]}$. This agrees with the results of the path integral calculations of [2,3], and predicts that the symmetry breaking ($\tilde{K} = 0$) occurs when $2\eta = \pi$ and not for $\eta = 0$ as was predicted using the states of equ(12).

Although I have presented this calculation as a straightforward variation-perturbation zero temperature theory result, a better way to proceed is to perform a variational unitary transformation on H, find a family of new Hamiltonians with the same spectrum as H and do a variational calculation with a limited basis to find the lowest free energy state at any temperature T. This defines a variationally determined unitary transformation, allows us to define a new H(0) and new perturbation V, upon which we can base a new perturbation theory calculation. This is the procedure followed in [4], to which the reader is referred. The result of this procedure is a derivation of the equation of motion for <P(t)>

$$<\ddot{P}(t)> + (2\tilde{K})^2 <P(t)> + (1 + \tilde{A})\,\tilde{\Gamma}\,<\dot{P}> = 0 \qquad\qquad 20.$$

where $\tilde{A} = \exp(-2\tilde{K}/kT)$ and $\tilde{\Gamma}$ is the relaxation term due to coupling of the transformed bath modes with the transformed tunneling system. In this picture, the tunneling particle has a distortion or cloud of quanta which it must "drag" through the barrier in order to tunnel to the other well. The fluctuations in the bath interfere with this process, and give rise to the damped harmonic oscillator equation of motion. For Ohmic coupling and for $2\eta > \pi$, the expression for $\tilde{\Gamma}$ is easily calculated. It predicts that at zero temperature, there is no tunneling at all, but as temperature is raised, the fluctuations in the bath allow tunneling to proceed. Thus, we have the highly unusual case of tunneling only occuring when incoherent thermal fluctuations open up an allowed channel.

This model has been extended to asymmetric wells and higher temperatures [4], and had been applied (previously) to exciton phonon coupled systems[9] and to ferroelectrics[10].

Acknowledgements: I would like to thank my friend, Bob Harris, of the University of California at Berkeley with whom all this work was done. This work was partially supported by a grant from the NSF.

References

[1] Quantum Mechanics, C. Cohen-Tannoudji, B. Diu and F. Laloe (Wiley-Interscience) 1977, page 455 et seq.
[2] A. Caldeira and A. Leggett, Phys. Rev. Lett **46**, 211 (1981); Ann. Phys. (NY), **149** 374 (1984); A. Leggett, S. Chakravarty, A. T. Dorsey, M. Fisher, A. Garg, W. Zwerger, Rev. Mod. Phys. **59**, 1, (1987)
[3] H. Grabert, U. Weiss, and P. Hanggi, Phys. Rev. Lett **52**, 2193 (1984); U. Weiss, H. Grabert, P. Hanggi and P. Riseborough, Phys. Rev **B35**, 9535 (1987)
[4] R. A. Harris and R. Silbey, J. Chem. Phys. **78** 7330 (1983); R. Silbey and R. A. Harris, J. Chem. Phys.**80**, 2615 (1984); ibid **83**, 1069 (1985) ; R. Silbey and R. A. Harris , J. Phys. Chem. **93**, 7062 (1989).
[5] E. Pollack, Phys. Rev **A33**, 4244 (1986); Chem. Phys. Lett. **127**, 178 (1986)
[6] Density Matrix Theory and Applications, K. Blum (Plenum, N.Y.) 1981
[7] Green's Functions for Solid State Physicists, S. Doniach and E. Sondheimer (Benjamin, N.Y.) 1974
[8] Many Body Physics, G.D. Mahan (Plenum, N.Y.) 1981
[9] D. Yarkony and R. Silbey, J. Chem. Phys. **65**, 1042 (1976); J.W. Allen and R. Silbey, Chem. Phys. 43, 341 (1979).
[10] A. Blumen and R. Silbey, J. Chem. Phys. **69**, 1072 (1978).

DYNAMICS OF QUANTUM PARTICLES:

COUPLED COHERENT AND INCOHERENT MOTION

P. Reineker, J. Köhler, A. Jayannavar, V. Kraus, H. Däubler

Abteilung Theoretische Physik, Universität Ulm
Albert–Einstein–Allee 11, 7900 Ulm, Germany

1 INTRODUCTION

Elementary excitation and transport properties in condensed matter are frequently described in terms of quasiparticles, such as electrons, plasmons, excitons, phonons, and polarons [1,2]. In the following we are mainly interested in these properties in connection with electrons and excitons [3–5] and if there is no need to differentiate between both kinds of particles, we denote them together as quantum particles. Also in organic materials, electrons naturally are of importance for the transport of charge, both in the case of ordinary and photoconductivity [6–8]. Excitons on the other hand play a role with respect to the energy transport in pure and doped molecular aggregates including photosynthetic systems.

The question whether the transport of these quantum particles occurs in a coherent or incoherent manner, in other words whether it has to be described by the Schrödinger [9] or by a master equation [10–12] has been a challenge both for theorists and experimentalists. The first unified theory describing the coupled coherent and incoherent motion and containing as limiting cases the purely coherent and the purely incoherent transport was developed in [13–16] using a stochastic model. A microscopic approach to this problem was given in [17,18]. A description of the coupled coherent and incoherent motion using generalized master equations was carried out in [19] and following papers. Related questions and problems are still under investigation in several fields: in the investigation of the self–trapping of electrons and excitons [20,21] and the formation of excimers, in connection with quantum tunneling in macroscopic systems [22–24], with tunneling systems in glasses [25,26], and in the transport of muons [27] and hydrogen [28] in metals. In the following presentation we shall concentrate on the transport of photoelectrons and excitons.

2 COHERENT, INCOHERENT, COUPLED COHERENT AND INCOHERENT MOTION OF QUANTUM PARTICLES

2.1 Mechanism of Exciton Transport

Organic materials are usually composed of rather large molecules, at least as compared to inorganic crystals. Fig. 2.1 shows the unit cell of an anthracene crystal, containing two identical, but differently oriented molecules. A single molecule is composed of three benzene rings. The physical properties of the molecule and of the crystal are strongly influenced by the

Large-Scale Molecular Systems, Edited by W. Gans *et al.*
Plenum Press, New York, 1991

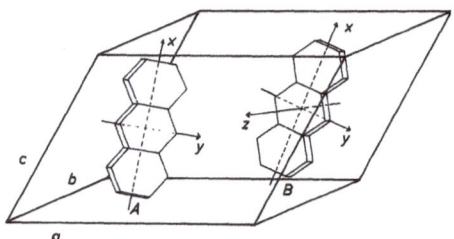

Fig. 2.1. Unit cell of anthracene

so–called π–electrons, whose wave functions are delocalized over the whole molecule. When the molecule is irradiated by light, an electron is excited into a higher state. If the spin orientation is not changed during the excitation, the excited state is a singlet state, in the other case a triplet state. Fig. 2.2 shows the energy level scheme of the molecule. The thick lines denote the electronic states, on the lefthand side for a triplet state, and on the righthand side for singlet states. The thin lines represent vibrational excitations. The life times of the higher excited states are rather short (10^{-12} s) as compared to those of the lowest excited states. Therefore in our model we assume that in addition to the ground state we have to consider only a single excited state, either in the triplet or in the singlet scheme.

In molecular crystals the interaction between the molecules is frequently determined by van der Waals forces, and thus rather weak. The simplest model for such a crystal is therefore the oriented lattice gas model, in which the molecules are arranged in the same way as in the crystal, but the interaction between them is completely neglected. Because of this missing interaction, the molecular excitation cannot be transferred to a neighbouring molecule. Therefore, if one is interested in the investigation of energy transport, a model as in Fig. 2.3 has to be considered, in which molecule n is excited. The energy transport occurs via a transfer matrix element $J_{m,n}$, which in the case of excitons has two contributions: a Coulomb–(Förster)–term (2.1), and an exchange–term (2.2):

$$J^c_{m,n} = \int \int dx\,dx' \, \varphi^*_{n,1}(x)\varphi^*_{m,0}(x')V(x,x')\varphi_{m,1}(x')\varphi_{n,0}(x) \tag{2.1}$$

$$J^e_{m,n} = \int \int dx\,dx' \, \varphi^*_{m,0}(x)\varphi^*_{n,1}(x')V(x,x')\varphi_{m,1}(x')\varphi_{n,0}(x) \tag{2.2}$$

In the Coulomb term (2.1) an electron at site m is deexcited and an electron at site n is excited, i.e. the electrons remain at the same molecule, but the energy is transferred from site m to site n. In contrast, the energy transfer described by the exchange–mechanism (2.2) occurs with an exchange of electrons at sites m and n. This requires an overlap of the wave functions at the two sites, and therefore this mechanism is usually shortrange. For singlet excitons both mechanisms contribute, whereas for triplet excitons the Coulomb term vanishes if the influence of the spin–orbit interaction can be neglected [29]. Therefore for triplet excitons the transfer matrix element usually is much smaller than for singlet excitons.

2.2 Coherent Versus Incoherent Motion in a Dimer

The limiting cases of the completely coherent and of the completely incoherent motion

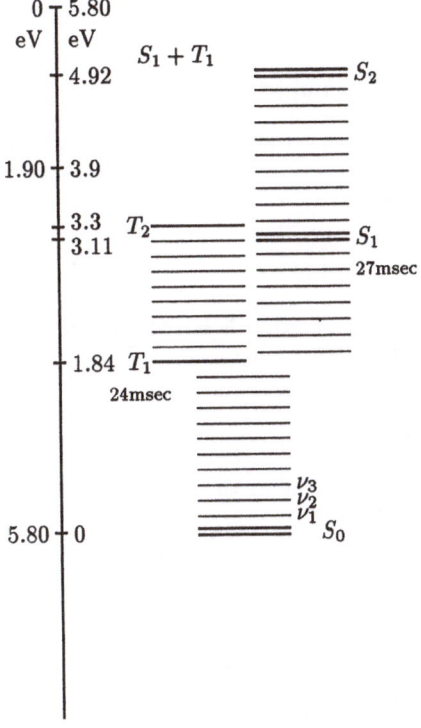

Fig. 2.2. Energy level scheme of anthracene

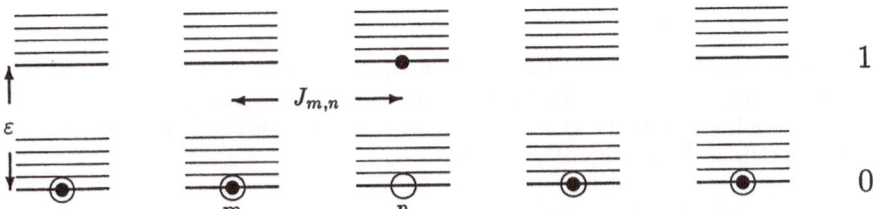

Fig. 2.3. Model for the description of the transport of quantum particles (ε: excitation energy, $J_{m,n}$: transfer matrix element)

can most easily be discussed for a dimer. Starting with the coherent limit, the Schrödinger equation is given by

$$i\dot{\psi} = H\psi. \tag{2.3}$$

In the Hamiltonian

$$H = \varepsilon(a_1^\dagger a_1 + a_2^\dagger a_2) + J(a_1^\dagger a_2 + a_2^\dagger a_1) \tag{2.4}$$

ε denotes the excitation energy at the two sites of the dimer and J the transfer matrix element. $a_i^\dagger$ and a_i are creation and annihilation operators for a quantum particle at site i. Denoting by φ_i, $i = 1,2$ the Hilbert space vector with the quantum particle at site i, $\psi(t)$ may be expanded in the following form:

$$\psi(t) = c_1\varphi_1 + c_2\varphi_2. \tag{2.5}$$

With the initial condition that the particle is initially sitting at site 1, i.e. $t = 0, c_1(0) = 1, c_2(0) = 0$, we obtain the following solution:

$$\underline{c}(t) = \begin{pmatrix} c_1(t) \\ c_2(t) \end{pmatrix} = \begin{pmatrix} \cos Jt \\ i\sin Jt \end{pmatrix}. \tag{2.6}$$

From this solution the probabilities of finding the quantum particle at site 1 and 2, respectively, of the dimer is given by eqs.(2.7,2.8):

$$p_1(t) = c_1^* c_1 = \cos^2 Jt = \frac{1}{2}(1 + \cos 2Jt) \tag{2.7}$$

$$p_2(t) = 1 - p_1(t) = \frac{1}{2}(1 - \cos 2Jt). \tag{2.8}$$

Phase relations between the two sites are determined from

$$c_1^* c_2 = i\cos Jt \sin Jt = \frac{i}{2}\sin 2Jt. \tag{2.9}$$

Obviously, both the occupation probabilities and phase relations oscillate with frequency $2J$.

Describing the particle motion by a Pauli master equation for the occupation probability P_i at site i, we have

$$\begin{aligned} \dot{P}_1 &= -2\gamma P_1 + 2\gamma P_2 \\ \dot{P}_2 &= 2\gamma P_1 - 2\gamma P_2. \end{aligned} \tag{2.10}$$

The occupation probability at site 1 decreases because particles move to site 2 with hopping rate 2γ, and it increases because with the same rate particles hop from site 2 to site 1. Starting with the same initial condition as above, i.e. the particle is initially localized at site 1, we obtain as solution of the rate equations (2.10)

$$P(t) = \begin{pmatrix} P_1(t) \\ P_2(t) \end{pmatrix} = \frac{1}{2}\begin{pmatrix} 1 + e^{-4\gamma t} \\ 1 - e^{-4\gamma t} \end{pmatrix}. \tag{2.11}$$

In this case the occupation probability at site 1 decays exponentially and phase relations do not exist at all.

2.3 Coherent Versus Incoherent Motion on a Linear Chain

The equation of motion in the coherent case is again the Schrödinger equation (2.3). In the case of nearest neighbour interaction the Hamiltonian reads

$$H = \sum_n J(a_{n+1}^\dagger a_n + a_n^\dagger a_{n+1}). \tag{2.12}$$

We expand the wave function in terms of states φ_n, describing a quantum particle localized at site n:

$$\psi(t) = \sum_n c_n(t)\varphi_n. \tag{2.13}$$

Assuming that the particle is initially sitting at the origin, its time development is determined by

$$c_n(t) = \frac{1}{N}\sum_k e^{-i2Jt\cos k}e^{-ikn} \rightarrow J_n(2Jt). \tag{2.14}$$

In the case of an infinite long chain the sum is replaced by an integral over the Brillouin zone and leads to a Bessel function . The occupation probabilities are then given by (2.15), and phase relations between sites n and n' by (2.16):

$$P_n(t) = c_n^* c_n = J_n^2(2Jt), \tag{2.15}$$

$$c_n^* c_{n'} = J_n(2Jt)J_{n'}(2Jt). \tag{2.16}$$

In the incoherent case the equation of motion is given by a Pauli master equation

$$\dot{P}_n = 2\gamma(P_{n+1} + P_{n-1} - 2P_n), \tag{2.17}$$

where the $P_n(t)$ are the occupation probabilities at site n and 2γ the hopping rates between neighbouring sites. Using again the initial condition that the particle is sitting at the origin at $t = 0$, i.e. $P_n(0) = \delta_{n,0}$, the solution is given by

$$P_n(t) = e^{-4\gamma t}\frac{1}{N}\sum_k e^{-4\gamma t\cos k}e^{-ikn} \rightarrow e^{-4\gamma t}I_n(4\gamma t). \tag{2.18}$$

$I_n(4\gamma t)$ on the right hand side is the modified Bessel function of order n. In arriving at this result, we have again assumed that the length of the chain is infinite and the sum over wave vectors can be replaced by an integral. In this incoherent description phase relations do not exist.

2.4 Description of the Coupled Coherent and Incoherent Motion of a Quantum Particle

The purely coherent motion [9] of quantum particles is conveniently described by the Schrödinger equation

$$i\dot{\psi} = H\psi. \tag{2.19}$$

Expanding the wave function in terms of states $|n\rangle$ describing a quantum particle at site n,

$$\psi(t) = \sum_n c_n(t)|n\rangle, \tag{2.20}$$

the probability of finding the particle at site n and phase relations between sites n and n' is given by (2.21,2.22), respectively:

$$\varrho_{nn}(t) = c_n^* c_n \tag{2.21}$$

$$\varrho_{nn'}(t) = c_n^* c_{n'}. \tag{2.22}$$

157

This purely coherent motion is disturbed by the phonons, under the influence of which phase relations decay in the course of time. If this decay occurs rapidly as compared to the transition between different sites, phase relations play no role, and the motion of the particle can be considered as purely incoherent. This purely incoherent motion [10–12] has been treated in the framework of the Pauli–Master–equation

$$\dot{\varrho}_{nn} = \sum_{n'} W_{n-n'}(\varrho_{n'n'} - \varrho_{nn}), \qquad (2.23)$$

where $W_{n-n'}$ are hopping rates between sites n and n'. In this description only occupation probabilities ϱ_{nn} can be calculated, phase relations play no role. The general case of a finite decay time of phase relations, the so–called coupled coherent and incoherent particle motion [13–15] is conveniently described with the help of a density operator. Its equation of motion can be written in the following form:

$$\dot{\varrho} = -iL\varrho. \qquad (2.24)$$

Here L is the Liouville operator, which in the case of the purely coherent particle motion is given by the commutator with the Hamiltonian. The diagonal matrix elements of ϱ

$$\varrho_{nn} = \langle n|\varrho|n\rangle \qquad (2.25)$$

are the probabilities of finding the particle at site n, the non–diagonal elements

$$\varrho_{nn'} = \langle n|\varrho|n'\rangle \qquad (2.26)$$

describe phase relations.

A different possibility to describe the coupled coherent and incoherent motion of quantum particles is given by the generalized master equation [30,31]. In the context of exciton motion this description has been introduced in [19] using a projection operator which projects down to diagonal elements of the density operator. Subsequently related techniques have been used in [32–34]. The generalized master equation reads

$$\dot{\varrho}_{nn}(t) = \int_0^t dt' \sum_{n'} W_{n-n'}(t') \left(\varrho_{n'n'}(t - t') - \varrho_{nn}(t - t')\right). \qquad (2.27)$$

Here $W_{n-n'}$ is a memory function, containing the information on the non–diagonal elements of the site representation of the density operator which have been projected out of the set of equations. Still another procedure for the description of the coupled coherent and incoherent exciton motion uses Green's functions [18].

3 INTERACTION OF A QUANTUM PARTICLE WITH VIBRATIONS: STOCHASTIC MODEL DESCRIPTION

The microscopic Hamiltonian for a quantum particle interacting with vibrational degrees of freedom is given by the following expression:

$$\begin{aligned} H = &\sum_n \varepsilon_n a_n^\dagger a_n + \sum_{n\neq n'} J_{n-n'} a_n^\dagger a_{n'} && \text{particle} \\ &+ \sum_n V_n(\{Q_i\}) a_n^\dagger a_n + \sum_{n\neq n'} V_{nn'}(\{Q_i\}) a_n^\dagger a_{n'} && \text{interaction} \\ &+ \sum_i \frac{1}{2}\left(P_i^2 + Q_i^2\right) && \text{vibration.} \end{aligned} \qquad (3.1)$$

Here ε_n is the energy of the particle at site n, $J_{n-n'}$ the transfer matrix element between sites n and n', and $V_n(\{Q_i\})$ and $V_{nn'}(\{Q_i\})$ describe the influence of the phonons on site energy and transfer matrix element, respectively. The operators $a_n^\dagger$ and a_n create and annihilate a quantum particle at site n, and P_i, Q_i are operators for the vibrational mode i. A general solution of the eigenvalue problem of this Hamiltonian was not possible up to now. Therefore one way to proceed is to replace the complicated Hamiltonian (3.1) by a simpler one, for example by treating the influence of the vibrations in a stochastic manner. In the treatment according to Haken, Strobl, and Reineker [13–15] the phonons give rise to fluctuations of the site energies, $h_n(t)$, and the transition Matrix elements, $h_{n,n'}(t)$. The Hamiltonian may then be written as

$$
\begin{aligned}
H = &\sum_n \varepsilon_n a_n^\dagger a_n + \sum_{n \neq n'} J_{n-n'} a_n^\dagger a_{n'} &&: H_0 \\
&+ \sum_n h_n(t) a_n^\dagger a_n + \sum_{n \neq n'} h_{nn'}(t) a_n^\dagger a_{n'} &&: H_1(t).
\end{aligned}
\tag{3.2}
$$

The stochastic matrix elements $h_n(t)$ and $h_{n,n'}(t)$ are assumed to be given by δ–correlated Markov processes with correlation functions

$$
\langle h_n(t) h_{n'}(t') \rangle = 2\gamma_0 \delta(t - t') \delta_{nn'}
\tag{3.3}
$$

and

$$
\langle h_{nn'}(t) h_{n''n'''}(t') \rangle = 2\gamma_{n-n'} \delta(t - t')(\delta_{nn''}\delta_{n'n'''} + \delta_{nn'''}\delta_{n'n''}).
\tag{3.4}
$$

In the following we shall investigate the transport properties of this Hamiltonian using the density matrix formalism.

4 TRANSPORT OF A QUANTUM PARTICLE UNDER THE INFLUENCE OF WHITE NOISE

4.1 Density Operator Equation

The total Hamiltonian is composed of a time independent part H_0, describing the coherent motion, and a stochastically time dependent part $H_1(t)$, taking into account the influence of the phonons:

$$
H = H_0 + H_1(t).
\tag{4.1}
$$

The equation of motion for the density operator becomes

$$
i\dot{\tilde{\varrho}} = [H, \tilde{\varrho}] = [H_0, \tilde{\varrho}] + [H_1(t), \tilde{\varrho}].
\tag{4.2}
$$

We introduce the average over the fluctuations, $\varrho = \langle \tilde{\varrho} \rangle$, and obtain the following equation of motion for the averaged density operator

$$
i\dot{\varrho} = [H_0, \varrho] + \langle [H_1(t), \tilde{\varrho}] \rangle = L\varrho.
\tag{4.3}
$$

The explicit evaluation of the average (see e.g. [35]) results in the following equation of motion, where $A^\times B = [A, B]$:

$$
\begin{aligned}
\dot{\varrho} = &-i[H_0, \varrho] - \sum_{n_1} \sum_{n_2} \gamma_{|n_1-n_2|} (b_{n_1}^\dagger b_{n_2})^\times (b_{n_2}^\dagger b_{n_1})^\times \varrho(t) \\
&- \sum_{n_1} \sum_{n_2} \overline{\gamma}_{n_1-n_2} (1 - \delta_{n_1 n_2})(b_{n_1}^\dagger b_{n_2})^\times (b_{n_1}^\dagger b_{n_2})^\times \varrho(t).
\end{aligned}
\tag{4.4}
$$

We represent the equation of motion in a basis $|m\rangle$, which describes a particle at site m, and arrive at the following equations for the time derivatives of the diagonal and nondiagonal elements of the density operator:

$$
\dot{\varrho}_{mm} = -i[H_0, \varrho]_{mm} + \sum_r 2\gamma_{m-r}(\varrho_{rr} - \varrho_{mm}),
\tag{4.5}
$$

$$
\dot{\varrho}_{mn} = -i[H_0, \varrho]_{mn} - 2\Gamma\varrho_{mn} + 2\gamma_{m-n}\varrho_{nm},
\tag{4.6}
$$

159

with

$$\Gamma = \sum_r \gamma_r. \tag{4.7}$$

4.2 Transport in a Dimer

In this subsection we consider the coupled coherent and incoherent motion of a quantum particle in a dimer (see Fig. 4.1). Both molecules have the same excitation energy ε, the transfer matrix element is denoted by J.

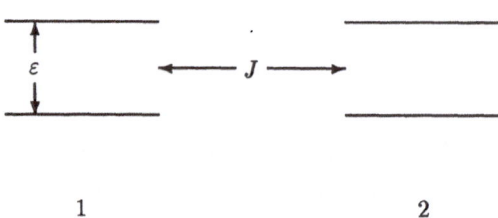

$$\begin{array}{cc} 1 & 2 \end{array}$$

Fig. 4.1. Dimer model for energy transfer

In the case of a dimer, the density operator is represented in the following way

$$\varrho = \sum_{m,n=1}^{2} |m\rangle \varrho_{mn} \langle n|. \tag{4.8}$$

The equations of motion for a diagonal and a non–diagonal matrix element are given by

$$\dot{\varrho}_{11} = -iJ(\varrho_{21} - \varrho_{12}) + 2\gamma_1(\varrho_{22} - \varrho_{11}) \tag{4.9}$$
$$\dot{\varrho}_{12} = -iJ(\varrho_{22} - \varrho_{11}) - 2(\gamma_0 + \gamma_1)\varrho_{12} + 2\gamma_1\varrho_{21}. \tag{4.10}$$

The equations of motion for the two remaining matrix elements are easily obtained because on account of normalization we have $\varrho_{22} = 1 - \varrho_{11}$ and because $\varrho_{12} = \varrho_{21}^*$. In the equations for the diagonal elements the incoherent part of the Hamiltonian gives rise to a contribution with the structure of a Pauli–master equation. The hopping rate is determined by the strength of the non–local fluctuations $2\gamma_1$. In the equations for the non–diagonal elements of ϱ the incoherent part of the Hamiltonian causes the decay of phase relations. This decay is determined both by the strengths of local and non–local fluctuations γ_0, and γ_1, respectively. Furthermore, the equations show that the coherent part of the Hamiltonian gives rise to a mixing of diagonal– and non–diagonal matrix elements.

The equations of motion (4.9,4.10) can be solved exactly. Assuming that the particle is sitting initially at the origin, i.e. $\varrho_{11}(0) = 1$, $\varrho_{22}(0) = \varrho_{12}(0) = \varrho_{21}(0) = 0$, we obtain for the time dependence of the probability of finding the particle at site 1

$$\varrho_{11}(t) = \frac{1}{2} + \left\{ \frac{1}{4}\left(1 + \frac{\gamma_0}{\sqrt{}}\right) e^{Rt} + \text{c.c.} \right\}. \tag{4.11}$$

The time dependence is determined by R, one of the eigenvalues of the density matrix equation, which is given by

$$R = -(\gamma_0 + 4\gamma_1) + i\sqrt{4J^2 - \gamma_0^2}. \tag{4.12}$$

For $\gamma_0 < 2J$ the square root is real and the occupation probability shows damped oscillations, whereas for $\gamma_0 > 2J$ the root becomes imaginary , the whole exponent negative, and the occupation probability decays exponentially. For $\gamma_0 = 2J$ we have a transition from the oscillatory to the purely exponentially decaying behaviour; therefore this value is considered as a transition from coherent to incoherent motion.

For the non–diagonal elements of the density operator, describing phase relations, we have a similar behaviour, i.e. for $\gamma_0 < 2J$ we have damped oscillations, for $\gamma_0 > 2J$ a purely exponential decay

$$\varrho_{12}(t) = \frac{i}{2\sqrt{4J^2 - \gamma_0^2}} e^{(\gamma_0 + 4\gamma_1)t} \sin\sqrt{4J^2 - \gamma_0^2}\ t. \tag{4.13}$$

4.3 Transport on a Linear Chain

The diagonal elements of the density matrix equation, i.e. the occupation probability, for the transport of a quantum particle on a linear chain is given by [35,36]

$$\varrho_{nn}(t) = \int_{-\pi}^{\pi} dk\, e^{ikn} a_{ks} \bar{\varrho}_0^{\ ks} e^{R_{ks}t} \tag{4.14}$$

$$+ \int_{-\pi}^{\pi} dk\, e^{ikn} \int_{0}^{\pi} dl\, \left(a_{kl}{}^s \varrho_0^{kl} + b_{kl}{}^a \varrho_0^{kl} \right) e^{R^{kl}t}. \tag{4.15}$$

Its time dependence is determined by two types of eigenvalues, a band eigenvalue which is complex

$$R^{kl} = -2\Gamma + i4H_1 \sin\frac{k}{2}\cos l \tag{4.16}$$

and an additional eigenvalue which is purely real

$$R^{ks} = -2\Gamma + \sqrt{\left(2\tilde{\tilde{\gamma}}_k\right)^2 - \left(4H_1 \sin\frac{k}{2}\right)^2}. \tag{4.17}$$

This additional eigenvalue exists only for those values of the wave vector k which satisfy

$$\left|2\tilde{\tilde{\gamma}}_k\right| \approx 2\gamma_0 > \left|4H_1 \sin\frac{k}{2}\right|. \tag{4.18}$$

Its limiting value $\bar{k}$ is determined from

$$\left|2\tilde{\tilde{\gamma}}_{\bar{k}}\right| \approx 2\gamma_0 = \left|4H_1 \sin\frac{\bar{k}}{2}\right|. \tag{4.19}$$

4.4 Mean Square Displacement on a Linear Chain

The mean square displacement, i.e. the second moment of the diagonal elements of the density operator, is defined by [38–40]

$$\dot{M}_2 = \sum_n n^2 \dot{\varrho}_{nn}. \tag{4.20}$$

Starting from this expression and using the equation of motion for the density operator, we arrive at the following closed set of equations:

$$\dot{M}_2 = 2\sum_m m^2 \gamma_m - \frac{1}{2}\sum_m m H_m V_m \tag{4.21}$$

$$\dot{V}_m = -2(\Gamma + \gamma_m)V_m - 4mH_m + \sum_{r(\neq -m)} 2r H_r W_{m+r} \tag{4.22}$$

$$\dot{W}_m = -2(\Gamma - \gamma_m)W_m. \tag{4.23}$$

161

V_m and W_m are lower moments of the density operator defined by

$$V_m = -\sum_n (2n+m) i (\varrho_{n+m,n} - \varrho_{n,n+m}) \qquad (4.24)$$

$$W_m = \sum_n (\varrho_{n+m,n} + \varrho_{n,n+m}). \qquad (4.25)$$

The set of equations (4.21,4.22,4.23) is easily solved and with the initial condition

$$\varrho_{n,m} = \delta_{n,m} \delta_{m,0} \qquad (4.26)$$

the mean square displacement as a function of time is given by

$$\langle n^2(t) \rangle = 2 \left(2\gamma_1 + \frac{J^2}{\Gamma + \gamma_1} \right) t + \frac{J^2}{(\Gamma + \gamma_1)^2} \left(e^{-2(\Gamma + \gamma_1)t} - 1 \right) \qquad (4.27)$$

$$\Gamma = \gamma_0 + 2\gamma_1. \qquad (4.28)$$

In writing down eqs.(4.27,4.28) we have restricted the expressions to nearest neighbour interaction only.

In the completely incoherent case the mean square displacement is given by

$$\langle n^2(t) \rangle = 2 \cdot 2\gamma_1 t, \qquad (4.29)$$

a result well known from hopping or diffusion theory. On the other hand by considering the purely coherent limit, the mean square displacement becomes

$$\langle n^2(t) \rangle = 2J^2 t^2, \qquad (4.30)$$

which describes just the free motion of a particle. In the limit of large times we arrive at

$$\langle n^2(t) \rangle = 2 \left(2\gamma_1 + \frac{J^2}{\Gamma + \gamma_1} \right) t - \frac{J^2}{(\Gamma + \gamma_1)^2}. \qquad (4.31)$$

For large times the time independent term on the right side may be neglected, the mean square displacement again becomes proportional to time, and the diffusion constant is given by $2\gamma_1 + J^2/(\Gamma + \gamma_1)$ [14]. It consists of two contributions: the first increases with increasing strength of the fluctuations, the second depends both on the coherent transfer matrix element and on the strength of the fluctuations and decreases with the latter.

The model presented in this section was the first to describe the interplay between coherent and incoherent transport, i.e. the coupled coherent and incoherent motion. Furthermore, within the framework of this model contact with a series of experiments, such as optical absorption, ESR and NMR, and measurements of the diffusion constant were possible [35,37–39,41–43]. On the other hand, the model definitely has some insufficiencies: it does not include the temperature explicitly; furthermore at lower temperatures the δ−correlation of the stochastic process may not be appropriate. The first problem has been discussed in several papers [17,18,44–50]. Some aspects of the second problem have been treated in recent papers [51–56] and the following section is essentially devoted to this question.

5 TRANSPORT OF A QUANTUM PARTICLE UNDER THE INFLUENCE OF COLOURED NOISE

5.1 Density Operator Description of the Transport of a Quantum Particle under the Influence of Dichotomic Coloured Noise

In this subsection we consider the transport of a quantum particle under the influence of dichotomic coloured noise [57]. Fig. 5.1 shows the model in an energy level scheme. As in Fig. 4.1 we have two sites 1 and 2, each with two energy levels a distance ε apart. The interaction between the two molecules is described by the transfer matrix element J. The

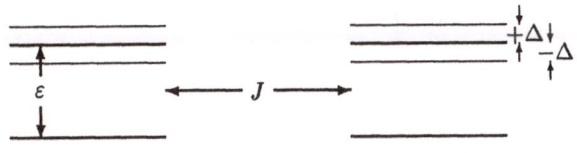

Fig. 5.1. Energy level scheme of a dimer in the case of a dichotomic stochastic process

influence of the phonons is modelled by a dichotomic stochastic process, which gives rise to fluctuations $\pm\Delta$ of the excitation energy. These fluctuations are indicated in Fig. 5.1 by the thin lines. The Hamiltonian of the model is given by

$$H = J(a_1^\dagger a_2 + a_2^\dagger a_1) + \sum_n h_n(t) a_n^\dagger a_n \qquad (5.1)$$

$$= H_0 + H_1(t). \qquad (5.2)$$

H_0 and $H_1(t)$ denote the coherent and the incoherent parts of the Hamiltonian. $h_n(t)$ describes the dichotomic stochastic process which is pictured schematically in Fig. 5.2. λ and λ^{-1} are the switching rate and average switching time, respectively, between the two possible energy values of the dichotomic process.

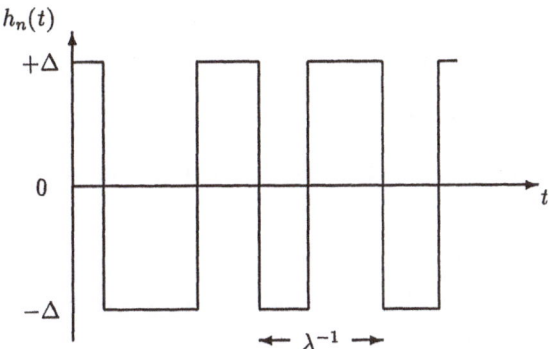

Fig. 5.2. Realization of a dichomoic stochastic process

As regards the stochastic processes, at each lattice site we assume that their mean values vanish

$$\langle h_n(t)\rangle = 0 \qquad (5.3)$$

and that their correlation functions are given by

$$\langle h_n(t) h_{n'}(t')\rangle = \delta_{nn'} \Delta^2 e^{-\lambda|t-t'|}, \qquad (5.4)$$

i.e. we assume that the energy fluctuations at different lattice sites are independent. In contrast to the Haken–Strobl model we now assume that the correlation function decays exponentially with finite decay constant λ. The amplitude of the energy fluctuations is Δ. In the dichotomic process multitime correlation functions are calculated according to

$$\langle h_n(t_1)h_n(t_2)\ldots h_n(t_n)\rangle = \langle h_n(t_1)h_n(t_2)\rangle\langle h_n(t_3)\ldots h_n(t_n)\rangle \tag{5.5}$$

with

$$t_1 > t_2 > \cdots > t_n. \tag{5.6}$$

The equation of motion for the density operator is given by

$$\dot{\varrho} = -i\left[H(t),\varrho\right]. \tag{5.7}$$

Because of the fluctuating part in the Hamiltonian the density operator in this equation is itself a fluctuating quantity. Finally we are interested in occupation probabilities and phase relations which are averaged over the fluctuations. Averaging (5.7) over the fluctuations we arrive at

$$\langle\dot{\varrho}_{11}\rangle = -iJ\langle\varrho_{21}\rangle + iJ\langle\varrho_{12}\rangle \tag{5.8}$$
$$\langle\dot{\varrho}_{22}\rangle = +iJ\langle\varrho_{21}\rangle - iJ\langle\varrho_{12}\rangle \tag{5.9}$$
$$\langle\dot{\varrho}_{12}\rangle = -iJ\langle\varrho_{22}\rangle + iJ\langle\varrho_{11}\rangle - i\langle h_1(t)\varrho_{12}\rangle + i\langle h_2(t)\varrho_{12}\rangle \tag{5.10}$$
$$\langle\dot{\varrho}_{21}\rangle = iJ\langle\varrho_{22}\rangle - iJ\langle\varrho_{11}\rangle + i\langle h_1(t)\varrho_{21}\rangle - i\langle h_2(t)\varrho_{21}\rangle. \tag{5.11}$$

In (5.10,5.11) new averages such as $\langle h_1(t)\varrho_{12}\rangle$ occur. The calculation of these quantities is carried out using a theorem by Shapiro and Loginov [58]:

$$\frac{d}{dt}\langle h(t)\phi_t[h]\rangle = \langle h(t)\frac{d}{dt}\phi_t[h]\rangle - \lambda\langle h(t)\phi_t[h]\rangle. \tag{5.12}$$

In this expression $h(t)$ is the stochastic process and $\phi_t[h]$ an arbitrary functional of it. With the help of this theorem we can derive an equation of motion for the new correlation functions:

$$\frac{d}{dt}\langle h_i(t)\varrho_{mn}\rangle = \langle h_i(t)\dot{\varrho}_{mn}\rangle - \lambda\langle h_i(t)\varrho_{mn}\rangle. \tag{5.13}$$

We insert the equation of motion for the density operator in the first term on the righthand side of (5.13) and arrive at correlation functions of the type $\langle h_i(t)h_k(t)\varrho_{mn}\rangle$. We apply the theorem another time to get an equation for this quantity

$$\frac{d}{dt}\langle h_i(t)h_k(t)\varrho_{mn}\rangle = \langle h_i(t)h_k(t)\dot{\varrho}_{mn}\rangle - 2\lambda\langle h_i(t)h_k(t)\varrho_{mn}\rangle. \tag{5.14}$$

With the help of (5.5) we now arrive at a closed set of linear differential equations which can be written in the following way:

$$\langle\dot{\varrho}(t)\rangle = L\langle\varrho(t)\rangle. \tag{5.15}$$

Here L is a 16–dimensional non–Hermitean matrix. With the ansatz

$$\varrho(t) = e^{Rt}\varrho \tag{5.16}$$

the solution of the system of differential equations is transformed into a non–hermitean eigenvalue problem:

$$R_i\varrho_i = L\varrho_i. \tag{5.17}$$

From the numerical solution we get eigenvectors ϱ_i and eigenvalues $R_i = \gamma_i + i\omega_i$, where the real part γ_i describes damping and the imaginary part ω_i oscillations. The general solution of the problem is given by a superposition of these eigensolutions and reads

$$\varrho(t) = \sum_{i=1}^{16} c_i\varrho_i e^{R_it}. \tag{5.18}$$

The coefficients c_i are determined from the initial condition. In the following figures we have assumed that the quantum particle is initially sitting at site 1 and that all non–diagonal elements of the density operator vanish, i.e. $\varrho_{n,n'}(0) = \delta_{n,n'}\delta_{n,1}$.

Fig. 5.3 shows the time dependence of the occupation probability for a quantum particle at site 1 for $J = \lambda = 1$ and various values of Δ. For $\Delta = 0$, the occupation probability oscillates without damping between the two sites. With increasing values of Δ this probability shows damped oscillations and for $\Delta > 2$ almost no oscillations are seen. The time dependence of $\langle \varrho_{12} \rangle$ describing phase relations between the two sites is pictured in Fig. 5.4. Also from this figure it is obvious that with increasing values of Δ the damping of the phase relations increases.

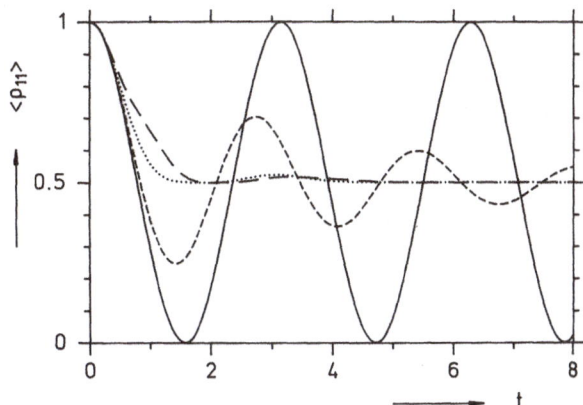

Fig. 5.3. Time dependence of $\langle \varrho_{11} \rangle$ for $J = 1, \lambda = 1$ and various values of Δ (solid line $\Delta = 0$, short dashed line $\Delta = 1$, dotted line $\Delta = 2$, long dashed line $\Delta = 3$)

In Fig. 5.5 $\langle \varepsilon_1 \varepsilon_2 \varrho_{22} \rangle$, a quantity which is characteristic for the coloured noise process and does not occur in the δ–correlated case, is shown as a function of time for various values of Δ. For vanishing values of the fluctuations (completely coherent motion) this quantity is identically zero. With increasing values of Δ we observe essentially a single peak of $\langle \varepsilon_1 \varepsilon_2 \varrho_{22} \rangle$ as a function of time. For times larger than the decay time of the peak one can show that the equations of the Haken–Strobl model can be rederived. Finally Fig. 5.6 shows again $\langle \varrho_{11} \rangle$, the occupation probability at site 1, this time, however, for various values of λ and $J = \Delta = 1$. The figure shows that with increasing values of λ the oscillatory behaviour of the occupation probability is stronger. This means that the influence of the fluctuations is averaged out and the particle motion approaches the coherent limit.

5.2 Transition from Coherent to Incoherent Motion

In the case of white noise (Haken-Strobl model) the transition from coherent to incoherent motion was determined by the eigenvalues

$$R_{3/4} = \gamma_0 \pm i\sqrt{4J^2 - \gamma_0^2}. \tag{5.19}$$

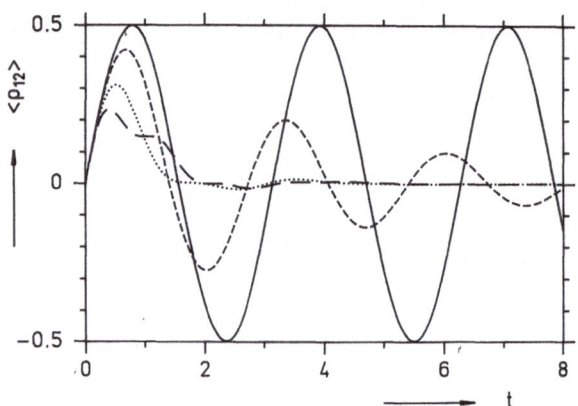

Fig. 5.4. Time dependence of $\langle \varrho_{12} \rangle$ for $J = 1, \lambda = 1$ and various values of Δ

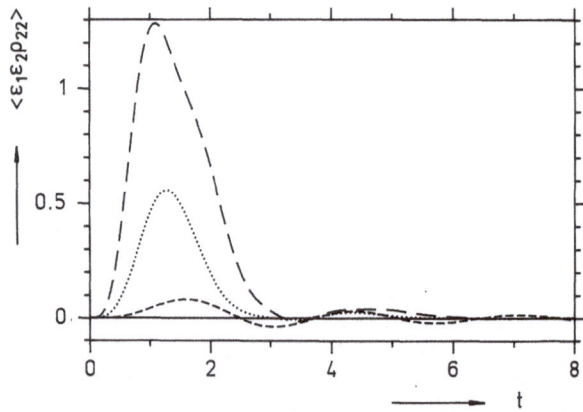

Fig. 5.5. Time dependence of $\langle \varepsilon_1 \varepsilon_2 \varrho_{22} \rangle$ for $J = 1, \lambda = 1$ and various values of Δ

In the coherent case ($\gamma_0 < 2J$) the eigenvalues are complex allowing for damped oscillations of the occupation probabilities, whereas in the incoherent case ($\gamma_0 > 2J$) a purely exponential decay is observed. To derive a criterion for this transition in the case of coloured noise, we consider the time integral of the correlation function. In the white noise case we have

$$\int_{-\infty}^{\infty} \langle h_n(t) h_n(t') \rangle \, dt = \int_{-\infty}^{\infty} 2\gamma_0 \delta(t - t') \, dt = 2\gamma_0 \tag{5.20}$$

and in the coloured noise case

$$\int_{-\infty}^{\infty} \langle h_n(t) h_n(t') \rangle \, dt = \int_{-\infty}^{\infty} \Delta^2 e^{-\lambda |t - t'|} \, dt = 2 \frac{\Delta^2}{\lambda}. \tag{5.21}$$

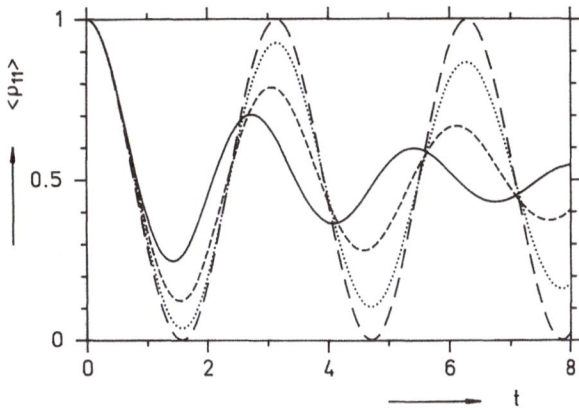

Fig. 5.6. Time dependence of $\langle \varrho_{11} \rangle$ for $J = 1, \Delta = 1$ and various values of λ

In the white noise case correlations decay infinitely fast, i.e. $\lambda \to \infty$. The comparison of the righthand sides of the two equations above shows that at the same time also the amplitude Δ has to approach infinity in order to keep $\gamma_0 = \Delta^2/\lambda$ constant. Using the criterion for the transition between coherent and incoherent motion in the white noise case, we have

$$\gamma_0 = \frac{\Delta^2}{\lambda} = 2J \tag{5.22}$$

or written in a different way

$$\frac{1}{\lambda} = 2 \frac{J}{\Delta^2}. \tag{5.23}$$

This equation is shown in Fig. 5.7 as the dashed straight line. The white noise case corresponds to the region close to the origin where coherent motion is represented by the area below the straight line and incoherent motion by the area above. On the other hand, when the fluctuations are purely static, i.e. $\lambda = 0$, the motion is also coherent. Therefore, we expect that somewhere above the straight line, a second transition from incoherent to coherent motion exists. In the white noise case the transition from coherent to incoherent motion was determined by all eigenvalues becoming purely real. In the case of coloured noise the situation is more complex: instead of only four eigenvalues as in the case of white noise, we have now 16. The numerical evaluation shows that 8 of these 16 eigenvalues are always purely real, whereas the other 8 occur in complex conjugate pairs, if the motion is definitely coherent. When the influence of the fluctuations becomes stronger two or four of these complex eigenvalues become also purely real. As a first approach to the investigation of the transition between coherent and incoherent motion in the case of coloured noise, we have therefore numerically determined the values of the parameters λ and Δ, for which the number of complex eigenvalues becomes smaller than 8. These parameter values are pictured in Fig. 5.7 by the full line. According to these preliminary results we denote the region I as coherent and region II as incoherent.

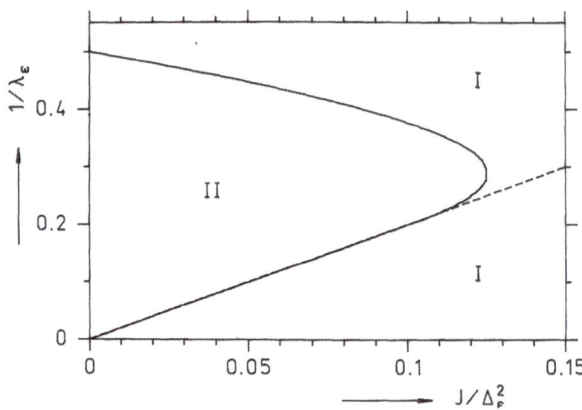

Fig. 5.7. Transition between coherent and incoherent motion of a quantum particle

6 MEAN SQUARE DISPLACEMENT OF A QUANTUM PARTICLE UNDER THE INFLUENCE OF CLOURED NOISE

6.1 Model and Equations of Motion

In this section we investigate the mean square displacement of a quantum particle under the influence of energy fluctuations with coloured noise [55,56]. The Hamiltonian is given by

$$H = \sum_{n,a} J a^{\dagger}_{n+a} a_n + \sum_n h_n(t) a^{\dagger}_n a_n = H_0 + H_1(t). \tag{6.1}$$

$a^{\dagger}_n$ and a_n are again creation and annihilation operators for a quantum particle at site n. J is the transfer matrix element between neighbouring sites, and $h_n(t)$ describes the energy fluctuations at site n with the properties defined in (5.3–5.6). The sums over n and a run over all unit cells of a crystal and its nearest neighbours.

The *equation of motion for the density operator* is given by (for the notation see (4.4); details of the following calculations are given in [55,56])

$$\dot{\varrho}(t) = -i(H^{\times}_0 + H^{\times}_1(t))\varrho(t). \tag{6.2}$$

Using the time development operator $U(t)$ and averaging over the fluctuations we write the formal solution of the density operator equation in the following form

$$\langle \varrho(t) \rangle = \langle U(t) \rangle \varrho(0) \equiv K(t) \varrho(0). \tag{6.3}$$

The averaged time development operator $K(t)$ is determined from the following integro–differential equation:

$$\dot{K}(t) = -iH_0^\times K(t) - \int_0^t d\tau \, \tilde{M}(t-\tau) K(\tau) \tag{6.4}$$

where $\tilde{M}(t)$ denotes its kernel.

With the help of the tetrade $K(t)$ the *mean square displacement* of the quantum particle is written as

$$\langle \hat{R}^2(s) \rangle = \sum_n n^2 \langle \varrho_{nn}(s) \rangle = \sum_n n^2 K_{nn00}(s). \tag{6.5}$$

After some analytical manipulation with the intention to bring a factor J^2 in front of the equation, we arrive at

$$\langle \hat{R}^2(s) \rangle = \frac{2J^2}{s^2} \sum_{a,a'} aa' \sum_n \hat{K}_{n,n+a,0,a'}(s). \tag{6.6}$$

With the definition of the contraction

$$\sum_n \hat{K}_{n,n+a,0,a'}(s) = \hat{\kappa}_{aa'}(s) \tag{6.7}$$

the mean square displacement is written as

$$\langle \hat{R}^2(s) \rangle = \frac{2J^2}{s^2} \sum_{a,a'} aa' \hat{\kappa}_{aa'}(s). \tag{6.8}$$

The contraction is determined from the following exact equation

$$\left\{ s + \hat{\tilde{\mu}}(s) \right\} \hat{\kappa}(s) = 1. \tag{6.9}$$

This equation has the structure of equations for Greens functions with $\hat{\tilde{\mu}}$ playing the role of the self–energy.

The following evaluation of $\hat{\tilde{\mu}}$ up to second order in J/Δ allows us to determine the mean square displacement up to fourth order in J/Δ. For a linear chain $\hat{\kappa}(s)$ and thus the mean square displacement can then be evaluated analytically in closed form without further approximations. For 2– and 3–dimensional crystals the mean square displacement is determined in a recursive manner. To that end we define the following expression

$$\hat{R}_n(s) = \frac{2J^2}{s^2} \sum_{a,a',a_1,\ldots,a_{2n}} aa' \hat{\kappa}_{a-a_1-\cdots-a_{2n},a'}. \tag{6.10}$$

The comparison with (6.8) shows that the mean square displacement is then obtained from

$$\hat{R}_0(s) = \langle \hat{R}^2(s) \rangle. \tag{6.11}$$

The *recurrence relation* for $\hat{R}_n(s)$ is obtained from

$$\frac{\varphi(J,s)}{z} \hat{R}_0(s) + \frac{2n+1}{\zeta_{n+1}} \chi(J,s) \hat{R}_n(s) + \frac{2n+1}{\zeta_{n+1}} \psi(J,s) \hat{R}_{n+1}(s) = \frac{2}{s^2} J^2 a^2. \tag{6.12}$$

In this equation $\varphi(J,s)$, $\chi(J,s)$, and $\psi(J,s)$ are well known functions [55,56] and z is the number of nearest neighbours. Subtracting two consecutive equations we arrive at the following tridiagonal recurrence relation

$$\sigma_n \chi(J,s) \hat{R}_n(s) - \left(\chi(J,s) - \sigma_n \psi(J,s) \right) \hat{R}_{n+1}(s) - \psi(J,s) \hat{R}_{n+2}(s) = 0 \tag{6.13}$$

which is written in the form of a continued fraction:

$$Y_n = \frac{\hat{R}_{n+1}}{\hat{R}_n} = \frac{\sigma_n \chi}{\chi - \sigma_n \psi + \psi Y_{n+1}}. \tag{6.14}$$

Here σ_n is defined via

$$\sigma_n = \frac{2n+1}{2n+3} \frac{\zeta_{n+2}}{\zeta_{n+1}} \tag{6.15}$$

and ζ_n has to be determined from

$$\zeta_n = \sum_{a_1,\ldots,a_{2n}} \delta(0, a_1 + a_2 + \cdots + a_{2n}) \tag{6.16}$$

and is the number of possibilities to return to the origin after $2n$ steps each to a nearest neighbour.

The *diffusion constant* can then be evaluated in Laplace space via the following limiting procedure

$$D = \lim_{s \to 0} \frac{s^2}{2} \langle \hat{R}^2(s) \rangle = \lim_{s \to 0} \frac{s^2}{2} \hat{R}_0(s). \tag{6.17}$$

In terms of continued fractions we have

$$D = za^2 J^2 \frac{1\ |}{|\phi + \chi +} \frac{\sigma_0 \chi \psi\ |}{|\chi - \sigma_0 \psi +} \frac{\sigma_1 \chi \psi\ |}{|\chi - \sigma_1 \psi +} \cdots \frac{\sigma_{N-2} \chi \psi\ |}{|\chi - \sigma_{N-2} \psi +} \cdots \tag{6.18}$$

Using a theorem by Pringsheim [59] one can show that the continued fraction converges if

$$\frac{\Delta^2}{J^2} > z(z+1), \tag{6.19}$$

i.e. for large amplitudes of the fluctuations (incoherent limit).

6.2 Results for the Diffusion Constant

In the following we shall discuss our results for the diffusion constant (6.18) and compare it to several results obtained recently in the literature. Kitahara and Haus [53] study diffusion on a linear chain within a stochastic model which reduces to the present one if the stochastics are dichotomic Markov processes. Their perturbation theory approach is applicable to fluctuations with short correlation times. Furthermore, because they take the stochastic part of the Hamiltonian as a perturbation, their approximation will fail for $\Delta \gg J$, except for extremely short correlation times. Thus the Kitahara–Haus result holds under the conditions

$$\frac{\lambda}{J} \gg 1 \quad , \qquad \frac{\Delta}{J} \ll 1 \tag{6.20}$$

and in our notation their result for the diffusion constant reads

$$D = a^2 J^2 \frac{\lambda}{\Delta^2} \left(1 + 3 \frac{J^2}{\lambda^2} \right). \tag{6.21}$$

Our treatment is valid, if $\Delta \gg J$. Therefore, one might conclude that it is not meaningful to compare our result to (6.21). However, as we claim that the second and fourth order (in J) terms of our result are exact, (6.21) must be included in it, containing only J^2 and J^4 terms. For that reason (6.21) yields a useful countercheck to our theory.

Exactly the same model as the one presented here was investigated by Inaba [54]. He assumes that all density matrix elements apart from the diagonal and nearest neighbour off-diagonal ones vanish. This assumption is justified, if the correlation times are short and the amplitude of the stochastic fluctuations is large, i.e.

$$\frac{\lambda}{J} \gg 1 \quad , \qquad \frac{\Delta}{J} \gg 1. \tag{6.22}$$

The assumptions become doubtful, if only one of these conditions is fulfilled. If the fluctuations are weak, the deterministic motion is not perturbed sufficiently to render the density matrix 'tridiagonal' in spite of the short correlation time. If, on the other hand, the correlation time is long, the coherence of the wave function is not destroyed within a few lattice constants in spite of the magnitude of the fluctuation amplitudes. Inaba obtains for the diffusion constant

$$D = \frac{z}{2}a^2 J^2 \left(\frac{\lambda}{\Delta^2} + \frac{1}{\lambda} - \frac{J^2}{\lambda^3} \right).$$

(6.23)

Our calculation has been carried out under the assumption

$$\frac{\Delta}{J} \gg 1.$$

(6.24)

To compare our result with (6.21) and (6.23) we take the second approximant of (6.18) and neglect the ψ term in the second denominator. The resulting expression is still exact up to fourth order in J:

$$D = \frac{z}{2}a^2 J^2 \left\{ \frac{\lambda}{\Delta^2} + \frac{1}{\lambda} + \frac{J^2}{\lambda^3} \left[-\frac{z+2}{4} + \left(z + \frac{\zeta_2}{3z} \right) \frac{\lambda^2}{\Delta^2} - \frac{27}{4} \frac{\lambda^2}{3\lambda^2 + 4\Delta^2} \right] \right\}.$$

(6.25)

In the limit

$$\Delta \to \infty \quad , \qquad \lambda \to \infty \quad , \qquad \frac{\Delta^2}{\lambda} = \gamma_0$$

(6.26)

the first term gives the diffusion constant according to Haken and Reineker [14,39] if only energy fluctuations are considered. Comparing the three results we find that (6.23) and (6.25) agree up to second order in J whereas in (6.21) $1/\lambda$ is neglected against λ/Δ^2. A comparison of the fourth order terms shows that for $z = 2$ (linear chain) the first term in formula (6.25) is identical to the J^4 term in Inaba's formula, the second to that of formula (6.21) ($\zeta_6 = 6$ for a linear chain); none of these formulae includes the third J^4 term of (6.25) stemming from four times correlation functions of the primary process. For large Δ, Inaba's and our formulae agree, because the second and third J^4 terms may be neglected.

For $z > 2$, only (6.23) and (6.25) can be compared. The formulae do not even agree in the case of large Δ — in Inaba's formula the diffusion constant is just proportional to z whereas ours contains terms $\propto z^2$. To make clear the origin of such a difference, we remark that in Inaba's treatment all the memory of jumps of the diffusing particle beyond nearest neighbours is lost; our result, however, contains the memory effects of closed paths consisting of four jumps (to a nearest neighbour each) in the approximation (6.25) and of even more in the full continued fraction.

To illustrate the difference between the three results we plot the reduced diffusion constant $D/(a^2 J)$ corresponding to (6.21,6.23,6.25) as a function of the reduced switching rate λ/J (Fig. 6.1). The picture shows clearly that approximation (6.21) becomes poor for small values of λ/J and that the other two approximations (6.23,6.25) agree pretty well for large values of Δ/J (linear chain case). Results (6.23) and (6.25) are also in good agreement over most of the λ/J range with numerical results plotted in the same figure which were obtained as a solution to the stochastic Schrödinger equation [60].

A further treatment of the same diffusion problem has been given by Inaba [60]. In a CPA calculation, diagrams describing backscattering of the diffusing particle are neglected. Therefore one expects that the diffusion constant obtained from this approximation will be too large. From Inaba's density matrix approach [54] one should obtain a too small diffusion constant in the case of large λ/J, because backscattering effects are overemphasized; for small λ/J, there is no simple way of prediction, because backscattering between nearest neighbours is fully taken into account, but backscattering after going to a next to nearest (or farther) neighbour is neglected. These conclusions are confirmed by Fig. 6.2 comparing the two solutions with our approximation (6.25). The CPA result is larger than ours, although at high switching rates the difference between the two curves is zero. The density matrix approximation yields a smaller diffusion constant than our theory for rapid fluctuations. In the case of slow fluctuations our result is smaller, indicating that more long range coherence leading to an increased backtransfer is taken into account.

In Fig. 6.3 we compare (6.25) and several approximants of the continued fraction solution (6.18) of the 1D case to the diffusion constant obtained from a direct solution of (6.9). For $\Delta/J = 1$ the latter solution does not exist over the whole range of λ/J, because the radicand of square roots contained in it becomes negative within a certain interval of λ/J values. In the same interval the continued fraction diverges. However, the figure also shows that for $\Delta/J \geq 1.5$ there is perfect convergence over the whole range of λ/J and that already the fifth approximant is a very good approximation to the total continued fraction.

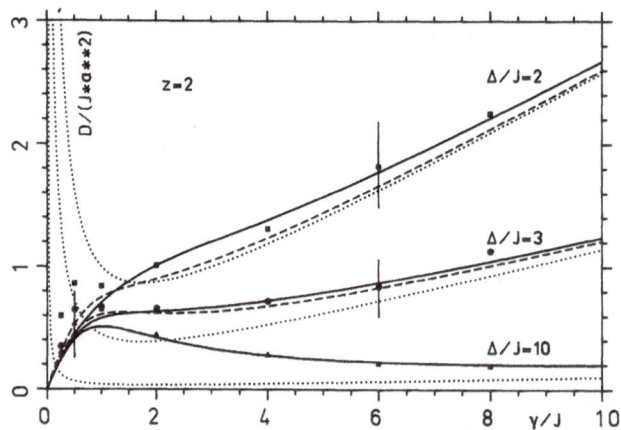

Fig. 6.1. Dependence of the diffusion constant on the switching rate for several fluctuation strengths Δ/J. Linear chain case. Dotted lines represent the Kitahara–Haus result [53], dashed ones Inaba's result [54], full lines show approximation (6.25) of the present theory. Single points are simulation results obtained by Inaba [60] (the largest error bar is given for each case).

We have calculated some approximants of the continued fraction (6.18) for several values of Δ/J and different lattices as a function of λ/J. The results of the calculations show that for sufficiently large Δ/J the continued fraction converges numerically within the whole range of switching rates. Examples of this behaviour are given in Fig. 6.4 where Inaba's approximation is plotted, too. The value of Δ/J, below which a region of non–convergence exists (in which our approximation fails) is increasing with the number of nearest neighbours; in the case of a linear chain we have a limiting value of $\Delta/J = 1.5$, for a fcc lattice $\Delta/J = 9$. These values are somewhat smaller than those estimated from (6.19).

We have seen that Inaba's formula is well behaved everywhere. One cannot conclude, however, from this nice analytical property that (6.23) is superior in the range where our expression diverges. The density matrix approximation breaks down there as well, because the assumptions necessary for its derivation are no longer valid. In fact, it is an advantage of our theory that the final result yields the possibility of estimating the range of its approximate validity. Basically, our theory is an unusual kind of perturbation scheme treating as perturbation the non–stochastic part of the Hamiltonian. For this reason, it should be a good approximation for sufficiently large Δ/J, no matter how small λ/J. On the other hand, for large λ/J taking into account only a few powers of J in the 'self–energy' is sufficient because of strong memory loss effects (the δ–correlated limit is approached).

The figures show that for a fixed value of Δ/J the continued fraction (6.18) goes to zero with $\lambda/J \to 0$. It may be meaningless in the case of small Δ/J, because there is a range without convergence between the regime of large λ/J where the theory is perfectly applicable and the limiting case we are interested in. But from the above considerations we conclude that in the case of large Δ/J it is allowed to interpret the behaviour of this limit as Anderson localization [61]. The figures, however, also show that even a small time dependence in the fluctuations results in a non vanishing diffusion constant and thus in a delocalization of the particle.

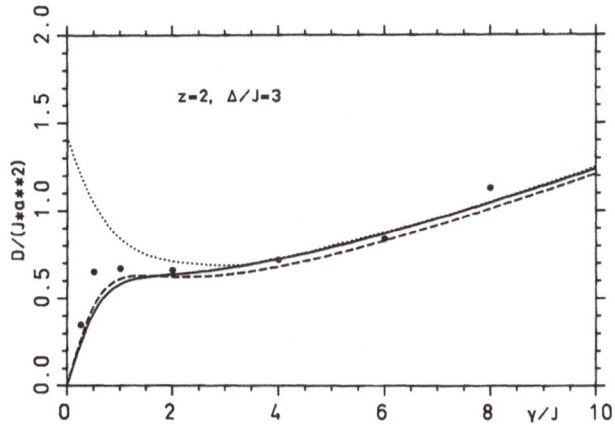

Fig. 6.2. Comparison of approximation (6.25) (full line) to Inaba's density matrix approach [54] (dashed line), a numerical CPA calculation [60] (dotted line), and a simulation of the stochastic Schrödinger equation [60] (single dots).

Another important feature of (6.18) is the fact that the diffusion constant is determined by all the numbers ζ_n of closed paths of given lengths $(2na)$ a particle can take starting out at some lattice point. But this means that the whole structure of the lattice enters the mathematical expression as soon as the stochastics is not δ correlated. This is a genuine memory effect; at a given time the particle 'remembers' all the loops it has already run through.

There is, however, no dramatic change in the absolute value of the diffusion constant. Fig. 6.4 shows that it is enhanced in comparison to the approximation $D = (1/2)za^2J^2(\lambda/\Delta^2)$ describing the 'corresponding white noise case' [60] or case of total memory loss [54]. For large values of λ/J this enhancement is somewhat larger than expected from Inaba's density matrix approximation and well described by the CPA result whereas for slow fluctuations only the

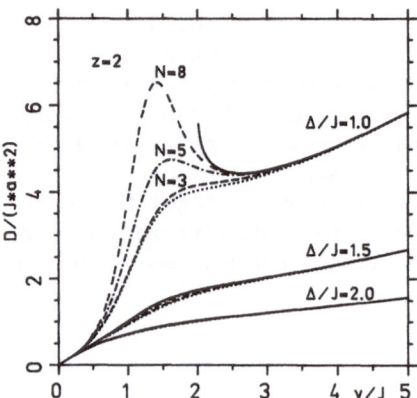

Fig. 6.3. Approximants of the continued fraction representing the diffusion constant as compared to the solution derived from (6.9) (full line) in the linear chain case. The dotted lines show approximation (6.25). For $\Delta/J > 1$ only the third and fifth approximants are plotted. For $\Delta/J = 1$ the solution on the basis of (6.9) does not exist over the whole range of λ/J.

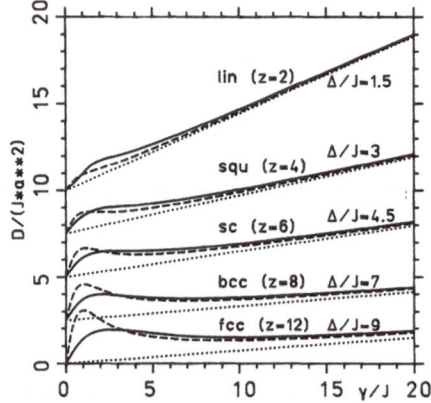

Fig. 6.4. Continued fraction solution (N=10, full lines) for several lattices and fluctuation strengths. The origin of each curve is shifted 2.5 units upwards with respect to the preceding one to avoid intersections. For reasons of comparison, Inaba's result (dashed lines) and the total memory loss case (dotted lines) [54] are plotted as well.

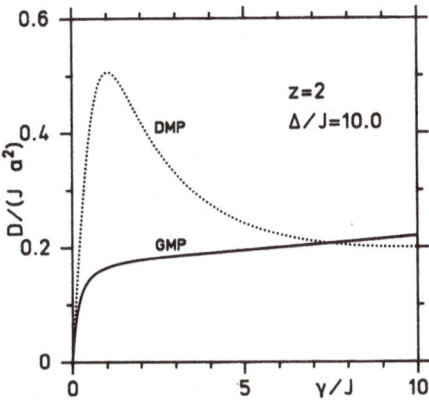

Fig. 6.5. Comparison of the linear chain results (10th approximant) for strong fluctuations as modeled by Gaussian (solid line) or by dichotomic (dotted line) Markov processes.

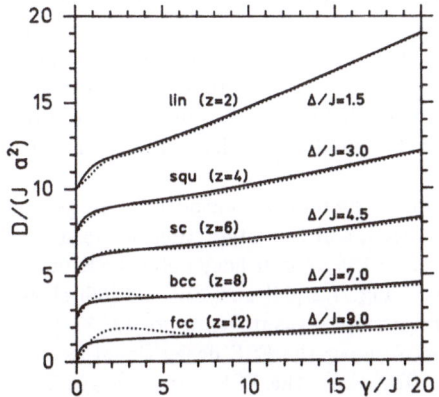

Fig. 6.6. Diffusion constant for several lattices and fluctuation strengths. Solid lines correspond to Gaussian, dotted to dichotomic processes (10th approximant given in each case). To avoid intersection the origin of each curve is shifted upward by 2.5 units with respect to its predecessor.

first of these approximations agrees qualitatively with our theory. The deviations become larger with increasing number of nearest neighbours.

To investigate the difference between the dichotomic and a Gaussian stochastic process, Kassner [62] has evaluated the diffusion constant when the local energy fluctuations are described by a Gaussian process. The Hamiltonian for the model is again given by (6.1). The average value and the two–time correlation function are given by (5.3,5.4). The calculation of multi–time correlation functions, however, becomes different. They are given by

$$\langle h_n(t_1)h_n(t_2)h_n(t_3)\ldots h_n(t_n)\rangle = \sum \langle h_n(t_1)h_n(t_2)\rangle \langle h_n(t_3)\ldots h_n(t_n)\rangle, \tag{6.27}$$

where the sum runs over all different pairings of the stochastic processes. Using the fact, that for Gaussian processes cumulants of higher than second order vanish, we have

$$\langle \exp\left\{-i\int_0^t d\tau\, h_n(\tau)\right\}\rangle = e^{-\theta\lambda t}\exp\left\{\theta\left(1-e^{-\lambda t}\right)\right\} \tag{6.28}$$

$$\theta = \frac{\Delta^2}{\lambda^2}. \tag{6.29}$$

The small change in the expression for the multi–time correlation functions results in rather involved equations for the diffusion constant. The numerical evaluation of the diffusion constant is shown in Fig. 6.4 and Fig. 6.5 [62]. Fig. 6.4 shows the reduced diffusion constant for a linear chain as a function of λ/J both for the dichotomic and for the Gaussian process. The diffusion constant of the Gaussian process increases continuously with increasing value of λ/J in contrast to the one of the dichotomic process which runs through a maximum. A similar, but quantitatively less distinct behaviour also holds for more complicated lattices as is seen in Fig. 6.6.

7 ESR AND OPTICAL LINESHAPES: INFLUENCE OF STATIC DISORDER

7.1 Influence of Static Disorder on ESR Line Shapes

The motivation for the investigations of this section stems both from experimental and theoretical results. In recent years a lot of investigations have been carried out to understand the transport of charge carriers in quasi–one–dimensional organic materials [63], especially in radical cation salts. Fig. 7.1 [64] shows the projection of the structure of the fluoranthene radical cation salt $(FA)_2PF_6$ parallel (a) and perpendicular (b) to the stack axis. The structure shows the stack of fluoranthene donor molecules and inbetween the inorganic acceptors.

One way of getting information on the transport of charge carriers is from the investigation of spin resonance. On account of the interaction between the spin of the charge carrier and the proton spins localized in the fluoranthene units, ESR properties are influenced by the motion of the charge carrier. Fig. 7.2 [65] and Fig. 7.3 [66] show electron spin echo (ESE) decay functions. Fig. 7.2 represents measurements for $(FA)_2^+[(SbF_6)_{1-x}(PF_6)_x]^-$, $x \approx 0.5$. The lefthand part of the figure shows the ESE decay signal as a function of τ, the righthand part as a function of $\tau^{3/2}$. Obviously, the ESE signal decays as $exp(-\tau^{3/2})$ for small times and purely exponentially for large times. In [65] this was interpreted as a transition from 1–dimensional to 3–dimensional motion of the particle. On the other hand the ESE in Fig. 7.3 decays in a purely exponential manner. Thus, from experimental results the question arises whether the ESE decay occurs according to

$$\exp(-(\gamma t)^{3/2}) \tag{7.1}$$

or as

$$\exp(-\overline{\gamma}t). \tag{7.2}$$

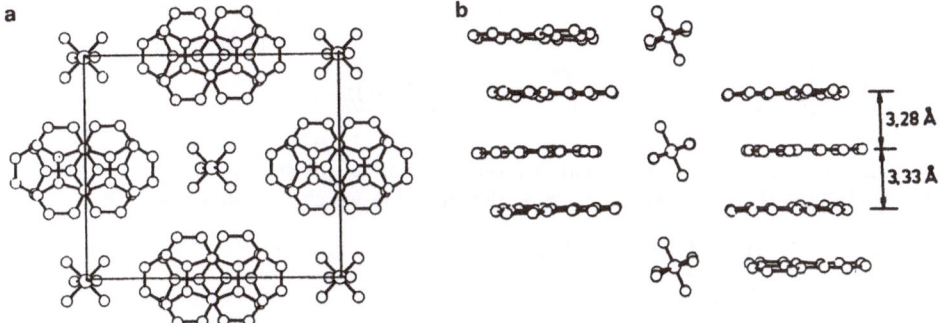

Fig. 7.1. Projection of the crystal structure parallel (a) and perpendicular (b) to the stack axis.[64]

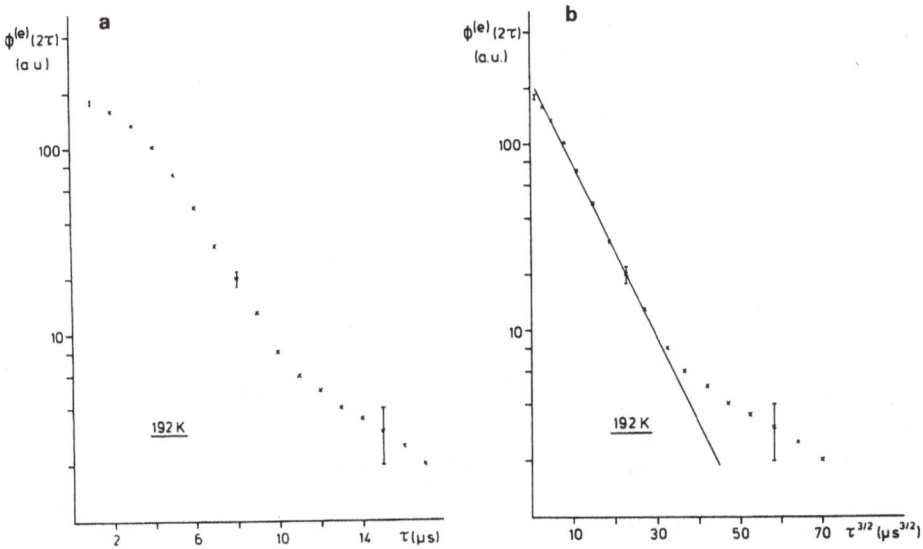

Fig. 7.2. ESE decay function plotted versus τ (a) and $\tau^{3/2}$ (b), respectively [65].

But a similar question comes up also from theoretical investigations of the free induction decay (FID) which for homogeneous systems contains the same information as the ESE decay for inhomogeneous ones. Analytical investigations in [67] arrived at the conclusion that the FID decay occurs according to (7.1), Monte–Carlo simulations carried out in [68] led to the conclusion that at least asymptotically the decay follows an exponential law.

Theory of ESR Line Shape and Free Induction Decay. From linear response theory the ESR line shape is given by the imaginary part of the magnetic susceptibility

$$\chi''(\omega) = \frac{1}{2N}\left(\frac{e}{m}\right)^2 (1 - e^{-\beta\omega})\,\mathrm{Re}\int_0^\infty dt\,e^{-i\omega t}\,\underbrace{\langle S^- S^+(t)\rangle}_{F(t)}. \qquad (7.3)$$

N is the number of unit cells, e and m are charge and mass of the charge carrier, $\beta = (kT)^{-1}$, and $F(t) = \langle S^- S^+(t)\rangle$ is the two time correlation function which describes the FID:

$$F(t) = \langle S^- S^+(t)\rangle = \int_{-\infty}^\infty dt\,e^{i\omega t}\,F(\omega). \qquad (7.4)$$

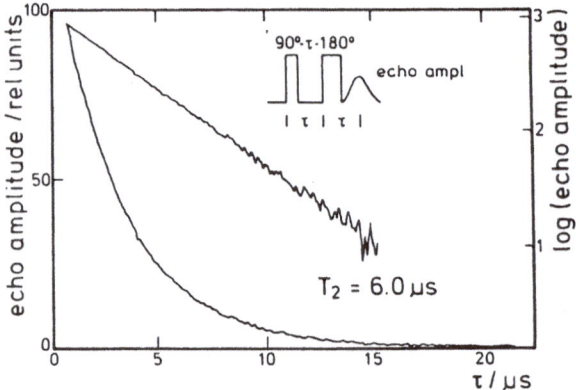

Fig. 7.3. ESE decay as a function of τ [66].

The comparison with (7.3) shows that — apart from constant factors — the ESR line shape is given by

$$F(\omega) = I(\omega) = \frac{1}{2\pi}\int_{-\infty}^\infty dt\,e^{-i\omega t}\,F(t) = \mathrm{Re}\frac{1}{\pi}\int_0^\infty dt\,e^{-i\omega t}\,F(t). \qquad (7.5)$$

Model Hamiltonian. The Hamiltonian of our model is represented by (7.6):

$$H = \underbrace{\sum_{n,n'} h_{n,n'}(t)\,a_n^\dagger a_{n'}}_{\text{incoherent electron transport}} + \underbrace{\omega_0 S^z}_{\text{Zeeman}} + \underbrace{H'}_{\text{interaction}} . \qquad (7.6)$$

The first part describes the incoherent part of the motion of the charge carrier (in the sense of Sec. 4). The next term is the Zeeman energy of the charge carrier spin in an external magnetic field. The last term finally describes the hyperfine structure interaction between the electron and proton spins. Explicitly, this term is given by

$$H' = A \sum_n \vec{S} \vec{I}_n a_n^\dagger a_n = A \sum_n (S^z I_n^z + S^+ I_n^- + S^- I_n^+) a_n^\dagger a_n, \tag{7.7}$$

which means that in our model calculation we have taken into account only the contact interaction. As regards the transport, we again assume that the fluctuating matrix element is given by a δ-correlated Gaussian Markov process with correlation functions

$$\langle h_{n,n\pm1}(t) h_{n,n\pm1}(t') \rangle = \langle h_{n,n\pm1}(t) h_{n\pm1,n}(t') \rangle = 2\gamma \, \delta(t - t'). \tag{7.8}$$

To complete the model description, we mention that lateron in the numerical evaluation we shall replace the quantum mechanical hyperfine structure interaction Hamiltonian by a simpler Hamiltonian with frozen proton spins, whose orientation along the z-direction is random. For this simplified model the Hamiltonian becomes

$$H' \to \sum_n \omega_n a_n^\dagger a_n S^z. \tag{7.9}$$

A pictorial representation of the model is given in Fig. 7.4.

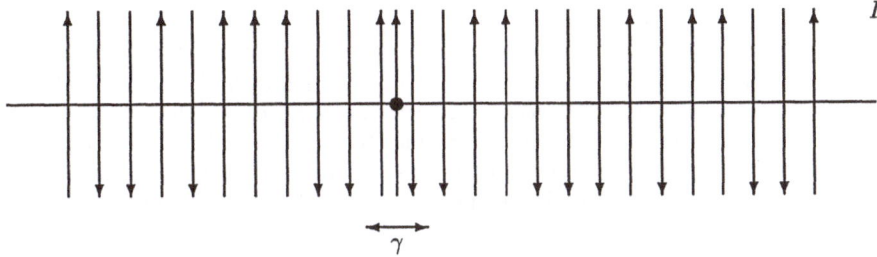

Fig. 7.4. Frozen random orientation of proton spins on a
linear chain. The dotted arrow denotes the spin
of the mobile particle.

Analytical Evaluation. In the framework of the Mori–formalism the equation of motion for the FID decay function is given by

$$\dot{F}(t) \;=\; i\Omega F(t) - \int_0^t dt' \, M(t - t') F(t') \tag{7.10}$$

$$\Omega \;=\; (S^+, LS^+) \cdot (S^+, S^+)^{-1} \tag{7.11}$$

$$M(t) \;=\; (iQL'S^+, \overrightarrow{T} \, e^{i \int_0^t d\tau \, QL(\tau)} \, iQL'S^+) \cdot (S^+, S^+)^{-1}. \tag{7.12}$$

In this equation Ω is the frequency, $M(t)$ the memory function, and L the Liouville operator. τ_M is the decay constant for the memory function, and τ_c the one for the free induction signal.

If the memory function is evaluated in Born–approximation, we arrive at

$$M(t) = e^{i\Omega t} \frac{A^2}{4} \langle a_0^\dagger a_0 \, \overrightarrow{T} \exp\left(i \int_0^t dt' L_M(t')\right) a_0^\dagger a_0 \rangle = e^{i\Omega t} \frac{A^2}{4} P(t). \tag{7.13}$$

The inspection of this result shows that $P(t)$ is the conditional probability of finding the particle at the origin at time t, if it was there at the initial time: $P(t) = P(r = 0, t; r = 0, t = 0)$. The analytical results [67,70,71] are summarized in the following.

If the decay time of the memory function is much larger than the decay time of the correlation function, i.e.

$$\tau_M \gg \tau_c \quad, \text{ or } \qquad \gamma/A \ll 1, \tag{7.14}$$

which describes the case of slow particle motion, the ESR line is inhomogeneously broadened with the inhomogeneous width determined by the hyperfine structure interaction. The homogeneous width of one component is determined by the life time of the charge carrier at a particular site, i.e. by the hopping rate γ.

In the opposite limit when the decay time of the memory function is much smaller than the decay time of the correlation function, i.e.

$$\tau_M \ll \tau_c \quad , \text{i.e.} \qquad \gamma/A \gg 1, \tag{7.15}$$

describing the case of fast particle motion, the FID decay function can be taken out of the integral in (7.10). In the interaction representation we finally arrive at

$$\dot{\tilde{F}}(t) = -\int_0^t dt' \, \tilde{M}(t')\tilde{F}(t) \longrightarrow \tilde{F}(t) = \exp -\int_0^t dt'(t-t')\tilde{M}(t'), \tag{7.16}$$

where the expression on the righthand side is the solution of the differential equation on the left. For sufficiently long times and taking into account that in the case considered the memory function decays fast, the second term in the exponent can be neglected and asymptotically we arrive at an exponential decay of the correlation function:

$$\tilde{F}(t) = \exp\left(-t\int_0^\infty dt' \, \tilde{M}(t')\right). \tag{7.17}$$

After this rather general consideration, let us investigate the influence of the dimensionality of the motion on the FID decay. In Sect. 4.3 we have seen that in the case of the purely incoherent motion the probability of finding the particle at lattice site n is given by the modified Bessel function of order n, if the particle was at the origin at the initial time. Therefore $P(t)$ from above is given by the modified Bessel function of order 0. We therefore have

$$\tilde{M}(t) = \frac{A^2}{4}e^{-\gamma t}I_0(4\gamma t) \overset{continuum}{\longrightarrow} (8\pi\gamma t)^{-1/2} \quad \text{algebraic decay} \tag{7.18}$$

where the expression on the righthand side describes the behaviour in the continuum limit. If this is inserted into (7.16), we get for the FID decay

$$\tilde{F}(t) = \exp\left(-(\Delta\omega_d t)^{3/2}\right) \qquad \text{dimension d: } 1. \tag{7.19}$$

In the case of independent random motion in the various space directions, the probability is given by the product of Bessel functions with a corresponding behaviour in the continuum limit. The calculation of the FID then gives

$$\tilde{F}(t) = \exp\left(-\Delta\omega_d t\right) \qquad \text{dimension d: } 2 \text{ and } 3. \tag{7.20}$$

Eq. (7.19) is just the result obtained in [67]. The evaluation of the ESR linewidth in dependence of the dimension results in

$$\Delta\omega_d = \frac{1}{\tau_c} = (\langle\omega_n^2\rangle^2/\gamma)^{1/3} \qquad \text{dimension d: } 1 \tag{7.21}$$

$$\Delta\omega_d = \frac{1}{\tau_c} = \langle\omega_n^2\rangle/\gamma \qquad \text{dimension d: } 2 \text{ and } 3. \tag{7.22}$$

Numerical Evaluation. Writing out explicitly (7.4), we get

$$F(t) = \sum_n \sum_{s=\uparrow\downarrow} \langle ns|S^-\langle\vec{T}\exp\{i\int_0^t d\tau \, L(\tau)\}S^+\rangle_{RW} \, \rho_0|ns\rangle. \tag{7.23}$$

The states $|ns\rangle$ describe a quantum particle at site n with spin s, ϱ_0 is the density operator at the initial time and the index RW means averaging over the random walk. Assuming as initial condition equal occupation probabilities at each lattice site, (7.23) may be written as follows:

$$F(t) = \frac{1}{N} \sum_n \langle \downarrow n | \langle \overleftarrow{T} \exp\{-i \int_0^t d\tau \, L(\tau)\} S^- \rangle_{RW} | n \uparrow \rangle \qquad (7.24)$$

$$= \frac{1}{N} \sum_n c_n(t). \qquad (7.25)$$

The functions $c_n(t)$ are defined by the last equality, and are to be determined from the following set of differential equations:

$$\dot{c}_n(t) = (i\omega_n - 4\gamma) c_n + 2\gamma (c_{n+1} + c_{n-1}). \qquad (7.26)$$

In matrix form this equation reads

$$\dot{\vec{c}} = \overleftrightarrow{A} \, \vec{c}. \qquad (7.27)$$

In this equation $\overleftrightarrow{A}$ is a matrix with random diagonal elements ω_n. After a Laplace transformation this set of differential equations is transformed into a set of algebraic ones. For motion of the particle in one dimension, this equation is tridiagonal and can easily be solved with the help of continued fractions. For 2– and 3–dimensional motion we have used multigrid methods [72].

Fig. 7.5 shows line shapes calculated for motion of the particle on a linear chain with 10^7 sites for dichotomic disorder of the local magnetic fields and various hopping rates, normalized to the strength of the local magnetic fields (strengths ± 1, i.e. $\sigma = 1$). For very small hopping rates as compared to the strength of the local magnetic field, i.e. $\gamma = 0.1$, we have two ESR lines with positions determined by the two values of the local magnetic fields. The width of the lines is given by the life time at the site and thus determined by the hopping rate γ. When γ becomes comparable to the distance in the line positions ($\gamma = 0.5$), the two lines merge into a single line. With increasing hopping rate ($\gamma = 10, 100$) the line narrows; the narrowing, however, is not $\propto \gamma^{-1}$ but given by (7.21). Furthermore, the comparison with the dashed Lorentzian line in the figure for $\gamma = 100$ shows that in this range of the hopping rates the line shape is not Lorentzian. Increasing γ further, the ESR line becomes structured. We shall show below that the origin of these structures are clusters in the dichotomic distribution of the spins. Increasing the hopping rate further to $\gamma = 10^{10}$, these structures are finally averaged out and we arrive at a Lorentzian line, whose width is $\propto \gamma^{-1}$.

Fig. 7.6 shows the FID signal for a chain with $N = 10^7$ sites, a hopping rate $\gamma = 1000$, and a Gaussian random distribution of the local magnetic fields. The curves are obtained by Fourier transforming line shapes obtained as described in connection with Fig. 7.5. In the left figure in the upper row, $F(t)$ as a function of t is represented and shows the decay of the FID signal. The figure on the right hand side shows $(-\ln F(t))/t^{3/2}$ as a function of t. In the lower row we have represented $F(t)$ in a logarithmic scale as a function of $t^{3/2}$ in the left figure, and as a function of t on the right hand side. The comparison with the dashed straight line shows that for short times the FID signal decays according to an $\exp(-(\gamma t)^{3/2})$ law and for large times as $\exp(-\bar{\gamma}t)$. Therefore our calculations show that we have a crossover between the two decay laws, and that the analytical result of [67] holds for short times whereas the simulation result of [68] is valid asymptotically. Furthermore our result also shows that the transition between the two decay laws can occur for purely 1–dimensional hopping and does not necessarily allow the conclusion that there is a transition from 1– to 3–dimensional motion.

From the analysis of the numerical line shapes we have derived the free induction decay time (maximum of the normalized line shape curve) as a function of the hopping rate for a chain with $N = 10^5$ sites. The plot of $\tau_c/\gamma^{1/3}$ as a function of γ in Fig. 7.7 on the lefthand side shows that for small values of γ the correlation time is proportional to $\gamma^{1/3}$ and thus the

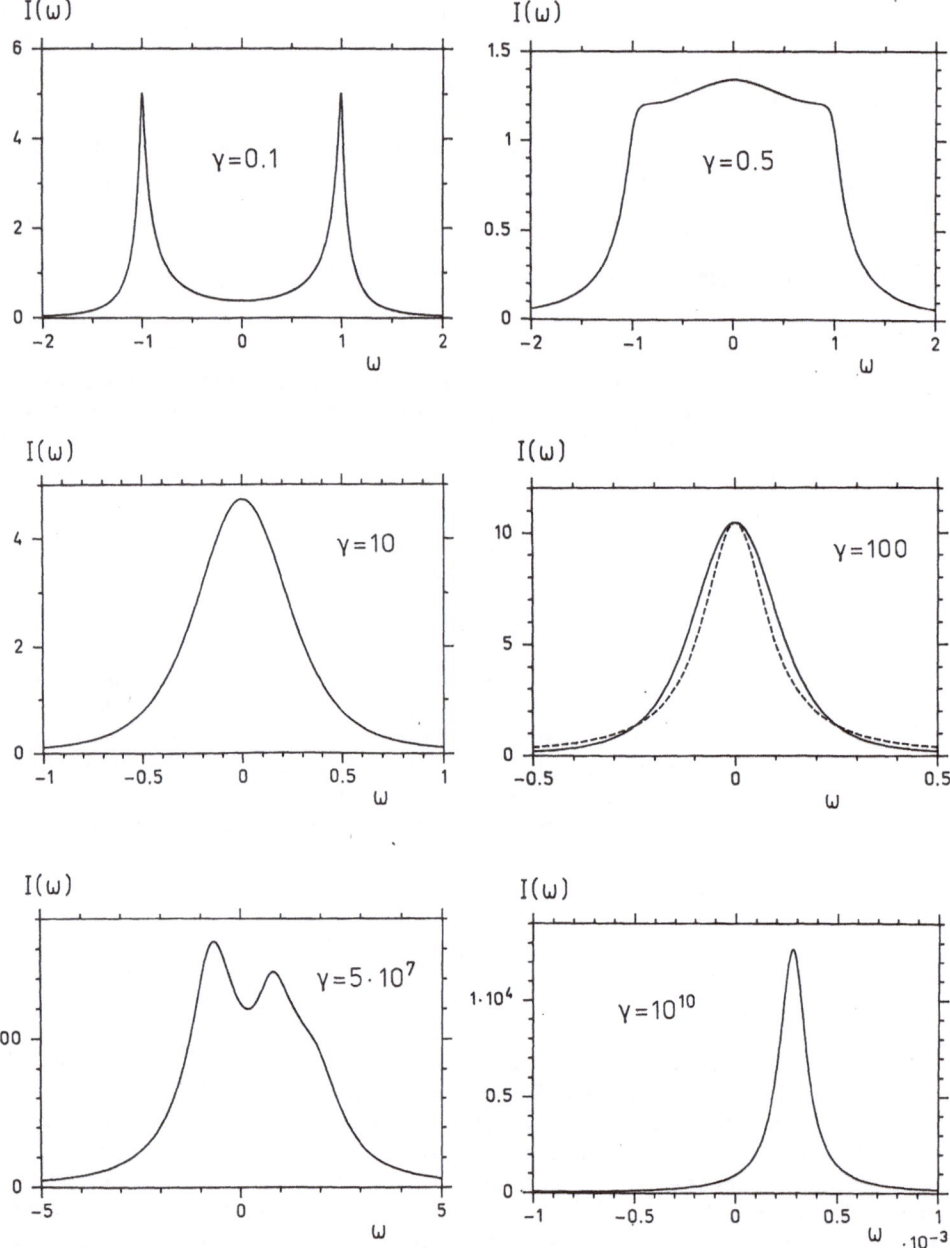

Fig. 7.5. Line shape as a function of the hopping rate γ for a 1–dimensional motion and dichotomic disorder ($\sigma = 1, N = 10^7$).

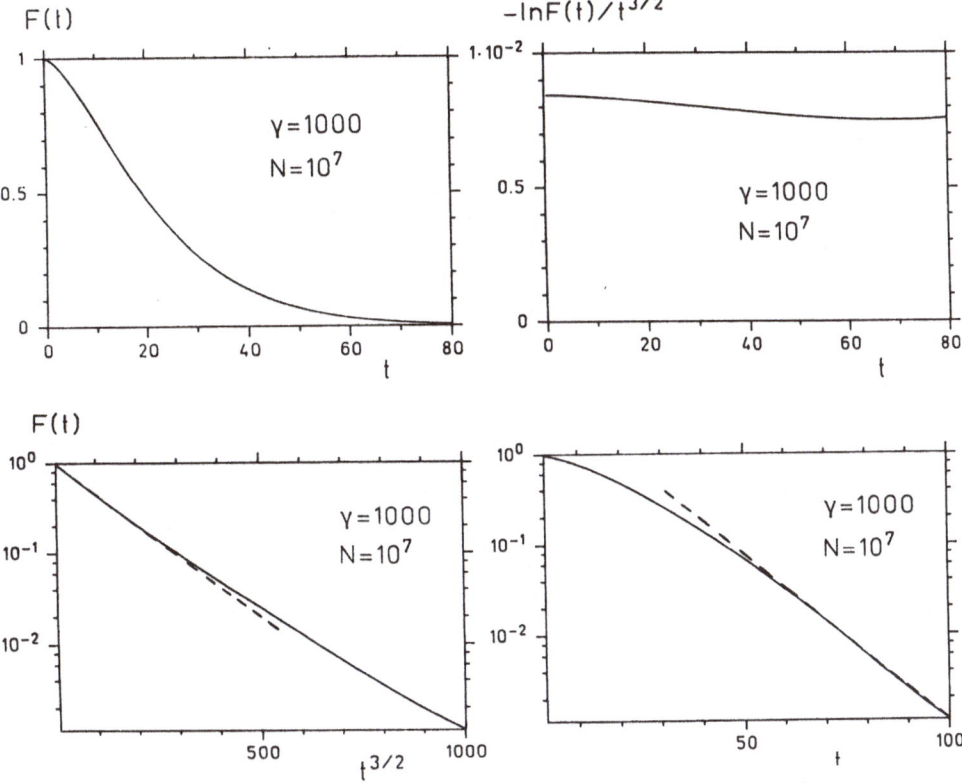

Fig. 7.6. Free induction decay $F(t)$ for Gaussian distributed local fields and 1-dimensional motion of the particle on a chain with 10^7 sites. Hopping rate $\gamma = 1000$ in units of the standard deviation of the local magnetic fields.

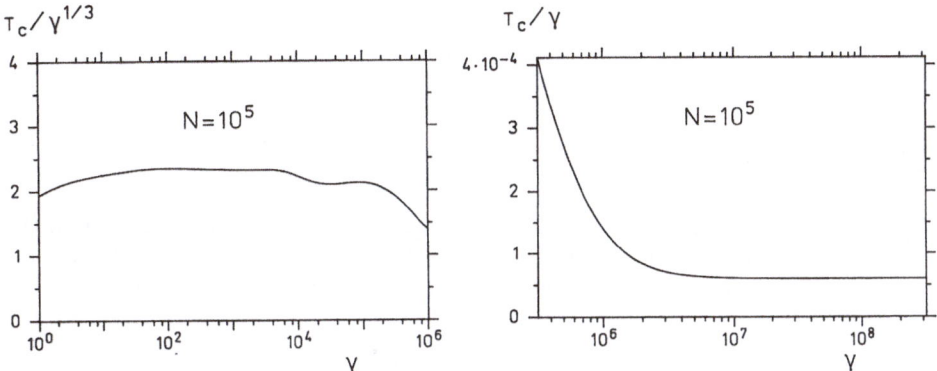

Fig. 7.7. Free induction decay time as a function of the hopping rate γ in 1 dimension.

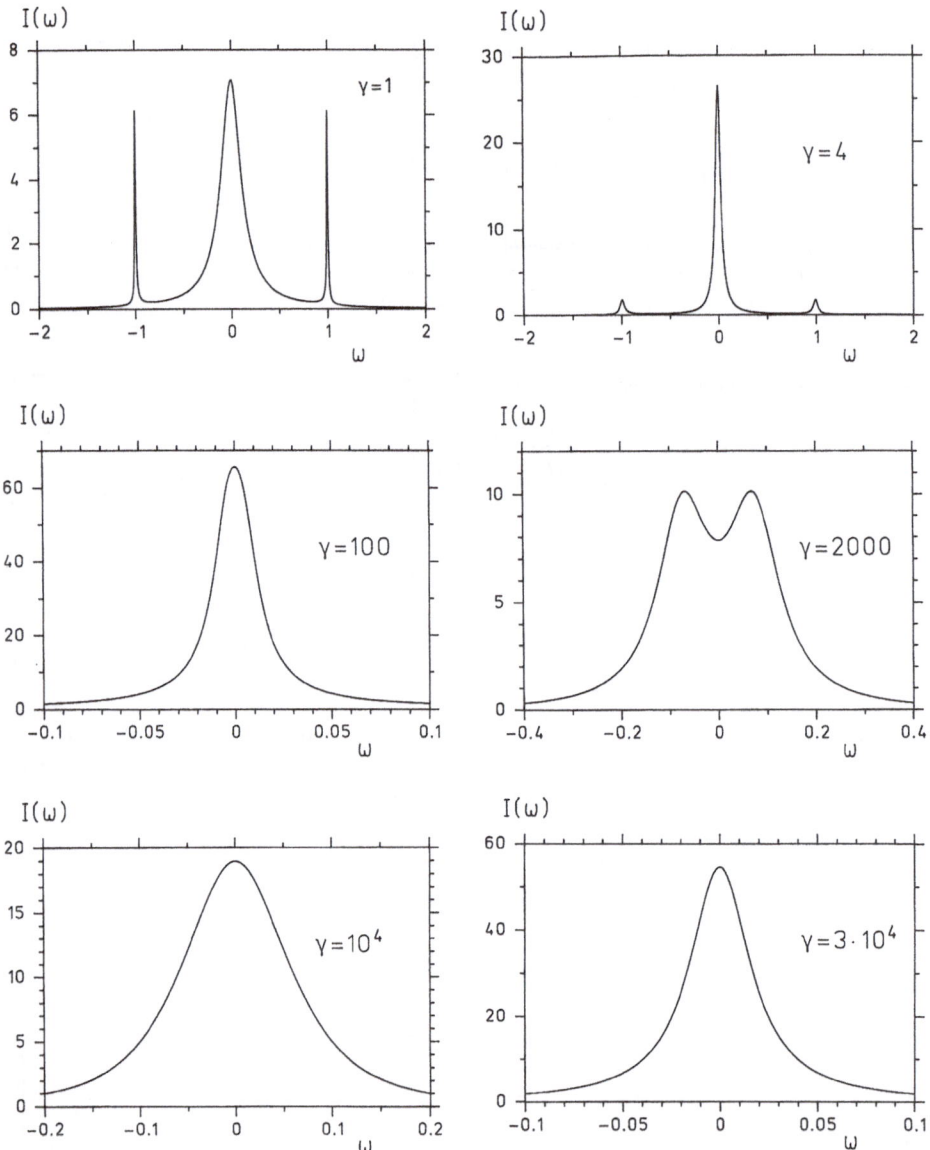

Fig. 7.8. ESR line shapes for a periodic arrangement of alternating local fields with two symmetrically positioned clusters of 50 sites with the local field pointing upwards and downwards, respectively.

linewidth is proportional to $\gamma^{-1/3}$. On the other hand from the figure on the righthand side we see that for large values of γ the correlation time is proportional to γ and the linewidth therefore proportional to γ^{-1}.

In Fig. 7.5 we have seen that for $\gamma = 5 \cdot 10^7$ the line shape becomes structured. To show that this structure stems from clusters of local magnetic fields oriented in a parallel manner, we consider a linear arrangement of 1000 sites with the local magnetic fields pointing up and down alternatively. In this periodic arrangement of local magnetic fields we have replaced in a symmetric manner two clusters of 50 up and down fields by two clusters with 50 up and 50 down fields, respectively. The line shape of this artificial cluster arrangement is shown in Fig. 7.8. For very small hopping rates ($\gamma = 1$) we have a central line stemming from the periodic arrangement of up and down fields. The two lines at $\omega = \pm 1$ have their origin in the two clusters of up and down fields, respectively.

After increasing the hopping rate to $\gamma = 4$, the central line shows motional narrowing and the two side lines are broadened because of the finite life time of the particle on the clusters. For $\gamma = 100$ the two cluster lines are broadened to such an extent that they can no more be identified in the total ESR line. Increasing the hopping rate further to $\gamma = 2000$ the exchange between the clusters and their surroundings becomes such fast that the particle can average out the local fields in the clusters and their surroundings. This gives rise to the structure in the ESR line. The last two graphs in Fig. 7.8 show that for still larger hopping rates the ESR line becomes symmetric around $\omega = 0$ and shows the usual motional narrowing.

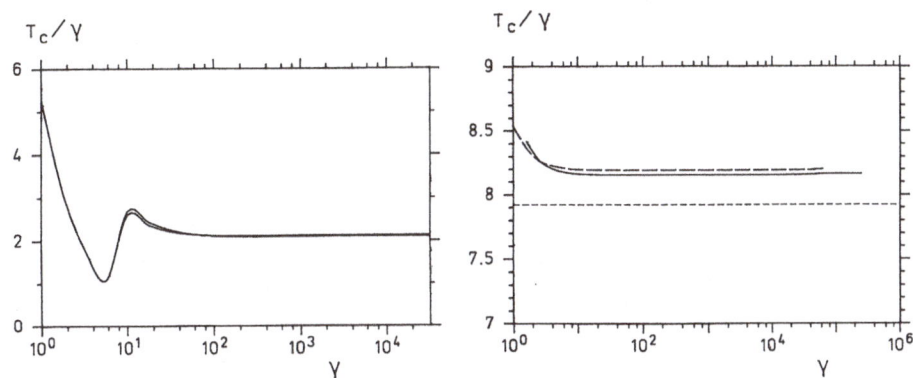

Fig. 7.9. τ_c/γ as a function of the hopping rate γ for a 300×300 lattice (left figure). The two curves correspond to two realizations of a Gaussian process. The figure on the righthand side shows the same quantity for a $28 \times 28 \times 28$ lattice. The full curve represents a Gaussian distribution, short dashes a dichotomic process, long dashes an analytical calculation in the framework of the Mori–formalism.

Fig. 7.9 shows preliminary numerical results for the normalized correlation time τ_c/γ for 2– and 3–dimensional lattices obtained [70] with the help of multi–grid methods [72]. From the figure for the 2–dimensional 300×300 lattice we see that for $\gamma > 100$ (the hopping rates are normalized to the standard deviation of the local field) $\tau_c \propto \gamma$, i.e. the line is motionally narrowed and Lorentzian. The two curves in the figure on the lefthand side correspond to two realizations of a Gaussian stochastic process. The figure on the right shows also τ_c/γ, however now for a $28 \times 28 \times 28$ lattice. In this case $\tau_c \propto \gamma$ as soon as $\gamma > 10$. The full and long–dashed curves correspond to a Gaussian and a dichotomic process, respectively. The short dashed curves are analytical results obtained from the Mori–formalism.

Fig. 7.10 gives a *summary* of this subsection. For small values of γ/A we have inhomogeneously broadened ESR lines, whose width is determined by the distribution of local magnetic fields modelling the hyperfine structure interaction. The width of a single component in the

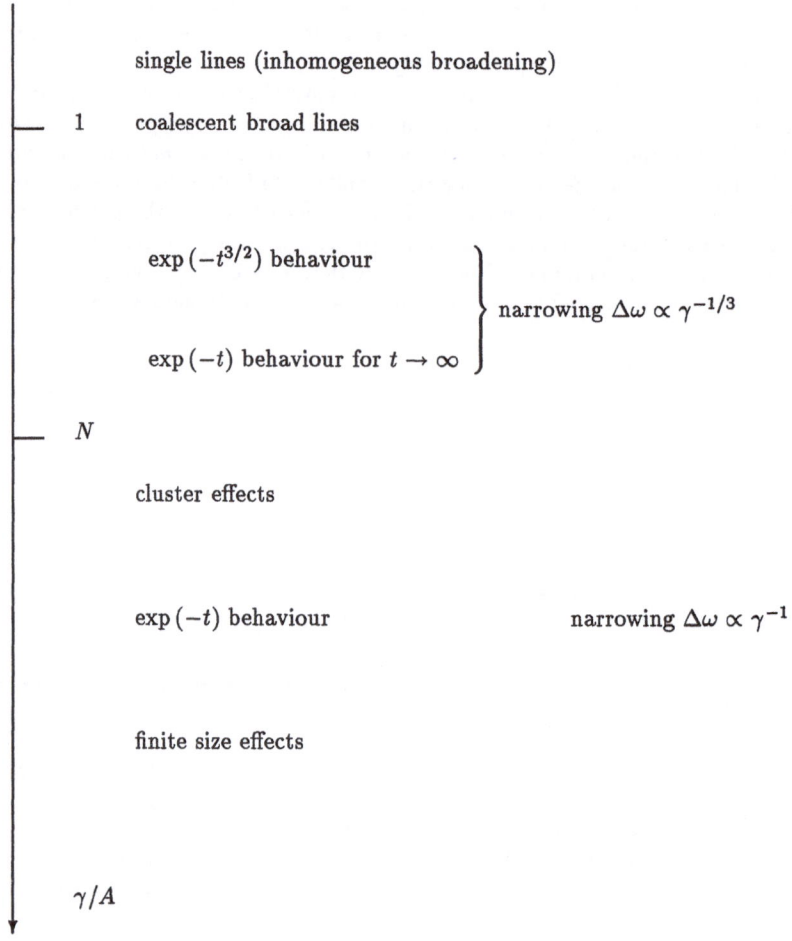

single lines (inhomogeneous broadening)

1   coalescent broad lines

$\exp\left(-t^{3/2}\right)$ behaviour

$\left.\begin{array}{l}\end{array}\right\}$ narrowing $\Delta\omega \propto \gamma^{-1/3}$

$\exp\left(-t\right)$ behaviour for $t \to \infty$

N

cluster effects

$\exp\left(-t\right)$ behaviour   narrowing $\Delta\omega \propto \gamma^{-1}$

finite size effects

γ/A

Fig. 7.10.   Summary of spin dynamics for a particle hopping on a linear chain.

inhomogeneous distribution is determined by the time interval the particle spends at a specific lattice site, i.e. by the hopping rate. For $\gamma/A \approx 1$ we have a coalescent broad line. In the following range of the hopping rate the FID is described by an $\exp(-t^{3/2})$ law, and asymptotically by an exponential function. The width of the ESR line is proportional to $\gamma^{-1/3}$. In the range $\gamma/A \approx N$, where N is the number of sites in the chain, we observe cluster effects. For still larger values of the normalized hopping rate, for which the particle probes the whole chain, the FID is described by an exponential function. The line shape is now Lorentzian because of finite size effects.

7.2 Influence of Static Disorder on Optical Line Shapes

Optical line shapes in solids at low temperatures are frequently asymmetric. For energies towards the band edge a steep decay of the line shape is observed, whereas the line shape is Lorentzian for energies towards the interior of the band. At higher temperatures the line shape becomes symmetric. This is clearly seen in Fig. 7.11, which shows the line shape of 1.4 dibromonaphthalene at various temperatures between 1.6K and 25K [73].

The reason for the asymmetry at low temperatures is static disorder caused by crystal defects, impurities, or a mixture of two kinds of atoms in the crystal. The disorder causes elastic scattering of the wave function of the quantum particle, which gives rise to quantum interference. These interference effects finally create the localized states of the particle, which show up in the steep tail of the line shape towards the band gap. At higher temperatures the quantum particles are inelastically scattered by phonons, the quantum interference is destroyed, the localization suppressed, and the line shape becomes symmetric.

Theoretical treatments of this problem have been carried out by various groups on the basis of the average t–matrix and the coherent potential approximation (ATA and CPA,

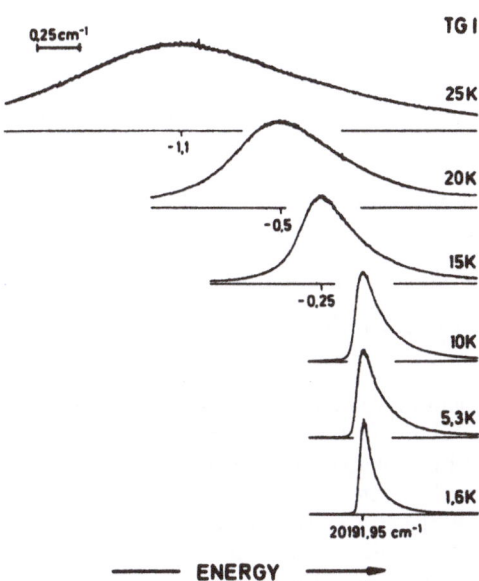

Fig. 7.11. Optical line shape of 1.4 dibromonaphtalene for various temperatures [73]

respectively) [73–75]. Numerical investigations are published in [76,77]. Our interest in this problem is twofold and different from the earlier investigations. Firstly, in disordered systems physical properties, e.g. the electrical resistance, may critically depend on the specific disorder configuration. The question arises, whether this is also true for optical spectra. Secondly, in binary disordered systems, e.g. in a mixed crystal of two atoms A and B, the density of states is heavily structured because of cluster formation. We shall investigate how these clusters influence the optical spectra of Frenkel excitons.

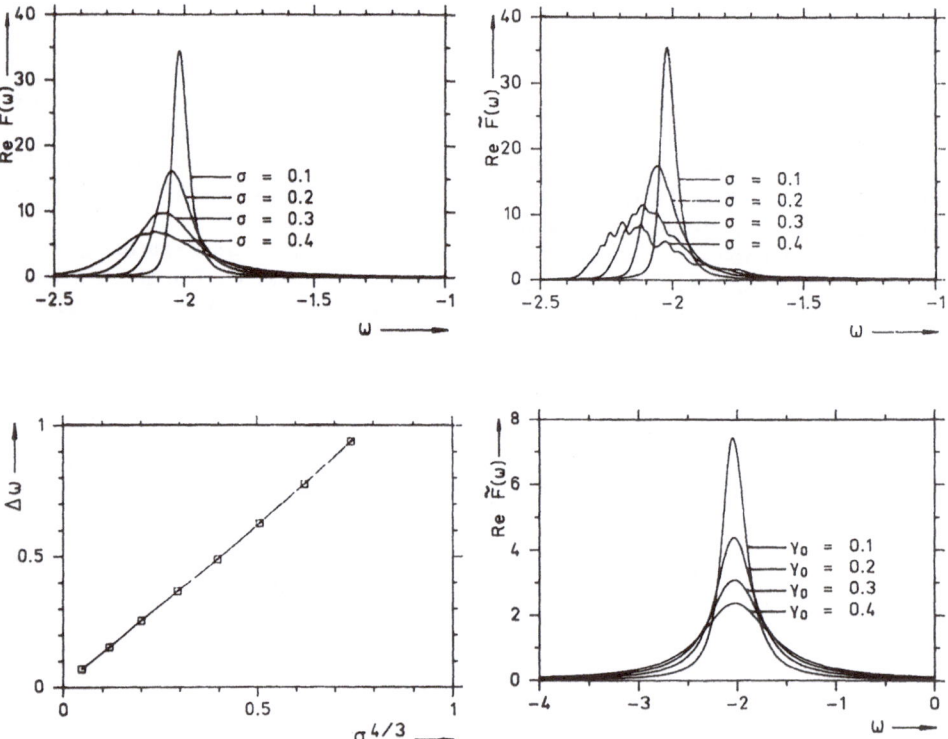

Fig. 7.12.   Optical line shapes for a linear chain of $N = 10^6$ sites, $\gamma_0 = 10^{-2}$ and various values of the standard deviation σ for Gaussian (a) and dichotomic (b) disorder.

Full width at half maximum as a function of $\sigma^{4/3}$ (c).

Optical line shape for Gaussian disorder, $\sigma = 0.2$, and various values of γ_0 (d).

The *Hamiltonian of the model* is given by

$$H = \sum_n (\epsilon_n + \epsilon'_n(t)) a_n^\dagger a_n + V \sum_{\langle n,m \rangle} a_n^\dagger a_m. \tag{7.28}$$

In this Hamiltonian $a_n^\dagger$ and a_n are creation and annihilation operators for an excitation at site n. ϵ_n are the static energy fluctuations caused by disorder, which in our calculation are assumed to be described by a Gaussian or dichotomic distribution. V describes the coherent transfer of the excitation between neighbouring sites. This description of the transport is completely different from the one in the previous section, which was incoherent hopping. Finally, $\epsilon'_n(t)$ introduces dynamical energy fluctuations in the model which slightly broaden the δ peaks in the line shape of a completely coherent treatment. In this way divergencies in the numerical calculation are avoided.

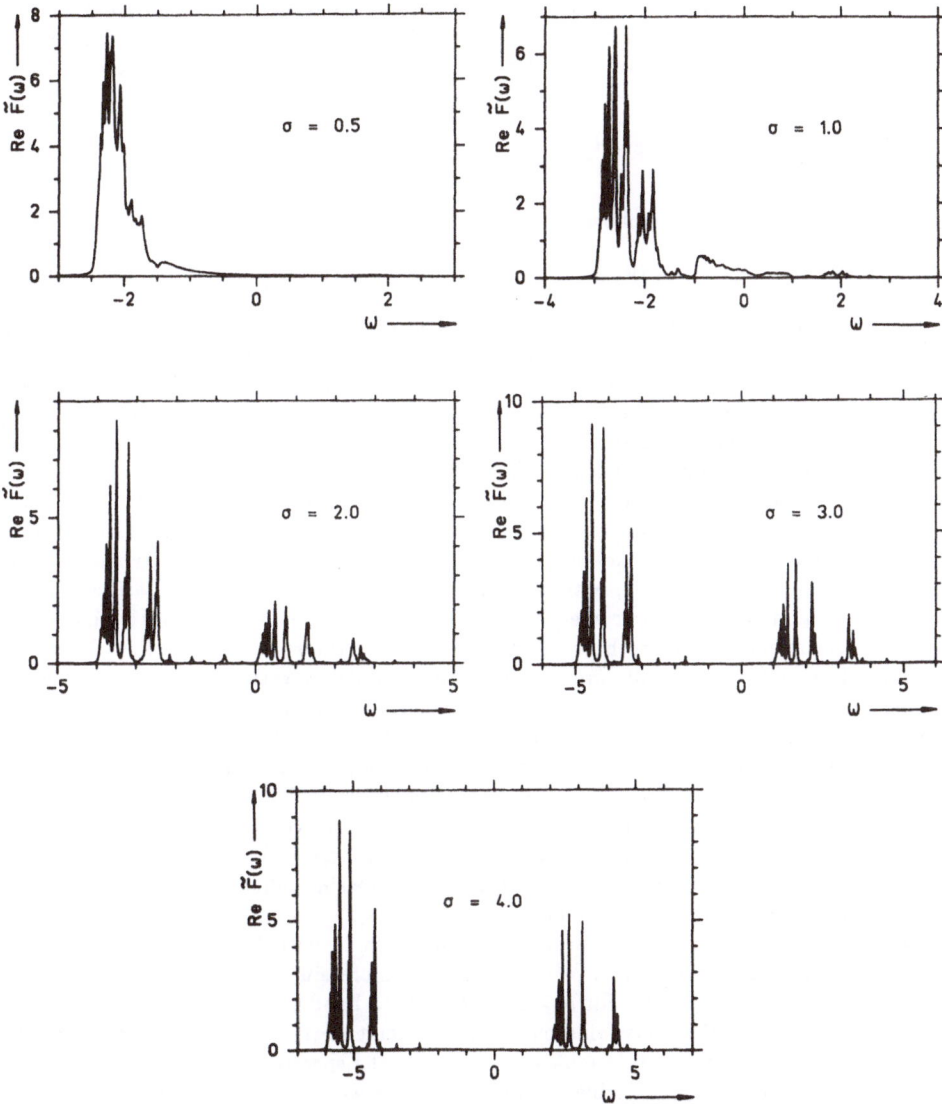

Fig. 7.13. Optical absorption spectra for binary (dichotomic) disorder and various values of the standard deviation σ. Band splitting is observed for $\sigma \geq 2$.

The optical line shape is given by the imaginary part of the dielectric susceptibility and in the framework of linear response theory we obtain

$$\chi''(\omega) = \sum_m \sum_n \mu^2 \, \text{Re} \int_0^\infty dt \, e^{-i\omega t} \langle\langle 0|a_m a_n^\dagger(t)|0\rangle\rangle, \qquad (7.29)$$

i.e. it is given by the half sided Fourier transform of the two–time correlation function of the optical dipole moment operator, which is expressed by the creation and annihilation operators for localized excitations. μ is the optical transition dipole moment matrix element. As in [77] we introduce $Z_m(t)$ as the n–sum over the correlation functions and arrive at the following expression for the line shape

$$\chi''(\omega) = \sum_m \mu^2 \, \text{Re} \int_0^\infty dt \, e^{-i\omega t} \bar{Z}_m(t). \qquad (7.30)$$

$\bar{Z}_m(t)$ is determined from the following set of equations [77]:

$$\dot{\bar{Z}}_m = (-i\epsilon_m - \gamma_0)\bar{Z}_m - iV(\bar{Z}_{m+1} + \bar{Z}_{m-1}). \qquad (7.31)$$

In this equations γ_0 is the strength of the dynamical energy fluctuations. The static energy fluctuations ϵ_m are taken from a pseudo random number generator. The structure of the equation is the same as the one of (7.26), however, the real hopping rate 2γ there is replaced here by the imaginary expression $-iV$ containing the coherent transfer matrix element. After a Laplace transformation we again arrive at a tridiagonal set of algebraic equations from which the line shape can be calculated with the help of continued fractions.

Fig. 7.12 (a) and (b) show optical line shapes for a linear chain of $N = 10^6$ sites, $\gamma_0 = 10^{-2}$, various values of the standard deviation σ, and Gaussian and dichotomic disorder, respectively. All quantities are measured in units of the coherent transfer matrix element V. The line shapes are asymmetric, with a steeper decay at the low energy side (band edge), and the maximum shifts to lower energies [74]. Obviously the asymmetry of the line shapes increases with increasing strength of the disorder. Furthermore, we see that in the case of Gaussian disorder the line shapes are smooth, whereas they are rather structured in the case of binary disorder. Fig. 7.12c represents the full line width at half maximum as a function of $\sigma^{4/3}$. The straight line behaviour indicates that the line width increases $\propto \sigma^{4/3}$. This behaviour agrees with numerical results obtained in [76] for smaller samples, and disagrees with analytical results based on ATA [74]. Recent analytical calculations based on CPA [75] also show this behaviour. Finally, Fig. 7.12d shows the influence of the strength of dynamic fluctuations on the optical line shape. We see that with increasing strength of the dynamic fluctuations, simulating increasing temperature, i.e. increasing influence of the phonons, the lines become broader and more symmetric.

In Fig. 7.13 we show optical spectra for a linear chain with binary disorder for higher values of the standard deviation σ. We see that with increasing values of σ the structure of the spectra becomes more rich. For $\sigma \geq 2$ we observe a band splitting. These structures in the spectra are the result of clustering effects and difficult to treat analytically. On the other hand, for each set of parameters we have investigated the spectra for several samples drawn out randomly from a given ensemble. In all these cases we have seen negligible fluctuations from sample to sample, indicating that the optical spectra in the case of Frenkel excitons are self–averaging.

ACKNOWLEDGEMENT

Two of the authors gratefully acknowledge financial support by the Volkswagenwerk Foundation (J. K.) and the Alexander von Humboldt Foundation (A. M. Jayannavar).

References

[1] H. Haken, "Quantenfeldtheorie des Festkörpers", B. G. Teubner, Stuttgart (1973)

[2] G. D. Mahan, "Many Particle Physics", Plenum Press, New York, London (1981)

[3] V. M. Kenkre, P. Reineker, "Exciton Dynamics in Molecular Crystals and Aggregates", Springer Tracts in Modern Physics 94, Springer, Berlin, Heidelberg, New York (1982)

[4] P. Reineker, H. Haken, H. C. Wolf, "Organic Molecular Aggregates", Springer Series in Solid–State Sciences 49, Springer, Berlin, Heidelberg, New York, Tokyo (1983)

[5] V. M. Kenkre, "Mathematical Methods for the Description of Energy Transfer", in: Energy Transfer Processes in Condensed Matter, ed. B. di Bartolo, NATO ASI Series B 114, Plenum Press, New York, London, (1983)

[6] M. Pope, Ch. E. Swenberg, "Electronic processes in organic crystals", Clarendon, Oxford (1982)

[7] J. Simon, J.–J. André, "Molecular Semiconductors", Springer, Berlin, Heidelberg, New York, Tokyo (1985)

[8] N. Karl, "Getting Beyond Impurity-Limited Transport in Organic Photoconductors", in: Proceedings of the International Conference on the Science and Technology of Defect Control in Semiconductors, ed. K. Sumino, Elsevier (1990)

[9] R. E. Merrifield, J. Chem. Phys. bf 28, 647 (1958)

[10] Th. Förster, Ann. Phys. (Leipzig), 2, 55 (1948)

[11] D.L. Dexter, R.S. Knox, "Excitons", Interscience, New York (1965)

[12] M. Trlifaj, Czech. J. Phys. 8, 510 (1958)

[13] H. Haken, G. Strobl, "Exact Treatment of Coherent and Incoherent Triplet Exciton Migration", in: The Triplet State, ed. A. Zahlan, Cambridge University Press, London (1967)

[14] H. Haken, P. Reineker, Z. Phys. 249, 253 (1972)

[15] H. Haken, G. Strobl, Z. Phys. 262, 135 (1973)

[16] P. Reineker, H. Haken, "The Coupled Coherent and Incoherent Motion of Frenkel Excitons in Molecular Crystals" in: Localization and Delocalization in Quantum Chemistry, Vol. 2, eds. O. Chalvet, R. Daudel, S. Diner, J. P. Malrieu, D. Reidel Publishing Company, Dortrecht–Holland, Boston–USA (1976)

[17] H. Haken, P. Reineker, "Comments on the Interaction of Excitons and Phonons", in: Excitons, Magnons and Phonons in Molecular Crystals, ed. A. Zahlan, Cambridge University Press, London (1968)

[18] M. Grover, R. Silbey, J. Chem. Phys. 52, 2099 (1970); 54, 4843 (1971)

[19] V. M. Kenkre, R. S. Knox, Phys. Rev. B 9, 5279 (1974); Phys. Rev. Lett. 33, 803 (1974)

[20] Y. Toyozawa, "Localization and Delocalization of an Exciton in the Phonon Field", in: Organic Molecular Aggregates, eds. P. Reineker, H. Haken, H. C. Wolf, Springer, Berlin, Heidelberg, New York, Tokyo (1983)

[21] E.I. Rashba, "Self-Trapping of Excitons", in: Excitons, eds. V.M. Agranovich, A.A. Maradudin, North Holland, Amsterdam, New York, Oxford (1982)

[22] A.O. Caldeira, A.J. Leggett, Phys. Rev. Lett. 46, 211 (1981)

[23] H. Grabert, U. Weiss, P. Hänggi, Phys. Rev. Lett. 52, 2193 (1984)

[24] P. Hänggi, J. Stat. Phys. 42, 105 (1986)

[25] P. W. Anderson, B. I. Halperin, C. M. Varma, Phil. Mag. 25, 1 (1972)

[26] P. Reineker, H. Morawitz, K. Kassner, Phys. Rev. B. 29, 4546 (1984)

[27] K.W. Kehr, K. Kitahara, J. Phys. Soc. Jpn. 56, 889 (1987)

[28] H. Wipf, A. Magerl, S.M. Shapiro, S.H. Satiga, W. Thomlinson, Phys. Rev. Lett. 46, 947 (1981)

[29] R. S. Knox, "Theory of Excitons", in: Solid State Physics Supplement, Vol. 5, eds. H. Ehrenreich, F. Seitz, D. Turnbull, 2nd print, Academic Press, New York (1972)

[30] S. Nakajima, Prog. Theor. Phys. 20, 948 (1958)

[31] R. Zwanzig, J. Chem. Phys. 33, 1338 (1960)

[32] P. Reineker, R. Kühne, Phys. Rev. B 21, 2448 (1980)

[33] V. Čápek, I. Barvík, J. Phys. C 18, 6149 (1985)

[34] I. Barvík, V. Čápek, Phys. Rev. B 40, 9973 (1989)

[35] P. Reineker, "Exciton Dynamics in Molecular Crystals and Aggregates; Stochastic Liouville Equation Approach: Coupled Coherent and Incoherent Motion, Optical Line Shapes, Magnetic Resonance Phenomena", in: Springer Tracts in Modern Physics, Vol. 94, ed. G. Höhler, Springer, Berlin, Heidelberg, New York (1982)

[36] P. Reineker, Phys. Lett. 44A, 429 (1973)

[37] P. Reineker, H. Haken, Z. Phys. 250, 300 (1972)

[38] P. Reineker, Phys. Lett 42A, 389 (1973)

[39] P. Reineker, Z. Phys. 261, 187 (1973)

[40] E. Schwarzer, H. Haken, Phys. Lett. 42A, 317 (1972)

[41] P. Reineker, phys. stat. sol. (b) 52, 439 (1972)

[42] P. Reineker, Z. Naturforsch. 29a, 282 (1974)

[43] P. Reineker, Phys. Rev. B 19, 1999 (1978)

[44] V. M. Kenkre, T. S. Rahman, Phys. Lett. 50A, 170 (1974)

[45] U. Schmid, P. Reineker, Mol. Physics 55, 77 (1985)

[46] J. Köhler, P. Reineker, Chem. Phys. 93, 209 (1985)

[47] R. Winkler, P. Reineker, Mol. Physics 60, 1283 (1987)

[48] K. Lindenberg, B. J. West, Phys. Rev. Lett. 51, 1370 (1983)

[49] S. Pabst, Diplom–Thesis, University of Ulm (1989)

[50] H. Däubler, Diplom–Thesis, University of Ulm (1989)

[51] A. Blumen, R. Silbey, J. Chem. Phys. 69, 3589 (1978)

[52] I. B. Rips, V. Čápek, phys. stat. sol. (b) 100, 451 (1980)

[53] K. Kitahara, J. W. Haus, Z. Phys. B 32, 419 (1979)

[54] Y. Inaba, J. Phys. Soc. Jap. 50, 2473 (1981)

[55] K. Kassner, P. Reineker, Z. Phys. B 59, 357 (1985)

[56] K. Kassner, P. Reineker, Z. Phys. B 60, 87 (1985)

[57] R. C. Bourret, U. Frisch, A. Pouquet, Physica 65, 303 (1973)

[58] V.E. Shapiro, V. M. Loginov, Physica 91A, 563 (1979)

[59] A. Pringsheim, "Über einige Konvergenzkriterien für Kettenbrüche mit komplexen Gliedern", Sitzungsberichte München 35, 359 (1905)

[60] Y. Inaba, Thesis, University of Tokyo (1982)

[61] P. W. Anderson, Phys. Rev. 109, 1492 (1958)

[62] K. Kassner, Z. Phys. B 70, 229 (1988)

[63] D. Jérome, L. G. Caron, "Low–Dimensional Conductors and Superconductors", NATO ASI Series B 155, Plenum, New York, London (1986)

[64] V. Enkelmann, B. S. Mora, Ch. Kröhnke, G. Wegner, Chem. Phys. 66, 303 (1982)

[65] J. Sigg, Th. Prisner, K. P. Dinse, H. Brunner, D. Schweitzer, K. H. Hausser, Phys. Rev. B 27, 5366 (1983)

[66] W. Stöcklein, B. Bail, M. Schwoerer, D. Singel, J. Schmidt, "Spin Resonance and Conductivity of Fluoranthenyl Radical Cation Salts", in: Organic Molecular Aggregates, eds. P. Reineker, H. Haken, H. C. Wolf, Springer, Berlin, Heidelberg, New York, Tokyo (1983)

[67] M. J. Hennessy, C. D. McElwee, P. M. Richards, Phys. Rev. B 7, 930 (1973)

[68] R. Czech, K. Kehr, Phys. Rev. B 34, 261 (1986)

[69] H. Mori, Prog. Theor. Phys. 34, 423 (1965)

[70] J. Köhler, Thesis, University of Ulm (1989)

[71] J. Köhler, P. Reineker, Chem. Phys., in print

[72] W. Hackbusch, "Multi-Grid Methods and Applications", Springer, Berlin, Heidelberg, New York (1985)

[73] H. Port, H. Nissler, R. Silbey, J. Chem. Phys. 87, 1994 (1987)

[74] J. Klafter, J. Jortner, J. Chem. Phys. 68, 1513 (1978)

[75] D. L. Huber, W. Y. Ching, Phys. Rev. B 39, 8652 (1989)

[76] M. Schreiber, Y. Toyozawa, J. Phys. Soc. Jpn. 51, 1528 (1982)

[77] J. Köhler, A. M. Jayannavar, P. Reineker, Z. Phys. B 75, 451 (1989)

CHAOTIC MOTION OF MOLECULAR CHAINS[1]

P. Reineker, R. G. Winkler, G. Siegert, G. Glatting

Abteilung für Theoretische Physik, Universität Ulm
Albert-Einstein-Allee 11, 7900 Ulm, Germany

1 INTRODUCTION

Elastic properties of metals and elastomers are rather different. In metals the maximum elastic deformation is about 0.2%, whereas in elastomers values up to 1500% are possible. While in metals in this range Hook's law holds, this is in general not true for elastomers. The comparison of the elastic moduli shows that it is of the order of 10^5MPa in metals and 1MPa in elastomers, it decreases with increasing temperature in metals but in contrast in elastomers an increase is observed. Metals under given strain elongate with increasing temperature whereas elastomers shorten. Beyond the limit of the elastic deformation, metals deform in a plastic manner while elastomers generally are destroyed. Finally, adiabatic extension of metals causes cooling but in elastomers the temperature raises. The basic reason for this different behaviour of metals and elastomers is the fact that in the first class of materials the elastic extension increases the internal energy whereas in the second group of materials it results in a decrease of entropy. This completely different behaviour originates from the difference in microscopic structure: the structural units in metals are atoms, while elastomers are composed of polymers, i.e. long molecular chains.

For the investigation of the stress–strain behaviour of macromolecules Kuhn [1] considered a socalled phantom chain, i.e. a chain of connected, freely orientable rigid segments. For the probability of finding an end–to–end distance $a = |\vec{a}|$, by counting the number of possible configurations he obtained a Gaussian distribution assuming that the number Z of segments is much larger than the segment length l and the end–to–end distance:

$$Z \gg 1, \qquad l \ll |\vec{a}| \ll Zl \tag{1}$$

$$W(Z,\vec{a})\,d^3a = \left(\frac{3}{2\pi Z l^2}\right)^{3/2} \exp\left\{-\frac{3a^2}{2Zl^2}\right\}\,d^3a \tag{2}$$

This Gaussian chain is frequently used for the description of macromolecules. A generalization of this theory for arbitrary end–to–end distances was carried out by Kuhn and Grün [2]:

$$Z \gg 1, \qquad |\vec{a}| \lesssim Zl \tag{3}$$

---

[1]Dedicated to Professor W. Pechhold on the occasion of his 60th birthday.

$$W(a)\,da = B\left(\frac{sh\beta}{\beta}\right)^Z \exp\left\{-\frac{\beta a}{l}\right\} a^2\,da \tag{4}$$

$$L(\beta) = \coth\beta - \frac{1}{\beta} = \frac{a}{Zl} \tag{5}$$

A probability distribution for all end–to–end distances and an arbitrary number of segments of the chain has been derived by Treloar [3]:

$$W(a)\,da = \frac{a}{2l^2}\frac{Z^{Z-2}}{(Z-2)!}\sum_{s=0}^{k}(-1)^s\frac{Z!}{s!(Z-s)!}\left(m-\frac{s}{Z}\right)^{Z-2}da \tag{6}$$

$$m = \frac{1}{2}\left(1-\frac{a}{Zl}\right), \qquad k < mZ \leq k+1 \tag{7}$$

Using the connection between probability and entropy or free energy, the force extension relation is given by the following equations:

Entropy $\qquad\qquad S = k\ln W = \text{const.} - kca^2 \tag{8}$

Free energy $\qquad\qquad\qquad dF = dU - T\,dS \tag{9}$

Force $\qquad\qquad f = \frac{dF}{da} = -T\frac{dS}{da} = 2kTc\,a \tag{10}$

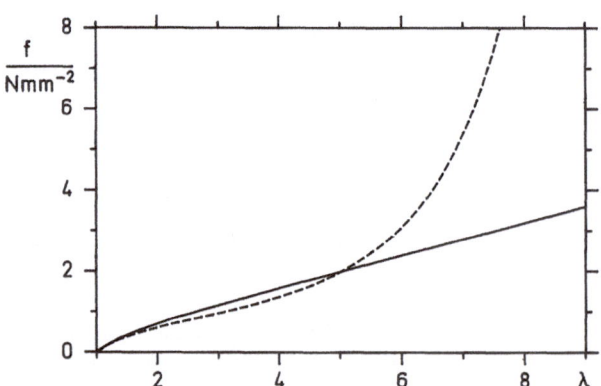

Fig. 1. Force–extension relation for an elastomer; dashed line: phenomenological van der Waals model of Kilian [28] (coincident with experiment); full line: statistical treatment of a Gaussian network

A first approach towards a dynamical treatment of the stress–strain relation was undertaken in [4,5]. More recently, Weiner [6] showed that the dynamical treatment of a chain consisting of masses and springs corresponds to the usual statistical treatment [7-15], if the limit of an infinite spring constant is considered.

The description of the elastic deformation of a polymeric network, connecting the chains by crosslinks, was based on the statistical treatment of the single chain and assuming an affine deformation down to microscopic length scales [9,16-18]. Fig. 1 shows the force-extension relation for a polymeric network. The unsatisfactory quality of the results obtained in this way shows up especially in the socalled Mooney–plot [19,20] (see Fig. 2). Both for large and small deformations there are large deviations between experimental and theoretical results. The deviations for large deformations have their origin in the finite size of the network which was not taken into account in the theory. To describe the deviations for small deformations, interactions between the chains in the form of topological constraints have been introduced, which are described by tubes [21-24] or slip–links [21,25,26]. In the constraint fluctuation model [27] fluctuations of the crosslinks are hindered by surrounding chains. In these models, the origin of the softening of the elastic force observed in the Mooney–plot stems from a reduction of the influence of the constraints with increasing deformation, i.e. by a relative increase of the number of conformations. However, up to now, none of the models can explain several experiments using the same set of parameters.

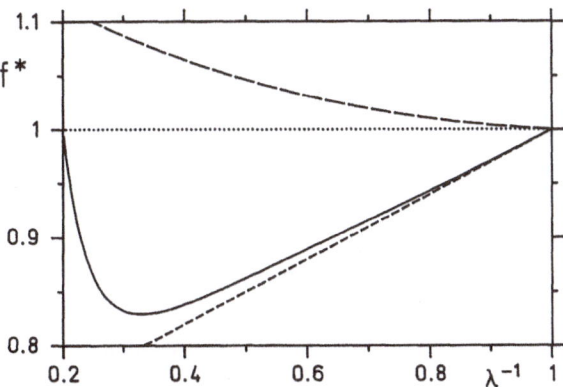

Fig. 2.   Mooney–plot, λ = relative extension, $f^* = f/(\lambda - \lambda^{-2})$; full line: van der Waals model [28]; long dashed line: inverse Langevin model [2]; short dashed line: Mooney–Rivlin model [9]; dots: Gaussian model [1];

In addition to these models there exist phenomenological descriptions, the socalled Mooney-Rivlin equation [9] and the van der Waals model of Kilian [28]. Both descriptions agree well with the experimental data for small deformations, the van der Waals model also in the range of large deformations. Unfortunately up to now there exists no connection between microscopic models and the phenomenological descriptions.

In section 2 we summarize the theoretical basis for our dynamical treatment of molecular chains. Section 3 discusses in detail the Hamiltonian dynamics for a chain of three mobile masses. In section 4 the force extension relation for chains of various lenghts moving in 2 and 3 dimensions is presented. In section 5 we consider fluctuations of energies and positions of several mass points on the chains. Section 6 investigates crosslinked chains and discusses the influence of restrictions on the motion of the chains. Finally section 7 gives a summary and concluding remarks.

2 DYNAMICAL TREATMENT OF A FREELY JOINTED CHAIN

2.1 Molecular dynamics for a chain with N mobile mass points

In the following we consider the molecular chain of Fig. 3 with $N + 2$ mass points connected by rigid rods of length l [29-35]. The end points of the chain are fixed a distance $|\vec{a}|$ apart, whereas the other N points of the chain are freely mobile and only subject to the constraints

$$|\vec{r}_i - \vec{r}_{i-1}| = l \ , \ i = 1, ..., N + 1 \tag{11}$$

These holonomous constraints are taken into account in the Lagrangian, which contains only kinetic energy and no potential energy, using the method of Lagrangian multipliers

$$L = \sum_{i=1}^{N} \frac{m_i}{2} \dot{\vec{r}}_i^{\,2} + \sum_{i=1}^{N+1} \frac{\bar{\lambda}_i}{2} \{(\vec{r}_{i-1} - \vec{r}_i)^2 - l^2\} \tag{12}$$

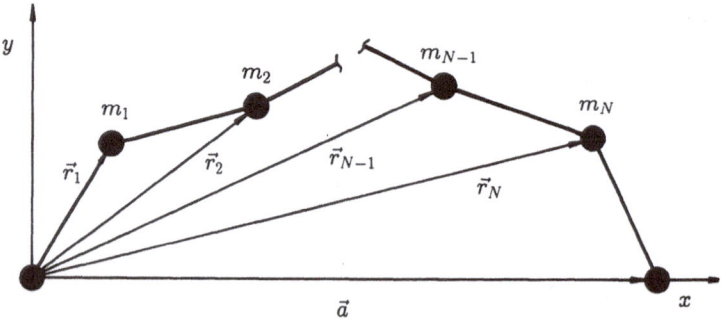

Fig. 3.   Model of a polymer chain

From the Euler–Lagrangian equation one obtains the following equation of motion

$$m_i \ddot{\vec{r}}_i = \bar{\lambda}_{i+1}(\vec{r}_i - \vec{r}_{i+1}) + \bar{\lambda}_i(\vec{r}_i - \vec{r}_{i-1}) \ , \ i = 1, ..., N \tag{13}$$

On the right side of this equation the forces of constraints generated by the geometrical restrictions enter. To simplify these equations, the site vector, the end–to–end distance, the Lagrangian multipliers, the time and the velocity are scaled in the following way:

$$\vec{r}_i(t) = l\vec{\rho}_i(\tau) \ , \ \vec{a} = l\vec{\alpha} \ , \ \bar{\lambda}_i(t) = \frac{E}{l^2} \lambda_i(\tau) \tag{14}$$

$$\tau = \sqrt{\frac{E}{ml^2}} \, t \ , \ \dot{\vec{r}}_i(t) = \sqrt{\frac{E}{m}} \, \dot{\vec{\rho}}_i(\tau) \tag{15}$$

E is the energy of the conservative system. Assuming that all masses of the chain are the same, we arrive at the following simplified form of the equations of motion:

$$\ddot{\vec{\rho}}_i(\tau) = \lambda_{i+1}(\vec{\rho}_i - \vec{\rho}_{i+1}) + \lambda_i(\vec{\rho}_i - \vec{\rho}_{i-1}) \tag{16}$$

The total energy of the chain is then given by

$$\frac{1}{2}\sum_{i=1}^{N}\dot{\vec{\rho}}_i{}^2 = 1 \qquad (17)$$

i.e. it is a conserved quantity of the system and normalized to 1. Obviously in the equations of motion of the chain only the number $(N+1)$ of the chain segments and the scaled end–to–end distance α occurs. Using in addition (11) we obtain a set of equations which allows to determine the positions of the mass points and the forces due to the constraints. Especially the force of constraint on the right end of the chain is given by

$$\vec{f}_{N+1} = \lambda_{N+1}(\vec{\rho}_N - \vec{\alpha}) \qquad (18)$$

We shall see that this force generally fluctuates heavily in time. The quantity, which is relevant for the stress–strain relation, is its time average

$$
\begin{aligned}
< \vec{F}_{N+1}(t) > \; &= \; \frac{E}{l} lim_{t\to\infty} \frac{1}{t} \int_0^t \lambda_{N+1}(\tau)(\vec{\rho}_N - \vec{\alpha})d\tau \\
&= \; \frac{E}{l} < \vec{f}_{N+1} > \qquad (19)
\end{aligned}
$$

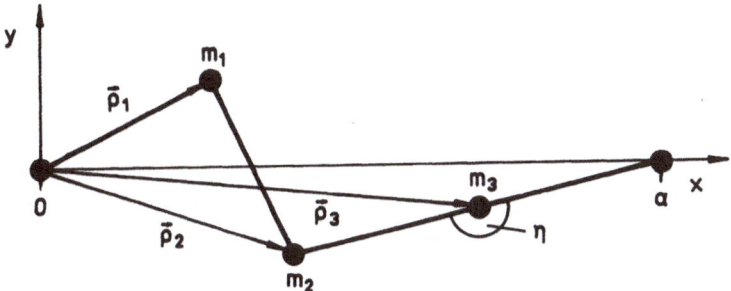

Fig. 4.   Definition of the Poincaré–Section

In this average the quantities mentioned below and the segment length l occur. The average force is independent of the masses of the chain because we have assumed all of them to be equal. For an ergodic system the energy E is given by the canonical expectation value of the kinetic energy; in this way we arrive at

$$< E > = \frac{1}{2}\left([d-1]N - 1\right) kT \qquad (20)$$

Here d is the dimension of the motion. Inserting this value of the energy, the average force becomes proportional to the temperature.

2.2 Stochastic dynamics treatment of the freely jointed molecular chain

In an alternative treatment we have considered the influence of friction and external fluctuations on the motion of the molecular chain. In this case the equations of motion have the form of Langevin equations and are given by

$$m_i \ddot{\vec{r}}_i = \bar{\lambda}_i(\vec{r}_i - \vec{r}_{i-1}) - \bar{\lambda}_{i+1}(\vec{r}_{i+1} - \vec{r}_i) - \bar{\gamma}m_i\dot{\vec{r}}_i + \vec{\Gamma}_i \;, i = 1, ..., N \qquad (21)$$

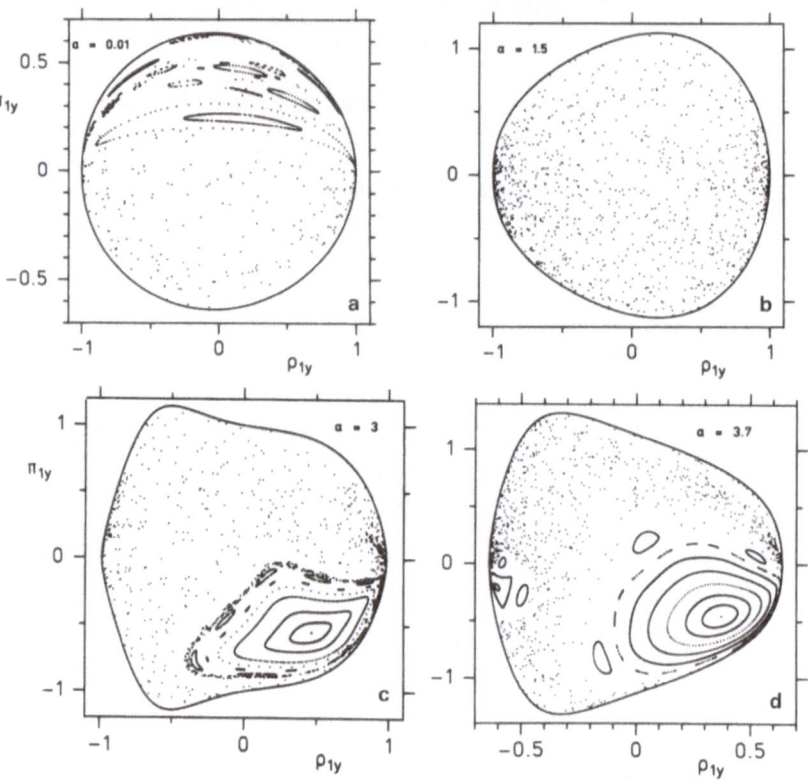

Fig. 5.   Poincaré–sections for various chain extensions: $\alpha = 0.01$ (a), 1.5 (b), 3.0 (c), 3.7 (d)

As usual we assume that the average value of the fluctuations vanishes, that they are δ–correlated and the fluctuation–dissipation theorem holds:

$$< \vec{\Gamma}_i > \, = \, 0 \tag{22}$$

$$< \Gamma_{i\nu}(t_2)\,\Gamma_{i\mu}(t_1) > \, = \, 2 k_B T \, \bar{\gamma} \, m_i \, \delta_{ij}\, \delta_{\nu\mu}\, \delta(t_2 - t_1) \tag{23}$$

3 HAMILTONIAN DYNAMICS FOR THREE MOBILE MASSPOINTS

3.1 Solution of the equations of motion

The system described by (16) together with the constraints can show deterministic chaos depending on the extension of the chain and on the initial conditions. This behaviour shows up even for short chains with $N = 3$ (see Fig. 3) mass points moving in two dimensions. In this case the equations of motion read

$$\ddot{\vec{\rho}}_1 = \lambda_1 \vec{\rho}_1 + \lambda_2(\vec{\rho}_1 - \vec{\rho}_2) \tag{24}$$

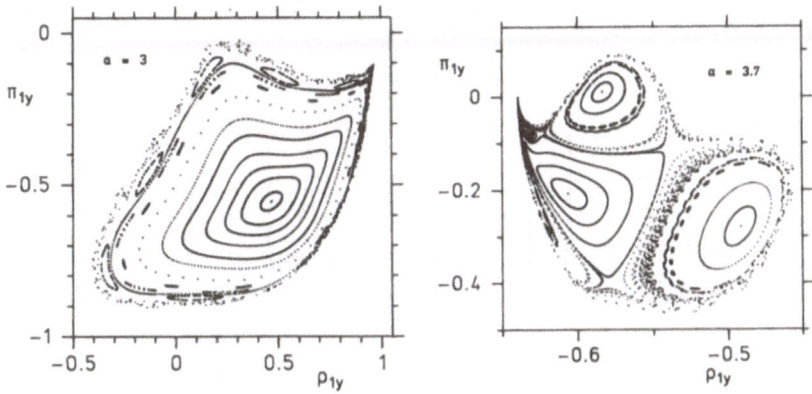

Fig. 6. Enlarged sections of the regular areas of Figs. 5c, d

$$\ddot{\vec{\rho}}_2 = \lambda_2(\vec{\rho}_2 - \vec{\rho}_1) + \lambda_3(\vec{\rho}_2 - \vec{\rho}_3) \tag{25}$$

$$\ddot{\vec{\rho}}_3 = \lambda_3(\vec{\rho}_3 - \vec{\rho}_2) + \lambda_4(\vec{\rho}_3 - \vec{\alpha}) \tag{26}$$

The Lagrangian parameter λ_1, for example, is given by the following equation

$$\lambda_1 = \frac{A}{B} \tag{27}$$

Here A and B are nonlinear functions of the velocities and site vectors of the mass points of the chain:

$$
\begin{aligned}
A = \ & \dot{\vec{\rho}}_1^{\,2}\{[(\vec{\rho}_1 - \vec{\rho}_2)(\vec{\rho}_3 - \vec{\rho}_2)]^2 - 2(2 - [(\vec{\rho}_3 - \vec{\rho}_2)(\vec{\rho}_3 - \vec{\alpha})]^2)\} \\
& + (\dot{\vec{\rho}}_1 - \dot{\vec{\rho}}_2)^2 \, [\vec{\rho}_1(\vec{\rho}_1 - \vec{\rho}_2)] \, (2 - [(\vec{\rho}_3 - \vec{\rho}_2)(\vec{\rho}_3 - \vec{\alpha})]^2 \\
& - (\dot{\vec{\rho}}_2 - \dot{\vec{\rho}}_3)^2 \, [\vec{\rho}_1(\vec{\rho}_1 - \vec{\rho}_2)] \, [(\vec{\rho}_1 - \vec{\rho}_2)(\vec{\rho}_3 - \vec{\rho}_2)] \\
& + \dot{\vec{\rho}}_3^{\,2} \, [\vec{\rho}_1(\vec{\rho}_1 - \vec{\rho}_2)][(\vec{\rho}_3 - \vec{\rho}_2)(\vec{\rho}_3 - \vec{\alpha})] \\
& \quad \times [(\vec{\rho}_3 - \vec{\rho}_2)(\vec{\rho}_1 - \vec{\rho}_2)]
\end{aligned}
\tag{28}
$$

$$
\begin{aligned}
B = \ & 4 - 2[(\vec{\rho}_3 - \vec{\rho}_2)(\vec{\rho}_3 - \vec{\alpha})]^2 - [(\vec{\rho}_1 - \vec{\rho}_2)(\vec{\rho}_3 - \vec{\rho}_2)]^2 \\
& - 2 \, [\vec{\rho}_1 \, (\vec{\rho}_1 - \vec{\rho}_3)]^2 \\
& + [\vec{\rho}_1(\vec{\rho}_1 - \vec{\rho}_2)]^2[(\vec{\rho}_3 - \vec{\rho}_2)(\vec{\rho}_3 - \vec{\alpha})]^2
\end{aligned}
\tag{29}
$$

The nonlinear system of differential equations was solved using a fourth order predictor-corrector method [36] and the Verlet–method [37]. The latter procedure is especially adapted for systems with constraints.

The system described by (24–26) has two degrees of freedoms and the energy as a global integral of motion. From the theory of dynamic systems [38] we know that because of the nonlinearities introduced by the constraints deterministic chaos can occur. In our case the ocurrence of regular or chaotic trajectories depends on the extension of the chain and on the initial conditions. Because of energy conservation the system moves on a 3-dimensional submanifold of phase space. With the help of a Poincaré–section the system is described in

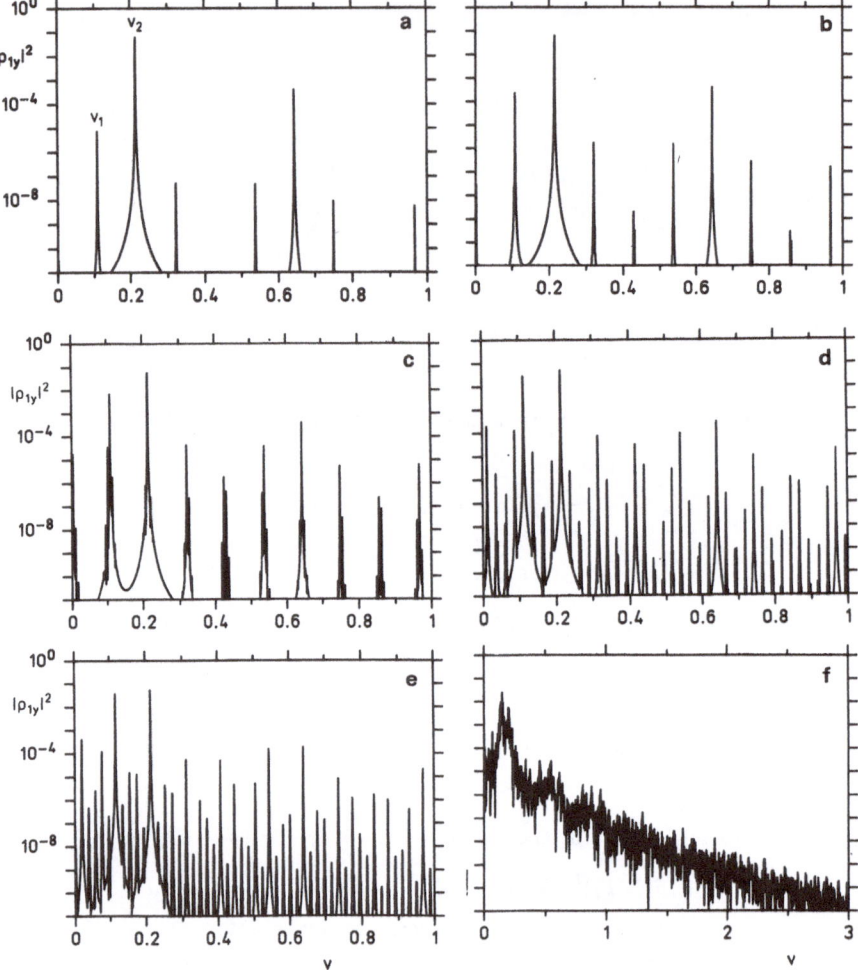

Fig. 7. Power spectrum of the coordinate ρ_{1y} of a chain with three mass points mobile in two dimensions and an end–to–end distance of $\alpha = 3$. Initial conditions (ρ_{1y}, π_{1y}): $(0.468, -0.55)$ (a); $(0.5, -0.525)$ (b); $(0.4, -0.4)$ (c); $(0.2, -0.25)$ (d); $(-0.285, -0.7)$ (e); $(0.0, 0.5)$ (f).

a two–dimensional plane. In the present case the section was carried out as shown in Fig. 4: the angle between segment 2 and 3 equals π and the y-component of the momentum of mass point 3, i.e. Π_{3y} has to be positive. As variables in the representation of the Poincaré–sections in Fig. 5 we have chosen the y–components of the momentum and site vector of the first mass point.

For a very small extension of the chain, Fig. 5a shows for initial conditions in its upper part regular trajectories, whereas the lower part represents a chaotic trajectory. In Fig. 5b for $\alpha = 1.5$ only chaotic trajectories have been found. With increasing extension of the chain ($\alpha = 3.0, 3.7$) more and more regular trajectories occur. The detailed investigation [29], however, shows that even for an almost streched chain ($\alpha = 3.99$) a considerable part of phase space is still occupied by chaotic trajectories. Figs. 6a and 6b present in an enlarged scale the regular ranges of Fig. 5c,d. They show nicely the sections of KAM–tori [38] as well as

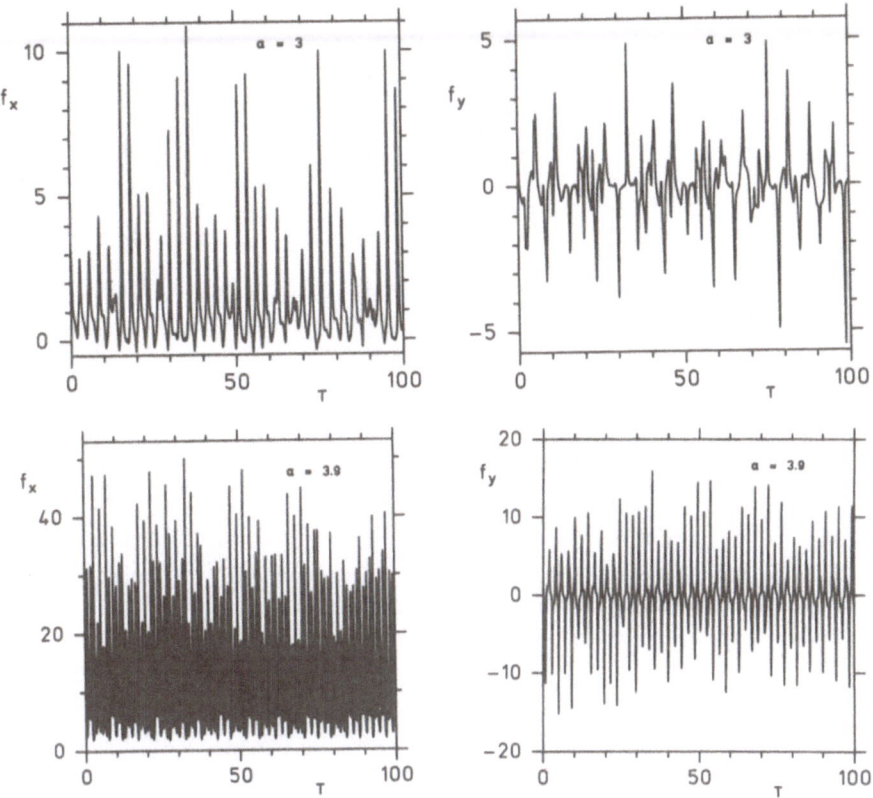

Fig. 8.   Time dependence of the x–and y–components of the force at the righthand
side endpoint for a chain with N = 3 masspoints mobile in two dimensions

elliptic and hyperbolic fixpoints which are expected according to the Poincaré–Birkhoff–theorem [38].

The Poincaré–section technique is difficult to generalize to chains with a larger number of
degrees of freedoms. In [30] this has been carried out for a chain with two mass points moving
in three dimensions using a finite slice technique [39]. However, in this case and also in cases
with a higher number of degrees of freedoms, the investigation of the power spectrum, i.e. the
square of the modulus of the Fourier transform of a coordinate, is a more convenient method.
For the purpose of comparison Fig. 7 shows such a power spectrum for the same situation
as in Fig. 5c. Figs. 7a–f correspond to different initial conditions where the motion becomes
more and more irregular. The continuous spectrum of Fig. 7f corresponds to a trajectory,
which starts in the middle of the chaotic range of Fig. 5c. Finally we mention that a further
method for the investigation of chaotic and regular trajectories, which has also been applied
in [29] uses the calculation of Lyapunov exponents [38].

3.2 Time dependence of the forces of constraints

Having determined the Lagrangian multipliers in the course of the numerical solution of
the system of differential equations, we can calculate the forces of constraints, especially the

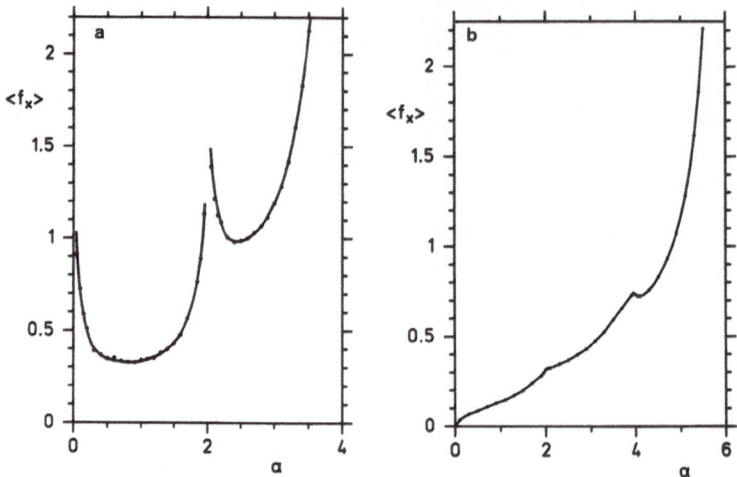

Fig. 9.   Averaged force–extension relation for chains mobile in two dimensions. N =
3 mass points (a); N = 5 mass points (b).

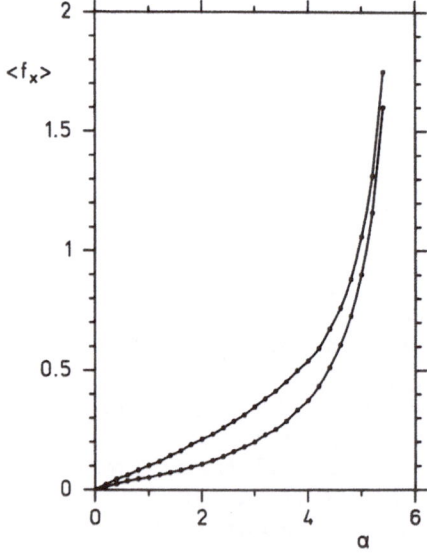

Fig. 10.   Averaged force–extension relation for chains mobile in three dimensions. Vanishing angular momentum: upper curve; large angular momenta: lower curve.

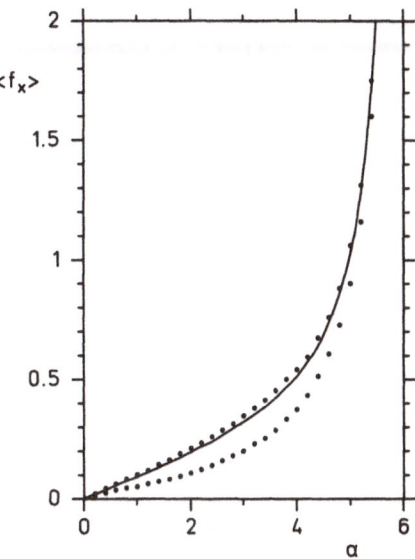

Fig. 11.   Comparison of the forces from the molecular dynamic model (dots) with the results obtaind within the stochastic dynamics calculation (full curve) for a chain with N = 5 mobile mass points.

one exerted on the end point of the chain. For the same chain as above, the components in $x-$ and $y-$direction of the force at the righthand side end point are represented in Fig. 8 for two different extensions $\alpha = 3.0$ and 3.9 . The figure shows that for the $y-$component of the force the number of positive and negative peaks is comparable and therefore its time average vanishes. In the case of the $x-$component the number of positive values prevails. Therefore one obtains a finite average value, which increases with increasing extension of the chain, because both the amplitude and the frequency of the force peaks increase.

4 FORCE EXTENSION RELATION FOR CHAINS OF VARIOUS LENGTHS MOVING IN TWO AND THREE DIMENSIONS

Fig. 9 shows the average value of the force as a function of the extension for chains with N = 3 (a) and N = 5 (b) mass points which move in two dimensions. It is obvious that with increasing extension the force generally increases and diverges when the extension approaches the length of the chain. The discontinuities of the force in the case of the shorter chain in Fig. 9a originate from special configurations of the almost non–extended and the approximately half–extended chain. The comparison with Fig. 9b shows that these discontinuities rapidly decrease with increasing chain length.

A still smoother behaviour of the force–extension curve is observed for chains with N = 5 mass points if the chain moves in three dimensions (see Fig. 10).

In the case of the three dimensional motion a further conservation law exists: the angular momentum around the $x-$axis (axis of extension) is constant. Apparently the value of the angular momentum influences the strength of the force; in Fig. 10 the curve above represents the force for vanishing angular momentum, the curve below belongs to large angular momenta.

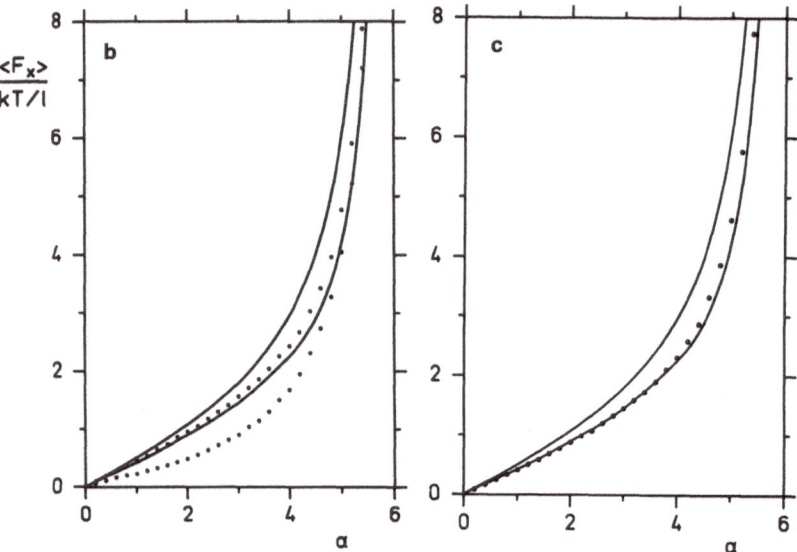

Fig. 11.   Comparison of the forces calculated within the dynamic model (dots above: vanishing angular momentum; dots below: large angular momenta) with those from the static models (curve above: inverse Langevin–function; curve below: force according to Treloar) for a chain with N = 5 mobile mass points (b). Comparison of the forces from the static models (full curves) with the stochastic dynamics simulation (dots) for a chain with N = 5 mobile mass points (c).

In Figs. 11 we compare the force extension relation obtained from the simulation of the molecular dynamics with other theoretical results. In Fig. 11a we compare the results of the molecular dynamics simulation with those of the stochastic dynamic calculation using Langevin equations. The stochastic dynamics calculation gives only a single force extension relation, because on account of the influence of the stochastic forces the angular momentum is no longer a conserved quantity. The calculations show that the force extension relation of the stochastic dynamics is close to the the one from molecular dynamics for vanishing angular momentum. Fig. 11b shows the comparison with the results from the static chain models: the upper full curve is calculated from the inverse Langevin-function (Kuhn and Grün [2]), the lower one gives the force according to the theory of Treloar [3]. The dots are as in Fig. 10 the result of the simulation of the molecular dynamics. For the short chains shown here with N = 5 mobile mass points the deviation between the various methods is considerable. Finally Fig. 11c compares the simulation of the stochastic dynamics (dots) with the results of the static chain models. Obviously, the result of the stochastic dynamics calculation agrees well with the one according to Treloar's theory.

The following figures compare the force–extension relations of the various chain models considered for chains with N = 30 mass points which move in three dimensions. Fig 12 gives the results from the molecular dynamics simulation, above for vanishing angular momentum, below for large angular momenta with a considerable difference between both.

Fig. 13a compares the result of the molecular dynamics simulation for vanishing angular momentum (dots) with the results of the static models (above: Langevin, below: Treloar). The

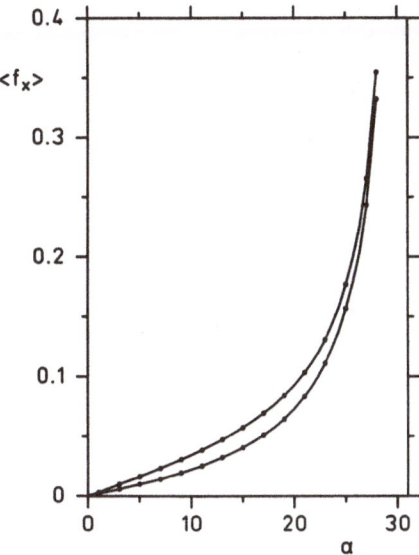

Fig. 12.   Averaged force-extension relation for a chain with N = 30 masspoints moving in three dimensions. Vanishing angular momentum: upper curve; large
angular momenta: lower curve.

deviations between the three curves is much smaller than the deviation between the two curves
in Fig. 12. Furthermore, for large extensions the molecular dynamics simulation seems to
agree well with the inverse Langevin–function description. In Fig. 13b, finally, we compare the
simulation of the molecular dynamics for vanishing angular momentum (full curve) with the
stochastic dynamics calculation (dots). The results of both calculations practically coincide.
This means that in the averaging over the distribution of the angular momenta, which is
implicit in the the stochastic dynamics calculation, small angular momenta must have a much
stronger weight than large ones.

5 FLUCTUATIONS OF THE ENERGY AND POSITIONS OF THE MASS-POINTS

In Fig. 14 we present the average values of the contributions to the kinetic energy for
the motion of various mass points parallel (x-direction) and perpendicular (y-direction) to the
streching direction of the chain as a function of its extension. Fig. 14a shows the situation for
a chain with N = 5 and Fig. 14b for one with N = 30 mobile mass points. The interesting point
is that with increasing extension of the chain the contribution to the energy in y-direction
increases whereas the one in x-direction decreases.

To investigate the influence of different masses we considered a chain with N = 5 mass
points which were allowed to move in three dimensions and where the mass of the point
in the middle of the chain was 100 times larger than the masses of the other points. The
numerical calculations showed that the force extension relation was not very much influenced
by the variation of the masses. The distribution of the kinetic energy, however, was influenced
considerably as may be seen from Fig. 15, which shows the average value for the first and
third mass points for motion along the x–(decreasing points) and y–directions (ascending

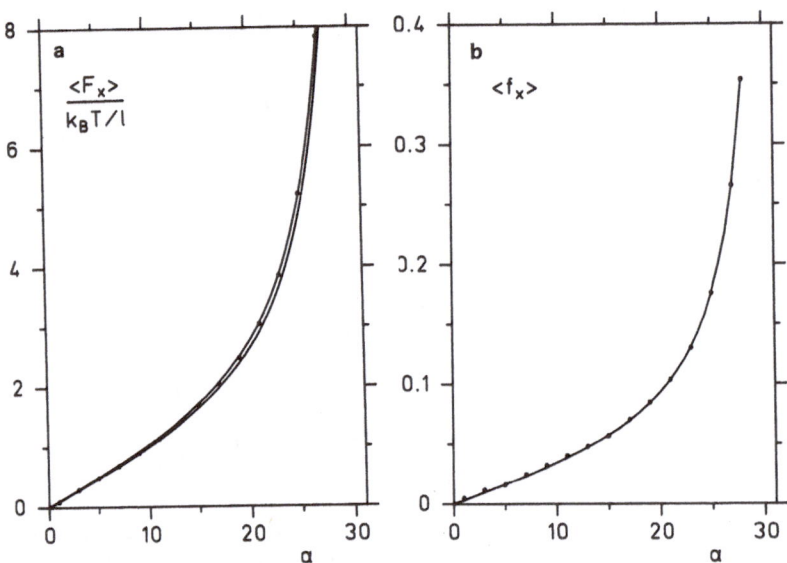

Fig. 13. Comparison of the forces according to the dynamic model with vanishing angular momentum (dots) with those of the static chain models (inverse Langevin–function: upper curve; force according to Treloar: lower curve) for a chain with $N = 30$ mobile mass points (a). Comparison of the forces calculated from the molecular dynamic model (full curve) with the results of the stochastic dynamics simulation (dots) for a chain with $N = 30$ mobile mass points (b).

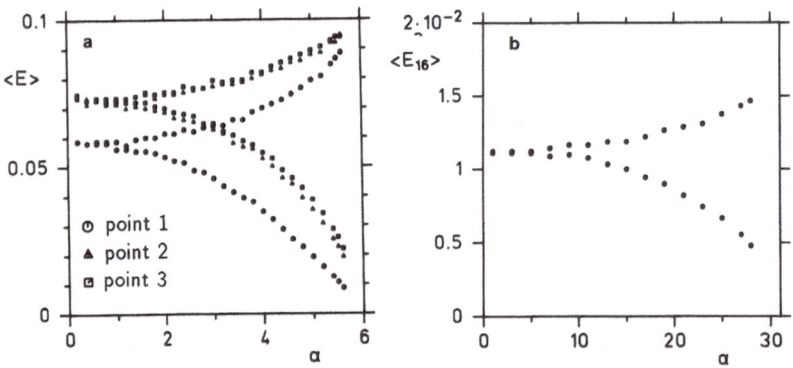

Fig. 14. Average value of the kinetic energy contribution of the masses 1, 2, and 3 for motion in x– (descending dots) and y–direction (ascending dots) for a chain with $N = 5$ mass points moving in three dimensions (a). Average value of the kinetic energy contribution of the 16th mass point of a chain with 30 masses moving in three dimensions. Motion in x– direction (descending dots); Motion in y–direction (ascending dots) (b).

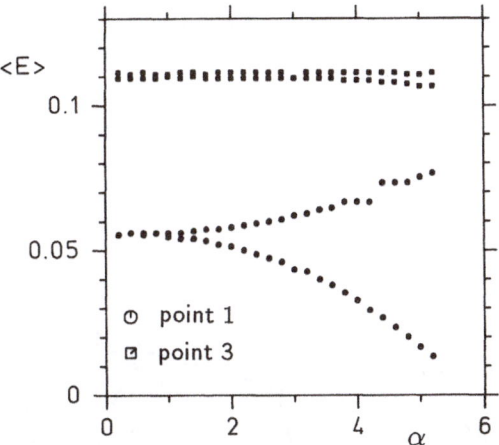

Fig. 15. Contributions of the motion in x– (descending points) and y–directions (ascending points) to the average value of the kinetic energy of the first and third mass point of a chain with N = 5 masses moving in three dimensions. The mass of the point in the middle of the chain is 100 times larger than that of the other points.

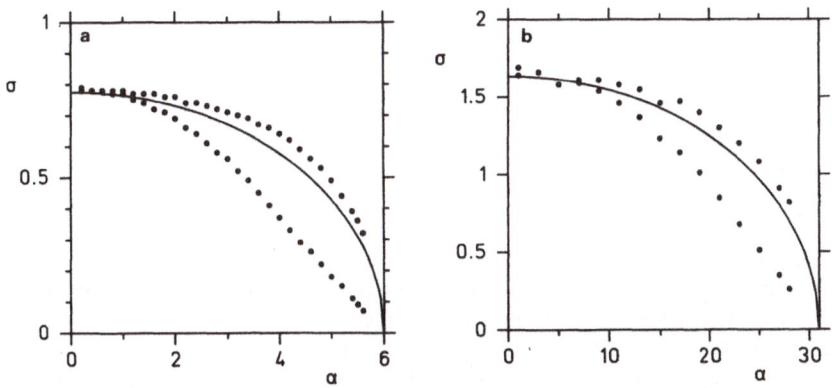

Fig. 16. Variance of the position of the middle point of a chain with N = 5 (a) and N = 30 (b) mass points moving in three dimensions for the x– (lower points) and y–direction (upper points). The full curve describes a Gaussian chain with fixed end points.

point), respectively. The comparison with Fig. 14a shows that the point in the middle has a considerably higher kinetic energy as compared to the case of a chain with equal masses.

Fig. 16 shows that the variance of the position of the mass point in the middle of a chain with N = 5 (a) and N = 30 (b) mass points, respectively, is smaller in the streching direction than perpendicular to it. Both variances become smaller with increasing stretching of the chain.

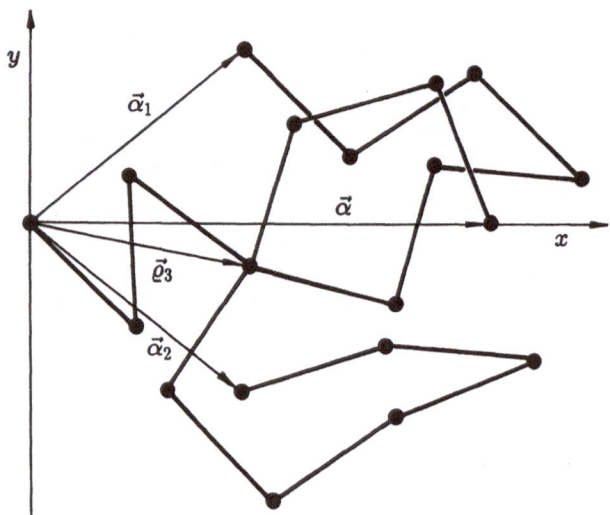

Fig. 17.   Model of a branched polymer chain.

6 HAMILTONIAN DYNAMICS FOR A CROSSLINKED CHAIN

6.1 Crosslinked chain

In the framework of the Hamiltonian dynamics we have also investigated the branched molecular chain shown in Fig. 17. Both the main chain and the two side chains are formed from N = 5 mobile mass points. The end–to–end distance of the main chain is again denoted by α. As regards the side chain, we have considered a symmetric situation with chain ends at $\alpha_{1,2} = (\alpha/2, \pm 2.5, 0)$ and an asymmetric one with chain ends at $\alpha_{1,2} = (0, \pm 2.5, 0)$.

The average value of the force is shown for both situations in Fig. 18. The lower curves are the force-length relations for the side chains in the asymmetric situation; the lowest curve represents the force component in the stretching direction, the almost horizontal one the component in y–direction. The two curves above give the force of the main chain in the stretching direction with the upper most curve representing the asymmetric situation and the curve below the symmetric one. The consideration of the mean values of the coordinates [29], which are not represented here, shows that in the non–symmetric situation the righthand side of the chain is, as expected, stretched more strongly than the lefthand one. Therefore, in the average the crosslink is situated to the left of the middle of the chain.

6.2 Additional constraints

To investigate the influence of additional geometric constraints on the force–extension relation, we considered a chain with N = 2 mass points which were mobile in three dimensions. The chain was included between walls perpendicular to the direction of extension and positioned at $x = 0$ and $x = \alpha$ [30]. As long as the extension of the chain was large enough to prevent the chain from getting in touch with the walls, the force–extension relation was the same as in the case without walls (see Fig. 19). This force, however, was reduced when after a reduction of the end–to–end distance the chain was able to get in contact either with the wall on the righthand side or with the one on the lefthand side, or with both. Furthermore, it is interesting that the force changes sign for a finite value of the end–to–end distance, which means that the equilibrium is characterized by a finite value of chain extension.

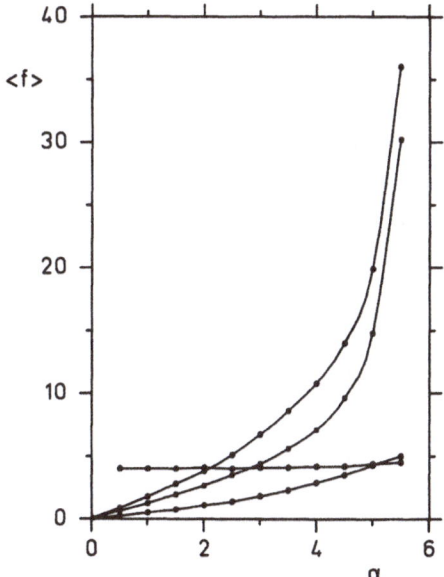

Fig. 18.   Average value of the forces in x– and y–direction as a function of the end–to–end distance for a branched system. Lowest curve: Force of a side chain in x–direction for $\alpha_{1,2} = (0,\pm2.5,0)$; horizontal curve: force of a side chain in y–direction for $\alpha_{1,2} = (0,\pm2.5,0)$; upmost curve: force of the main chain in x–direction for $\alpha_{1,2} = (0,\pm2.5,0)$; curve below of it: force of the main chain in x–direction for $\alpha_{1,2} = (\alpha/2,\pm2.5,0)$.

In a further calculation for the branched polymer, together with the extension of the main chain also the side chains were streched reducing the possibilities for the motion of the main chain. Also in this case the force of constraint at the end point of the chain was weaker as compared to the case without additional restrictions.

7 CONCLUDING REMARKS

The investigations presented in this contribution are carried out in order to give a different approach to a microscopic understanding of the stress–strain behaviour of polymeric networks. To this end we have considered non–branched and branched molecular chains with rigid segments using molecular and stochastic dynamics simulations. The results have been compared with static statistical mechanics calculations. Using various methods in the framework of Hamiltonian dynamic calculations it was shown that the motion of the chains shows molecular chaos. We have shown that the various models and methods of treatments yield different results for short chains. For long enough chains the various results seem to approach each other. From the calculations the divergent Langevin–behaviour of the force is obtained if the extension of the chain approaches its total length. However, the Mooney weakening was not observed in all of these calculations. Furthermore, we investigated the distribution of the kinetic energy along the chain and fluctuations of the positions of the mass points also under the influence of different masses. First investigations of chains subject to additional constraints show that they result in a reduction of the force. Calculations along these lines will be the subject of future investigations.

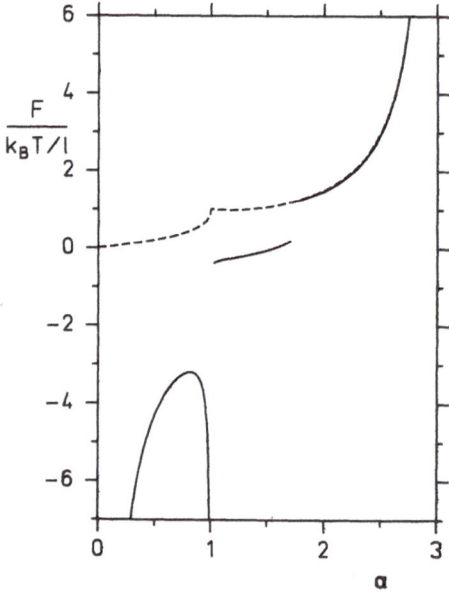

Fig. 19.   Force of a dynamic chain with N = 2 mass points moving in three dimensions. Dashed line: force of the chain not inclosed between walls. Full line: force of the chain inclosed between walls at x = 0 and x = α.

ACKNOWLEDGEMENT

This research project is part of the Sonderforschungsbereich 239. The support by the Deutsche Forschungsgemeinschaft is gratefully acknowledged.

References

[1] W. Kuhn, Kolloid–Z. **68**:2 (1934)

[2] W. Kuhn, F. Grün, Kolloid–Z. **101**:248 (1942)

[3] L. R. G. Treloar, Trans. Faraday Soc. **42**:77 (1949)

[4] K. H. Meyer, G. von Susich, E. Valkó, Kolloid–Z. **59**:208 (1932)

[5] F. H. Müller, Kolloid–Z. **95**:139 (1941)

[6] J. H. Weiner,Macromolecules **15**:542 (1982)

[7] M. V. Volkenstein, "Configurational Statistics of Polymeric Chains", John Wiley & Sons, New York (1963)

[8] P. J. Flory, "Statistical Mechanics of Polymeric Chains", John Wiley & Sons, New York (1969)

[9] L. R. G. Treloar, "The Physics of Rubber Elasticity", Clarendon Press, Oxford (1975)

[10] P.–G. de Gennes, "Scaling Concepts in Polymer Physics", Cornell University Press, Ithaca (1979)

[11] J. H. Weiner, "Statistical Mechanics of Elasticity", John Wiley & Sons, New York (1983)

[12] M. Doi, S. F. Edwards, "The Theory of Polymer Dynamics", Clarendon Press, Oxford (1986)

[13] J. des Cloizeaux, G. Jannink, "Les Polymères en Solution: leur Modélisation et leur Structure", Les Edition de Physique, Les Ulis (1987)

[14] K. F. Freed, "Renormalization Group Theory of Macromolecules", John Wiley & Sons, New York (1987)

[15] K. F. Freed, Adv. Chem. Phys. **22**:1 (1972)

[16] W. Kuhn, Kolloid–Z. **76**:258 (1936)

[17] H. M. James, E.Guth, J. Chem. Phys. **11**:455 (1943)

[18] R. T. Deam, S. F. Edwards, Phil. Trans. Roy. Soc. Lond. **A280**:317 (1976)

[19] M. Mooney, J. Appl. Phys. **11**:582 (1940)

[20] R. S. Rivlin, Trans. Roy. Soc. (London) **A241**:379 (1948)

[21] M. Doi, S. F. Edwards, J. Chem. Soc. Faraday Trans. II **74**:1802 (1978)

[22] G. Marrucci, Rheol. Acta **18**:193 (1979)

[23] W. W. Graessley, Adv. Polym. Sci., **46**:67 (1982)

[24] M. Gottlieb, R. J. Gaylord, Polymer, **24**:1644 (1983)

[25] R. C. Ball, M. Doi, S. F. Edwards, M. Warner Polymer **22**:1009 (1981)

[26] S. F. Edwards, Th. Vilgis, Polymer **26**:101 (1986)

[27] P. J. Flory, B. Erman, Macromolecules **15**:800 (1982)

[28] H. G. Kilian Polymer **22**:208 (1981)

[29] R. G. Winkler, "Untersuchungen zum statistischen und molekulardynamischen Verhalten von Polymerketten", Ph.D. Thesis, University of Ulm (1989)

[30] G. Siegert, "Deterministisches Chaos in der Dynamik von Molekülketten" Diplom Thesis, University of Ulm (1989)

[31] R. G. Winkler, P. Reineker, M. Schreiber, Europhys. Lett **8**:493 (1989)

[32] R. G. Winkler, P. Reineker, M. Schreiber, Entropy Elastic Forces of Chain Molecules, in: "Molecular Basis of Polymer Networks", A. Baumgärtner, C. E. Picot, eds., Springer, Berlin, Heidelberg (1989)

[33] P. Reineker, R. G. Winkler, Phys. Lett. **A141**:264 (1989)

[34] R. G. Winkler, P. Reineker, Makromol. Chem., Makromol. Symp., **30**:215 (1989)

[35] P. Reineker, R. G. Winkler, Progr. Colloid & Polymer Sci. **80**:101 (1989)

[36] C. D. Conte, D. de Boor, "Elementary Numerical Analysis", Mc Graw Hill, Kogakusha, Ltd. Tokyo (1972)

[37] L. Verlet, Phys. Rev. **159**:98 (1967)

[38] A. J. Lichtenberg, M. A. Liebermann, "Regular and Stochastic Motion", Springer, New York (1983)

[39] M. Hénon, Numerical exploration of Hamiltonian Systems, in: "Chaotic Behaviour of Deterministic Systems," G. Iooss, R.H.G. Helleman, R. Stora, eds., North-Holland Publishing Company, Amsterdam (1983)

COMPLEX SURFACE GEOMETRY IN NANO-STRUCTURE SOLIDS:

FRACTAL VERSUS BERNAL-TYPE MODELS

Peter Pfeifer

Department of Physics and Astronomy
University of Missouri
Columbia, MO 65211, USA

David Avnir and Dina Farin

Department of Organic Chemistry and
F. Haber Research Center for Molecular Dynamics
The Hebrew University of Jerusalem
Jerusalem 91904, Israel

1. INTRODUCTION

In this chapter we present a discussion of a controversy that has recently arisen and that has been witnessed at this conference in the lectures by J. Klafter and by one of us (P.P.). The controversy concerns the question whether the fractal concept, as found to be highly successful in a large number of surface science problems (see Refs. 1-3 for recent surveys), is applicable to a certain class of porous silicas as reported by us and others,[4-11] or whether these silicas are more appropriately described by traditional random-packed sphere models as proposed by Drake, Levitz, and Klafter (DLK).[12-16] The fractal model asserts that the surface geometry scales with dimension $D \approx 3$ and thus is highly disordered, from atomic length scales upward. The random-packed sphere model, to which we shall refer as Bernal-type model because of its similarity to the familiar Bernal model for liquids,[17-19] describes the solid as a random assembly, more or less close-packed, of hard spheres with fixed diameter. A typical value of the sphere diameter, as proposed by DLK, is 70 Å. Their model thus asserts that the surface is smooth and scales with dimension $D = 2$ from atomic lengths up to the sphere diameter. The question then, which of the two models describes the structure of the silicas more adequately, is a rather specialized one and may seem of limited interest. But DLK have elevated it to an issue from which they wish to conclude that "the situation with surface fractals is less clear and still controversial,"[15] as if to build a general case against the fractal concept in surface science. We therefore believe it is important to clarify the issue, to assess the two models in an unbiased way, and to obtain, from the analysis, guidelines for similar studies of other systems.

Elsewhere[20] we have performed such an analysis based on a comprehensive comparison of DLK's experimental data with our data (molecular tiling, surface area from N_2 adsorption, pore-size distribution, electron microscopy, electronic energy transfer, small-angle X-ray scattering), including a discussion of possible experimental uncertainties. The conclusion was:

(i) if one corrects DLK's data for an elementary, but far-reaching calculational error, their experimental data and ours are in full *agreement* (within respective experimental error bars and to the extent that the two data sets are comparable); (ii) if one tries to explain the data in terms of a Bernal-type model, *no consistent* interpretation is possible; (iii) if one assumes a fractal model instead, very *consistently* a fractal dimension $D \approx 3$ over a range of length scales of about 4 - 10 Å is obtained for the silica surfaces in question.

Actually, the electronic energy transfer experiments discussed at length in that analysis suggest that the fractal regime extends up to 100 Å rather than ending near 10 Å; but such an extended regime seemed to be in conflict with the small-angle scattering data. According to the Porod law in small-angle scattering, the scattering curves indicate a smooth surface, $D = 2$, at length scales above 15 Å. In the present paper we show that this Porod behavior can be explained equally well by a fractal model, with $D = 3$ in the entire regime 4 - 100 Å, and that the model is consistent with all other data.

This will resolve the controversy in a rather complete fashion: The fractal model offers a fully consistent description of the surface in terms of the single dimension $D \approx 3$ over length scales 4 - 100 Å, whereas DLK's model offers a picture rich in contradictions. This outcome makes the case also interesting from the viewpoint of inverse problems, i.e., from the viewpoint of interpreting one and the same set of experimental data in terms of inequivalent structural models. The two models are not easy to distinguish by small-angle scattering because they give essentially the same scattering curves. This is remarkable because, to the best of our knowledge, no scatterer shape has been known so far that is intrinsically irregular and yet scatters like a smooth surface. With respect to all other data in the regime 4 - 100 Å, the two models are trivially distinguishable, however. We demonstrate this by identifying the consistency tests that DLK have failed to perform for their interpretation, each of which would have signaled the inadequacy of their model at an early stage.

The plan of the paper is as follows. In Section 2 we present the salient geometric features of the two models. Section 3 is devoted to the consistency tests. The fractal model that reproduces the experimental small-angle scattering data is introduced in Section 4. It consists of interconnected clusters, each with $D = 3$ up to 100 Å and with a nominal diameter in the neighborhood of 270 Å, such that the intercluster correlation function is of the Percus-Yevick type familiar from theory of liquids. Qualitatively the Percus-Yevick correlation function is similar to the interparticle correlation function in the Bernal model. Thus a Bernal-type model does apply to the silicas at issue, but with *large* particles carrying a *fractal* surface instead of the small nonfractal spheres considered by DLK. Final conclusions are drawn in Section 5.

2. THE GEOMETRY OF THE MODELS

A solid with a fractal surface is characterized by three parameters: the dimension D ($2 \leq D \leq 3$), the inner cutoff u_1, and the outer cutoff u_2. It has the property that if the surface is covered with molecules of diameter u, the number $N(u)$ of molecules needed to form a monolayer (tiling) scales as

$$N(u) \propto u^{-D} \qquad\qquad (u_1 < u < u_2). \qquad\qquad (1a)$$

Equivalently, the surface area $A(u)$ of the solid as measured by a tiling of the surface with these molecules, and the volume $V(u)$ outside the solid that is inaccessible to these molecules, scale as

$$A(u) \propto u^{2-D} \qquad\qquad (u_1 < u < u_2), \qquad\qquad (1b)$$
$$dV(u)/du \propto u^{2-D} \qquad\qquad (u_1 < u < u_2). \qquad\qquad (1c)$$

These relations express that the surface becomes increasingly inaccessible, due to geometric heterogeneity of the surface, to molecules of increasing size u. The area $A(u)$ is called apparent

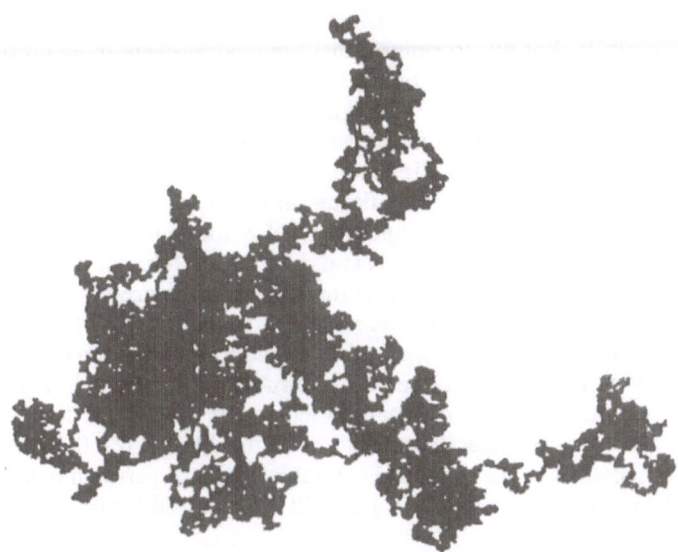

Fig. 1. Trail of a random walk in the plane as a cross-sectional model of a surface with D = 3. The surface may be thought of as constructed by translating the trail out of the figure plane. The walk fills some regions of the plane compactly (dimension two of the trail) and leaves other regions empty, so that the surface fills space in a *non-uniform* way and contains pores of all sizes. If we assign a length of 1 Å to each step of the walk, the diameter of the trail shown (largest distance between two points on the trail) equals 410 Å, and the lengths u_1, u_2 between which the surface obeys the scaling relations (1) with D = 3 are approximately 4 Å and 100 Å. The walk consisted of 250000 steps and visited 31447 distinct sites.

surface area and V(u) is the cumulative volume of pores with diameter less than u. Apart from factors of order unity, the three functions are related by $A(u) \approx u^2 N(u)$ and $dV/du \approx -u\, dA/du$. So Eqs. (1b,c) are elementary consequences of (1a). The lengths u_1 and u_2 define the length-scale regime within which the surface heterogeneity is described by the dimension D. Above u_2 (or below u_1 if the inner cutoff is distinct from an atomic scale) the surface may scale with a different dimension, or may not scale at all. For the sake of definiteness, we assume that the surface giving rise to (1) is self-similar, as opposed to merely self-affine. This assumption is motivated by the fact that the silicas under consideration are all formed under isotropic conditions (polymerization, drying, etc.) and therefore have no preferred direction. But the assumption is not critical because Eqs. (1) hold also for a self-affine surface if the necessary distinction between local and global dimension is made.[21]

The porous silicas we are concerned with are called Si-40, Si-60, and Si-100 because they have a nominal average pore diameter of 40, 60, and 100 Å (for more specifics, including manufacturers, see Refs. 5-15). Si-60 is the most extensively studied silica of all with regard to fractal surface properties; and it is for this system that we have proposed a dimension of D ≈ 3, beginning at $u_1 \approx 4$ Å. Now a surface with D = 3 is usually called space-filling because it winds back and forth so much that it visits essentially a region of nonzero volume. This has led some researchers, perhaps even DLK,[15] to expect that a surface with D = 3 fills space in a way that leaves no room for pores into which molecules could adsorb and give rise to the power laws $N(u) \propto u^{-3}$, $A(u) \propto u^{-1}$, and $dV/du \propto u^{-1}$ over an appreciable range of diameters, as implied by (1). The random-walk model in Fig. 1 gives a counterexample[21] to this expectation. It shows that a surface with D = 3 can easily be space-filling and at the same time allow external probes, such as molecules or X rays, to detect this space-filling property.

We are not suggesting, of course, that the silicas under consideration really have a structure related to random walks. But we do suggest that the random-walk model provides a useful picture of what a silica surface which obeys (1) with D = 3 may qualitatively look like. It follows that such a surface is very different from that in a zeolite-like solid: the surface of a zeolite with channel width w and wall thickness w' is three-dimensional, too, at length scales above max(w,w'); but it has pores of size w only and hence cannot give (1) with D = 3. A zeolite is an example of an uniformly space-filling surface not at issue here.

The random-packed sphere model proposed by DLK for these silicas is illustrated in Fig. 2. It is characterized by the sphere radius R and the packing fraction η (ratio of the volume occupied by spheres to the total volume). If the spheres were arranged in a regular close-packed assembly (hexagonal or face-centered cubic), η would be 0.74. If the spheres were poured into a container with irregular walls, they would form a random close-packed assembly with $\eta = 0.64$.[18,19] This random close-packed structure is the Bernal model of a liquid, in which the spheres represent atoms or small molecules. It models the liquid at its point of highest density. More general random packings, with tunable $0 < \eta < 0.74$, may be obtained from sample configurations of the canonical ensemble (classical partition function) for a hard-sphere fluid with number density $3\eta/(4\pi R^3)$, as will be discussed further in Section 4.

Fig. 2. Random-packed sphere model of a porous solid. The spheres form a contiguous network such that the ratio of occupied volume to empty volume, locally averaged, is constant throughout the system. The ratio is $\eta/(1-\eta)$, where η is the packing fraction and $1-\eta$ is the porosity. In DLK's model for Si-60, the sphere diameter equals 70 Å and $\eta = 0.33$. With this assignment, the length scales in Figs. 1 and 2 are the same.

Examples where such random packings have been used before to model the structure of porous solids are compacted powders of spherical silica particles (R ≈ 20 Å) and compressed aerogels (R ≈ 10 Å).[22] In the application to porous solids, one would like to relate the sphere radius R and the packing fraction η to the average number z of contacts between a sphere and its neighbors (coordination number), and to the average radius ρ of interstitial pores between the spheres. The commonly used empirical relations for this purpose are[22]

$$z = 2\, e^{2.4\,\eta} \tag{2}$$

$$\rho = [1 - 0.12\,(z - 4)]R \qquad (z \leq 10). \tag{3}$$

They have been obtained from fits to the respective values for regular packings on various lattices. The pore radius in (3) is taken to be the radius of the largest sphere that fits into a typical cavity of the packing. Eqs. (2,3), approximate as they are, are important because they link the model to experimentally measurable quantities. [DLK's electron micrograph[12,14,15] of Si-60 shows beads of diameter 2000 - 4000 Å and does not resolve any smaller-scale structure; so the proposed sphere diameter of 70 Å (2R) for Si-60 is not a quantity that can be directly measured.] Thus from Eqs. (2,3) and experimental values for η and ρ one can calculate the sphere radius R. Alternatively, if η and some apparent surface area A(u) have been measured, one may calculate the radius R from Eq. (2) and[22]

$$A(u) = (3V_0/R)\left[1 - \frac{z}{4}\frac{u}{R}\right]. \tag{4}$$

Here u is the diameter of the molecule used to tile the surface and V_0 is the volume occupied by the spheres. The factor 3/R is the surface to volume ratio of a single sphere; and the term $-zu/(4R)$ in [...] is the first term in the Taylor expansion that corrects for the inaccessible surface area near the points of contact of the spheres. In this lowest order, the expansion is independent of the details of the packing geometry. When the sphere radius is large compared to the size of the molecule, R >> u, the correction becomes negligible of course.

Eqs. (2-4) may be considered as counterpart of Eqs. (1b,c) of the fractal model: (1b, 4) describe the inaccessibility of parts of the surface in terms apparent surface areas, and (1c, 3) describe the inaccessibility in terms of pores of certain sizes. In their content, the two models are very different, however. Eq. (4) states that inaccessibility is a small, perturbative effect at most; whereas Eq. (1b) treats inaccessibility as a dominant, recurring phenomenon. Eq. (3) states that there is essentially only one relevant pore size in the system; whereas Eq. (1c) says that pores of all sizes (between u_1 and u_2) are present.

3. THE MODELS VIS-A-VIS THE EXPERIMENTAL DATA

We now proceed to analyze the two models in Section 2 with respect to their consistency with the experimental data.

Consistency Test # 1: Tilings with Molecules of Different Sizes. A summary of all tiling experiments performed by us and others (excluding DLK) on Si-60 can be found in Fig. 1 of Ref. 11. The probe molecules were N_2, six branched alkanols, and six non-branched alkanes. The number of molecules needed to form a monolayer, N, was determined from standard adsorption isotherm analyses. The size of the molecules was expressed in terms of cross-section area σ of the molecule, obtained from the liquid density of the adsorptive (guidelines for reliable σ values are described in Refs. 20, 24-26). Thus if the surface is fractal, one should observe $N \propto \sigma^{-D/2}$ (variant of Eq. (1a)). The experimental data obey such a scaling relation with remarkable accuracy. In fact, all thirteen data points fall onto a single curve, $N \propto \sigma^{-3/2}$, even though the alkane experiments were done on a sample from a different manufacturer. This data collapse, with data from chemically diverse probe molecules and from different samples, is a strong test of the scaling hypothesis (1). The separate D values and σ ranges from the alkanol and alkane tilings are given in the first line of Table 1.

Similar tiling experiments have been carried out by DLK. Their D value from alkane data agrees with our value of 3 within error bars. But DLK have asserted repeatedly that their tilings with alkanols lead to an absurd D value between 5 and 6.[12,15] This assertion is based on a miscalculation of the slope in their log N vs. log σ data, as we discovered recently.[20] The data in Ref. 15 yields one half of the slope asserted by DLK. So in the second line of Table 1 we collect what we believe are the best available results from their experiments. It shows that their results are entirely consistent, within the available accuracy, with our results.

There is substantial scatter in DLK's data, as evidenced by the error estimates for D in Table 1, and to a lesser degree also in the alkane data of Ref. 11. This contrast to the small

Table 1. D values from the relation $N \propto \sigma^{-D/2}$ as observed on Si-60 and on a reference silica with smooth surface, CPG-75.

Surface	Tiling with alkanols	Tiling with alkanes
Si-60 (Refs. 5,11,23)	$D = 2.97 \pm 0.02$ ($\sigma = 18\text{-}42$ Å^2) [a,b]	$D = 3.0 \pm 0.2$ ($\sigma = 17\text{-}53$ Å^2) [b]
Si-60 (Refs. 12,15)	$D = 2.7 \pm 0.5$ ($\sigma = 16\text{-}35$ Å^2) [c,b]	$D = 3.4 \pm 0.4$ ($\sigma = 17\text{-}42$ Å^2) [d]
CPG-75 (Refs. 11,23)	–	$D = 2.09 \pm 0.08$ ($\sigma = 17\text{-}59$ Å^2) [b]

[a] If the data point for N_2 ($\sigma = 16$ Å^2) is included, the same D is obtained.
[b] The error bar for D is the standard deviation.
[c] D is corrected for the error made by DLK (see text). The data points used are from Fig. 6 in Ref. 15 (σ from liquid densities) and include N_2.
[d] Quoted from Ref. 12.

scatter in our alkanol data can be understood as follows. We chose branched alkanols to ensure that each molecule in the series interacts in the same way with the surface (hydrogen bonding of the alcohol hydroxyl group to the surface silanol groups); and to ensure sphericity of the molecules, thus eliminating all questions of orientation on the surface. The series should be optimal for making the N-σ relation to depend on surface geometry only, and we interpret the small standard deviation of the corresponding D value as confirming this. DLK's alkanol series, on the other hand, includes also linear higher members and thus is likely to contain effects of variable orientation of the molecules on the surface. The same thing holds for all alkane series in Table 1 which consisted of linear members only.

From this viewpoint, the alkanes may be regarded as particularly unsuitable to probe the surface geometry. The fact that they do surprisingly well is perhaps the best indicator that the observed power laws are dominated by geometric accessibility effects, and not by specific chemical interactions as proposed by DLK.[15] To amplify this point, Table 1 also includes the result of alkane tiling experiments on a controlled-pore glass, CPG-75. This is a silica with an average pore diameter of 75 Å (comparable to Si-60) and with a surface that is widely believed to be smooth at molecular scales. Tiling of this surface gives $D \approx 2$ in agreement with such smoothness. Were the N-σ relation controlled by chemical factors, then CPG-75 and Si-60 would have to behave the same because their chemistry is the same. Conversely: were Si-60 smooth at molecular scales, as DLK's random-packed sphere model implies, then CPG-75 and Si-60 would have to behave the same on geometric and chemical grounds.

The inconsistency of DLK's sphere model with the tiling data can be made explicit even further. Eq. (1b) and the value $D \approx 3$ for Si-60 from Table 1 imply that the effective surface area A(u) decreases by a factor of two when the molecular diameter is increased from u = 3.8 Å for N_2 by a factor of two. For this decrease to be described by the sphere model, Eq. (4), the sphere radius would have to be R = 3zu/4, which is 10.4 Å if we assume an average coordination number of z = 4 (see below). This differs by more than a factor of three from DLK's value of R = 35 Å. Even worse, in the next paragraph we shall see that Eq. (4) would no longer be valid if R were anywhere near 10 Å.

Altogether then, the tiling experiments strongly support the fractal model. A conservative estimate of the resulting parameters is: $D = 3.0 \pm 0.2$, $u_1 = 3.8$ Å (diameter of N_2), and $u_2 >$ 10.3 Å (diameter of n-hexane). To try to explain the data in terms of the sphere model, with or without special chemical assumptions, leads to violent contradictions.

Consistency Test # 2: Tiling with N_2 Molecules Only. Here we investigate to what extent the random-packed sphere model is at least consistent with the N_2 tiling data. This is motivated by DLK's assertion that the sphere model conforms with the N_2 data, porosity $1-\eta$, and average pore radius ρ of the sample.[14] If this assertion is correct, one can use Eq. (4) to determine the sphere radius R from the surface area measured with N_2, and check whether the correction term $-zu/(4R)$ in Eq. (4) is small as it should be for consistency of the model. Moreover, the so obtained radius R should also agree with the R value that can be calculated from Eqs. (2,3).

Table 2 shows the results of the test. The most striking feature is that for Si-40 there exists no R value that would satisfy Eq. (4). The resulting quadratic equation for R has complex roots. This means that the spheres, in order to generate a surface area of 768 m², would have to be so small that the correction term $-zu/(4R)$ becomes large and formally counteracts any gain in area resulting from small R (Fig. 3). The spheres must be so small, and hence the corrections for inaccessible area so large, that (4) is no longer valid. But if (4) fails, then the sphere model altogether fails, because the basic premise of the model is that the surface is smooth at molecular scales and thus is fully accessible to molecules like N_2. Equivalently: if (4) fails, then pore diameters all the way down to $u = 3.8$ Å matter and ρ is no longer the only relevant pore size–which is precisely what the fractal model says.

One might argue that the sphere model could be saved by including higher-order correction terms in (4). This is not possible because the higher terms in the Taylor expansion of A(u) are non-universal; they depend on hopelessly many details of how the spheres are packed and how the excluded area is defined.[27] In one particular way of counting the excluded area (Ref. 22, p. 485), the next higher term in (4) is

$$-\frac{z}{16}\left(\frac{u}{R}\right)^2 . \tag{5}$$

Since this term is negative, its inclusion in (4) would make the no-solution dilemma only worse (Fig. 3).

For Si-60, which is the system of main interest in this paper, Eq. (4) does have a solution R when the data in Table 2 is used. But it is close to the situation of Si-40. Indeed, the condition for Eq. (4) to have a solution can be expressed as $z \leq 3V_0/(uA(u))$, which for Si-60 gives $z \leq 9.3$. Thus it is only the low coordination number z from Eq. (2) that saves Si-60 from having no solution. Since the coordination number does not affect the structure of the inaccessible region at the point of contact of two spheres, it follows that Si-60 has a surface similarly inaccessible to N_2 as it would have if Eq. (4) had no solution.

Table 2. Sphere radii R calculated from Eqs. (2-4) for the silicas Si-40, Si-60, and Si-100. The first three lines are the data from Ref. 14. A(u) is the surface area measured with N_2 ($u = 3.8$ Å) for a sample of 1 g mass and $V_0 = 0.46$ cm³. In the fifth line, R is the larger of the two roots.

	Si-40	Si-60	Si-100
A(u) [m²]	768	391	281
ρ [Å]	18	35	60
η	0.41	0.33	0.27
z from Eq. (2)	5.4	4.4	3.8
R from Eq. (4) [Å]	no solution	30	45
R from Eq. (3) [Å]	21	37	59
R from Ref. 14 [Å]	18	35	60

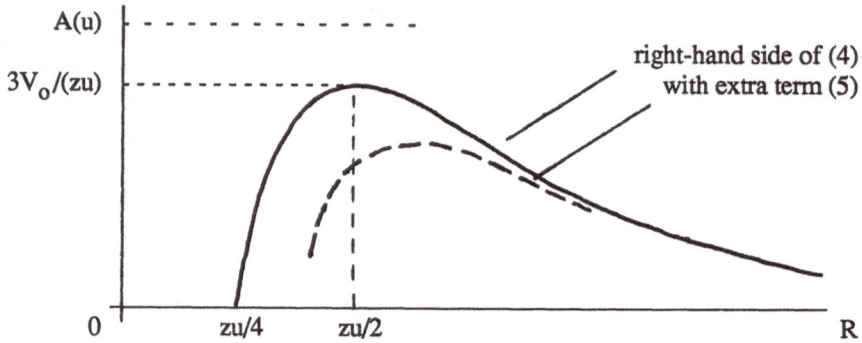

Fig. 3. The no-solution case of Eq. (4). It occurs whenever A(u) > $3V_0/(zu)$.

Other consistency checks give no better results. For Si-60 and Si-100, the value of the correction term zu/(4R) in Eq. (4) equals 0.14 and 0.08, respectively. This amounts to 14 % and 8 % of the surface area being nominally inaccessible to N_2 molecules and indicates all by itself that the surface is not smooth at molecular scales. The agreement between R values calculated from Eq. (4) and from Eq. (3) is within 30 % only. The R values from Ref. 14 correspond to the assumption R = ρ. It is interesting to note that DLK, in their original analysis[12] of the data discussed here, did estimate the size of the term zu/(4R), too. But they presented only the result for Si-100 (< 0.05) and remarked that the correction "would be more important for Si-40 and Si-60". It therefore may be that DLK are not unaware of some of the problems described above.

<u>Consistency Test # 3: Electronic Energy Transfer</u>. The energy transfer experiments exploit the fact that when an electronically excited donor molecule on the surface transfers energy to neighboring acceptor molecules on the surface by dipole-dipole interaction, the rate of transfer depends on the number of   acceptors within a given distance from the donor. On a surface with fractal dimension D, this number grows as distance to the power D with increasing distance and leads to the following survival probability P(t) of the excited donor as a function of time t:[7,28]

$$P(t) = \exp[-t/\tau - \gamma (t/\tau)^{D/6}] \tag{6}$$

$$\gamma = \frac{M}{N(u_1)} \frac{2}{D} \Gamma(1-D/6) (u_2^*/u_1)^D . \tag{7}$$

Here τ is the fluorescence life time of the donor, M is the number of acceptors on the surface, $N(u_1)$ is the number of reference molecules of diameter u_1 needed to tile the surface, u_2^* is twice the Förster radius of the transfer process, and Γ is the gamma function. If P(t) follows the law (6) for times $0 < t < t_{max}$, the surface is D-dimensional over the range of lengths[14]

$$u_1 < u < (t_{max}/\tau)^{1/6} u_2^* . \tag{8}$$

The meaning of u is the same as in Eqs. (1): the process samples points on the surface at distance u/2 from the donor and thus a region of diameter u. Experimentally, P(t) is obtained from measurement of the fluorescence decay. Representative values are $t_{max}/\tau \approx 5$ and $u_2^* \approx$ 100 - 150 Å, so that energy transfer experiments permit to probe the surface geometry over a considerably larger length range than the tiling experiments.

The reference molecule in (7) is N_2 and originates from using (1a) to eliminate tiling with acceptor molecules in favor of tiling with N_2. The quantities τ, M, u_2^*, u_1, and $N(u_1)$ can all be determined experimentally. Thus when Eqs. (6,7) are used to fit the measured fluorescence decay, D is the only adjustable parameter. This makes the analysis a particularly sensitive test

Table 3. D values from Eqs. (6,7) as observed on Si-60 under various conditions (from Ref. 10). Error bars are those estimated in Ref. 7. Each value was determined at several acceptor concentrations, resulting in variations less than 0.05 for D. The last column gives the length range estimated from Eq. (8) and data in Refs. 7,8,10.

Donor-acceptor pair	D	u_2* [Å]	range [Å]
fluorescein, diethyloxydicarbo-cyanineiodide; index-matched	2.71 ± 0.10	82	4-120
rhodamine 6G, malachite green; index-matched	2.71 ± 0.10	104	4-130
rhodamine 6G, malachite green; air	2.78 ± 0.10	122	4-140
rhodamine B, malachite green; air	2.82 ± 0.10	162	4-190

of surface fractality or nonfractality. Additional testing grounds are provided by varying the acceptor concentration (M) and the donor-acceptor pair (u_2*). Such experiments have been carried out in extenso by Huppert et al. and us.[6-8,10] An overview can be found in Ref. 10; a detailed comparison with similar experiments by DLK[13,14] is given in Ref. 20.

Table 3 shows that the energy transfer experiments are fully consistent with the fractal model. Within error bars, the D values agree with those from the tiling experiments and extend the previous scaling regime, 4 - 10 Å, to approximately 4 - 100 Å (conservative estimate). The fractal model says that the surface has no structural element that would single out a particular length scale in that interval. In agreement with this, the energy transfer data show no indication of any crossover from (6) to some over behavior, neither as the acceptor concentration is varied nor as the Förster radius is varied.

If the surface had the structure implied by the sphere model or cylindrical-pore model,[14] a distinct crossover would have to occur. Specifically, for a sample with $\rho = R$ as conceived by DLK, the fluorescence decay would have to follow (6) with D = 2 when the process samples only points on the surface at distance less than 2R from the donor. According to (8), such short-distance sampling occurs when

$$t_{max}/\tau < (4R/u_2*)^6 , \tag{9}$$

which can be achieved by making the acceptor concentration high (small t_{max}) or by making u_2* small. With R = 35 Å for Si-60 (Table 2), the right-hand side of (9) takes the value 25, 6, and 2 for the first three donor-acceptor pairs in Table 3. This means that the time scale in those experiments was such that a decay with D = 2 in (6) should have been observed if the sphere model were correct. Thus the sphere model is in gross violation of the results in Table 3.

The corresponding results by DLK[14] for Si-60 are conflicting and difficult to assess. Their experimental conditions are similar to the second case in Table 3. DLK fit (6) to their fluorescence curves with both D and γ as adjustable parameters, and obtain D = 2.2 ± 0.2 independently of the acceptor concentration. Even at the very lowest concentration, for which they estimate the length range (8) to go up to 190 Å,[13] they still find D = 2.14 ± 0.03. In terms of the sphere model, this means that the sphere and pore diameter would have to exceed 95 Å which does not agree with the estimates in Table 2. Moreover, the γ values obtained by DLK show a concentration dependence vastly different from what the sphere model predicts.[14] With so many internal and external problems in DLK's results, we find it hard to accept their energy-transfer analysis as definitive.

4. THE SCATTERING MODEL

The analysis in the previous section shows that the fractal model, with $D \approx 3$ over a length range of 4 - 100 Å, provides a very consistent description of the surface structure of Si-60 and is, in fact, the only consistent description that has been put forth so far. We are now going to show, by means of an explicit model calculation, that this description is also consistent with the small-angle X-ray scattering data for Si-60.

The experimental scattering curve for our sample is reproduced in Fig. 4. It agrees with the scattering curve for Si-60 measured by Sinha and DLK[12,15,16] (their curve extends to even smaller angles). We write the scattered intensity I as function of the magnitude q of the scattered wave vector, $q = 4\pi\lambda^{-1} \sin(\theta/2)$, where λ is the X-ray wave length and θ is the scattering angle. The two basic features of the experimental curve are the power law

$$I(q) \propto q^{-4} \tag{10}$$

for 0.08 Å$^{-1} < q < 0.4$ Å^{-1}; and the plateau, or weak local maximum, at $q \approx 0.03$ Å^{-1}. The curve continues to rise at smaller q values[12] so that the plateau may be thought of as a washed-out peak superimposed on a power law. The power law (10) is called Porod law, which states that a scatterer with smooth surface gives rise to (10) when $2\pi/q$ varies in the length-scale regime at which surface is smooth.[29] For a scatterer with fractal surface of di-

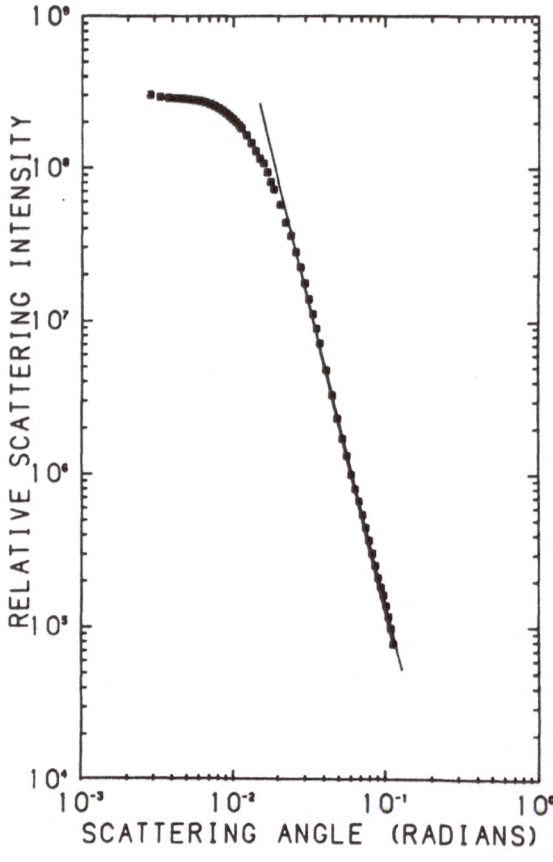

Fig. 4. Small-angle X-ray scattering curve for Si-60 (from Ref. 6). The wave length is 1.54 Å. At these small angles, 0.1 rad corresponds to a q value of 0.408 Å^{-1}. The solid line is the least-square fit with slope -3.95 ± 0.06, giving the power law (10).

mension D, on the other hand, the scattered intensity obeys the power law[30-32]

$$I(q) \propto q^{D-6} \tag{11}$$

when $2\pi/q$ varies in the fractal regime. For a smooth surface, $D = 2$, the fractal law (11) leads back to the Porod law as it should. Thus the problem to be solved is: if the surface of Si-60 has $D \approx 3$, why does it scatter as q^{-4} and not as q^{-3} ?

The problem is nontrivial because the power law (11), being the leading term of an asymptotic expansion for large q ($q \gg L^{-1}$ where L is the scatterer diameter), may have a vanishing prefactor in the limit $D \rightarrow 3$.[30] In this case, higher-order terms become the leading ones and mask the three-dimensional behavior. This scenario was assumed in Ref. 6, together with the assumption that the next term in (11) is of Porod form, q^{-4}. But the limit $D \rightarrow 3$ may also be taken in such a way that the prefactor in (11) does not vanish.[31] The two scenarios correspond to the two ways in which a three-dimensional surface can be space-filling, namely uniformly or non-uniformly (Section 2). That a non-uniformly space-filling surface gives a non-vanishing prefactor in (11) can easily be seen from Fig. 1. The presence of pores of all sizes gives rise to all the fluctuations in the electron density needed for a nonzero term q^{-3}. This is the case at issue here.

Our model is follows. It assumes Si-60 to consist of particles (clusters) with diameter L, each with a surface area distribution given by Eq. (1b) with $D = 3$, $u_1 = 4$ Å, and $u_2 = 100$ Å. The diameter of a particle is defined as the diameter of the smallest circumscribing sphere. The particles are assumed to form a random-packed assembly (interconnected network) described by the pair distribution function of a hard-sphere fluid with sphere diameter L and packing fraction η (both L and η are unspecified as yet). The pair distribution function gives the probability of finding a second particle at a given distance from some reference particle. The motivation for the model is that the "peak" at $q \approx q_0 = 0.03$ Å^{-1} in the scattering curve indicates the presence of structural units with a definite length scale. The units are the particles of diameter L. The existence of a definite length scale (related to L as we shall see) is also suggested by the peak at 70 Å in the pore-diameter distribution for Si-60,[12,14,15] if one trusts that the cylindrical-pore model underlying DLK's distribution gives a reasonable estimate of the dominant pore diameter. The least biased model then is to assume that the particles are distributed randomly, subject to the condition that they do not overlap and form a connected network of some density. This is precisely what the hard-sphere fluid model does. A similar model has been used previously for small-angle scattering from Vycor glass and related systems,[33] except that there the structure of the particles was very different.

As is well-known, the scattered intensity in scattering from liquids is the product of the structure factor of the liquid (Fourier transform of the pair distribution function) and the atomic form factor (scattered intensity from a single atom).[34] In the same way, the intensity in our model is given by the product

$$I(q) = N_0 S(q) I_0(q) \tag{12}$$

where N_0 is the number of particles in the system, $S(q)$ is the structure factor of the hard-sphere fluid, and $I_0(q)$ is the scattered intensity from a single particle. Thus $I_0(q)$ is the analog of the atomic form factor in liquid scattering. However, it refers to a particle with a highly irregular surface. In the Percus-Yevick approximation, $S(q)$ has the analytic form[35]

$$S(q) = \left\{ 1 - 24\eta \, (qL)^{-3} \sum_{i=1}^{3} \alpha_i \, f_i(qL) \right\}^{-1} \tag{13}$$

$$\alpha_1 = -(1 + 2\eta)^2 \, (1 - \eta)^{-4} \tag{14a}$$
$$\alpha_2 = 6\eta \, (1 + \eta/2)^2 \, (1 - \eta)^{-4} \tag{14b}$$
$$\alpha_3 = (\eta/2) \, \alpha_1 \tag{14c}$$

225

$$f_1(y) = \sin y - y \cos y \tag{15a}$$

$$f_2(y) = y^{-1}\{-2 + 2y \sin y + (2 - y^2) \cos y\} \tag{15b}$$

$$f_3(y) = y^{-3}\{24 + (-24y + 4y^3) \sin y + (-24 + 12y^2 - y^4) \cos y\}. \tag{15c}$$

It has the familiar peaks of the liquid structure factor (short-range order), with higher-order maxima becoming rapidly weak in intensity. The position and intensity of the peaks depend on L and η through Eqs. (13,14). The structure factor (13) does not depend on temperature, however, even though it derives from the canonical ensemble at nonzero temperature. This is a pecularity of the Percus-Yevick approximation as applied to the hard-sphere model. The exact structure factor for the hard-sphere system does depend on temperature (crystalline state at zero temperature); likewise the Percus-Yevick approximation does give a temperature-dependent pair distribution function for interactions other than the hard-sphere potential. For our present purposes, this temperature independence is welcome because it gives a model for S(q) with only two instead of three parameters.

The intensity $I_0(q)$ for a single particle can be calculated from the general framework developed in Ref. 32. There it was shown that for arbitrary scatterer shape the intensity, after spherical averaging, is given by

$$I_0(q) = 4\pi I_{el} n^2 \Omega \int_0^\infty \left\{ 1 - (2\Omega r)^{-1} \int_0^r v(r')dr' + O(rv(r)) \right\} \frac{\sin qr}{qr} r^2 dr, \tag{16}$$

where I_{el} is the scattered intensity per electron, n is the electron density of the scatterer (constant), Ω is the volume of the scatterer, and v(r) is the volume of all points inside the scatterer whose distance from the surface is less than or equal to r. The expression {...} as a function of r is the correlation function of the scatterer and says that the leading contribution, for small r, is entirely determined by the volume v(r). By the definition of our model for the particle, the apparent surface area A(u) satisfies

$$A(u) = A(u_1) \begin{cases} 1 & \text{for } 0 < u \le u_1 \\ u_1/u & \text{for } u_1 < u < u_2 \\ u_1/u_2 & \text{for } u_2 \le u < \infty \end{cases} \tag{17}$$

where we formally assume that the surface is smooth at length scales below u_1 and above u_2. Thus the behavior (1b) with D = 3 is described by the part $A(u_1)u_1/u$ in (17). Note that here A(u) refers to the surface area, as measured by tiling with molecules of size u, of a single particle rather than of the entire Si-60 system. The volume v(r) needed for (16) can be calculated from (17) by

$$v(r) = \int_0^r A(u)du = A(u_1)u_1 \begin{cases} r/u_1 & \text{for } 0 \le r \le u_1 \\ 1 + \ln(r/u_1) & \text{for } u_1 < r < u_2 \\ \ln(u_2/u_1) + r/u_2 & \text{for } u_2 \le r < \infty. \end{cases} \tag{18}$$

In fact, in a mathematically rigorous definition of the area A(u), the function A(u) is equal to the derivative, with respect to u, of the volume of all points outside the solid whose distance from the surface is less than or equal to u.[21] Thus all that we have assumed in (18) is that the volume within distance u from the surface is the same on the outside of the solid as on the inside. The integration of (18) and the subsequent Fourier transform (16) are somewhat tedious but straightforward to perform. In order to get convergence when the the higher-order term O(rv(r)) in (16) is neglected, one introduces an extra factor $e^{-\varepsilon r}$ in (16) and lets $\varepsilon \to 0$ in the end. The final result is

$$I_0(q) = 2\pi I_{el} n^2 A(u_1) u_1 q^{-3} \left\{ \frac{1 - \cos(qu_1)}{qu_1} + si(qu_2) - si(qu_1) + \frac{\cos(qu_2)}{qu_2} \right\} \qquad (19)$$

where $si(.)$ is the sine integral.[36] As expected from Eq. (11) and $D = 3$, the intensity (19) varies as q^{-3} in the fractal regime $u_1 < 2\pi/q < u_2$. The dominant term in the fractal regime is $-si(qu_1)$ and gives $\{...\} = \pi/2 + O(qu_1)$. Similarly, it easy to verify that (19) reduces to the Porod law (10) when $2\pi/q$ varies below u_1 and above u_2 where the surface is smooth. The numerical evaluation of (19) gives a monotonically decreasing function, showing that the oscillatory terms in $\{...\}$ are all suppressed by the power-law behavior.

Eqs. (12-15) and (19) give the complete scattering behavior of the model. For the packing fraction in $S(q)$, we choose $\eta = 0.70$. This is considerably higher than the value of 0.33 in DLK's sphere model (Table 2). Qualitatively, the argument for a higher η is that much of the porosity of 0.67 measured by DLK comes from micro- and mesopores in the fractal regime of each of the particles (intraparticle porosity) rather than from interstitial pores between the particles (interparticle porosity). So the interparticle porosity must be much lower than 0.67, whence a packing fraction much higher than 0.33. Quantitatively, the value $\eta = 0.70$ is obtained by choosing η so that the average interparticle pore diameter agrees with the average pore diameter of 70 Å measured by DLK. Indeed, having chosen η, we fix L so that the position of the first peak in the structure factor $S(q)$ agrees with the position q_0 of the peak in the experimental scattering curve. This gives $L = 8.0/q_0 = 267$ Å for $\eta = 0.70$. From Eqs. (2,3) one

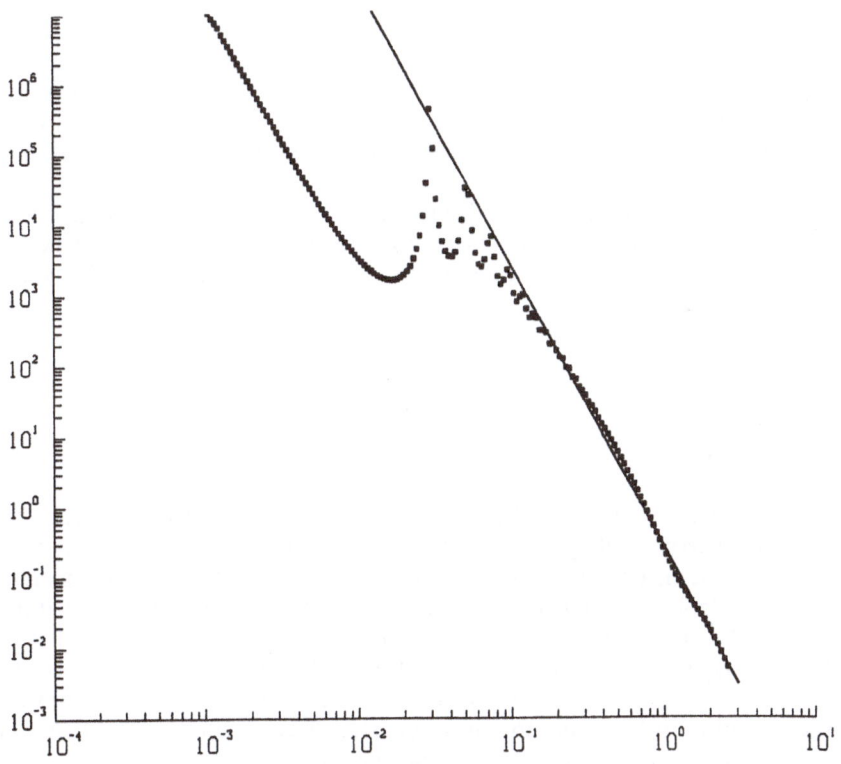

Fig. 5. Calculated scattering curve from Eqs. (12-15, 19) with $\eta = 0.70$, $L = 267$ Å, $u_1 = 4$ Å, and $u_2 = 100$ Å. The horizontal axis is q in units of Å^{-1} and the vertical axis is the intensity in arbitrary units. The solid line is the power law $I(q) \propto q^{-4}$ fitted to the calculated curve.

then gets $z \approx 10$ and $2\rho \approx 70$ Å as desired. The total scattered intensity resulting from these parameter values is shown in Fig. 5.

At first sight, the resemblance between the calculated and the experimental curve may seem poor. The calculated curve has a sharp peak at $q = 0.03$ Å^{-1} which originates from the first peak in the Percus-Yevick structure factor. By contrast, the peak at $q = 0.03$ Å^{-1} in the experimental curve is broad and plateau-like. The calculated curve has subsidiary maxima to the right of the main peak (originating from the higher-order maxima in the structure factor) for which there exists no counterpart in the experimental curve. However, if one ignores these differences for a moment and considers the maxima as representative of the calculated intensity, the outcome in Fig. 5 is remarkable. The scattered intensity follows a nearly perfect Porod law, Eq. (10), over the entire fractal regime. Thus the model does not give a q^{-3} behavior, as might have been anticipated from Eq. (19), but a q^{-4} behavior as observed experimentally. The reason is simple. The structure factor (13) is constant at very large q values and develops oscillations of increasing amplitude as q decreases. When multiplied with $I_0(q)$, these oscillations make the total intensity grow faster than q^{-3} for decreasing q, resulting in an effective q^{-4} behavior. We note that such a change of exponent appears to be be specific of the case where the single-particle intensity $I_0(q)$ grows as q^{-3}. No change of exponent occurs for $I_0(q) \propto q^{-D}$ with $D \approx 2$.[33] On the other hand, there exist experimental examples where a q^{-3} power law for the measured intensity has been observed.[37] So the change of exponent in the present model is due to the special form of both $I_0(q)$ and $S(q)$.

The model is clearly crude from the viewpoint of its assuming that all particles have the same fixed diameter L. A refined model would allow for some distribution of particle diameters, which in very good approximation can be incorporated by averaging Eq. (12) over the diameter distribution. This would broaden the main peak and wash out all subsidiary maxima in the calculated intensity,[38-39] and hence would make Fig. 5 resemble the experimental scattering curve much more closely. We have deliberately not introduced such refinements in order to exhibit the "masking mechanism," i.e. the mechanism by which the q^{-3} scattering law for a fractal three-dimensional surface may be transformed into a q^{-4} behavior, in its simplest form.

By construction, the model is consistent with all the experimental data discussed in Section 3. This then completes the final test of the fractal model for Si-60, namely that it is consistent also with the small-angle scattering data. A lesson important much beyond Si-60 is that the model gives a counterexample to the "Porod paradigm" which assumes the power law (10) to be a proof of surface smoothness.

5. CONCLUSION

Motivated by a series of papers by Drake, Levitz, and Klafter we have presented an in-depth analysis of two competing, inequivalent models for the structure of porous silica Si-60. We have shown that the random-packed sphere model proposed by those authors is inconsistent with the available experimental data across the board. We have shown that the fractal model, on the other hand, provides a fully coherent account of all data. The resulting estimate for the fractal dimension of the surface is $D = 3.0 \pm 0.2$ for the length-scale interval 4 - 100 Å. The model asserts that the three-dimensional nature of the surface is of non-uniform type. We find no justification for referring to our analysis as "fractal dilemma."[15] If one cares for dilemmas, one can find them in the papers of those authors.

Acknowledgement is made to the donors of the Petroleum Research Fund, administered by the American Chemical Society; to the US-Israel Binational Foundation; to the Wolfson Foundation; and to the Valatzi-Pikovski Foundation for partial support of this work.

REFERENCES

1. D. Avnir, ed., "The Fractal Approach to Heterogeneous Chemistry," Wiley, Chichester (1989).
2. A.J. Hurd, ed., "Fractals, Selected Reprints," American Association of Physics Teachers, College Park, MD (1989).
3. M. Graetzel and J. Weber, eds., "Fractal Structures, Fundamentals and Applications in Chemistry," *New J. Chem.* **14**, 185-254 (1990).
4. D. Avnir and P. Pfeifer, *New J. Chem.* **7**, 71 (1983).
5. D. Farin, A. Volpert, and D. Avnir, *J. Am. Chem. Soc.* **107**, 3368 (1985).
6. D. Rojanski, D. Huppert, H.D. Bale, X. Dacai, P.W. Schmidt, D. Farin, A. Seri-Levy, and D. Avnir, *Phys. Rev. Lett.* **56**, 2505 (1986).
7. D. Pines-Rojanski, D. Huppert, and D. Avnir, *Chem. Phys. Lett.* **139**, 109 (1987).
8. D. Pines, D. Huppert, and D. Avnir, *J. Chem. Phys.* **89**, 1177 (1988).
9. P.W. Schmidt, A. Höhr, H.B. Neumann, H. Kaiser, D. Avnir, and J.S. Lin, *J. Chem. Phys.* **90**, 5016 (1989).
10. D. Pines and D. Huppert, *Isr. J. Chem.* **29**, 473 (1989).
11. D. Farin and D. Avnir, in: Ref. 1, pp. 271-293.
12. J.M. Drake, P. Levitz, and S. Sinha, in: "Better Ceramics Through Chemistry," C.J. Brinker, D.E. Clark, D.R. Ulrich, eds., *Mat. Res. Symp. Proc.* **73**, 305 (1986).
13. P. Levitz and J.M. Drake, *Phys. Rev. Lett.* **58**, 686 (1987).
14. P. Levitz, J.M. Drake, and J. Klafter, *J. Chem. Phys.* **89**, 5224 (1988).
15. J.M. Drake, P. Levitz, and J. Klafter, *New J. Chem.* **14**, 77 (1990).
16. J.M. Drake and J. Klafter, *Physics Today*, May 1990, p. 46.
17. J.D. Bernal, *Nature* **183**, 141 (1959).
18. J.M. Ziman, "Models of Disorder," Cambridge Univ. Press, Cambridge (1979), p. 77.
19. R. Zallen, "The Physics of Amorphous Solids," Wiley, New York (1983), p. 49.
20. D. Avnir, D. Farin, and P. Pfeifer, submitted.
21. P. Pfeifer and M. Obert, in: Ref. 1, pp. 11-43.
22. R.K. Iler, "The Chemistry of Silica," Wiley, New York (1979), Chap. 5.
23. S.V. Christensen and H. Topsøe, Haldor Topsøe Co., Denmark, did the alkane experiments. Personal communication (1987).
24. P. Pfeifer and D. Avnir, *J. Chem. Phys.* **79**, 3558 (1983).
25. D. Avnir, D. Farin, and P. Pfeifer, *J. Chem. Phys.* **79**, 3566 (1983).
26. A.Y. Meyer, *J. Comp. Chem.* **7**, 144 (1986).
27. P. Pfeifer, unpublished.
28. J. Klafter and A. Blumen, *J. Chem. Phys.* **80**, 875 (1984).
29. A. Guinier and G. Fournet, "Small-Angle Scattering of X-Rays," Wiley, New York (1955).
30. H.D. Bale and P.W. Schmidt, *Phys. Rev. Lett.* **53**, 596 (1984).
31. P. Pfeifer and P.W. Schmidt, *Phys. Rev. Lett.* **60**, 1345 (1988).
32. P. Pfeifer, *Springer Ser. Surf. Sci.* **10**, 283 (1988).
33. A. Höhr, H.B. Neumann, P.W. Schmidt, P. Pfeifer, and D. Avnir, *Phys. Rev. B* **38**, 1462 (1988).
34. See e.g. Ref. 18, p. 122.
35. N.W. Ashcroft and J. Lekner, *Phys. Rev.* **145**, 83 (1966).
36. M. Abramowitz and I.A. Stegun, "Handbook of Mathematical Functions," Dover, New York (1965), p. 232.
37. P.Z. Wong, J. Howard, and J.S. Lin, *Phys. Rev. Lett.* **57**, 637 (1986).
38. P.W. Schmidt and M. Kalliat, *J. Appl. Cryst.* **17**, 27 (1984).
39. W.L. Griffith, R. Triolo, and A.L. Compere, *Phys. Rev. A* **35**, 2200 (1987).

Aggregation Phenomena

M.Kolb

Laboratoire de Chimie Théorique

Ecole Normale Supérieure

46, allée d'Italie, 69364 Lyon Cedex 07

Introduction

Growth processes lead to many beautiful and intriguing structures in nature. Here we wish to describe methods to characterize and model such processes. Only recently has one begun to investigate growth phenomena in a systematic fashion. Theoretical investigations are difficult as the processes are far from equilibrium and frequently the structures are disordered. Recent progress came from two sides: 1) Scaling and the notion of fractals provide efficient ways to describe aggregation structures in a quantitative way - analogous to the scaling approach to critical phenomena - and 2) Large scale simulations of basic irreversible processes determine the presumably universal properties which are measured in experiments.

The first step in describing disordered systems consists of giving a geometrical characterization of the objects. As a second step one tries to establish a relationship between the structure and the dynamical process responsible for the growth. The principal experimental techniques in aggregation phenomena are electron microscopy and scattering experiments (light, X-rays and neutrons). The theoretical task is to determine which aspects are relevant for for a specific structure to appear, and which aspects can be ignored. Modeling, paired with knowledge drawn from from the study of equilibrium phase transitions, have permitted to classify many aggregating systems.

Here, a short survey is given of different types of aggregation mechanisms and their experimental realizations. Examples from physics, chemistry and biology will be cited. The list of phenomena and theories presented is by no means exhaustive; for details the reader will be referred to the literature listed below. A growing number of comprehensive reviews and books have appeared in the past years[1-27].

First, basic experimental results of aggregation phenomena will be summarized. They include irreversible colloidal

Large-Scale Molecular Systems, Edited by W. Gans *et al.*
Plenum Press, New York, 1991

aggregation and gelation phenomena, macroscopic aggregation, electrodeposition, dielectric breakdown, and properties of porous media. More esoteric examples are cloud formation and the size and distribution of galactic dust particles.

A brief discussion is given of the notion of a fractal. This concept has been (re-)discovered in the last few years in many different fields. A few examples will be cited stressing the most fundamental property of a fractal: its dilation symmetry. This property appears most clearly in the recursive scheme of mathematical examples. Other properties include selfaffinity, renormalization and anomalous physical properties of fractals. Experimentally, it is often the unusual physical properties that evidence the underlying fractal structure. A more general way to characterize a fractal is in terms of multifractal spectra.

Some of the basic irreversible growth processes will be introduced among the many models that have been conceived to describe the different experimental situations. The role of the irreversibility for the growth processes and the relation to equilibrium systems is discussed.

Finally, two recent calculations describing structure formation mechanisms will be sketched: 1) Irreversible shear induced aggregation and 2) Interpretation of structures developing during late stage spinodal decomposition in terms of percolation theory.

Experimental realizations of aggregation processes

Disordered growth processes surround us everywhere. It is easy to appreciate the beauty of a tree or the impressive patterns of lightning, but it is quite difficult to accurately describe these objects.

An early attempt to classify the geometry of natural phenomena has been given by D'Arcy Thompson in 'On growth and form'[1]. This work compiles a wealth of information on different structures observed in nature and advances explanations for the origin of the different shapes encountered. However, it only considers 'weakly' disordered structures. Strongly disordered irreversible systems have been studied quantitatively in the last decade; these studies rely heavily on numerical simulations.

Aggregation and growth as we understand it here can for the most part be thought of as an irreversible process between particles or aggregates of particles. Nevertheless, the range of application is not limited to particle aggregation processes because formal analogies permit one to use the same models in different contexts. Two fundamentally different aggregation processes will be distinguished: the cluster-cluster processes and the particle-cluster processes. The cluster-cluster processes are growth processes where objects of comparable size aggregate, they are hierarchical: two clusters of size M form a cluster of size 2M, then two such objects generate a 4M cluster, etc. The second class of processes is asymmetric: one cluster of size M>>1 grows by accreting individual particles of size 1. The size of the cluster and the size of the particle being different, there are two scales in this process. The cluster-cluster process can easily be renormalized (from 2M→4M to M→2M), but this is not possible in an analogous way for the particle-cluster process. Therefore the two types of growth have different properties.

There are several prototype models for both types of growth mechanisms. These models are rather simplified. The hope is that the scaling and fractal properties of more realistic models are unchanged, analogous to the properties of the Ising model for ferromagnetism.

The basic cluster-cluster aggregation process is observed in many experiments of colloidal aggregation. Metallic aggregates have been analyzed in the literature for well over twenty years. Many aspects of the aggregation process have been studied in the 60's, for example the conditions under which a protective agent prevents aggregation[28]. However, the geometry of the structures was not considered.

Micrograph pictures of aerosols lead to similar patterns, suggesting that the underlying mechanism is the same. The images of U_3O_8 aggregated in an explosion chamber[29] shows disordered, stringy objects. Again, the purpose of this and related early studies was to investigate under what conditions aggregation takes place.

Inspired by the theory of critical phenomena of the 70's an interpretation in terms of scaling laws was proposed for aggregated metal particles (i.e. iron)[30]. The analysis of the density-density correlation function revealed that such aggregation processes might be a kinetic version of critical phenomena. This experiment found that the density correlation function $C(r)$ falls off as r^{-A} with the same exponent $A \cong 0.28$ for several different metals suggesting that A is universal, (depending on the mechanism, but not on material details).

The visual similarity between aerosol and colloid aggregation was confirmed quantitatively by accurate measurements on gold colloids[31]. The experiments were done by direct analysis of transmission electron micrographs. The conditions in this system are particularly favorable as gold aggregates grow to be very large and yield micrographs with a high contrast. In general, the limit of the aggregation process is reached when the clusters reach a size such that they restructure , break or sediment - the last case limits the colloidal gold experiment. The elementary particles of this process are colloids of about 7.5 nm in diameter which aggregate to form clusters of up to μ-size. The analysis of the micrograph shows that this is a process of the cluster-cluster type: all aggregates are roughly of the same size. The aggregation conditions in the aequous solution are such that the clusters grow rapidly, implying that aggregates stick instantaneously whenever they touch each other. The analysis of the mass M as a function of the radius R indicates a power law behaviour. The relation $M \sim R^D$ is found to hold with $D \cong 1.75$. The non standard mass-volume relation is the signature of a fractal. In support of the universal fractal interpretation is the consistency of the exponents D of the colloids and the exponent A measured for the aerosols. For isotropic and homogeneous fractals they are expected to satisfy $A+D=2$. The variation of the (average) cluster mass with time also follows a power law.

The aggregation of gold colloids can be modified by varying the number of neutralized surface charges. This changes the probability that two clusters stick on contact. If the charges are only partially neutralized the aggregation process takes several hours and the structures that develop are more compact[32]. The exponent D increases to $D \cong 2.10$. Now the reactivity of the chemical bonds determines the growth in contrast to the previous case where the diffusion dominated the process. In addition, the cluster size distribution changes: for fast aggregation it is bell shaped with a well defined maximum (which grows with time) whereas for slow aggregation it decreases monotonically. The temporal evolution of the cluster size for the slow process is better described by an exponential rather than by a power law.

There are other methods to demonstrate the selfsimilar fractal aspect of colloidal aggregates. The direct method used for the gold colloids provides a very detailed analysis, but is restricted to two dimensions. For ramified structures with D<2 this works well but for fractals with D>2 one cannot determine D from projections. In this case scattering experiments (light, x-rays or neutrons) are well suited to determine the fractal properties. For silica aggregates this technique has been used extensively[33]. The scattering intensity I as a function of the wave vector q follows a power law relation for a fractal, $I \sim q^{-D}$, in the 1/q range between the particle size and the cluster size. Data in double logarithmic plots at different stages of the aggregation process show that the range of the power law gradually increases with time, indicating that the aggregates grow bigger and the fractal range increases.

Silica chemistry is extremely complicated and may produce a rich variety of phenomena depending on the conditions of preparation and aggregation. So far growth processes have been studied at low initial particle concentrations where no gelation takes place. Increasing the concentration will change the aggregation mechanism and necessarily leads to structures which are no longer fractal. However they may have a fractally rough surface. This behaviour also shows up in scattering experiments[34]. Scattering off a fractally rough surface leads to power law scaling of the form $I \sim q^{6-D_s}$, where $2<D_s<3$ is the fractal exponent of the surface. The distinction between volume and surface fractals can in principle be identified from the slope in I vs. q plots: it is less than three for volume fractals and larger than three for surface fractals. The silica structures in these aggregation experiments are more or less rigid. Under certain conditions and over long periods of time they are observed to restructure. As a consequence the fractal properties measured in scattering experiments change.

It is possible to study two dimensional aggregation by confining the silica to an interface[35]. The clusters are far more ramified in two dimensions than in three dimensions – the fractal exponent is smaller. Long range electrostatic forces probably influence this aggregation process.

Another class of experiments is aggregation of polymer latex spheres under rapid aggregation conditions[36,37]. Optical measurements using a flow cell accurately determine the number of clusters of a given (small) size. This method allows for much better statistics than direct microscopy. The data suggest that fractal aggregation can be modeled by a kinetic equation of the Smoluchowski type, provided the kernel is modified to account for the fractal nature of the aggregates.

Red blood cells have a tendency to form rouleaux when their flow is slowed or stopped. The presence of these rouleaux are known to influence the physiological properties of the blood. In contrast to previous growth processes, this system is at equilibrium. Kinetic equations with appropriate rate constants for the formation and branching are able to model the average number of cells in a rouleau and the number of branching points[38].

Calibrated polystyrene spheres can aggregate and then crystallize during restructuring[39]. Two dimensional aggregation is induced by an ac field between two charged plates. The resulting structures are ordered at short distances and are fractal globally. The measured exponents are those of diffusive aggregation at high particle density. The crystalline order can be destroyed by increasing the driving frequency. This melting occurs via a hexatic phase.

Macroscopic aggregation experiments show the subtle interplay between random and hydrodynamic effects. A realization of such a process is an assembly of randomly distributed wax balls on a water surface, under a shearing motion[40]. The balls aggregate to form ramified clusters in two dimensions. Depending on the shear velocity several different regimes have been observed.

Aggregates with an open fluffy structure appear to be an important component of the interplanetary dust cloud. There is evidence that such particles influence the properties of the zodiacal light[41].

Let us now consider the second group of aggregation processes: particle-cluster aggregation. The most direct application (though it was not the first discovered) is electrodeposition. It can, under certain conditions, be modeled directly by diffusion limited particle aggregation. A commonly observed example are the fragile deposits developing on the terminals of a car battery. Electrodeposition of zink from an electrolyte onto a point-like electrode yields ramified structures whose density correlation function was determined by digitalizing the growing pattern[42]. In order to see the fractal structures it is necessary that the current is sufficiently small such that the growth is controlled by a diffusion process. The density correlation function clearly shows a power law behaviour. Larger currents lead to drift and to more compact objects.

There are several situations where the growth is described by the same equation even though no particles aggregate. A beautiful example governed by the same law is fluid-fluid displacement[43]. Suppose a fluid with a low viscosity displaces another fluid with a very high viscosity. If the surface tension can be neglected, structures similar to those in electrodeposition will appear as the basic growth equations are the same. The growing structure is sensitive to the properties of the fluids. Using less miscible fluids increases the surface tension, resulting in different patterns. The pressure is another parameter used to control the type of growth in fluid instabilities. The transition from dendritic growth, a regular but unstable growth process, to noisy aggregation of the DLA type has been observed and shown to be dependent on anisotropy[44].

The range of phenomena can be extended by including snowflakes, which are more symmetric patterns than dendrites. Thousands of different examples exist which are all perfectly regular, but with details which depend on the thermodynamics in a subtle way[45].

A different example of a pattern which forms according to the same equation is the displacement of a fluid through a porous medium. For the invasion of a system of ducts by an immiscible fluid the emerging pattern depends on the flow rate. At low flow rate the pattern is random (invasion percolation), at higher flow rate it becomes Laplacian (diffusion limited)[46]. The fractal exponents for the two cases are different. Patterns of the same type arise naturally in porous media. Metal oxides enclosed in shale or limestone clearly show a similarly irregular structure[6].

Another class of phenomena where fractals are observed is porous rock. The growth mechanism is not clearly identified. The properties are extracted experimentally by coating the materials with monolayers of different sized molecules, or by scattering experiments[47]. In the best examples scaling is observed over two orders of magnitude in length.

A beautiful phenomenon closely related to the particle-cluster aggregation model is dielectric breakdown. This problem has been studied over 200 years ago by Lichtenberg[48]. A similar pattern was investigated more recently for its fractal properties. Two dimensional discharge patterns (lightning) have a structure typical of fractal growth[49]. The analysis of the branching structure of the breakdown pattern confirms that this process is also described by the Laplacian growth model. The process of dielectric breakdown is microscopically very complicated. For the present description one assumes that a breakdown occurs with a probability proportional to the local electric field. More general dependencies may be considered which result in different fractal patterns. The experimental results are consistent with the simplest hypothesis.

There are other irreversible growth processes that lead to fractal structures. Thin metal film deposits in the context of a metal-insulator transition generate clusters of the type random percolation[50]. The portion of the surface that determines the conductivity corresponds to the backbone of the percolation cluster.

Biological growth has been modeled by particle-cluster growth. A fungus can have an astounding variety of surface shapes depending on its cell structure and on nutrient conditions[51]. The observed patterns are not fractal, but have an irregular surface structure with a possibly nontrivial scaling behaviour.

Another example of fractal structure formation are the shapes of clouds. The fluid mechanical description of this phenomenon is extremely complicated. Empirically scaling behaviour is observed over strikingly wide range of a few to over 10^3 km in perimeter[52]. A log-log plot of the area vs. the perimeter follows a straight line with an exponent 4/3.

In all the examples presented the structures show a fractal behaviour over a finite range. Usually there are natural bounds to the scaling regime, as for example the particle size as a lower cutoff and the mechanical stability of the clusters as an upper cutoff. To determine the fractal properties reliably the latter must be substantially larger than the former.

Fractals

There are many different aspects to the notion of fractals[12,13]. In its simplest form a fractal describes geometrical scale invariance. It can also be viewed as nontrivial dimensional analysis and as such has been used for a long time. There are many examples for nontrivial scaling. Consider the metabolic rate of different animals: the heat produced by a body as a function of the body weight. A simple argument would assert that the metabolic rate B is proportional to the surface area A and hence proportional to $W^{2/3}$, where W denotes the weight. Experimental data for different animals shows that B vs. W indeed follows a power law (a straight line in a log-log plot), but with a different exponent[53]. Another example from the animal world is the oxygen conductance and the pore length (=shell thickness) of birds' eggs as a function of the egg weight. These parameter are relevant to the breathing of the eggs during incubation. Both quantities follow a power law that can be estimated from simple geometrical considerations. Naively one would expect that the pore length scales as $W^{1/3}$ and the conductance as the total pore area (=number of pores times area per pore) divided by the pore length, or $(W^{2/3})^2/W^{1/3}=W$. For this case the experimental data is in reasonable agreement with the prediction of dimensional analysis.

Scaling expresses a symmetry property. Analogous to translational resp. rotational invariance, the dimensional analysis express invariance under dilation symmetry. The utility of scaling theory is in showing deviations from simple dimensional analysis, indicating more subtle symmetries. Fractals capture the essential aspects of such symmetry properties in a geometrically transparent way. Their power is well illustrated in the algorithms used to generate trees[15]: even the simplest rules yield remarkably realistic patterns.

Classical examples of fractals are hierarchical models such as the Koch curve, proposed by H. von Koch in 1904 to describe the shape of snowflakes. The construction is as follows: An equilateral triangle is refined by replacing each side of length $a_0=1$ by four segments of length $a_1=1/3$. Then such a side is itself replaced by four segments of length $a_2=1/9$, etc. By continuing this process ad infinitum a continuous curve is obtained which is nowhere differentiable. It is bounded in space but its length is infinite. At each step, the side length decreases by 1/3 and their number increases by a factor of 4. The total length tends to infinity as $L_n=3(4/3)^n$ as n, the number of iterations, goes to infinity. The fractal dimension D in the relation $L_n \sim a_n^{-D}$ yields $D=\log4/\log3 \cong 1.26$. Linear objects satisfy this relation with D=1 and two-dimensional objects with D=2. The Koch snowflake defines an object of dimension D>1 starting from a linear object. One can do the opposite by decimating a two-dimensional object to make it fractal. A well known example is the Sierpinski gadget. Here one starts from the filled equilateral triangle of side length $a_0=1$. It can be divided naturally in four triangles of side length 1/2. The center triangle of side length 1/2 is removed. The next iteration consists in removing the center triangle of each of the remaining three triangles of side length $a_1=1/2$, etc. The area $A \sim (3/4)^n$ expressed in terms of $a_n=(1/2)^n$ decreases as $A_n \sim a_n^{-D}$ where $D=\log3/\log2 \cong 1.58$. For this case the area becomes infinitesimally small in contrast to the length of the curve becoming infinite in the previous example. These examples generate connected objects in a two-dimensional euclidean space and hence their fractal dimensions necessarily lie in the range $1 \leq D \leq 2$.

Dusts of particles need not be connected, and their distribution in space may be fractal, as illustrated by the cantor set. This construction consists of removing the center third of the line (0,1) at step 1 and the center third of the remaining intervals at subsequent steps. The fractal dimension of the mass ($\sim 2^{-n}$) vs. the interval length ($\sim 3^{-n}$) is $D=\log2/\log3 \cong 0.63<1$. This construction is embedded in a one dimensional euclidean space.

In order to describe in the simplest fashion the fractal clusters that are observed in aggregation phenomena, the following algorithm has been proposed[21]. Start with a seed particle at step 0. At step 1 replace the seed by an ensemble of five particles forming a cross (the seed plus its four nearest neighbors on a square lattice). On a hypercubic lattice in d dimensions the analogous rule replaces the particle by 2d+1 particles. The next step consists in replacing each of the five particles by a cross, etc. These clusters have a fractal dimension close to the one found for the diffusion limited cluster aggregation process in two and three dimensions. For d=2 the mass increases by a factor 5, the linear size of the cluster by a factor 3, at each iteration. The mass vs. radius relation defines D, $M \sim R^D$, with $D=\log5/\log3 \cong 1.40$. In d dimensions one finds $D=\log(2d+1)/\log3$. The major difference with real aggregation clusters is that these clusters are totally deterministic.

A prominent example of a random fractal is the path of a brownian particle. The end-to-end distance r obeys the law $<r^2> \sim N$ where N is the number of steps and an average over different realizations has been performed. This leads to $N \sim r^D$ with D=2. Hence a random walk fills space for d=2 and is fractal in d>2. Interestingly, the hull of a random walk in two dimensions is a curve with a fractal dimension $D=4/3$[13]. Such constructions resemble coastlines on an aerial photograph. A related random fractal describing the statistics of linear polymers is the self avoiding walk. The constraint of self avoidance swells the trajectory in a way that the fractal

238

dimension lowers from D=2 to D=4/3 in two dimensions and from D=2 to D≡5/3 in three dimensions. Above four dimensions D is unchanged; the self avoidance becomes irrelevant.

The iterative fractal constructions presented are the inverse of the renormalization group - instead of decimating and rescaling one decorates in a scale invariant way. Conversely the renormalization group can be applied to calculate the properties of fractal aggregates.

Complicated fractal structures can also been generated by very simple iterations in the complex plane. A well studied example is the iteration of the function $f(z)=z^2+c$, where c is a constant[13]. Starting with a seed z_0 one forms $z_1=f(z_0)$, $z_2=f(z_1)$, ... The points z_0 which for a given large n satisfy $|z_n|<2$ form a complicated fractal structure. Instead of asking which points z_0 remain bounded upon iteration for a fixed parameter c one may ask for which values of c does the mapping remain bounded for $z_0=0$. The resulting picture is also of extraordinary complexity.

For fractals encountered in physical situations there are a number of different methods to calculate D. Even though they are not rigorously equivalent, they usually give consistent results. Three methods prove to be particularly useful. The first consists of placing boxes of increasing size L around a particle of the aggregate. The fractal exponent then derives from the $M\sim L^D$ relation as long as L does not exceed the radius R of the aggregate. The second method is to calculate the density correlations, the probability that a site at a distance r from a particle of the aggregate is occupied; it behaves like r^{D-d} for $1\ll r\ll R$. Because of finite cluster size corrections this yields a measured D value that is usually higher than the one measured by the box counting method. In scattering experiments the measured intensity is the Fourier transform of the density correlation function. Another method is sometimes used to measure coastlines. In this method one uses rulers of variable length 1 and marches on the surface in steps of size 1. One deduces D from $N\sim l^{-D}$, where N is the number of times the ruler of length 1 has to be applied. This approach, although in principle similar to the box counting, does not always give the same result. For example, if there are bottle necks hiding large interior surface areas the fractal properties may be different. This is an important consideration for adsorption experiments[23]. Numerous rough surfaces (in porous materials) have been investigated in this way. Often they are surface fractals; viewed as a massive object they are non fractal but the surface has a nontrivial scaling behaviour.

In some cases fractals are not selfsimilar but nevertheless scale invariant.The next more complicated possibility is the notion of a self-affine fractal, which is scale invariant if it is rescaled differently in different directions. An example is a random walk trajectory as a function of space and time. In order to rescale it in an invariant fashion time has to be changed by b^2 while rescaling space by b.

How many independent exponents are required to characterize a physical fractal? It is obvious that the fractal dimension cannot determine a fractal uniquely - different constructions may yield fractals with the same D. Thus D is the most important, but not the only scaling index. There are other exponents related to physical properties as for example chemical reactions on fractal networks or dynamical properties of fractals.

A more general method to characterize a fractal is in terms of multifractal spectra. The utility of this approach has been demonstrated in dynamical systems as well as in aggregation phenomena. Consider the Cantor set studied above. Associate with the original interval (level 0) a probability unity. At level 1 distribute the probability evenly among the two intervals retained, $p=1/2$ each. At the next level the same procedure is repeated by dividing the probability for each interval among its sub-intervals. One is now interested in the averages $Z(q,\tau,l)=\Sigma p_i^q/l_i^\tau$ where the sum is over all intervals i, p_i and l_i are the measure resp. the length of interval i and τ is an exponent adjusted in such way that $Z(q,\tau,l)$ neither goes to zero nor diverges as the l_i go to zero. With other words one assumes that Σp_i^q scales as $l^{\tau(q)}$. Defining $\tau(q)=(q-1)D_q$ one makes contact with the fractal dimension previously defined, $D=D_0$, and defines more general fractal exponents for different q-values. For the uniform cantor set just defined, D_q can be calculated explicitly. Noting that $(Z(q,\tau,l))^2=Z(q,\tau,l)=1$ one finds that $D_q=\log2/\log3$ for all q. With other words all the powers of the measure scale with the same type of singularity and D_q does not depend on q.

There is another way of describing the scaling of $\tau(q)$[54]. Suppose that a given p_i^q behaves as $l_i^{\alpha q}$ where α is the singularity strength characteristic for a given interval. Furthermore assume that the number of intervals with a singularity in an interval $d\alpha$ at α be $l^{-f(\alpha)}d\alpha$. l is a typical (maximum) scale of the l_i. Then $\Sigma p_i^q=\int d\alpha\, l^{-f(\alpha)}\, l^{\alpha q}$. Using a saddle point approximation, the $\tau(q)$ of the moments can be related to the singularity spectrum $f(\alpha)$. The relation is $\alpha(q)=d\tau(q)/dq$ and $f(\alpha)=\tau(q)-q\alpha$. For the uniform Cantor set, there is a single singularity with $\alpha=D_0$ and $f(\alpha)=D_0$. Let us mention a more complicated example: a two scale cantor set. Instead of subdividing an interval in equal parts and distributing the probability evenly, the two intervals at each step shall have length $l_1=0.4$ and $l_2=0.25$ and the measures associated with it shall be $p_1=0.6$ and $p_2=0.4$. In this case the relation between τ and q are given by the equation $p_1^q/l_1^\tau+p_2^q/l_2^\tau=1$. Solving this equation (numerically) yields for D_q a q-dependance and for $f(\alpha)$ a bell shaped curve with its maximum the fractal dimension $D=D_0$. An analogous procedure to generate a fractal measure can be used in two dimensions. The generalized concept of multifractal spectra is useful in characterizing time series of chaotic systems and for describing aggregation clusters. It is particularly relevant to diffusive particle aggregation as the harmonic measure directly determines the growth of the cluster[20,21].

Basic Growth processes

Much of the progress in the field of aggregation phenomena in the past years is due to modelization of the different growth mechanisms by numerical simulations. The approach consists in reducing the complexity of physical aggregation phenomena to the minimum while conserving the physically relevant ingredients. Often this means considering idealized kinetics on lattices. Particle-cluster aggregation processes and cluster-cluster aggregation processes will be discussed.

We start with the particle-cluster models, where a single cluster grows by accumulating many individual particles. The simplest model is the Eden model, introduced to describe statistical properties of cellular growth[55]. Despite its simplicity, this model has many non-trivial features. It does not lead to fractal structures, but as it is far from equilibrium, its surface has a non-trivial scaling structure.

Let's define how one grows an Eden cluster. Place a seed particle on a square lattice. The particle has four nearest neighbours. Pick one of these neighbours at random and place a particle on it to form a dimer. Pick one of the six nearest neighbours of the dimer at random and place another particle on it. The process is now continued by adding new particles at the surface of the growing cluster, one at the time. The position of the new particle is selected at random among the available surface sites at the moment when it is deposited. The result of this iteration process is a connected cluster with an irregular surface.

Eden clusters are not only non-fractal, but totally dense: deep inside not a single site remains unoccupied. There are variants of this model with the same asymptotic properties but with varying rates of convergence towards the asymptotic shape. There is another type of modification that changes the interior density of the cluster but does not change the scaling aspects: growth is only allowed at the outer surface. This physically motivated change in the rules leaves holes in the structure: whenever a particle blocks a bay, it turns it into a lake. Attempts have been made to formulate and to solve this model analytically[56]. The theoretical predictions and the simulations of the surface properties agree well.

Certain aspects of the Eden model clearly show differences between irreversible kinetic processes and critical phenomena. One example is the role of the lattice on which the growth occurs. While lattice effects have been shown not to influence the scaling behaviour in equilibrium systems, large Eden clusters have shapes that are related to the underlying lattice. Notably a square lattice cluster which appears round at first will, for larger sizes, become diamond shaped. A similar effect, but even more pronounced, is observed in the DLA model discussed below. The Eden model can be formulated off-lattice in which case no anisotropy develops.

The most important particle-cluster aggregation model is the Witten-Sander model or diffusion limited aggregation (DLA)[57]. It has a rich and complex structure and is observed in numerous experimental situations, where noise and instability determine disordered structures. The model is defined as follows. Place a seed particle at the origin of a square lattice. Place another particle randomly at a distance R_0 from the seed on a lattice site. The second particle now performs a random walk on the lattice until it hits a site next to the seed. At that moment this particle stops and becomes part of the cluster. In order to prevent the particle from diffusing too far away a cutoff is placed at, say, $3R_0$: whenever the particle attempts to go farther it is removed and a new particle is placed at radius R_0. The process just described for particle number 2 is repeated indefinitely. The particles are diffused one at a time until they stick to the cluster. In practice, the radius R_0 is always chosen in such a way as to be larger than the actual cluster radius R. This guarantees that the particles diffuse from the outside towards the cluster and fully feel the influence of the screening due to the branches. DLA clusters are much more ramified than Eden clusters because the particles preferentially stick to the outer tips.

Analyzing mass versus radius, $R \sim M^{1/D}$ shows that D=1.7 in two dimensions, indicating a fractal structure. The same type of analysis was also performed on other lattices and in higher dimensions. The conclusion of all these simulations is that DLA clusters are selfsimilar fractals with a well defined fractal dimension that is independent of the lattice type but varies with dimension d: D=1.7, 2.5, 3.3, 4.2 and 5.3 for spatial dimensions d=2, 3, 4, 5 and 6. Note that there is a steady increase of D with d indicating that it never saturates. This is in contrast to equilibrium models like random walks or percolation where there is an upper critical dimension above which free field behaviour is observed with D=constant. The qualitative understanding is that for DLA the tips screen the newly diffusing particles from penetrating deep inside. If D were constant, for high d there would not be enough tips to effectively screen - leading to more compact clusters. Thus D must grow along with d. A more rigorous treatment indicates that D approaches d-1 as d goes to infinity. Theoretical attempts have tried to relate D to d by scaling arguments. A numerically successful formula is $D=(d^2+1)/(d+1)$; it has been derived in several different ways. Off-lattice simulations have been performed as well as for DLA. The trajectory of the diffuser, the cluster, or both can be defined in continuous space and yield fractal properties that are close but not identical to the lattice simulation results.

An interesting modification to DLA is the same model with a finite sticking probability. The growth rule is the same as above with the sole exception that the particle does not stick to the cluster the first time it touches it, but only sticks with a probability p<1 and continues to diffuse otherwise. During diffusion, the particle cannot, however, step on the cluster. For a given p, the particle gets stuck on the cluster after, on average, 1/p attempted collisions with it. Once it is stuck, it never detaches itself from it anymore. With other words there is no restructuring. In the limit when p goes to zero, all sites become equally accessible and the model reduces to the Eden model. For small but finite p there is a crossover from Eden-like to DLA-like behaviour. This can be understood by associating a length ξ with p. ξ is the distance a diffuser travels, on average, between the moment when it first touches the cluster until it sticks to it. This distance is (roughly) independent of the cluster size R. The modified DLA process then behaves Eden-like when R<<ξ and DLA-like when R>>ξ.

Another variant of DLA consists in replacing the diffusive path of the incoming particles by a straight-line path. The clusters grown this way are compact, possibly with logarithmic corrections. Again a crossover between ballistic and diffusive regimes occurs if the trajectories are diffusive but with a ballistically large step length. The ballistic mechanism prevails for small clusters and the diffusive mechanism dominates the late stages of the growth.

The peculiarities of the kinetic critical phenomena and the subtle structures associated with it become apparent when one considers DLA with anisotropic diffusion or with anisotropic sticking probabilities[10]. Diffusion on a lattice is anisotropic if a particle has a higher probability to jump north and south than to jump east and west. In order to describe this structure two different scaling indices are necessary to characterize the aggregate horizontally and vertically, respectively.

The physical importance of diffusion limited aggregation stems from a formal analogy. The model as described above is equivalent to a Laplacian growth pattern of the following form: suppose a cluster has grown to a certain size. To add the next particle iterate the following rules. First solve the Laplace equation $\Delta\phi=0$ subject to the boundary condition $\phi=0$ on the cluster and $\phi=$constant far from the cluster. Then select a surface site i with probability $p_i \sim \nabla_\perp \phi$. The first part is deterministic, the second part stochastic. Many of the applications are Laplacian problems perturbed by noise - as described in the second rule.

Lets now consider the second type of growth processes: cluster-cluster aggregation[58,59]. The basic model is the following: consider a box of side length L (imagine the process on a square lattice in two dimensions, generalizations are straightforward). Place a number N_0 of particles or monomers randomly in this box (for the time being the density N_0/L^2 is assumed to be low). Now let all the particles diffuse at the same time and independently of each other. Whenever two of them touch each other (nearest neighbours on the lattice) they stick to each other. From now on they diffuse as a dimer, but independently of the other clusters. The dimer can again collide with other monomer or dimers, forming larger clusters according to the same rules. The important point is that the clusters are formed irreversibly, they never break apart. Hence invariably larger and larger clusters develop. The clusters may perform a rotational diffusion as well. Besides the spatial dimension d and the initial concentration an additional parameter enters the model: the diffusion constant and in particular its dependence on the cluster size (usually a small cluster diffuses faster than a large one). This is accounted for by a parameter α in $D(M) \sim M^\alpha$, the diffusion constant or jumping probability of a cluster of mass M, per unit time. Negative α are physical for colloidal diffusion. Under these conditions the diffusive cluster-cluster aggregation or clustering of cluster process leads to fractal structures.

The fractal dimension of this process is lower than that for diffusive particle-cluster aggregation. This is a consequence of the fact that the process, for α negative, takes place in a hierarchical fashion. Roughly speaking monomers aggregate to dimers, dimers to tetramers, etc. A typical collision consists of a diffusional approach of two equal-sized clusters, which obviously have a much lower chance to interpenetrate than an individual particle and a large cluster. The dimension D measured in the simulations is D=1.4, 1.8 and 2.0 in d=2, 3 and 4. Incidentally, one can make the model purely hierarchical, by diffusing two M—mers to form one 2M—mer, etc. The result for D is the same to within numerical accuracy.

While D clearly depends on d, the dependence of D on α is more subtle. For $\alpha < 0$ the simulations show almost no dependence on it, along with the observation that the size distribution of the clusters is always monodisperse. This means that there is a typical cluster size M^* that characterizes all the moment of the size distribution. The number of clusters deviating appreciably from M^* decreases exponentially both for sizes much larger and much smaller than M^*. Above a critical value α_c the size distribution changes: while the large clusters continue to grow rapidly, a large number of small clusters remains present at all times. The process changes to a particle—cluster type process and the fractal dimension changes accordingly. This fact illustrates how closely the kinetics and the structural properties are interconnected.

One other parameter can be varied in this model: the initial particle concentration. The results presented so far are valid in the low concentration regime where the distance between the clusters is larger than the linear size of a cluster. As the clusters grow this condition gradually fails — a direct consequence of their fractal structure. If one continues the growth process, it becomes a correlated cluster aggregation process with an effective fractal dimension that increases with time. Finally the clusters are practically dense in space and a kinetic gelation process sets in leading to compact structures with fractally rough surfaces[60,61]. Hence true scaling is only observed at monomer densities close to zero and close to one, inbetween there is a crossover towards the high density regime.

Many other aspects of the cluster process that have been studied. The clusters are internally isotropic and have an overall shape with a finite ratio between small and large axis, independent of their size. Lattice and off—lattice simulations yield similar results for the fractal properties; the latter have the advantage that rotational aspects can be considered. Diffusional rotation of the type encountered in colloidal aggregation is weak and does not affect the fractal properties.

Interesting extensions of this model include mechanisms for restructuring and reversible bonding between the aggregates. In physical systems , bending and stretching may change the aggregation process when the clusters are large.

The last model that shall be discussed is chemical clustering[62,63]. The growth rule of diffusive cluster aggregation is simply modified by allowing the clusters to collide (but never to overlap) many times before sticking. A sticking probability p determines the probability to form a bond per collision. A crossover is observed between pure chemical clustering, where two cluster explore each others surface uniformly before sticking irreversibly to one another, and diffusive clustering, where the bond forms at the first contact. Experimentally the pure chemical case is easily realized, as often 10^4 collisions are needed before the clusters stick. In experiments the crossover can be controlled by chemically varying the bonding probability.

The simulations indicate that for chemical clustering D=1.56

and D=2.02 in two and three dimensions. Because of the low sticking probability the size distribution of the diffusive regime is irrelevant for the chemical process. Nevertheless geometry and kinetics are intimately related in this process. The size of the clusters as a function of time grows exponentially and the size distribution is mildly polydisperse.

The models presented and their properties as determined by simulations show how even for kinetic processes universal laws can be found that are independent of detailed microscopic properties. However, in general the irreversibility of the growth process requires a more subtle analysis than in equilibrium systems.

Shear induced aggregation

An extension of the models used to describe aggregating particles to systems that are under external constraints such as shear or applied fields are of great practical importance. For example, the rheological properties of concentrated colloidal suspensions depend on the structure of the colloidal aggregates present. The viscosity as a function of shear rate and its time dependence or thixotropy is a phenomenon influenced by the aggregates present in solution. The relation is a complicated one as the aggregate structure influences the shearing flow and is influenced by it at the same time. Experimentally the question of how aggregates behave under shear has been studied by scattering techniques. The fractal properties of carbon black dispersions in a mineral oil has been shown to depend on the floc concentration and the shear rate[64]. Theoretically, it has been argued that the viscosity of a flocculated suspension depends on the fractal properties of the flocs. This is the first step in treating the coupled system of aggregates in a shear flow[65].

Generally one may distinguish between the rheological properties of a flocculated suspension in equilibrium, where the stationary floc structure depends on the shear rate, and the far from equilibrium irreversible aggregation process under shear where the flocs grow steadily. Here we will address the second type of process. As a motivation one can imagine that by controlled shear conditions a particular type of aggregation mechanism can be favored which in turn determines a particular floc structure. A model for aggregation in the presence of shear – or more precisely *by* shear – will be defined and explored in numerical simulations.

A colloidal aggregation process usually takes place when diffusing flocs grow by an irreversible sticking process whenever they collide. Under shear additional collisions are induced by the shearing motion. Under the assumption that the growth is strictly irreversible, the brownian motion becomes negligible and the shearing motion dominates upward from a certain floc size. In colloidal systems this occurs when the flocs have a size around one micron.

In the spirit of scaling and universality we will first introduce and study a pure shear model and later discuss the effects of superposing a diffusive motion. The model is defined as follows: N particles of unit radius are placed at random in a square (or hypercubic) box of linear dimension equal to L.

245

The particles do not overlap. A particle at position (x,y) moves in the direction of the positive x-axis with velocity $v_x(y)=y/L$. When two particles collide they stick irreversibly to each other and form a rigid dimer. This cluster moves with a velocity $v=(v_1+v_2)/2$. The cluster also rotates around its center of mass with a corresponding angular velocity. When two clusters collide they stick at the point of first contact and the new cluster moves with the flow velocity of its center of mass. Its angular motion is determined by the average angular velocity of its constituent particles around the center of mass. The cluster rotates as a rigid body. These rules completely define the aggregation process.

The fact that all the particles feel the unperturbed flow field is clearly an oversimplification. A more realistic model would treat the central 'core' of a cluster differently from its (fractal) 'arms'.

As the aggregation process is irreversible it will continue until a single cluster is left or, at low concentration, until no trajectories overlap any longer. In practice we use periodic boundary conditions in the x-direction and a moving periodic boundary in the y-direction: for example, a particle that leaves the system at the bottom reenters it at the top at a position $x=v_{max}t$ with $v_{max}=1$, where t is the time.

This shear induced aggregation process is, apart from the initial condition, completely deterministic. It is only due to the shearing motion that aggregation occurs – hence the name shear-induced aggregation. Aside from the spatial dimension there is only a single parameter in the problem: the initial particle concentration. The aggregation process is qualitatively different at low and at high concentration. The results are now summarized.

Simulations have been performed for $N=500$ to $N=5000$ for $L=200$, in two dimensions. The time steps are chosen sufficiently small that the particles move at most a distance of order unity. At low concentration the clusters are fractal and appear spherical despite the shearing motion. There is a high degree of polydispersity: few large clusters and numerous monomers coexist. At high concentration growth is more rapid and there is a clear anisotropy of the clusters. They also have a preferred orientation in space. The structure of large clusters is no longer fractal.

The fractal dimension at low density has been extracted from the mass-radius relation for the largest cluster, $M \sim R^D$. The measured value is $D=1.48(5)$. The increase of the cluster mass with time can also be fitted to a power law form, with an exponent 3.7. In contrast, the data at high concentration suggests that the mass of large clusters diverges in a finite time. The cluster size distribution is polydisperse both for low and for high initial density.

Inhomogeneous spinodal decomposition

The late stage growth during spinodal decomposition[66-68] is characterized by a growth law $R \sim t^a$ where R is the linear size of the domains and t is the time. Theory predicts a=1/3, both in two and in three dimensions[69,70]. Most simulations and experiments measure a exponents in the range 0.15 to 0.25, below the theoretical value. The discrepancy is attributed to surface diffusion effects[71].

Density fluctuations, temperature variations and boundary effects are present in any phase separation experiment. A recent experiment has shown that the interactions between the phase separating system and the boundaries lead to measurable effects[72]. This motivates us to consider spinodal decomposition for spatially inhomogeneous systems. We show that the dynamics of the domain walls can be described in terms of gradient percolation[73]. More precisely, the change of the fractal properties of the percolation hull is directly related to the coarsening of the domains during phase separation. This connection provides an explanation for fractal properties measured for metal deposits in polyimide films[74,75].

The precise model studied is interacting diffusion in a concentration gradient[76,77]. The simulations permit to draw the following conclusions: initially inhomogeneous density profiles do not change with time. As an example consider particles distributed randomly (T=∞) but with a linearly decreasing probability in the x-direction, from $c_1=0.60$ at x=0 to $c_2=0.40$ at $x=L_x$. In the y-direction the system is homogeneous. At time t=0, the system is quenched to a temperature $T<T_c$ such that c_1 and c_2 are both in the unstable region of the phase diagram. The only observable effect are boundary related fluctuations - they grow with time. As another example, a source has been placed at x=0 and a sink at x=L (always maintaining a concentration c=1 and c=0, respectively). As initial profile a step function was chosen, with $c_1=0.55$ for $x \leq L_x/2$ and $c_2=0.45$ for $x>L_x/2$. During spinodal decomposition the step function profile is conserved at all times. Close to the source and the sink, and near the discontinuity at $L_x/2$, there are again boundary related fluctuations which increase with time, both in magnitude and in width.

For a more quantitative characterization of the boundary effects the structure factor for the *average* density in the x-direction was calculated. It scales like the correlation function structure factor (with an effective exponent a≅0.20-0.22) showing that the density fluctuations induced by the boundary is governed by the same growth law as the coarsening process itself.

We now use the invariance of the density profiles to interpret spinodal decomposition in terms of gradient percolation theory. In gradient percolation the external perimeter (hull) of the percolation cluster is fractal over a range which is determined by the concentration gradient ∇. This result implies that many interfaces encountered in experiments should be fractal. The following scaling relations hold in two dimensions: the width σ of the hull and the density n_f of particles on the hull ($n_f=N_f/L_y$ where N_f is total number of particles on the hull) depend on the concentration gradient $\nabla=-R_0 dc/dx$ as

$$(\sigma/R_0) \sim \nabla^{-\nu/(1+\nu)}$$

$$n_f \sim \nabla^{-1/(1+\nu)}$$

where $\nu=4/3$ is the correlation length exponent and R_0 is the lower cutoff (the lattice spacing) and $c(x)$ is the average density in the x-direction. The fractal aspect of the hull then is described by $n_f \sim (\sigma/R_0)^{D_H-1}$ with D_H the corresponding fractal dimension.

We propose to interpret the domain structure during phase separation as a random assembly of droplets of size R ($R \gg R_0$). Knowing that the initial density profile does not change with time, the change of the scaling properties of the percolation hull is determined entirely by the change of the length R. The properties of σ and of n_f can be calculated by replacing the elementary length R_0 by R and by postulating that the droplets are smooth on a scale between R_0 and R. Then the effective gradient (in terms of the concentration of droplets of size R) decreases with time, $\nabla = (c_1-c_2)/(L_x R_0/R)$. With $R \sim t^a$ one gets

$$\sigma \sim t^z$$

$$n_f \sim t^{-z}$$

with $z=a/(1+\nu)$. Simulation data for σ and n_f were analyzed as a function of time. Two definitions of the hull were used[78]: the first hull (nearest neighbor hull) consists of all the sites of the percolation cluster which form the outer surface; the second hull (next nearest neighbor hull) excludes those sites of the first hull which are connected to the outside only by narrow openings ('bays'). Both hulls converge to the same limit and σ as well as $1/n_f$ scale with $z \cong 0.09$. For comparison, using the theoretical values $a=1/3$ and $\nu=4/3$ yields $z=1/7$, larger than the values measured in the simulations. However, with the effective value $a \cong 0.20$ which is measured in simulations over comparable time ranges one obtains $z \cong 0.09$ in agreement with the exponent determined directly from the percolation hull.

Gradient percolation is an efficient method to determine the percolation threshold c_p. It is instructive to monitor how it changes during phase separation. Our simulations show that c_p decreases rapidly from the initial value $c_p=0.593$ (site percolation on a square lattice) to $c_p=0.5$. The latter value is expected for a system for which the lattice structure is irrelevant. For random droplets of size $R \gg R_0$ one expects that lattice effects disappear. Hence the measured $c_p=0.5$ supports the random droplet assumption.

Finally the results obtained here are compared with experiments of metal interdiffusion in polyimide films[74,75]. In these experiments a film is prepared in such a way that the concentration of the metal particles varies across the film. Provided the domain structure is the result of a phase separation process as described above, one may interpret the domain structure in terms of gradient percolation. The two dimensional section of the film may then be described in terms of two dimensional gradient percolation - despite the three dimensional geometry of the experimental setup. The fractal dimension of the hull was calculated from micrographs. Its value $D_H=1.73(4)$ agrees with percolation theory, supporting our line of reasoning.

References

1   D'Arcy W.Thompson, 'On growth and form', Cambridge 1917/1961
2   'Kinetics of aggregation and gelation', F.Family and D.P.Landau (eds.), North Holland (1984)
3   'On growth and form', H.E.Stanley and N.Ostrowski (eds.), Matinus Nijhoff, Dortrecht (1985)
4   'Scaling phenomena in disordered systems', R.Pynn and A.Skjeltorp (eds.), NATO ASI series B, Vol. 133, Plenum (1985)
5   'Physics of finely divided matter', N.Boccara and M.Daoud (eds.),Springer Proceedings in Physics Vol.5, Springer (1985)
6   'Aggregation', Film of growth processes, M.Kolb (author), FU Berlin (producer), 1985
7   H.J.Herrmann, Phys.Rept. 136, 153 (1986)
8   'The physics of structure formation', W. Güttinger (ed.), Springer Series in Synergetics (1987)
9   'Random fluctuations and pattern growth: Experiments and models', H.E.Stanley and N.Ostrowsky (eds.), NATO ASI Vol. 157, Kluwer Academic 1988
10 P.Meakin, in 'Phase transition and critical phenomena',C.Domb and J.Lebowitz (eds.), Vol. 12, Academic 1989
11 'Statistical physics' (STATPHYS 16), H.E.Stanley (ed.), North Holland 1986
12 B.B.Mandelbrot, 'Les objets fractals' (Flammarion 1975);
13 B.B.Mandelbrot, 'The fractal geometry of nature' (Freemann 1982)
14 'Système de création d'images', Minis et Micros 219, p. 57 (1984)
15 I. Stewart, 'Les fractals' (Belin, Bande Dessinée)
16 M. Berger, 'Atelier fractal', MIB, La Chaux-de Fonds (1986)
17 'Fractals in physics', L.Pietronero and E.Tosatti (eds.), Elsevier (1986)
18 H.U.Peitgen and P.H.Richter, 'The beauty of fractals' Springer, 1986
19 H.U.Peitgen and D.Saupe (eds.) 'Fractal images', Springer, 1988
20 J.Feder, 'Fractals', Plenum, 1988
21 T.Vicsek, 'Fractal growth phenomena', World Scientific, 1989
22 'Fractals in physics', A.Aharony and J.Feder (eds.), North Holland 1989
23 D.Avnir (ed.), 'Fractal approach to heterogeneous chemistry', Wiley, 1989
24 J.F.Gouyet, 'Physique et structures fractales' Cours d'option', 1989
25 B.Sapoval, 'Les fractales', Additech Paris, 1990
26 A. le Méhauté, 'Les géométries fractales', Hermès 1990
27 'Fractals in the physical sciences', H.Takayasu, Manchester Univ. Press 1990
28 H.Thiele and H.S. von Levern, J.Coll.Sci. 20, 679 (1965)
29 F.G.Karioris and B.R.Fish, J.Coll. Sci. 17, 155 (1962)
30 S.R.Forrest and T.A.Witten, J.Phys. A12,L109 (1979)
31 D.A.Weitz and M.Oliveria, Phys.Rev.Lett. 52, 1433 (1984)
32 D.A.Weitz, M.Y.Lin and C.J.Sandroff, Surf. Sci. 158, 147 (1985)
33 D.W.Schaefer, J.E.Martin, P.Wiltzius and D.S.Cannell, Phys.Rev.Lett. 52, 2371 (1984)

34 K.D.Keefer, Mat. Res. Soc.Proc. 73,295 (1986)

35 A.Hurd and D.W.Schaefer, Phys.Rev.Lett. 54, 1043 (1985)

36 M.Broide, Thesis, MIT, Cambridge

37 J.Cahill, P.G.Cummins, E.J.Staples and L.Thoson, unpublished

38 F.W.Wiegel and A.S.Perelson, J.Stat.Phys. 29, 813 (1982)

39 P.Richetti, J.Prost and P.Barois, J.Physique Lett. 45, 1137 (1984)

40 C.Camoins and R.Blanc, J.Physique Lett. 46, 67 (1985)

41 R.H.Giese, K.Weiss, R.H.Zerull and T.Ono, Astron. Astrophys. 65, 265 (1978)

42 M.Matsushita, M.Sano, Y.Hayakawa, H.Honjo and Y.Sawada, Phys.Rev.Lett. 53, 286 (1984)

43 J.Nittmann, G.Daccord and H.E.Stanley, Nature 314, 141 (1985)

44 Y.Sawada, Physica 140A, 134 (1986)

45 J.S.Langer, Rev.Mod.Phys. 52,1 (1980)

46 R.Lenormand, C.Zarkone and A.Sarr, J.Fluid.Mech. 135, 337 (1983)

47 H.D.Bale and P.W.Schmidt, Phys.Rev.Lett. 53, 596 (1984)

48 Physikalische Blätter, Cover photograph, 1986

49 L.Niemeyer, L.Pietronero and H.J.Wiesmann, Phys.Rev.Lett. 52, 1033 (1984)

50 G.Deutscher, A.Kapitulnik and M.Rappaport, in 'Percolation structures and processes', G.Deutscher, R.Zallen and J.Adler (eds.), Ann. Isr. Phys. Soc. (1983)

51 H.Stiller, private communication

52 S.Lovejoy and B.B.Mandelbrot, Tellus 37A, 209 (1985)

53 G.B.West, Los Alamos Science 11, 2 (1984)

54 T.C.Halsey, M.H.Jensen, L.P.Kadanoff, I.Procaccia and B.I.Shraiman, Phys.Rev.A33, 1411 (1986)

55 M.Eden, in 'Proc of 4. Berkeley Symp. on Math. Statistics and Probabilities', F.Neyman (ed.), UC Press (1961)

56 M.Kardar, G.Parisi and V.C.Zhang, Phys.Rev.Lett. 56, (1986)

57 T.A.Witten and L.M.Sander, Phys.Rev.Lett. 37, 1400 (1981)

58 P.Meakin, Phys.Rev.Lett. 51, 1119 (1983)

59 M.Kolb, R.Botet and R.Jullien, Phys.Rev.Lett. 51, 1123 (1983)

60 M.Kolb and H.J.Herrmann, J.Phys. A18, L435 (1985)

61 H.J.Herrmann and M.Kolb, J.Phys. A19, L1027 (1986)

62 M.Kolb and R.Jullien, J.Physique Lett. 45, 2977 (1984)

63 W.D.Brown and R.C.Ball, J.Phys. A18, L517 (1985)

64 J.A.Helsen and J.Texeira, Coll. Poly. Sci. 264, 619 (1986)

65 P.Mills, J. de Physique Lett. 46, L301 (1985)

66 J.S.Langer, Ann. Phys. (N.Y.) 41,108 (1967); ibid. 54, 258 (1969); ibid. 65, 53 (1971)

67 J.D.Gunton, M.San Miguel, and P.S.Sahni, in 'Phase transitions and critical phenomena, Vol. 8, C.Domb and J.L.Lebowitz, eds. (Academic 1983)

68 H.Furukawa, Adv. Phys. 34, 703 (1985)

69 I.M.Lifshitz and V.V.Slyozov, J. Phys. Chem. Solids 19, 35 (1961)

70 C.Wagner, Z. Electrochem. 65, 581 (1961)

71 D.A.Huse, Phys.Rev. B34, 7845 (1986)

72 P.Guenoun, 'Journées de physique statistique', Janvier 1989, Paris (France)

73 M.Kolb, T.Gobron, J-F.Gouyet and B.Sapoval, Europhys.Lett. 11, 601 (1990)

74 S.Mazur and S.Reich, J.Phys.Chem. 90, 1365 (1986)

75 R.P.Wool, B.-L.Yuan and O.J.Mcgarel, Poly.Eng.Sci. 29, 1340 (1989)

76 B.Sapoval, M.Rosso and J.-F.Gouyet, J.Physique Lett. 46, 149 (1985)

77 M.Kolb, J.-F.Gouyet and B.Sapoval, Europhys. Lett. 3, 33 (1987)

78 T.Grossmann and A.Aharony, J.Phys. A19, L745 (1986); ibid. A20, L1193 (1987)

ELECTRODEPOSITS: PHENOMENOLOGY AND THEORY

Werner Gans

Freie Universität Berlin, Institut für physikalische und

theoretische Chemie, Takustr. 3, D–1000 Berlin 33, FRG

1 Introduction

The process of electrodeposition belongs to the large class of growth and aggregation processes in nature. Biological growth processes were first considered in a systematic way at the
beginning of this century by *D'Arcy Wentworth Thompson* in his famous book *"On Growth
and Form"* [1]. It became evident that different growth conditions yield a rich phenomenology
of patterns, and that these growth conditions should somehow be related to the particular
pattern formed. Growth in biological systems is, to a large extent, governed by the underlying
genetic code; in physical and chemical systems, there is no such thing as a genetic code, therefore in these systems the resulting pattern depends much more on the external conditions
under which it is formed, a fact which makes physical growth processes much more amenable
to experimental investigation.

Physical phenomena, in which growth phenomena take place, are:

- formation of colloids and aerosols

- solidification of substances from their undercooled melt or from oversaturated vapour
  (snow flakes)

- electrodeposits from aqueous and ionic solutions

- multiphase flow in fluids ("viscous fingering", Saffman–Taylor fingers)

- sputtering process on surfaces

- lightning, dielectric breakdown

- gelation in polymer systems

This list of examples is by no means exhaustive; the reader who wants to see pictures of
the patterns formed in the various experiments is referred to the review article by *Kessler et
al.* [2].

Large-Scale Molecular Systems, Edited by W. Gans *et al.*
Plenum Press, New York, 1991

It seems that the formation of flat or spherical surfaces is rather the exception; on the other hand there are many different kinds of irregular growth patterns, which look qualitatively different, but which used to be very difficult to classify in a quantitative manner. Since the time the concept of the *fractal dimension* [3,4,5] has entered physics, the situation has improved. Many experiments were designed to produce fractal patterns, in order to understand more about these highly irregular structures and their growth conditions. Therefore, we will start with a brief account of old and new experiments in electrochemistry relating to the problem of pattern formation.

2   Experiments on Electrodeposition

Experiments investigating the nature of dendritic growth processes during electrodeposition go back to at least 1935 [6], but the first quantitative work seems to be [7]. The authors considered the electrolytic growth on silver spheres of dendrites from the silver + silver nitrate system in liquid sodium nitrate + potassium nitrate. They found a critical current density i_{crit} for dendrite formation, which is proportional to the concentration of Ag^+; i_{crit} corresponds to an overpotential of about 3 mV. The velocity of dendrite growth at a given potential was found to be constant with time, fastest growth occurred for dendrites initiated at high overpotentials, slowest growth for those initiated at low overpotentials. Also some branching was observed, unfortunately the article does not contain any photographs of the patterns. Assuming diffusion control (which, under the experimental conditions, is indeed the case), one gets for the current density i

$$i = \frac{(zF)^2}{RT} \frac{Dc_\infty}{\delta} \, \eta,$$

(1)

where z = charge number of the ion,
F = Faraday constant,
D = diffusion constant,
c_∞ = concentration far away from the electrode,
δ = effective thickness of the diffusion layer in the steady state,
η = overpotential.

Equation (1) is a linearized version of the *Butler–Volmer* equation (*cf.* sec. 3.2, eq. (23)). In their theoretical treatment, the authors establish formulas for the growth rate of the tips of the dendrites as a function of the overpotential for high and for low exchange current densities, surface energy effects on the overpotential are also included. The comparison with experimental growth rates is done by assuming "reasonable" parameters and yields agreement within a factor of 2 or 3.

Another experimental study of dendritic growth is by *Faust* [8] for Ag, Cu and Ni in molten salt baths. He, too, found a lower limit of the current density for dendrites to develop. At high current densities he indicates the appearance of "polycrystalline and spongy forms" without further investigating them, and it is precisely this morphology which will interest us next. For two reviews on the phenomenology of electrodeposition (and dissolution) of metals up to around 1980, the reader is referred to [9] and [10].

In 1984, *Brady* and *Ball* [11] reported the fractal growth of copper electrodeposits from aqueous solutions of $CuSO_4$ (concentration: 0.01 or 0.02 M) in the presence of Na_2SO_4 as inert electrolyte and 1.4 wt% polyethylene oxide to increase viscosity. The cations were deposited on an cathode of cylindrical shape; the (constant) voltage applied was in the range of 0.3–0.8 V; changing the instantaneous voltage did not change the instantaneous current significantly, so that the authors could be sure that the process was limited by diffusion of the Cu^{2+} ions. The product of the electrodeposition, of which two micrographs are shown at different length scales, exhibits a very rough and spongy structure. The fractal dimension $\bar{d}$ of the deposit was determined via the scaling of the density ρ with the radius (the *Smoluchowski radius*) R_s,

$$\rho \propto R_s^{\bar{d}-3},$$ (2)

which is equivalent to the so-called mass–radius relation [12]

$$N \propto R^{\bar{d}}$$ (3)

for N mass points within a radius R.

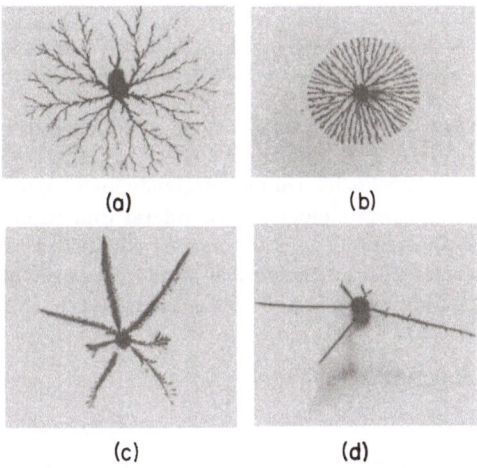

(a)   (b)

(c)   (d)

Figure 1. Electrochemical deposition of zinc. The patterns (a) – (d) correspond to the points a – d in Fig. 2. (a) 'diffusion–limited' aggregation, (b) dense radial, (c) dendritic, (d) needle crystal in the extreme dendritic limit .

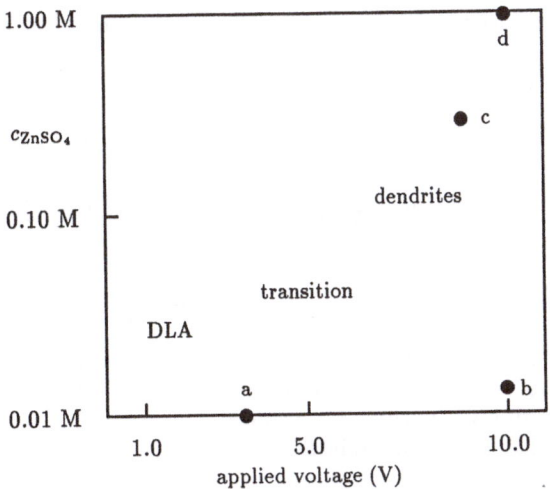

Figure 2. Observed regions of pattern formation in electrochemical deposition of Zn.

The mean dimension is reported to be $\bar{d} = 2.433 \pm 0.03$, in a range of 40 nm to 400,000 nm, a value which agrees well with earlier computer predictions for aggregates from diffusion–limited aggregation (DLA) [13], which gave $\bar{d} = 2.495 \pm 0.06$.

In 1986, two groups came up with a series of experiments on the electrodeposition of zinc from aqueous solutions of $ZnSO_4$ [14,15]. The most striking difference in experimental conditions, compared to the experimental work presented so far, is the absence of an *inert electrolyte* (like Na_2SO_4 in [11]). Before discussing this point, let us have a look at their experimental findings: Figure 1 shows different patterns obtained during electrodeposition, while Figure 2 locates these patterns in a concentration versus (applied) voltage diagram [14] (the results obtained in [15] are similar).

The data in Fig. 2 are a summary of about 250 growth conditions, and the solid lines suggest a separation of different regimes in the process of electrodeposition.

Further recent experiments with main emphasis on producing fractal electrodeposits were done e.g. by *Sagués et al.* [16] (*cf.* also these Proceedings) and *Hibbert* and *Melrose* [17]. Especially the latter authors, who let copper electrodeposits grow in paper support from acid solutions, produced wild morphologies with fractal dimensions in the range 2.0 – 1.7 (determination by image analysis). The experimental conditions of this largely exploratory work cannot be easily translated into theoretical models, its principal relevance may be the successful production of highly ramified and dense structures, both tree–like and radial.

3 Theoretical Approach

The theory of the growth of electrodeposits has much in common with the theory of solidification from an undercooled melt; in both cases, a field, which can be controlled from the exterior, is responsible for the process: the electric field in the one case, the temperature field in the other. There are, however, also significant differences, which are mainly due to the presence of long–range forces in the electrochemical case. So before considering electrodeposition, it seems appropriate to present a brief account to the theory of solidification; an enormous body of work has been devoted to this problem during the last 15 years.

3.1 Crystallization

During the process of crystallization, the dynamical field is the temperature field T, and in principle, the differential equations and boundary conditions governing the process can be written down quite easily. They are:

1. the heat transport equation

$$\frac{\partial T}{\partial t} = D_i \nabla^2 T, \qquad (4)$$

   D is the heat diffusion coefficient, index i is liquid or solid phase;

2. conservation of energy

$$[D_{sol}c_{p_{sol}}(\nabla T)_{sol} - D_{liq}c_{p_{liq}}(\nabla T)_{liq}] \cdot \mathbf{n} = Lv_n, \qquad (5)$$

   c_p = Heat capacity per volume of the solid or liquid phase,
   $\mathbf{n}$ = local normal vector at the interface, directed into the liquid phase,
   L = latent heat of crystallization per volume,
   v_n = local normal velocity of the interface;

3. undercooling of the liquid far away from the solid phase:

$$T_{|x|\to\infty} = T_m - \Delta, \qquad (6)$$

T_m = melting temperature,
Δ = undercooling;

4. an equation taking into account the surface tension of the solid phase

$$T_{int} = T_m(1 - d_0\kappa),\qquad(7)$$

T_{int} = temperature at the interface
$d_0 = \gamma/L$ = capillary length, where γ = surface tension
κ = inverse mean radius of the curvature of the interface.

Equation (7) is better known as *Gibbs–Thomson* relation, it determines the temperature at the interface. The applicability of the *Gibbs–Thomson* relation presupposes at least local thermodynamic equilibrium, a condition which is hardly fulfilled during growth processes; on the other hand, this is the simplest way of incorporating the effect of surface tension into the relevant set of equations.

The fundamental equations (4) – (7) can be transformed into an equivalent set of integral equations. For this one has to take into account that an element of the interface at position $\mathbf{x}'$ at time t' is the source of an amount $Lv_n(\mathbf{x}',t')/c_p$ of temperature, and that the temperature at any other space–time point is the background temperature T_∞ plus the superposition of sources at all previous interface points, propagated by the diffusion Green's function (*cf.* [2]).

$$T(\mathbf{x},t) = T_\infty + \int dS' \int_{-\infty}^{t} dt' G(\mathbf{x}-\mathbf{x}',t-t')Lv_n(\mathbf{x}',t')/c_p,\qquad(8)$$

where S' is the d–dimensional surface element and

$$G(\mathbf{x},t) = (4\pi Dt)^{-d/2} \exp[-\mathbf{x}^2/4Dt].\qquad(9)$$

This transformation does not, however, reduce the difficulties of the full set of equations; for performing the integration over t', one needs to know the full history of the problem. For illustrative purposes, we will reduce the problem to a two–dimensional interface propagating with constant shape and velocity $v\mathbf{z}$. The interface is then parametrized as $\mathbf{x}' = (x', z(x')+vt')$; taking into account the boundary condition (7) on the left–hand side of equation (8), changing variables from arclength s' to x' ($v_n ds' = vdx'$) and performing the t' integral, one gets:

$$\tilde{\Delta} - \tilde{d}_0\kappa = \frac{v}{2\pi D} \int_{-\infty}^{\infty} dx'[v(z(x') - z(x))/2D]\, K_0\left\{\frac{v}{2D}\sqrt{[(x - x')^2 + (z(x) - z(x'))^2]}\right\},\qquad(10)$$

where K_0 is a Bessel function, $\tilde{\Delta} = (T_m - T_\infty)c_p/L$ is the dimensionless undercooling, and $\tilde{d}_0 = d_0 T_m c_p/L$ is the capillary length.

Usually one is not interested in the temperature profile, but rather in the shape of the interface solid/liquid during the process. If $z = \zeta(x,t)$ denotes points on the interface, then the curvature κ can be expressed in the following way:

$$\kappa\,(\zeta) = -\frac{\partial^2\zeta/\partial x^2}{[1 + (\partial\zeta/\partial x)^2]^{3/2}},\qquad(11)$$

which makes eq. (10) highly nonlinear.

Before showing some special solutions of the general equations, the effect of surface tension on the process should be discussed qualitatively. Regions of the interface protruding into the liquid phase are defined to have positive curvature. In this case the foremost atoms or molecules have fewer neighbours than the surface atoms on a flat interface and therefore have a tendency to leave the solid phase; macroscopically this results in a local decrease of the melting temperature. On the other hand, in the case of a negative curvature, the local melting temperature will be increased – this is what the *Gibbs–Thomson* relation (7) contains. A more refined treatment will be given instantly, when we consider the stability of flat interfaces during the solidification process.

The simplest solution of equations (4) – (7) is a plane interface moving with constant velocity v in the z–direction; the temperature in the solid is simply T_m and constant, and choosing a moving frame of reference, $\xi = z - vt$, one calculates the time–independent temperature field $T(\xi)$ in the liquid phase. In the moving frame of reference, the diffusion equation simplifies to

$$-v\frac{\partial T}{\partial \xi} = D\frac{\partial^2 T}{\partial \xi^2},\tag{12}$$

with boundary conditions $T(0) = T_m$ and $-c_p DT'(0) = Lv$. The solution is then

$$T(\xi) = T_m - \frac{L}{c_p}(1 - \exp(-v\xi/D)).\tag{13}$$

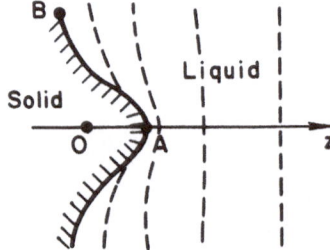

Figure 3. Schematic representation of the *Mullins–Sekerka* instability ([19])

This solution, however, exhibits some strange features, namely: the velocity is *arbitrary* and the *unique* undercooling is given by $\Delta = L/c_p$. The last fact is simply a consequence of heat conservation in the particular case of a flat interface, but the arbitrariness of the velocity renders these solutions somewhat "unphysical".

One further important point is that the interface is completely unstable with respect to small perturbations, e.g. wiggles imposed on the surface. This is the famous *Mullins–Sekerka* instability [18], the analysis of which clearly shows the interplay of the role of surface tension and heat transport near the interface. Qualitatively, this picture can be explained as follows (see fig. 3, and [19]). Purely thermal considerations suggest that the release of heat at point A is easier than at point B, because the isotherms are closer to each other at A; therefore A will grow unstably. On the other hand, the surface tension will reduce the temperature at A so that there will be a heat flow from B to A and consequently a tendency for A to melt back.

Mathematically, one can discuss this instability by using an unperturbed solution of eq. (8), $z(x,t) = vt$ with $\kappa = 0$ and $ds' = dx'$, for which one gets $\tilde{\Delta} = 1$ (or $\Delta = L/c_p$), as previously, then perturbing the interface by putting $z = vt + \delta(x,t)$. Evaluating (8) on this perturbed interface and expanding everything to first order in δ, expressing δ in terms of Fourier modes, $\delta = \delta_0(\omega t + ikx)$, and finally taking into account the non–zero curvature, $\kappa = -\delta'' = k^2\delta$ (to order δ), one arrives at a dispersion relation:

$$\omega = \frac{v}{l}[A^{1/2}\,(1 - l\tilde{d}_0 k^2) - 1], \tag{14}$$

where $l = 2D/v$ is the diffusion length, $\omega_0 = v^2/4D$ and $A = 1 + \omega/\omega_0 + (kl)^2$. Under the assumption that the temperature field relaxes rapidly compared to the movement of the interface (quasistatic approximation) and restricting oneself to short wavelengths ($kl \gg 1$), one obtains a simplified dispersion relation:

$$\omega = v|k|(1 - l\tilde{d}_0 k^2). \tag{15}$$

One can see that in the absence of surface tension, $\tilde{d}_0 = 0$, the system is completely unstable, whereas for nonzero surface tension, the growth rate is positive only for a limited band of 'unstable' wavenumbers.

The steady–state diffusion equation (12), for which the planar solution was obtained in the rather special case of $\Delta = L/c_p$, admits a whole family of steady-state, shape preserving solutions in the case of vanishing surface tension, $\tilde{d}_0 = 0$, and for $\Delta < L/c_p$. To guess the shape of these solutions, let us note that eq. (12) is separable in parabolic coordinates. The solutions are parabolas or paraboloids in two resp. three dimensions, the parabolas were first derived by Ivantsov [20] in 1947. Detailed descriptions of these "needle–crystals" can be found in [19] or [21]. Here we only want to mention that although recalling the shapes of morphologies observed in nature, these solutions are unphysical, because the undercooling depends only on the *product* of growth velocity v and tip radius ρ, whereas it is experimentally observed that the undercooling uniquely determines ρ and v.

Due to the enormous difficulties encountered in treating the full problem of solidification with its nonlinearities and also the built–in nonlocality, so–called *local models* have been developed which treat the interface like a string (in the two–dimensional problem) and which take account of effects like surface tension and the *Mullins–Sekerka* instability in an ad hoc manner. The best– known of these models are the *geometrical model* [22] and the *boundary layer model* [23]. Considerable progress has been made with these models in understanding the mechanisms of dendritic growth, one of the latest achievements concerns the prediction of dendritic growth rates for comparison with experiments [24].

After this short review of the basic theory of crystallization, we will now turn to the even more intricate problem of electrodeposition or electrocrystallization, as it is sometimes called.

3.2 Electrodeposition

As in the case of crystallization from an undercooled melt, the relevant set of equations can be written down fairly easily. One of the distinctions is that primarily not energy (heat), but matter (ions in the solution) is transported. This ionic transport can occur by

- diffusion,

- migration along the electric field,

- convection.

Convection does not seem to be important in unstirred electrochemical cells, so we will disregard it completely in the following. The effects of convection on mass transport in electrochemical cells can be found in [25].

The dynamical variables are:

- the species deposited,

- all other species,

- the electric field.

The basic equations are then:

$$\frac{\partial c_i}{\partial t} = -\nabla \cdot \mathbf{j}_i, \tag{16}$$

where

$$\mathbf{j}_i = -\frac{D c_i}{kT} \nabla \mu_i \tag{17}$$

is the flux of the i-th dissolved species, and μ_i denotes the *electrochemical potential* of the i-th disolved species of concentration c_i and charge q_i,

$$\mu_i = kT \ln c_i + q_i \phi. \tag{18}$$

Let c_1 denote the concentration of the species deposited, then in the cathode (pure metal) $c_1 = 1$. Finally we have *Poisson*'s law,

$$-\nabla^2 \phi = \frac{4\pi}{\epsilon V_s} \sum_i q_i c_i, \tag{19}$$

where V_s = mean molecular volume of the solution, and ϵ = dielectric constant of the solution (or of water, for simplicity). To simplify the set of equations a bit, one can now (i) linearize the equations arond the concentration far away from the electrodes, (ii) assume local charge neutrality. Assumption (ii) is justified if the length scales of the deposit are very much larger than the *Debye* screening length

$$l_D^2 = \frac{4\pi kT \epsilon V_s}{\sum q_i^2 c_{i\infty}}, \tag{20}$$

which is usually of the order of magnitude of several Å, and if one considers the situation in the bulk of the solution, outside of the electric double layer (*Helmholtz* layer), whose depth is several Å, too. Generally speaking, the situation regarding electroneutrality a few Å off the electrode is harmless for not too dilute solutions and for those with a supporting electrolyte [26].

If again species 1 denotes the electroactive species, we have the following boundary conditions:

$$\mathbf{n} \cdot \mathbf{j}_i = 0 \quad \text{for} \quad i \neq 1 \tag{21}$$

and

$$\mathbf{n} \cdot \mathbf{j}_1 = -v_n(1 - c_1), \tag{22}$$

where $\mathbf{n}$ denotes the normal on the deposit, and v_n the normal growth velocity. The last equation follows from the charge neutrality requirement.

We already noted that the process of crystallization is far from thermodynamic equilibrium; here the situation is even more complicated, and one has to consider explicitly the kinetics of charge transfer at the cathode and the influence of (possibly) existing large concentration gradients at the interface electrode/solution. In the macroscopic theory, the effects of the charge transfer process and the distribution of ions on the current density and on the dynamics of electrodeposition are usually incorporated in the *Butler–Vollmer* equation. The net reaction rate r is expressed in the following way:

$$r = r_0\{\exp[\eta(1 - \beta)] - \exp(-\eta\beta)\}, \tag{23}$$

where β is a dimensionless (electrode–dependent) symmetry parameter between 0 and 1, r_0 is equal to the reaction rate in either direction, when no net reaction takes place, and η is the (dimensionless) overpotential,

$$\eta = \frac{q_1}{kT} (\Delta\phi - \Delta\phi_e). \tag{24}$$

Here $\Delta\phi$ means the potential difference across the double layer, and $\Delta\phi_e$ is the equilibrium overpotential, at which no net reaction takes place ("$r_0 = r$").

Generally, the main factors which govern the current and potential distributions over an electrode are [27]:

- the geometry of the system,

- the conductivity of the solution and electrodes,

- the activation overpotential (depending on the kinetics of the electrode reaction),

- the transport or concentration overpotential (caused by the concentration differences between the electrode–solution interface and the bulk solution).

So the importance of overpotential effects are quite obvious for the process of electrodeposition. An expression for $\Delta\phi_e$ can be found by using the fact that a planar electrode is in equilibrium with a saturated solution at zero potential drop; if one also considers the influence of surface tension, one gets the following expression:

$$q_1\Delta\phi_e = kT \ln (c_1/c_{sat}) - \gamma\kappa V_m, \tag{25}$$

where V_m is the atomic volume of the metal of the cathode, γ = surface tension, and κ = curvature of the interface.

At this stage, one can introduce simplifying assumptions (*cf.* [2]), which, in some sense, do away with the complexity of the system completely. Firstly, the bulk equation for ion 1 is reduced to $\nabla^2\mu_1 = 0$, which is the quasistatic approximation to the diffusion equation; secondly, if the growth rate is very much smaller than r_0, the *Butler–Vollmer*-equation can tentatively be replaced by $\eta = 0$, i.e. the influence of the overpotential is neglected completely. Then the problem can be reformulated as a kind of crystallization with the dimensionless "undercooling" $\tilde{\Delta}$:

$$\tilde{\Delta} = \frac{q_1 c_{1,\infty}}{kT} \Delta\phi_{ext}, \tag{26}$$

with $\Delta\phi_{ext}$ = externally applied voltage.

Although it is claimed in [2] that the last equation at least qualitatively describes the shape of lines in the "phase diagram" of [15] separating the different structures, one must ask the question, whether the simplifying assumptions mentioned above do not destroy the *observed* physical conditions of growth for dendritic and spongy morphologies. An experimental study from 1969 of dendritic electrocrystallization of zinc [28] shows that dendritic growth is observed at overpotentials more negative than −75 to −85 mV, in fact up to −160 mV. Above and below these limiting values of overpotential, a sponge (maybe of different texture, but with a fractal dimension?) was formed. Remarkably, silver dendrites begin to form at much lower overpotentials, at around −3 mV. On the other hand, it seems that not very much is known about the kind of overpotential involved (see above), the influence of surface tension, however, is acknowledged in [28]. Therefore it seems necessary to have further experimental material (following the best traditions of electrochemistry!) and to devise more realistic theoretical models taking into account the various effects present in the "simple" system *ionic solution and two electrodes.*

These models should, for example, take into account the different overpotentials arising for various electrode profiles. If the thickness of the diffusion layer is large compared to the mean height of the profile, the concentration overpotential is smaller on the crests and counteracts the action of the activation potential, the latter tending to equalize the current density at the interface. On the other hand, if the diffusion layer is thin compared to the height of the profile (this includes the case of plane electrodes and of electrodes with macroscopic crests), the diffusion layer closely follows the profile. In this case the concentration overpotential acts in the same direction as the activation potential, namely, it tends to make the current distribution more uniform and thus to improve the "throwing power" (as electrochemists would say) on the macroscopic scale [27].

It would be interesting to have a look at the interplay of these two overpotentials (and possibly also of the influence of surface tension) and their effect on the interfacial current distribution (the so-called tertiary current distribution) in the case of a fractal electrode surface. The relation between electrodeposition and diffusion–limited aggregation for secondary current distribution (activation overpotential only) was considered recently in [29], the electrochemical system at the basis of the model does not, however, fulfill the condition of diffusion–limited aggregation, as it is understood in electrochemistry (*cf.* [11]) with the solution containing a high concentration of supporting electrolyte.

So, in the end it seems that many questions are still open in the macroscopic theory of electrodeposition, even for the "pure case", i.e. solutions without organic additives and without the host of other empirical methods and tricks of trade used in this technologically important branch of physical chemistry.

References

[1] Thompson, d'A.W.: *On Growth and Form.* Cambridge University Press, Cambridge 1917/1961

[2] Kessler, D.A., J. Koplik, H. Levine, Adv. Physics **37**, 255–339 (1988)

[3] Mandelbrot, B.B.: *The Fractal Geometry of Nature.* Freeman, San Francisco, 1982

[4] Feder, J.: *Fractals.* Plenum Press, New York, 1988

[5] Vicsek, T.: *Fractal Growth Phenomena.* World Scientific, Singapore, 1989

[6] Papapetrou, A., Z.Krist. **92**, 89 (1935)

[7] Barton, J.L., J.O'M. Bockris, Proc.Royal Soc. (London) **A268**, 485–505 (1962)

[8] Faust, J.W., J.Cryst.Growth **3**, **4**, 433–435 (1968)

[9] Budewski, E.B., in: Conway, B.E., J.O'M. Bockris, E. Yeager, S.U.M. Khan, R.E. White (eds.) *"Comprehensive Treatise of Electrochemistry"* Vol. 7, 399–450; Plenum Press, New York 1983

[10] Despić, A.R., in: Conway, B.E., J.O'M. Bockris, E. Yeager, S.U.M. Khan, R.E. White (eds.) *"Comprehensive Treatise of Electrochemistry"* Vol. 7, 451–528; Plenum Press, New York 1983

[11] Brady, R.M., R.C. Ball, Nature **309**, 225–229 (1984)

[12] Kolb, M., these Proceedings

[13] Witten, T.A., L.M. Sander, Phys.Rev. B**27**, 5686–5697 (1983)

[14] Grier, D., E. Ben–Jacob, R. Clarke, L.M. Sander, Phys.Rev. Lett. **56**, 1264–1267 (1986)

[15] Sawada, Y., A. Dogherty, J.P. Gollub, Phys.Rev. Lett. **56**, 1260–1263 (1986)

[16] Sagués, F., F. Mas, M. Vilarrasa, J.M. Costa, J. Electroanal. Chem. **278**, 351–360 (1990)

[17] Hibbert, D.B., J.R. Melrose, Phys.Rev. A**38**, 1036–1048 (1988)

[18] Mullins, W.W., R.F. Sekerka, J. appl. Phys. **34**, 323 (1963); J. appl. Phys. **35**, 444 (1964)

[19] Langer, J.S., Rev.Mod.Phys. **52**, 1–28 (1980)

[20] Ivantsov, G.P., Dokl. Akad. Nauk SSSR **58**, 567 (1947)

[21] Langer, J.S., *"Lectures in the Theory of Pattern Formation"* in: J. Souletie, J. Vannimenus and R. Stora (eds.) *"Chance and Matter"*, Elsevier, Amsterdam, 1987

[22] Brower, R.C., D.A. Kessler, J. Koplik, H. Levine, Phys.Rev. Lett. **51**, 1111 (1983), Phys.Rev. A, **29**, 1335 (1984), Scripta metall. **18**, 463 (1984)

[23] Ben–Jacob, E., N. Goldenfeld, J.S. Langer, G. Schön, Phys. Rev.Lett. **51**, 1930 (1983), Phys.Rev. A**29**, 330 (1984)

[24] Barbieri, A., J.S. Langer, Phys.Rev. A**39**, 5314–5697 (1989)

[25] Ibl, N., O. Dossenbach, in: E. Yeager, J.O'M. Bockris, B.E. Conway, S. Sarangapani (eds.) *"Comprehensive Treatise of Electrochemistry"* Vol. 6, 133–237; Plenum Press, New York 1983

[26] Ibl, N., in: E. Yeager, J.O'M. Bockris, B.E. Conway, S. Sarangapani (eds.) *"Comprehensive Treatise of Electrochemistry"* Vol. 6, 1–63; Plenum Press, New York 1983

[27] Ibl, N., in: E. Yeager, J.O'M. Bockris, B.E. Conway, S. Sarangapani (eds.) *"Comprehensive Treatise of Electrochemistry"* Vol. 6, 239–315; Plenum Press, New York 1983

[28] Diggle, J.W., A.R. Despić, J.O'M. Bockris, J.Electrochemical Society **116**, 1503–1514 (1969)

[29] Halsey, T.C., H. Leibig, J.Chem.Phys. **92**, 3756–3767 (1989)

SCREENING IN ELECTROLYTES AND IN POLYMER SOLUTIONS :

THE CHARGE STRUCTURE FUNCTION

Gérard Jannink

Laboratoire Léon Brillouin (CEA-CNRS)
CEN-Saclay
91191 Gif-sur-Yvette cedex, France

I. INTRODUCTION

In 1923, P. Debye[1] and E. Hückel introduced screening in order to study concentration effects on the structure of ionic aqueous solutions. This concept is very usefull in general, to account for long range effects. It was used in particular by S.F. Edwards[2] in order to determine the structure of neutral polymer solutions at finite concentrations.

We discuss :

a) the measurement of the screening length ξ associated to the screening effect
b) the variation of ξ with concentration and temperature

Our aim is to examine in greater detail the <u>charge</u> structure function in electrolyte solutions. This function can now be measured directly. Developments are therefore expected, in charged as well in neutral systems.

II. ELECTROLYTE SOLUTION

We have learned from Debye[1] and Hückel that the significant length in an electrolyte of ionic concentration $\mathbb{C} = \mathbb{C}^- + \mathbb{C}^+$ is the screening length

$$\xi_D = \frac{1}{\sqrt{\mathbb{C}\ \ell_B}} \tag{1}$$

rather than the average neighbour distance

$$\delta = \frac{1}{\mathbb{C}^{1/3}} \tag{2}$$

The quantity ℓ_B (or Bjerrum length) is equal to $e^2/\epsilon\ kT$, where e is the charge and ϵ the dielectric constant.

When a negative point charge is introduced in the aqueous solution,

Large-Scale Molecular Systems, Edited by W. Gans *et al.*
Plenum Press, New York, 1991

this charge induces a compensating "charge" distribution $C_z(r) = C^-(r) - C^+(r)$.

The electric field $\varphi(r)$ satisfies the Poisson relation

$$\nabla^2 \varphi(r) = -4\pi \{e\delta(r) + e \langle C_z(r) \rangle\} \tag{3}$$

where $e\delta(r)$ is the bare response function of the point charge.

The charge distribution (or polarization charge) creates a field

$$\langle C_z(r) \rangle = \frac{2}{3} C \frac{\varphi(r)}{\epsilon} \tag{4}$$

Inserting into (3), using (1), we obtain in reciprocal space

$$\tilde{\varphi}(q) = \frac{4\pi e}{q^2 + \xi_D^{-2}} \tag{5}$$

The singularity at the origin is now removed and the bare Coulomb interaction

$$V(r) = \frac{e^2}{r} \tag{6}$$

is screened by the polarization charge

$$V(r) = e\varphi(r) = \frac{e^2 \exp(-r/\xi_D)}{r} \tag{7}$$

The electrostatic energy U_e is proportional to $e^2/\epsilon\xi_D$ rather than to $e^2/\epsilon\delta$: U_e scales like $C^{1/2}$ and vanishes at high temperature.

The experimental determination of the screeening length consists in measuring the osmotic coefficient

$$f = \frac{\pi}{\pi_0} \tag{8}$$

which is the ratio between effective and ideal osmotic pressures. This ratio is related to the electric interaction energy. For the dilute case

$$f = 1 - \frac{e^2 \xi_D^{-1}}{6\epsilon \, kT} \tag{9}$$

III. POLYMER SOLUTION

In 1966, S.F. Edwards[2] introduced the concept of screening in the study of neutral polymer solutions, to describe the effects of polymer concentration on the monomer-monomer correlation function. De Gennes[3] pointed out the similarity of this approach and the previous one, if the roles of bare response and bare interaction are reversed.

In the limit of an infinite chain, the monomer-monomer correlation function is proportional to $C/2\pi r$, where $C = Cs$ (C is the chain concentration and S the chain equivalent Brownian area). This result is derived from the Debye form function of a "Brownian" polymer chain :

$$h(y) = \frac{2}{y^2} (e^{-y} - 1 + y) \simeq \frac{1}{y} \ , \ y = q^2(S/2) \gg 1 \tag{10}$$

For $S \rightarrow \infty$, the asymptotic part of the form function is singular at the origin $(1/q^2)$. Screening removes this singularity.

A. <u>Screening due to repulsive interaction</u>

Let us consider the effect of an external potential, U_{ext}. This potential induces a perturbation δC_q in the monomer concentration, (where C_q is the Fourier transform of

$$C(\vec{r}) = \sum_a \int_0^S ds \left\langle \delta(\vec{r} - \vec{r}_a(s)) \right\rangle \tag{11}$$

From the linear response theory, we obtain

$$\delta C_q = -\chi_o(q) \ U_{ext} \tag{12}$$

where

$$\chi_o(q) = \beta \int d^3r \ e^{iqr} \left\{ \langle C(0)C(r) \rangle C^2 \right\} = CS\beta h(q^2 R_G^2) \tag{13}$$

is the bare response function. This function has the behaviour predicted in (10), and in the limit $S \rightarrow \infty$, it has the same singularity as the bare Coulomb potential. The point like interaction b screenes the bare response function. The induced internal energy is :

$$U_{int} = \frac{b \ \delta C_q}{\beta} \tag{14}$$

The modified linear response is :

$$\delta C_q = -\chi_o(q)\{U_{int} + U_{ext}\}$$
$$= -\chi(q) \ U_{ext} \tag{15}$$

As a result, for $qR_G > 1$

$$H(q) = \frac{4C}{q^2 + 4CbS} \tag{16}$$

which can also be derived directly by diagrammatic expansion, in the simple tree approximation. In this approximation, the screening length is

$$\xi_E = \frac{1}{2} \frac{1}{\sqrt{CbS}} \tag{17}$$

Expression (16) does not hold true for "good" solutions, with strong repulsive interactions. However, the significant length in a polymer solution is, in the case of chain overlap (des Cloizeaux[4])

$$\xi_K = (C \ X^D)^{-(d-D)} \tag{18}$$

rather than the average distance δ (equation (2)).

Here $D = 1/\nu$ is the fractal dimension of the polymer chain, and

$$X^2 = \lim_{C \to 0} \frac{R^2}{3}.$$

Scaling implies that there can be only one significant length. There is experimental evidence for the existence of a screening length ξ_e. The latter has to obey the relation :

$$\xi_e \simeq \Gamma \, \xi_R \tag{19}$$

where Γ is a universal constant.

a) The variation of ξ_e with concentration and temperature has been observed experimentally. The results agree with prediction (18). The data show in particular that in good solutions, the interaction b varies as $1/T$ (see figure 1). They also prove the scaling hypothesis, which assumes that the scaling variables are defined at zero concentration.

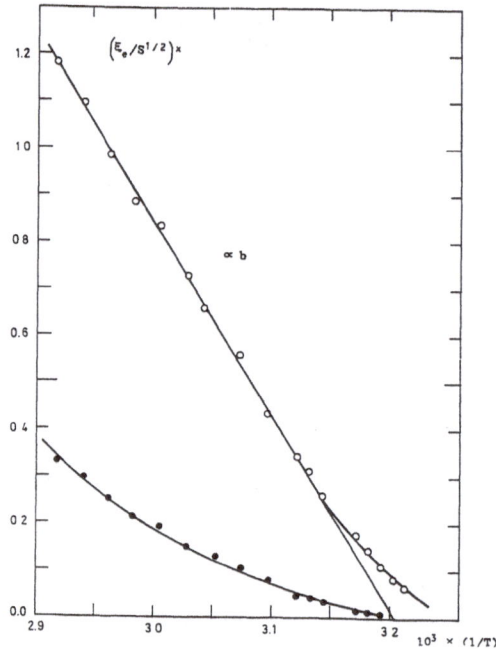

Figure 1. Decrease of the screening length with rising temperature in good solutions. The concentration is fixed at 25 g/ℓ. Plot of $(\xi_e/S)^X$ as a function of $1/T$ (reference (4)).

    ○ $x = -\dfrac{d-D}{2-D}$

Here $(\xi_e/S)^X$ is proportional to the interaction b.

    ● $x = -\dfrac{d-D}{2-D} \times \dfrac{d-D}{D} \times 2$

This representation of the data implies that the scaling variable is the size R^2 of the chain, at the concentration $C = 25$ g/ℓ. This hypothesis is not correct.
The top curve suggests that the correct scaling variable is the size X of the chain at zero concentration.

b) The universal constant Γ can be evaluated theoretically. The method of des Cloizeaux[4] consists in calculating ξ_e in dimension 4, and as a function of $\epsilon = 4\text{-}d$. In the random phase approximation, valid for $\epsilon = 0$

$$\xi_e = \frac{1}{2} \frac{1}{\sqrt{b\mathbb{C}s}} \tag{20}$$

We use the identities

$$b = (2\pi)^{d/2} \, z \, s^{(d/2)\text{-}2} \tag{21}$$

and the development

$$z = g + g^2 \, (8 + (\text{-}2\text{-}4\ln^2)) + \ldots \simeq g \tag{22}$$

We write

$$\xi_e = \frac{1}{2} \, (2\pi)^{\text{-}(4\text{-}\epsilon)/4} \, (g\mathbb{C}s^{(1\text{-}\epsilon)/2})^{\text{-}1/2}$$

We let successively $g \to g*$ ($z \to \infty$) and $\epsilon \to 0$ (note that $g(\epsilon = 0) = 0$). The result is

$$\xi_e = \frac{1}{4\pi} \frac{1}{\sqrt{g*}} \, (\mathbb{C}s)^{\text{-}1/2}$$

Returning to the case $d = 3$ ($\epsilon = 1$), we have

$$\xi_e = \frac{1}{4\pi} \frac{1}{\sqrt{g*}} \, (\mathbb{C}x^D)^{\text{-}\frac{1}{d\text{-}D}}$$

hence

$$\Gamma = \frac{1}{4\pi} \frac{1}{\sqrt{g*}} = 0.18 \tag{23}$$

The method of Broseta[5] et al. consists in deriving ξ_e from the osmotic compressibility

$$H(\vec{0}) = \frac{d\text{-}D}{d} \frac{1}{F_\infty} \xi_\kappa^d \tag{24}$$

Using (16) we write

$$H(\vec{0}) = \frac{4 \, \xi_\kappa}{4 \frac{d}{d\text{-}D} F_\infty \xi_\kappa^{\text{-}2}}$$

We find

$$\Gamma = \frac{1}{2} \frac{1}{\sqrt{\frac{4d}{d\text{-}D} F_\infty}} \simeq 0.21 \tag{25}$$

The discrepancy with (23) shows that the exact calculation of structure functions is still a difficult task. The experimental value is $\Gamma = 0.19\pm0.01$.

B. Screening due to chain ends

The singularity $(1/q^2)$ is naturally removed, when the polymer chain is finite

$$\frac{1}{q^2} \rightarrow \frac{1}{q^2 + \dfrac{1}{R^2}}$$

In general there is a competition between finite end and repulsive interaction effects. The screening length corresponds to a pole in the complex q plane. Des Cloizeaux[4] showed that the pole closest to the real plane is given by :

$$\xi_{er} = 0.371 \sqrt{s} \tag{26}$$

(in the limit of infinite repulsive interaction). However, simple extrapolation of the structure function yields ξ_e.

An interesting application of the screening effect due to finite ends, has been developed by Benoit[6] and collaborators, who studied transesterification reactions in copolymer melts. This results can be summarized as follows. At time $t = 0$, polyester H and polyester D form a liquid melt. The scattered intensity I is

$$\frac{\mathcal{N}(b_H - b_D)^2}{I} = \frac{1}{xN_D h_D} + \frac{1}{1-x)N_H h_H} \tag{27}$$

where $\mathcal{N}$ is the total number of monomers and x the deuterated fraction. At time 0+ the melt is heated up : the polyester undergo scission and recombination. During the chemical process $(qR > 1)$

$$\frac{\mathcal{N}(b_H - b_D)^2}{I} = \frac{q^2 R^2 / N}{12x(1-x)} + Z(t) \tag{28}$$

where $Z(t)$ is a pure number related to the screening length

$$Z(t) = Z_o + \frac{1}{2x(1-x)} [1 - \exp(-t/\tau)]$$

where $\tau = \dfrac{2}{kN_T}$ is the rate of scission. The effect of the chemical reaction is to make block copolymers. The main contribution of the screening to the correlation function is now due to the ends of the block sequences, in addition to the natural ends of the chain.

Thus the screening concepts is used here to follow the structure of a system undergoing chemical reactions toward a state of greater homogeneity.

270

IV. CHARGE STRUCTURE FUNCTION

We now consider the concentration fluctuations of a species $\mathcal{A}$, whose average concentration is $\mathbb{C}_{\mathcal{A}}$. The partial structure function $H_{\mathcal{A}\mathcal{A}}(q)$ is defined by

$$H_{\mathcal{A}\mathcal{A}}(q) = \int d^3r \; e^{iq \cdot r} \left\{ \frac{\langle \mathbb{C}_{\mathcal{A}}(0)\mathbb{C}_{\mathcal{A}}(r) \rangle}{\mathbb{C}_{\mathcal{A}}^2} - 1 \right\} \tag{29}$$

and the partial scattered intensity is

$$I_{\mathcal{A}} = b_{\mathcal{A}}^2 \; \mathbb{C}_{\mathcal{A}}^2 \; H_{\mathcal{A}\mathcal{A}}(q) \tag{30}$$

We are interested in the particular case where $\mathcal{A}$ represents the charge $(\mathcal{A} \rightarrow Z)$

$$\mathbb{C}_Z(r) = \mathbb{C}^+(r) - \mathbb{C}^-(r)$$

On the average $\mathbb{C}_Z = 0$, because of electrical overall neutrality. We can write formally :

$$\mathbb{C}_Z^2 \; H_{ZZ} = \int d^3r \; e^{iq \cdot r} \left\langle (\mathbb{C}^+(r) - \mathbb{C}^-(r))(\mathbb{C}^+(0) - \mathbb{C}^-(0)) \right\rangle$$

$$= (\mathbb{C}^+)^2 \; H^{++}(q) + (\mathbb{C}^-)^2 \; H^{--}(q) - 2\mathbb{C}^+\mathbb{C}^- \; H^{+-}(q). \tag{31}$$

The charge structure function obeys characteristic sum rules.

Combining the linear responses to external force and electric fields we have[7] :

$$\left. \begin{array}{l} i\vec{q} \cdot \vec{\mathcal{D}}(q) = 4\pi e \\[2mm] i\vec{q} \cdot \vec{\mathcal{E}}(q) = 4\pi \; \{e + e\mathbb{C}_Z(q)\} \end{array} \right\} \tag{32}$$

$$\epsilon(q) = \mathcal{D}(q)/\mathcal{E}(q) \tag{33}$$

$$\frac{1}{\epsilon(q)} = 1 - \frac{4\pi \; e^2\beta\mathbb{C}}{q^2} \left(\mathbb{C}_Z^2 \; H_{zz}(q) \right) \tag{34}$$

Perfect screening implies :

$$\lim_{q \to 0} \frac{1}{\epsilon(q)} = 0 \tag{35}$$

Debye Hückel screening implies

$$\lim_{q \to 0} \epsilon(q) = 1 + \frac{4\pi \; \mathbb{C}e^2\beta}{q^2} \tag{36}$$

As a result

$$\mathbb{C}_Z^2 H_{zz}(q) = \frac{q^2}{q^2 + \xi_D^{-2}} \simeq q^2\xi_D \quad (\text{for } q \; \xi_D < 1) \tag{37}$$

Stillinger[8] and Lovett propose a different derivation of the sum rules (35) and (37).

The essential consequence of screening is to built up a structure of the charge distribution.

In neutral systems, similar correlations exist, but at zero average contrast. Consider a block copolymer made of two equal sequences ($\alpha = 1,2$) which are chemically identical and which differ only by their isotopic composition (deuteration). In the melt state the scattered intensity reads[9]

$$I = (b_1 - b_2)^2 \; \frac{C_1^2 H_{11} \, C_2^2 H_{22} - C_1^2 C_2^2 H_{12}^2}{C_1^2 H_{11} + C_2^2 H_{22} + 2C_1 C_2 H_{12}}$$

$$\propto q^2 \, \frac{R^2}{6} \quad , \quad qR < 1$$

In the case of a dilute solution (and $qR < 1$)

$$I = b^2(x) - \frac{q^2}{3} \sum_{\alpha=1}^{2} \sum_{\beta=1}^{2} \frac{1}{2} b_\alpha(x) \, b_\beta(x) \, \frac{N_\alpha}{N} \frac{N_\beta}{N} R_{\alpha\beta}^2 \qquad (38)$$

where

$$R_{\alpha\beta}^2 = \frac{1}{N_\alpha N_\beta} \sum_{j \in \alpha} \sum_{\ell \in \beta} \left\langle (r_j - r_\ell)^2 \right\rangle$$

and

$$b(x) = \sum_\alpha N_\alpha b_\alpha / N$$

At zero average contrast, $b(x_0) = 0$, $b_1(x_0) = -b_2(x_0)$.

The coefficient of interest is then

$$b_1^2(x_0) \left[R_{11}^2 + R_{22}^2 - 2 R_{12}^2 \right] = -b_1^2(x_0) \, R_{G1G2}^2$$

where R_{G1G2}^2 is the square average distance between the centers of the sequences. The result is :

$$\frac{I}{b_1^2(x_0)} = \frac{q^2 \, R_{G1G2}^2}{3} \qquad (39)$$

Comparing (37) and (39) we note that electrostatic interaction and chain correlations have similar effects on the structure functions, at least in the lower q range.

We may now combine electrostatic and chain like correlations, as found in polyelectrolytes. For the case of a single polyion, J.M. Victor[9] obtained :

$$\mathbb{C}_z^2 \; H_{zz} = \frac{q^2}{q^2 + \xi_D^{-2}} \frac{\left(1 + \frac{q^2 S}{q^2 + \xi_D^{-2}} \, h(q) \right)}{2} \qquad (40)$$

The charge structure function is characterized here by a maximum. The question arises as to whether the dielectric response function $\epsilon(q)$ also follows a similar pattern.

V. CONCLUSION

We have examined screening effects in electrolytes and in (neutral) polymer solutions. The observed similarities leads us to compare dielectric and density susceptibilities in multicomponent systems. Scattering experiments zero average contrasts are proposed, in order to determine the "charge" structure and its screening.

REFERENCES

1. P. Debye and E. Hückel, Physik. Zeit. 9:185 (1923).
2. S.F. Edwards, Proc. Phys. Soc. (London) 88:265 (1966).
3. G. Jannink and P.G. de Gennes, J. Chem. Phys. 48:2260 (1968).
4. J. des Cloizeaux et G. Jannink, "Les Polymères en Solution : leur Modélisation et leur Structure", Ed. de Physique, Les Ulis. (1987).
5. D. Broseta, L. Leibler, A. Lapp and C. Strazielle, Europhys. Lett. 2:733 (1986).
6. H. Benoit, E.W. Fisher, Zackmann, Polymer, 30:379 (1989).
7. D. Pines and P. Nozières, "The Theory of Quantum Liquids", Benjamin Inc., New-York (1966).
8. F. Stillinger and R. Lovett, J. Chem. Phys. 49:1991 (1968).
9. L. Leibler and H. Benoit, Polymer 22:195 (1980).
10. J.M. Victor, Thèse, Paris (1988).

IN SEARCH OF SCALING LAWS IN POROUS SILICA GELS

J. M. Drake*, P. Levitz** and J. Klafter***

*Exxon Research and Engineering Company
 Clinton Township, Annandale, NJ 08801, USA

**CNRS, CRSOCI, 47051-Orleans, Cedex 2, France

***School of Chemistry, Tel Aviv University
   Tel Aviv 69978, Israel

INTRODUCTION

The role of surface morphology in adsorption processes has been a long standing problem. When adsorption takes place on the interior surface of porous materials (i.e. silica gel, alumina oxide, phase separated glasses, etc.), describing the adsorption process through the thermodynamics of the adsorbates/surface interactions can be complex, making it difficult to interpret the chemical behavior of these systems. These types of porous materials are used in both separation and/or catalytic processes throughout the chemical industry. The selection of a porous material for a specific application is done in part by understanding the interplay between the morphological features of the pore interface and the adsorption and transport properties of adsorbates confined within these pore networks[1-3].

It has recently been claimed that the interfaces of many porous materials, including certain silica gels, are fractally rough[4], thereby requiring that the general morphology of the surface is independent of the scale at which the surface is probed. This claim is based on the premise that the number of identical molecules needed to form a monolayer (N_m) on a fractal surface should scale with the adsorption cross-section of the adsorbates (σ) as:

$$N_m \propto \sigma^{-D/2},$$ (1)

where D is the fractal dimension[4]. Eq.(1) constitutes a general scaling relationship which holds for all molecules independent of their specific interactions with the surface. The only relevant parameter is σ. Confirmations for such a scaling relationship have been reported by analyzing adsorption data from the literature[4,5]. The idea of fractal roughness of pore interfaces appeared to introduce an intriguing simplification into the description of the structure of these complex materials and therefore attracted a lot of attention in related fields[6-8]. If applicable, Eq.(1) offers an important tool to predict the surface area "seen" by molecules of different sizes σ. It is exactly this possibility that led us to critically evaluate the

Large-Scale Molecular Systems, Edited by W. Gans *et al.*
Plenum Press, New York, 1991

applicability of the fractal concept to pore interface of silica gel. At the same time we addressed the more general question: can the morphology of a surface be probed uniquely through the adsorption process? We found no evidence for fractal interfaces in these porous silica gels.

ADSORPTION FROM BINARY SOLUTIONS

Our focus in this contribution is on low surface capacities which correspond to submonolayers. This adsorption range was originally proposed for fractal surface analysis[4,5]. In order to confirm the applicability of Eq.(1) one needs an *accurate measure* of N_m, and a *reliable value* of σ. Obtaining accurate and reliable measures of these parameters is by no means an easy task.

We have investigated a series of porous silica gels, Si-40, Si-60 and Si-100 (E. Merk, Darmstadt, Germany), which have been claimed to have fractal surfaces. The reader is referred to refs. 1,3,9 for a detailed characterization of these materials. We have performed adsorption measurements on all three silicas using alcohol in toluene solutions (methanol, ethanol, 1-propanol, 2-propanol, 1-butanol, 2-butanol and tertiary amyl alcohol) in an attempt to reproduce previous results reported by Hoffman et al.[10] and by Pfeifer et al.[11].

Adsorption from binary solutions is a complicated equilibrium process[12-14]. Since there are no vacancies in either the surface layer or bulk solution, only substitution of the molecules of one component by those of the other is possible at the interface. Therefore, there is no direct method to follow the adsorption of a single component from the liquid phase onto the surface. Characteristic of adsorption from solutions are excess (net) isotherms of the type exhibited in Fig. (1). This is the excess isotherm of ethanol in toluene adsorbed on Si-60. The very low surface coverages were measured by analyzing the equilibrium alcohol solutions by gas chromatography (GC) and the high coverages by solution refractive index changes (RI) . The excess adsorption isotherms are not very instructive by themselves and one can, making certain assumptions, resolve them into individual isotherms as discussed in refs. 12-14. The resolved isotherms which correspond to the system in Fig. (1) are shown in Fig. (2). At very low coverages the resolved alcohol isotherms overlap with the excess isotherms. Determining monolayer coverages from either the excess or the individual isotherms is problematic especially at higher coverages because the adsorption process is dominated by adsorbate-adsorbate interactions. One is limited then to analyzing the low surface coverage portion of the isotherm to obtain an estimate of N_m. The analysis at low coverages is done using a Langmuir type model which leads to the Everett expression:

$$N_2^s = \frac{N_m bC_s}{1+bC_s} , \qquad (2)$$

where C_s is the equilibrium alcohol solution concentration in moles per liter N_2^s is the surface concentration of alcohol per unit mass of silica and b is given by the equilibrium constant (K) for adsorption normalized by solvent concentration. How low should the coverages be to confidently use Eq.(2) is not clear. In Fig.(3) we fit the very low coverage portion of the excess isotherm to Eq.(2), using N_m and b as adjustable parameters to obtain the best fit to the isotherm data.

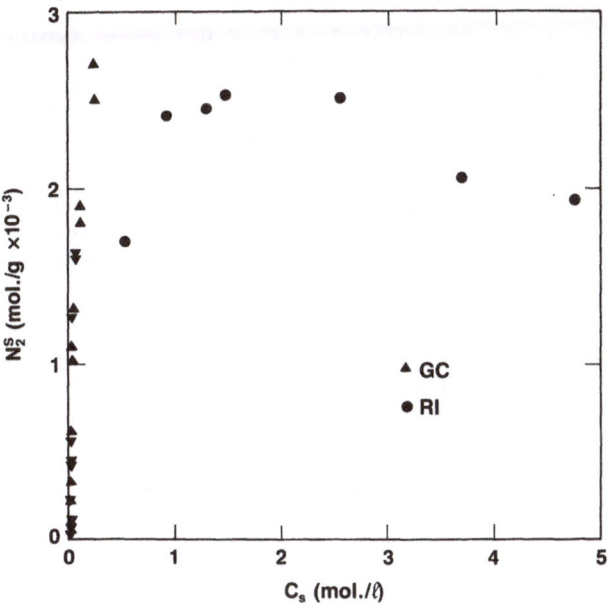

Fig. 1. The excess isotherm of ethanol/toluene on Si-60.

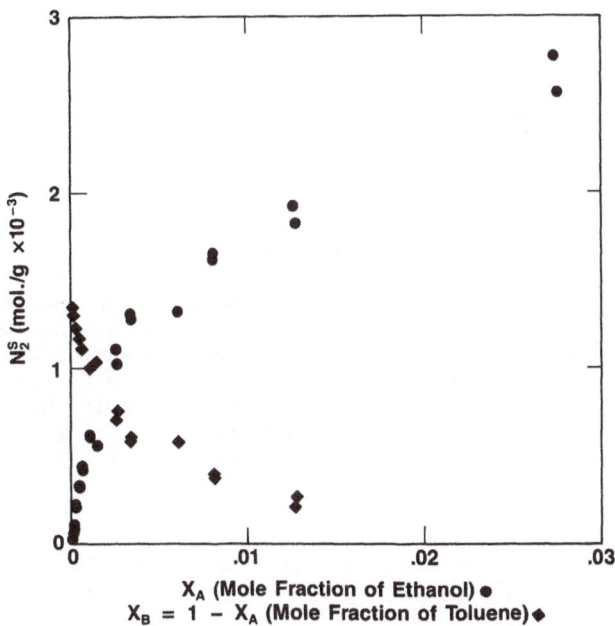

Fig. 2. Individual (resolved) isotherms of ethanol and toluene on Si-60.

Note that we analyze the GC data at, C_s, 0.3 $\frac{moles}{L}$ and that a high density of data points are accumulated at $C_s < 0.1 \frac{moles}{L}$ (which corresponds
to an alcohol mole fraction of less than 0.01!). The solid line in
Fig. (3) is the best fit of Eq.(2) to the data using a non-linear least
squares optimization. The residuals show a strong systematic variation
as the fit evolves from low coverage. This systematic variation suggests that the adsorption process itself changes over the concentration
range measured. This then demonstrates that the extrapolation of the
adsorption isotherm to measure N_m is unreliable and leads to large
errors in N_m. This is seen clearly in Fig. (4), where we show two
values of N_m for 1-propanol, 2-propanol, 1-butanol and t-amyl alcohol
using two different ranges of alcohol solution concentrations.

As for the other parameter needed for the scaling in Eq.(1), the
actual adsorption cross-section for each adsorbate is very difficult to
establish unambiguously. There is little agreement in the literature
about the σ values for a given homologous series of molecules[15] and it
is not clear how meaningful the concept is considering the various
possible orientations of the adsorbed molecules on the surface[16] and
the likely dependence of σ on surface coverage.

Because both N_m and σ are uncertain, adsorption is generally not
reliable in probing surface morphology and should not be applied to
verify fractal surface behavior. Furthermore the range over which the
σ vary is quite limited (16-48Å^2) as is the range of N_m (~ factor 4),
which create an additional obstacle in justifying the fractal scaling
relationship. Fig. (4) presents N_m, vs. liquid density values of σ
(used in ref. [11]) for the alcohols on Si-60. A general trend is
evident in the figure, but no scaling, as suggested in Eq.(1), is
observed. Monolayer coverages depend not only on adsorbate sizes, but
also on their interactions with the interface.

SMALL ANGLE SCATTERING

The most straightforward way to demonstrate fractal roughness is
through small angle scattering of x-rays or neutrons (SAXS, SANS)[17,18].
It has been demonstrated that for fractal surfaces,

$$I(Q) \sim Q^{D-6} \quad , \tag{3}$$

where $I(Q)$ is the scattered intensity as a function of momentum
transfer Q. For smooth surfaces $D = 2$ and one obtains the Porod law
Q^{-4}. For fractals where $2 \leq D < 3$ the exponent ranges between -3 and
-4. The insert in Fig. (5) shows SAXS data for Si-40, Si-60 and
Si-100. On the scale of Q that correspond to the surfaces simple Porod
behavior, Q^{-4}, is observed for length larger than 10 Å, namely for *our
purposes these surfaces are smooth*. No self-similarity has been
observed. It is interesting to note that the scattering curves can be
scaled in Q and in intensity in the following way[3,18,19]:

$$I(Q) = I_0(\xi)F(\xi Q), \tag{4}$$

where F is a structure function and $I_0(\xi)$ and ξ are the appropriate
parameters for each silica (Fig. (5)). The value of ξ turns out to be
given approximately by the mean pore size R_p. This observation is
basically consistent with the picture which originates from the electron microscopy results. The scaling in Fig. (5) demonstrates the
similar (but not self-similar!) morphological features of different

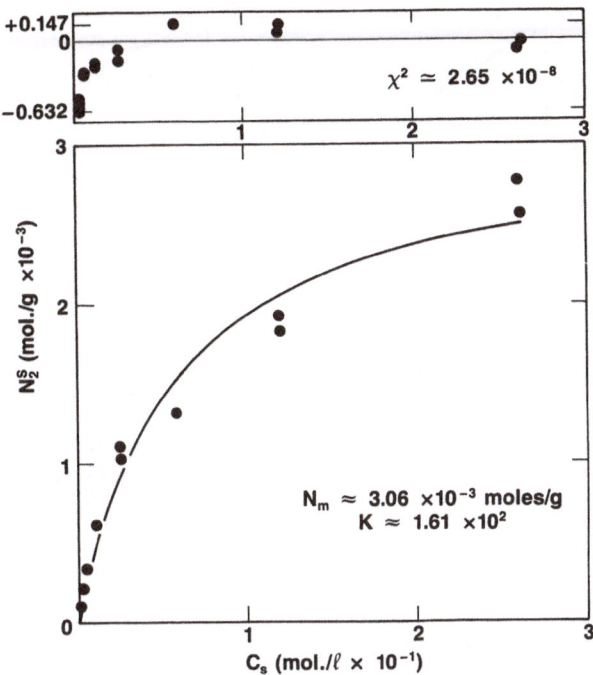

Fig. 3. Determination of N_m from the low coverage adsorption isotherm of ethanol from toluene on Si-60. The solid line is the fit to Eq. (2).

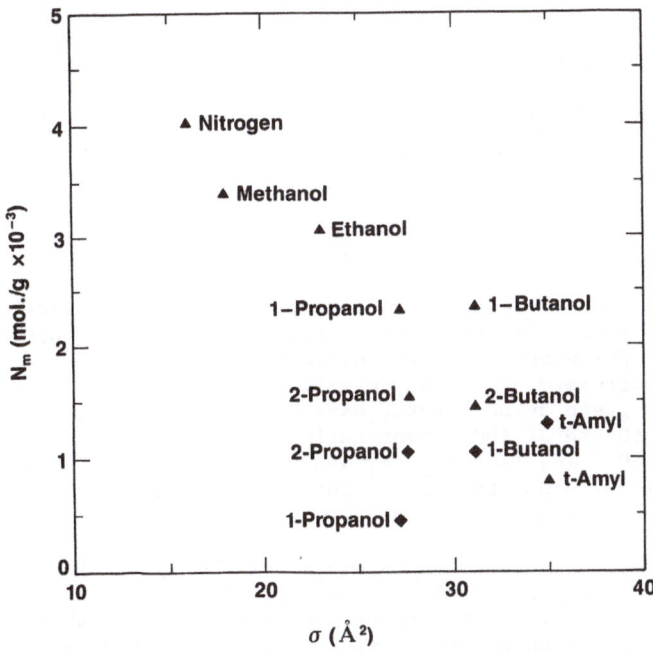

Fig. 4. Monolayer capacities, N_m, vs. cross-sections, σ, for a homologous series of alcohols on Si-60.

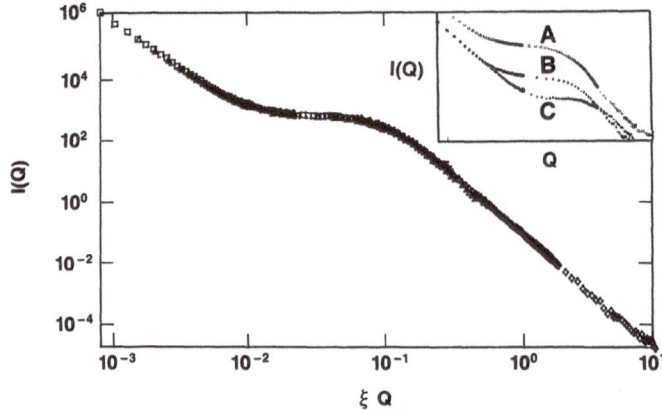

Fig. 5. The scaled behavior of small-angle x-ray scattering for Si-40,
Si-60 and Si-4000. The insert shows scattering intensities
I(Q) for (A) Si-100, (B) Si-60 and (C) Si-40.

silicas of mean pore sizes as small Si-40 and as large as Si-4000. In
the latter case there is no argument about the smoothness of the
interface[18]. Although the surfaces appear smooth the adsorption
results show richness and complexity which must originate from the
details of the specific interactions between the adsorbate and surface.

DIRECT ENERGY TRANSFER

Another measurement which may, if done on fractals, probe directly
the fractal dimension is long range dipolar energy transfer of the
Forster type among dye molecules[20,21]. The basic idea behind the
energy transfer measurements is to tag a surface with a random
distribution of donor and acceptor molecules at concentrations low
enough (again by studying the adsorption isotherms) for only a one step
transfer of the initially excited donor to the acceptors. One then
monitors the fluorescence of the donors which should follow the
pattern[21]:

$$\Phi(t) = \exp[-At^{D/6}], \tag{5}$$

where A depends on the concentration and detailed donor-acceptor
interaction. Energy transfer probes length scales up to approximately
100 Å[3,9,22]. For a more detailed discussion see refs. 3 and 9.
Experiments performed on the three silicas, with rhodamine 6G as donors
and malachite green as acceptors, have been fit by Eq. (5) with
D ~ 2 for Si-60 and Si-100 demonstrating again smooth surfaces and
D ~ 3 for Si-40. The latter case, when corroborated with SAXS results
reveals the local connectedness of the pore network. Again no
self-similar behavior has been observed.

In conclusion, adsorption is not a reliable probe for surface
morphology. Clearly it should not be applied when attempting to verify
surface self-similarity. In the case of silica gels both SAXS and
energy transfer probes do not provide evidence for fractal surface
structure, which has been erroneously claimed before. All this does
not rule out, of course, the existence of other fractal surfaces.
However, each case should be carefully checked.

REFERENCES

1. J. M. Drake and J. Klafter, <u>Physics Today</u> 43:46 (1990).
2. <u>Molecular Dynamics in Restricted Geometries</u>, ed. J. Klafter and J. M. Drake, John Wiley, New York (1989).
3. J. M. Drake, P. Levitz, S. K. Sinha and J. Klafter, <u>Chem. Phys.</u> 128:199-207 (1988).
4. D. Avnir, D. Farin, and P. Peiffer, <u>Nature</u> 308:261-263 (1984).
5. D. Avnir and D. Farin, <u>New J. Chem.</u> 14:197 (1990).
6. R. Kopelman, <u>Science</u>, 241:1620-1626 (1988).
7. A. Katz and A. Thompson, <u>Phys. Rev. Lett.</u> 54:1325-1328 (1985).
8. H. D. Bale and P. W. Schmidt, <u>Phys. Rev. Lett.</u> 53:596-599 (1984).
9. P. Levitz, J. M. Drake, and J. Klafter, <u>J. Chem. Phys.</u> 89:5224-5236 (1988).
10. R. L. Hoffman, D. G. McConnel, G. R. List and C. D. Evans, <u>Science</u> 90:550-551 (1967).
11. P. Pfeifer, D. Avnir, and D. Karin, <u>J. Stat. Phys.</u> 36:699-716 (1984).
12. J. Oscik, <u>Adsorption</u> (Ellis Horwood, Chichester) (1982).
13. J. J. Kipling, <u>Adsorption from Solution of Nonelectrolytes</u>, Academic Press, London (1965).
14. S. Schay and L. G. Nagy, <u>J. of Colloid and Interface Sci.</u> 38:302-311 (1972).
15. A. L. McClellan and H. F. Harnsberger, <u>J. Colloid. and Interface Sci.</u>, 23:577-599 (1967).
16. J. M. Drake, P. Levitz and J. Klafter, <u>New J. Chem.</u> 14:77 (1990).
17. S. B. Ross, D. M. Smith, A. J. Hurd and D. W. Schaefer, <u>Langmuir</u> 4:977 (1988).
18. P. W. Schmidt, <u>in</u>: "Characterization of Porous Solids," K. K. Unger, J. Rouquerol, K. S. W. Sing and H. Karl, ed., Elsevier, Amsterdam (1988).
19. J. M. Drake, P. Levitz, and S. Sinha, <u>Mat. Res. Soc. Symp. Proc.</u> 73:305 (1986).
20. T. Forster, Z. <u>Naturforsch</u> 4A:321 (1949).
21. J. Klafter and A. Blumen, <u>J. Chem. Phys.</u> 80:874 (1984).
22. P. Levitz and J. M. Drake, <u>Phys. Rev. Lett.</u> 58:686 (1987).

STOCHASTIC ASPECTS IN REACTION KINETICS

Alexander Blumen and Horst Schnörer

Physics Institute and BIMF
University of Bayreuth
P.O. Box 101251
D-8580 Bayreuth
Federal Republic of Germany

I. INTRODUCTION

Randomness occurs in many areas of modern physics. For example, in the past decade interest has turned increasingly towards the investigation of amorphous solids such as polymers and glasses. Furthermore randomness can occur in many forms. Thus, transport properties of spatially random systems (mixed crystals, alloys) are triggered by a distribution of microscopic (site-to-site) transfer rates (temporal disorder) and by different interactions with the surroundings (energetic disorder).

That glasses and polymers are substances with enormous potential in the applied sciences is self-evident; that the same materials display features making them one of the favorite testing-grounds for new theoretical concepts is not so well-known. In the authors' opinion this underscores the fact that progress in natural sciences occurs historically. We mean by this that formerly disjoint fields are suddenly perceived as related, and thereafter develop in parallel. Thus in materials science theoretical modeling and chemical synthesis are, in fact, much more intimately linked than what one might naively expect.

An area in which randomness is central is that of reactions under diffusion-limited conditions; this is a classical topic of physical chemistry.[1,2] In this field recently developed analytical methods and the increased computing power have revealed intriguing, unexpected features.[3-8] These investigations have rendered evident many qualitative

Large-Scale Molecular Systems, Edited by W. Gans *et al.*
Plenum Press, New York, 1991

departures from the accepted Smoluchowski-type decay laws, when reactions in confined geometries (as found in low-dimensional systems or in porous media) are studied.

In this article we center on the A+B→0 and A+B→B reactions, both in regular spaces and also on fractals. In most cases we will model the dynamics through random walks, and let the disorder be represented through its spatial aspect (i.e. through fractals). More sophisticated ways for modeling disorder (continuous-time random walks, CTRW, or hierarchical energetic structures such as ultrametric spaces, UMS) are beyond the scope of this work; we refer the interested reader to other review articles.[9-11]

Basic to our understanding of disorder is the temporal evolution of the systems under investigation. Here we will review reactions of widespread use, such as pseudo-unimolecular and bimolecular decays. The analysis of actually observed relaxation patterns will reveal that long-time decays in disordered materials seldom have kinetical chemical counterparts; the reason behind this is that the kinetical scheme implicitly assumes a 'well-stirred' chemical reactor. Most disordered media lack such an external homogenization; this leads then to interesting deviations from the kinetic picture.

Let us note that many methods used to probe disordered materials can be viewed as being generalized chemical reactions. Thus the capture of an electron by a scavenger is a reaction in which (at the microscopic level) one free electron and one active scavenger disappear. The same holds for the electron-hole recombination in glasses: at each recombination step one electron and one hole are annihilated. Also the sensitized energy transfer in glasses (by which the energy is exchanged between different molecular species) may be modeled as a reaction which changes the numbers of excited molecules of different kinds. Triplet - triplet annihilation in mixed molecular solids provides another example: here two triplets are annihilated and one excitation in the singlet state appears; the whole process is followed by the delayed fluorescence of the singlet. Furthermore, chemical analogs appear in several models for dielectric relaxation; these are based on the assumption that local strains relax only when free defects (or volumes) which roam through the glass happen to be in their vicinity. Evidently, the reactions involved here are of A+B→0, A+B→B or A+A→0 type.

This article is structured as follows: In Sec. II we focus on the direct energy transfer between donors and acceptors, which may be interpreted as an A+B→B reaction with immobile reactants. Sec. III is devoted to bimolecular annihilation reactions of A+B→0 type, where A and B are again immobile. In Sec IV and V we consider the diffusion-controlled reactions A+B→B and A+B→0, respectively, where A and B react on contact after undergoing diffusive motion. We summarize our findings in Sec. VI.

II. THE DIRECT ENERGY TRANSFER (A+B→B)

In this section we focus on the kinetics of the A+B→B reaction when the reactants are immobile and use the lines of approach developed by us in Refs. 12 and 13. A major example of this kind of reaction is provided by the relaxation of an excited donor (A) due to the direct energy transfer to randomly distributed acceptors (B). The direct energy transfer from an excited donor molecule to acceptors embedded in condensed media is of considerable interest; starting with the work of Förster[14] the problem has been extensively studied, both theoretically[15-20] and experimentally.[17,21,22] Here we summarize the results which obtain both when the systems are homogeneous and infinite and also when limitations occur (restricted geometries and situations where the acceptor density varies from place to place).

II.1. General Theory

We start from the general formalism for the relaxation of an excited donor located at site $\vec{r}_0$ due to the direct energy transfer to acceptors that occupy some of the sites $\vec{r}_i$ of a given structure. Here the important feature is that the transfer rates $w(\vec{r})$ depend on the mutual distance r between each donor-acceptor pair. Thus the probability of the decay of the donor due to the presence of an acceptor at $\vec{r}_i$ is:

$$f(t,\vec{r}_i,\vec{r}_0) = \exp[-tw(\vec{r}_i-\vec{r}_0)] \quad . \tag{2.1}$$

Furthermore, in a very good approximation the acceptors act independently; this means that they contribute multiplicatively to the decay. Let $g(j)$ be the probability of having j acceptors at one site. The decay of the donor is given by[12]

$$\Phi(t,\vec{r}_0) = \prod_{\vec{r}_i} \left\{ \sum_j g(j)[f(t,\vec{r}_i,\vec{r}_0)]^j \right\} \quad , \tag{2.2}$$

where the product extends over all the sites of the underlying structure, with the exception of the donor location $\vec{r}_0$.

Consider the case of a substitutional occupancy of sites by acceptors with probability p, namely a binomial distribution:

$$g(j) = (1-p)\delta_{0,j} + p\delta_{1,j} \quad , \tag{2.3}$$

where each site contains either, with probability 1-p, no acceptor or, with probability p, exactly one acceptor and $\delta_{i,j}$ is the Kronecker delta. Inserting Eq. (2.3) into Eq. (2.2), one obtains[12,18,19]

$$\Phi(t,\vec{r}_0) = \prod_{\vec{r}_i} \left\{ 1 - p + p \exp[-tw(\vec{r}_i - \vec{r}_0)] \right\} \quad , \tag{2.4}$$

which for a small acceptor concentration, p≪1, may be approximated by[12,19]

$$\Phi(t,\vec{r}_0) = \prod_{\vec{r}_i} \exp\left[- p \left\{ 1 - \exp[-tw(\vec{r}_i - \vec{r}_0)] \right\} \right]$$

$$= \exp\left[- p \sum_{\vec{r}_i} \left\{ 1 - \exp[-tw(\vec{r}_i - \vec{r}_0)] \right\} \right] \quad . \tag{2.5}$$

In the following also cases in which the donor is located on the surface of pores or on fractals will be considered. The decay laws Eqs. (2.4) and (2.5) depend explicitly on the location $\vec{r}_0$ of the donor. In general (when the donor itself is randomly located) one has to average these decay laws with respect to $\vec{r}_0$.

Starting point of the investigations is Eq. (2.5), which contains the basic ingredients of the decay. By introducing a site density function $\rho_0(\vec{r})$, one can transform the sum in Eq. (2.5) to the usual integral form. One writes:

$$\rho_0(\vec{r}) = \sum_{\vec{r}_i} \delta(\vec{r} - \vec{r}_i) \quad , \tag{2.6}$$

where the index 0 in $\rho_0(\vec{r})$ acts as a reminder that $\vec{r}_0$ is excluded from the sum on the right-hand side. With $\rho_0(\vec{r})$ one obtains:

$$\Phi(t,\vec{r}_0) = \exp\left[- p \int d\vec{r}\, \rho_0(\vec{r}) \left\{ 1 - \exp[-tw(\vec{r}_i - \vec{r}_0)] \right\} \right] \quad . \tag{2.7}$$

286

From Eq. (2.7) one now arrives at the Förster-type decays by taking $\rho_0(\vec{r}) \equiv \rho = $ const. and extending the integration over the whole space:

$$\Phi(t,\vec{r}_0) = \exp\left[- p\rho \int d\vec{r} \left\{ 1 - \exp[-tw(\vec{r}_i-\vec{r}_0)] \right\} \right] . \tag{2.8}$$

To obtain the particular decay patterns which correspond to the situation at hand it is now only necessary to specify the interaction $w(\vec{r})$ and to perform the integration in Eq. (2.8). Often encountered forms for $w(\vec{r})$ are:

$$w(\vec{r}) = \alpha r^{-s} \tag{2.9}$$

for isotropic multipolar interactions,[14] and

$$w(\vec{r}) = \alpha e^{-\kappa r} \tag{2.10}$$

for isotropic exchange interactions.[15] The parameter s in Eq. (2.9) equals 6 for dipole-dipole, 8 for dipole-quadrupole and 10 for quadrupole-quadrupole interactions, and the parameter κ in Eq. (2.10) is a measure for the range of the exchange interaction. Evidently, Eqs. (2.9) and (2.10) are only approximations to realistic transfer rates, which may have a considerably more complex structure.[23] Inserting now Eq. (2.9) into Eq. (2.8) leads to

$$\Phi(t) = \exp(-At^{-d/s}) . \tag{2.11}$$

Here d is the spatial dimension of the underlying lattice and A is a time-independent constant

$$A = V_d p\rho\Gamma(1-d/s)\alpha^{d/s} , \tag{2.12}$$

where V_d is the volume of the unit sphere in d dimensions and $\Gamma(z)$ is the Euler-gamma function. One should note that Eq. (2.11) has a stretched-exponential form.

Considering now exchange interactions, and inserting Eq. (2.10) into Eq. (2.8) gives

$$\Phi(t) = \exp[-Bg_d(\alpha t)] , \tag{2.13}$$

where B is again a time-independent constant

$$B = V_d p\rho\kappa^{-d} \tag{2.14}$$

and $g_d(z)$ is an analytical function of z (see Ref. 24 for details). For longer times, $\alpha t \gg 1$, one has[24]

$$g_d(\alpha t) \simeq \ln^d(\alpha t) \tag{2.15}$$

and one obtains from Eq. (2.13)

$$\Phi(t) = \exp[-B\ln^d(\alpha t)] \ . \tag{2.16}$$

Expression (2.17) is an example for an exponential-logarithmic decay.[10]

II.2. Restricted Geometries

In this subsection we display the role of confined geometries on the direct energy transfer. A main result of this subsection consists in the appearance of crossover effects in the decay patterns.[25,26]

The role of confined geometries on the decay is amenable to an analytical study by introducing a site-density function $\rho_0(\vec{r})$. As first example consider a donor positioned on the surface of a pore. To simplify matters, for the donor-acceptor transfer the multipolar interaction, Eq. (2.9), will be used. Thus one has to evaluate the following integral, which appears in the exponent of Eq. (2.7):[25]

$$I(t,\vec{r}_0) = \int d\vec{r} \ \rho_0(\vec{r}) \left\{ 1 - \exp[-tw(\vec{r}_i - \vec{r}_0)] \right\} . \tag{2.17}$$

The geometry enters in Eq. (2.17) through a judicious choice of $\rho_0(\vec{r})$. Thus, say for a spherical pore, one may take $\rho_0(\vec{r})$ as being constant inside the sphere and zero outside. Hence let us concentrate on a spherical pore of radius R and calculate the energy decay of an excited donor placed on the surface of the sphere, say at $\vec{R}_0 = (0,0,R)$ on the z axis. In order to find a simple expression for $\rho_0(\vec{r})$, it is convenient[25] to use spherical coordinates centered at $\vec{R}_0$. Since $w(\vec{r})$ of Eq. (2.9) depends only on r, one needs the r dependence of $\rho_0(\vec{r})$. For this we follow Ref. 25 and evaluate the volume comprised between two spherical shells centered at $\vec{R}_0$ and belonging to the sphere:

$$\rho_0(r)dr = \rho r^2 dr \int_0^{2\pi} d\psi \int_0^{\theta_{max}} \sin\theta \ d\theta = 2\pi r^2 \left(1 - \frac{r}{2R}\right) \rho \ dr \tag{2.18}$$

with $\theta_{max} = \arccos(r/2R)$. Inserting Eq. (2.18) into Eq. (2.17), one obtains

$$I(t,\vec{R}_0) = 2\pi\rho \int_0^{2R} dr\ r^2[1-\exp(-t\alpha r^{-s})] - \frac{\pi\rho}{R}\int_0^{2R} dr\ r^3[1-\exp(-t\alpha r^{-s})]\ .\quad (2.19)$$

For physical interactions $(s\geq 6)$ both integrals on the right-hand side of Eq. (2.19) exist. If $R\to\infty$, one can neglect the second term, and then:

$$\Phi(t,\vec{R}_0) = \exp[-(2\pi/3)p\rho\Gamma(1-3/s)(\alpha t)^{3/s}]\ .\qquad (2.20)$$

This result is very similar to the Förster decay in an infinite three-dimensional space:[19]

$$\Phi(t) = \exp[-(4\pi/3)p\rho\Gamma(1-3/s)(\alpha t)^{3/s}]\ .\qquad (2.21)$$

The difference by a factor of 2 in the exponents of Eqs. (2.20) and (2.21) is due to the fact that the donor on the surface of the sphere sees, as $R\to\infty$, only one-half of the space occupied by acceptors.

Again following Ref. 25 we consider the corrections due to the finite volume. At long times $t\gg(2R)^s/\alpha$, the $\exp(-t\alpha r^{-s})$ terms in Eq. (2.19) may be neglected and

$$I(t,R_0) = 2\pi\rho(2R)^3/3 - (\pi\rho/R)(2R)^4/4 = (4\pi/3)\rho R^3\ .\qquad (2.22)$$

The decay according to Eq. (2.7) is for large t

$$\Phi(t) = \exp[-p\rho(4\pi/3)R^3]\qquad (2.23)$$

and is hence a nonzero, time-independent value. The decay is given in terms of the mean number N of acceptors inside the spherical volume $V=4\pi R^3/3$, that is, $N=p\rho V$. Stated differently, the probability that the donor does not decay at all, $\exp(-N)$, is simply that of not finding any acceptor inside V. This result follows in general fashion from an expression based on the Poisson distribution, see Ref. 13 for details. At long times t, for any finite volume V such that $tw(\vec{r})\gg 1$ for all $\vec{r}$ inside V, one can neglect $\exp(-tw)$ in Eq. (2.17) and thus $I(t,\vec{r}_0)=\rho V$.

The short-time behavior of Eq. (2.19) may be found[25] along lines similar to those leading to Eq. (2.20). For small times one has:

$$I(t,R_0) = \frac{2\pi\rho}{3}(\alpha t)^{3/s}\Gamma(1-3/s) - \frac{\rho}{4R}(\alpha t)^{4/s}\Gamma(1-4/s) + \ldots\ ,\qquad (2.24)$$

that is the Förster decay for the half-space derived for $R\to\infty$ [Eq. (2.20)] corrected so as to obtain slower decay at moderate times $t < (2R)^s/\alpha$. Due to the shape of the restricted space, there is a crossover in the donor decay law from a three-dimensional Förster-type behavior $[t \ll (2R)^s/\alpha]$ to a time-independent limit $[t \gg (2R)^s/\alpha]$. The geometry slows down the donor relaxation. It should be noted that the calculation does not explicitly include the excitation's lifetime.

The next example represents the effects of a cylindrical volume on the energy transfer process, and starts from an infinite cylinder of radius R. The acceptors are distributed within the volume, while the donor is located on the surface of the cylinder at the origin $\vec{0} = (0,0,0)$. The axis of the cylinder lies at a distance R from the origin along the z direction. The site density function $\rho_0(\vec{r})$ for this case is

$$\rho_0(\vec{r})\,drdz = 2r\rho \arccos(r/2R)\,dr\,dz \ . \tag{2.25}$$

Inserting Eq. (2.25) into Eq. (2.17) leads to:[25]

$$I(t,\vec{0}) = 4 \int_0^{2R} dr \int_0^{\infty} dz \ r\rho \arccos\left(\frac{r}{2R}\right) \{1 - \exp[-t\alpha(r^2 + z^2)^{-s/2}]\} \ . \tag{2.26}$$

In the cylindrical case, as in the spherical-volume example, at short times $t \ll R^s/\alpha$ one recovers the Förster-type decay [Eq. (2.21)], where the donor sees only half of the space occupied by the acceptors. For the long-time decay, $t \gg R^s/\alpha$, Eq. (2.26) yields the Förster decay corresponding to one dimension:

$$\Phi(t,0) = \exp[-\rho\rho 2\pi R^2 \Gamma(1-1/s)(\alpha t)^{1/s}] \ . \tag{2.27}$$

Thus, due to the cylindrical geometry, the decay shows a crossover between a three-dimensional and a one-dimensional Förster-type relaxation. Here again the geometrical restriction induces a slower relaxation of the donor. Other examples that deal with molecules adsorbed on the surface of spheres and cylinders can be found in Refs. 20 and 27.

II.3. Fractals

Another type of decay that can be obtained by focusing on $\rho_0(\vec{r})$ is the direct transfer to acceptors randomly distributed on fractals. Fractal structures are self-similar and display dilatational symmetry; their site density $\rho_0(\vec{r})$ can be defined in terms of the Euclidean dimension d of the embedding space and of the Hausdorff fractal dimension $\bar{d}$ $(\bar{d} \le d)$:[28]

$$\rho_0(\vec{r}) = Ar^{\overline{d}-d} \ . \tag{2.28}$$

Here A is a proportionality constant. Many stochastically disordered structures, such as percolation clusters at criticality, linear and branched polymers, aggregates constructed by diffusion-limited growth, and various porous materials have been viewed as being fractal. However, fractals need not be stochastic. A well-known example are the deterministically built Sierpinski gaskets.[28] The Hausdorff fractal dimension of a gasket embedded in the d-dimensional Euclidean space is $\overline{d}=\ln(d+1)/\ln2$. In Fig. 2.1 we present a two-dimensional Sierpinski gasket for which $\overline{d}=\ln3/\ln2$.

Now let us display the relaxation due to acceptors randomly distributed on fractals. Inserting Eq. (2.28) into Eq. (2.7), one obtains:[29]

$$\Phi(t) = \exp\left[-4\pi pA \int dr \ r^{\overline{d}-1} \left\{ 1 - \exp[-tw(r)] \right\} \right] , \tag{2.29}$$

an expression independent of $\vec{r}_0$. For multipolar interactions [Eq. (2.9)], one then has:[29]

$$\Phi(t) = \exp(-pBt^{-\overline{d}/s}) , \tag{2.30}$$

where B is time-independent. This is an extension of the results for

Fig. 2.1. Sierpinski gasket embedded in two-dimensional space.

Euclidean dimensions d to fractal dimensions $\bar{d}$. It is interesting to note that on fractal structures the decay law due to direct energy transfer depends on the fractal dimension $\bar{d}$ only. This result differs from the indirect energy transfer, which involves the migration of the energy over the donor subsystem; in this case a different dimension, the spectral (fracton) dimension $\tilde{d}$,[30,31,10] enters. This case will be discussed in section IV. It should be noted that similar extensions of the decay laws from regular to fractal structures have been derived also for interactions mediated by exchange. Inserting Eq. (2.10) into Eq. (2.29), one has:[10,12,29]

$$\Phi(t) = \exp(-pC \ln^{\bar{d}} t) , \qquad (2.31)$$

where C is time-independent. The decay in Eq. (2.31) is again exponential-logarithmic, eg. Eq. (2.16).

As a final point we emphasize that a crossover in the decay laws is to be expected due to length scale limitations of the fractal behavior. As an example consider an infinite cluster above the percolation threshold. Here an upper cutoff exists (the correlation length ℓ); at larger length scales the system appears homogeneous. Let now ℓ denote in general the upper length limit for self-similar behavior and $w(\ell)$ the corresponding transfer rate [Eq. (2.9) or (2.10)]. Then for times such that $w(\ell)t \ll 1$, we expect the relaxation to follow Eq. (2.30) or (2.31). At longer times, such that $w(\ell)t \gg 1$, a regular relaxation (corresponding to Euclidean dimension d) is to be seen.

Thus the direct energy transfer reflects the nature of the underlying fractal structure and provides some insight into the characterization of disordered media. In this way Eq. (2.30) has been extensively applied in the analysis of the direct energy transfer in various complex systems,[32-34] such as polymers, silica gels, Vycor glass, vesicles, microemulsions, and zeolites.

In summary, we have presented examples for the influence of restricted geometries and of density fluctuations on the direct energy transfer. Here the decay laws reflect directly the spatial restrictions on the donors' and acceptors' positions; the decay often shows richer patterns than the forms obtained in regular, infinite structures. We note that all pore models studied show slower relaxations (when compared to the regular, infinite structures) and crossover effects.

III. BIMOLECULAR ANNIHILATION REACTIONS (A+B→0):
THE CASE OF IMMOBILE REACTANTS

In the preceding chapter, where we studied the A+B→B reaction, the number and positions of B particles did not change during the course of the reaction; therefore the A particles could be regarded as acting independently. This is quite different in the case of the A+B→0 reaction, where the number of B particles also changes during the course of the reaction; the reaction introduces thus correlations between the different particles present in the sample.

Bimolecular reactions between like (A+A→products) and unlike particles (A+B→products) have been extensively investigated during the last few years.[5-7,35-57] Most of the work centered on diffusion-limited reactions, where the particles react on contact after undergoing diffusion. It took a long time to realize that the usual kinetic reaction schemes are only approximate; this is so because such schemes do not reflect the spatial aspects accurately.[5,6,38] Thus, theories which treat only average particle densities cannot describe the problem properly, since spatial correlations between the reactants have to be taken into account. Although several attempts have been made to find an analytical approach to the general problem, up to now only a few exact solutions for special cases are known, which center on one-dimensional systems.[36,43,51,53]

In this section we consider the A+B→0 reaction and concentrate on immobile particles, which interact via exchange. We are interested in the occurrence of spatial correlations; these appear most vividly when the particles are immobile, since diffusion works against cluster formation.

An analytical approach to bimolecular reactions, which takes into account spatial correlations through a many-body formalism has recently been developed by Kuzovkov and Kotomin.[49] They used the Kirkwood superposition approximation[60,61] to decouple the three-body correlation functions; this approach reduces the infinite set of equations to a finite set of coupled integro-differential equations. Such equations can be solved numerically. However, since no analytical estimation for the validity of the Kirkwood approximation exists,[60] the results have to be compared to computer simulations of the full problem, in order to assess their validity.

In this section we perform this comparison for the reaction A+B→0 with immobile reactants and we show that the Kirkwood approximation works very well even in this extreme case; we find a very good agreement both for the global reactant concentrations and for the AB-correlations.

III.1. Analytical Approach

If the spatial distribution of particles is homogeneous at all times the concentrations $n_A(t)$ and $n_B(t)$ obey the following equations:[10]

$$\frac{\partial n_A(t)}{\partial t} = \frac{\partial n_B(t)}{\partial t} = - k \, n_A(t) \, n_B(t) \quad . \tag{3.1}$$

In Eqs. (3.1) the constant k is the reaction rate. Let us denote the initial concentrations by n_A^0 and n_B^0. For $n_A^0 = n_B^0$ the solution of Eq. (3.1) is:

$$n_A(t) = \frac{n_A^0}{1 + n_A^0 k t} \quad . \tag{3.2}$$

Asymptotically one has thus $n_A(t) \sim t^{-1}$. On the other hand, for $n_A^0 < n_B^0$ and setting $\Delta = n_B^0 - n_A^0$ one has:

$$n_A(t) = \frac{\Delta}{(n_B^0/n_A^0)\exp(\Delta k t) - 1} \quad . \tag{3.3}$$

Thus for large t the concentration of the minority species decays exponentially in time, $n_A(t) \sim \exp(-\Delta k t)$.

However, due to the discrete nature of the particles, spatial fluctuations in the initial distributions can occur. Then clusters of like particles (i.e. regions where only one kind of particles is present) develop in the course of the reaction. This leads to a slowing down of the overall decay, since mainly particles at the border of such clusters are able to react. In this case the decay law depends on the specific conditions of the reaction and on the restrictions imposed on the space in which the reaction takes place.

In the following we consider randomly distributed A- and B-particles, which react via exchange, $w(\vec{r}) = w(r) = w_0\exp(-r/r_0)$. Here w is the probability rate that one A and one B molecule situated at a distance r react with each other; r_0 denotes a constant, which is a measure of the range of the interaction. The exponential form is typical for exchange-dominated reactions in solids, such as scavenging or the recombination of electrons and holes via tunneling.[10] In the analysis of recombination kinetics we follow the approach of Kuzovkov and Kotomin.[49] A complete derivation of the analytical equations is given elsewhere.[49,57] Here we only summarize the main ideas and present the way leading to the basic equations.

Starting point of the analysis is an infinite set of differential equations for the many-body densities $\rho_{m,m'}$. Let $n_{A(B)}(\vec{r},t)$ be the (microscopic) particle concentrations, and consider a situation where m A-particles are placed at $\vec{r}_1,\ldots,\vec{r}_m$ and m' B-particles are placed at $\vec{r}'_1,\ldots,\vec{r}'_{m'}$. From this one introduces the many-point densities $\rho_{m,m'}$ as ensemble averaged products of reactant concentrations:

$$\rho_{m,m'} = \left\langle \prod_{i=1}^{m} n_A(\vec{r}_i,t) \prod_{j=1}^{m'} n_B(\vec{r}'_j,t) \right\rangle . \qquad (3.4)$$

Thus one has for the one- and for the two-body densities:

$$\rho_{1,0} = \langle n_A(\vec{r},t) \rangle \equiv n_A(t)$$
$$\rho_{0,1} = \langle n_B(\vec{r},t) \rangle \equiv n_B(t) \qquad (3.5)$$

and

$$\rho_{2,0} = \langle n_A(\vec{r}_1,t)\, n_A(\vec{r}_2,t) \rangle \equiv n_A^2(t)\, X_A(\vec{r}_1-\vec{r}_2,t)$$
$$\rho_{0,2} = \langle n_B(\vec{r}'_1,t)\, n_B(\vec{r}'_2,t) \rangle \equiv n_B^2(t)\, X_B(\vec{r}'_1-\vec{r}'_2,t) \qquad (3.6)$$
$$\rho_{1,1} = \langle n_A(\vec{r}_1,t)\, n_B(\vec{r}'_1,t) \rangle \equiv n_A(t)\, n_B(t)\, Y(\vec{r}_1-\vec{r}'_1,t) ,$$

where we have introduced the global particle concentrations $n_{A(B)}(t)$ and the pair correlation functions $X_{A(B)}(\vec{r},t)$ and $Y(\vec{r},t)$.

Now, the following infinite system of coupled differential equations for the many-body densities $\rho_{m,m'}$ holds:[49]

$$\frac{\partial}{\partial t}\, \rho_{m,m'} = -\sum_{i=1}^{m}\sum_{j=1}^{m'} w(\vec{r}_i - \vec{r}'_j)\,\rho_{m,m'}$$

$$-\sum_{i=1}^{m}\int w(\vec{r}_i - \vec{r}'_{m'+1})\,\rho_{m,m'+1}\,d\vec{r}'_{m'+1}$$

$$-\sum_{j=1}^{m'}\int w(\vec{r}'_j - \vec{r}_{m+1})\,\rho_{m+1,m'}\,d\vec{r}_{m+1} \qquad . \qquad (3.7)$$

The contributions to the decay of $\rho_{m,m'}$ may be readily understood as follows: The first term accounts for reactions between any pair of A- and B-particles from the (m,m') set, whereas the second and the third term involve reactions of one of the particles of the original (m,m')-set with a particle of opposite type from outside the (m,m')-set.

In order to decouple the infinite set of equations, one expresses three-body densities in terms of two-body densities according to:

$$\langle n_1\, n_2\, n_3\rangle = \frac{\langle n_1\, n_2\rangle\, \langle n_2\, n_3\rangle\, \langle n_3\, n_1\rangle}{\langle n_1\rangle\, \langle n_2\rangle\, \langle n_3\rangle} \qquad . \qquad (3.8)$$

This approximation is due to Kirkwood and is often called superposition approximation. Its range of validity, however, cannot be well-assessed by analytical methods, but has to be checked by comparison to experiments or simulations.

Using Eq. (3.8) in conjunction to Eq. (3.7) leads (after some calculations, see Refs. 49 and 57) to the following set of coupled differential equations:

$$\frac{\partial n_A(t)}{\partial t} = \frac{\partial n_B(t)}{\partial t} = -\,n_A(t)\,n_B(t)\int w(\vec{r}')\,Y(\vec{r}',t)\,d^d r' \qquad (3.9)$$

$$\frac{\partial \ln Y(r,t)}{\partial t} = -w(\vec{r}) - \int w(\vec{r}')Y(\vec{r}',t)\Big\{ n_A(t)\Big[X_A(\vec{r}-\vec{r}',t)-1\Big] +$$

$$n_B(t)\Big[X_B(\vec{r}-\vec{r}',t)-1\Big]\Big\}\, d^d r' \qquad (3.10)$$

$$\frac{\partial \ln X_{A,B}(\vec{r},t)}{\partial t} = -\,2\,n_{B,A}(t)\int w(\vec{r}')Y(\vec{r}',t)\Big[Y(\vec{r}-\vec{r}',t)-1\Big]\,d^d r' \qquad . \qquad (3.11)$$

A simple way to visualize, say Eq. (3.10), is to observe that a given AB pair may disappear $[\partial(n_A n_B Y)/\partial t]$ either through direct annihilation $[-wn_A n_B Y]$ or by having one of the partners annihilate with a third particle, whose position distribution is given in the Kirkwood approximation by a $n_A^2 n_B YYX_A-$ or $n_A n_B^2 YYX_B-$type product. The -1 in the square brackets stems from the $\partial n_{A,B}/\partial t$ terms in the $\partial(n_A n_B Y)/\partial t$ expression, where use was made of Eq. (3.9).

The starting point both for the simulations and also for the analytical development according to Eqs. (3.9) – (3.11) is an uncorrelated distribution of A and B particles with initial densities $n_{A,B}(0) = n_{A,B}^0$ and $Y(\vec{r},0) \equiv X_{A,B}(\vec{r},0) \equiv 1$. Evidently, in the course of the reaction one has at all times:

$$\lim_{r\to\infty} X(\vec{r},t) \;=\; \lim_{r\to\infty} Y(\vec{r},t) \;=\; 1 \;\;. \tag{3.12}$$

In the following we use dimensionless variables and measure distances in units of r_0, concentrations in units of r_0^{-d} and time in units of w_0^{-1}. The reaction rate takes now the form $w(\vec{r}) = \exp(-r)$. Furthermore we define the reaction radius ξ through $w(\xi)t = 1$; ξ measures the time-dependent range of the interaction. For exchange-type interaction one has $\xi = r_0 \ln(w_0 t)$, which in dimensionless variables reads $\xi = \ln t$.

III.2. Comparison Between the Simulations and the Analytical Approach

Here we present the results of numerical investigations of the annihilation process A+B→0 on one- and two-dimensional lattices. Starting point for the simulations are random distributions of A- and B-particles. The reaction takes place either on a linear chain or on a square lattice of 1000 x 1000 sites. In both cases periodic boundary conditions are implemented. The random mutual annihilation of unlike particles is simulated through a minimal process method:[62] At each reaction step from all AB-pairs one pair is selected randomly, according to its reaction rate; the time increment τ corresponding to the step is computed as $\tau = - (\ln T)/R$, with R being the sum of the rates of all AB-pairs present and T being a random number from a homogeneous distribution in the unit interval. The time is measured in units of w_0^{-1}, and the interaction parameter r_0 is set to $r_0 = 5$ (in units of the lattice spacing), which corresponds to typical exchange interactions in organic molecular crystals.

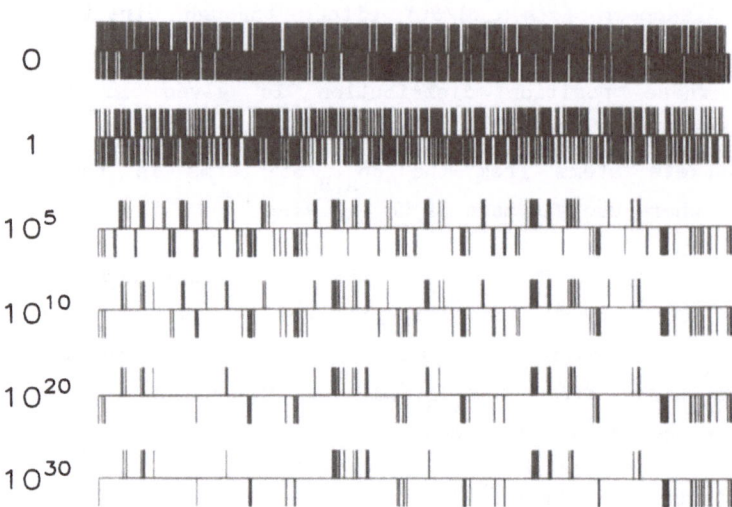

Fig. 3.1. Distribution of A- and B-particles (initially 1000 of each kind on a chain with L=10000 sites) during a numerical simulation. Each vertical line represents one A (up) or B (down) particle. [From Ref. 56]

Figure 3.1 shows the build-up of correlations after starting with 1000 A- and 1000 B-particles on a linear chain of length L = 10000. Already at $t=10^5$ correlation regions are clearly visible, a finding comparable for instance to the results for diffusion-limited reactions under pulsed excitation,[6,10,37] and also for steady-state situations.[40,50]

In Fig. 3.2 we present the distributions of A- and B-particles at different times for a reaction which takes place on a square lattice. Note that at t = 10 the particle distribution still looks quite random. On the other hand, already at $t = 10^5$ the evolution of clusters of like particles becomes clearly visible; this effect gets enhanced at later times. On a cautionary note, we remark that in our simulations the typical cluster sizes stayed well below the size of the lattice at all investigated times. The findings in Fig. 3.2 parallel the results for the one-dimensional case, where (with r_0 = 5) cluster formation also becomes visible from $t = 10^5$ on. Comparison to the findings for diffusion-limited reactions[6] shows now that the clusters are more pronounced when the reactants are immobile. This can be understood by noting that diffusion stirs the system and works against the tendency to form well-separated clusters.

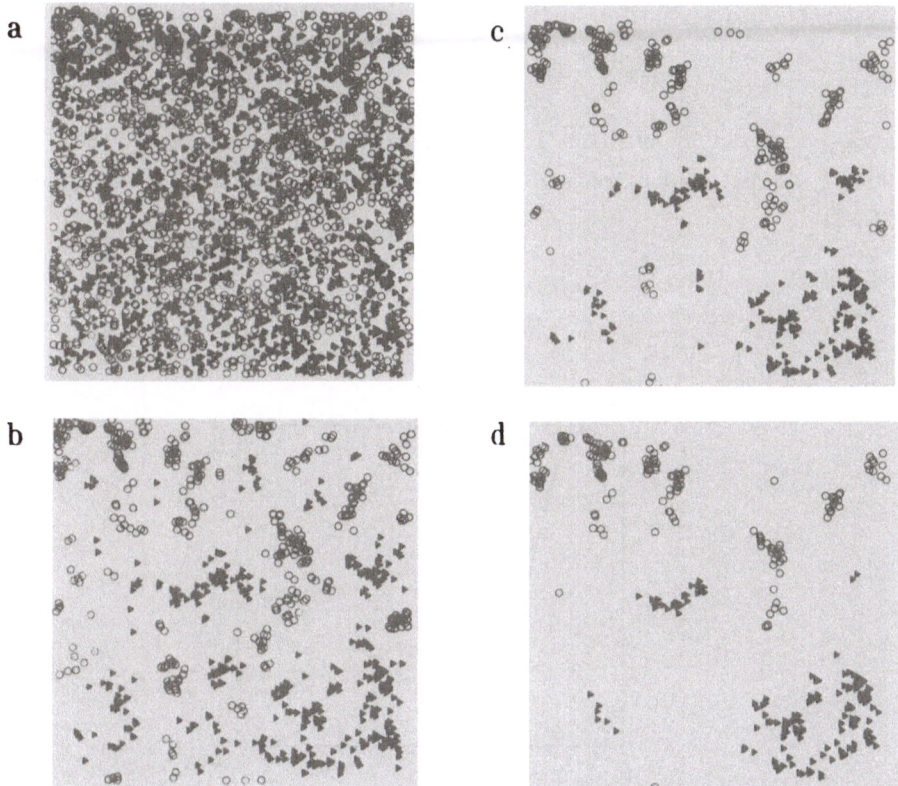

Fig. 3.2. Spatial distribution of A- and B-particles during a reaction process with $r_0 = 5$ on a 1000×1000 square lattice (with periodic boundary conditions) after starting with 10^4 particles of each kind. The situations correspond to the following times: (a) $t = 10$, (b) $t = 10^5$, (c) $t = 10^{10}$ and (d) $t = 10^{15}$. [From Ref. 57]

Equal initial concentrations of A and B. First we consider the case of equal initial concentrations, $n_A^0 = n_B^0$. In this case one has $n_A(t) = n_B(t) \equiv n(t)$ and $X_A(\vec{r},t) = X_B(\vec{r},t) \equiv X(\vec{r},t)$. Fig. 3.3 displays the decay of the concentration of A-(or B-)particles (curve a) on a square lattice; the decay was obtained by averaging over 20 different initial configurations ($r_0=5$, $n_A^0 = n_B^0 = 0.25$). The dotted lines indicate $\langle n(t) \rangle \pm s(t)$, where $s(t)$ is the standard deviation of the 20 individual curves, i.e. $s^2(t) = \langle n^2(t) \rangle - \langle n(t) \rangle^2$. Asymptotically, the simulations follow very closely a $[\ln(w_0 t)]^{-1}$ behavior,[57] as verified by us by replotting the data of Fig. 3.3.

Generally, on d-dimensional lattices the concentration $n(t)$ follows the asymptotic behavior

$$n(t) \sim [\ln(w_0 t)]^{-d/2} \sim \xi^{-d/2} \ , \qquad\qquad (3.13)$$

which was verified by us also for a one-dimensional lattice[56] and for a Sierpinski gasket, where instead of d the fractal dimension $\bar{d}$ enters into Eq. (3.13).[59]

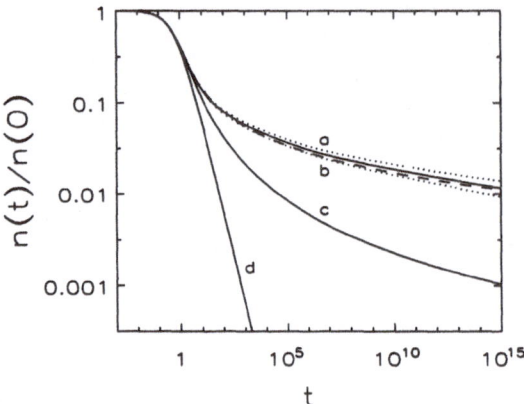

Fig. 3.3. Decay of the reactant concentration obtained from (a) averaging over 20 numerical simulations with initially 10^4 particles of each kind on a 1000 × 1000 square lattice and with interaction range $r_0 = 5$ (the dotted lines indicate the standard deviation of the data); (b) numerical evaluation of Eqs. (3.9) – (3.11) for d = 2 and n(0) = 0.25 (dashed line); (c) a raw approximation which neglects correlations between like particles ($X(\vec{r},t) \equiv 1$); (d) the solution of the simple kinetic scheme [Eq. (3.14)]. [From Ref. 57]

A very good description of this decay is provided by Eqs. (3.9) – (3.11). Here the numerical evaluation involved the discretization of the time and space variables. To improve the accuracy, for each discretized time interval the derivatives were expressed as symmetric functions of the interval's boundaries. This leads to a nonlinear set of iteration functions, which converge very fast under a quasilinearization procedure.[63]

The decay resulting from the numerical evaluation of Eqs. (3.9) – (3.11) with the appropriate parameters ($n_A^0 = n_B^0 = 0.25$, $d = 2$) is also given in Fig. 3.3 (curve b). One observes that over the whole range investigated simulation and numerical solution agree exceedingly well within the statistical errors of the simulation. Moreover, the asymptotic behavior which follows from Eqs. (3.9) - (3.11) also obeys a $[\ln(w_0 t)]^{-1}$-behavior.

By contrast, setting $X(r,t) \equiv 1$ in Eq. (3.10), which is equivalent to neglecting correlations between like particles, leads to a very different behavior; this is given in Fig. 3.3 as curve c. Here simulation and analytical result agree only in the very short-time domain ($t < 10$); furthermore, at longer times a $[\ln(w_0 t)]^{-d}$-pattern dominates.[57] At longer times the formation of clusters causes the simulation curve a to decay more slowly than curve c.

In Fig. 3.3 we also plotted the solution of the simple kinetic approach (curve d), which implies neglect of all spatial inhomogeneities ($X(r,t) \equiv 1$ and $Y(r,t) \equiv 1$). In this case the solution of Eq. (3.9) is

$$n(t) = \frac{n(0)}{1 + 2\pi r_0^2 n(0) w_0 t} . \qquad (3.14)$$

This approximation is only valid up to $t^* \approx 1$, as shown in Fig. 3.3. Then the assumption of spatial homogeneity breaks down.

We conclude that for an equal number of reactants of both kinds the superposition approximation leads to a very good description of complex reaction kinetics over two orders of magnitude in the decay.

A further test of the superposition approximation is provided by comparing the correlation functions $X(\vec{r},t)$ and $Y(\vec{r},t)$ to the simulation data. Here we calculated these functions from Eqs. (3.10) and (3.11) for the considered set of parameters ($n(0) = 0.25$, $d = 2$). In Fig. 3.4 we compare $X(\vec{r},t)$ and $Y(\vec{r},t)$ to the simulation results, obtained, as before, by averaging over 20 realizations of the reaction process. The scatter of the simulated data was reduced by averaging over radius intervals of length $6r_0$ (30 lattice constants). Thus the statistical error of the data is comparable to (or less than) the height of the symbols and decreases with increasing distance r, since the number of lattice sites with distance in the interval (r, r + dr) is proportional to r^{d-1} dr.

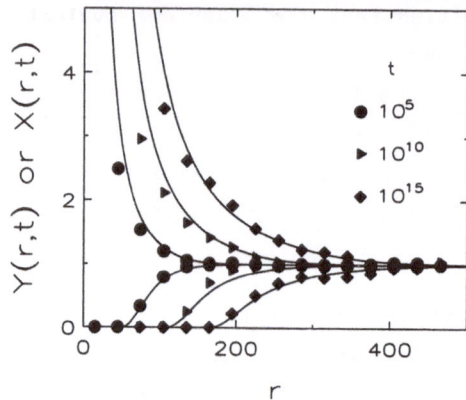

Fig. 3.4. Correlation functions for like (X) and unlike (Y) particles, calculated numerically from Eqs. (3.9) - (3.11) with d = 2 and n(0) = 0.25 (solid lines). The symbols mark the results from numerical simulations: each value is an average over 20 realizations of the process, smoothed out over an interval of $6r_0$. The reaction took place on a 1000 × 1000 square lattice, having initially 10^4 particles of each kind, which react via exchange ($r_0 = 5$). [From Ref. 57]

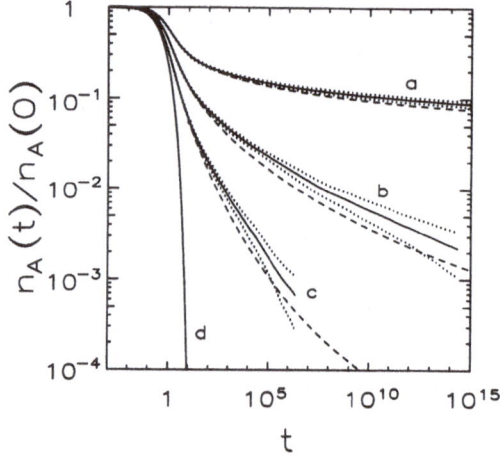

Fig. 3.5. Decay of the concentration $n_A(t)$ in d=1 resulting from direct simulations of the reaction (solid curves a-c), the corresponding numerical evaluation of Eqs. (3.9)-(3.11) (dashed lines) or an even simpler approximation, neglecting correlations between like particles (curve d). The dotted lines indicate the standard deviation of the simulated curves, each of which being the average over 10 realizations of the reaction process. The initial particle concentrations are $n_A^0 = 0.5$ (a-d) and $n_B^0 = 0.5$ (a), 0.75 (b) and 1 (c,d).

Again, simulation data and calculated correlation functions agree well in the major part of the r-range. Slight deviations occur only in the small-r regime, where the simulation data for the correlations between like particles lie somewhat below the $X(\vec{r},t)$ curves. On the other hand, the $Y(\vec{r},t)$ curves, which determine the decay of concentration via Eq. (3.9), fit well in the whole r-regime. Taking the value of r, where $Y(\vec{r},t)$ begins to deviate from zero, as the typical cluster size, Fig. 3.4 also reflects the logarithmic time dependence of this quantity, which turns out to mirror rather closely the previously defined reaction radius ξ.

Unequal reactant concentrations. In Figs. 3.5 and 3.6 we compare the simulation results for the case $n_A^0 < n_B^0$ to the outcomes of Eqs. (3.9) – (3.11) in one and in two dimensions. Here the initial number of A particles was 10^4, whereas the numbers chosen for the B particles were 10^4, 1.5×10^4 and 2×10^4 in d = 1 and 10^4, 1.25×10^4 and 1.5×10^4 in d = 2. The solid curves a-c in Figs. 3.5 and 3.6 are the results of the simulations, each curve being averaged over 10 realizations of the annihilation process. The curves are given as a function of the different initial concentrations of the B particles. In the figures the dotted lines indicate the standard deviation of the simulation data, whereas the dashed lines are the corresponding numerical solutions of Eqs. (3.9) – (3.11).

For comparison we also present in Figs. 3.5 and 3.6 the case of equal initial concentrations, $n_A^0 = n_B^0$, curves a; as noted before, for these the agreement between simulation and numerical evaluation of Eqs. (3.9) – (3.11) is very good and both sets of data follow asymptotically Eq. (3.13).

In the case $n_A^0 < n_B^0$ (curves b and c) Eqs. (3.9) – (3.11) lead to forms for which the agreement is less impressive. Furthermore the results of the two methods deviate more and more when the difference in the A and B concentrations increases.

Nonetheless, the agreement between simulation and the analytical approach, based on the superposition approximation is still quite good, when one takes into consideration other possible approximate schemes. Thus one may envisage a simpler approximation than Eqs. (3.9) – (3.11), in which correlations between like particles are neglected; this corresponds to $X_{A,B}(\vec{r},t) \equiv 1$. In this case one has from Eq. (3.10):

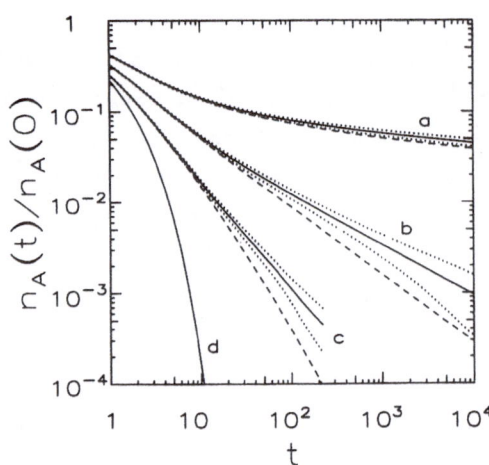

Fig. 3.6. Same as Fig. 3.5 but for d=2. Here the initial concentrations are $n_A^0 = 0.25$ (a-d) and $n_B^0 = 0.25$ (a), 0.3125 (b) and 0.375 (c,d).

$$Y(\vec{r}, t) = \exp[-w(\vec{r})t] \ . \tag{3.15}$$

Now only Eq. (3.9), where $Y(\vec{r}, t)$ is given by Eq. (3.15), has to be evaluated numerically. The outcome of this lower level approximation is indicated in Figs. 3.5 and 3.6; the results are the curves indicated by d, which were calculated for the same parameters as curves c. Compared to this simpler scheme, the results of the superposition approximation fit the simulation data much better.

On the other hand, this positive result raises the question of possible extensions to our system of Eqs. (3.9) – (3.11) by including higher order terms. From our experience we view this as being a hard task, which also would be very expensive in terms of computer time. Further improvements to the approach presented here, obtained by extending the analytical scheme to include further, higher order correlation functions seem to us at this stage prohibitive.

Interestingly, the simulation curves (and also the corresponding analytical curves) of Figs. 3.5 and 3.6 for the case $n_A^0 < n_B^0$ fit quite well the asymptotic behavior

$$n_A(t) \ \sim \ \exp\left[- \left(\sqrt{n_B^0} - \sqrt{n_A^0} \right) \times \left(\gamma \ \ln \ t \right)^{d/2} \right] \ . \tag{3.16}$$

This may readily be seen in the two-dimensional case, Fig. 3.6, where (since d = 2) one would expect from Eq. (3.16) a power-law dependence for $n_A(t)$, which is really fulfilled.

Why are in the case $n_A^0 < n_B^0$ the deviations between simulation and analytical approach larger than those for $n_A^0 = n_B^0$? Some insight into this problem may be obtained by analyzing the correlation functions. Thus we have plotted in Fig. 3.7 the correlation functions which correspond to curve b of Fig. 3.6. Here the solid and dashed curves are the solutions of Eqs. (3.9)-(3.11), whereas the symbols indicate the corresponding simulation results. Although the correlation functions $Y(\vec{r}, t)$ for pairs of unlike particles agree very well, there are significant deviations for the correlations between like particles, $X_A(\vec{r}, t)$ and $X_B(\vec{r}, t)$. This implies that at a given instant of time the predicted size of clusters is greater than what the simulations show, i.e. the functions $X_{A,B}(\vec{r}, t)$ of the analysis overestimate the formation of clusters of like particles. This becomes most obvious in the case of $X_A(\vec{r}, t)$, the correlation function for the minority species. Whereas for $n_A^0 = n_B^0$ X_A and X_B are identical and grow moderately with increasing time (see Fig. 3.4), here

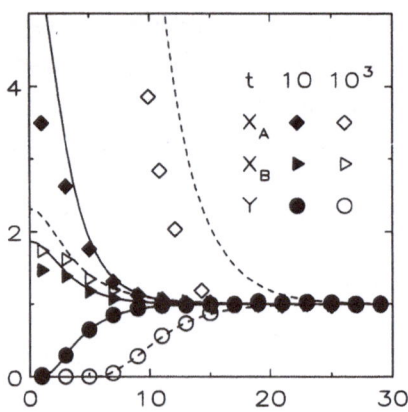

Fig. 3.7. Correlation functions for like (X_A, X_B) and unlike particles (Y) in d=2, calculated numerically from Eqs. (3.9)-(3.11) (solid or dashed curves) or obtained from direct simulations of the reaction (symbols), each value averaged over 10 realizations and .over a distance of $2r_0$ ($n_A^0 = 0.25$, $n_B^0 = 0.3125$, $r_0 = 5$). Solid lines and filled symbols belong to t = 10, whereas dashed lines and open symbols belong to $t = 10^3$.

X_A grows very fast, possibly overstretching the limits of validity of the superposition approximation.

IV. TRAPPING AND TARGET MODELS: THE A+B→B REACTIONS

After having studied in the last sections static relaxation kinetics, in which the reaction partners did not move, we now turn our attention to relaxation mechanisms, in which a series of steps (mostly randomly taken) is necessary to bring the particles together, which then react on contact. Typical for such behavior are generalized diffusion models, which for discrete geometrical structures correspond to random walks.[9,64-66] In this section we follow closely our approaches of Refs. 8 and 10. In the simplest models one has one A and several B particles, and the A particle is annihilated at the encounter of a B particle. Depending on which of the species performs the motion one distinguishes between the trapping model[9,37,65-70] (only the A moves), the target (scavenging) model[71-74] (only B move) and the moving targets[38] (both species move). We have to evaluate the survival probability of the A particle averaged over all possible realizations of particle distributions and motions.

We start with the simpler, the target model. We assume that the B molecules do not interact with each other, and we focus on the fate of a single A particle, taken to be at the origin of the coordinate system. Several possibilities may now be envisaged in creating the initial distribution of B particles on the remaining lattice sites. In Refs. 74 and 75 we have presented the survival probabilities for B particles which follow binomial, multinomial and Poisson distributions. Here we will exemplify the procedure using the last one, for which the occupancy of a site is taken to be distributed as

$$g(j) = e^{-p} \frac{p^j}{j!} \quad , \qquad (4.1)$$

where $g(j)$ is the normalized probability of having j B particles at one site and p is their average number density.

Typically, $F_m(\vec{r})$ denotes the probability that a random walker starting from $\vec{r}$ reaches the origin $\vec{0}$ for the first time in the mth step. For regular lattices, because of the symmetry of the walk $F_m(\vec{r})$ is also the first-passage time from $\vec{0}$ to $\vec{r}$, as defined by Montroll and Weiss.[76] In general, let $H_n(\vec{r})$ be the probability that a first passage from $\vec{r}$ to $\vec{0}$ occurred in the first n steps. Then:

$$H_n(\vec{r}) = \sum_{m=1}^{n} F_m(\vec{r}) \qquad (4.2)$$

The probability therefore that a walker from $\vec{r}$ did *not* reach $\vec{0}$ in the first n steps is thus:

$$\Phi_n(\vec{r}) = 1 - H_n(\vec{r}) \quad . \tag{4.3}$$

Using Eq. (4.3) the survival probability Φ_n of the A molecule is obtained by appropriately weighting products of the $\Phi_n(\vec{r})$ functions:

$$\Phi_n = \prod_{\vec{r}} \left\{ \sum_j g(j) \, [\Phi_n(\vec{r})]^j \right\} \quad . \tag{4.4}$$

Here the product extends over all structure sites, with the exception of the origin. Inserting now Eqs. (4.1) and (4.3) into Eq. (4.4) leads to:

$$\Phi_n = \prod_{\vec{r}} \left\{ \sum_j e^{-p} \frac{[p\Phi_n(\vec{r})]^j}{j!} \right\}$$

$$= \prod_{\vec{r}} \exp[-p + p\,\Phi_n(\vec{r})] = \exp\left[-p \sum_{\vec{r}} H_n(\vec{r}) \right] \quad . \tag{4.5}$$

For regular lattices (4.5) may be further simplified since, according to Eqs. (III.2) and (III.3) of Ref. 76, one has:

$$\sum_{\vec{r}} F_m(\vec{r}) = S_m - S_{m-1} \quad (m > 1). \tag{4.6}$$

Here S_m is the mean number of distinct sites visited by a random walker in m steps, and $S_0 = 1$.

As a reminder, for regular lattices the S_m are well-known. For a particular realization of the random walk on the structure let R_n denote the number of distinct sites visited in n steps, and set $R_0 = 1$. Here the starting point is irrelevant, but it may matter if fractals are considered. The first two moments of R_n are

$$S_n = \langle R_n \rangle \qquad \text{(mean of } R_n) \tag{4.7}$$

and

$$\sigma_n^2 = \langle R_n^2 \rangle - \langle R_n \rangle^2 \qquad \text{(variance of } R_n). \tag{4.8}$$

Now, for regular lattices one has for not too small n:[9]

$$d=1: \quad S_n = a_1\sqrt{n} + a_2/\sqrt{n} + \ldots \qquad \sigma_n^2 \sim 4\left(\ln 2 - \frac{2}{\pi}\right) n \tag{4.9}$$

308

$$d=2: \quad S_n = \frac{a_1 n}{\ln(a_2 n)} + \dots \qquad\qquad \sigma_n^2 \sim \frac{n^2}{\ln^4 n} \qquad\qquad (4.10)$$

$$d=3: \quad S_n = a_1 n + a_2 \sqrt{n} + \dots \qquad\qquad \sigma_n^2 \sim n \ln n \ , \qquad\qquad (4.11)$$

where the a_i are constants which depend on the lattices structure. Introducing now Equations (4.2) and (4.6) into (4.5), one has exactly

$$\Phi_n = \exp[-p(S_n - 1)] \ . \qquad\qquad (4.12)$$

This simple, very useful formula holds approximately even for fractals, under certain homogeneity conditions. We point out that the use of fractals as an underlying space for dynamical processes has been stressed by Alexander and Orbach in a by-now classic paper.[30] A fundamental point which emerged from their analysis is that the density of states of a fractal is connected to an additional parameter, called 'fracton'[30] or spectral[31] dimension $\tilde{d}$, which is in general distinct from $\bar{d}$, the fractal dimension.

In Refs. 77 and 78 a series of random walks on Sierpinski gaskets embedded in Euclidean spaces of dimensions d = 2, 3, 4, and 6 was simulated, the spectral dimensions being $\tilde{d}$ = 1.36, 1.55, 1.65 and 1.77, respectively. For the first two moments, S_n and σ_n^2 one finds that the relations

$$S_n \simeq a n^{\tilde{d}/2} \qquad (\tilde{d} < 2) \qquad\qquad (4.13)$$

and

$$\sigma_n^2 \simeq b n^{\tilde{d}} \qquad (\tilde{d} < 2) \qquad\qquad (4.14)$$

are well obeyed in the range investigated (see Figs. 1 and 2 of Ref. 78). Furthermore, for Sierpinski gaskets the inhomogeneity is minor, and one has, to a very good approximation,

$$\Phi_n \simeq \exp(-pn^{\tilde{d}/2}) \qquad (\tilde{d} < 2) \qquad\qquad (4.15)$$

and

$$\Phi_n \simeq \exp(-pn) \qquad (\tilde{d} > 2) \qquad\qquad (4.16)$$

Let us now turn to the trapping problem and focus on the fate of a single A molecule which performs a random walk over the lattice. The B molecules are assumed to be distributed randomly, with probability p over the structure. For simplicity, one lets at each step the walker move, with equal probability, to one of its neighboring sites. The reaction is assumed to occur instantaneously when the A molecule lands on a site occupied by a B molecule. Thus the B molecules act as traps.

In order to obtain the survival law of the A molecules, an average over all random-walk realizations and all distributions of B molecules is needed. As in Refs. 67 and 79 one can start by focusing on R_n, the number of distinct sites visited in n steps by a particular walk. For the same realization of the walk one lets $\tilde{F}_n$ denote the probability that trapping has not occurred up to the nth step:

$$\tilde{F}_n = (1-p)^{R_n-1} \tag{4.17}$$

assuming the origin of the walk not to be a trap. The measurable survival probability after n steps is $\tilde{\Phi}_n$, the average of $\tilde{F}_n$ over all realizations of random walks[37,67,69] and over all starting points:

$$\tilde{\Phi}_n = \langle \tilde{F}_n \rangle = \langle (1-p)^{R_n-1} \rangle \; . \tag{4.18}$$

Introducing $\lambda = -\ln(1-p)$, Eq. (4.18) turns into:

$$\tilde{\Phi}_n = e^\lambda \langle e^{-\lambda R_n} \rangle \; , \tag{4.19}$$

which allows the following expansion in the cumulants $K_{j,n}$ of the distribution of R_n:

$$\tilde{\Phi}_n = e^\lambda \exp\left[\sum_{j=1}^\infty K_{j,n} \frac{(-\lambda)^j}{j!} \right] \; . \tag{4.20}$$

As an example, $K_{1,n} = S_n$ and $K_{2,n} = \sigma_n^2$.

Apart from the case d=1, for which both Φ_n and the distribution R_n are known, in general not much information is available on R_n. Truncating the sum in Eq. (4.20) one obtains:

$$\tilde{\Phi}_{N,n} = e^\lambda \exp\left[\sum_{j=1}^N K_{j,n} \frac{(-\lambda)^j}{j!} \right] \; . \tag{4.21}$$

The quality of this approximation depends very much on the underlying lattice and on the value of N, see Ref. 10 for details.

Remarkably, in d=1, one is in the fortunate position that an exact analytic solution for trapping is known.[9,80] This expression allows to determine an asymptotic expansion for $\tilde{\Phi}_n$, valid for large n, or, equivalently, for large t in continuous time.[80-82] The leading term of this expansion is:[82]

$$\tilde{\Phi}_n \sim \lambda n^{1/2} \exp\left[-\frac{3}{2}(\pi\lambda)^{2/3} n^{1/3}\right].$$ (4.22)

This is a special case of a more general law, proven by Donsker and Varadhan[4] for arbitrary d:

$$\ln \tilde{\Phi}_n \sim -C\lambda^{2/(d+2)} n^{d/(d+2)}.$$ (4.23)

Eq. (4.23) was investigated by several groups;[3,83,84] and also extended to the fractal domain.[85] As a rule, however, the convergence of the decay laws to the asymptotic Donsker-Varadhan form, Eq. (4.25), is in general quite slow,[3,85-87] so that for dimensions d≥2 and not too large concentrations of traps this form is more of theoretical than of practical interest.[85]

V. THE DIFFUSION-LIMITED A+B→0 REACTION

In this section we focus on the diffusion-limited A+B→0 reaction. This reaction contrasts to the trapping and to the target problem, where we could consider the A particles to be independent from each other. Here, however, the A and B particles, even if they were initially independently distributed, get correlated due to the reaction. This effect was already mentioned in Sec. III. The correlations stem from the fact that the annihilation of each AB-pair changes the distribution of B particles seen by the remaining A particles. Thus collective phenomena have to be taken into account.

The main arguments for most of the analytical approaches to this type of reaction are based on fluctuation statistics. For an initially equal number of A and of B particles randomly situated in space one has asymptotically:[5-7,10,37]

$$n_A(t) \sim t^{-d/4} \qquad (d<4),$$ (5.1)

where d denotes the spatial dimension. For fractals one has as an evident extension $n_A(t) \sim t^{-\tilde{d}/4}$, as shown in Refs. 37 and 88. Note that the classical kinetic scheme predicts only the time dependence $n_A(t) \sim t^{-1}$ (see Sec. III.1.).

Now let us focus on the (heuristic) arguments which lead to Eq. (5.1). The picture is that during time t a diffusing particle moves in a volume V which is of the order of ℓ_D^d, where $\ell_D = (Dt)^{1/2}$ is the diffusion length and D is the diffusion coefficient. The number of surviving particles at time t is then taken to be the difference in the number of particles of opposite kind in volumes of size $V \sim (Dt)^{d/2}$. Evidently, the underlying idea is that during time t a complete, pairwise annihilation of particles takes place in each volume V. For an equal number of A and B particles in the total reaction volume Ω this procedure yields Eq. (5.1); clearly, the result is due mainly to fluctuations in the actual number of particles in the subregions V of Ω.[5,6,37]

What happens, however, for an unequal number of A and B particles, say when A is in the minority? For this case Kang and Redner[37] advanced the asymptotic form:

$$ n_A(t) \sim \exp\left[- \text{const.} \times \left(\sqrt{n_B(0)} - \sqrt{n_A(0)} \right) t^{d/4} \right], \qquad (5.2) $$

based on the following argument: In the $n_A(0) = n_B(0)$ case Eq. (5.1) may be obtained from the formal kinetic equation [see Eq. (3.1)]

$$ \frac{d}{dt}n_A(t) = - k(t)\, n_A(t)\, n_B(t) , \qquad (5.3) $$

when one sets $k(t) \sim t^{d/4-1}$. With this reaction rate the solution of Eq. (5.3) in the case $n_A(0) < n_B(0)$ is Eq. (5.2). Furthermore, the analysis by Kang and Redner showed that Eq. (5.2) agrees quite well with computer simulations in one and two dimensions.

On the other hand, however, as also discussed in Ref. 37, the use of Eq. (5.2) becomes problematic in dimensions higher than two: Thus Eq. (5.2) predicts a decay which is faster than that found for trapping, i.e. for the A+B→B reaction where the A particles are mobile and the B particles immobile: Asymptotically the decay due to trapping follows[3,4]

$$ n_A(t) \sim \exp(-\text{const.} \times t^{d/(d+2)}) \qquad (5.4) $$

[see also Eq. (4.23)]. This point makes already clear that a strict derivation of the correct asymptotic decay for the diffusion-limited A+B→0 reaction when an unequal number of A and B particles is present is still lacking.

V.1. Fluctuation Statistics

In this section we will analyze the behavior of the $A+B \to 0$ reaction using a statistical approach; we will consider both cases, $n_A(0) = n_B(0)$ and $n_A(0) < n_B(0)$. Our analysis predicts forms which for $n_A(0) < n_B(0)$ differ from Eq. (5.2), but which (surprisingly at first sight) also agree very well with the computer simulations over several orders of magnitude in decay. It is certainly astonishing to find that the same set of data can be fitted to two different decay forms. This shows that one must be very careful when drawing conclusions about asymptotical decay forms based only on computer simulations.

In the following we will give a rough outline of our approach (for details see Ref. 89). We take a volume V consisting of N sites on which we distribute the A and B particles randomly and independently, according to their respective probabilities p_A and p_B ($p_A \neq 0 \neq p_B$ and $p_A \neq 1 \neq p_B$). Let m denote the difference in the number of particles: $m = n_A - n_B$. Then for large N the probability P(m) that the particle difference equals m is well-approximated by a Gaussian:

$$P(m) = \frac{1}{\sqrt{2\pi}\,\sigma_m} \exp\left[-\frac{(m - \bar{m})^2}{2\sigma_m^2} \right] \quad . \tag{5.4}$$

P(m) is centered at $\bar{m} = (p_A - p_B)N = -\Delta N$ and has the variance $\sigma_m^2 = wN$, where $\Delta = p_B - p_A$ and $w = p_A + p_B$.

When the reaction is turned on, the A and B particles will annihilate each other, so that in the long run only particles of the majority species survive. The determination of M(N), the average number of A particles surviving the annihilation process in the volume V, is now straightforward. One has:

$$M(N) = \sum_{m > 0} m\, P(m) \approx \int_0^\infty m\, P(m)\, dm \quad . \tag{5.5}$$

For a Gaussian distribution P(m) like that of Eq. (5.4) the average M(N) is:

$$M(N) = \frac{\sigma_m}{\sqrt{2\pi}} \exp\left(-\frac{\bar{m}^2}{2\sigma_m^2} \right) + \frac{\bar{m}}{2} \operatorname{erfc}\left(-\frac{\bar{m}}{\sqrt{2}\,\sigma_m} \right)$$

$$= \frac{\sigma_m}{\sqrt{2\pi}} e^{-z^2} \left[1 - \sqrt{\pi}\, z\, e^{z^2} \operatorname{erfc}(z) \right] \quad . \tag{5.6}$$

313

Here we set $z = -\overline{m}/(2\sigma_m^2)^{1/2} = \Delta(N/2w)^{1/2}$, and erfc(z) denotes the complementary error function, Eq. (7.1.2) of Ref. 90. Finally, the average concentration $n_A(N)$ of surviving A particles is:

$$n_A(N) = \frac{M(N)}{N} = \frac{\sigma_m}{\sqrt{2\pi}\,N}\,e^{-z^2}\left[1 - \sqrt{\pi}z\,e^{z^2}\mathrm{erfc}(z)\right].$$ (5.7)

For $p_A \neq p_B$ we now divide both sides of Eq. (5.7) by $\Delta = p_B - p_A$ and obtain:

$$\frac{n_A(N)}{\Delta} \equiv F(z) = \frac{1}{2\sqrt{\pi}\,z}\,e^{-z^2}\left[1 - \sqrt{\pi}\,z\,e^{z^2}\mathrm{erfc}(z)\right].$$ (5.8)

Eq. (5.8) is very interesting, since it states that for $\Delta \neq 0$ the average concentration of A particles surviving the annihilation reaction A+B→0 on N sites is only a function of Δ and z. This means that plotting $n_A(N)/\Delta$ versus z should yield the same curve for all initial concentrations. We will later demonstrate through simulation calculations that this scaling relation indeed holds.

We now turn our attention to some special cases of Eqs. (5.7) and (5.8). For equal initial concentrations, $p_A = p_B$, one has $\Delta = 0$ and hence also z=0. With $\sigma_m = wN$ one obtains from Eq. (5.7):

$$n_A(N) = \sqrt{\frac{w}{2\pi N}}\,.$$ (5.9)

As stated above, for diffusing particles $N \sim (Dt)^{d/2}$. Eq. (5.9) then leads to the well-known decay law $n_A(t) \sim t^{-d/4}$ (with d<4), Eq. (5.1).

In the case of $p_A < p_B$ one has z>0. Consider now the two limiting cases z≪1 and z≫1 (which means N≪N_0 or N≫N_0, respectively, with $N_0 = 2w/\Delta^2$). For z≪1 we get from Eq. (5.8):

$$F(z) \sim \frac{1}{2\sqrt{\pi}\,z}\,,$$ (5.10)

from which we recover for $n_A(N)$ the asymptotic form given in Eq. (5.9). For z≫1 we make use of the asymptotic series, Eq. (7.1.23) of Ref. 90:

$$\sqrt{\pi}\,z\,e^{z^2}\mathrm{erfc}(z) \sim 1 - \frac{1}{2z^2} + \dots\,.$$ (5.11)

Thus we obtain:

$$F(z) \sim \frac{1}{4\sqrt{\pi}\,z^3}\,e^{-z^2}\,,$$ (5.12)

which means

$$n_A(N) \sim \frac{\Delta}{4\sqrt{\pi}} \left(\frac{\Delta^2}{2w} N \right)^{-\frac{3}{2}} \exp\left(-\frac{\Delta^2}{2w} N \right) \qquad (5.13)$$

Therefore we expect a crossover from an algebraic decay for $z \ll 1$ ($N \ll N_0$) to an exponential form for $z \gg 1$ ($N \gg N_0$).

V.2. Computer Simulations

In this subsection we present results from computer simulations of the diffusion-limited $A+B \to 0$ reaction. We keep the B particles fixed and allow only the A (minority) particles to move. Here we remark that somewhat different results are found when both types of particles move. This feature is reminiscent of the $A+B \to B$ reaction, see Sec. IV, where one encounters different kinetics for the trapping problem, in which only the A particles move, and for the target problem, where only the B particles move.[10]

In our simulations each site is initially occupied by an A particle (probability p_A) or by a B particle (probability p_B), with $p_A + p_B \leq 1$ (trinomial distribution). Under these conditions the distribution $P(m)$ of the particle difference m can also be approximated by a Gaussian,[89] as given in Eq. (5.4), now with $\sigma_m^2 = (p_A + p_B - \Delta^2)N$, i.e. $w = p_A + p_B - \Delta^2$. The underlying lattices are either linear chains, consisting of 10^5 to 10^6 sites, or two-dimensional 1000×1000 square lattices, for which we implement in both cases periodic boundary conditions. At each step of the process an A particle is chosen at random and is moved to one of its neighboring lattice sites. The A and B particles annihilate at first encounter. The time increment for each step is equal to the reciprocal of the actual number of A particles. Thus during a time unit each A particle performs on the average one step.

In order to compare the simulations' outcomes to the analytical expressions given in terms of the volume V, we have to relate V to the time t. The heuristic argument mentioned before relates V to $\ell_D^d = (Dt)^{d/2}$, with a proportionality constant to be determined. In our simulations the A particles perform a symmetric random walk and thus results from the theory of random walks[9] can be used. Hence, in one dimension the volume V visited by a random walker is equal to the span of the random walk, which in 1d coincides with the number of distinct sites visited by the walker. As in Sec. IV, the average of the last quantity is denoted by S_ν, where ν is the number of steps. In one dimension one has exactly:[9,76]

$$S_\nu = \sqrt{\frac{8}{\pi} \nu} \ . \tag{5.14}$$

Since in the simulations each A particle performs on the average one step per time unit, one finds for the average volume V visited during time t the expression $V = (8t/\pi)^{1/2}$ in 1d.

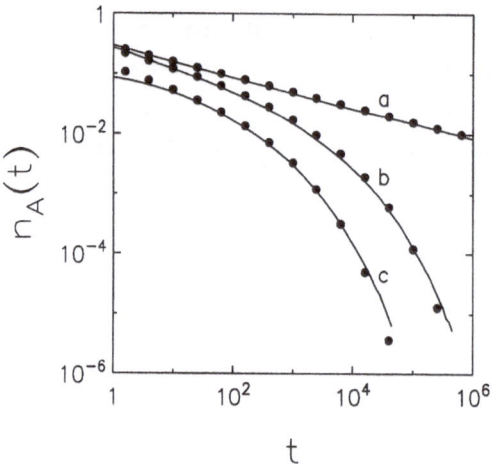

Fig. 5.1. Results of simulations in 1d (solid lines) and the corresponding analytical curves (dots), stemming from Eq. (5.7) with $N = (8t/\pi)^{1/2}$ and $w = p_A + p_B - \Delta^2$. The initial concentrations are (a) $p_A = p_B = 0.4$, (b) $p_A = 0.4$ and $p_B = 0.5$ and (c) $p_A = 0.1$ and $p_B = 0.2$, respectively. The linear chain consisted of 10^5 sites for curve a and of 10^6 sites for curves b and c. Each of the curves b and c was averaged over 10 realizations of the reaction.

We have performed simulations for various initial concentrations. In Fig. 5.1 we compare the outcomes of some of these simulations in d=1 (solid lines) to the corresponding results of the fluctuation statistics (crosses), obtained from Eqs. (5.7) and (5.8) by taking, according to Eq. (5.14), $N = V = (8t/\pi)^{1/2}$. Firstly, for equal initial concentrations

($p_A = p_B = 0.4$, curve a) we recover the well-known $t^{-1/4}$-law, a fact in perfect agreement with Eq. (5.9). Furthermore, also for the case $p_A < p_B$ (curves b and c) our theory based on fluctuation statistics agrees very well with the simulation results. Thus, just by using the correct initial concentrations (and without any parameters) the agreement between simulations and analytical results, Eq. (5.7), is very good. The slight deviations in the small-t regime are due to the discreteness of the lattice and to the fact that the Gaussian approximation for P(m) is valid only for large N.

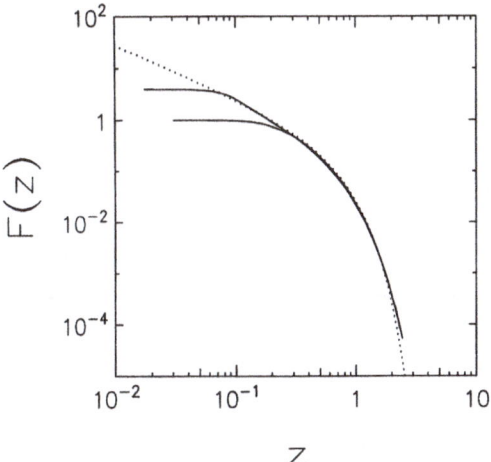

Fig. 5.2. Curves b and c of Fig. 5.1 (solid lines), rescaled with respect to $z = \Delta(8t/\pi)^{1/4}/(2w)^{1/2}$ (see text). The dashed line indicates $F(z)$, Eq. (5.8).

A further stringent test of our analytical approach is provided by the scaling law, Eq. (5.8). In Fig. 5.2 we therefore rescaled the simulation results, curves b and c of Fig. 5.1, and plotted $n_A(t)/\Delta$ versus $z = \Delta(8t/\pi)^{1/4}/(2w)^{1/2}$. The result is that (after an initial regime) all curves fall together, a fact which demonstrates scaling. Moreover Fig. 5.2 shows that our analytical expression, the function $F(z)$, describes very well the simulated decay over 4 orders of magnitude.

In Fig. 5.2 we have indicated the function F(z) by a dashed line. A detailed analysis[89] shows that the asymptotic form of F(z) for $z \gg 1$, Eq. (5.12), approximates F(z) reasonably well only for $F(z) < 10^{-3}$.

We infer that our theoretical approach based on fluctuation statistics describes very well the simulated kinetics of the $A+B \to 0$ reaction over 4 orders of magnitude in concentration decay. Interestingly, the simulations for $p_A < p_B$ (Fig. 5.1) also fit quite well the behavior predicted by Kang and Redner [see Eq. (5.2)]. This finding shows that one must proceed with much care when extracting asymptotic dependencies from computer simulations.

We are aware of the fact that our theory based on fluctuation statistics does not hold asymptotically when the B particles are immobile. This can be seen by noticing that with $N \sim t^{d/2}$ the theory predicts an asymptotic decay which is faster than that of the $A+B \to B$ trapping reaction. (A similar problem arises with Eq. (5.2) in dimensions $d>2$, as mentioned above.[37]) We infer that, although Eq. (5.7) provides a reasonable approximation over the experimentally relevant part of the decay, it may not always lead to a correct asymptotical form.

In the long-time regime the picture of fluctuation kinetics can be complemented by features which follow some ideas of Ref. 82, as we now show. For B immobile and $p_A < p_B$ at very large times one finds only single A particles (or small clusters of A particles) in large holes among the fixed B particles, holes which have been created during the earlier stages of the reaction. At this point in time the A clusters are well-separated from each other, and the situation is similar to that encountered for the $A+B \to B$ trapping reaction, see Sec. IV, but with a different distribution of hole sizes. Instead of putting in Eq. (5.7) N equal to the average volume S_ν explored by a random walker in ν steps, we now average $n_A(N)$ over the distribution of the span of a random walk and get the asymptotic result (for details see Ref. 89):

$$\bar{n}_A(t) = \sqrt{\frac{8}{3}} \, \frac{\Delta}{\pi^2} \, \exp\left[-\frac{3}{2} \left(\frac{\Delta^4 \pi^2}{4w^2} \, t \right)^{\frac{1}{3}} \right] . \tag{5.15}$$

Thus for large times we expect the simulations to follow Eq. (5.15) rather than Eq. (5.7), with $N=S_\nu$.

In Fig. 5.3 the results of an extensive simulation (solid curve) are depicted. Starting conditions are 10^7 A and 2×10^7 B particles, distributed on a linear chain of 10^8 sites. The results are the average over 4 realizations of the reaction. This allows us to monitor about 6 decades in the concentration decay. The dotted line is the analytical curve corresponding to the decay based on fluctuation statistics, Eq. (5.7) with $N=(8t/\pi)^{1/2}$, whereas the dashed line results from Eq. (5.15). Fig. 5.3 shows that during the early stages of the reaction the simulations are well-described by simple fluctuation statistics, whereas at later times the simulation curve approaches more and more the behavior of Eq. (5.15).

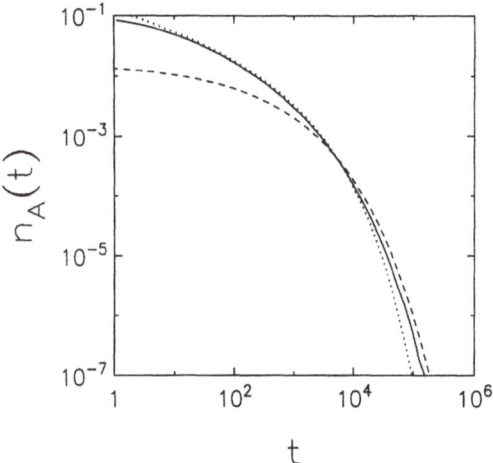

Fig. 5.3. Results of the simulation of the A+B-->0 reaction (solid curve) on a linear chain with 10^8 sites. The initial concentrations are p_A=0.1 and p_B=0.2. The dotted line represents Eq. (5.7), whereas the dashed line results from Eq. (5.15).

Summarizing, the concentration decay for the diffusion-limited A+B$\rightarrow$0 reaction, where the majority particles (B) are at rest, is (at early stages of the reaction) well-described by a theory based on simple fluctuation statistics, whereas for longer times a crossover to the asymptotic form $\rho_A(t) \propto \exp(-\text{const.}\times t^{1/3})$, Eq. (5.15), is observed in one dimension.

VI. CONCLUSIONS

In this work we have summarized recent results concerning the reaction kinetics for processes of the type A+B→products. For a spatially homogeneous distribution of particles at all times ('well-stirred reactor') the chemical kinetic scheme predicts exponential decay patterns for pseudo-unimolecular (A+B→B or A+B→0 with $A_0 < B_0$) or a 1/t form for bimolecular reactions (A+B→0 with $A_0 = B_0$).

As we have shown, deviations from such relaxation behaviors are widespread and may arise due to different kinds of randomness. Here we studied the effects on the kinetics of restrictions on the space in which the reaction takes place. Furthermore we highlighted the role played by spatial fluctuations on the reactants' decay. We considered both the case of immobile particles reacting through exchange or through multipolar interactions and also the case of diffusing particles which react on contact.

Only for a few special cases are exact analytical results available. Most problems, however, can be treated by approximate methods only. We have presented different analytical approaches to various reaction types and we have compared the analytical results to the outcomes of computer simulations. This helped us to assess the validity of several analytical approximations. The more sophisticated theories are in good agreement with the simulations at short and medium times, i.e. in the time regime relevant to most experiments.

In the problems studied here we have encountered decay patterns which reach from stretched-exponential over exponential-logarithmic to power-law forms. It is not easy to distinguish experimentally between the different decay patterns, since for this one needs measurements which cover several orders of magnitude in time and concentration decay. For similar reasons the exact long-time asymptotics is also difficult to establish.

In real systems the situation may be even more complex than in the models analyzed here. Thus, one often encounters different kinds of randomness simultaneously. In order to have a good starting point for analytical predictions, the main challenge is to find adequate model systems. For this a close cooperation between theory and experiment is crucial.

ACKNOWLEDGEMENTS

The authors are indebted to Dr. M. Drake, Prof. D. Haarer, Prof. J. Klafter, Prof. V. Kuzovkov, Dr. I.M. Sokolov and Dr. G. Zumofen for their cooperation on many of the topics discussed here. Stimulating discussions with G.H. Köhler and S. Luding are gratefully acknowledged. This work was supported by the Deutsche Forschungsgemeinschaft (SFB 213) and by the Fonds der Chemischen Industrie (grant of an IRIS workstation).

REFERENCES

1. S. Chandrasekhar, *Rev.Mod.Phys.* 15, 1 (1943)
2. D.F. Calef and J.M. Deutsch, *A.Rev.Phys.Chem.* 34, 493 (1983)
3. B.Ya. Balagurov and V.G. Vaks, *Zh.Eksp.Teor.Fiz.* 65, 1939 (1973) [*Sov.Phys. JETP* 38, 968 (1974)]
4. M.D. Donsker and S.R.S. Varadhan, *Commun.Pure Appl.Math.* 28, 525 (1975)
5. A.A. Ovchinnikov and Ya.B. Zeldovich, *Chem.Phys.* 28, 215 (1978)
6. D. Toussaint and F. Wilczek, *J.Chem.Phys.* 78, 2642 (1983)
7. G. Zumofen, A. Blumen and J. Klafter, *J.Chem.Phys.* 82, 3198 (1985)
8. A. Blumen and G.H. Köhler, *Proc.R.Soc.Lond.A* 423, 189 (1989)
9. G.H. Weiss and R.J. Rubin, *Adv.Chem.Phys.* 52, 363 (1983)
10. A. Blumen, J. Klafter and G. Zumofen, *in:* "Optical Spectroscopy of Glasses", I. Zschokke, ed., Reidel, Dordrecht (1986) p. 199
11. A. Blumen, G. Zumofen and J. Klafter, *in:* "Fractals, Quasicrystals, Knots and Algebraic Quantum Mechanics", A.Amann, ed., Kluwer Academic Publishers (1988) p. 21
12. A. Blumen, J. Klafter and G. Zumofen, *J.Chem.Phys.* 84, 1397 (1986)
13. J. Klafter, A. Blumen and J.M. Drake, *in:* "Molecular Dynamics in Restricted Geometries", J. Klafter and J.M. Drake, eds., Wiley, New York (1989) p. 1
14. T. Förster, *Z.Naturforsch.A* 4, 321 (1949)
15. D.L. Dexter, *J.Chem.Phys.* 21, 836 (1953)
16. M. Inokuti and F. Hirayama, *J.Chem.Phys.* 43, 1978 (1965)
17. D.L. Huber, *in:* "Laser Spectroscopy of Solids", W.M. Yen and P.M. Selzer, eds., Springer, New York (1981) p. 85
18. A. Blumen and J. Manz, *J.Chem.Phys.* 71, 4694 (1979)
19. A. Blumen, *Nuovo Cimento B* 63, 50 (1981)
20. J. Klafter and A. Blumen, *in:* "Energy Transfer Processes in Condensed Matter", B. DiBartolo, ed., Plenum, New York (1984) p. 621
21. D. Rehm and K.B. Eisenthal, *Chem.Phys.Lett.* 9, 387 (1971)
22. G. Porter and C.J. Tredwell, *Chem.Phys.Lett.* 56, 278 (1978)
23. I.M. Rozman, *Izv.Akad.Nauk SSSR Ser.Fiz.* 36, 922 (1972) [*Bull.Acad. Sci.USSR Phys.Ser.* 36, 833 (1972)]
24. A. Blumen, *J.Chem.Phys.* 72, 2632 (1980)
25. G. Zumofen, A. Blumen and J. Klafter, *J.Chem.Phys.* 84, 6679 (1986)
26. C.L. Yang, P. Evesque and M.A. El-Sayed, *J.Phys.Chem.* 89, 3442 (1985)
27. P. Levitz, J.M. Drake and J. Klafter, *J.Chem.Phys.* 89, 5224 (1988)
28. B.B. Mandelbrot, "The Fractal Geometry of Nature", W.H. Freeman, San Francisco (1982)
29. J. Klafter and A. Blumen, *J.Chem.Phys.* 80, 875 (1984)
30. S. Alexander and R. Orbach, *J.Phys.Lett. (Paris)* 43, L625 (1982)
31. R. Rammal and G. Toulouse, *J.Phys.Lett. (Paris)* 44, L13 (1983)
32. A. Takami and N. Mataga, *J.Phys.Chem.* 91, 618 (1987)
33. N. Tamai, T. Yamazaki, I. Yamazaki and N. Mataga, *Springer Ser.Chem. Phys.* 46 (1986)

34. Y. Lin, M.C. Nelson and D.M. Hanson, *J.Chem.Phys.* 86, 158 (1987)
35. P.G. de Gennes, *J.Chem.Phys.* 76, 3316; 3322 (1982)
36. D.C. Torney and H.M. McConnell, *J.Phys.Chem.* 87, 1941 (1983)
37. K. Kang and S. Redner, *Phys.Rev.Lett.* 52, 955 (1984); *Phys.Rev.A* 32, 435 (1985)
38. A. Blumen, G. Zumofen and J. Klafter, *J.Phys. (Paris)* 46, Colloque C7, 3 (1985)
39. V.P. Zhdanov, *Sov.Phys.Sol.St.* 27, 453 (1985)
40. L.W. Anacker and R. Kopelman, *Phys.Rev.Lett.* 58, 289 (1987)
41. P. Argyrakis and R. Kopelman, *J.Phys.Chem.* 91, 2699 (1987)
42. D. Ben-Avraham, *Phil.Mag.B* 56, 1015 (1987)
43. A.A. Lushnikov, *Phys.Lett.A* 120, 135 (1987)
44. Y.-C. Zhang, *Phys.Rev.Lett.* 59, 1726 (1987)
45. D. Ben-Avraham and C.R. Doering, *Phys.Rev.A* 37, 5007 (1988)
46. M. Bramson and J.L. Lebowitz, *Phys.Rev.Lett.* 61, 2397 (1988)
47. C.R. Doering and D. Ben-Avraham, *Phys.Rev.A* 38, 3035 (1988)
48. R. Kopelman, *Science* 241, 1620 (1988)
49. V. Kuzovkov and E. Kotomin, *Rep.Prog.Phys.* 51, 1479 (1988)
50. K. Lindenberg, B.J. West and R. Kopelman, *Phys.Rev.Lett.* 60, 1777 (1988)
51. J.L. Spouge, *Phys.Rev.Lett.* 60, 871 (1988)
52. P. Argyrakis and R. Kopelman, *J.Phys.Chem.* 93, 225 (1989)
53. C.R. Doering and D. Ben-Avraham, *Phys.Rev.Lett.* 62, 2563 (1989)
54. G.H. Weiss, R. Kopelman and S. Havlin, *Phys.Rev.A* 39, 466 (1989)
55. B.J. West, R. Kopelman and K. Lindenberg, *J.Stat.Phys.* 54, 1429 (1989)
56. H. Schnörer, V. Kuzovkov and A. Blumen, *Phys.Rev.Lett.* 63, 805 (1989)
57. H. Schnörer, V. Kuzovkov and A. Blumen, *J.Chem.Phys.* 92, 2310 (1990)
58. H. Schnörer, V. Kuzovkov and A. Blumen, *submitted*
59. H. Schnörer, S. Luding and A. Blumen, *in:* "Fractal 90", J. Henriques, ed., Fundacao Gulbenkian, Lisbon (1990)
60. R.Balescu, "Equilibrium and Nonequilibrium Statistical Mechanics", Wiley, New York (1975) p. 247
61. J.G. Kirkwood, *J.Chem.Phys.* 3, 300 (1935)
62. P. Hanusse and A. Blanche, *J.Chem.Phys.* 74, 6148 (1981)
63. E. Angel and R. Bellman, "Dynamic Programming and Partial Differential Equations, Academic Press, New York (1972) ch. 11
64. M.N. Barber and B.W. Ninham, "Random and Restricted Walks", Gordon and Breach, New York (1970)
65. Papers presented at the Symposium on Random Walks, *J.Stat.Phys.* 30, No.2 (1983)
66. M.F. Shlesinger and B.J. West, "Random Walks and their Applications in the Physical and Biological Sciences", Amer.Inst.Phys., New York (1984)
67. G. Zumofen and A. Blumen, *Chem.Phys.Lett.* 88, 63 (1982)
68. H.E. Stanley, K. Kang, S. Redner and R.L. Blumberg, *Phys.Rev.Lett.* 51, 1223 (1983)
69. R. Kopelman, P.W. Klymko, J.S. Newhouse and L.W. Anacker, *Phys.Rev.B* 29, 3747 (1984)
70. J. Klafter, A. Blumen and G. Zumofen, *J.Phys.Chem.* 87, 191 (1983)
71. E.W. Montroll and J.T. Bendler, *J.Stat.Phys.* 34, 129 (1984)
72. M.F. Shlesinger and E.W. Montroll, *Proc.Natl.Acad.Sci. USA* 81, 1280 (1984)
73. S. Redner and K. Kang, *J.Phys.A* 17, L451 (1984)
74. A. Blumen, G. Zumofen and J. Klafter, *Phys.Rev.B* 30, 5379 (1984)
75. G. Zumofen, A. Blumen and J. Klafter, *in:* "Structure and Dynamics of Molecular Systems", R. Daudel, J.P. Korb, J.P. Lemaistre and J. Maruani, eds., Reidel, Dordrecht (1985) p. 87
76. E.W. Montroll and G.H. Weiss, *J.Math.Phys.* 6, 167 (1965)
77. A. Blumen, J. Klafter and G. Zumofen, *Phys.Rev.B* 28, 6112 (1983)
78. J. Klafter, A. Blumen and G. Zumofen, *J.Stat.Phys.* 36, 561 (1984)

79. A. Blumen, G. Zumofen and J. Klafter, *in:* "Structure and Dynamics of Molecular Systems", R. Daudel, J.P. Korb, J.P. Lemaistre and J. Maruani, eds., Reidel, Dordrecht (1985) p. 71

80. B. Movaghar, G. Sauer, D. Würtz and D.L. Huber, *Sol.St.Comm.* 39, 1179 (1981); *J.Stat.Phys.* 27, 473 (1982)

81. S. Redner and K. Kang, *Phys.Rev.Lett.* 51, 1729 (1983)

82. J.K. Anlauf, *Phys.Rev.Lett.* 52, 1845 (1984)

83. P. Grassberger and I. Procaccia, *J.Chem.Phys.* 77, 628 (1982)

84. R.F. Kayser and J.B. Hubbard, *Phys.Rev.Lett.* 51, 6281 (1982)

85. J. Klafter, G. Zumofen and A. Blumen, *J.Phys.Lett. (Paris)* 45, L49 (1984)

86. G. Zumofen, A. Blumen and J. Klafter, *J.Phys.A* 17, L479 (1984)

87. S. Havlin, M. Dishon, J.E. Kiefer and G.H. Weiss, *Phys.Rev.Lett.* 53, 407 (1984)

88. P. Meakin and H.E. Stanley, *J.Phys.A* 17, L173 (1984)

89. H. Schnörer, I.M. Sokolov and A. Blumen, *submitted*

90. M. Abramowitz and I.A. Stegun, "Handbook of Mathematical Functions", Dover, New York (1972)

PART B

The Seminars

COHERENCE AND QUANTUM MECHANICS

Stephan Zanzinger

Institut für Theoretische Physik
Auf der Morgenstelle 14
D–7400 Tübingen

1 Introduction

Coherent superposability of pure states is one characteristic feature of traditional quantum mechanics. It is one of those points, which forces the departure from classical physics and the use of Hilbert space quantum mechanics. If one want to use the algebraic approach to quantum mechanics, two problems arise. The first one is how to formulate superposability of states if there is no Hilbert space, and the second one is how to handle mixed states, which are the only states relevant to certain thermodynamic representations.

One solution of these problems is the use of the GNS–representation, where a suitable Hilbert space is constructed. This is practicable, but has certain drawbacks. A first minor disadvantage is the nonuniqueness of the vectorstate status of a state. This could be solved by the use of irreducible representations. But then one looses the power of that notion concerning macroscopically different states and the second problem remains unsolved. The major objection is sophisticated and a mainly aesthetic question. In the algebraic approach the GNS–Hilbert space is a derived concept and therefore *it should not be used for such fundamental investigations.*

For that reason we advocate an abstract point of view, which focuses on the convex structure of the state space, induced by the statistical character of quantum mechanics. This has the further advantage that the basic notions are so general, that our definitions can also be applied to every fundamental approach having the concept of states. For detailed analysis we use more special structures, induced by the algebraic properties of the observables, discovered by [6, 5] . On the other hand their results show that the observables are predetermined by the structure of the state space, which is deeply connected with the relation of superposability used in the sequel.

We'll use the notion introduced by [11], which is different from the more popular one due to Varadarajan [4] used in the context of quantum logic [1]. It has the advantage that it takes a relative measure of mixedness into account, e.g. it allows only pure states to be superpositions of pure states, and distinguishes between coherent superpositions and statistical mixtures. This relation although originally defined in the W^*-algebraic context we'll investigate in the more general algebraic environment of Jordan algebras and demonstrate how it can be used to characterize C^*-algebras. By doing this an other well–known concept of quantum mechanics, the concept of transition probability, is generalized to this algebraic framework.

---

[1] for a extensive discussion of the superposition principle and the other quantum principles cf. [9]

2 General Setting of Jordan Algebraic Theories

In this paper the *observables* are represented by a Jordan subalgebra $\mathcal{A}$ of $B(\mathcal{H})_{sa}$ the selfadjoint operators on the complex Hilbertspace $\mathcal{H}$, i.e. $\mathcal{A}$ is a norm closed subspace of $B(\mathcal{H})_{sa}$, which is also closed under the special Jordan product:

$$A \circ B = \tfrac{1}{2}(AB + BA) \qquad A, B \in B(\mathcal{H})_{sa}$$

where the multiplication on the left hand side of the above equation is the usual composition of operators. It is worth mentioning that this product can be defined using only the operation of squaring, which is connected with the forming of variances and is therefore essential for a statistical theory:

$$A \circ B = \tfrac{1}{2}[(A + B)^2 - A^2 - B^2]$$

The Jordan product is commutative and distributive, but in general *not* associative, reflecting the noncommutativity of the ordinary product in $B(\mathcal{H})$. On the motivation for the use of Jordanalgebras in quantum mechanics see the fundamental paper of Jordan, von Neumann and Wigner [8] and the book of Emch [1]. Equip $\mathcal{A}$ with the order generated by the positive cone of squares, i.e. operators with positive spectrum, and let the identity be a member of $\mathcal{A}$.

States are represented in this context by normalized, positive, linear functionals on $\mathcal{A}$. Denote by $\mathcal{K}$ the set of all states. $\mathcal{K}$ is a convex, w^*-compact set and is therefore by the Krein–Milman–Theorem generated by its pure states. The physical operation of statistical mixing of states is represented by a convex combination of the elements of $\mathcal{K}$.

Further consider also the bidual of $\mathcal{A}$, $\mathcal{A}^{**} = A^b(\mathcal{K})$, which consists of the bounded affine functionals on $\mathcal{K}$, and is the completion of $\mathcal{A}$ in the weak operator topology. It can be proved ([12] Theorem 1.4.) that it is also a Jordan algebra, a so–called JW–Algebra, the Jordan counterpart of von Neumann Algebras, with $\mathcal{K}$ as normal state space. For the theory of Jordan algebras see [2].

As a paradigmatic example of these structures remember that in traditional Hilbert space quantum mechanics, with superselection rules these structures are:

$$A^b(\mathcal{K}) = \bigoplus_{i \in I} B(\mathcal{H}_i)_{sa} \qquad \text{and} \qquad \mathcal{K} = \bigoplus_{i \in I} \mathcal{T}_1^+(\mathcal{H}_i)$$

Here the $\mathcal{H}_i$ are the coherent subspaces, where the unrestricted Dirac superposition principle holds and $\mathcal{T}_1^+(\mathcal{H}_i)$ are the density operators on $\mathcal{H}_i$. Superselection rules were introduced, because we want to exhibit in this example how the notion of (non)superposability can be used to distinguish macroscopically different states.

3 The Coherence Relation

In order to state the notion of superposability we'll need first some fundamental structures in the state space $\mathcal{K}$.

Definition 3.1

(a) $F \subseteq \mathcal{K}$ is called property (norm–closed face) if the following holds:

   (i)   F is norm–closed

  (ii)   F is convex (invariant under statistical mixing)

 (iii)   F is invariant under statistical decomposition:

$$F \ni \omega = \lambda\psi + (1 - \lambda)\chi \;\; \& \;\; \psi, \chi \in \mathcal{K} \implies \psi, \chi \in F$$

The set of all properties is denoted by $\mathcal{E}$.

(b) A property C is called classical *(split face) if $C^{\perp} := \bigcup\{F \in \mathcal{K}, F \cap C = \emptyset\}$ is a property and every state is the* unique *statistical mixture of a state of C and one of $C^{\perp}$. Let $\mathcal{F}$ be the set of all classical properties.*

It is now easy to see that if E, F are (classical) properties then E∩F is also a (classical) property and this can be extended to the intersection of an arbitrary number of properties. This allows now to define the supremum of a family of properties $(F_i)_{i \in I}$, denoted by $\bigvee_{i \in I} F_i$, as least property containing all the F_i, and for a state ω E_{ω} (C_{ω}) the (classical) property generated by ω, as the least (classical) property containing ω. In our paradigmatic example E_{ω} for a state ω represented by the density matrix ρ_{ω} consists of all those density matrices, whose images are contained in the image of ρ_{ω}. This is one facette of the fundamental order isomorphism between properties and the closed linear span of the union of images of its members. In this isomorphism the superselection sectors are the images of the classical properties. Further define the so–called orthocomplement [2] $F^{\perp} = \{\omega, \forall \psi \in F : \|\omega - \psi\| = 2\}$, call E orthogonal to F $(E \perp F)$ iff $E \subseteq F^{\perp}$ and denote by $\wedge$ the usual settheoretical meet.

It must be stressed, that most of the fundamental concepts should be formulated on this level, because there the structure is more transparent. Compatibility of properties $E, F \in \mathcal{E}$ (ECF), for example, is expressed by the formula: $E = (E \wedge F) \vee (E \wedge F^{\perp})$. This is a certain kind of distributivity, because iff in a lattice all elements are compatible, the lattice is Boolean[3]. Further is orthomodularity characterized by the symmetry of the above relation (for the lattice–theoretical background cf. [3]).

Proposition 3.1 $\langle \mathcal{E}, \vee, \wedge, \subseteq, \perp \rangle$ *is a complete, orthomodular lattice with $\mathcal{K}$ as greatest and $\emptyset$ as least elements. $\mathcal{F}$ is the centre of $\mathcal{E}$, i.e. the Boolean sublattice, whose elements are compatible with every property.*

PROOF: The main idea of the proof is, that according to [7] (Lemma 4.3.) $\mathcal{E}$ is isomorphic in the lattice theoretic sense to the lattice of projections of $A^b(\mathcal{K})$. The isomorphismus works as follows:
Let E be a property and then let P_E[4] be the least projection $P \in A^b(\mathcal{K})$ with $\omega(P) = 1$ for all ω in E and for a given projection $P \in A^b(\mathcal{K})$ is $E_P = \{\omega \in \mathcal{K}, \omega(P) = 1\}$ the associated property.
Also the following formula is valid: $P_{E^{\perp}} = 1 - P_E$.
This projection lattice shares almost all features with the projection lattice of a von Neumann algebra [2] and one of them is orthomodularity. The image of $\mathcal{F}$ under this isomorphism are the central projections. ∎

This means that $\mathcal{E}$ is a so called quantum logic, with $\mathcal{F}$ as classical part. A further immediate consequence of the above definitions is, that the atoms of $\mathcal{E}$, i.e. the least properties greater than $\emptyset$, are precisely those with one element, which is then automatically pure.

Definition 3.2 (Coherent superposition) *Let ω, ψ, φ states. We call φ a coherent superposition of ω and ψ $(K(\omega, \psi, \varphi))$ if the following holds:*

(i) $E_{\omega} \vee E_{\psi} = E_{\omega} \vee E_{\varphi} = E_{\psi} \vee E_{\varphi}$
(ii) $E_{\omega} \wedge E_{\psi} = E_{\omega} \wedge E_{\varphi} = E_{\psi} \wedge E_{\varphi} = \emptyset$

Further call ω, ψ coherently superposable $(K(\omega, \psi))$ if there is a state χ with $K(\omega, \psi, \chi)$.

---

[2]under the above isomorphism this is the orthogonal subspace to the subspace representing E
[3]In the projection lattice of von Neumann algebras compatibility coincides with commutability
[4]P_E is called support projection of E

This extends the usual notion of coherent superposability, because by an easy application of the correspondence between properties and kernels in our example a pure state $\varphi = |\varphi><\varphi|$ with $|\varphi> \in \mathcal{H}_i$ is a coherent superposition of the states ω, ψ iff ω, ψ are vector states in $\mathcal{H}_i$ and $|\varphi>$ is in the linear span of $|\omega>$ and $|\psi>$. This is also valid in the general case:

Proposition 3.2 *Let $\omega, \psi \in \mathcal{K}$ be pure states. $K(\omega, \psi, \varphi)$ holds for some state $\varphi \in \mathcal{K}$ iff φ is pure and $\varphi \in E_{\omega,\psi} := E_\omega \vee E_\psi = E_{\frac{1}{2}(\omega+\psi)}$.*

PROOF: By (i) of Definition 3.2 it is necessary that $\varphi \in E_{\omega,\psi}$. Using the covering property[5] of $\mathcal{E}$, proved by the isomorphy mentioned in the proof of Prop. 3.1, E_φ is all of $E_{\omega,\psi}$ iff φ is not pure, i.e. (i) holds. (ii) of Definition 3.2 is always true for different pure states ω, ψ, φ. ∎

Corollary 3.3 *The pure states ω, ψ are coherently superposable iff $C_\omega = C_\psi$, i.e. iff they are macroscopically identical.*

PROOF: If $C_\omega = C_\psi$ then there is another pure state $\varphi \in E_{\omega,\psi}$ and by Proposition 3.2 $K(\omega, \psi, \varphi)$. If $C_\omega \neq C_\psi$ then $C_\omega \wedge C_\psi = \emptyset$ and by the unique decomposition property $E_{\omega,\psi} = [\omega, \psi] = \{\lambda\omega + (1-\lambda)\psi, \lambda \in [0,1]\}$ and contains no additional pure state. ∎

The property $E_{\omega,\psi}$ for pure states ω, ψ plays now an important role in the sequel. It is used in [6, 5] as a tool to characterize state spaces of JB-Algebras and C^*−algebras among other compact, convex sets. One of their results is the following proposition.

Proposition 3.4 *$E_{\omega,\psi}$ for pure states ω, ψ is affinely homeomorphic to a ball $\mathcal{B}$ in some real Hilbertspace $\mathcal{H}$, called n–ball, where n is the dimension of $\mathcal{H}$. If $C_\omega \neq C_\psi$ then $E_{\omega,\psi}$ degenerates to the line segment $[\omega, \psi]$.*

PROOF: cf. [6] Theorem 3.11. ∎

This is the so–called Hilbert ball property. Now the occuring cases can further be discriminated, using the dimension n of the Hilbert space $\mathcal{H}$.

Proposition 3.5 *Let all be as in Proposition 3.4. If $C_\omega = C_\psi$ then C_ω is affinely homeomorphic (norm topology) to the state space of the following Jordan algebras C ,i.e. $C = A^b(C_\omega)$ is a direct summand of $A^b(\mathcal{K})$:*

   (i)   *n=2: $\mathcal{B} = B(\mathcal{G})_s$*
         the symmetric, bounded operators on a real Hilbert space $\mathcal{G}$.
   (ii) *n=3: $\mathcal{B} = B(\mathcal{G})_{sa}$*
         the selfadjoint, bounded operators on a complex Hilbert space $\mathcal{G}$.
   (iii) *n=5: $\mathcal{B} = B(\mathcal{G})_{sa}$*
         the selfadjoint, bounded operators on a quaternionic Hilbert space $\mathcal{G}$.
   (iv) *a n–dimensional spinfactor.*

Here is n, as in Proposition 3.4, the dimensionality of the ball homeomorphic to $E_{\omega,\psi}$.

PROOF: Cf. [5] Proposition 3.2. ∎

The 3–ball–property, i.e. 3.5 (ii) holds for every pair of macroscopically identical, different, pure states, characterizes together with orientability C^*−algebras among JB-algebras ([5] Theorem 8.4.). As a corollary one gets combining 3.2 and 3.4, that the space of all coherent superpositions of pure ω, ψ, viewed under the above mentioned homeomorphism is the surface of the n-ball $\mathcal{B}$.

---

[5]If $\varphi \in \mathcal{K}$ is pure ,$\varphi \notin E \in \mathcal{E}$, then there is no property F with $E < F < E_\varphi \vee E$

Now the investigations of Mielnik [10], concerning transition probabilities on general convex sets can be applied. In his nomenclature the n–balls are the transition probability spaces S(2,n). The transition probability between pure states ω, ψ is defined as:

$$p(\omega, \psi) := x^2$$

Here x is the geometrical distance between ω and $\psi^{\perp}$ [6] the antipodal point of ψ on the n–ball $E_{\omega, \psi}$, with radius 1/2. In our example this is exactly $|<\psi|\omega>|^2$, the usual transition probability of vector states. Another representation for the transition probability arises through the isomorphism mentioned in Proposition 3.1: $p(\omega, \psi) = \omega(P_{E_{\psi}})$. Mielnik ([10] Section 5) established also numerical criteria, i.e. a hierarchy of inequalities between transition probabilities, which discriminate the occuring cases. For instance we have the following propositions.

Proposition 3.6 *Let* $\omega, \psi, \varphi, \chi \in E_{\omega, \psi} \cong \mathcal{B}$ *be pure states, where each state is the coherent superposition of two of the others. Further let the transition probability between each two different states be constant and equal to p. Then these states can be realized in traditional quantum mechanics, i.e.* $\mathcal{B}$ *is a 3-ball iff* $p = \frac{1}{\sqrt{3}}$. *If* $0 < p \leq \frac{1}{\sqrt{3}}$ *then* $\mathcal{B}$ *can be a n-ball with* $n > 3$. *The remaining case* $1 > p > \frac{1}{\sqrt{3}}$ *is not attainable in Jordan algebraic quantum mechanics.*

PROOF: That is elementary geometry, i.e. the pure geometrical problem to place an tetraeder in such a way in a three dimensional ball, that the corners lie on the surface. The other statements follow from the same considerations in higher dimensions. ∎

Proposition 3.7 *Let* $\omega_1, ..., \omega_n \in \mathcal{B}$ *be pure states, where each state is the coherent superposition of two of the others. Assume also, that* $p(\omega_i, \omega_k) = \frac{1}{2}$ *for* $i \neq k$. *Then n is the minimal dimension of* $\mathcal{H}$ *in which* $\mathcal{B}$ *can be embedded, i.e.* $\mathcal{B}$ *is at least a n-ball.*

PROOF: This is a consequence of the fact, that $p(\varphi, \psi) = \frac{1}{2}$ means, that the images of φ, ψ are orthogonal in $\mathcal{H}$. ∎

References

[1] EMCH,G.G. *Algebraic Methods in Statistical Mechanics and Quantum Field Theory* Wiley New York 1971

[2] HANCHE–OLSEN,H. STØRMER,E. *Jordan Operator Algebras* Pitman Boston 1984

[3] KALMBACH,G. *Orthomodular Lattices* Academic Press London 1983

[4] VARADARAJAN,V.S. *Geometry of Quantum Theory I* Van Nostrand New York 1968

[5] ALFSEN,E.M. HANCHE–OLSEN,H. SHULTZ,F.W. State Spaces of C^*-Algebras *Acta Mathematica* **144** (1980) pp. 267-305

[6] ALFSEN,E.M. SHULTZ,F.W. State Spaces of Jordan Algebras *Acta Mathematica* **140** (1978) pp. 155-190

[7] EDWARDS,C.M. RÜTTIMANN,G.T. On the Facial Structure of the Unit Balls in a GL-Space and its Dual *Mathematical Proceedings of the Cambridge Philosophical Society* **98** (1985) pp. 305-322

[8] JORDAN,P. VON NEUMANN,J. WIGNER,E. On an Algebraic Generalization of Quantum Mechanic Formalism *Annals of Mathematics* **35** (1934) pp. 29-34

[9] LAHTI,P.J. Characterization of Quantum Logics *International Journal of Theoretical Physics* **19** (1980) pp. 905-923

[10] MIELNIK,B. Geometry of Quantum States *Communications of Mathematical Physics* **9** (1968) pp. 55-80

---

[6]$\{\psi^{\perp}\} = E_{\omega, \psi} \wedge \{\psi\}^{\perp}$

[11] RAGGIO,G. RIECKERS,A. Coherence and Incompatibility in W^*-Algebraic Quantum Theory *International Journal of Theoretical Physics* **22** (1983) pp. 267-291

[12] SHULTZ,F.W. On Normed Jordan Algebras which are Banach Dual Spaces *Journal of Functional Analysis* **31** (1979) pp. 360-376

THE DYNAMICAL GENERATION OF MACROSCOPIC COHERENT LIGHT

Reinhard Honegger

Institut für Theoretische Physik
Auf der Morgenstelle 14
D–7400 Tübingen

1 Introduction

In quantum optics for the theory of the laser commonly only are calculated photon rate equations and the coherence properties of the emitted light are only discussed by the mean and mean-square photon numbers in the single mode approach of the quantized electromagnetic field. In terms of operator algebraic quantum statistical mechanics the present work gives a detailed microscopic description how the coherent laser light is generated during the time evolution in dependence upon the macroscopic preparation of the radiating atoms or molecules. For this we study the extended Dicke model, a very large system of (infinitely many) two-level atoms interacting with the radiation field quantized in the whole euclidean space $\mathbb{R}^3$.

The infinitely many atoms we index by $\mathbb{N}$, but e.g. one may assume a spatial lattice. For each two-level atom we have the complex 2×2-matrices M_2 as the algebra of observables, and hence for the whole of the atoms we take the inductive limit C^*-algebra $\mathcal{A} := \bigotimes_{n \in \mathbb{N}} M_2$. If $\sigma^1, \sigma^2, \sigma^3$ and $\sigma^\pm = \sigma^1 \pm i\sigma^2$ are the usual Pauli spin matrices let us put $s_n^k := \mathbb{1} \otimes \cdots \otimes \mathbb{1} \otimes \frac{1}{2}\sigma^k \otimes \mathbb{1} \cdots \in \mathcal{A}$ with $k \in \{1, 2, 3, -, +\}$ for the spin-1/2 description of the n-th atom, where $\frac{1}{2}\sigma^k$ appears at the n-th factor.

For simplicity for the electromagnetic field only one direction of polarization is considered. As C^*-algebra of the photons we use the Weyl algebra $\mathcal{W}(E)$ over an appropriate testfunction space E dense in $\mathsf{L}^2(\mathbb{R}^3)$ and which will be specified later. The one-photon hamiltonian is given by $\sqrt{-\Delta}$ with the usual Laplacian $\Delta = \frac{\partial^2}{\partial x^2} + \frac{\partial^2}{\partial y^2} + \frac{\partial^2}{\partial z^2}$ in $\mathbb{R}^3$, and we assume E to be invariant under the group $(e^{it\sqrt{-\Delta}})_{t \in \mathbb{R}}$. For small times $t \in \mathbb{R}$, the photon system is supposed to have no macroscopic part. Therefore we restrict ourselves to the Fock representation $(\Pi_{\mathcal{F}}, \mathcal{F}_+(\mathsf{L}^2(\mathbb{R}^3)))$ of $\mathcal{W}(E)$. By $a_{\mathcal{F}}(g)$, $a_{\mathcal{F}}^*(g)$ and $W_{\mathcal{F}}(g)$ we denote the (smeared) annihilation, creation and Weyl operators in the Fock space $\mathcal{F}_+(\mathsf{L}^2(\mathbb{R}^3))$ over $\mathsf{L}^2(\mathbb{R}^3)$. In the Fock representation the free photon Heisenberg dynamics is given by $e^{itd\Gamma(\sqrt{-\Delta})} W_{\mathcal{F}}(g) e^{-itd\Gamma(\sqrt{-\Delta})} = W_{\mathcal{F}}(e^{it\sqrt{-\Delta}}g)$, where $d\Gamma(\sqrt{-\Delta})$ is the second quantization of $\sqrt{-\Delta}$, which in momentum space formally is expressed by $\int_{\mathbb{R}^3} |\mathbf{k}| \, a_{\mathbf{k}}^* a_{\mathbf{k}} \, d^3\mathbf{k}$. Note, with the Fock vacuum vector $\Omega_{\mathcal{F}}$ the Fock representation is just the GNS-representation of the Fock state $\omega_{\mathcal{F}}(W(g)) = \exp\{-1/4 \|g\|^2\} \; \forall g \in E$, where $W(g)$, $g \in E$, denote the Weyl operators in $\mathcal{W}(E)$.

After having choosen some physically suitable representation $(\Pi_a, \mathcal{H}_a)$ of $\mathcal{A}$ we set up the local hamiltonians — local, that is for the atoms in $\Lambda \subset \mathbb{N}$, $|\Lambda| < \infty$ — in the standard manner (rotating wave approximation, cf. e.g. [1])

$$H_\Lambda = \sum_{n \in \Lambda} \varepsilon \, \Pi_a(s_n^3) \otimes \mathbb{1}_{\mathcal{F}} + \mathbb{1}_a \otimes d\Gamma(\sqrt{-\Delta}) + \frac{1}{|\Lambda|} \sum_{n \in \Lambda} \Big(\Pi_a(s_n^-) \otimes a_{\mathcal{F}}^*(f) + \Pi_a(s_n^+) \otimes a_{\mathcal{F}}(f) \Big).$$

Large-Scale Molecular Systems, Edited by W. Gans *et al.*
Plenum Press, New York, 1991

Here $\varepsilon > 0$ denotes the energy difference of the two levels of each atom, and the values $\hat{f}(\mathbf{k})$, where $\hat{f}$ is the Fourier transform of $f \in \mathsf{L}^2(\mathbb{R}^3)$, are the coupling constants between the atomic system and the mode $\mathbf{k}$ of the field. The coupling is taken as $|\Lambda|^{-1}$, because we desire a finite rate of radiation in the limit $\Lambda \to \mathbb{N}$.

In the representation $\Pi_a \otimes \Pi_{\mathcal{F}}$ of $\mathcal{A} \otimes \mathcal{W}(E)$ we show that the limiting Heisenberg dynamics $\lim_{\Lambda \to \infty} e^{itH_\Lambda} Z e^{-itH_\Lambda} = e^{itH} Z e^{-itH}$, $Z \in \Pi_a \otimes \Pi_{\mathcal{F}}(\mathcal{A} \otimes \mathcal{W}(E))$, $t \in \mathbb{R}$, exists and deduce its generator H. We find that the limiting hamiltonian H is integrable and obtain an explicit closed form for the unitary time evolution group $(e^{itH})_{t \in \mathbb{R}}$. For an arbitrary state ϱ on $\mathcal{A} \otimes \mathcal{W}(E)$, which is normal to $\Pi_a \otimes \Pi_{\mathcal{F}}$ — ϱ can be regarded as a density operator on $\mathcal{H}_a \otimes \mathcal{F}_+(\mathsf{L}^2(\mathbb{R}^3))$ —, we consider its time evolved restriction to the Weyl algebra $\mathcal{W}(E)$

$$\omega_t^\varrho(X) \quad := \quad \mathrm{tr}_{\mathcal{H}_a \otimes \mathcal{F}_+} \left[e^{-itH} \varrho\, e^{itH} \left(\mathbb{1} \otimes \Pi_{\mathcal{F}}(X) \right) \right], \qquad X \in \mathcal{W}(E), \;\; t \in \mathbb{R}. \tag{1}$$

For very large times $t > 0$, ω_t^ϱ will be shown to become a (fully) coherent state on $\mathcal{W}(E)$. The coherence properties will be discussed in their operator algebraic version [2] [3], which is an extension of Glauber's original definition [4] and obtained by a smearing procedure with one-photon testfunctions. This extended coherence condition naturally allows to consider more general cases than coherent states, which are normal to the Fock representation, whereas Glauber only investigated the latter. And indeed, here in the Dicke model it depends on the coupling function f, if the generated coherent light is of microscopic or macroscopic nature, that is, if the appearing coherent states are normal resp. disjoint to the Fock representation. As we will see, the coherent states occuring for large times are characterized by a linear form $L : E \to \mathbb{C}$, which is additively composed by two very different terms. The first one represents the resonance ($\|\mathbf{k}\| = \varepsilon$) between the radiation field and the two levels of each atom, whereas the other term picks up in the surroundings of the resonant modes $\mathbf{k}$. Because in physical applications the coupling function f is uniquely determined by the wave functions of the two energy levels of the type of atoms or molecules under consideration, in general the linear form L will be unbounded (unbounded with respect to the scalar product topology of the pre-Hilbert space $E \subseteq \mathsf{L}^2(\mathbb{R}^3)$). As a consequence our coherent states are not realizable by density operators on Fock space. In fact, for unbounded L there are so much emitted photons, that ω_t^ϱ leaves Fock space for very large times $t > 0$. More exactly, ω_t^ϱ no longer remains normal to the Fock representation, when the time t goes to infinity. For more details see Section 5.

In [1] for the same problem also a limit $\Lambda \to \mathbb{N}$ is done, but along a sequence of eigenstates of the total angular momentum of the spins in Λ, $|\Lambda| < \infty$, and, since only the radiation field is considered, by averaging out these local atomic states. Regarded from the atomic part this limit seems very artificial and is far from being done in a representation of $\mathcal{A}$. There two macroscopic parameters of the atomic system, the global cooperation and excitation number, are choosen and then in the restriction to the radiation field the limit $\Lambda \to \mathbb{N}$ is carried through to obtain just these two parameters in the limit. Contrary to [1] we consider here the infinite atomic system in the only correct way, namely as a mean field quantum lattice system, and the above two macroscopic parameters are choosen with the selection of the representation Π_a. So, for $\Lambda \to \mathbb{N}$ we have the correct dynamical description of the total system and not only, as is put on in a somewhat ad hoc manner in [1], in the restriction to the photons and some kind of collective behaviour of the atoms. However, in [1] for $t \to \infty$ the same Cesàro averaged states on $\mathcal{W}(E)$ are obtained. We remark that both works [1] inspired the present investigation.

In the present manuscript for $(\Pi_a, \mathcal{H}_a)$ we take the GNS-representations of some permutation invariant states on $\mathcal{A}$, each of which has a sharp cooperation number μ and a sharp degree of excitation γ. Because of the tensor product structure of $\mathcal{H}_a = \mathcal{H}_0 \otimes \mathsf{L}^2([0, 2\pi[, \mathrm{d}\vartheta/(2\pi))$, we are in a situation from which can be seen what is going on in [1]. But one may ask, if it is possible to take larger representations of the atomic system, perhaps the largest one, the partially universal representation Π_a^p associated with the folium F_a^p generated by all permutation invariant states on $\mathcal{A}$. The answer is: "Yes". Indeed, it turns out to be possible working with Π_a^ρ. However, the mathematical techniques are much more complicated, especially the direct integral method used in Section 3 has to be replaced by the spectral integral, which is more general and more difficult to handle with (integration of non-commutative operator-valued functions with respect to projection-valued measures). For large times the states ω_t^ρ become coherent too, but coherent in first order and in general not fully coherent. Moreover it depends on the variance and the sharpness of the preparation of the macroscopic value $\beta = \sqrt{(\mu/2)^2 - (\gamma - 1/2)^2}$ of the atomic system at $t = 0$, how stringent the higher order coherence properties are satisfied. In other words: Due to this general formalism one is able to calculate explicitly the sharpness of the coherence in each order of the emitted light (after long waiting) in dependence from the macroscopic preparation of the atomic system at $t = 0$ by the experimentalist. However, if the macroscopic parameters μ and γ are sharp, the present results suffice and always fully coherent light is obtained. Still more general, the macroscopic ordering inherent in the atomic system, partially by preparation at $t = 0$ and partially by the system itself, one refinds in the formulae of the light states which are build up during the time evolution. Especially one aspect are the discussed coherence properties. In the here presented material in detail this effect is visible in Section 4.

For the photon system one also may take other representations than the Fock one. In the experiment these are due to macroscopic preparations of the light at $t = 0$. As mentioned above, if only a few photons are present at $t = 0$, the Fock representation is the correct one.

After this outlook let us return to the present work and proceed as follows. In Section 2 the mean field model briefly is discussed, the considered representations Π_a are specified and for $\Lambda \to \mathbb{N}$ the existence of the limiting dynamics is ensured, which is written in its closed form in Section 3. For large times $t > 0$ the states ω_t^ρ are investigated in Section 4, and the last Section is devoted to the emitted coherent light. We mention, all result are stated without proofs.

2 The Infinite Atom Limit

For choosing the representation $(\Pi_a, \mathcal{H}_a)$ of $\mathcal{A}$ let us define the mean field operators $J_\Lambda^k := \frac{1}{|\Lambda|} \sum_{n \in \Lambda} s_n^k$, $k \in \{1, 2, 3, -, +\}$, and the square $(\vec{J}_\Lambda)^2 := (J_\Lambda^1)^2 + (J_\Lambda^2)^2 + (J_\Lambda^3)^2$ for each $\Lambda \subset \mathbb{N}$ with $|\Lambda| < \infty$. Together with the local hamiltonians $A_\Lambda := \varepsilon |\Lambda| J_\Lambda^3$ and the associated local Heisenberg dynamics $\alpha_t^\Lambda(.) := e^{itA_\Lambda} . e^{-itA_\Lambda}$ on $\mathcal{A}$ we are in the situation of a simple mean field quantum lattice system. Using the identification of $K_{1/2} := \{x \in \mathbb{R}^3 \mid \|x\| \leq 1/2\}$ with the state space of M_2, there is a well known affine bijection p from the probability measures $M_+^1(K_{1/2})$ on $K_{1/2}$ onto the permutation invariant states $S^p(\mathcal{A})$ on $\mathcal{A}$. Thereby the point measures are just the elements of the extreme boundary of $S^p(\mathcal{A})$. Further, there is a unique C^*-dynamics $(\alpha_t)_{t \in \mathbb{R}}$ on $\mathcal{A}$ such that $\alpha_t(X) = \|.\| - \lim_{\Lambda \to \infty} \alpha_t^\Lambda(X) \ \forall X \in \mathcal{A} \ \forall t \in \mathbb{R}$. The associated Schrödinger dynamics $(\alpha_t^*)_{t \in \mathbb{R}}$ then is given on $S^p(\mathcal{A})$ by the flow $(\varphi_t)_{t \in \mathbb{R}}$ on $K_{1/2}$, which is just the rotation around the z-axis with velocity ε: $\varphi_t(z, r, \vartheta) = (z, r, \vartheta + \varepsilon t)$, where we have choosen polar coordinates $x_1 = r \cos(\vartheta)$, $x_2 = r \sin(\vartheta)$, $x_3 = z$, $\vartheta \in [0, 2\pi[$, and $p^{-1}(\alpha_t^*(\omega)) = p^{-1}(\omega) \circ \varphi_t^{-1} \ \forall \omega \in S^p(\mathcal{A})$. In the partially universal representation Π_a^p of the folium F_a^p generated by $S^p(\mathcal{A})$ the limits s–$\lim_{\Lambda \to \infty} \Pi_a^p(J_\Lambda^k)$ exist in the strong operator topology.

Obviously the operators $|\Lambda| J_\Lambda^k$ and $|\Lambda|^2 (\vec{J}_\Lambda)^2$ describe the total angular momentum of the spins in Λ, $|\Lambda| < \infty$. Now for a state ω on $\mathcal{A}$ the local cooperation numbers $\mu_\Lambda^\omega \in [0, 1]$ are defined by $\omega(|\Lambda|^2 (\vec{J}_\Lambda)^2) =: \mu_\Lambda^\omega |\Lambda|/2 \, (\mu_\Lambda^\omega |\Lambda|/2 + 1)$ and the proportion $\gamma_\Lambda^\omega \in [0, 1]$ of excited atoms in Λ by $\omega(|\Lambda| J_\Lambda^3) =: \gamma_\Lambda^\omega |\Lambda| - |\Lambda|/2$. Because of the Cauchy-Schwarz inequality we obtain $|\gamma_\Lambda^\omega - 1/2|^2 \leq (\omega(J_\Lambda^3))^2 \leq \omega((J_\Lambda^3)^2) \leq \omega((\vec{J}_\Lambda)^2) = \mu_\Lambda^\omega/2 \, (\mu_\Lambda^\omega/2 + 1/|\Lambda|)$.

Now consider fixed $\mu, \gamma \in [0, 1]$ with $|\gamma - 1/2| \leq \mu/2$ and the associated circle line $T := \{x \in K_{1/2} \mid x_1^2 + x_2^2 = \beta^2, \ x_3 = \gamma - 1/2\}$ with the radius $\beta := \sqrt{(\mu/2)^2 - (\gamma - 1/2)^2}$. For $\omega \in p(M_+^1(T))$ we have $\lim_{\Lambda \to \infty} \mu_\Lambda^\omega = \mu$ and $\lim_{\Lambda \to \infty} \gamma_\Lambda^\omega = \gamma$. In this sense μ and γ are interpreted as the global cooperation and excitation numbers.

In the subrepresentation of Π_a^p associated with the subfolium of F_a^p generated by $p(M_+^1(T))$ the C^*-dynamics $(\alpha_t)_{t \in \mathbb{R}}$ extends to a W^*-dynamics, and hence would be an appropriate candidate for Π_a. However for simlicity and because there is a simple unitary implementation for $(\alpha_t)_{t \in \mathbb{R}}$ we use here as Π_a the GNS-representation $(\Pi_a, \mathcal{H}_a, \Omega_a)$ associated with the state $\omega_a = p(\lambda)$, where λ is the normalized Lebesgue measure on T. Note, if $\rho \in M_+^1(T)$ is absolutely continuous to λ, then $p(\rho)$ is normal to Π_a. With the above mentioned parametrization of T by the phase angles $\vartheta \in [0, 2\pi[$ $\mathcal{H}_a$ can be realized as $\mathcal{H}_0 \otimes L^2([0, 2\pi[, d\vartheta/(2\pi))$ and $\mathcal{M}_a = \Pi_a(\mathcal{A})'' = \mathcal{M}_0 \bar{\otimes} L^\infty([0, 2\pi[, d\vartheta/(2\pi))$ with a separabel $\mathcal{H}_0$ and a von Neumann algebra $\mathcal{M}_0$ acting on $\mathcal{H}_0$. Because of $\omega_a \circ \alpha_t = \omega_a \; \forall t \in \mathbb{R}$ by standard arguments there is a selfadjoint A on $\mathcal{H}_a$ with $\Pi_a(\alpha_t(X)) = e^{itA}\Pi_a(X) e^{-itA} \; \forall X \in \mathcal{A} \; \forall t \in \mathbb{R}$. Moreover we have

$$\text{s-}\lim_{\Lambda \to \infty} \Pi_a(J_\Lambda^\pm) = 1_{\mathcal{H}_0} \otimes \beta e^{\pm i\vartheta} =: B_\pm = B_\mp^*, \qquad e^{itA} B_\pm \, e^{-itA} = e^{\pm i\varepsilon t} B_\pm. \qquad (2)$$

With the operators J_Λ^k the local hamiltonians ($\Lambda \subset \mathbb{N}$, $|\Lambda| < \infty$) of the interacting system write $H_\Lambda = \Pi_a(A_\Lambda) \otimes 1_{\mathcal{F}} + 1_a \otimes d\Gamma(\sqrt{-\Delta}) + \Pi_a(J_\Lambda^-) \otimes a_{\mathcal{F}}^*(f) + \Pi_a(J_\Lambda^+) \otimes a_{\mathcal{F}}(f)$, which are selfadjoint by a Kato-Rellich argument. From the above formulae it is now clear, that for $\Lambda \to \mathbb{N}$ the local operators H_Λ in some sense should converge to the operator formally given by $H = A \otimes 1_{\mathcal{F}} + 1_a \otimes d\Gamma(\sqrt{-\Delta}) + B_- \otimes a_{\mathcal{F}}^*(f) + B_+ \otimes a_{\mathcal{F}}(f)$. Indeed by perturbation theoretical methods [5] and the Trotter product formula one rigorously deduces:

Theorem 1 H is essentially selfadjoint in $\mathcal{H}_a \otimes \mathcal{F}_+(L^2(\mathbb{R}^3))$ and $\text{s-}\lim_{\Lambda \to \infty} e^{itH_\Lambda} Z \, e^{-itH_\Lambda} = e^{itH} Z \, e^{-itH} \; \forall Z \in \Pi_a \otimes \Pi_{\mathcal{F}}(\mathcal{A} \otimes \mathcal{W}(E)) \; \forall t \in \mathbb{R}$ in the strong operator topology.

3 Integration of the Limiting Hamiltonian

Because of the simple relations (2) by perturbation techniques it is possible to calculate an explicit expression for e^{itH}. Before the formulation of the result we rewrite the Hilbert space $\mathcal{H}_a \otimes \mathcal{F}_+(L^2(\mathbb{R}^3))$ in a direct integral

$$\begin{aligned}
\mathcal{H}_a \otimes \mathcal{F}_+(L^2(\mathbb{R}^3)) &= \mathcal{H}_0 \otimes L^2([0, 2\pi[, d\vartheta/(2\pi)) \otimes \mathcal{F}_+(L^2(\mathbb{R}^3)) \\
&= \mathcal{H}_0 \otimes \int_{[0,2\pi[}^\oplus \mathcal{F}_+(L^2(\mathbb{R}^3)) \, \frac{d\vartheta}{2\pi} \, .
\end{aligned}$$

Theorem 2 We have

$$e^{itH} = \left[1_{\mathcal{H}_0} \otimes \int_{[0,2\pi[}^\oplus e^{i\kappa(t)} \, W_{\mathcal{F}}(f(t, \vartheta)) \, \frac{d\vartheta}{2\pi} \right] \left[e^{itA} \otimes e^{itd\Gamma(\sqrt{-\Delta})} \right], \qquad (3)$$

where $\kappa(t) = -\beta^2 \Im \langle f, \Psi_t(\sqrt{-\Delta} - \varepsilon)f \rangle$ and $f(t, \vartheta) = \sqrt{2}\, \beta e^{-i\vartheta} \, \Theta_t(\sqrt{-\Delta} - \varepsilon)f$ with $\Psi_t(y) = (e^{ity} - 1 - ity)/(iy)^2$ and $\Theta_t(y) = (e^{ity} - 1)/(iy)$ for $y \in \mathbb{R}$.

Another method for proving Theorem 2 is by cocycle theory. There one starts with the right hand side $(=: V_t)$ from (3) and determines cocycles for the functions $t \mapsto \kappa(t)$ and $(t, \vartheta) \mapsto f(t, \vartheta)$ such that V_t, $t \in \mathbb{R}$, becomes a unitary group. For special solutions of the cocycle equations one is able to derive the generator (cf. [1]).

4 The Photon Field at Large Times

For each $t \in \mathbb{R}$ let $h_t := i \int_0^t e^{it(\varepsilon - \sqrt{-\Delta})\tau} f \, d\tau \in L^2(\mathbb{R}^3)$. Then $L_t(g) := \langle h_t, g \rangle$, $g \in E$, defines a bounded linear form on E. Observing $i \lim_{t \to \infty} \int_0^t e^{i(x-\varepsilon)\tau} \, d\tau = i\pi\delta(x-\varepsilon) + \text{pv}\frac{1}{x-\varepsilon} =: \delta^\dagger(x-\varepsilon)$, with the (unitary) Fourier transformation, $g \mapsto \hat{g}$, on $L^2(\mathbb{R}^3)$ we obtain for $\hat{f}, \hat{g} \in L^2(\mathbb{R}^3)$

continuously differentiable in a neighborhood of the ε-sphere $S_\varepsilon := \{\mathbf{k} \in \mathbb{R}^3 \mid \|\mathbf{k}\| = \varepsilon\}$, the limiting linear form $L = \lim_{t \to \infty} L_t$

$$
\begin{aligned}
L(g) &= \lim_{t \to \infty} L_t(g) = -i \lim_{t \to \infty} \int_0^t \langle e^{it(\varepsilon - \sqrt{-\Delta})\tau} f, g \rangle \, d\tau \\
&= -i \lim_{t \to \infty} \int_{\mathbb{R}^3} \overline{\widehat{f}(\mathbf{k})} \, \widehat{g}(\mathbf{k}) \left(\int_0^t e^{it(\|\mathbf{k}\| - \varepsilon)\tau} \, d\tau \right) d^3\mathbf{k} \\
&= -i\pi \int_{S_\varepsilon} \overline{\widehat{f}(\mathbf{k})} \, \widehat{g}(\mathbf{k}) \, dS(\mathbf{k}) \;-\; \mathrm{pv} \int_{\mathbb{R}^3} \frac{\overline{\widehat{f}(\mathbf{k})} \, \widehat{g}(\mathbf{k})}{\|\mathbf{k}\| - \varepsilon} \, d^3\mathbf{k} \, .
\end{aligned}
$$

The prerequisites, for which the above calculation holds, now allow to specify the testfunction space E:

$$
E \subseteq E_{\max} := \left\{ g \in \mathsf{L}^2(\mathbb{R}^3) \mid \widehat{g} \text{ is } C^1 \text{ in a neighborhood of } S_\varepsilon \right\},
$$

and the coupling function f should be an element of $E_{\max}$.

Now for an arbitrary density operator ϱ on $\mathcal{H}_a \otimes \mathcal{F}_+(\mathsf{L}^2(\mathbb{R}^3))$ we are interested how the state ω_t^ϱ on $\mathcal{W}(E)$ from (1) looks like for very large times $t > 0$. Remembering the flow restricted to T, $\varphi_t|_T(\vartheta) = \vartheta - \varepsilon t$, we have the following result.

Theorem 3 *To each density operator ϱ on $\mathcal{H}_a \otimes \mathcal{F}_+(\mathsf{L}^2(\mathbb{R}^3))$ there corresponds a measure $\nu^\varrho \in M_+^1([0, 2\pi[)$ absolutely continuous with respect to $d\vartheta/(2\pi)$ such that*

$$
\text{weak*-} \lim_{t \to \infty} \left(\omega_t^\varrho - \widetilde{\omega}_t^\varrho \right) = 0 \, ,
$$

where $\widetilde{\omega}_t^\varrho(W(g)) := \omega_{\mathcal{F}}(W(g)) \int_{[0,2\pi[} \exp\{i\sqrt{2}\,\Re(e^{-i\vartheta}\beta L(g))\} \, d(\nu^\varrho \circ \varphi_{-t})(\vartheta) \;\; \forall g \in E$.
Moreover, averaging over the period of the flow φ_t one arrives at

$$
\text{weak*-} \lim_{t \to \infty} \frac{\varepsilon}{2\pi} \int_t^{t + 2\pi/\varepsilon} \omega_\tau^\varrho \, d\tau = \overline{\omega}
$$

with $\overline{\omega}(W(g)) := \omega_{\mathcal{F}}(W(g)) \int_{[0,2\pi[} \exp\{i\sqrt{2}\,\Re(e^{-i\vartheta}\beta L(g))\} \, d\vartheta/(2\pi) \;\; \forall g \in E$.

We remark: If $\varrho = \sum_n \gamma_n |\phi_n\rangle\langle\phi_n|$ is the spectral decomposition of ϱ, then with $\phi_n = \int_{[0,2\pi[}^{\oplus} \phi_n(\vartheta) \, d\vartheta/(2\pi) \in \int_{[0,2\pi[}^{\oplus} \mathcal{H}_0 \otimes \mathcal{F}_+(\mathsf{L}^2(\mathbb{R}^3)) \, d\vartheta/(2\pi)$ the measure ν^ϱ calculates as $d\nu^\varrho(\vartheta) = \left(\sum_n \gamma_n \|\phi_n(\vartheta)\|^2 \right) d\vartheta/(2\pi)$.

Obviously, as is seen in Theorem 3, the macroscopic ordering of the atomic system, which is expressed by the phase angles $\vartheta \in [0, 2\pi[$, for large times $t > 0$ forces an ordering upon the generated light, which appears as a phase integral in the states $\widetilde{\omega}_t^\varrho$ and $\overline{\omega}$ and leads to the coherence properties stated in the next section. Note also the appearance of the classical flow, φ_t, in $\widetilde{\omega}_t^\varrho$.

5 Microscopic and Macroscopic Coherent Light

First we examine the linear form $L : E \to \mathbb{C}$. If $\widehat{f}|_{S_\varepsilon} \neq 0$, then L is unbounded. However for $\widehat{f}|_{S_\varepsilon} = 0$ there are two cases. First, for $f \notin \mathcal{D}((\sqrt{-\Delta} - \varepsilon)^{-1})$ the principle value, $\mathrm{pv} \int \ldots$, is proper and hence L is unbounded. Second, for $f \in \mathcal{D}((\sqrt{-\Delta} - \varepsilon)^{-1})$, i.e. $\int_{\mathbb{R}^3} \frac{|\widehat{f}(\mathbf{k})|^2}{\|\mathbf{k}\| - \varepsilon|^2} d^3\mathbf{k} < \infty$, and which e.g. is fulfilled by $\widehat{f}(\mathbf{k}) = 0$ in a neighborhood of S_ε, then L is a bounded linear form on E: $L(g) = \langle (\sqrt{-\Delta} - \varepsilon)^{-1} f, g \rangle \;\; \forall g \in E$. Also there exists an integral kernel for L: $L(g) = \int_{\mathbb{R}^3} \overline{L(x)} g(x) \, d^3x$, where $L(x) = -i \int_{\mathbb{R}^3} f(y) \frac{\exp\{i\sqrt{2\varepsilon}\,\|x-y\|\}}{2\pi\|x-y\|} d^3y$.

With the linear form βL each state $\omega \in \{\widetilde{\omega}_t^\varrho \mid \varrho, t \text{ arbitrary}\} \bigcup \{\overline{\omega}\}$ satisfies the factorization condition for a fully coherent state

$$
\omega\left(a_\omega^*(g_1) \cdots a_\omega^*(g_m) \, a_\omega(g_{m+1}) \cdots a_\omega(g_{2m}) \right) = \beta^{2m} \, L(g_1) \cdots L(g_m) \, \overline{L(g_{m+1})} \cdots \overline{L(g_{2m})} \tag{4}
$$

for all $g_1, \ldots, g_{2m} \in E$ and all $m \in \mathbb{N}$. Hence by [2] or [3] ω is a (fully) coherent state. In addition $\overline{\omega}$ is also gauge-invariant, which arises from the averaging procedure over the period of the flow. However, here it is possible to state the original formal factorization property for coherence from Glauber with the annihilation and creation operators at each point in $\mathbb{R}^3$. In position space one gets with the integral kernel $x \mapsto L(x)$

$$\omega\left(a^*_{x_1} \cdots a^*_{x_m} \, a_{x_{m+1}} \cdots a_{x_{2m}}\right) = \beta^{2m} \, \overline{L(x_1)} \cdots \overline{L(x_m)} \, L(x_{m+1}) \cdots L(x_{2m})$$

and in momentum space using the generalized function $y \mapsto \delta^\dagger(y - \varepsilon)$

$$\omega\left(a^*_{\mathbf{k}_1} \cdots a^*_{\mathbf{k}_m} \, a_{\mathbf{k}_{m+1}} \cdots a_{\mathbf{k}_{2m}}\right) = \beta^{2m} \, \delta^\dagger(\|\mathbf{k}_1\| - \varepsilon) \cdots \delta^\dagger(\|\mathbf{k}_{2m}\| - \varepsilon).$$

Suppose $\beta \neq 0$. For bounded L the states $\widetilde{\omega}^\varrho_t$ and $\overline{\omega}$ are normal to the Fock representation and hence microscopic coherent states (because of the finiteness of the expectation value of the number operator). For unbounded L the mean number of photons is strictly infinite, which ensures $\widetilde{\omega}^\varrho_t$ and $\overline{\omega}$ to be macroscopic coherent states on $\mathcal{W}(E)$ not realizable by density operators on Fock space and disjoint to the Fock state $\omega_{\mathcal{F}}$ [2], [3]. The statistical correlations inherent in the factorization property (4) extend for unbounded L over so many photons that macroscopic classical features are generated, especially a macroscopic phase observable is displayed. The collective phenomenon is expressed by an additional classical field, which arises by just one additional macroscopic mode uniquely determined by the unbounded L (for more details see [3]).

For $\widetilde{\omega}^\varrho_t$ and $\overline{\omega}$ it is possible to deduce the photon density, more exactly, how the photons are distributed througout the space $\mathbb{R}^3$. These states are locally normal to the Fock representation, that is, for each bounded $U \subset \mathbb{R}^3$ there is a density operator $\rho^{t,\varrho}_U$ on the Fock space $\mathcal{F}_+(\mathrm{L}^2(U))$ such that $\widetilde{\omega}^\varrho_t(W(g)) = \mathrm{tr}_{\mathcal{F}_+}\left[\rho^{t,\varrho}_U \, W_{\mathcal{F}}(g)\right]$ for every $g \in E$ with $\mathrm{supp}(g) \subseteq U$. The number expectation value — N_U denotes the number operator in $\mathcal{F}_+(\mathrm{L}^2(U))$ — calculates to be $\mathrm{tr}_{\mathcal{F}_+}\left[\rho^{t,\varrho}_U N_U\right] = 1/2\,\beta^2 \int_U |L(x)|^2 \, d^3x$, which is independent from ϱ and t and the same for $\overline{\omega}$. Thus $x \in \mathbb{R}^3 \mapsto 1/2\,\beta^2 \, |L(x)|^2$ is interpreted as the photon density (cf. also [1]).

In the simple case, when no photon is present at time $t = 0$, one can explicitly calculate the expected number of photons emitted up to time $t > 0$. The initial state now is given by $\varrho = \varrho_a \otimes |\Omega_{\mathcal{F}}\rangle\langle\Omega_{\mathcal{F}}|$ for an arbitrary density operator ϱ_a on $\mathcal{H}_a$, and one obtains with the number operator N in $\mathcal{F}_+(\mathrm{L}^2(\mathbb{R}^3))$ the equation $\omega^\varrho_t(N) = \beta^2 \, \|h_t\|^2 \; \forall t \in \mathbb{R}$. We turn now to an estimate of $t \in \mathbb{R} \mapsto \|h_t\|^2$ in the case of $\widehat{f}|_{S_\varepsilon} \neq 0$: there is a function $\phi : [0, \infty[\mapsto \mathbb{R}$ and $c_1, c_2 > 0$ such that $|\phi(t)| \leq c_1 + c_2\sqrt{t}$ for $t \geq 1$ and $\|h_t\|^2 = 2\pi\left(\int_{S_\varepsilon} |\widehat{f}(\mathbf{k})|^2 dS(\mathbf{k})\right) t + \phi(t)$. Consequently the number of emitted photons increases for large times in a linear rate. As is seen this radiation rate is maximal for $\beta = 1$, or equivalently for $\mu = 1$ and $\gamma = 1/2$. This effect in the literature on quantum optics is known as superradiance.

References

[1] E.B. Davies, *The Infinite Atom Dicke Maser Model I, II*, Commun. math. Phys. **33**, 187 (1973); **34**, 237 (1973)

[2] R. Honegger, and A. Rapp, *General Glauber Coherent States on the Weyl Algebra and their Phase Integrals*, Preprint, Tübingen (1990)

[3] R. Honegger, and A. Rieckers, *The General Form of non-Fock Coherent Boson States*, Publ. RIMS, Kyoto Univ. **26** (1990)

[4] R.J. Glauber, *The Quantum Theory of Optical Coherence*, Phys. Rev. **130**, 2529 (1963)

[5] R. Honegger, *Unbounded Perturbations of Boson Equilibrium States in their GNS-Representations*, Helv. Phys. Acta **63**, 139 (1990)

MACROSCOPICALLY INHOMOGENEOUS
BOSE–EINSTEIN CONDENSATION

Jochen Hertle

Institut für Theoretische Physik
Auf der Morgenstelle 14
D–7400 Tübingen

1 Introduction

Supra–fluid ^{4}He is commonly regarded as an example for Bose Einstein condensation, since the particles are almost free of interaction. The first approach towards the theoretical description in operator–algebraic terms was done by Araki and Woods 1963, [3]. To recover the classical observables of condensate it is necessary to calculate the limiting Gibbs state. This was first done by Lewis and Pulè 1973, [4]. Other investigations including external potentials followed ([5], [6]), the limiting Gibbs state however was only calculated for special potentials. The purpose of this paper is to handle general potentials and to introduce an explicit macroscopically inhomogeneous formalism. This work is an improvement and continuation of [11].

2 The Thermodynamic Limit for Bosons in a Vessel

In general, we take an arbitrary open, bounded subset Λ_0 of the Euclidean space $\mathbb{R}^3$ for the vessel. Later we will add some regularity conditions on the boundary.

It is the purpose to perform the thermodynamic limit in a scaling where the vessel Λ_0 is kept fixed and the Bosons are located at the coordinates $x \in \Lambda_0$. The idea is when looking through a magnifying glass at the point x we will see the ordinary quantum mechanics for Bosons. Therefore, we need a description of the one–particle system, which is independent of the relative size of the system compared to the vessel.

2.1 The One–Particle Hamiltonian

As one–particle pre–Hilbert space we take $\mathsf{L}^2_c(\mathbb{R}^3) := \{f \in \mathsf{L}^2(\mathbb{R}^3) \mid f \text{ has compact support }\}$. This space is having a quasi–local structure in the following sense:
$\mathsf{L}^2_c(\mathbb{R}^3) = \bigcup_\Lambda \{\mathsf{L}^2(\Lambda) \mid \Lambda \text{ bounded and open }\}$. We wish to locate Bosons, described with functions $f \in \mathsf{L}^2_c(\mathbb{R}^3)$, near an arbitrary point $x \in \Lambda_0$. This is done by the following unitaries:

Definition 2.1 *For each $x \in \Lambda_0$ and each $\lambda \geq 1$ let $U_{\lambda x}$ be the linear transformation*

$$U_{\lambda x} : \mathsf{L}^2_c(\mathbb{R}^3) \longrightarrow \mathsf{L}^2_c(\mathbb{R}^3) \quad , \quad (U_{\lambda x}f)(y) = \lambda^{\frac{3}{2}} f(\lambda(y - x)) \quad a.e.$$

These operators have the following properties:

1. All $U_{\lambda x}$ are unitaries and $(U_{\lambda x}^* f)(y) = \lambda^{-\frac{3}{2}} f(y/\lambda + x)$ a.e.

2. $L_c^2(\mathbb{R}^3)$ is invariant under each $U_{\lambda x}$.

3. Given any neighborhood O_x of x, we can find a $\lambda_f \geq 1$, so that the support of $U_{\lambda x} f$ is contained in O_x for all $\lambda \geq \lambda_f$.

These properties follow by simple evaluations.

As we don't want to change the one–particle description, we also must transform the observables so that the mean values are independent of x and λ. If we want to get the value $-\frac{\hbar^2}{2m} \langle f, \Delta f \rangle$ for the kinetic energy of the Boson f, we have to take $-\frac{\hbar^2}{2m} \lambda^{-2} \Delta$ to get the equation

$$-\frac{\hbar^2}{2m} \langle f, \Delta f \rangle = -\frac{\hbar^2}{2m} \left\langle U_{\lambda x} f, \lambda^{-2} \Delta U_{\lambda x} f \right\rangle .$$

Now it is important, that the description of Bosons in a vessel requires boundary conditions for the Laplacian $-\Delta$. Also an external potential V should be allowed. This leads to the following

Definition 2.2 *As one–particle Hamiltonian for the located Bosons $U_{\lambda x} f$ we take $H^\lambda :=$ $-\frac{\hbar^2}{2m} \lambda^{-2} \Delta_{\Lambda_0} + V$, where $-\Delta_{\Lambda_0}$ is the Laplacian with Dirichlet boundary condition in the vessel Λ_0 and V a continuous function on $\overline{\Lambda_0}$ with $\min\{V(x) \mid x \in \overline{\Lambda_0}\} = 0$.*

It can be shown that H^λ has purely discrete spectrum and $e^{-\beta H^\lambda}$ is trace–class on $L^2(\Lambda_0)$. Further the ground–state energy $\varepsilon_0(\lambda)$ satisfies $\varepsilon_0(\lambda) > 0$ $\forall \lambda \geq 1$ and $\lim_{\lambda \to \infty} \varepsilon_n(\lambda) = 0$ for all eigen–values $\varepsilon_n(\lambda)$ of H^λ.

2.2   The Second Quantization

The many–particle system is properly described by a density operator ρ^λ in the symmetric Fock space $\mathcal{F}_+(L^2(\Lambda_0))$. Its state is uniquely determined by the following conditions:

1. $\mathrm{tr}\left[\rho^\lambda N_{\mathcal{F}}\right] = N$ (number of particles)

2. $\mathrm{tr}\left[\rho^\lambda H_{\mathcal{F}}^\lambda\right] = U$ (total inner energy)

3. $-\mathrm{tr}\left[\rho^\lambda \ln(\rho^\lambda)\right]$ be maximal.

Here it is $N_{\mathcal{F}}$ the number operator in Fock space, $H_{\mathcal{F}}^\lambda$ the second quantization of the one–particle Hamiltonian H^λ, which can be expressed by $H_{\mathcal{F}}^\lambda = \sum_{n=0}^{\infty} \varepsilon_n(\lambda) a^*(\Phi_n^\lambda) a(\Phi_n^\lambda)$, where $a^*(\Phi_n^\lambda)$ and $a(\Phi_n^\lambda)$ are the creation and annihilation operators, smeared with the one–particle eigen–functions of H^λ, corresponding to the eigen–values $\varepsilon_n(\lambda)$. $\mathrm{tr}\,[.]$ stands for evaluating the trace in Fock space.

It is well known, that ρ^λ is given by the expression

$$\rho^\lambda = e^{-\beta(H_{\mathcal{F}}^\lambda - \mu N_{\mathcal{F}})} \Big/ \mathrm{tr}\left[e^{-\beta(H_{\mathcal{F}}^\lambda - \mu N_{\mathcal{F}})}\right] ,$$

where $\beta = (K_B T)^{-1}$ is the inverse temperature, K_B the Boltzmann factor and μ the chemical potential. For our proposes it is important to note, that μ is uniquely determined by the condition 1., i.e. $\mu = \mu(N, \lambda)$. The states ρ^λ are leading to the the grand canonical Gibbs states on the Weyl algebra in the Fock representation.

2.3 The Algebraic Formulation

An introduction to this topic is given in [1]. The quasi–local Weyl–algebra $\mathcal{W}(\mathsf{L}_c^2(\mathbb{R}^3))$ is generated by the operators $W(f) = e^{\frac{i}{\sqrt{2}}(a^*(f)+a(f))}$, $f \in \mathsf{L}_c^2(\mathbb{R}^3)$, satisfying the following relations (Weyl form of CCR):

1. $W^*(f) = W(-f)$

2. $W(f) \cdot W(g) = e^{-\frac{i}{2}\Im\langle f,g\rangle} \cdot W(f + g)$.

$\Im\langle .,.\rangle$ denotes the imaginary part of the inner product in $\mathsf{L}_c^2(\mathbb{R}^3)$, which is assumed to be linear in the second factor. Our grand canonical density operator ρ^λ defines a quasi–free gauge–invariant state on $\mathcal{W}(\mathsf{L}_c^2(\mathbb{R}^3))$, given by

$$\omega_\mu^\lambda\left(W(U_{\lambda x}f)\right) :=$$
$$\mathrm{tr}\left[\rho^\lambda W(U_{\lambda x}f)\right] = \exp\left\{-\tfrac{1}{4}\|f\|^2 - \tfrac{1}{2}\left\langle U_{\lambda x}f, e^{-\beta(H^\lambda-\mu)}(\mathbb{1}-e^{-\beta(H^\lambda-\mu)})^{-1}U_{\lambda x}f \right\rangle\right\},$$

see also [1, Proposition 5.2.28]. In the exponent occurs the two–point function

$$\omega_\mu^\lambda\left(a^*(U_{\lambda x}f)a(U_{\lambda x}f)\right) = \left\langle U_{\lambda x}f, e^{-\beta(H^\lambda-\mu)}(\mathbb{1}-e^{-\beta(H^\lambda-\mu)})^{-1}U_{\lambda x}f \right\rangle .$$

It is sufficient to calculate the limit of this expression to get the limiting Gibbs state.

3 The Limiting Gibbs States

First we must restrict the class of vessels Λ_0, demanding the vessel to have the $\mathsf{N}^{0,1}$–property, c.f. [2, Definition 2.4]. This ensures, that we can introduce locally Lipschitz continuous co-ordinate functions for the boundary. $\mathsf{N}^{0,1}$–sets may have points and edges like polygons etc.. The normalized eigen–functions of the local one–particle Hamiltonian can be choosen to be continuous and then they will be unique up to a phase.

The class of allowed potentials must be restricted, too. If $\mathcal{O}$ denotes the interior of the minimum set $\{x \in \overline{\Lambda}_0 \mid V(x) = 0\}$, we also demand $\mathcal{O}$ to have the $\mathsf{N}^{0,1}$–property as well as $\overline{\mathcal{O}} = \{x \in \overline{\Lambda}_0 \mid V(x) = 0\}$. For simplicity we assume $\mathcal{O}$ to be arc–wise connected, but it is not essential.

3.1 The limit at fixed particle density

In performing the thermodynamic limit, we want to keep the (scaled) particle density at a constant level, i.e. we choose a constant $\overline{n} > 0$ with

$$|\Lambda_0|^{-1}\lambda^{-3}\omega_\mu^\lambda(N_\mathcal{F}) = \overline{n} \qquad \forall\lambda \geq 1 . \tag{1}$$

As the left–hand side depends on λ, we have to vary $\mu = \mu(\lambda, \overline{n})$, so that equation (1) holds.

Proposition 3.1 *For each $\lambda \geq 1$ equation (1) determines $\mu(\lambda, \overline{n}) \in]-\infty, \varepsilon_0(\lambda)[$ uniquely. Let*

$$n_c := |\Lambda_0|^{-1}(2\pi m^{-1}\beta\hbar^2)^{-\frac{3}{2}}\int_{\Lambda_0}\sum_{k=1}^\infty k^{-\frac{3}{2}}e^{-k\beta V(x)}\,d^3x \tag{2}$$

be the critical density. Then it is $\lim_{\lambda\to\infty}\mu(\lambda, \overline{n}) = 0$ for $\overline{n} \geq n_c$ and $\lim_{\lambda\to\infty}\mu(\lambda, \overline{n}) = \overline{\mu} < 0$ exists for $\overline{n} < n_c$.

Our proof has been done with similar techniques as in [5]. Now we present the central result:

Theorem 3.2 (The limits of the two–point functions) *Let be $f, g \in L^2_c(\mathbb{R}^3)$, $x, x' \in \Lambda_0$ and $\bar{n} < n_c$. Then*

$$\lim_{\lambda \to \infty} \omega^\lambda_{\mu(\lambda,\bar{n})} \left(a^*(U_{\lambda x}f)a(U_{\lambda x'}g) \right) = \begin{cases} 0 & \text{for } x \neq x' \\ \sum\limits_{k=1}^{\infty} e^{-k\beta(V(x)-\bar{\mu})} \langle g, e^{\frac{k\beta\hbar^2}{2m}\Delta} f \rangle & \text{for } x = x' \end{cases}$$

Above the critical density, $\bar{n} \geq n_c$, the limits are given by

$$\begin{cases} |\Lambda_0| (\bar{n} - n_c)\overline{\Phi_0(x')\hat{g}(0)}\Phi_0(x)\hat{f}(0) & \text{for } x \neq x' \\ |\Lambda_0| (\bar{n} - n_c)\overline{\Phi_0(x)\hat{g}(0)}\Phi_0(x)\hat{f}(0) + \sum\limits_{k=1}^{\infty} e^{-k\beta V(x)} \langle g, e^{\frac{k\beta\hbar^2}{2m}\Delta} f \rangle & \text{for } x = x' \end{cases}$$

$-\Delta$ denotes the Laplacian in whole $\mathbb{R}^3$, $\hat{f}$ the Fourier transform of f and Φ_0 is a continuous normalized ground state eigen–function of $-\Delta_{\mathcal{O}}$, the Laplacian with Dirichlet boundary condition in the minimum set $\mathcal{O}$. As mentioned above this function is unique up to a phase.

SOME ASPECTS CONCERNING THE PROOF: The following statements had to be shown:

1. $\langle U_{\lambda x}f, e^{-\beta H^\lambda} U_{\lambda x}g \rangle \xrightarrow{\lambda \to \infty} \langle f, e^{-\frac{\beta \hbar^2}{2m}(-\Delta + V(x))}g \rangle \quad \forall \beta > 0$.

2. $\mathrm{tr}\left[U^*_{\lambda x} e^{-\beta H^\lambda} U_{\lambda x} \right]_{L^2(\Omega)} \leq |\Omega| (2\pi m^{-1}\beta\hbar^2)^{-\frac{3}{2}}$ for all open, bounded $\Omega \subset \mathbb{R}^3$.

3. $\lim\limits_{\lambda \to \infty} \lambda^2 \varepsilon_n(\lambda) = \varepsilon_n$ for all $n \in \mathbb{N}$, where ε_n are the eigen–values of $-\Delta_{\mathcal{O}}$.

4. For the corresponding continuous eigen–functions: $\lim\limits_{\lambda \to \infty} \Phi^\lambda_n = \Phi_n$ uniformly on Λ_0.

The proof of 2. was done using the following scale of estimations:

$$0 \ll U^*_{\lambda x} e^{-\beta H^\lambda} U_{\lambda x} \ll U^*_{\lambda x} e^{\frac{\beta\hbar^2}{2m}\Delta_{\Lambda_0}} U_{\lambda x} \ll e^{\frac{\beta\hbar^2}{2m}\Delta} ,$$

where "$0 \ll A$" denotes the positivity preserving property of A. The proof of the latter two statements requires a very general Feynman–Kac formula, which can be obtained with the help of [9]. Also techniques from [5] were applied.

3.2   Discussion of the theorem

Below the critical density there is no off–diagonal long range order ($x \neq x'$) and the limiting sesquilinear form yields a closable quadratic form, belonging to the operator $e^{-\beta(-\frac{\hbar^2}{2m}\Delta + V(x) - \bar{\mu})}(\mathbb{1} - e^{-\beta(-\frac{\hbar^2}{2m}\Delta + V(x) - \bar{\mu})})^{-1}$. The limiting Gibbs state is given by

$$\omega^x_{\bar{\mu}}\left(W(f) \right) = \exp\left\{ -\frac{1}{4}\|f\|^2 - \frac{1}{2}\sum_{k=1}^{\infty} e^{-k\beta(V(x)-\bar{\mu})} \langle f, e^{\frac{k\beta\hbar^2}{2m}\Delta} f \rangle \right\} .$$

According to [8] it is a factor state and hence there are no classical observables to describe condensation. The Bosons located at $x \in \Lambda_0$ "see" the constant potential $V(x)$. This constant potential together with $\bar{\mu}$ provides an effective chemical potential $(\bar{\mu} - V(x))$. Hence the particle density is high where the potential is low. Further investigations yield a kind of barometric formula for Bosons, [12]. For the general theory of quasilocal algebras, the following statement is interesting to note:

Proposition 3.3 $\omega^x_{\overline{\mu}}$ *is locally normal to Fock space, i.e. on each subalgebra $\mathcal{W}(L^2(\Lambda))$ of $\mathcal{W}(L^2_c(\mathbb{R}^3))$ (Λ open and bounded) the state is given by a density operator on Fock space. Explicitly we have for $A \in \mathcal{W}(L^2(\Lambda))$*

$$\omega^x_{\overline{\mu}}(A) = \mathrm{tr}\left[A\,\Gamma\left(P_\Lambda e^{-\beta(-\frac{\hbar^2}{2m}\Delta + V(x) - \overline{\mu})} P_\Lambda \right) \right] \Big/ \mathrm{tr}\left[\Gamma\left(P_\Lambda e^{-\beta(-\frac{\hbar^2}{2m}\Delta + V(x) - \overline{\mu})} P_\Lambda \right) \right] .$$

Γ *denotes the second quantization in Segal's notation and P_Λ is the projection onto $L^2(\Lambda)$.*

PROOF: Most important $P_\Lambda e^{-\beta(-\frac{\hbar^2}{2m}\Delta + V(x) - \overline{\mu})} P_\Lambda$ is trace–class on $L^2(\mathbb{R}^3)$, which follows from a theorem of Mercer. Now use [1, Prop. 5.2.27 and 5.2.28]. ∎

Even more interesting is the situation well above critical density. The limiting Gibbs state is given by

$$\omega^x_{\overline{n}}\left(W(f) \right) = \exp\left\{ -\tfrac{1}{4}\|f\|^2 - \tfrac{1}{2}|\Lambda_0|\,(\overline{n} - n_c)\left|\Phi(x)\hat{f}(0)\right|^2 - \tfrac{1}{2}\sum_{k=1}^{\infty} e^{-k\beta V(x)}\left\langle f, e^{\frac{k\beta\hbar^2}{2m}\Delta} f \right\rangle \right\} .$$

There is off–diagonal long range order in connection with the occurence of a singular quadratic form, which vanishes for $x \notin \overline{\mathcal{O}}$ ($\Phi(x) \neq 0$ only for $x \in \mathcal{O}$!). Thus the condensate is only present in $\mathcal{O}$ and different points in $\mathcal{O}$ are statistically correlated. Also for the condensate we have generalized coherence of first order (compare [10]), as for $x \neq x'$ the limit of the two–point function factorizes into a product of linear forms $L_x(f) = (|\Lambda_0|\,(\overline{n} - n_c))^{\frac{1}{2}}\Phi(x)\hat{f}(0)$.

If we want to know, whether there are classical observables for the condensate, we have to construct the GNS–representation.

4   The GNS–Representation

Below critical density we have the factor states $\omega^x_{\overline{\mu}}$ with $\overline{\mu} < 0$ and $x \in \Lambda_0$. The corresponding GNS–representation is well known, compare [7] or [3]. As this representation is contained in the GNS–representation above the critical density, let's have a look at that case. The result is:

Theorem 4.1
Let be $x \in \mathcal{O}$, $T_x := e^{-\beta(-\frac{\hbar^2}{2m}\Delta + V(x))}(\mathbb{1} - e^{-\beta(-\frac{\hbar^2}{2m}\Delta + V(x))})^{-1}$, $c_{\overline{n}} := \sqrt{|\Lambda_0|\,(\overline{n} - n_c)}$ and J any anti–linear involution on $L^2(\mathbb{R}^3)$ with $J = J^*$. Further let be $\Phi_0(x) = |\Phi_0(x)|\,e^{i\theta(x)}$ with an arbitrary continuous function θ. Then the GNS–representation of $\omega^x_{\overline{n}}$ is given by:

- $\mathcal{H}^x_{\overline{n}} = \mathcal{F}_+(L^2(\mathbb{R}^3)) \otimes \mathcal{F}_+(L^2(\mathbb{R}^3)) \otimes L^2(\mathbb{R}^2, (2\pi)^{-1}e^{-R}dRd\varphi)$
  (reconstructed Hilbert space)

- $\Omega^x_{\overline{n}} = \Omega_{\mathcal{F}} \otimes \Omega_{\mathcal{F}} \otimes 1$ *(cyclic vector)*

- $\Pi^x_{\overline{n}}(\,W(f)\,) = W_{\mathcal{F}}((\mathbb{1} + T_x)^{\frac{1}{2}}f) \otimes W_{\mathcal{F}}(J(T_x)^{\frac{1}{2}}f) \otimes e^{ic_{\overline{n}}|\Phi_0(x)|\sqrt{2R}\,\Re\{e^{-i(\varphi+\theta(x))}\hat{f}(0)\}}$
  (represented Weyl operators)

- $a^{x*}_{\overline{n}}(f) = a^*_{\mathcal{F}}((\mathbb{1}+T_x)^{\frac{1}{2}}f)\otimes\mathbb{1}\otimes\mathbb{1} + \mathbb{1}\otimes a_{\mathcal{F}}(J(T_x)^{\frac{1}{2}}f)\otimes\mathbb{1} + \mathbb{1}\otimes\mathbb{1}\otimes c_{\overline{n}}\,|\Phi_0(x)|\,\sqrt{R}\hat{f}(0)e^{-i(\varphi+\theta(x))}$
  (creation operator)

- $a^x_{\overline{n}}(f) = a_{\mathcal{F}}((\mathbb{1}+T_x)^{\frac{1}{2}}f)\otimes\mathbb{1}\otimes\mathbb{1} + \mathbb{1}\otimes a^*_{\mathcal{F}}(J(T_x)^{\frac{1}{2}}f)\otimes\mathbb{1} + \mathbb{1}\otimes\mathbb{1}\otimes c_{\overline{n}}\,|\Phi_0(x)|\,\sqrt{R}\overline{\hat{f}(0)}e^{i(\varphi+\theta(x))}$
  (annihilation operator)

- $\mathcal{Z}^x_{\overline{n}} = \mathbb{1} \otimes \mathbb{1} \otimes L^\infty(\mathbb{R}^2, (2\pi)^{-1}e^{-R}dRd\varphi)$ *(the center)*

The index $\mathcal{F}$ denotes the quantities in the Fock space, $\Re(z)$ means the real part of the complex number z. For $x \notin \mathcal{O}$ the third factor and therefore the center becomes trivial.

The construction was explicitly done in [8], [12]. In the case $x \in \mathcal{O}$ we have an additional factor belonging to a classical statistical description. It has two additional degrees of freedom, belonging to the coordinates R and φ. It can be shown, that the multiplication operator with R is approximated by local particle density operators, [12]. Therefore we interpret R as the classical particle density of the condensate. The variable φ is connected with the gauge transformation and the infinite system exhibits breaking of gauge–invariance, as the central decomposition of $\omega_{\frac{x}{n}}$ runs over φ. It is an additional observable for the condensate which has no classical analogue. Due to the inhomogeneous formalism we see that arbitrary continuous distributions of phase over the condensate are possible $(\theta(x))$. These inhomogeneous distributions generate supra–fluid currents even in thermodynamic equilibrium.

4.1 Discussion

The GNS–representation below critical density just contains factor one and two. The additional third factor above critical density is interpreted as an extra phase so that we are dealing with a two–fluid model. The condensate is located in the absolute minimum $\mathcal{O}$. In consequence a system having condensate in local minima of the external potentials cannot be in equilibrium and a flow towards the absolute minimum takes place. This is a simple explanation of the ^{4}He–film–effect. More detailed explanation requires non–equilibrium models.

References

[1] O. BRATTELI AND D. W. ROBINSON: *Operator Algebras and Quantum Statistical Mechanics II*. Springer, 1981

[2] J. WLOKA: *Partielle Differentialgleichungen*. Teubner, Stuttgart, 1982

[3] H. ARAKI AND E. J. WOODS: *Representations of the CCR, describing a non–relativistic infinite Bose Gas*. J. Math. Phys. **4**, 637 – 662, 1963

[4] J. T. LEWIS AND J. V. PULÈ: *The Equilibrium States of the Free Boson Gas*. Commun. math. Phys. **36**, 1 – 18, 1974

[5] E. B. DAVIES: *The ideal Boson Gas in a Weak External Potential*. Commun. math. Phys. **30**, 229 – 247, 1973

[6] J. V. PULÈ: *The free Boson Gas in a Weak External Potential*. J. Math. Phys. **24(1)**, 138 ff, 1983

[7] J. MANUCEAU AND A. VERBEURE: *Quasi–Free States of the CCR–Algebra and Bogoliubov Transformations*. Commun. math. Phys. **9**, 293 – 302, 1968

[8] R. HONEGGER: *Decomposition of positive Sesquilinear Forms and the Central Decomposition of Gauge–Invariant Quasi–free States on the Weyl–Algebra*. Z. Naturforsch. **45a**, 17 – 28, 1990

[9] R. K. GETOOR: *Markov Operators and their associated Semi–Groups*. Pacific J.Math. **9**, 449 – 472, 1959

[10] R. HONEGGER AND A. RIECKERS: *The General Form of non–Fock Coherent Boson States*. To appear in Publ. RIMS, Kyoto Univ.

[11] R. BRENDLE: *Einstein Condensation in a Macroskopic Field*. Z. Naturforsch. **40a**, 1189 – 1198, 1985

[12] J. HERTLE: *Macroskopisch inhomogene Gleichgewichtszustände von Bosonen in äußeren Feldern*. Diplomarbeit Univ. Tübingen, 1989

Generalized Squeezing of Boson States

Gerhart Schroff

Institut für Theoretische Physik
Auf der Morgenstelle 14
D–7400 Tübingen

1 Introduction

The goals in quantum optics are often very different from those in general quantum field theory. The goal of quantum field theory is to describe the dynamics of interacting particles. In quantum optics, however, correlations play a fundamental role, especially in the concept of optical coherence [1]. That means one is interested in quantum correlations at delayed coincidence detection of photons [2]. These experiments still suggest the need for an adequate theoretical approach.

The present investigation is devoted largely to introducing a new approach for these fundamental problems in optics. We do this by introducing states called α_n–localized states and by giving an interpretation for the detection process of photons, based on these α_n–localized states. We further define special α_n–localized states, the so called coherent states in a space-time-region $\mathcal{X}$ and coherent states in $\mathcal{X}$ which have a further property which will allow a direct identification with detector setups called n-detector-arrangements. The most important difference to earlier approaches is, perhaps, that in general the states describe the whole system and not the fields themselves. For this reason one get a better insight into quantum noise phenomena.

As in this context squeezing is an important effect [5], even because requiring field quantization for explanation, we study these squeezed states. To generate squeezed states we use special *–automorphisms, called squeezing automorphisms and minimal squeezing automorphisms. By means of these squeezed states we outline how to use the formalism and discuss several consequences.

A convenient mathematical structure to describe quantized systems is provided by C^*–algebras. See for example [3]. In our context the convenient C^*–algebra is the so called Weyl-algebra introduced in [4]. In contrast to [3] we are not limited to the free electromagnetic fields. This is a consequence of how we interpret the photon-detection-process, based on the α_n–localized states.

2 Operator Algebraic Approach

First of all we have to show how to describe quantum correlations within a rigorous mathematical approach. To do this it is necessary to make some initial assumptions which will have to be verified in particular models. Before doing this we establish our notation and recall some facts we need.

Let $\mathcal{E}$ be a (complex) pre-Hilbert space with inner product $\langle .,. \rangle$ (linear in the second factor) and $\mathcal{W}(\mathcal{E})$ the Weyl-algebra over $\mathcal{E}$, generated by Weyl elements W(f), f∈$\mathcal{E}$ ([4]). In the GNS-representation $(\mathcal{H}_\omega, \pi_\omega, \Omega_\omega)$ of a regular state ω on $\mathcal{W}(\mathcal{E})$, the existence of selfadjoint operators $\phi_\omega(f)$ such that $\pi_\omega(W(tf)) = exp(it\phi_\omega(f)) \forall t \in \mathbb{R}$ is ensured, via the theorem of Stone. Using the $\phi_\omega(f)$ one constructs the so called annihilation and creation operators associated with ω

$$a_\omega(f) = \tfrac{1}{\sqrt{2}}(\phi_\omega(f) + \phi_\omega(if)) \quad , \quad a_\omega^*(f) = \tfrac{1}{\sqrt{2}}(\phi_\omega(f) - i\phi_\omega(if)) \quad .$$

They are densely defined, closed, it is $a_\omega^*(f) = (a_\omega(f))^*$, $f \to a_\omega(f)$ is antilinear and $f \to a_\omega^*(f)$ is linear and they fulfill the canonical commutation relations CCR; for proof see [4]. For analytic states ω the cyclic vector Ω_ω is in the domain of each polynomial of the operators $\phi_\omega(f)$. This leads to the very natural extension of ω

$$\omega(A \, \phi_\omega(f_1) \cdots \phi_\omega(f_n)) = \langle \, \Omega_\omega, \, \pi_\omega(A) \, \phi_\omega(f_1) \cdots \phi_\omega(f_n) \, \Omega_\omega \, \rangle \quad , \quad A \in \mathcal{W}(\mathcal{E}) \,.$$

Let us start now.

Assumption 2.1 *Let $\mathcal{F}$ be the space of all functions $f : \mathbb{R}^4 \to \mathbb{C}^3$. Let $\mathcal{E} \subseteq \mathcal{F}$ be a pre-Hilbert space with the inner product $\langle .,. \rangle$, generated by the solutions of inhomogeneous wave equations, i.e. inhomogeneous, hyperbolic, partial differential equations. The elements can then be interpreted as electric and magnetic fields or as their potentials. The assumption is now, that all questions connected with quantum correlations at delayed coincidence detection processes of photons can be adequately formulated and analyzed with a so selected pre-Hilbert space $\mathcal{E}$.*

The following definition gives an example :

Definition 2.1 *$\mathcal{F}$ as in Assumption 2.1. Let S be a linear operator in $\mathcal{F}$, $\mathcal{D}(S) = \mathcal{F}, \Omega \subseteq \mathbb{R}^4$ and χ_Ω be the characteristic function of the set Ω. Obviously*

$$\mathcal{E}_S^\Omega = \{ f \in \mathcal{F} ; \; f = h + Sh , \; \Box f = g\chi_\Omega , \; h, g \in \mathcal{F} \}$$

is a pre-Hilbert space. $\Box$ is the D'Alembert Operator.

$\mathcal{E}_S^\Omega$ is a linear subspace of $\mathcal{F}$ for every set $\Omega \subseteq \mathbb{R}^4$ and every linear operator S. These particular pre-Hilbert spaces $\mathcal{E}_S^\Omega$ allow a straightforward translation of earlier discussions of squeezing, (see [5]), into our rigorous mathematical approach (see [6] and paragraph 4 in the present paper). For better understanding of the further discussion it is necessary to talk about detectors ([6]). As every photodetector is ample in space and time we have to choose a point of it to make clear what is meant with: A detector D is at point $x \in \mathbb{R}^4$, where x is a point in space-time. With that knowledge we can formulate the second assumption. Before doing this we define the α_n-localized states.

Definition 2.2 *Let $\mathcal{P}_i$ be n disjoint subsets of $\mathcal{E}$. Further let Ω_i, $i = 1, \cdots, n$, be some regions in space–time $\mathbb{R}^4$. Then for a measured correlation $0 \leq \alpha_n \leq 1$ we define the analytic state ω to be α_n–localized, if the following conditions hold:*

(i) $\forall f_1, ..., f_n, f_i \in \mathcal{P}_i :$ $\omega(a_\omega^*(f_1) \cdots a_\omega^*(f_n) a_\omega(f_n) \cdots a_\omega(f_1)) = \alpha_n$ *and*

(ii) $\forall f_1, ..., f_n \in \mathcal{E}$ *with* $f_i = 0$ *on* $\bigcup_{1 \leq i \leq n} \Omega_i :$ $\omega(a_\omega^*(f_1) \cdots a_\omega^*(f_n) a_\omega(f_n) \cdots a_\omega(f_1)) = 0.$

346

Assumption 2.2 $\omega, \Omega_i, \alpha_n$ *as in Definition 2.2.* Ω_i *are the space-time-regions of the detectors* D_i.

The assumption is, that the magnitude α_n *relative to* ω *can always be interpreted as photon-detection-expectation-values from the n-detector-arrangement* $D_1, \ldots, D_n$.

Every fixed arrangement of n detectors $D_1, \ldots, D_n$ is called a n-detector-arrangement. It should be mentioned that we use the following notation: ω is a α_∞–localized state, if for every $n \in \mathbb{N}$ the state ω is an α_n–localized state in the sense of Definition 2.2.

As coherent states play an essential role in quantum optics we give the following definitions.

Definition 2.3 ω *as in Definition 2.2,* $x_i \in \Omega_i$, $0 \le \alpha_i \le 1$, *for all* $i \in \mathbb{N}$. *Let* $\mathcal{P}_i$ *be disjoint subsets of* $\mathcal{E}$ *for all* $i \in \mathbb{N}$. *A state* ω *is called* α_∞–*coherent in the points* x *of the space-time-region* $\mathcal{X} = \{x_i, i \in \mathbb{N}\}$ *if there is a linear form* $L : \mathcal{E} \to \mathbb{C}$ *such that*

$$\omega\left(a_\omega^*(f_1) \cdots a_\omega^*(f_n) a_\omega(g_k) \cdots a_\omega(g_1)\right) = L(f_1) \cdots L(f_n)\overline{L(g_k)} \cdots \overline{L(g_1)} \tag{1}$$

for all $k, n \in \mathbb{N}$, $k = n$, *for all* $f_1 \in \mathcal{P}_1, ..., f_n \in \mathcal{P}_n, g_1 \in \mathcal{P}_1, ..., g_k \in \mathcal{P}_k$ *and if* ω *is* α_∞–*localized* .

ω *is called* (α_m)–*coherent in* $\mathcal{X}$ *if condition (1) is valid for all* $n = k \le m$ *and if* ω *is* α_n–*localized for all* $n \le m$. x_i *are the well choosen points of the detectors* D_i.

Definition 2.4 *Let* $0 \le \lambda_i \le 1$, $1 \le i \le m \in \mathbb{N}$, *m fixed and* $L : \mathcal{E} \to \mathbb{C}$ *a linear form. A linear form* L *is called* (λ_m)–*localized, if:*

For all $f \in \bigcup_{i \le m} \mathcal{P}_i$ *exists* $\lambda \in \{\lambda_i\}$ *with : there exists* $(f_n)_{n \in \mathbb{N}}$ *with* $f_n(x) = 0, \forall n \in \mathbb{N}$,

for all $x \in \cup_{i \le m} \Omega_i$ *and* $f_n \xrightarrow{\|\cdot\|} f$ *on* $\mathbb{R}^4 \setminus (\cup_{i \le m} \Omega_i)$ *such that* $\lim_{n \to \infty} |L(f - f_n)|^2 = \lambda$

If L *is for every* $m \in \mathbb{N}$ *a* (λ_m)–*localized linear form, we call* L *a* (λ_∞)–*localized linear form.*

Let ω be a state, α_∞–coherent in $\mathcal{X}$, determined by the linear form L. If L is also a (λ_∞)–localized linear form such that

$$\forall m \in \mathbb{N} : \alpha_m = \Pi_{i \le m} \lambda_i \tag{2}$$

then it is always possible to identify the linear form L for every $m \in \mathbb{N}$, m fixed, with the detectors $D_1, ..., D_m$. This is not the case if the linear form L does not possess the property (2). This is an important difference! That means L describes qualities of the whole system in contrast to a linear form possessing property (2) which could always be understood as a description of the detectors. It should be clear, that if L could be understood as a description of the detectors, we have immediately a prescription for measurement (see [6] and paragraph 4 in the present paper). For the further discussion we assume that equation (1) in Definition (2.3) is valid for all $f_1, ..., f_n, g_1, ..., g_k \in \mathcal{E}$!

3 Squeezed States

In the context of quantum noise phenomena squeezing is an important effect (see [5]). These earlier considerations can be readily translated into our rigorous mathematical approach with some important and essential differences. First of all we shall define what we want to understand by squeezing.

Definition 3.1 *Let ω be an analytic state. A *-automorphism τ on $\mathcal{W}(\mathcal{E})$, $\tau(W(f)) = W(Tf)$ $\forall f \in \mathcal{E}$, $T \in \mathcal{S}(\mathcal{E})$, the set of all symplectic operators on $\mathcal{E}$, is called a squeezing-automorphism relative to ω if the following condition is fulfilled:*

There exist elements $f, g \in \mathcal{E}$, $f, g \neq 0$, and a state ω, such that

$$\exists h \in \{f, g\} : \operatorname{Var}_{\omega_\tau} (\phi_{\omega_\tau}(h)) < \tfrac{1}{2} \mid \omega_\tau ([\phi_{\omega_\tau}(f), \phi_{\omega_\tau}(g)]) \mid \ .$$

$\omega_\tau = \omega \circ \tau$ and $\operatorname{Var}_\omega (\pi_\omega(A)) = \omega\left((\pi_\omega(A))^2\right) - \omega(\pi_\omega(A))^2$, $\forall A \in \mathcal{W}(\mathcal{E})$.

Thus a squeezing-automorphism τ is a special Bogoliubov-Transfomation.

Definition 3.2 *Let f, g and ω as in Definition 3.1. A squeezing-automorphism τ is called minimal if*

$$\operatorname{Var}_{\omega_\tau} (\phi_{\omega_\tau}(f)) \operatorname{Var}_{\omega_\tau} (\phi_{\omega_\tau}(g)) = \tfrac{1}{4} \mid \omega_\tau ([\phi_{\omega_\tau}(f), \phi_{\omega_\tau}(g)]) \mid^2 \ . \tag{3}$$

The following operator T gives an example:

Let $\mathcal{E}^n$ be a n-dimensional subspace of $\mathcal{E}$ with ONB $\{e_i, 1 \leq i \leq n\}$. Obviously

$$T : \mathcal{E}^n \longrightarrow \mathcal{E}^n, \ f = \Sigma_{i \leq n} c_i e_i \longrightarrow \Sigma_{i \leq n} \lambda_{\mu_i, \nu_i}(c_i) e_i \ ;$$

$$\lambda_{\mu_i, \nu_i} : \mathbb{C} \longrightarrow \mathbb{C}, \ z \longrightarrow \mu_i z - \nu_i \overline{z} \ , \ \mid \mu_i \mid^2 - \mid \nu_i \mid^2 = 1 \ , \ \mu_i, \nu_i \in \mathbb{C} \ , \ \mu_i, \nu_i \neq 0$$

is a symplectic operator on $\mathcal{E}^n$. T depends on the ONB! It can be easily demonstrated that T determines a squeezing-automorphism τ relative to (α_∞)-coherent and quasi-free states in a space-time-region $\mathcal{X}$. By using this operator it is simple to transfer earlier discussions of squeezing (see [7]) to our rigorous mathematical approach. In Definition 2.3 one requirement for a state to be a α_∞-coherent state in $\mathcal{X}$ is, that ω satisfies the factorization property (1). There is nothing said about the case $k \neq n$. In this case that is if the factorization property (1) is true for all $k, n \in \mathbb{N}$ it can be shown [8] that this is equivalent to the fact that ω is an α_∞-coherent and quasi-free state in $\mathcal{X}$. By $\mathcal{S}_L(\mathcal{W}(\mathcal{E}))$ we denote the set of all C^∞ states which fullfill property (1) for a given linear form L.

Theorem 3.1 *τ as in Definition 3.1 and let $\omega \in \mathcal{S}_L(\mathcal{W}(\mathcal{E}))$ be a quasi-free state, $T \in \mathcal{S}(\mathcal{E})$. The following conditions are equivalent:*

(i) τ is a minimal squeezing $-$ automorphism relative to ω ,

(ii) $\exists f, g \in \mathcal{E}$, $f, g \neq 0$: $Re(\langle Tf, Tg \rangle) = 0$ and $\exists c \in \mathbb{C}$: $Tf = cTg$.

PROOF: Since ω is quasi-free, it follows that ω_τ is quasi-free, because τ is a Bogoliubov Transformation. With that we get 1.): $\operatorname{Var}_{\omega_\tau}(\phi_{\omega_\tau}(f)) = \tfrac{1}{2} \parallel Tf \parallel^2$ $\forall f \in \mathcal{E}$. It is easy to show that $(\mathcal{H}_\omega, \pi_\omega \circ \tau, \Omega_\omega)$ is the GNS-representation of ω_τ. It follows immediately 2.): $\mid \omega_\tau ([\phi_{\omega_\tau}(f), \phi_{\omega_\tau}(g)]) \mid^2 = \mid i \operatorname{Im}(\langle f, g \rangle) \mid^2$ $\forall f, g \in \mathcal{E}$. For $f, g \in \mathcal{E}$, $f, g \neq 0$ it follows : $\parallel f \parallel^2 \parallel g \parallel^2 = \mid \operatorname{Im}(\langle f, g \rangle) \mid^2 \Longleftrightarrow \{Re(\langle f, g \rangle) = 0 \text{ and } f, g \text{ are l.d.} \}$. By this and 1.) and 2.) the assertion follows. ∎

Let L be an unbounded linear form on $\mathcal{E}$. The GNS-representation of a state $\omega \in \mathcal{S}_L(\mathcal{W}(\mathcal{E}))$ is

$$
\begin{aligned}
\mathcal{H}_\omega &= \mathcal{F}_+(\mathcal{E}) \otimes \mathcal{H}_\nu \\
\Omega_\omega &= \Omega_F \otimes w \\
\Pi_\omega(W(f)) &= \Pi_F \otimes W_\nu(L(f))
\end{aligned}
$$

$\forall f \in \mathcal{E}$ (see for proof [9]). $W_\nu(\alpha) = exp\left\{\frac{i}{\sqrt{2}}(\alpha \nu + \overline{\alpha} \nu^*)\right\}$, $\alpha \in \mathbb{C}$, ν is a unitary operator on the Hilbert space $\mathcal{H}_\nu$. The unitary operator ν is clearly defined by ω. w is the cyclic vector of the C^*–algebra $\mathcal{A}_\nu$ generated by the $W_\nu(\alpha)$, $\alpha \in \mathbb{C}$, and $\phi_\omega(f) = \phi_F(f) \otimes 1\!\!1_\nu + 1\!\!1_F \otimes \phi_\nu(L(f))$, with $\phi_\nu(\alpha) = \frac{1}{\sqrt{2}}(\alpha \nu + \overline{\alpha} \nu^*)$, $\alpha \in \mathbb{C}$. $\mathcal{F}_+(\mathcal{E})$ is the Bose-Fock space and $\pi_F(W(f))$ are the Fock-Weyl operators.

Theorem 3.2 *Let L be a unbounded linear form, $S \in \mathcal{S}(\mathcal{E})$ and τ as in Definition 3.1, $\omega \in \mathcal{S}_L(\mathcal{W}(\mathcal{E}))$. The following conditions are equivalent:*

(i) τ *is a minimal squeezing – automorphism relative to* ω,

(ii) $\exists f, g \in \mathcal{E}$, $f, g \neq 0$:

$$
\begin{aligned}
&(1) \quad Re\left(\langle Tf, Tg \rangle\right) = 0 \\
&(2) \quad \exists c \in \mathbb{C} : Tf = c\,Tg \\
&(3) \quad \mathrm{Var}_\nu\left(\phi_\nu(L(Tf))\right) = \mathrm{Var}_\nu\left(\phi_\nu(L(Tg))\right) = 0.
\end{aligned}
$$

$\mathrm{Var}_\nu\left(\phi_\nu(\alpha)\right) = \langle w, \phi_\nu(\alpha)^2 w \rangle - \langle w, \phi_\nu(\alpha) w \rangle^2$, $\alpha \in \mathbb{C}$.

PROOF : One easily checks that for $f, g \in \mathcal{E}$, $\omega \in \mathcal{S}_L(\mathcal{W}(\mathcal{E}))$:

$$
\begin{aligned}
\mathrm{Var}_\omega\left(\phi_\omega(f)\right) \mathrm{Var}_\omega\left(\phi_\omega(g)\right) =\ & \mathrm{Var}_F\left(\phi_F(f)\right) \mathrm{Var}_F\left(\phi_F(g)\right) \\
&+ \mathrm{Var}_F\left(\phi_F(f)\right) \mathrm{Var}_\nu\left(\phi_\nu(L(g))\right) \\
&+ \mathrm{Var}_\nu\left(\phi_\nu(L(f))\right) \mathrm{Var}_F\left(\phi_F(g)\right) \\
&+ \mathrm{Var}_\nu\left(\phi_\nu(L(f))\right) \mathrm{Var}_\nu\left(\phi_\nu(L(g))\right) \quad (4)
\end{aligned}
$$

$$
\begin{aligned}
\mathrm{Var}_F\left(\phi_F(f)\right) \mathrm{Var}_F\left(\phi_F(g)\right) &\geq \tfrac{1}{4} |\, \omega\left([\phi_F(f), \phi_F(g)]\right)\,|^2 = \tfrac{1}{4} |\, Im\langle f, g \rangle\,|^2 \\
&= \tfrac{1}{4} |\, \omega\left([\phi_\omega(f), \phi_\omega(g)]\right)\,|^2 . \quad (5)
\end{aligned}
$$

It is shown in Theorem 3.1 that $(\mathcal{H}_\omega, \pi_\omega \circ \tau, \Omega_\omega)$ is the GNS-representation of $\omega_\tau = \omega \circ \tau$, $\omega \in \mathcal{S}(\mathcal{W}(\mathcal{E}))$. " $\Longrightarrow$ " : As τ is a squeezing-automorphism, there exists elements $f, g \in \mathcal{E}$, $f, g \neq 0$, such that equation (3) is valid. With $\mathrm{Var}_{F_{\circ \tau}}\left(\phi_{F_{\circ \tau}}(f)\right) = \tfrac{1}{2} \| Tf \|^2$ $\forall f \in \mathcal{E}$ and (4), (5) by using ω_τ instead of ω (ii) follows. " $\Longleftarrow$ " : As $(ii)(1) - (ii)(3)$ are valid, it is easy to check the statement. ∎

4 Squeezed States and Photon-Detection-Prozesses

During the last few years several laboratories have performed experiments in which the recorded noise level in homodyne detection has clearly been reduced (see for example [10] and [11]). But it is still a generally unsolved problem how these noise reductions can be identified with the squeezing of the electromagnetic field. In order to do this it is necessary to determine the quantum limit of the field fluctuations, often called the vacuum fluctuations.

The most natural way in describing experimental setups like this is to describe the accessible magnitudes. In this context that means to describe expectation values of photon-detection-processes. This can be easily done within the approach developed in this paper, because in contrast to earlier approaches to quantum optics these photon-detection-expectation-values are accessible in a very explicit way. For this reason one gets a better insight into quantum noise phenomena and quantum noise reduction phenomena.

To make the connection between earlier discussions of squeezing and our approach, let us consider an ordinary homodyne detector arrangement. In ordinary homodyne detection the signal is combined,via a beam splitter, with intense local oscillator light oscillating at the carrier frequency of the signal light. This light is then directed to a photon-detector. In earlier approaches this experimental setup has been understood like this: The signal delivered by the photon-detector results from the constructive or destructive interference of the signal light with a local oscillator light at the photon-detector. See for example [12]. The natural description of an ordinary homodyne detector in our approach reads to: The photon-detector illuminated by a intensive local oscillator has to be understood as a new detector. With this arrangement, photon-detector plus intensive local oscillator, one can now analyse the signal light. This description of an ordinary homodyne detector is carried out in [6]. Doing this by using the pre-Hilbert space $\mathcal{E}_{S}^{\Omega}$ allows a straightforward translation of earlier discussions of squeezing. It should be noted that the approach introduced in this paper is not limited to ordinary homodyne detectors.

References

[1] Roy J. Glauber , Phy. Rev., Vol. **130**, No.6, 2529 (1963)

[2] R.Hanbury Brown and R.Q.Twiss , Nature **177**,27 (1956)

[3] C.A.Hurst , *Quantum theory of the free electromagnetic field.* Symmetries in Science II, Plenum Publishing Corporation, 1986.

[4] O.Bratteli and D.W.Robinson : *Operator Algebras and Quantum Statistical Mechanics II,* Ch.5.2 . Springer, 1981.

[5] R.L.Loudon and P.L.Knight , Journal of Modern Optics, 1987 , Vol. **34** , Nos.6/7, 707-759.

[6] G.Schroff : Diplomarbeit, Tübingen, 1990.

[7] Horace P. Yuen : *Two-photon coherent states of the radiation field* , Phy. Rev. A , Vol. **13** , No. 6, 1976.

[8] R.Honegger, A.Rapp, *General Glauber coherent states on the Weyl algebra and their phase integrals.* Preprint, Tübingen, 1989.

[9] R.Honegger and A.Rieckers, *The General Form of non-Fock Coherent Boson States.* Publ. RIMS, Kyoto Univ., 26 (1990).

[10] R.E.Slusher, L.W.Hollberg and B.Yurke, J.F.Valley , Phy. Rev. Let., Vol. **55**, No.22 (1985).

[11] Ling-An Wu, H.J.Kimble, J.L.Hall, Huifa Wu, Phy. Rev. Let., Vol. **57**, No.20 (1986).

[12] Bernard Yurke , Phy. Rev. A , Vol. **32** , No.1 (1985).

EQUILIBRIUM STATES OF LONG RANGE INTERACTING QUANTUM LATTICE SYSTEMS

Thomas Gerisch

Institut für Theoretische Physik
Auf der Morgenstelle 14
D–7400 Tübingen

1 Introduction

At this place some thermodynamic aspects of a quantum lattice in thermic equilibrium will be handled. At the center of the discussion stand different variational principles and their connection by generalized Legendre transformation. To convey this program, one restricts to a completely homogeneous lattice that will be realized by chosing the set of permutation invariant states $S^P(\mathcal{A})$ as an adequate thermodynamic state space [1]. The generalization of polynomial interactions [2, 3, 1] to equivalence classes of approximately symmetric sequences [4] becomes apparent as a fruitful concept. This large class of possible thermodynamic systems is to be seen as a natural class of mean field interactions. To clarify the behavior of the thermodynamic functionals internal energy, entropy and free energy in the thermodynamic limit, the full generality of [4] is not exhausted, especially only finite–dimensional algebras on a lattice point are treated and the absolute entropy is used in contrast to the relative one. The absolute properties allow a direct relation to the measurable expectation values of a macroscopic thermodynamic system. Starting with the principle of minimal free energy density for equilibrium states, the convex duality between free energy and entropy density is proved (Theorem 3.1). With a suited restriction of the variation to extremal permutation invariant states, the duality properties are modified in a way which distinguishes between an interaction in the lattice and an external field (Theorem 3.3). The rigorous treatment uses mainly results of convex analysis [5, 6]. Some ideas can be found in the investigations, presented in [7, 8] for short range interactions on translation invariant systems and the case of quadratic mean field interactions on a permutation invariant lattice in [2]. In contrast to [7], the homogeneity simplifies the calculations. The introduction of the so–called subgradient allows a geometrical interpretation of states, minimizing the free energy functional. The formalism is suited to treat non–trivial phase transitions of first and second order, [2, 3]. A discussion of models presented, e.g. in [9] seems to be possible and will be worked out in future.

2 Quantum Statistical and Thermodynamic Foundations

The quantum lattice is described by a infinite countable index set $\mathcal{I}$. At each point $i \in \mathcal{I}$ of the lattice, there is a finite dimensional algebra $\mathcal{B}_i \cong \mathcal{B} \cong M_n(\mathbb{C})$ containing the physical properties of the lattice point, e.g. a particle with spin $\frac{1}{2}$. Describing a finite region $\Lambda \in \mathcal{L} := \{\Lambda' \subset \mathcal{I} | \Lambda'$ contains only a finite number $|\Lambda'|$ of lattice points$\}$, the local algebra $\mathcal{A}_\Lambda := \otimes_{i \in \Lambda} \mathcal{B}_i$ is used. With the C^*-inductive limit [10] one obtains the quasi local algebra $\mathcal{A} := \otimes_{i \in \mathcal{I}} \mathcal{B}_i$. The local algebras $\mathcal{A}_\Lambda$ are canonically embedded in $\mathcal{A}$.

The physical states $S(A)$ are the positive linear functionals on A with $\langle \omega; \mathbb{1} \rangle = 1$ for all $\omega \in S(A)$. With intent to treat a homogeneous lattice, it is sufficient to consider only the set of permutation invariant states $S^P(A)$. This is the set of all $\omega \in S(A)$ with $\langle \omega; P_\sigma x \rangle = \langle \omega; x \rangle$ for all $x \in A$ and P_σ is the automorphismn, that permutes the lattice points of a finite region; this is the linear continuous extension of $P_\sigma \otimes_{i \in \mathcal{I}} x_i := \otimes_{i \in \mathcal{I}} x_{\sigma(i)}$, with σ as a permutation of a finite region in $\mathcal{L}$ and $x_i \neq \mathbb{1}$ only for a finite number of i. $S^P(A)$ is a Bauer simplex in $S(A)$ with extremal boundary $\partial_e S^P(A) \cong S(B)$; that means: for each state $\omega \in S^P(A)$ there exists a unique positive normed regular Borel measure μ_ω' with support in $\partial_e S^P(A)$ and

$$\omega = \int_{\partial_e S^P(A)} \omega' d\mu_\omega'(\omega') = \int_{S(B)} \Pi_\varphi d\mu_\omega(\varphi), \tag{2.1}$$

with the product state Π_φ as the associated element in $\partial_e S^P(A)$ for all $\varphi \in S(B)$ and $\mu_\omega := \mu' \circ \Pi$, [11]. This decomposition is the *central decomposition* of the permutation invariant state ω, [2]. The support of this measure gives the *pure phase states*.

Let us begin with the definition of the considered class of thermodynamic models. These are families of local Hamiltons $H_\Lambda \in A_\Lambda$ for all finite regions Λ. These families must be specified in a more concrete way, e.g. by the set of the *(approximately) symmetric sequences*. The definition of these sequences is given in [4] and says, that an (approximately) symmetric sequence in A is a family of local Hamiltonian densities $h = (\frac{H_\Lambda}{|\Lambda|})_{\Lambda \in \mathcal{L}}$ with: H_Λ is invariant under permutations of the region Λ and if $|\Lambda| = |\Omega|$ for $\Lambda, \Omega \in \mathcal{L}$, for all bijections $P : \Lambda \to \Omega$, there is $H_{P\Lambda} = H_\Omega$. The crucial condition for a symmetric sequence is, that $\frac{H_{\Lambda \cup \Omega}}{|\Lambda| + |\Omega|} = sym \frac{H_\Lambda}{|\Lambda|}$, for all $\Lambda, \Omega \in \mathcal{L}$ with $\Lambda \cap \Omega = \emptyset$ and $|\Lambda|$ is greater than a fixed $n_0 \in \mathbb{N}$. $sym_{\Lambda \cup \Omega}$ means the usual symmetrization of a tensor in $A_{\Lambda \cup \Omega}$. The set of all symmetric sequences in A is denoted with $\mathcal{Y}$. A straight forward extension of this condition is the approximate symmetry. This means, that $h = (\frac{H_\Lambda}{|\Lambda|})_{\Lambda \in \mathcal{L}}$ is approximately symmetric, if $\forall \varepsilon > 0 \ \exists h' = (\frac{H'_\Lambda}{|\Lambda|})_{\Lambda \in \mathcal{L}} \in \mathcal{Y}$ $\exists n_0 \in \mathbb{N} \ \forall \Lambda \in \mathcal{L}, |\Lambda| > n_0 \ \frac{\|H_\Lambda - H'_\Lambda\|}{|\Lambda|} < \varepsilon$. The set of all approximately symmetric sequences will be denoted as $\tilde{\mathcal{Y}}$.

$\tilde{\mathcal{Y}}$ is strictly related with the continuous functions on $S(B)$ (in the usual norm, since B is a finite–dimensional algebra). That is: $\forall \varphi \in S(B)$ there exists the limit $\lim_{\Lambda \in \mathcal{L}} \frac{\langle \Pi_\varphi; H_\Lambda \rangle}{|\Lambda|}$. Using this convergence it is possible to define a surjective map j by

$$j : \tilde{\mathcal{Y}} \longrightarrow C(S(B), \mathbb{C}), h = \left(\frac{H_\Lambda}{|\Lambda|}\right)_{\Lambda \in \mathcal{L}} \longrightarrow \left[j(h) \right](\varphi) := \lim_{\Lambda \in \mathcal{L}} \frac{\langle \Pi_\varphi; H_\Lambda \rangle}{|\Lambda|}, \quad \forall \varphi \in S(B). \tag{2.2}$$

In [4], it is proved, that $j(h) \in C(S(B), \mathbb{C}) \ \forall h \in \tilde{\mathcal{Y}}$ and j is surjective. If one only wants to treat possible equilibrium states, it is sufficient to use the equivalence classes of all the approximately symmetric sequences h with the same continuous function $j(h)$. These equivalence classes can be indexed by an element $h \in C(S(B), \mathbb{C})$. If the convex analysis is developed, it is enough, to use the equivalence classes of approximately symmetric sequences with H_Λ selfadjoint for all $\Lambda \in \mathcal{L}$. This set is given by $C(S(B), \mathbb{R})$.

In this context we use the notation of approximately symmetric sequences; this is not absolutely necessary, since it is enough to use all sequences of local Hamiltonians, where the limit $\lim_{\Lambda \in \mathcal{L}} \frac{\langle \Pi_\varphi; H_\Lambda \rangle}{|\Lambda|} =: \left[j(h) \right](\varphi)$ exists for all $\varphi \in S(B)$ and $j(h) \in C(S(B), \mathbb{C})$. The use of the approximately symmetric sequences is necessary in [4], to handle the limiting Gibbs states; in our case it can be considered as the reason, why it is allowed to work with the continuous functions on $C(S(B), \mathbb{C})$. After these fundamental considerations, it is possible to deal with internal energy, entropy and free energy as functionals on $S^P(A)$:

Proposition 2.1 *The following three limits exist* $\forall \omega \in S^P(A)$, *fixed* $h \in C(S(B), \mathbb{R})$ *and* $(\frac{H_\Lambda}{|\Lambda|})_{\Lambda \in \mathcal{L}} \in j^{-1}(\{h\})$:

(i)
$$\lim_{\Lambda \in \mathcal{L}} \frac{\langle \omega; H_\Lambda \rangle}{|\Lambda|} =: u(h, \omega), \tag{2.3}$$

(ii)
$$- \lim_{\Lambda \in \mathcal{L}} \frac{\mathrm{tr}_\Lambda(\varrho_\Lambda^\omega \ln \varrho_\Lambda^\omega)}{|\Lambda|} =: s(\omega), \tag{2.4}$$

with ϱ_Λ^ω as the density matrices, resulting from the restriction of $\omega \in \mathcal{S}^P(\mathcal{A})$ to a state on $\mathcal{A}_\Lambda$.

(iii)
$$\lim_{\Lambda \in \mathcal{L}} \frac{1}{|\Lambda|} \left(\langle \omega; H_\Lambda \rangle + \frac{1}{\beta} \mathrm{tr}_\Lambda(\varrho_\Lambda^\omega \ln \varrho_\Lambda^\omega) \right) = u(h, \omega) - \frac{1}{\beta} s(\omega) =: f(\beta, h, \omega) \tag{2.5}$$

with $\beta \in (0, +\infty)$.

The three thermodynamic functionals are affine, w^*-continuous functionals on $\mathcal{S}^P(\mathcal{A})$ with values in $\mathbb{R}$ and describe the energy, entropy and free energy density of the quantum lattice system.

Proof:
Statement (i) is a consequence of (2.2) and the Lebesgue convergence theorem. (ii) is proved in [1]. Affinity in (i) is clear by construction and in (ii), use [12, Theorem 6.2.25]. The w^*-continuity follows with the central decomposition. $\square$

Note that these limits are treated in [4] in a more general manner, by use of the relative entropy. In this context, the absolute functionals are chosen to stress the thermodynamic character.

In the (2.3), the energy density was calculated for a special Hamiltonian $(H_\Lambda)_{\Lambda \in \mathcal{L}}$. Now, $u(h, .) : \mathcal{S}^P(\mathcal{A}) \longrightarrow \mathbb{R}$ is extended to the functional

$$\langle .; . \rangle : \mathcal{S}^P(\mathcal{A}) \times \mathcal{C}(\mathcal{S}(\mathcal{B}), \mathbb{R}) \longrightarrow \mathbb{R}$$
$$(\omega, h) \longrightarrow \langle \omega; h \rangle := u(h, \omega) = \int_{\mathcal{S}(\mathcal{B})} u(h, \Pi_\varphi) d\mu_\omega(\varphi), \tag{2.6}$$

on $\mathcal{S}^P(\mathcal{A}) \times \mathcal{C}(\mathcal{S}(\mathcal{B}), \mathbb{R})$. μ_ω is the unique decomposition of ω in states of $\partial_e \mathcal{S}^P(\mathcal{A})$. With this extension, $\langle .; . \rangle$ is affine in the first argument and linear in the second. Since $u(h, .)$ is continuous, the definition of $\langle .; . \rangle$ can be extended to the integration over all regular Borel measures μ on $\mathcal{S}(\mathcal{B})$ and this is just the dual $\mathcal{C}(\mathcal{S}(\mathcal{B}), \mathbb{C})^*$ of $\mathcal{C}(\mathcal{S}(\mathcal{B}), \mathbb{C})$ by the Riesz representation theorem.

Further, the entropy density functional will be extended as a functional on $\mathcal{C}(\mathcal{S}(\mathcal{B}), \mathbb{C})^*$. Note, that the entropy is a continuous functional on $\partial_e \mathcal{S}^P(\mathcal{A})$, since $\mathcal{B}$ is finite dimensional (see [2]) and just as in (2.6) we have by affinity for $\omega \in \mathcal{S}^P(\mathcal{A})$ with central measure μ_ω: $s(\omega) = \int_{\mathcal{S}(\mathcal{B})} s(\Pi_\varphi) d\mu_\omega(\varphi)$. This definition of the entropy will be extended to $s : \mathcal{C}(\mathcal{S}(\mathcal{B}), \mathbb{C})^* \longrightarrow \overline{\mathbb{R}}$ by continuation of (2.4) with $s(\alpha) := -\infty$ if $\alpha \in \mathcal{C}(\mathcal{S}(\mathcal{B}), \mathbb{C})^*$ but not in $\mathcal{S}^P(\mathcal{A})$. This extension has only technical character and allows to use the notation of convex analysis in a straight forward way. The value $-\infty$ states in this context, that there is no information available for the entropy, if α is not a state.

3 Convex Analysis over Equilibrium States

The set of all equilibrium states $S(\beta, h)$ at inverse temperature β for the thermodynamic system $h \in \mathcal{C}(\mathcal{S}(\mathcal{B}), \mathbb{R})$ is given by the set of all states $\omega \in \mathcal{S}^P(\mathcal{A})$ minimizing the free energy density $f(\beta, h, .)$, that is, $\omega \in \mathcal{S}^P(\mathcal{A})$ fulfills

$$f(\beta, h, \omega) = \inf\{f(\beta, h, \omega') | \omega' \in \mathcal{S}^P(\mathcal{A})\} =: f(\beta, h). \tag{3.1}$$

$f(\beta, h)$ is finite, since

$$\begin{aligned} f(\beta, h, \omega) &= u(h, \omega) - \frac{1}{\beta} s(\omega) = \int_{\mathcal{S}(\mathcal{B})} \left[u(h, \Pi_\varphi) - \frac{1}{\beta} s(\Pi_\varphi) \right] d\mu_\omega(\varphi) \\ &\geq \inf\{f(\beta, h, \Pi_\varphi) | \varphi \in \mathcal{S}(\mathcal{B})\}, \end{aligned} \tag{3.2}$$

and $f(\beta, h, .) : \mathcal{S}(\mathcal{B}) \longrightarrow \mathbb{R}$ is a continuous function on a compact set, which takes its infimum in $\mathcal{S}(\mathcal{B})$.

Now, the minimal free energy of a system h is interpreted as a functional on $C(\mathcal{S}(\mathcal{B}), \mathbb{R})$, that is $f(\beta, .) : C(\mathcal{S}(\mathcal{B}), \mathbb{R}) \longrightarrow \mathbb{R}, h \longrightarrow f(\beta, h)$. Then, the following relation is valid:

$$
\begin{aligned}
f(\beta, h) &= \inf\{f(\beta, h, \omega) | \omega \in \mathcal{S}^{\mathcal{P}}(\mathcal{A})\} = \inf\{\langle \omega; h \rangle - \frac{1}{\beta} s(\omega) | \omega \in \mathcal{S}^{\mathcal{P}}(\mathcal{A})\} \\
&= \inf\{\langle \alpha; h \rangle - \frac{1}{\beta} s(\alpha) | \alpha \in C(\mathcal{S}(\mathcal{B}), \mathbb{R})^*\} = \left[\frac{1}{\beta} s(.) \right]^* (h),
\end{aligned}
\tag{3.3}
$$

that is the *concave conjugate function* to the entropy density s. The spaces $C(\mathcal{S}(\mathcal{B}), \mathbb{R})$ with norm topology and $C(\mathcal{S}(\mathcal{B}), \mathbb{R})^*$ with the w^* - topology are a pairing in the sense of [5]. Therefore a second conjugation leads back to the entropy density, since s and $f(\beta, .)$ are both upper semicontinuous concave functions, [5, Theorem 5].

Theorem 3.1 : *Consider the entropy density functional* $s : C(\mathcal{S}(\mathcal{B}), \mathbb{R})^* \to \mathbb{R}$ *in the extension given in Section 2 and the minimal free energy functional* $f(\beta, .) : C(\mathcal{S}(\mathcal{B}), \mathbb{R}) \to \mathbb{R}$ *introduced in (3.1). Then the two functionals are related by concave conjugation (cf. (3.3)), that is:*

(i)
$$
f(\beta, h) = \left[\frac{1}{\beta} s(.) \right]^* (h),
\tag{3.4}
$$

(ii)
$$
s(\omega) = \beta \left[f(\beta, .) \right]^* (\omega).
\tag{3.5}
$$

This means, that there exists a unique one–to–one relation between free energy and entropy density. The transformation is given by concave duality (or the so called *Fenchel–transformation*, see [5]), which is a generalization of the *Legendre–transformation*. This is an important result, since it reproduces in a much more general formalism the duality of free energy and entropy in usual thermodynamic formalism. Further, for that reason the considered Hamiltonians are a natural class of interactions for a homogeneous quantum lattice.

Starting with the minimizing condition for the free energy, we can give a geometrical interpretation for equilibrium states. This is expressed in the term of the *subgradient* of a concave function (see [5]). For each $\omega \in S(\beta, h)$ and $\forall h' \in C(\mathcal{S}(\mathcal{B}), \mathbb{R})$, we have:

$$
\begin{aligned}
f(\beta, h) &= f(\beta, h, \omega) = \langle \omega; h \rangle - \frac{1}{\beta} s(\omega) \\
&= \langle \omega; h - h' \rangle + \langle \omega; h' \rangle - \frac{1}{\beta} s(\omega) \geq \langle \omega; h - h' \rangle + f(\beta, h').
\end{aligned}
\tag{3.6}
$$

(3.6) is just the condition, that $\omega \in \partial f(\beta, h)$, the set of all subgradients to $f(\beta, .)$ in $h \in C(\mathcal{S}(\mathcal{B}), \mathbb{R})$. The other way round, take a linear functional $\alpha \in \partial f(\beta, h)$. By definition of s as a functional on $C(\mathcal{S}(\mathcal{B}), \mathbb{R})$ it follows at once, that $\alpha \in \mathcal{S}^{\mathcal{P}}(\mathcal{A})$. Now, one can use the fact that $\alpha \in \partial f(\beta, h) \Longleftrightarrow h \in \partial \frac{1}{\beta} s(\alpha)$:

$$
\begin{aligned}
f(\beta, h, \alpha) &= \langle \alpha; h \rangle - \frac{1}{\beta} s(\alpha) \leq \langle \alpha; h \rangle - \langle \alpha - \omega; h \rangle - \frac{1}{\beta} s(\omega) \\
&= \langle \omega; h \rangle - \frac{1}{\beta} s(\omega) \qquad \forall \omega \in \mathcal{S}^{\mathcal{P}}(\mathcal{A}),
\end{aligned}
\tag{3.7}
$$

but this is the minimizing condition for a state $\omega \in S(\beta, h)$; thus the following proposition is valid:

Proposition 3.2 *For all* $h \in C(\mathcal{S}(\mathcal{B}), \mathbb{R})$, *the set of equilibrium states* $S(\beta, h)$ *is given by* $\partial f(\beta, h)$, *the subdifferential of* $f(\beta, .)$ *in* $h \in C(\mathcal{S}(\mathcal{B}), \mathbb{R})$.

Now start with a given system h in the thermic equilibrium; if it is possible to vary the Hamiltonian of this system in the neighborhood of h, one can learn something about the equilibrium states. But in general, this is not the physical problem, since the variety of the system can never be the complete neighborhood of h in $C(\mathcal{S}(\mathcal{B}), \mathbb{R})$. On the other hand, if the system h is fixed exactly, a small variety of h could change the set of equilibrium states in an almost arbitrary way. Therefore one has to look for a 'smaller' variation of the Hamiltonian. This leads to a variational principle on an finite–dimensional affine subspace of $C(S(\mathcal{B}), \mathbb{R})$, which contains in some sense more information about the thermodynamic system and can reproduce information about a suitable homogeneous mean field, describing an experimental situation.

Therefore, a special basis of description is introduced in the following way: choose $n_\mathcal{B} := n^2$ selfadjoint elements $\hat{e} := (\hat{e}_1, ..., \hat{e}_{n_\mathcal{B}})$ in $\mathcal{B}$, being a basis of operators in $\mathcal{B}$. Then define for a fixed *interaction* $w \in C(S(\mathcal{B}), \mathbb{R})$ and ($\frac{W_\Lambda}{|\Lambda|}$)$_{\Lambda \in \mathcal{L}} \in j^{-1}(\{w\})$ the local Hamiltonians

$$H_\Lambda : \mathbb{R}^{n_\mathcal{B}} \longrightarrow A_\Lambda, \varepsilon \longrightarrow H_\Lambda(\varepsilon) := |\Lambda| \sum_{i=1}^{n_\mathcal{B}} \varepsilon_i m_\Lambda(\hat{e}_i) + W_\Lambda, \tag{3.8}$$

with $m_\Lambda(x) := \frac{1}{|\Lambda|} \sum_{k \in \Lambda} 1 \otimes ... \otimes 1 \otimes x \otimes 1 \otimes ....$ and the x appears at the k-th place in the tensor. The associated equivalence class of thermodynamic systems is given by:

$$h(\varepsilon) = \sum_{i=1}^{n_\mathcal{B}} \varphi(\hat{e}_i)\varepsilon_i + \Big[j(w) \Big](\varphi) := \langle \varphi(\hat{e}); \varepsilon \rangle + w(\varphi(\hat{e})). \tag{3.9}$$

An interpretation of this Hamiltonian can be given in the following way: w describes all effective interactions of the mean field in the quantum lattice. This interaction is fixed and assumed to be independent of temperature β and ε. The vector ε describes an external parameter, e.g. if $\mathcal{B} = M_2(\mathbb{C})$ and $\hat{e}$ are chosen to be the Pauli matrices, ε is a magnetic field up to some constants. Therefore take the point of view, that ε can be varied (experimentally) in an arbitrary manner. Again, the condition for equilibrium states is used and one calculates:

$$f(\beta, \varepsilon) := f(\beta, h(\varepsilon)) = \inf\{\langle \omega; h(\varepsilon) \rangle - \frac{1}{\beta}s(\omega) | \omega \in S^P(\mathcal{A})\}$$

$$= \inf\{\langle \varphi(\hat{e}); \varepsilon \rangle + w(\varphi(\hat{e})) - \frac{1}{\beta}s(\Pi_\varphi) | \varphi \in S(\mathcal{B})\}. \tag{3.10}$$

With

$$\sigma(\beta, .) : \mathcal{B}^* \longrightarrow \overline{\mathbb{R}}, \quad \sigma(\beta, \varphi) := \begin{cases} \frac{1}{\beta}s(\Pi_\varphi) - w(\varphi) & \varphi \in S(\mathcal{B}) \\ -\infty & \text{else} \end{cases}, \tag{3.11}$$

conclude that $f(\beta, \varepsilon) = [\sigma(\beta, .)]^*(\varepsilon)$ and therefore all results in [6, 5] can be used. In contrast to Theorem 3.1, the variation is now reduced to the finite–dimensional case (since $\mathcal{B}^*$ is isomorphic to $\mathbb{R}^{n_\mathcal{B}}$), but the one–to–one relation between $f(\beta, .)$ and $\sigma(\beta, .)$ is no longer valid.

Theorem 3.3 *Take the functionals* $f(\beta, .) : \mathbb{R}^{n_\mathcal{B}} \longrightarrow \mathbb{R}$ *and* $\sigma(\beta, .) : \mathcal{B}^* \longrightarrow \mathbb{R}$. *It holds:*

(i)
$$f(\beta, \varepsilon) = \Big[\sigma(\beta, .)\Big]^*(\varepsilon) \qquad \forall \varepsilon \in \mathbb{R}^{n_\mathcal{B}}, \tag{3.12}$$

(ii) $\Big[f(\beta, .)\Big]^*(\varphi) = \Big[\sigma(\beta, .)\Big]^{**}(\varphi) = \Big[\text{cl co}\sigma(\beta, .)\Big](\varphi) =: \hat{\sigma}(\beta, \varphi), \qquad \forall \varphi \in \mathcal{B}^*, \tag{3.13}$

with co as the concave hull of an arbitrary function and cl as the closure of a function defined as in [6, 5]. In this case the closure equals to the upper semicontinuous hull.

(iii)
$$f(\beta, \varepsilon) = \Big[\hat{\sigma}(\beta, .)\Big]^*(\varepsilon). \tag{3.14}$$

Proof:
(i) is derived in (3.10). (ii) is a result in [5, Theorem 5]. (iii) is a consequence from the concavity of $f(\beta, .)$ as the conjugate of $\sigma(\beta, .)$. Since $f(\beta, \varepsilon)$ is finite for all $\varepsilon \in \mathbb{R}^{n_\mathcal{B}}$, $f(\beta, .)$ is a continuous function on $\mathbb{R}^{n_\mathcal{B}}$ (see [6, Section 10]) and consequently $f(\beta, .) = \text{cl co}f(\beta, .)$. $\square$

Now, a finite–dimensional variation principle is established, where the parameter, which is varied has an explicit interpretation. The Fenchel–transform of the minimal free energy appears as a functional of the entropy, modified by a term of the interaction in the lattice.

This additional part of the interaction is a consequence of the specialized variation, only over extremal permutation invariant states, which kills a part of the information about the entropy appearing in Theorem 3.1.

Just as in Theorem 3.2 a geometrical structure can be established in terms of the subgradient. With the equilibrium condition follows:

$$M(\beta,\varepsilon) = \{\varphi \in \mathcal{S}(\mathcal{B}) | f(\beta,\varepsilon,\Pi_\varphi) = f(\beta,\varepsilon)\} \subseteq \partial f(\beta,\varepsilon) \qquad \forall \varepsilon \in \mathbb{R}^{n_B}. \qquad (3.15)$$

Note, that in general, there is no equality between theses sets, but there exist many relations among them, see [2] in the case of quadratic interactions.

At this place, we concentrate on (3.13) in Theorem **3.3** and its consequences. We have interpreted the function $\hat{\sigma}(\beta,.)$ as a modified entropy density functional, created by the specialized variation, only over extremal permutation invariant states. On the other hand, one can also see a term carrying information about the effective mean field interaction, which is modified by a part of the entropy. This point of view allows to get information about the interaction in a given lattice if the minimal free energy functional is given (— maybe experimentally). The Fenchel–transformation, then produces $\hat{\sigma}(\beta,.)$; starting with $\hat{\sigma}(\beta,.)$ one gets the interaction in eliminating the entropy part (which is fixed if the quantum lattice is fixed).

There are two qualitatively different cases: either $f(\beta,.)$ is differentiable or not. In the first case the interaction is uniquely determined. The other case, that $f(\beta,.)$ is not differentiable is more interesting, since this situation has relevant consequences in the case of a first order phase transition, see [2]. Here the interaction is not exactly determined, since there exists a set $\mathcal{X} \subseteq \mathcal{S}(\mathcal{B})$, with $\hat{\sigma}(\beta,.)|_\mathcal{X}$ is affine and $\dim \mathcal{X} = \dim \mathcal{S}(\mathcal{B})$ as subsets in $\mathbb{R}^{n_B}$. Exactly in $\mathcal{X}$, the interaction has a large variety and there exist almost arbitrary continuous interactions with same minimal free energy functional and phase structure. This is a consequence of $\hat{\sigma}(\beta,.) := \mathrm{cl}\, \mathrm{co}\, \sigma(\beta,.)$.

Acknowledgements

The author is indebted to Prof. Dr. A. Rieckers for helpful discussions, especially clarifying the role of the variational principles.

References

[1] Th. Gerisch and A. Rieckers, *The Quantum Statistical Free Energy Minimum Principle for Multi-Lattice Mean Field Theories,* Preprint Tübingen 1990

[2] A. Rieckers and H. - J. Volkert, *Variational Principles and Equilibrium States for a Class of Long-Range Interacting Quantum Lattice Systems,* Preprint Tübingen 1988

[3] H. - J. Volkert and A. Rieckers, *Equilibrium States and Phase Transitions of some FCC-Multi-Lattice Systems,* Preprint Tübingen 1988

[4] G. A. Raggio and R. F. Werner, *Quantum statistical mechanics of general mean field systems,* Helvetica Physica Acta, Vol. **62**(1989) 980-1003

[5] R. T. Rockafellar, *Conjugate Duality and Optimization,* Society for Industrial and Applied Mathematics, Philadelphia (1974)

[6] R. T. Rockafellar, *Convex Analysis,* Princeton University Press (1970)

[7] R. B. Israel, *Convexity in the Theory of Lattice Gases,* Princeton University Press (1979)

[8] G. L. Sewell, *Quantum Theory of Collective Phenomena,* Clarendon Press, Oxford, (1986)

[9] J.-C. Tolédano and P. Tolédano, *The Landau Theory of Phase Transitions,* World Scientific Publishing, Singapore - New Jersey - Hong Kong, (1987)

[10] S. Sakai, *C*-Algebras and W*-Algebras,* Springer - Verlag, Berlin, (1971)

[11] E. Størmer, *J. Funct. Anal.,* **3**, 48 (1969)

[12] O. Bratteli and D. W. Robinson *Operator Algebras and Quantum Statistical Mechanics II,* Springer Verlag, New York - Heidelberg - Berlin (1981)

ENVIRONMENT AND SYMMETRY BREAKING

IN QUANTUM FIELD THEORY

E. Graziano and G. Vitiello

Dipartimento di Fisica dell'Università
I-84100 Salerno, Italia

In this communication we want to discuss the dynamical effects of the coupling between a quantum system and the environment in the framework of Quantum Field Theory (QFT).

Actually, in realistic conditions any microscopic system is always embedded in some environment and we observe that the "environment" should not necessarily be thought as substantially different from the system. One may consider a local region of the system as embedded in the remaining part of the system and consider this last one as the "environment" surrounding the former one; or else, one may consider certain degrees of freedom of the system acting as environment for other degrees of freedom; such a view may be useful for example in cosmological applications where no environment is outside the Universe[1] .

Due to the system-environment coupling, any fluctuation or change in the state of the system (or environment) may produce some reaction field. In a perturbative approach weak reaction fields may be neglected in lower order approximations.

A first part of the discussion presented here is devoted to show that, under some circumstances, even very tiny reaction fields may play a crucial role in driving the system into an asymmetric ground state (spontaneous breakdown of symmetry). As a consequence one should not completely rely on a perturbative analysis which indeed may not display relevant dynamical features. Weak reaction fields may be of great relevance in many body physics as well as in high energy physics; also, they play a relevant role in chemistry and in biology. As an example, it is well known that in Quantum Mechanics pyramidal molecules like NH_3 AsH_3 and PH_3 show tunnelling instability in the semiclassical limit caused by very small perturbations due to system-environment coupling[2-5].

The second part of our discussion concerns systems whose dynamic coupling with the environment is not explicitly known; in many cases the environment effects reveal themselves through the changes in the vacuum expectation values of some relevant fields (order parameters). Different values of the order parameter represent different realizations of the same basic dynamics; these realizations correspond to different, i.e. unitarily inequivalent, representations of the canonical commutation relations (CCR); in some cases we may say that they represent different phases of the system. The occurrence of (infinitely many) unitarily inequivalent representations of the CCR is a characteristic feature of systems with infinite degrees of freedom which are studied in QFT. The foliation of the states of the system in unitarily inequivalent representations of the CCR finds a rigorous treatment in the operator algebra approach to QFT[6]. For sake of shortness, we refer to the existing literature on the operator algebra formalism and we will not discuss it in this paper. In our discussion we need not only a Lagrangian (quantum operator) equation describing the system evolution in a given representation but we also need a c-number equation describing the order parameter variations (evolution) under the influences of the environment. This last evolution may be thought as a trajectory or path in the "space of the representations of CCR"; in other words, as an evolution across unitarily inequivalent representations of the CCR, which therefore cannot be Lagrangian: in this sense it corresponds to non-unitary evolution. We may have in this way a description of phase transition and of non-equilibrium processes. We observe also that variations in the order parameter reflect themselves in changes in the vacuum structure (condensate) whose knowledge is relevant in problems such as the early Universe physics, bubble formation, extended objects (as vortices, monopoles, etc.), and in general in problems facing spontaneous formations of structures. In these cases the changing of the order parameter crucially depend upon physical quantities, like temperature, density, etc...

Let us consider now the problem of weak reaction fields inducing spontaneous breakdown of symmetry. We use the functional integral techniques in order to have a model independent analysis. We may summarize our results as follows[7] :

-Even very tiny perturbations arising from the system-environment interaction (reaction field) may trigger the breakdown of symmetry thus producing long range correlations in the system.
-The system order parameter does not depend on the details of the environment state.
-Finite volume effects set a threshold above which the intensity of the reaction field must be in order to have symmetry breakdown.
-The symmetry pattern and the coherence of the ground state and the classical limit are controlled by the Inönü-Wigner group contraction of the Lagrangian symmetry group.

We always refer to continuous internal symmetry groups in this paper. We consider, for simplicity, a spin operator system $S^{(i)}$, i=1,2,3, with the SU(2) algebra $[S^{(i)}, S^{(j)}] = i\epsilon_{ijk} S^{(k)}$. The spin density operators $S^{(i)}(x)$ may be thought as products of basic (Heisenberg) spin doublet fields $\psi(x) = (\psi_\uparrow, \psi_\downarrow)$. The system Lagrangian is given in terms of the quantum fields $\psi(x)$ and is

assumed to be invariant under SU(2) transformations
$$\psi(x) \to \psi'(x) = g\psi(x), \quad g \in SU(2) \quad : \quad L[\psi(x)] = L[\psi'(x)].$$
In the path-integral formalism the generating functional is

$$W[J] = \frac{1}{N} \int [D\psi][D\psi^{*}] \exp(i \int d^4 x (L[\psi(x)] +$$

$$+ J^{(+)}(x) S_{\psi}^{(-)}(x) + J^{(-)}(x) S_{\psi}^{(+)}(x) - i\lambda S_{\psi}^{(3)}(x))), \qquad (1a)$$

$$N = \int [D\psi][D\psi^{*}] \exp\left(i \int d^4 x (L[\psi(x)] - i\lambda S_{\psi}^{(3)}(x)) \right) \qquad (1b)$$

The λ term in eq.(1) denotes the coupling of the reaction field λ (assumed to be the x-independent average of the induced magnetic polarization in the environment and collinear to $S^{(3)}(x)$) with the system polarization. By deriving the Ward-Takahashi identities from eq.(1), we can show[7,8] that the "magnetization" $<S^{(3)}(x)>_{\lambda} = M(\lambda)$ is non vanishing in the $\lambda \to 0$ limit provided a gapless mode ($\omega(p=0)$) exists. In such a case we also show that the value $M = M(\lambda)|_{\lambda=0}$ does not depend on the intensity of the reaction field λ.

Then our conclusion is that for vanishingly small value of the perturbing reaction field a non-zero magnetization is possible only when long range correlations (the gapless modes) may be dynamically created: these correlations acting as cooperative effect are the well known Goldstone bosons of spontaneously broken symmetry theories[9]. We observe that since the reaction field is assumed to arise from the system fluctuations, the λ term in eq.(1) may be viewed as a non-linear self-coupling term.

In the case of finite (but large) volume $V = 1/\eta$ of the system we find that a non-zero magnetization still exists and is independent of λ provided $\lambda \gg \omega(p=\eta) \cdot c$ (with c a theory parameter)[7]. In the infinite volume limit the Goldstone bosons are massless and therefore their condensation in the ground state does not contribute to its energy. For small but not zero η, however, the "effective mass" contribution $\omega(\eta)$ must be taken into account in the ground state energy, besides the kinetic energy. The splitting between the lowest eigenvalues becomes thus larger as an effect of the perturbing reaction field.

It is known[10,11] that the boson condensation is controlled by the Inönü-Wigner group contraction of the Lagrangian symmetry group. In the case considered above SU(2) contracts to E(2). It has also been observed[10] that the occurrence of group contraction in spontaneously broken symmetry theories signals a transition from quantum to classical regime. This suggest to us that the reaction field effects above discussed, by inducing spontaneous breakdown of symmetry, actually drive the system to a "classical limit" regime. This can be seen by introducing the effective Lagrangian[12,13] $L_{\epsilon}(\xi, M)$ where ξ denotes the set of asymptotic fields and $\epsilon = (\sqrt{V})^{-1}$ or $\epsilon = (\sqrt{N})^{-1}$ is the group contraction parameter: $L(\psi) \to L_{\epsilon}(\xi, M) = \epsilon^{-2} L(\xi, M)$. Use of L_{ϵ} in eq.(1) gives the exponent $i(\epsilon^{-2}/\hbar) \int d^4 x \ L(\xi, M)$. We then see that the group contraction limit ($\epsilon \to 0$) acts as the classical limit $\hbar \to 0$. Notice that we always keep $\hbar \neq 0$ in this paper.

We thus conclude that small reaction fields from the environment may trigger the spontaneous breakdown of symmetry with the formation of ordered patterns in the system. The long range correlations set in through the group contraction which regulates the boson condensation and the macroscopic behaviour (classical limit) of the system. As a consequence, by considering that any microscopic system is always embedded in some environment we also conclude, with ref.[14], that "symmetry breaking is a universal phenomenon under semiclassical conditions".

As a final comment on this part, we may observe that the quantum measure process may be understood as producing the wave function localization (i.e. symmetry breaking) and perhaps the interaction with the apparatus amounts to switching on of long range interaction fields.

Let us now consider systems with variable order parameter. Let us study a simple example for sake of clarity and shortness. We assume the Lagrangian for the quantum field $\hat{\phi}(x)$, $x = (\vec{x}, t)$:

$$L = \frac{1}{2}\partial_\mu\hat{\phi}\partial^\mu\hat{\phi} - \frac{1}{2}\mu^2\hat{\phi}^2 - \frac{1}{4}\lambda\hat{\phi}^4, \quad \lambda > 0 \tag{2}$$

$$(\nabla^2 - \partial_t^2 - \mu^2)\hat{\phi} = \lambda\hat{\phi}^3 \tag{3}$$

We assume that a non-zero vacuum expectation value $v(x, \sigma)$ exists:

$$_v<0|\hat{\phi}(x)|0>_v = v(x, \sigma). \tag{4}$$

σ is a physically relevant parameter, e.g. temperature, upon which the order parameter v may depend. The ground states $|0>_v$ are labelled by the values taken by v and in the infinite volume limit $|0>_v$ is orthogonal to $|0>_{v'}$ with $v \neq v'$: they are ground states belonging to unitarily non-equivalent representations of the CCR. Variations of $v(x, \sigma)$ induce thus transitions across the unitarily inequivalent representations of the CCR. Our problem is finding an "evolution" law for $v(x, \sigma)$. We assume $\sigma = \beta = 1/(kT)$ and use the Bogoliubov inequality[15] in the variational approach:

$$F \leq F_1 = F_0 + <H - H_0>_0, \tag{5}$$

where F denotes the free energy for the quantum system $\hat{\phi}$, $F = -kT\ln Tr[\exp(-\beta H)]$; H_0 is the trial Hamiltonian which may be derived from the trial Lagrangian

$$L = \frac{1}{2}\partial_\mu\hat{\rho}\partial^\mu\hat{\rho} - \frac{1}{2}\mu_0^2\hat{\rho}^2 \quad .$$

$<\cdot>_0$ denotes statistical average with respect to H_0 . Use of eq. (4) to compute (5) gives[16,17]:

$$F_1(v(x, \beta)) = \int d^3x[(\partial_0 v)^2 - L_{eff}(v(x, \beta), \partial_\mu v(x, \beta))] \tag{6a}$$

$$L_{eff} = \frac{1}{2}\partial_\mu v\partial^\mu v - U_{cl}(v) - F_0(\mu_0^2) + \frac{3}{4}\lambda <\rho^2>_0^2 \tag{6b}$$

$$U_{cl}(v) = \frac{1}{2}\mu^2 v^2 + \frac{1}{4}\lambda v^4 \qquad (6c)$$

Eq. (6) is a generalization of the Ginzburg-Landau functional. We then see that the free energy for the quantum system $\hat{\phi}$ plays the role of the functional energy for the classical field $v(x,\beta)$. We can derive from eq.(6) the Euler-Lagrangian equation for $v(x,\beta)$ by least-action-principle:

$$(\nabla^2 - \partial_t^2 + m^2)v(x,\beta) = \lambda v^3(x,\beta) \qquad (7a)$$

$$m^2 = \lambda v^2(\beta) = |\mu^2| - 3\lambda <\rho^2>_0 \qquad (7b)$$

$$\mu_0^2 = \mu^2 + 3\lambda v^2(x,\beta) + 3\lambda <\rho^2>_0 \qquad (7c)$$

We observe that m^2 in eq.(7) has the "wrong" sign: eq. (7a) describes indeed non-unitary evolution in the "space of the representations of CCR"; in other words, the classical field $v(x,\beta)$ evolves through trajectories or paths according to a least-action-principle in the space of the folia of the quantum system states. Such an evolution is a classical one and char- acterizes the macroscopic behaviour of the quantum system. L in eq. (2) controls the quantum dynamics of $\hat{\phi}$ in each folium (for each value of $v(x,\beta)$); L_{eff} in eq.(6) controls the classical evolution of $v(x,\beta)$ through the folia. As shown, the relation between the two evolutions is by means of the free energy for $\hat{\phi}$ and eq. (7b) is a consistency relation between the "inner" dynamics described by L and the macroscopic evolution controlled by L_{eff}.

The above mentioned labelling of the representations by the order parameter values may be rigorously performed in the GNS construction in the C*-algebra approach to QFT.

We also note that in a non-equilibrium dynamics β in general may depend on time.

Application of the above scheme has been worked out in the problem of quantization of free matter field in the presence of a classical gravitational background[18], for the description of unstable states in QFT[19], in gauge theory models with the vortex and monopole solutions[16,20] and for a formulation of the basic dynamics for the living matter[21].

References

1. W. H. Zurek, Quantum measurements and the environment induced transition from quantum to classical, NSF-ITP- 88-197 (1989).
2. G. Jona-Lasinio, F. Martinelli and E. Scoppola, Phys. Rep. 77, 313 (1981); Mathematics+Physics, Lectures on recent results Vol. II, L. Streit. (Ed.) World Scientific Publ. Singapore (1986).
3. G. Jona-Lasinio, F. Martinelli and E. Scoppola, Comm. Math. Phys. 80, 223 (1981).
4. S. Graffi, V. Grecchi and G. Jona-Lasinio, J. Phys. A: Math. Gen. 17, 2935 (1984).

5.   P. Claverie and G. Jona-Lasinio, Phys. Rev. A 33, 2245 (1986).

6.   O. Brattelli and D. W. Robinson, Operator Algebras and Quantum Statistical Mechanics, Springer, Berlin, 1981.

7.   E. Celeghini, E. Graziano and G. Vitiello, Phys. Lett. A 145, 1   (1990).

8.   M. N. Shah, H. Umezawa and G. Vitiello, Phys. Rev. 10 B, 4724 (1974) .

9.   J. Goldstone, Nuovo Cimento 19, 154 (1961).
     J. Goldstone, A.Salam and S. Weinberg, Phys. Rev. 127, 965 (1962).

10. C. De Concini and G. Vitiello, Nucl. Phys. 116 B, 141 (1976).

11. H. Umezawa, H. Matsumoto and M. Tachiki, Thermo Field Dynamics and Condensate States, North Holland Publ. Comp. (1982).

12. S. Coleman, J. Wess and B. Zumino, Phys. Rev. 177, 2239 (1969);
     C. G. Callan Jr., S. Coleman, J. Wess and B. Zumino, Phys. Rev. 177, 2247 (1969).

13. E. Celeghini, P. Magnollay, M. Tarlini and G. Vitiello, Phys. Lett. 162 B, 133 (1985).

14. G. Jona-Lasinio and P. Claverie, Prog. Theor. Phys. Suppl. 86, 54 (1986).

15. R. P. Feynman, Statistical Mechanics (Benjamin, Reading MA, 1972).

16. R. Mańka, J. Kuczynski and G. Vitiello, Nucl. Phys. 276 B, 533 (1986).

17. E. Del Giudice, R. Mańka, M. Milani and G. Vitiello, Phys. Lett. 206 B, 661 (1988).

18. M. Martellini, P. Sodano and G. Vitiello, Nuovo Cimento 48 A, 341 (1978).

19. S. De Filippo and G. Vitiello, Lett. Nuovo Cimento 19, 92 (1977).

20. R. Mańka and G. Vitiello, Annals of Phys. (N.Y.), in print.

21. E. Del Giudice, S. Doglia, M. Milani and G, Vitiello, in Biological Coherence and Response to External Stimuli, H. Fröhlich ed., Springer, Berlin, 1988

GAUGE CHEMISTRY

Jan C.A. Boeyens

Department of Chemistry, University of the
Witwatersrand, Johannesburg, South Africa

ABSTRACT. Gauge theory describes interactions in terms of internal symmetries that reflect an intrinsic property of space time. It has revolutionized particle physics, but still awaits application in chemistry. It affects free particles through the quantum potential and mediates their response to chemical environments. Simple applications are outlined and the origin of the quantum potential in the vacuum is discussed.

1. INTRODUCTION

The mechanical description of chemical systems mandates an awkward choice. At the sub–atomic level the correct description is only provided by quantum mechanics, which is commonly assumed to contain standard classical mechanics, the logical choice for ordinary macrosystems, as a special case. This creates the problem that transition from one system to the other is neither obvious nor smooth. The quantum/classical limit is defined at $h \longrightarrow 0$ and requires a switch in logic from a non–Boolean to a Boolean system [1]. In practice this leaves an undefined grey area that cannot be consistently demarcated nor assigned to either domain. This situation is in striking contrast to the choice required between classical and relativistic mechanics for astronomically large systems, or as $v \longrightarrow c$. In this case the transition is smooth, and classical systems emerge as special cases in relativistic mechanics with a well–defined correction factor; e.g. the mass increase with velocity is

$$m' = m_0 \left(1 - v^2/c^2\right)^{-\frac{1}{2}}$$

The choice of mechanics becomes a matter of convenience and dictated by the accuracy required.

Large-Scale Molecular Systems, Edited by W. Gans *et al.*
Plenum Press, New York, 1991

It seems that classical mechanics is defined as a subset of both quantum mechanics and of relativistic mechanics. To further complicate matters, relativistic systems are strictly local whereas correlated quantum systems in causal reality are not [2]. A smooth merging of quantum into classical mechanics seems to be the element missing from the present formula. Noting that h is constant and that quantum effects become important only at the sub–atomic level, the transition from classical to quantum descriptions occurs as h/m ——→ ∞, m ——→ 0. One has the interesting sequence that dictates the choice of mechanics:

$$QM \xleftarrow{\text{m} \to 0} CM \xrightarrow{\text{m} \to \infty} RM,$$

which suggests that the mechanics of moderate things breaks down for the very large and for the very small. It is tempting to generalize that the special effects at both ends of the scale are due to non–Euclidean features of space–time. Special relativity defines the condition when classical mechanics is replaced by the relativistic as v ——→ c, m ——→ ∞. The physical meaning of this condition is provided by General relativity in terms of a geometrical distortion of space–time in the vicinity of large masses. An equivalent argument would seek for a General quantum theory that links the Special quantum condition, m ——→ 0, to some other geometrical distortion of space–time. This distortion was first inferred by Weyl [3] in an attempt to unify gravitational and electromagnetic effects, suggesting that a vector transplantation in non–Euclidean space requires a gauge transformation to compensate for a distortion of the linear measurement scale.

2. GAUGE THEORY [4]

Vector transplantation in an arbitrary coordinate system is formulated in differential form as

$$d\xi^\alpha = \Gamma^\alpha_{\beta\gamma} dx^\beta \xi^\gamma$$

where $\Gamma^\alpha_{\beta\gamma}$ are the symmetric connections of the manifold, ξ^α are the components of the vector and dx^β the local displacement vector. In terms of the metric tensor $g_{\mu\nu}$ the length of the vector ξ^α is defined at each point of the manifold as

$$l^2 = \| \xi \|^2 = g_{\mu\nu} \xi^\alpha \xi^\beta$$

In Riemannian space the length of this vector remains unchanged under the transplantation law. It is a basic assumption that physical laws are invariant under Poincaré transformations, i.e., Lorentz transformations and space–time translation.

To accommodate electromagnetic variables in addition to the gravitational, as geometric features of the manifold it was necessary for Weyl [3] to postulate a differential geometry that allows for transplantation with non–constant l. If the vector ξ^α is interpreted as the length of a measuring rod it means that this length could change under displacement. It was assumed that the increment in length is proportional to the length itself and a linear homogeneous function of the displacement vector. Hence

$$dl = (\phi_\beta dx^\beta)l$$

where the covariant vector ϕ_β plays the role of the connection $\Gamma^\alpha{}_{\beta\gamma}$.

For a time–independent gauge vector ϕ_0, like an electrostatic field, the gauge transformation after a time x^o yields

$$l = l_0 \exp[\int_0^{x^o} \phi_0 \, dx^o] = l_0 \exp(\phi_0 x^o)$$

The idea, presented as a distortion of length almost fatally discredited the model, which however survived in the form that identifies the modified gauge with the vector potential of the electromagnetic field. However, only gauge invariant quantities can be obtained by direct measurement [5]. As an electron interacts with a field that changes from point–to–point, gauge invariance of its measurable properties is maintained by the exchange of virtual photons with the field. This local gauge transformation follows the internal U(1) symmetry–group operations which describe the interaction between charged particles by photon exchange and charge conservation. In addition to Poincaré invariance, internal symmetries, related to space–time–independent transformation of particle states, therefore appear.

The parallel with a quantum–mechanical wavefunction that transplants unchanged, except for a phase factor, was duly noticed [6], but the description of field–free quantum phenomena as a gauge transformation has never been completed, although an appropriate quantity that could act as the gauge potential has in fact been identified by Bohm [7]. It is now postulated that this quantum potential, which exists as a property of the vacuum, represents the gauge potential for free quantum particles (electrons). The gauge vector then is

$$\phi = -\frac{h^2 \nabla^2 \psi}{8\pi^2 m \psi}$$

where ψ is the time–independent wave function. The gauge transformation consists of

$$d\Psi = \psi \exp\{k\phi dt\}$$

i.e. $\quad \Psi = \psi \exp\{-k(h^2\nabla^2\psi/8\pi^2 m\psi)t\}$

where k is some scale factor. The term in brackets represents the kinetic energy of the free electron and by choosing $k = -2\pi i/h$ one has the familiar time–dependence

$$\Psi = \psi \exp\left(-2\pi i Et/h\right)$$

Gauge invariance therefore automatically introduces the complex phase, which is the most distinctive feature of quantum theory and emphasized by Schrödinger even before the formulation of wave mechanics [8]. It is emphasized that this last expression is valid only in the empty vacuum, for an isolated electron. Any charge or potential field perturbs the field–free quantum potential and hence the free–particle wavefunction. It is, for instance, easy to show [4,6] how the requirement of gauge invariance, superimposed on the classical orbit of an electron circling a proton, directly produces the Bohr quantum conditions, choosing a scale factor of $i\alpha$, the complex fine–structure constant. It follows that the hydrogenic wavefunctions are defined only in this central coulombic gauge field. As the gauge potential changes the electronic wavefunctions must follow.

3. CHEMICAL GAUGE THEORY

Since all chemical interactions are electromagnetic in nature it appears almost essential to discuss these interactions only within the formalism of U(1) gauge theory of the Maxwell field [9], briefly summarized below.

In the most general case, a set of matter fields $\{\phi_i(x)\}$ that furnishes a representation of the group U(1), generated by the charges Q_i of the field is gauged according to

$$\phi(x) \longrightarrow \phi(x)e^{-iQ\omega}$$

where ω is an arbitrary real constant. The gauge field

$$A^\mu(x) \longrightarrow A^\mu(x) + \partial^\mu \omega(x)$$

ensures gauge invariance of the Lagrangian, L, under the phase transformations

$$\phi \longrightarrow \phi' = \phi e^{-i\xi}$$
$$\phi* \longrightarrow (\phi*)' = \phi* e^{i\xi}$$

which describe a group of unitary 1×1 matrices called U(1) in terms of the continuous rotation parameter ξ. In relativistic field theory, $L = \partial_\mu \phi* \partial^\mu \phi - V(\phi\phi*)$, where $V = m^2\phi\phi* + \lambda(\phi\phi*) + \text{const}$. The Lagrangian is also invariant under interchange of ϕ and $\phi*$, known as charge conjugation. The degrees of freedom associated with ϕ and $\phi*$ are also called "particle" and "anti–particle".

These principles should find ready application in a number of problems familiar to chemists.

3.1 Ligand fields

The well–known ligand–field method has all the ingredients of a chemical gauge theory. It uses the symmetry of the environment to define the bonding around a central atom. The spherical symmetry of the free atom is broken by the environment and lowered to match the symmetry of the environment. In actual application however, the elements of gauge theory are used ad hoc and not applied consistently. It is typically assumed [10] that only the angle–dependence of the atomic orbitals,

$$\psi = R(r) \cdot \theta\,(\theta) \cdot \Phi\,(\phi)$$

is affected as the degeneracy is lifted by the environment, to give

$$\Phi' = \Phi e^{i\alpha} = e^{im(\phi + \alpha)}$$

for rotation through α. But now, in order to assess relative orbital energies there is reverted back to the original degenerate set ϕ_m, and effectively only the radial wavefunctions are modified qualitatively.

To be consistent the gauge transformation should act on the complete wavefunction and by the charge distribution of the ligand shell, quantitatively define the transformed molecular orbitals. As a first approximation one examines a spherically symmetrical uniform surface charge density q_s of radius σ. The electrostatic potential, V is constant for $r < \sigma$ and

$$V(0) = \frac{1}{4\pi\epsilon_0} \int_s \frac{q_s}{\sigma}\, ds = \frac{q_s S}{4\pi\epsilon_0 \sigma}$$

The field $E = -\mathrm{grad}\ V = 0$. A surface charge of n electrons has a gauge effect on an electron in the sphere like

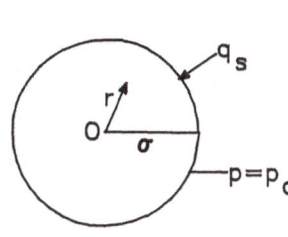

$$\psi(r) = \psi_0(r)\, \exp\{-\frac{2\pi i}{hc} \int_r^{\sigma} \frac{ne^2}{4\pi\epsilon_0\sigma}\, dr\}$$

Substitute $\alpha = e^2/2hc\epsilon_0 = 1/137$

$$\psi(r) = \psi_0(r)\, \exp\{-n i\alpha(1 - r/\sigma)\}$$

where $\psi_0(r) = |\psi(r)|$.

The phase, p, changes sign when

$$2ni\alpha(1 - r/\sigma) = 2\pi i l, \ l = \text{integer},$$

i.e.   at $r = \sigma(1 - \pi l/n\alpha)$

For $l = 0$, the phase changes sign at $r = \sigma$;

$\quad l = 1, \qquad\qquad\qquad\qquad$ at $n = \pi/\alpha \simeq 431$

As the phase changes sign, $\psi e^{-i\beta} \longrightarrow \psi^* e^{i\beta}$, and the wavefunction now describes positronic rather than electronic states. This implies that the electron is confined to the Faraday cage by diffraction, as in a Brillouin zone. Note that the probability density, $\psi^*\psi$ is gauge invariant as required.

A new boundary condition is therefore imposed on the electronic wavefunction by the environment. Any environment of discrete ligand charges thus generates appropriate wavefunctions on a central atom according to symmetry. In general, new functions like $\exp(-ar^p)$, $p > 1$, should result from simple exponential forms.

3.2   Realistic Wavefunctions

In view of the foregoing the use of atomic wavefunctions in molecular environments is a mathematical convenience only, and without physical meaning. Even for an atom in a chemically interactive environment the gauge field for the electron is different from the isolated–atom gauge. This is symmetry breaking. As the environment changes, the internal symmetry breaks spontaneously. Pure quantum states of the isolated atom disappear to be replaced by a mixture of states imposed by the environment, like a wave packet becoming increasingly localized, behaving more classically [11].

The outer electrons on atoms in a crowded, interacting environment are therefore more realistically described in terms of wavefunctions defined by the gauge field of the environment. This leads to the remarkable conclusion that, to a first approximation, these wavefunctions are independent of atomic type and a function of atomic size only. Since any laboratory sample consists of such an enormous number of particles, all with outer shields of electrons, all chemical environments are statistically comparable and can be considered to constitute equivalent chemical gauge fields. Small wonder that chemical systems are poorly described by the wave mechanics of pure systems [12] such as isolated free–atom wavefunctions, or even linear combinations thereof. Free–atom wavefunctions are of the form, $\exp(-ar)$, where r measures separation between electron and nucleus, and extends to infinity. Chemically realistic wavefunctions however, should resemble classical wave packets, in well–defined regions of space. In fact, the wavefunction

$$\psi(\mathbf{r}) = (3\gamma/4\pi n)^{\frac{1}{3}} \, (1/\sigma) \, \exp[-(\mathbf{r}/\sigma)^p], \, p >> 1,$$

assumed valid for all valence electrons, is known [13] to be remarkably effective in reproducing molecular properties like bond parameters, energies and force constants for any pair of atoms, in simple calculations. The parameter γ is nearly constant for all atoms, n is the principal quantum number and σ a characteristic atomic radius.

4. CONCLUSION

The only aspect of the argument left unexplored is the origin of the quantum potential, which supports the entire edifice. This quest inevitably leads to the geometry and topology of space–time. It was recently proposed [14] that the quantum potential appears naturally as a property of the vacuum, defined as an interface between enantiomeric regions of space–time.

This interface exists in a closed universe with a topology like the double cover of a non–orientable manifold. The involution that operates in a time dimension creates opposite chiralities and charge conjugation across the interface. Like the phase boundary in a liquid bilayer this vacuum interface creates a potential gradient, which significantly perturbs the behaviour of small particles. More massive particles are not seriously influenced by the interfacial gradient, but in turn, they have an effect on the vector field. It becomes necessary to recognize that space–time is non–Euclidean and to describe vector transplantation in terms of affine connections as in general relativity. The interfacial potential creates a gauge field and the interaction of an electron with its environment becomes a question of gauge invariance.

REFERENCES

[1]   D. Finkelstein (1965). The logic of quantum physics. Trans. N.Y. Acad. Sci., 25, 621–628.

[2]   J.S. Bell (1987). Speakable and Unspeakable in Quantum Mechanics. University Press, Cambridge.

[3]   H. Weyl (1918). Gravitation and Elektrizität. Sitzber. Preuss. Akad. Wiss. Berlin, 465–480.

[4]   R. Adler, M. Bazin, M. Schiffer (1965). Introduction to General Relativity. McGraw–Hill, New York.

[5]   W. Pauli (1941) Relativistic field theories of elementary particles. Rev. Mod. Phys. 13, 203–232.

[6]   F. London (1927). Quantenmechanische Deutung der Theorie von Weyl. Z. Physik, **42**, 375–389.

[7]   D. Bohm (1952). A suggested interpretation of the quantum theory in terms of "hidden" variables. I. Phys. Rev. **85**, 166–179.

[8]   C.N. Yang (1987). Square root of minus one, complex phases and Erwin Schrödinger; in C.W. Kilmister (ed) Schrödinger. Centenary celebration of a polymath. Cambridge University Press.

[9]   K. Huang (1982). Quarks, Leptons & Gauge Fields. World Scientific. Singapore.

[10]   F.A. Cotton (1963). Chemical Applications of Group Theory. Wiley. New York.

[11]   S. Clough (1989). Gauge Theory — A new outlook on solid state dynamics. Adv. Mater. 296–298.

[12]   H. Primas (1981). Chemistry, quantum mechanics and reductionism—perspectives in theoretical chemistry, Lecture Notes in Chemistry, Springer, Berlin.

[13]   J.C.A. Boeyens (1985). Molecular mechanics and the structure hypothesis. Struct. and Bonding, **63**, 65–101, and references therein.

[14]   J.C.A. Boeyens (1990). The geometry of quantum events. S. Afr. J. Sci., submitted for publication.

COHERENCE EFFECTS IN EXCITATION TRANSFER:

APPLICATION TO HEXAGONAL PHOTOSYNTHETIC UNIT

Ivan   Barvík

Institute of Physics
Charles University
Ke Karlovu 5, 121 16 Prague, Czechoslovakia

METHODS

An energy captured in an antenna system (AS) of   the photosynthetic unit (PSU) is transferred to a reaction centre (RC) in a form of localized Frenkel excitons.

Pearlstein and Zuber (1986) discussed unability of an incoherent picture which uses the Förster mechanism for exciton transfer to explain a regularity of cyclic antenna systems. An incorporation of coherent and intermadiate regimes was suggested.

Our aim is to show   the time dependence of the exciton site-occupation probabilities $P_n(t)$ in hexagonal PSU and follow consequences of different kinds of exciton transfer regimes in presence of a trap.

The character of the exciton motion (coherent or incoherent) is determined, besides an electronic intermolecular interaction, also by an exciton-bath (phonons) interaction.

In the convolution Generalized Master Equations method (GME) (Kenkre, 1982) a transition from the coherent regime to the incoherent one is well pronounced in the time dependence of the memory functions $w_{mn}(t)$ (MFs) entering convolution GME for the site occupation probabilities $P_n(t)$ (relevant but reduced information)

$$\partial/\partial t P_m(t) = \sum_n \int_0^t [w_{mn}(t-\tau)P_n(\tau) - w_{nm}(t-\tau)P_m(\tau)] d\tau \qquad (1)$$

The memory functions w_{mn} in (1) are complicated functions of hamiltonian matrix elements and temperature and, in most cases, cannot be found explicitly.

In our recent papers we succeeded in calculation of the long time asymptotics of the MFs for the exciton interacting locally with phonons (bath) (Čápek and Barvík, 1985,1987, Szöcs and Barvík, 1988, Barvík and Čápek, 1989). We demonstrated a two-channel character of MFs and determined (Szöcs and Barvík, 1988) a role of parameters like temperature, strength of the exciton-phonon interaction etc. on the character of the exciton motion in a linear chain.

On the other hand MFs are known only for some simple examples. We used a general analytical expression for coherent memory functions in finite systems (Barvik and Szocs, 1983) (where a destructive role of a site-energy difference for the coherent regime in a linear chain was demonstrated) for obtaining MFs for hexagonal PSU.

Excitation energy is transferred through AS to RC. In our treatment RC is modelled as a sink. Analysing carefully the time dependence of $P_a(t)$ Čápek and Szocs pointed (1984) the necessity of "transformation" of MFs. Absence of such a transformation could lead to completely erroneous results (Kenkre, 1982).

GME method can be connected with Stochastic Liouville Equations method (SLE) (Reineker, 1982), which is also able to describe the coherent as well as the incoherent regime and which is in this respect comparable with GME.

The basic equations of the SLE method read

$$i\hbar\partial/\partial t \rho_{mn} = \lfloor H_0, \rho \rfloor_{mn} - i\hbar\delta_{mn}\sum_p 2\gamma_{m-p} (\rho_{mm}-\rho_{pp}) - \tag{2}$$

$$-i\hbar(1-\delta_{mn})\lfloor (\Gamma_m+\Gamma_n)\ \rho_{mn} - 2\bar{\gamma}_{m-n}\rho_{nm}\rfloor$$

$$\Gamma_m = \sum_n \gamma_{m-n}$$

Here ρ_{mn} is the density matrix for the exciton system, $P_a = \rho_{mm}$, H_0 relates to the exciton system unperturbed by the exciton-phonon coupling. The off-diagonal parameters γ describe the phonon induced hopping of excitons, the Γ's control the decay of the off-diagonal elements $\rho_{mn} (m \neq n)$ which parallels the loss of coherence of the exciton motion.

We were able (Čápek and Barvík, 1985, Szocs and Barvík, 1988) to connect phenomenological parameters in SLE with microscopic parameters entering the Hamiltonian for an initially relaxed lattice.

A generalization of SLE to energetically heterogeneous systems was given in (Čápek, 1985, Čápek and Szocs, 1985).

We have chosen (Szocs and Barvík, 1986, Nedbal and Szocs, 1986) for our futher calculation the SLE method owing to its capability of describing the coherent regime as well as the incoherent one and its proper description of the role of the trap as suggested in (Čápek and Szocs, 1984).

MODEL

The Hamiltonian H_0 relates to the unperturbed exciton system (six antenna globulas on hexagon with one reaction centre) and includes only a nearst neighbour interaction J.

We concentrate on the coherence effects in the exciton transfer and their consequences. We include in our model the transfer through the quasicoherent channel (Szocs and Barvík, 1988) and we neglect a strongly temperature dependent (Čápek and Barvík, 1985, Čápek and Barvík, 1987, Szocs and Barvík, 1988) incoherent channel (γ_{mn} for m=n). The coherent memory functions are then damped with factor $\Gamma = \gamma_0$.

SLE (2) is completed by the "sink" term

$$-\sigma\ (\delta_{m7} + \delta_{n7})\ \rho_{mn}\ /2,$$

which describes the capture of the exciton in RC by the help of a trapping rate σ.

RESULTS

We show on following figures some examples (the time scale is given by $J/\hbar = 1$ psec^{-1}) of our extended computer modelling (Barvík, 1989, 1990) which was devoted to the calculation of the time development of the occupation probabilities $P_n(t)$.

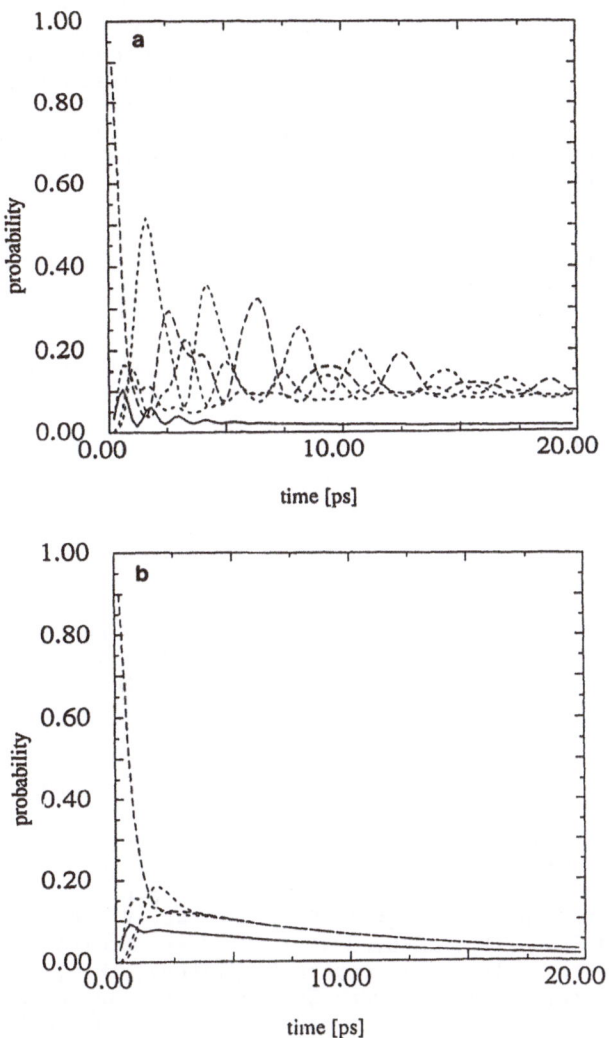

Fig 1. Occupation probabilities $P_n(t)$;
$J/\hbar=1.$, $\sigma=1.$ and $\gamma_s=0.1$ (a) or 1. (b) (in psec^{-1});
- - - $P_1(t)$ ······· $P_{2\text{-}6}(t)$ (AS) ——— $P_7(t)$ (RC)

For the initial condition corresponding to the excitation of one of antenna globulas at t=0, $P_n(0)=\delta_{n1}$ we demonstrate the loss of coherence $\Gamma=\gamma_s = 0.1$ (a) and 1.0 (b) with the trapping rate $\sigma = 1.0$ in units of $J/\hbar$.

CONCLUSIONS

It is very often argued , that after the initial period of several psec the occupation probabilities should reach a homogenous distribution in PSU and one can consequently use PME for their further time development. As it is seen from our results on Fig.1 this conclusion is for presumably realistic higher degree of coherence in exciton transfer unjustified. The oscillatory behaviour may be preserved in the antenna system for long times.

The coherence effects in the sence of time oscillations are preserved for shorter times with increasing incoherence, requiring the ultra-short time optical measurements for their discovery. The oscillatory character of $P_n(t)$ is smeared (averaged) in antenna and reaction center probabilities (Barvík, 1990) in the case of homogenous excitation of either AS, RC or PSU. The most probable oportunity to discover the oscillatory coherence effects are seen in the case of the initial excitation of RC. On the other hand the non oscillatory behaviour of RC does not rule out the coherent or quasicoherent regime in AS.

The second important result is the effective decoupling of RC from the AS with increasing the trapping rate σ. The excitation remains stored in AS avoiding the transfer to RC.

The obtained results are the first step in our understanding of coherence effects in hexagonal cyclic PSU. A more complete model would require further steps in a more complete set of input parameters for SLE method which includes:
   a) site-energy distribution in PSU
   b) hopping assisted channel ($\gamma_{ij} = 0$)

References

BARVÍK, I., 1990, submitted in J. theor. Biology.

BARVÍK, I., and SZÖCS, V., 1983, Coherent memory functions for finite molecular systems, Czech. J. Phys. B 33, 802:809.

BARVÍK, I., and ČÁPEK, V., 1989, Towards memory effects for electrons and phonons on nonrigid lattices, Phys. Rev. B 40, 9973:9976.

ČÁPEK, V., and BARVÍK, I., 1985, Memory functions for the electron-phonon system, J. Phys.C 18, 6149:6156.

ČÁPEK, V., and BARVÍK, I., 1987, Exciton memory functions in initially unrelaxed lattices, J. Phys. C 20, 1459:64.

ČÁPEK, V., and SZOCS, V., 1984, Phys. stat. sol. (b) 125, K137:140.

KENKRE, V. M., 1982, The master equation approach, in: "Exciton Dynamics in Molecular Crystals and Aggregates," G. Höhler, ed., Springer Tracts in Modern Physics, vol. 94., Springer-Verlag, Berlin.

NEDBAL, L., and SZÖCS, V., 1986, How long does exciton motion in the photosynthetic unit remain coherent, J. theor. Biol. 120, 411:418.

PEARLSTEIN, R., M., ZUBER H., 1986, Exciton state and energy transfer in bacterial membranes, in: Antennas and Reaction Centers of Photosynthetic Bacteria, M. E. Michel-Beyerle, ed., Springer Series in Chemical Physics, vol. 42., Springer-Verlag, Berlin.

REINEKER, P., 1982, Stochastic Liouville equation approach, "Exciton Dynamics in Molecular Crystals and Aggregates," G. Höhler, ed., Springer Tracts in Modern Physics, vol. 94., Springer-Verlag, Berlin.

SZÖCS, V., and BARVÍK, I., 1986, Description of excitation transfer and damping in the primary processes of bio-(chemical) systems, J. theor. Biol. 122, 179:186.

SZÖCS, V., and BARVÍK, I., 1988, Coupled coherent and incoherent motion of an excitation on a periodic linear chain, J. Phys. C21, 1533:1541.

RANDOM MATRIX THEORY AND ANDERSON LOCALISATION

Spiros Evangelou

Physics   Department
Division of Theoretical Physics
University of Ioannina
Ioannina   451 10 GREECE

ABSTRACT

The random statistical ensembles of short-ranged tight-binding lattice Hamiltonian matrices are introduced and studied in connection with the Anderson metal-insulator transition. The short and long range spectral fluctuation-correlation properties are examined for systems classified into three symmetry universality classes. In the metallic regime for two and three dimensions the results are consistent with the properties of level repulsion and spectral rigidity known from the Gaussian random matrix ensembles of Wigner-Dyson. In the insulating regime the spectrum is instead uncorrelated. The eigenvalue statistics is directly related to the universal conductance fluctuations observed in small metallic (mesoscopic) systems. The corresponding statistical behaviour of the wave functions is discussed and found to exhibit multifractal scaling at the transition. Similarities of the Anderson transition with what is known as the transition to "quantum chaos" are also discussed.

AN INTRODUCTION TO THE SCALING THEORY OF LOCALISATION AND THE PHYSICS OF MESOSCOPIC SYSTEMS

The eigenfunctions of non-interacting electrons in periodic solids are Bloch waves characterised by a wavevector k and the corresponding eigenenergies assembly into bands $E(k)$ forming a continuous spectrum. A small amount of disorder introduces elastic scattering after the electrons have travelled a characteristic distance which equals the mean free path l, up to which the wave functions retain their phase. By increasing the disorder l shortens and the d.c. conductivity σ decreases accordingly since from a Drude formula $\sigma \propto l$. Anderson[1] showed that in the presence of disorder the amplitude of the wave functions may also decay with an average localization length ξ leading to zero d.c. conductivity σ. Consequently for a fixed amount of randomness mobility edges E_c appear in the band separating extended states, which retain their spatial amplitude on the average ($\sigma \neq 0$), from non-conducting localized states which decay ($\sigma = 0$), usually exponentially. A metal-insulator transition is therefore expected when the Fermi energy E_f crosses E_c.

Large-Scale Molecular Systems, Edited by W. Gans *et al.*
Plenum Press, New York, 1991

The conductance G(L) for a good metallic hypercubic sample of linear size L and volume L^d from Ohm's law takes the form

$$G(L) = \sigma \cdot L^{d-2}, \tag{1}$$

where the proportionality coefficient σ is the conductivity of the sample. The L^{d-2} dependence arises from the fact that G(L) is proportional to the cross-section of the sample in the direction of the current and inversely proportional to its length. Following an early work by Thouless[2] if we combine Eq.(1) with the Einstein formula for the conductivity

$$\sigma = e^2 \cdot D \cdot \rho, \tag{2}$$

where D is the diffusion coefficient and ρ is the density of states at the Fermi level we obtain

$$G(L) = (e^2/h) \cdot E_T(L)/\Delta(L). \tag{3}$$

The characteristic energy scale E_T of the system, also known as Thouless energy, is inversely proportional to the time the electron takes to diffuse through the sample, that is

$$E_T(L) = h \cdot D/ L^2 \tag{4}$$

and the mean energy separation Δ has the L-dependence

$$\Delta(L) = [\rho \cdot L^d]^{-1}. \tag{5}$$

Then from Eq.(3) follows that E_T/Δ measures the number of levels N in an energy band of width E_T, that is

$$G(L) = (e^2/h) \cdot N(E_T) \tag{6}$$

with $e^2/h \approx 26 K\Omega^{-1}$. In the insulating phase G(L) decreases with increasing L and the Anderson transition occurs for a critical value G_c.

The simplest method to understand the Anderson transition is to use the dimensionless conductance in units of e^2/h, or Thouless number $g(L)=G(L)/(e^2/h)$, via a one-parameter scaling theory scheme[3] which rests on the calculation of the beta Gell-Mann renormalisation group function

$$\beta(g(L))=\partial \ln g(L)/\partial \ln L. \tag{7}$$

Knowledge of β as a function of g(L) and given $g(L_0)$ for some length scale L_0 permit the evaluation of g(L) at any other length scale L. Under certain assumptions (e.g. monotonicity) β interpolates smoothly between the metallic $(g \to \infty)$ limit $\beta \approx d-2$ from Eq.(1) and the insulating $(g \to 0)$ limit where $\beta \approx \ln g$. Positive β implies a g which increases with L and in the infinite L limit $g \to \infty$ (metal). Negative β implies decreasing g with L and eventually $g \to 0$ which leads to an insulator. When β has a zero this is fixed point indicating the presence of a metal-insulator transition. Because the next correction to the β function for large g is negative ($\beta=d-2-2 \cdot a \cdot (1/g)+...$) and since is expected to be negative for small g (ß≈lng) it never crosses the g-axis for d≤2. Therefore, a metal-insulator transition can be concluded only for three or higher dimensions. However in the presence of spin-dependent terms such as spin-orbit coupling or an applied strong magnetic field the 1/g correction term is positive and allows an Anderson transition even in d=2[4,5].

Theoretical studies of localisation are usually performed on a statistical ensemble and is expected that values measured for a given sample should correspond to the mean values calculated on the ensemble. In recent years it has become clear that fluctuation phenomena around the mean values are also very important. A new field has recently appeared in the subject of disordered systems which concerns quantum coherence transport phenomena in the diffusive elastic scattering regime, which at $T=0^0K$ is realized for systems with linear size L much greater than the mean free path l and much smaller than the localization length ξ. For $T\neq0^0K$ the system sizes must be smaller than a characteristic phase breaking length which is due to inelastic scattering. These phenomena are realised in small disordered metallic (mesoscopic) samples[6] and the related mesoscopic physics is characterized by large and universal properties of the statistics of fluctuations for the physical quantities. For example conductance fluctuations occur, reproducible for a given sample, which are described by a universal variance

$$<(\delta G)^2> = <(G-<G>)^2> \approx (e^2/h)^2. \tag{8}$$

This result is derived[6,7] for the weak localisation diffusive regime but holds for all mesoscopic systems and implies (if we use Eq.(1)) that the relative conductance fluctuations are

$$\delta G/<G> \propto L^{2-d}, \tag{9}$$

that is much larger than the normally expected statistical fluctuations $(\propto L^{-d/2})$ if d<4. These fluctuation effects are experimentally observed for very small samples at low T, in order to make inelastic scattering unimportant. At $T=1^0K$ samples with size of the order of 1μm are required and the observations are made on a single sample by varying an external magnetic field. It is clear that in order to describe theoretically the mesoscopic systems we must go beyond the average value or macroscopic description, that is the variance and even the higher moments of distributions must be considered. In this paper we consider the spectral and wave function distributions and study their statistical properties.

THE TIGHT-BINDING RANDOM MATRIX ENSEMBLES

An electron in a random potential is usually studied within a statistical ensemble of random one-electron matrix Hamiltonians, each corresponding to a specific microscopic configuration of disorder. The tight-binding approximation is used in an orthogonalised lattice-site basis representation including a spin-1/2 dependence. The diagonal matrix elements correspond to the site energies and are independent random variables chosen from a flat probability distribution of width W which describes the strength disorder. The off-diagonal hopping matrix elements, for every pair of nearest-neighbour sites, are represented by 2x2 complex matrices in spinor space. They take the form

$$\mathbf{V}=t^0.\mathbf{I}+i.\mu.(t^x.\sigma_x +t^y.\sigma_y +t^z.\sigma_z) , \tag{10}$$

where $\mathbf{I},\sigma_x,\sigma_y,\sigma_z$ are the identity and Pauli spin matrices forming a complete basis set. In the rest we choose $t^0=1$ and the t^x, t^y, t^z as independent random variables from a uniform probability distribution defined on the interval [-1/2,

1/2]. In such a way we can also describe the case of the random spin-orbit coupling whose strength is denoted by μ. The metal-insulator transition occurs at $W_c \approx 16.5$ when $\mu=0$ in d=3 and at $W_c \approx 7.0$ when $\mu=1$, in $d=2^5$. For $W>W_c$ all states are localized and the conductivity is zero, while for $W<W_c$ mobility edges appear in the band separating localised states near the edges from extended states near the band center.

Depending on the symmetry a classification scheme is possible for the tight-binding random matrix ensembles(TBME): the orthogonal ensemble(OE) in the case of a random potential, the unitary ensemble(UE) when a magnetic field breaking the time-reversal invariance is added and the symplectic ensemble(SE) when spin-orbit coupling is also present. Localisation and mobility edges occur in all three-dimensional tight-binding ensembles and in two-dimensions only for the symplectic and possibly for the unitary universality classes. This distinction shares many similarities with the symmetry classification of the Gaussian random matrix ensembles introduced by Wigner and Dyson[8]. It is important to note that the Wigner-Dyson ensembles consist of matrices with all their elements being independent Gaussian random variables whereas for the electronic models the matrix ensembles consists of short-ranged and sparse random matrices, that is with most of their matrix elements far from the main diagonal being zero. The Gaussian ensembles correspond to the high-dimensionality $(d \to \infty)$ limit of the TBME and are always metallic. As a consequence, it turns out that the metallic phase is well approximated, in any dimension, by the well-known results from the Gaussian random matrix theories[8]. It will be further shown that these results are ultimately responsible for the universal conductance fluctuations. In the rest we study the $OE(\mu=0)$ and the $SE(\mu \neq 0)$.

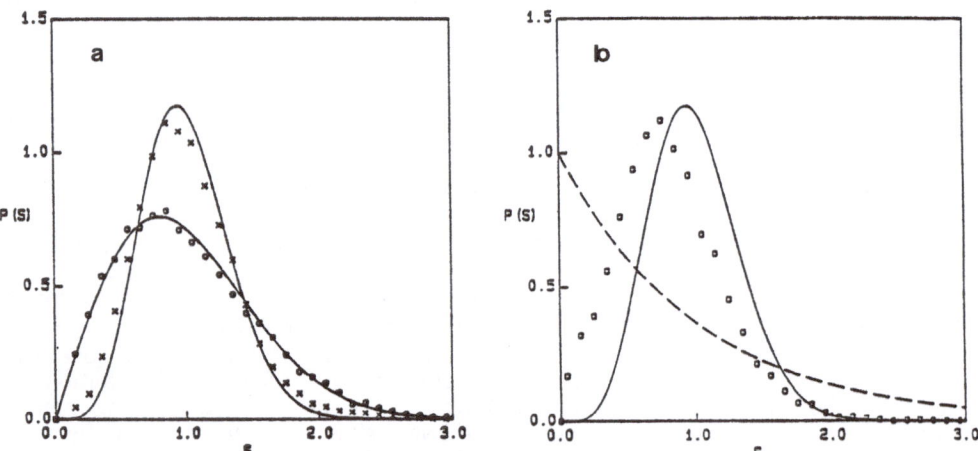

Fig.1(a). The calculated level-spacing distribution function without spin-orbit coupling W=4, $\mu=0$ (circles) and with spin-orbit coupling W=1, $\mu=1$ (crosses) in d=2 for the disordered metallic regime. The horizontal line is in units of the local mean-level spacing and the solid curves are the Wigner-Dyson formulae for the OE (linear repulsion, $\beta=1$) and SE (quadratic repulsion, $\beta=4$). They read $P_{OE}(S)=S\cdot(\pi/2)\cdot\exp(-\pi S^2/4)$ and $P_{SE}(S)=S^4\cdot(2^{18}/3^6\pi^3)\cdot\exp(-64S^2/9\pi^2)$.

1(b). As in Fig.1(a) but at the mobility edge (W=W_c=7, $\mu=1$). The circles are the numerical results and the lines are the corresponding metallic and insulating limits.

Spectral fluctuations

We have studied the most common spectral fluctuation measures for the TBME which are the nearest-level spacing distribution function P(S) and the Δ_3-statistics[8], which measures higher order level correlations. Firstly we consider the distribution function P(S) of the nearest-energy-level spacings $S_n=E_{n+1}-E_n$ measured in units of the local mean level spacing Δ. The details of the calculations can be found in Ref.9 but the results are displayed in Fig.1(a). It can be seen that, at least approximately, they follow the Wigner-Dyson[9-11] statistics characterised by the appropriate universality class index β which takes the values 1 and 4 for the OE and SE, respectively. We also studied the critical level statistics. The results shown in Fig.1(b). indicate that around the mobility edge an intermediate law for P(S) is obtained. On the other hand, the insulating phase is described by a simple Poisson law $P(S)=\exp(-S)$. The Anderson transition in this respect resembles the quantum chaotic transition[12].

It can be shown that the fluctuations for the density of states are partially responsible for the conductance fluctuations[10]. Although the averaged density of states is insensitive to the Anderson transition its higher moments are not. Using Eq.(6) we can obtain the averaged conductance for samples of linear size length L. Equivalently we can study fluctuations around $<N(E_T)>$, the averaged number of energy-levels in a band of width E_T. E_T can be numerically estimated by the energy shift due to variation in the boundary conditions of the sample[2]. The quantity of interest is the variance

$$<(\delta G)^2> = (e^2/h) \cdot <(\delta N(E_T))^2>. \tag{11}$$

Alternatively we may consider the $<\Delta_3>$-statistics which measures long range correlations. The numerical results are shown in Fig.2. For the correlated spectrum in the mesoscopic regime they agree well with the

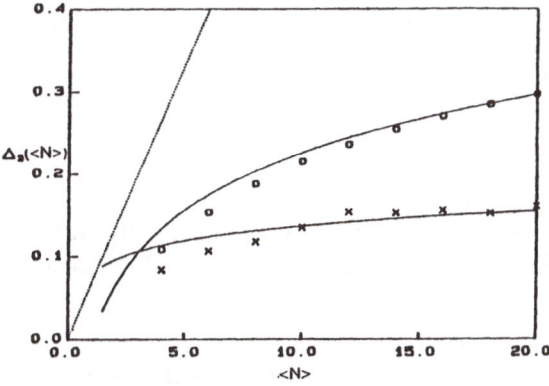

Fig.2. The calculated Δ_3-statistics for the same parameters as in Fig.1(a).

corresponding analytical expressions of Dyson[8] $<\Delta_3(<N>)>=(1/\pi^2)\cdot\ln(<N>)-0.007$ and $<\Delta_3(<N>)>=(1/4\pi^2)\cdot[\ln(4\pi<N>)+\gamma+1+(\pi^2-18)/8]$, $\gamma=0.57721$. This implies that the value of the variance $<(\delta N)^2>$ is of order one. In the insulating phase $<\Delta_3>=<N>/15$ and the spectrum is uncorrelated. The conductance fluctuations in this case dramatically increase since the variance $<(\delta N(E_T))^2>$ is of order $<N>$[10] as expected from normal statistical considerations. It was noted by Altshuler[10], who made more detailed studies, that this relationship between the mean and the variance also holds at the critical point. We saw that although the mesoscopic fluctuations in the energy-level density are very small (of order one) they can account for the large mesoscopic conductance fluctuations. Moreover due to the validity of the random matrix theory for the disordered metallic (mesoscopic) systems a very striking, experimentally accessible result is immediatelly extracted: the magnitude of the variance for the conductance is reduced by a factor of $\beta(=1,2,4)$ when changing universality class[13].

Wave function fluctuations

A simple way to characterize numerically the wave functions is to evaluate the the Haudorff dimension and estimate the degree of localization, for example by computing the Lyapunov exponents. A method which generalizes all these approaches by using the full information is by means of the D_q scaling exponents and the corresponding α-$f(\alpha)$ spectrum of scaling indices[14,15]. We utilise the fact that one again has to deal with distributions rather than simple averages. The different moments for the distribution of the wave function amplitude are considered and shown to scale in different ways at the modility edge and the infinite set of non-simply related scaling exponents D_q is obtained. There is numerical evidence[16] favouring the fact that the wave function at the mobility edge of disordered systems can be described in terms of multifractal measures, as it was done for mobility edges in incommensurate systems[17,18].

DISCUSSION

In this article we presented a fluctuation-correlation study of the spectra and wave functions in disordered lattice electronic systems in connection with the phenomenon of Anderson localisation via the introduction of the TBME. We give numerical evidence in favour of the Wigner-Dyson random matrix theory description for the disordered metallic phase. Consequently a simple expanation of the mesoscopic conductance fluctuations follows. Our results also point towards a connection between the Anderson transition and the problem of quantum dynamics in classically chaotic systems[12]. Clearly further work is needed to complete the present picture of fluctuation properties at the metallic and insulating phases as well as at the Anderson transition and explore the consequences for quantum transport. The methods and the results briefly presented here may serve as a guide towards the resolution of this important problem.

ACNOWLEDGMENTS: This work was supported by Research Grants from NATO and the Greek Secretariat of Science and Technology (Π.EN.E.Δ.).

REFERENCES

1. P.W.Anderson, Phys. Rev. 109, 1492 (1958)
2. D.J.Thouless , Phys.Rep. 13 93 [1974] and D.J.Thouless, in "Ill-Condensed Matter, Les Houches 1978 Session XXXI, Eds R Balian, R. Maynard and G. Toulouse [North Holland, Amsterdam, 1979]
3. E.Abrahams, P.W. Anderson, D.C. Licciardello and T.V.Ramakrishnan, Phys.Rev.Lett.42, 673, (1979)
4. S. Hikami, A.I.Larkin and Y.Nagaoka, Physics Reports 67, 15 (1980)
5. S.N.Evangelou and T.A.L.Ziman, Phys.C20, L235 (1987)
6. P.A.Lee and A.D.Stone, Phys. Rev. Lett 55, 1622 (1985)
7. B. L. Altshuler, JETP Lett 41, 648 (1985)
8. C.E.Porter, in "Statistical Theories of Spectra: Fluctuations", (Academic Press, New York,1965)
9. S.N.Evangelou, Phys.Rev.B.39, 12895 (1989) and References therein

10. B.L.Altshuler, Proc. 18th Int. Conf. on Low Temperature Phys.Kyoto Jap. Journal of Appl.Phys. Supp. 26, 1938 (1987) and References therein
11. U.Sivan and Y.Imry, Phys.Rev.B35, 6074 (1987)
12. B.Echardt, Phys.Rep. 163, 205 (1988)
13. J.-L.Pichard, N.Zanon, Y.Imry and A.D.Stone, J.Phys.France 51, 587 (1990)
14. T.C.Halsey, M.H.Jensen, L.P.Kadanoff, I.Procaccia and B.Shairman, Phys. Rev. A33, 1141 (1986)
15. H.E. Stanley and P. Meakin Nature 335, 405 (1988)
16. S.N.Evangelou, J.Phys.A.: Math. and Gen. (to appear)
17. S.N.Evangelou, J.Phys. C20, L295, (1987)
18. S.N.Evangelou in "Disordered Systems and New Materials", Eds N.Kirov, M.Borissov and A.Vavrek, World Scientific, p783 (1989)

LOCAL MOMENTS AND THE LOCALIZATION OF ELECTRONS IN LIQUIDS

David E. Logan

Oxford University
Physical Chemistry Laboratory
South Parks Road
Oxford OX1 3QZ

An understanding of the electronic properties of liquids is a goal of considerable current interest, see e.g. [1,2]. The conventional theoretical methods of crystalline solid state physics must be transcended, with explicit account taken of both the disorder and local structure characteristic of the liquid state. In addition, an independent electron description is not generally adequate to account for observed properties, and the effects of electron–electron interactions require consideration.

A topologically disordered generalization of the familiar Hubbard model is considered, with off–diagonal disorder in the one–electron transfer matrix elements arising from the disorder in the atomic centre of mass positions. The latter is determined by a canonical distribution for particles interacting via an appropriately chosen classical pair potential, the inherent disorder and structural characteristics of a liquid thus being introduced. The effects of electron correlation are treated at a generalized Hartree–Fock level. Following Cyrot [3], local magnetic moments are recognized as the first effect of electron correlation and are therefore introduced from the outset. A simplified description is employed in which the resultant spin disorder is caricatured by a self–consistently determined one–body local moment probability distribution. This maps the problem approximately onto a random Anderson model with both off–diagonal and uncorrelated, but self–consistently determined, site–diagonal disorder.

A discussion is then given of predicted electrical and magnetic properties of the system, and their evolution with number density and both structural and electronic parameters of relevance, as determined via techniques based on averaged Green functions and, for localization, a probabilistically based approach. Particular attention is given to the metal–insulator transition due to Anderson localization of the HF pseudoparticle states. It is argued that electrical and magnetic properties of the system in the insulating and dirty metallic domains should be correlated, for within the description employed the metal–insulator transition is due to an interplay between the localizing effects of disorder and the effects of electron correlation which, in giving rise to local magnetic moments, lead to a pseudogap in the density of states in which the Fermi level lies and the insulator–metal transition occurs.

The theory is used to interpret bulk experiments [2] on expanded fluid alkali elements produced by expansion along the liquid gas coexistence curve, with particular emphasis on the behaviour of, and observed correlation between, the density dependences of the d.c. conductivity and the paramagnetic susceptibility from the low density insulating domain, across the insulator–metal transition occurring around the critical density, ρ_{crit}, and through the dirty metallic regime up to densities around $2\rho_{crit}$ where the metal becomes 'clean'. The broad picture which emerges is similar to one of the possible scenarios suggested by Freyland [4] to explain the alkali experiments, and specific implications of the theory are shown to agree rather well with observation. Although most attention is given to the extensively studied case of Cs, inferences one can make about the behaviour of Rb appear to accord with experiment.

Full details will be given in a forthcoming publication [5].

References

1. Stratt, R.M., 1990, <u>Ann. Rev. Phys. Chem.</u>, <u>41</u>.
2. Freyland, W. and Hensel, F., 1985, in "The Metallic and Nonmetallic States of Matter", P.P. Edwards and C.N.R. Rao eds., Taylor and Francis, London.
3. Cyrot, M., 1972. <u>Phil. Mag.</u>, <u>25</u>, 1031.
4. Freyland, W., 1981, <u>Commun. Solid State Phys.</u> , <u>10</u>, 1.
5. Logan, D.E., 1990, Submitted for publication.

How Universal is the Scaling Theory of Localization ?

Michael Schreiber

Institut für Physikalische Chemie, Johannes-Gutenberg-Universität
Welderweg 11, D-6500 Mainz, F.R.Germany

The numerical implementation of the one-parameter scaling theory of
localization is reviewed for the Anderson model of disordered solids. A
finite-size scaling procedure is used to derive the 3D localization
length and d.c.-conductivity from the raw data computed for quasi-1D
systems by the strip-and-bar method. While a common scaling function
can be unambiguously obtained for different distributions of the
diagonal disorder in the Anderson model, discrepancies appear between
the box and the Gaussian distribution with regard to the derived
critical exponents. To discuss these effects, new results are presented
for a triangular distribution, and a new method for the computation of
the critical exponents is introduced, which yields larger values than
previously obtained.

1. INTRODUCTION

The problem of localization of electronic states in disordered systems
is centered on the distinction between extended and localized states.[1] While
the former contribute to transport even in the limit of vanishing
temperature, the latter do not. For a given (Fermi) energy, a critical
disorder W_c (or, vice versa, for a given disorder, a critical energy E_c)
separates the conducting regime from the non-conducting. Describing the
respective metal-insulator transition as a second-order phase transition one
expects the inverse d.c.-conductivity σ^{-1} and the localization length λ to
diverge according to power laws

$$\sigma^{-1} \propto (W_c-W)^{-s} \quad \text{and} \quad \lambda \propto (W-W_c)^{-\nu} \tag{1}$$

$$\text{or} \quad \sigma^{-1} \propto (E_c-E)^{-s'} \quad \text{and} \quad \lambda \propto (E-E_c)^{-\nu'}. \tag{2}$$

In principal, all four critical exponents could be different. Using scaling
arguments,[2,3] however, it is possible to equate these exponents.

Large-Scale Molecular Systems, Edited by W. Gans *et al.*
Plenum Press, New York, 1991

It is an essential point of the one-parameter scaling hypothesis[3] that a characteristic quantity g exists which describes the disordered system irrespective of the values of the parameters E and W. Originally the conductance g itself has been taken as this "scaling" variable. It is assumed to be characteristic in the following way: If the size of a given system is altered, the change in the behaviour of the system is governed by a function $\beta(g)$ which depends explicitly only on g, but not on, e.g., the values of E and W which determine g. If one accepts the assumptions of the scaling hypothesis, then the critical exponents should not be contingent upon the way how the metal-insulator transition is approached, i.e., $s = s'$ and $\nu = \nu'$.

The numerical investigation of the problem of localization is commonly based on an evaluation of the Anderson Hamiltonian

$$H = \sum_{n} | n > \varepsilon_n < n | + \sum_{m,n}^{n.n.} | m > < n | \tag{3}$$

which describes a regular lattice in tight-binding approximation with nearest neighbour transfer only (taken as the unit of energy) and a random distribution of site energies ε_n. The numerical implementation[4] of the scaling ideas in this model does not start from g, but uses a different scaling variable Λ; this raises no difficulties if a one-to-one relation between Λ and g can be postulated. Λ is defined as the ratio of the localization length λ_M and the width M of a very long strip or bar:

$$\Lambda = \lambda_M / M \tag{4}$$

As all wave functions in a 1D system with finite disorder are localized, such a quantity can always be defined for the investigated quasi-1D systems. A finite-size scaling procedure is then employed to extrapolate towards infinite width: $M \rightarrow \infty$, thus the 3D systems can be reached. The respective scaling behaviour of Λ is described by a numerically obtained scaling relation

$$\Lambda = f\left(\xi(E,W) / M \right) \tag{5}$$

which depends explicitly on a scaling parameter ξ, but only implicitly on E and W. This characteristic length ξ can easily be identified as the 3D localization length λ in the insulating case and can also be associated[4] with the inverse d.c.-conductivity σ^{-1}. If the scaling behaviour is the same on both sides of the transition (as it should be) the critical exponents s and ν are identical.

The actual size of the exponents has been determined from the non-linear σ-model:[5] In lowest order the ε-expansion with $\varepsilon = d-2$ yielded[5]

$$s = \nu = 1 \tag{6}$$

in agreement with a diagrammatic expansion starting from the metallic limit.[6] This result should be universally valid for a certain "symmetry class" of the underlying Anderson model. In particular, eq. (6) should hold for all 3D systems in the orthogonal case of non-interacting electrons without magnetic fields. In this case it is also clear from the ε-expansion that $d = 2$ is the limiting dimension in which all states are localized even for very small disorder. Breaking the time-reversal invariance under the influence of a magnetic field or magnetic impurities, i.e., in the unitary universality class, delocalized states appear already in 2D systems, yielding[7] $s = 1/2$. In the symplectic class, which is appropriate for the description of systems with spin-orbit coupling, a tendency towards weak antilocalization can be observed.[8]

The present investigation, however, is restricted to the orthogonal case. Here, the results were expected not to depend on the probability distribution of the diagonal matrix elements in the Anderson Hamiltonian.[9] But serious doubts have arisen about the universality of the scaling theory: An investigation[10] of the critical behaviour around the Anderson transition has shown that the critical exponents on both sides of the transition do indeed agree, but deviate considerably for different distributions. A detailed analysis[11] of highly accurate data yielded $s = \nu = 1.4 \pm 0.2$ for the uniform and $s = \nu = 0.9 \pm 0.3$ for the normal distribution. The latter result is still in agreement with eq. (6), but the error bars of the former are small enough to exclude an agreement.

On the other hand, as the evaluation of the effective Lagrangian in the non-linear σ-model, which lead to eq. (6), is based on an ensemble averaging over the random potentials in (3), taking only lowest order terms into account, the result is strictly valid only for the normal distribution.[12] For other distributions, in particular the uniform distribution with its discontinuities, higher moments may increase the importance of higher order terms, thus yielding deviating indices.

It is the purpose of this presentation to review the numerical results obtained for the uniform and the normal distributions and to complement them by an investigation of the Anderson model with random site energies distributed according to a triangular distribution. We have chosen the triangular distribution, because as a convolution of two uniform distributions it is the first step towards the normal distribution (which can be obtained by an infinite sum of uniformly distributed random numbers). It is therefore reasonable to expect that important features of both limits (the uniform and the normal) are somehow reflected. Another extreme case which we shall investigate in the future is of course given by the Lorentzian distribution with its infinite second moment.

2. COMPUTATION OF THE RAW DATA

As mentioned above, we employ the Anderson Hamiltonian (3) with a box (or uniform)

$$P(\varepsilon) = W^{-1} \theta(W/2 - |\varepsilon|), \tag{7}$$

a Gaussian (or normal)

$$P(\varepsilon) = \sqrt{6/\pi} \, / W \, \exp(-6\varepsilon^2/W^2), \tag{8}$$

or a triangular distribution

$$P(\varepsilon) = (\sqrt{2} - 2|\varepsilon| \, / \, W) \, / \, W \;\; \theta(W/\sqrt{2} - |\varepsilon|), \tag{9}$$

where θ denotes the usual step function, and the normalized distribution functions are scaled in such a way that the second moments are identical. It is our purpose to describe the development of the wave functions through a long bar of cross-section $M * M$ sites k. The Schrödinger equation corresponding to the Hamiltonian (3) is written as an initial value problem.[4] Starting on the first slice with M^2 orthogonal states j, which are described by their expansion coefficients in an $M^2 * M^2$ matrix $A_{jk}(1)$, one can then determine the wave functions on the $(L+1)$-th slice recursively from the transfer matrix equation

$$\begin{bmatrix} A(L+1) \\ A(L) \end{bmatrix} = T_L \begin{bmatrix} A(L) \\ A(L-1) \end{bmatrix}, \quad T_L = \begin{bmatrix} E-H_M(L) & -1 \\ 1 & 0 \end{bmatrix} \tag{10}$$

where $H_M(L)$ comprises the matrix elements on the L-th layer. The transfer matrix of the whole bar

$$\mathcal{T}_L = \prod_{L=1}^{L} T_L \tag{11}$$

diverges exponentially, but a limiting matrix

$$T = \lim_{L\to\infty} (\mathcal{T}_L \mathcal{T}_L^+)^{1/2L} \tag{12}$$

exists, because $\mathcal{T}_L$ satisfies the theorem of Oseledec. As T_L and therefore T is a symplectic matrix, its eigenvalues appear pairwise and can be written as $exp(\tau_i)$ and $exp(-\tau_i)$. τ_i are the Lyapunov exponents which characterize how the initial states "drift apart" exponentially. The inverse values reflect the different characteristic length scales. The largest length (i.e., the inverse of the smallest positive Lyapunov exponent) measures the largest possible extension of a state at a given energy, i.e., the localization length λ_M.

In this way, a large amount of data for a variety of parameters E, W, and M has been accumulated in the last years.[13] In Fig. 1 the recently obtained data for the triangular distribution are presented. The cross-section has been limited by $M = 13$ as usual in our calculations, because the amount of computation increases as M^7 at least. It should be noted, however, that systems with cross-sections up to $M * M = 6 * 6$ have always shown strong finite-size effects, which can easily be identified by

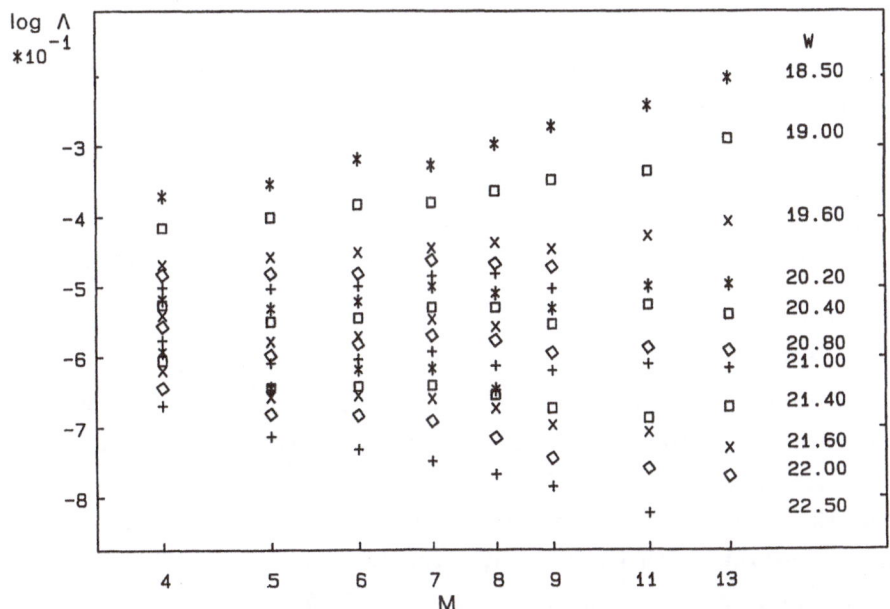

Fig.1. Raw data of the localization length $\lambda_M/M = \Lambda$ versus M for several disorders W of the triangular distribution in the band centre ($E = 0$).

changing the employed periodic boundary conditions into anti-periodic or non-periodic. For larger energies this problem is certain to occur up to $M =$ 9 and may be severe for even larger systems.[11] A very careful analysis is therefore necessary to exclude misleading data points. The length of our systems has been extended until the recursively determined relative error of λ_M was below 1%. Although the error bars of data for different M or W still overlap (see Fig.1), a further improvement is again restricted by the available computer resources, because the error enters the computation time quadratically. Unfortunately, the smallest (positive) Lyapunov exponent is the one most sensitive to a loss of significance due to rounding errors so that special care has to be taken with respect to the accuracy of the accumulated transfer matrix (11), otherwise a fast convergence towards wrong values of λ_M occurs.

The raw data in Fig. 1 can now be used to distinguish localized and extended states: if λ_M/M decreases with increasing M, the state is already more or less confined within the bar, i.e., it is localized. If, on the other hand, the obtained localization length increases faster than M, the state grows more and more towards an extended state in the infinite system. Accordingly the topmost data in Fig. 1 belong to extended states, the data at the bottom to localized states.

Disregarding the data for $M < 6$ because of finite-size effects, one can determine the metal-insulator transition to occur at $W_c = 20.5 \pm 0.5$. This result is close to the critical value obtained for the normal distribution $W_c = 20.95 \pm 0.1$, but not to the uniform case $W_c = 16.5 \pm 0.1$. It should be noted that the different values of the critical disorder cannot be interpreted as contradicting the universality of the scaling theory, because it is by no means clear that the second moments of the various distributions should be used to equate the distributions according to eqs. (7)-(9).

3. FINITE SIZE SCALING

As discussed in the introduction we assume that $\Lambda = \lambda_M/M$ is the relevant scaling variable and try to rescale the data for various W to obtain a common scaling relation (5). On the logarithmic scale of Fig. 1 changing the scale of M^{-1} via a factor simply means a horizontal shift by an amount commensurate with the scaling parameter $\xi(W)$. By a mean least squares fit of all available data for $E = 0$ it is indeed possible to determine the scaling curve shown in Fig. 2. The existence of a common scaling curve for different distributions has been taken[10] as a corroboration of the universality of the scaling theory.

390

A closer investigation of the critical region around the metal-insulator transitions which is given by the divergence of ξ in Fig. 2 shows, however, that deviations between the distributions are just not visible on the large scale of Fig. 2. For example, the critical value of Λ_c is given by 0.64 for the box[10], 0.6 for the Gaussian[11], and 0.58 for the triangular distribution (see Fig. 1), for energies at the band centre. For larger energies smaller values of Λ_c could be determined.[11]

4. THE CRITICAL BEHAVIOUR

The critical behaviour of the localization length and the inverse d.c.-conductivity is characterized by the critical exponents. It is straightforward but numerically very difficult to determine s and ν directly from the singularity of $\xi(W)$ by fitting the data to the power laws in eq. 1. The overall functional dependence of $\xi(W)$ can be easily obtained (see Fig. 4 of Ref. 10) from the mean least squares fit producing Fig. 2. But the values of $\xi(W)$ close to the transition are rather inaccurate, because the scale factor cannot be precisely determined for those data which are nearly undistinguishable within the error bars. The computational procedure will lead to a rounding of the singularity. The obvious alternative is to exclude data close to the singularity from the determination of the critical exponents. But then it is no longer sure, whether the data still belong to the critical regime.

To circumvent these problems we examine the raw data directly.

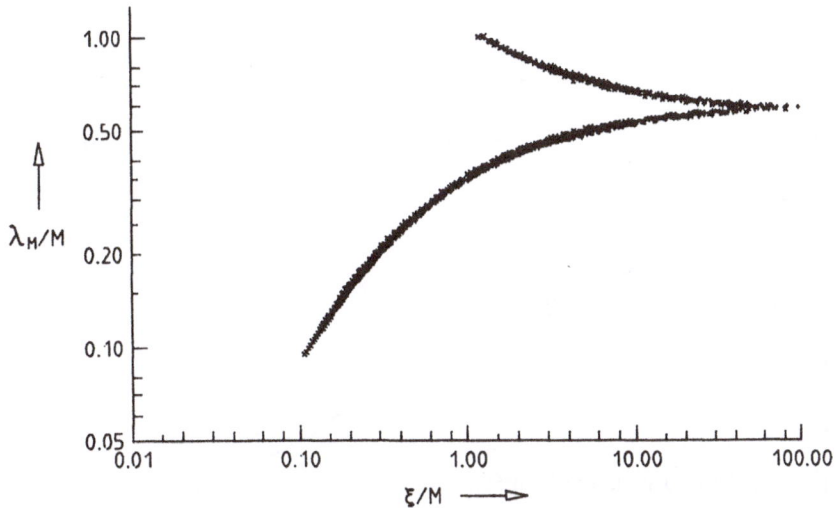

Fig. 2. Scaling relation for the Anderson model derived from the raw data obtained for the box and Gaussian distribution (see Fig. 2 of Ref. 10) including the data of Fig. 1 for the triangular distribution.

Requiring the existence of the scaling relation (5), one can expand the logarithmic derivative of Λ, the so-called χ-function

$$\chi (\ln \Lambda) = \frac{d \ln \Lambda}{d \ln M} \qquad (13)$$

which corresponds to the above-mentioned β-function into a Taylor series around the critical point $\chi = 0$. Considering this series only to first order, the differential equation

$$\chi = \frac{d \ln \Lambda}{d \ln M} = \chi' \, (\ln \Lambda - \ln \Lambda_c) \qquad \text{with} \qquad \chi' = \frac{d \, \chi}{d \ln \Lambda} \bigg|_{\Lambda_c} \qquad (14)$$

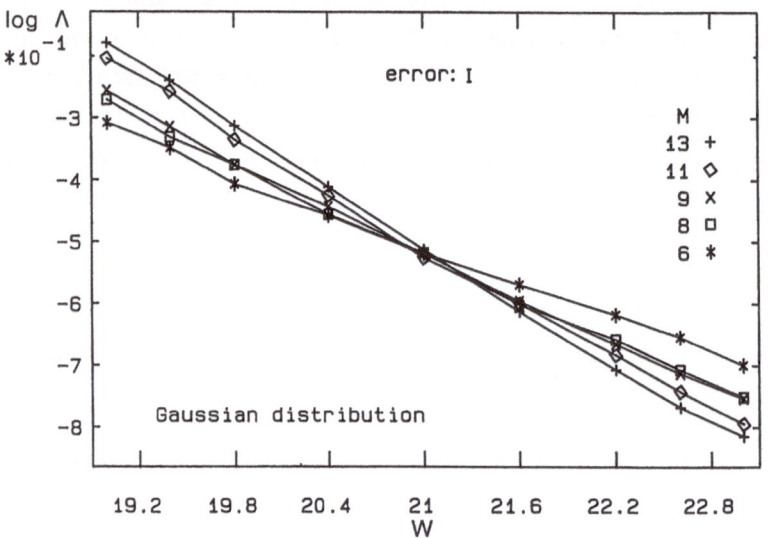

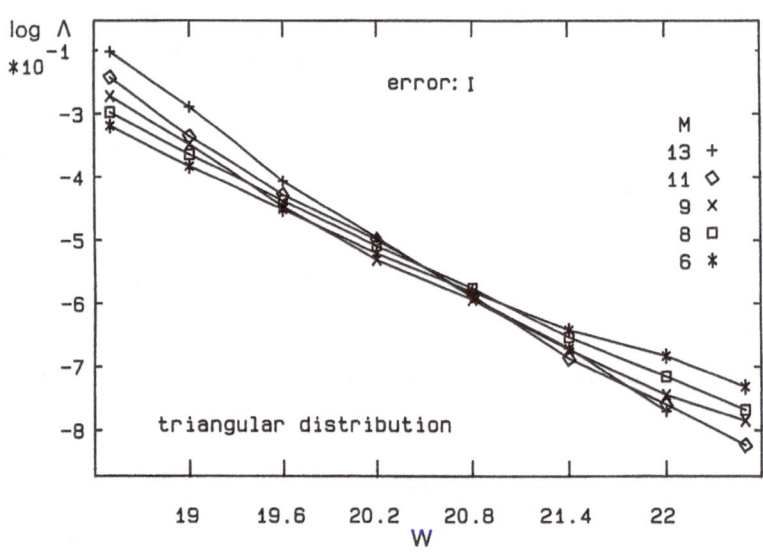

Fig. 3.

Dependence of the raw data on disorder for the Gaussian and the triangular distribution at $E = 0$, for various cross-sections $M * M$ of the quasi-1D bar. Connecting lines are for guiding the eye only.

Table 1. Parameters of the mean least squares fit of eq. (15) to the raw data. N is the number of values taken into account. For the determination of $\tilde{\chi}'$ see text.

Distribution	disorder W	N	Λ_c	c	W_c	$1/\chi'$	$1/\tilde{\chi}'$
uniform	15.5 - 17.5	168	0.581	-0.040	16.54	1.53	1.61
triangular	18.5 - 22.0	60	0.580	-0.031	20.50	1.52	1.66
normal	20.0 - 22.0	168	0.595	-0.018	21.05	0.14	1.18

results, which can be solved easily. Another expansion up to first order with respect to W yields the final result

$$\ln \Lambda = \ln \Lambda_c + c \, M^{\chi'} (W - W_c). \tag{15}$$

Here the analyticity of Λ with respect to M and W has been assumed, which implies the equality of the exponents on both sides of the transition. According to the scaling hypothesis (5), Λ should depend only on the ratio of ξ and M. Therefore the factor $(W - W_c)$ in eq. (15) must be proportional to $\xi^{-\chi'}$. According to eq. (1) we are thus able to derive $\nu^{-1} = s^{-1} = \chi'$.

Fig. 3 shows the dependence (15) for selected data out of our collection. We have previously[10,11] fitted interpolating polynomials to the data for each M. The intersection of all curves then directly yields the values of W_c and Λ_c. Comparing the steepness of the curves at W_c for different M, the exponent χ' can be determined. To avoid problems with establishing the intersection and to reduce the number of fit parameters, we now fit eq. (15) to all data simultaneously. With respect to the error analysis it is noteworthy that for a given exponent χ' the non-linear fit can be rewritten to yield a linear problem. We then vary the exponent χ' to optimize the fit. The results are presented in Table 1.

The computed statistical errors were less than 0.12 %, 1 %, and 0.05 % for Λ_c, c, and W_c, respectively. It is known, however, that systematic errors are considerably more significant in this analysis.[10,11] We have therefore tried several other fits. In particular, we have introduced a second-order term with respect to $W - W_c$ into eq. (15), because it is obvious from Fig. 3 that straight lines cannot be the best fit. It is important that this term does not contribute to the steepness at the critical point so that the determination of the critical exponent is not altered. The obtained exponents $\tilde{\chi}'$ are slightly smaller than χ' (see the last column of table 1). With this quadratic fit, a larger interval of disorder values could be taken into account, resulting in another slight increase. On the other hand, to further eliminate possible finite-size effects, we also excluded the smallest values of $M = 6$ and $M = 7$, achieving a 10% decrease of the critical exponents in agreement with previous investigations.[11] An exclusion of large M-values yielded only smaller changes. In all cases we observed that the fit parameters Λ_c and W_c varied very little.

In summary, our estimates of the critical exponents are 1.55 ± 0.15, 1.55 ± 0.15, and 1.15 ± 0.15 for the uniform, triangular, and normal distribution, respectively. These results are somewhat larger but still consistent with the previous investigation.[11] In spite of the large systematic errors the results show conclusively that uniform and normal distributions lead to different critical behaviour. That the triangular distribution does not differ from the box distribution in this behaviour can be attributed to the fact that the effects of the discontinuities of the derivative of $P(\varepsilon)$ are as severe as discontinuities in $P(\varepsilon)$ itself. In contrast, W_c is influenced most strongly by the density of states.[14] But in this respect the triangular and Gaussian distribution are similar, because the central peaks yield a high density in the centre of the band so that a rather large disorder is necessary to localize the states around $E = 0$.

In conclusion, although the scaling hypothesis has once more been corroborated, it could also be shown that the critical behaviour depends significantly on the kind of disorder which enters the Anderson Hamiltonian as a random potential. Consequently one-parameter scaling cannot be established universally.

REFERENCES

1   P.W. Anderson, Phys. Rev. B **109**, 1492 (1958)
2   F.J. Wegner, Z. Phys. B **25**, 327 (1976)
3   E. Abrahams, P.W. Anderson, D.C. Licciardello, T.V. Ramakrishnan, Phys. Rev. Letters **42**, 673 (1979)
4   A. MacKinnon, B. Kramer, Phys. Rev. Letters **47**, 1546 (1981); Z. Phys. B **53**, 1 (1983)
5   F. Wegner, in: *Localization, Interaction and Transport Phenomena*, Eds. B. Kramer, G. Bergmann, Y. Bruynseaede, Springer Ser. Sol. St. Sci. **61**, 99 (1985)
6   D. Vollhardt, P. Wölfle, Phys. Rev. B **22**, 4666 (1980); Phys. Rev. Letters **48**, 699 (1982)
7   R. Oppermann, K. Jüngling, Phys. Lett. **76**, 449 (1980); Z. Phys. B **38**, 93 (1980)
8   A. MacKinnon, in: *Localization, Interaction and Transport Phenomena*, Eds. B. Kramer, G. Bergmann. Y. Bruynseaede, Springer Ser. Sol. St. Sci. **61**, 90 (1985)
9   F. Wegner, private communications
10 M. Schreiber, B. Kramer, A. MacKinnon, Phys. Scripta T **25**, 67 (1988)

11 B. Kramer, K. Broderix, A. MacKinnon, M. Schreiber, Physica A **161** (1990), in print

12 R. Oppermann, private communications

13 see Ref. 10, 11 and the references therein

14 B. Bulka, M. Schreiber, B. Kramer, Z. Phys. B **66**, 21 (1987)

DIMERS, REPULSIONS, AND THE ABSENCE OF LOCALISATION

Philip Phillips[1], H.-L. Wu[1], and David Dunlap[2]

[1]Department of Chemistry, Rm. 6-223 Massachusetts Institute of Technology
    Cambridge, Mass. 02139
[2]Dept. Phys. Univ. New Mexico, Alberquerque, New Mexico 87131

Abstract: We review here two models we have recently proposed which do not conform to the standard view that disorder precludes long range transport in 1-dimension. The two models we consider are the repulsive binary alloy and the random dimer model. We show that the mean-square displacement of an initially-localised particle in either of these models will grow in time as $t^{\frac{3}{2}}$. Transport obtains in both models because $\sqrt{N}$ of the electronic states are unscattered by the disorder. The relevance of these models to the conductivity of Fibonacci semiconductor lattices and polyaniline is discussed.

1. Introduction

The Anderson model for site-diagonal disorder has proven to be of fundamental importance in our understanding of the role disorder plays in insulator-metal transitions in a wide range of materials, most notably Si:P[1,2]. A well-known result of the Anderson model for site-energy disorder is the vanishing of the diffusion constant of an initially localised particle for any amount of disorder in one and two dimensions. In this lecture we shall consider two models[3] which do not conform to this view. Both models we consider can be thought of as variations of a tight-binding description of the standard binary alloy. Let us first introduce the random dimer model. Consider a binary alloy consisting of two types of species a and b with corresponding site energies ε_a and ε_b, respectively. The random dimer model is simply a binary alloy in which at least one of the site energies is assigned at random to pairs of lattice sites, that is, two sites in succession. Let V be the constant nearest-neighbour matrix element. We show by numerical simulation that the mean-square displacement of an initially localised particle in the random dimer model will grow as $t^{\frac{3}{2}}$ provided that $-2V < \varepsilon_a - \varepsilon_b < 2V$. Diffusion is shown to obtain when $\varepsilon_a - \varepsilon_b = \pm 2V$. In all other cases, the particle remains localised at long times. The random dimer model is shown to be relevant to transport in the conducting polymer polyaniline. The other model we consider is, as we will see, the dual of the random dimer model. In this model, a and b are assigned at random to the lattice sites with the constraint that the b's, for example, do not occur on consecutive lattice sites. Such a situation would arise naturally in the standard binary alloy if strong repulsions existed between the b's. Let V_a be the nearest-neighbour matrix element connecting two sites assigned the energy ε_a and V_c the matrix element connecting neighbouring sites with energy ε_a and ε_b. We show that the repulsive binary alloy possesses a surprising localisation-delocalisation transition when $V_a |\varepsilon_a - \varepsilon_b| \leq 2 |V_a^2 - V_c^2|$ regardless of the spatial dimension. In one dimension when the strict inequality holds, superdiffusion obtains with the mean-square displacement growing asymptotically in time as $t^{\frac{3}{2}}$. Diffusion occurs when the equality is satisfied and localisation otherwise. We show that the repulsive binary alloy predicts that transmission resonances should be experimentally observable in Fibonacci semiconductor lattices.

2. Random Dimer Model

To determine the dynamics of an electron in the random dimer model (RDM), we numerically integrated the equations of motion

Large-Scale Molecular Systems, Edited by W. Gans *et al.*
Plenum Press, New York, 1991

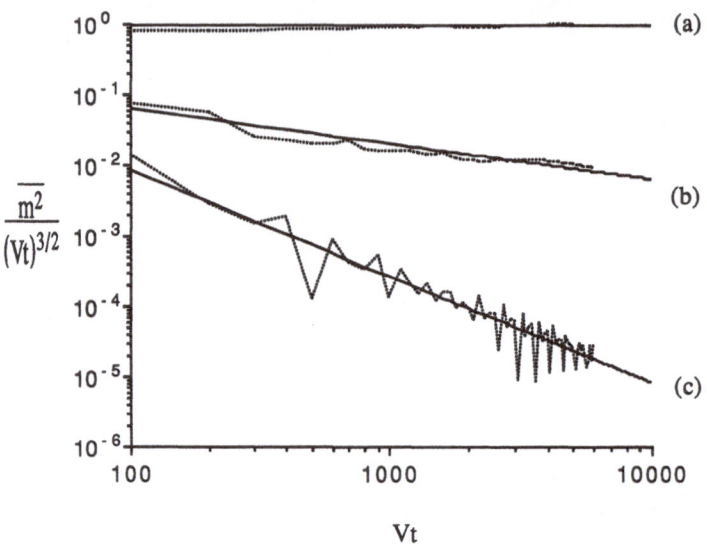

Fig. 1 Results from numerical simulations of the mean-square displacement of an initially
localized particle for a disordered 1-dimensional lattice with the disorder
determined by the three points (a), (b) and (c) .

$$i\dot{c}_n = \varepsilon_n c_n + V(c_{n+1} + c_{n-1}) \tag{1}$$

for the site amplitudes $C_n(t)$ and calculated the mean-square displacement

$$\overline{m^2} = \sum_m m^2 |c_m|^2 \tag{2}$$

Our calculations were performed on a self-expanding chain with the localised initial condition $C_0(t=0) =1$. The self-expanding chain was used to minimise end effects. The results from several random samples are shown in Fig. 1. The site energies were chosen from the bi-valued distribution $\varepsilon_n = \varepsilon_a$ and $\varepsilon_n = \varepsilon_b$ with equal probability. Fig. 1 compares the mean-square displacement in the random dimer model for three different values of $\varepsilon_a - \varepsilon_b$: a) $\varepsilon_a - \varepsilon_b = V$, b) $\varepsilon_a - \varepsilon_b = 2V$, and c) $\varepsilon_a - \varepsilon_b = 3V$. The dotted curves are the data from single numerical simulations, whereas least-square fits of the data with an expression of the form $\overline{m^2} = A(Vt)^b$ are represented by solid lines. The constant A was allowed to vary, while the exponent b was set to 3/2, 1 and 0 for the cases (a), (b), and (c) respectively. The subsequent straight-line fit through the data in each of these cases indicates that the transport properties are as advertised, namely superdiffusive for $-2V < \varepsilon_a - \varepsilon_b < 2V$, diffusive for $\varepsilon_a - \varepsilon_b = \pm 2V$ and localised otherwise.

Probably the simplest way to understand our results is to consider an otherwise ordered lattice with a single dimer defect. Let us place the dimer on sites 0 and 1. We assign the energy ε_a to all the sites except sites 0 and 1. Let the energy of sites 0 and 1 be ε_b. A constant nearest neighbour matrix element V mediates transport between the sites. We first show that $\sqrt{N}$ of the electronic states are unscattered by the dimer impurity. Then, we construct explicitly the unscattered states in a lattice containing randomly placed dimers. To proceed we calculate the reflection and transmission coefficients through the dimer impurity. Let us write the site amplitudes as $C_n = e^{ikn} + Re^{-ikn}$ for $n \leq -1$ and $C_n = Te^{ikn}$ for $n \geq 1$ where R and T are the reflection and transmission amplitudes, respectively. From the eigenvalue equation for sites -1 and 1, it follows that

$C_0 = 1 + R = T(\varepsilon_- e^{-ik} + V)/V$ with $\varepsilon_- = \varepsilon_a - \varepsilon_b$. Substitution of this result into the eigenvalue equation for site 0 results in the closed expression

$$|R|^2 = \frac{\varepsilon_-^2 (\varepsilon_- + 2V\cos k)^2}{(\varepsilon_-^2 - 2V^2 \sin^2 k + 2V\varepsilon_- \cos k)^2 + V^2 (V\sin 2k + 2\varepsilon_- \sin k)^2} \tag{3}$$

for the reflection probability. The reflection coefficient vanishes then when $\varepsilon_a - \varepsilon_b = -2V\cos k$ or equivalently when $-2V \leq \varepsilon_a - \varepsilon_b \leq 2V$. The location in the parent ordered band of the perfectly transmitted electronic state corrsponds to the wave vector $k_0 = \cos^{-1}[(\varepsilon_b - \varepsilon_a)/2V]$. Of course, no transport would obtain if only a single electronic state remained unscattered. To determine the total number of states that behave in this fashion, we expand R around k_0. To lowest order we find that in the vicinity of k_0, $|R|^2 \sim (\Delta k)^2$ where $\Delta k = k - k_0$. Consider now a crystal containing a certain fraction of randomly-placed dimer impurities. Electronic states in the vicinity of k_0 will be reflected with a probability proportional to $(\Delta k)^2$. The time between scattering events τ is inversely proportional to the reflection probability. As a result, in the random system, the mean free path $\lambda = \langle \text{velocity} \rangle \tau \sim 1/(\Delta k)^2$ in the vicinity of k_0[4]. Let $\Delta k = \Delta N / 2\pi N$. Upon equating the mean free path to the length of the system (N), we find that the total number (ΔN) of states whose mean free path is equal to the system size scales as $\Delta N = \sqrt{N}$. Because the mean free path ~ localisation length in 1 dimension, we find that the total number of states whose localisation lengths diverge as $\sqrt{N}$. Consequently, in the random dimer model $\sqrt{N}$ of the electronic states remain extended over the total length of the system. Such states move through the crystal ballistically with a constant group velocity $\langle v(k) \rangle$ except when they are located at the bottom or the top of the band where the velocity vanishes. Because all the other electronic states are localised, the diffusion constant is determined simply by integrating $v(k)\lambda(k)$ over the width of k-states that participate in the transport. The upper limit of the integration is then proportional to the total fraction of unscattered states or $1/\sqrt{N}$ and $\lambda(k) \sim N$. In the case when the velocity is a nonzero constant, we obtain that the $D \sim \sqrt{N}$. Because the states which contribute to transport traverse the length of the system with a constant velocity, t and N can be interchanged or $D \sim t^{\frac{1}{2}}$. Consequently the mean-square displacement grows as $t^{\frac{3}{2}}$. At the bottom or the top of the band where the group velocity vanishes, $v(k) \sim k$ and $D \sim O(1)$.

We now construct explicitly the unscattered states. In absence of any dimer impurities, the eigenstates are simply Bloch states of the form e^{ikn}. When the dimer impurities are present because the eigenstates in the vicinity of k_0 have unit transmission, it must be the case that these states are still of the Bloch form. These states can be constructed as follows. Consider the single dimer impurity case discussed earlier. The dimer impurity is located on sites 0 and 1. The unscattered state must be of the form e^{ikn} for $n \leq 0$ and $e^{ikn+i\Omega}$ for $n \geq 1$. That is, the only difference between the electron wavefunction before and after it has interacted with the impurity is that its phase changes by Ω. There is no reflected component. To determine Ω, we consider the eigenvalue equation, $E - \varepsilon_b = V(e^{ik+i\Omega} + e^{-ik})$, for site 0. The non-trivial solution occurs when $E - \varepsilon_b = 0$. Because E is the energy of the ordered band, $\varepsilon_a + 2V\cos k$, the vanishing of $E - \varepsilon_b = 0$ corresponds to the condition $-2V \leq \varepsilon_a - \varepsilon_b \leq 2V$. In this case, $\Omega = -2k + \pi$. Consequently, the Bloch state that satisfies the Schroedinger equation is e^{ikn} for $n \leq 0$ and $-e^{ik(n-2)}$ for $n \geq 1$ provided that $-2V \leq \varepsilon_a - \varepsilon_b \leq 2V$. The wavefunction on the second atom of the dimer is the negative of the a-type atom located on site -1. The unscattered state then is odd with respect to reflection around the first atom of the dimer. If another dimer were located on sites 2 and 3, the corresponding perfectly transmitted wave would be $\ldots e^{-2ik}, e^{-ik}, 1, -e^{-ik}, -1, e^{-ik}, 1, e^{ik}, e^{2ik}, \ldots$ provided of course that $-2V \leq \varepsilon_a - \varepsilon_b \leq 2V$. Such states can always be constructed regardless of the number of dimer impurities that are placed at random in the lattice. It is straightforward to verify that similar states cannot be constructed for single impurities in the absence of off-diagonal disorder.

An immediate application of the random dimer model is the conducting polymer polyaniline. Polyaniline refers to the general class of compounds that are formed from a series of nitrogen connected six-membered carbon rings of benzenoid or quinoid character[5]. Oxidation of the fully reduced form of the polymer results in a disordered bi-polaron lattice in which neighbouring nitrogen atoms are positively charged[5]. In the polaron latticer, then there are two kinds of nitrogen atoms: neutral N-H groups and positively charged N-H sites. The latter occurs in pairs. As a result, the conducting form of polyaniline is a random dimer model. In a forthcoming paper, we show that the experimentally observed non-ohmic dependence of the resistance on the electric field can be explained by the random dimer model[6].

3. Repulsive Binary Alloy

As in the random dimer model, we determine the transport properties of the repulsive binary alloy by computing the reflection and transmission coefficients on an otherwise ordered infinite 1-dimensional lattice with a single b-type defect. Let us place the defect with energy ε_b on site 0. All sites other than site 0 are assigned the energy ε_a. The matrix element V_a connects all nearest-neighbour sites except sites (-1,0) and (0,1) which are connected by V_c. We note at the outset that the sites immediately to the left and right of a single b-type defect will be connected to the defect with matrix element V_c. Consequently, the matrix elements in the repulsive binary alloy appear in pairs. In some sense then, this model is the dual of the random dimer model[3] discussed above. We now show that $\sqrt{N}$ of the electronic states have unit transmission coefficients through the single defect. Using the method discussed above for the random dimer model, we obtain

$$|R|^2 = \frac{(V_a\varepsilon_- - 2(V_a^2 - V_c^2)\cos k)^2}{(V_a\varepsilon_- - 2(V_a^2 - 2V_c^2)\cos k)(V_a\varepsilon_- + 2V_a^2\cos k) + 4V_c^2} \tag{4}$$

for the reflection probability, where $\varepsilon_- = \varepsilon_a - \varepsilon_b$ in the repulsive binary alloy. We see then that the reflection probability vanishes when $V_a|\varepsilon_a - \varepsilon_b| \leq 2\left|V_a^2 - V_c^2\right|$. The wavevector of the perfectly-transmitted wave is $k_0 = \cos^{-1}(\varepsilon_- V_a / 2(V_a^2 - V_c^2))$. Similarly, $|R|^2 \sim (\Delta k)^2$ in the repulsive binary alloy. As in the random dimer model then $\sqrt{N}$ of the electronic states are unscattered by the disorder. Our results are consistent then with the recent work of Flores[7] who has also shown that $\sqrt{N}$ of the electronic states in the dilute binary alloy have divergent localisation lengths. Based on the $\sqrt{N}$ unscattered states, it is straightforward to show that the transport properties of the repulsive binary alloy are identical to those of the random dimer model.

We now determine what the perfectly-transmitted states look like. We know they must be of the Bloch form. Consider the single defect problem discussed above with the defect located on site 0. The solution for the perfectly transmitted state must be of the form $C_n = e^{ikn}$ $n \leq -1$, $a_0 e^{i\Omega}$ for site 0, and $e^{ikn + i\Omega}$, $1 \leq n$. From the eigenvalue equation for site 0, it follows that $a_0 = V_a / V_c$ and $\Omega = 0$ provided that $V_a|\varepsilon_a - \varepsilon_b| \leq 2\left|V_a^2 - V_c^2\right|$. Consequently in the repulsive binary alloy, no phase change results when $\sqrt{N}$ of the electronic states scatter from the impurities; only the amplitude changes at each impurity. For example, if b-type impurities were located on sites 0,2,-2, the unscattered state would be $\ldots e^{3ik}, V_a/V_c e^{2ik}, e^{ik}, V_a/V_c, e^{ik}, V_a/V_c e^{2ik} \ldots$. In the random dimer model the unscattered states change phase at each impurity with no change in the amplitude.

Physical systems of considerable experimental interest that are special cases of the repulsive binary alloy are Fibonacci lattices formed from two types of materials such as GaAs and AlAs[8]. Let a=GaAs and b=AlAs. A Fibonacci sequence of a and b will of the form a, ab, aba, abaab, ... where the nth entry is a product of the (n-1)st and (n-2)nd. As is evident the b's will always appear singly. Hence, Fibonacci lattices of this sort are a subset of the repulsive binary alloy. Such systems have been fabricated experimentally[8] by molecular beam epitaxy because of the general interest in quasiperiodic order in crystals[9]. Previous tight-binding treatments[10] of Fibonacci lattices have failed to predict the resonances we have obtained here because they have treated either the site energies or matrix elements in a Fibonacci sequence. In the actual experimental system, the physical arrangement of the site energies and matrix elements will correspond to the sequence in the repulsive binary alloy, with the a-b site matrix elements appearing in pairs and the b site energies singly. Experimental confirmation of the resonances we predict here can be obtained by successively doping a Fibonacci heterostructure until the Fermi level coincides with the position of the perfectly unscattered states. Partial confirmation of these results are the resonances which have been found in Kronig-Penney models of Fibonacci lattices[11].

References

1. P. W. Anderson, Phys. Rev. **109**, 1492 (1958).
2. F. Wegner, Z. Phys. B **25**, 327 (1976); E. Abrahams, P. W. Abrahams, P. W. Anderson, D. C. Licciardello, and T. V. Ramakrishnan, Phys. Rev. Lett. **42**, 673 (1979); A. MacKinnon and B. Kramer, Phys. Rev. Lett. **47**, 1546 (1981).
3. D. H. Dunlap, H.-L. Wu, and P. Phillips, submitted, Phys. Rev. Lett.; H.-L. Wu, Phys. Rev. B. submitted; See also D. H. Dunlap, K. Kundu, and P. Phillips, Phys. Rev. B. **40**, 10999 (1989); D. H. Dunlap and P. Phillips, J. Chem. Phys. **92**, 6093 (1990).

4. S. Feng, private communication.
5. F. Zuo, M. Angelopoulos, A. G. MacDiarmid, and A. J. Epstein, Phys. Rev. B **36**, 3475 (1987).
6. H. -L. Wu and P. Phillips, in preparation.
7. J. Flores, J. Phys. Cond. Matt. **1**, 8471 (1989).
8. R. Clarke, T. Moustakas, K. Bajema, D. Grier, W. dos Passos, and R. Merlin, Superlattices and microstructures B4, 371 (1988).
9. D. Shechtman, I. Belch, D. Gratias, and J. W. Cahn, Phys. Rev. Lett. **53**, 1951 (1984).
10. M. Kohmoto, L. P. Kadanoff, and C. Tang, Phys. Rev. Lett. **50,** 1870 (1983); T. Odagaki and L. Friedman, Solid State Comm. **57,** 915 (1986).
11. J. Kollar and A. Suto, Phys. Lett. A. **117,** 203 (1986); F. Laruelle and B. Etienne, Phys. Rev. B. **37,** 4816 (1988).

ELECTRONIC EXCITATIONS IN POLYSILANES: FRENKEL EXCITONS OF A DISORDERED CHAIN [+]

A. Tilgner[*], H.P. Trommsdorff[*], J.M. Zeigler[**], and R.M. Hochstrasser[°]

[*]Laboratoire de Spectrométrie Physique, associé au C.N.R.S.
Université Joseph Fourier Grenoble I
B.P. 87, 3402 St. Martin d'Hères cedex, France
[**]Silchemy, 2208 Lester Dr. NE, Albuquerque, New Mexico 87112, U.S.A.
[°]Department of Chemistry, University of Pennsylvania
Philadelphia, PA 19104-6323, U.S.A.

INTRODUCTION

The soluble, disubstituted polysilanes, $(SiXY)_n$, present a class of novel materials with interesting electronic, optical, conformational and photochemical properties,[1] useful for a number of applications (photoresists, photoconductors, non-linear optical media). The lowest energy electronic states correspond to $\sigma\sigma^*$ excitations of the polymer backbone, delocalized over several tens of monomer units, and give rise to a strong absorption in the near UV. Strong electronic conformational coupling manifests itself in phase transitions, observed in solid films as well as liquid solutions and accompanied by large shifts of the near UV absorption band.[2] Many issues that were, and still are, a matter of debate for π-conjugated polymers also arise for the σ-conjugated polysilanes. Among these are: Of what type is the elementary excitation? What limits excitation delocalization? Is the phase transition an inter- or intramolecular process? The present paper focuses on the second question and discusses the spectra of poly (di-n-hexyl) silane (PDHS). At liquid helium temperatures the absorption spectrum in glassy solutions consists of a band at ca. 350 nm (as compared to 320 nm in room temperature solutions), which is only 2.5 nm broad and composed of still narrower lines (1-4 cm^{-1}), as revealed by spectral hole-burning.[3,10,11] This spectrum is successfully simulated by numerical modelling, describing the excitation as a Frenkel exciton of a disordered linear chain.

MODELS

The nearest neighbor exchange interaction is commonly accepted to be in the range 1-2 eV as suggested by quantum mechanical calculations and measurements of the ionization potential[4], so that long range multipolar interactions can be neglected. A good description for the backbone excitation is therefore given by the linear chain Hamiltonian:

$$H = \Sigma E_n\{|n><n|\} - \beta_{n,n+1}\{|n><n+1| + |n+1><n|\}$$

where E_n is the transition energy of site n and $\beta_{n,n+1}$ the exchange interaction between sites n and n+1. Disorder in the chain accounts for excitation delocalization lengths less than the whole polymer strand. Two extreme models have previously been proposed to treat disordered chains:
1) The independent segment model assumes that a few defects are distributed at random along the chain. They act as infinite barriers and therefore localize excitations on mutually

decoupled segments. The segments are taken as defect free linear chains of finite length. This model has successfully been applied to describe optical properties of π-conjugated polymers[5].

2) The wormlike chain model assumes a Gaussian distribution (standard deviation σ) of site energies and /or exchange interactions, again neglecting any correlation. This model was proposed for polydiacetylene[6] and has recently been subject to theoretical investigations for its conformational properties[7], but to the best of our knowledge, the consequences of this type of disorder for optical properties of these polymers have never been explicitly addressed.

For comparison, calculations have been performed for both models. The algorithm given in ref. 8 was used for the first model, and straightforward diagonalization of 3000 randomly generated tridiagonal matrices of dimension 200 yielded the results for the second model. Only diagonal (site energy) disorder was introduced as calculations with offdiagonal disorder gave comparable results. At low temperature, the homogeneous line width is small as compared to the whole absorption line and vibrational bands carry an insignificant fraction only of the total intensity,[11] so that the spectrum of a single segment or state is adequately modelled by a δ-function. Figure 1 shows a comparison of absorption spectra calculated within the first and second model (1b and 1c respectively) with the experimental spectrum (1a). In 1c, the disorder parameter σ was adjusted so that the oscillator strength of states at the red edge of the absorption equals the value determined from measurements of the fluorescence lifetime and yield, i.e. corresponding to a chromophore consisting of ca. 20 monomer units[9]. This requires σ/β to equal 1/20, giving rise to an absorption line of width 0.024β. Delocalization therefore significantly narrows the site energy distribution. In the first model the same linewidth is obtained for an average segment length of 100. The striking asymmetry of the lineshape in this model (1b) stems from the predominance of short segments in the length distribution. This distribution monotonically decreases with segment length (see fig. 7 of ref. 8) thereby giving rise to a large absorption tail extending to higher energies, whereas the rare long segments accumulate at low energy, $E_0-2\beta$. As the longest segments have the largest transition moment the maximum of the optical absorption is located at energies only slightly higher than $E_0-2\beta$.

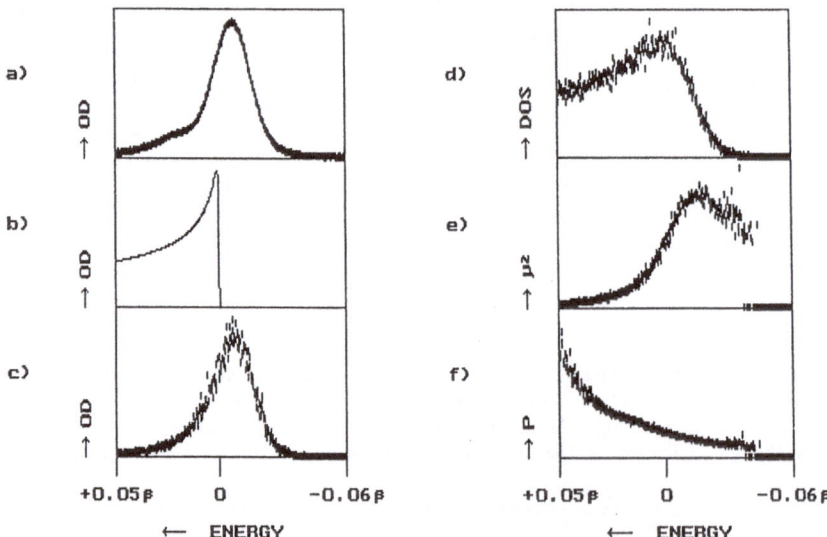

Fig. 1. Measured (a), and simulated (b,c) absorption spectra of PDHS in low temperature glasses. Density of states (d), average oscillator strength per state (e), and participation ratio (f) calculated according to the wormlike chain model. All spectra and distributions are drawn on the same reduced energy scale and the bottom of the infinite perfect chain is taken as the zero of energy.

404

The diluted disorder of the second model, on the other hand, yields a density of states similar to that of an ideal chain, but the divergences (for an infinite chain) at the band edges have been changed into smooth maxima (fig. 1d). As energy decreases, starting from the band center, the oscillator strength per eigenstate (fig. 1e) first increases since the number of knots in the wavefunctions decreases. At lower energy, this trend is overwhelmed by the increasing degree of localization which tends to reduce transition moments and the oscillator strength per state drops again. Fig. 1f shows this behavior of localization quantified by the participation ratio, defined as the inverse of the sum over all sites of the wavefunctions amplitude at a site to the fourth power: The low energy states are localized, states at the band center (not contained in the figure) extend over the whole chain and localization sets in again at the high energy band edge. In the whole range of disorder investigated ($\sigma/\beta = 1/50$ to $1/3$), the shapes of all the distributions shown in fig. 1c-f are, after readjustment of scales, independent of the value of σ/β (within the noise left by insufficient statistics).

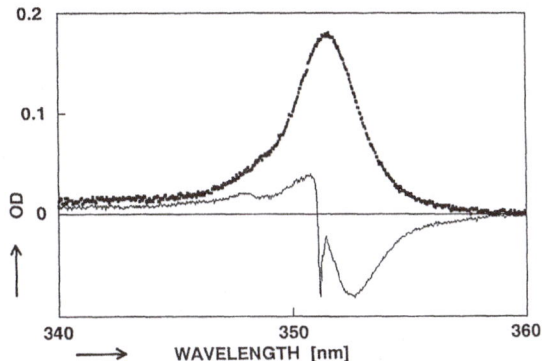

Fig. 2. Absorption and difference spectra after holeburning of PDHS in 3-methyl-pentane glass at 1.5 K. The narrow hole (negative spike) marks the position of the burning laser. Note that the optical density scale is the same for the two spectra.

The delocalization discussed above relates, in a time dependent picture, to "coherent transport" properties. Coupling to vibrations allows for incoherent transfer of the excitation between eigenstates. In the first model, the transfer between adjacent segments has to rely on multipolar interactions, which is not the case for transfer between the spatially overlapping wavefunctions, orthogonal at the chosen nuclear configuration, of the second model. Whatever the mechanism, it is expected that the residence time of the excitation on a segment/state is inversely proportional to the number of available acceptor states.

EXPERIMENTAL OBSERVATIONS

Comparison with experimental data (see Fig. 1, and ref. 10 and 11 for further details and figures) clearly favours the wormlike chain model as a suitable description of poly (di-n-hexyl) silane in low temperature glasses. The experimental and calculated absorption spectra are virtually indistinguishable, as the small hump at high energies can be assigned to a weak vibronic transition.[11]

Polysilanes are photochemically unstable and degrade in room temperature solutions by chain scission with high quantum yield. In the low temperature glasses photolysis occurs with a quantum yield of about 10^{-5}. This photosensitivity offers a convenient tool to trace the excitation flow: After spectrally selective excitation by a laser, photodamage of the polymer occurs either in the region initially excited or excitation transfer to an energetically lower lying state leads to damage in a spectrally and spatially different region. Difference spectra (absorption after irradiation minus absorption of undamaged samples) therefore show a sharp hole at the laser frequency and a diminution of absorption at lower energies corresponding to states burnt away after energy transfer. The photoproducts are shortened chains whose absorption appears at frequencies higher than that of the exciting laser. The width of the narrow hole becomes temperature independent at low temperatures and is therefore likely to be determined by the lifetime of the state initially excited by the laser. The main contribution to the decay is energy transfer to lower energy acceptor states. The concentration of potential acceptor states is proportional to the fraction of states located at wavelengths longer than that of the laser. When this fraction is determined according to the wormlike chain model from Fig. 1d, it is found that the increase of holewidth as a function of the concentration of acceptor states is approximately linear. This behavior is consistent with the transfer rate being, on average, independent of the energy of the available acceptor states.

The experimentally observed shift of fluorescence w.r. to absorption can entirely be attributed to energy transfer and becomes very small (≤ 10 cm^{-1}) when the excitation lies at the low energy edge of the absorption: No significant structural relaxation occurs (e.g. polaron formation) on the timescale of the fluorescence lifetime (200ps).

The influence of site energy correlation and bimodal energy distributions will be investigated numerically in a near future, whereas on the experimental side, hole burning at variable temperatures will establish the impact of vibrational coupling on the homogeneous linewidth in an extended excitonic system.

+This research was supported by NSF-DMR-8519059, by Sandia National Laboratories, supported by the U.S. Department of Energy under contract number DEAC04-76-DP00789, and by the Conseil Régional de la Région Rhône-Alpes (# 900.3/1308V5655). R.M.H. and H.P.T. are grateful to NATO for support of their joint collaborative research on polysilanes.[3,10,11] A.T. acknowledges a stipend from the Dr. Otto Röhm Foundation.

REFERENCES

1. For reviews see: R.D. Miller and J. Michl, Chem. Rev. 89(1989)1359; J. Michl, J.W. Downing, T. Karatsu, K.A. Klingensmith, G.M. Wallraff, R.D. Miller, in Inorganic and Organometallic Polymers, eds. M. Zeldin et al., ACS Symposium Series 36 (1987) 61; J.M. Zeigler Synth. Met. 28 (1989) C581

2. L.A. Harrah, J.M. Zeigler, J. Poly. Sci., Poly. Lett. Ed. 23 (1985) 209, and Macromolecules 20 (1987) 601; P. Trefonas, J.R. Damewood, R. West, R.D. Miller, Organometallics 4 (1985) 1318

3. H.P. Trommsdorff, J.M. Zeigler, R.M. Hochstrasser J. Chem. Phys. 89 (1988) 4440

4. H. Takeda, H. Teramae, N. Matsumoto, J. Amer. Chem. Soc. 108 (1986) 8186; A. Herman, B. Dreczewski, W. Wojnowski, Chem. Phys. 98 (1986) 7413

5. K.S. Schweizer, J. Chem. Phys. 85 (1986) 1156,1176

6. G. Wenz, M.A. Müller, M. Schmidt, G. Wegner Macromolecules 17 (1984) 837

7. G. Rossi, R.R. Chance, R. Silbey, J. Chem. Phys. 90 (1989) 7594

8. K.S. Schweizer, J. Chem. Phys. 85 (1986) 4181

9. Y.R. Kim, M. Lee, J.R.G. Thorne, R.M. Hochstrasser, J.M. Zeigler, Chem. Phys. Lett. 145 (1988) 75; J.R.G. Thorne, R.M. Hochstrasser, J.M. Zeigler, J. Phys. Chem. 1988, 92, 4275

10. A. Tilgner, H.P. Trommsdorff, J.M. Zeigler, R.M. Hochstrasser, J. Lum. 45 (1990) in press

11. A. Tilgner, J.P. Pique, H.P. Trommsdorff, J.M. Zeigler, R.M. Hochstrasser, Polym. Prepr. in press

THE STUDY OF ENERGY CHANGE

IN RADICAL SYSTEMS

Emine CEBE* and Mustafa CEBE**

Uludağ University
Faculty of Science
*Department of Physics
**Department of Chemistry
Bursa-Turkey

INTRODUCTION

A number of theoretical investigations have been developed in order to describe the kinetics of a homogenous electron self exchange reaction:

$$R + R^{\mp} \xrightleftharpoons{k_{hom}} R^{\mp} + R \tag{1}$$

and heterogenus

$$R + e^{-} \xrightleftharpoons{k_{het}} R^{\mp} \tag{2}$$

The method of Marcus is mostly used to compare experimental results of radical reaction kinetic, presuming that the inner quantum states of the system are equilibrated. Thus the homogenous rate constant k is given by equation (3).

$$k = KZexp(-\Delta G^*/RT) \tag{3}$$

K is called the transmission factor and is K=1 for the adiabatic processes. Z is collision parameter of spherical molecules[2] ΔG^* is a fonction of the free enthalpy ΔG^0 and the reorganization energy λ of the system. In the rate expression the activation energy ΔG^* is a sum of the both reorganization energy terms: one of these terms is the inner sphere reorganization energy λ_I describing the changes in bond angles between the two redox sites R and $R^{+/-}$ and the other the outer sphere reorganization energy λ_0 depending on the changes in solvent orientation by means of the charge transfer[3]

$$\Delta G = 1/4 \ (\lambda_I + \lambda_0) \tag{4}$$

In order to compare experimental rate constans and activation energies with the theoretical values obtained by using eqs.(1) and (2) it is necessary to have good models for the calculations of λ_I and λ_0 besides good models for the preexponential factors.

Large-Scale Molecular Systems, Edited by W. Gans *et al.*
Plenum Press, New York, 1991

The reaction activation energy ΔG^* is dependent on the type of the rate equation. According to eqs.(1) and (2) the following relationships can be written respectively:[3,4]

$$k_{hom} = K_{hom} Z \exp(-\Delta G^*_{hom}/RT) \tag{5}$$

and for heterogenous,

$$k_{het} = K_{het} Z \exp(-\Delta G^*_{het}/RT) \tag{6}$$

For homogenous electron transfer, the standart free reaction enthalpy is taken as zero and no Coulomb work term W_R between the two reactans in eqn(1). Thus it has to be considered ($W_R = 0$).

To calculate for λ_0 two models are mostly used. One is the Born-Based continuum model introduced by Marcus[5] reviewed and improved by Dogonadze[6]. The other model is called by the cavity model or Kirkwood and Westheimer. This second model developed by Brunschwig and Sutin is recently applied to electron transfer reactions. All these models predict a solvent dependence of the rate constant through the solvent parameter. γ is given by

$$\gamma = (1/n^2 - 1/\varepsilon) \tag{7}$$

n is the refractive index and ε the dielectric constant of the solvent. Such a solvent dependence is very important on the reaction activation energy ΔG^* and is recently investigated[6] for several organic electron transfer reactions.

An important problem is to get exact λ_I-values since knowledge of bond length of the molecule in both redox sites is necessary. Most of the authors dealing with organic electron transfer reactions neglect λ_I, assuming it is small. One of the reasons of neglecting λ_I may be the uncertainties in the calculations caused by the unknown bond lengths changes between the two redox sites.

λ_I is sometimes very much important especially for inorganic electron transfer reactions. It is determined in this investigation how great problem to find is exactly the values of λ_I for some organic radical systems. The aim of this study is to find experimental λ_I-values obtained from crystallographic data and them to compare with theoretical calculated ones obtained from bond-length/bond order relations and to improve these relations.

THEORY

If only the electron exchange between the redox couple of a substance is discussed. $\Delta G^0 = 0$; Coulomb work term for reactant W_R and for radical product W_P can be accepted same ($W_R = W_P = W$). Then the activation energy ΔG^* is given by

$$\Delta G^* = W + \lambda/4 \tag{8}$$

λ is composed of inner and outer reorganization energy:

$$\lambda = \lambda_I + \lambda_0 \tag{9}$$

With a harmonic approximation λ_I can be expressed by the force constants f of reactant (R) and product (P) and the change of atomic distances Δq_{r-s} during the reaction:

$$\lambda_1 = \sum_{n=1}^{m} \frac{f_{r-s}^{R} \cdot f_{r-s}^{P}}{f_{r-s}^{R} + f_{r-s}^{P}} \ (\Delta q_{r-s})^2 \qquad (10)$$

f_{r-s}^{R} : force constants of the bond between the atoms r and s in the reactant R.

f_{r-s}^{P} : force constants of the bond between the atoms r and s in the radical product P

Δq_{r-s} : the difference in equilibrium length of the same chemical bond in both states (in reactant and product states)

q_{r-s}^{R} : the bond-length between the atoms r and s in the reactant R.

q_{r-s}^{P} : the bond-length between the atoms r and s in the radical product P.

The difference of the bond length Δq_{r-s} is given by

$$\Delta q_{r-s} = q_{r-s}^{R} - q_{r-s}^{P} \qquad (11)$$

The general electron transfer reaction can be given on the anthracene system as in fig 1.

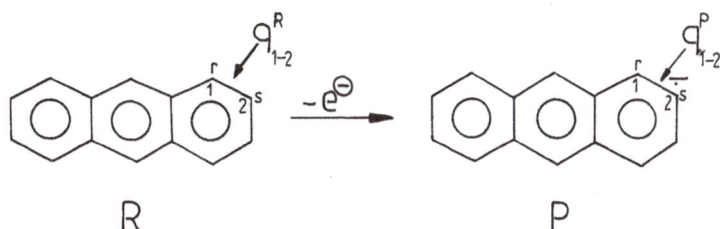

Fig. 1 . The electron transfer reaction of the anthracene $(R \xrightarrow{-e} P^{-})$

There exist several type bond-length/bond-order relations. According to Coulson the following equation can be written.

$$q_{r-s} = q_s - \frac{q_s - q_d}{1 + k \ (1-p_{rs}^{x})} \qquad (12)$$

q_s is the distance of the single bond, q_d the double bond distance and k a constant. The bond order p_{rs}^{x} is defined

$$p_{rs} = \sum_{j=1}^{N} b_j \, a_{jr} \, a_{js} \qquad (13)$$

b_j is the number of occupation by electrons (b_j = 0,1 or 2) and a_{jr} is the coefficient of atomic orbital contributed by atom r in j. molecular orbital.[6,7]

According to Pauling the following relation is valid for the chemical bond distance q_{rs} between the atoms r and s.

$$q_{rs} = q_s - 0.71.\log(p_{rs}^{\pi} + 1) \qquad (14)$$

The other relationship for heterogenous bonds is given by Gordy[8]

$$q_{rs} = \left(\frac{a}{p_{rs} - b} \right)^{1/2} \qquad (15)$$

a and b in equation are constants for different bonds[8]. The parameters a and b for the different bonds are given as,

Bond	$a.10^{-4}$ cm^2	b
C-C	6.80	-1.71
C-N	6.48	-2.00
N-O	4.98	-1.45
C-S	11.90	-2.59

METHOD AND RESULTS

In this investigation two different moleculs-radical radical systems are studied. These are the couple of N-methlylphenazin (Nmp/Nmp+) and N-methylphenotiazine (N-mePZT/N-mePZT+).

The force constants f_j which can be obtained by the Gordy relations are shown in equation (16)

$$f_j = g \, (1 + p_{r-s}^{\pi}) \left(\frac{X_r \, X_s}{q_{rs}^2} \right)^{3/4} + h \qquad (16)$$

The constants g and h are $1.67.10^{-13}$ Nm$^{5/2}$ and 30 N/m respectively. X_r and X_s are the electronegativity indices of the atoms r and s.

To determine the experimental bond-lenghts are used the results of x-ray both of the couples Nmp/Nmp+[9,10] and N-mepZT/N-mepZT+[10,11]. For two couples are calculated the experimental bond-length and λ_i(exp) by using the equations (15) and (16). For the theoretical calculations are used Hückel molecular Orbital theory (HMO)[12] modifed by POL-1 method.[13] The obtained results are given as an example for the couple N-mePZT/N-mePZT+ in table 1a and 1b. The model of radical cation is shown in fig.2. The results showed that the electronic parameters for the molecule-radical couple are different.

r-s	p^N_{r-s}	q^N_{r-s} (A°)	p^R_{r-s}	q^R_{r-s} (A°)	q^N_{r-s}(theo.) (A°)	f^N_{r-s}(theo) (N/m)	f^N_{r-s}(exp.) (N/m)
1-2 12-13	.6676	1.386	.6331	1.348	1.419	681.318	704.642
5-6 8-9	.6208	1.400	.5711	1.348	1.429	656.449	675.898
2-3 11-12	.6445	1.390	.6770	1.375	1.424	669.211	692.588
4-5 9-10	.6610	1.387	.7008	1.375	1.420	677.789	701.246
6-7 7-8	.3943	1.430[26]	.4963	1.430	1.382	679.756	647.113
1-14 13-14	.2486	1.820[26]	.3625	1.870	1.761	382.792	365.693
3-4 10-11	.6668	1.377	.6225	1.387	1.419	680.890	710.940
1-6 8-13	.5862	1.399	.5268	1.397	1.436	638.293	662.787

r-s	q^R_{r-s}(theo.) (A°)	f^R_{r-s}(theo.) (N/m)	f^R_{r-s}(exp) (N/m)	λ_n(theo.) $\times 10^{21}$	λ_n(exp) $\times 10^{21}$	Δq(theo.) $\times 10^2$(A°)	Δq(exp) $\times 10^2$(A°)
1-2 12-13	1.426	662.950	718.818	0.1792	5.1382	-0.7303	3.8
5-6 8-9	1.440	630.431	692.667	0.3738	9.2501	-1.0781	5.2
2-3 11-12	1.417	686.356	716.603	0.1550	0.7924	0.6762	1.5
4-5 9-10	1.412	699.176	726.347	0.2377	0.5138	0.8311	1.2
6-7 7-8	1.361	742.946	692.258	1.4634	0	2.0304	0
1-14 13-14	1.735	423.511	381.723	1.3130	4.6692	2.5555	-5.0
3-4 10-11	1.428	657.346	685.687	0.2956	0.3490	-0.9401	-1.0
1-6 8-13	1.449	607.582	640.399	0.5358	0.0130	-1.3119	0.2

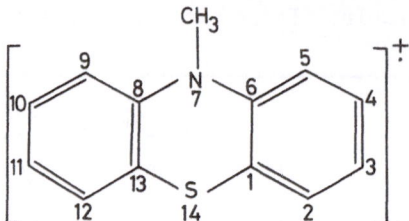

Fig.2. The order of the atoms on the radical cation of N-mepZT[+]

CONCLUSION

The molecular structures of two system are very similar but the inner reorganization energie λ_I are different from each other. λ_I(exp) and λ_I(theo.) are found 4.445 kj/mol and 3.301 kj/mol respectively for the couple Nmp/Nmp[+] and λ_I(exp)=5.482 kj/mol and λ_I(theo.)=4.954 kj/mol for the system N-mepZT/ N-mepZT[+] The experimental values of inner reorganization energies are greater than the theoretical values. These values generally are valid to calculate the free activation energies ΔG^* by using the equation (4).

REFERENCES

1) R.A. Marcus. Annu. Rev. Phys. Chem. 15: 155 (1964).
2) N. Sutin, B.S. Brunschwig and J.R. Winkler, Pure Appl. Chem. 60: 1817 (1988).
3) R. R. Dogonadze. A. M. Kusnetzov and T.A. Marsagishivili, Electrochim. Acta. 25: 1 (1980).
4) N. Sutin, Prog. Inorg. Chem. 30: 441 (1983).
5) J. G. Kirkwood and F. H. Westheimer, J. Chem. Phys., 6: 506 (1938).
6) R. A. Marcus and N. Sutin, Biochem. Biophys. Acta, 811: 265 (1985).
7) G. Grampp and W. Jaenicke. Ber. Bunsenges. Phys. Chem. 88: 325 (1984).
8) G. Grampp, W. Jaenicke and W. Harrer, J. Chem. Soc. Faraday Trans. I. 83: 161 (1987).
9) S. C. Chu. Shirley, Acta Cryst. B30: 2489 (1974).
10) H. Kobayashi, Bull. Chem. Soc. Japan, 46: 2945 (1973).
11) C. J. Fritchie. Acta Cryst. 20: 892 (1966).
12) F. H. Herbstein, G. M. Schmidt, Acta Cryst.. 8: 399 (1955).

DECAY FROM AN INITIAL UNSTABLE STATE

IN A CHEMICAL REACTION MODEL

Sergio Ciuchi° and Bernardo Spagnolo*

°Dipartimento di Fisica
Università La Sapienza
Piazzale A. Moro 2
00185 Roma, Italy

*Dipartimento di Energetica ed
Applicazioni di Fisica
Viale delle Scienze
90128 Palermo, Italy

INTRODUCTION

Non linear autocatalytic chemical reaction models have been developed in order to study the main features of dissipative structures[1].

A model of different physical systems, such as autocatalytic chemical reactions, liquid crystals, dye lasers, etc..., which undergo phase transitions far from equilibrium, is given in terms of stochastic differential equations (SDE)[2]:

$$dx = [ax - bx^{\gamma}] dt + x\sqrt{\varepsilon} dW \tag{1}$$

where $W(t)$ is the wiener process with the usual statistical properties:

$$< dW > = 0 , < dW(t) dW(t') > = \delta(t-t')dt dt'$$

For $\gamma = n$ the eq.(1) is associated with the kinetics of the autocatalytic steps in chemical reactions of the following type (Schlögl model):

$$A + X \rightleftharpoons nX$$

$$B + X \longrightarrow C$$

These SDE (1) are characterized by the multiplicative fluctuations, which arise due to the randomness of the environment to which the system is coupled (here the fluctuations of the A and

B species). The noise enters the system dynamics via its coupling to the state variable.

Although the importance of the multiplicative noise in the context of electronic oscillators has been known for many years it is only recently that its importance for non equilibrium systems has been realized, in fact the effects of multiplicative noise are far less intuitive than those produced by additive noise[3] .

Moreover, simple random multiplicative processes have been used as a prototype of multifractal statistics[4].

In the present work we obtain exact analytic results for the time-average process defined as:

$$\Phi(t) = \frac{1}{t} \int_0^t x(t') dt' \qquad (2)$$

particularly we obtain the generating function of the process $\Phi(t)$ and the analytic expression of the moments of the process $z(t)$:

$$z(t) = g \int_0^t m(0) e^{\delta t' + \sqrt{\varepsilon} W(t')} dt'$$

which is related to the real process of SDE (1).

Some of our results are in agreement with theoretical results obtained, within a different approach, independently by Suzuki et al[5]. and by Brenig and Banai[6].

THE LINEAR MODEL AND THE TIME AVERAGE PROCESS

A generalization of eq.(1) is the following SDE:

$$d\varphi_i = \left[\sum_j J_{ij} \varphi_j + k\varphi_j - g\varphi_i^2 \right] dt + \sqrt{\varepsilon} \, dW_i \varphi \qquad (3)$$
$$i = 1,...N$$

where J_{ij} is the interaction matrix. With the position:

$$\varphi_i = \frac{\varphi_i^L}{y_i} \qquad (4)$$

we obtain a linear SDE for the process φ_i^L, with solution:

$$\varphi_i^L = \exp\left[\left(k - \frac{e}{2} \right) t + \sqrt{e} \, W_i(t) \right] \sum_j \left[e^{\vec{J}} \right]_{ij} \varphi_j(0) \qquad (5)$$

where we used the Ito calculus. We consider now long-range interaction and homogeneous state: $J_{ij} = J/N$, $m(0) = \varphi_i^L(0)$ with:

$$m(t) = \frac{1}{N} \sum_j \varphi_j^L(t)$$

414

and

$$\left[e^{\mathbf{J}t}\right]_{ij} = \delta_{ij} + \frac{1}{N}(e^{Jt} - 1)$$

so we have:

$$\varphi_i^L = m(0) e^{\delta t + \sqrt{\varepsilon} \, W_i(t)} \tag{6}$$

where: $\delta = k - \varepsilon/2 + j$

The mean value and the relative variance of φ_i^L are divergent for long time scale, this means that even a very small noise (i.e. $\varepsilon \ll k$) will give to amplification of fluctuations in the long time regime.

We expect therefore for the process φ_i, during the decay from the initial unstable state, an anomalous fluctuation phenomenon, that is rare large fluctuations which do not scale with the inverse system size.

The real process is related to the linear one by the expression:

$$\varphi_i = \frac{\varphi_i^L}{1 + g \int_0^t \varphi_i^L(t')dt'} \tag{7}$$

Previous work of ref.s[5],[6] have been devoted to the study of the statistical properties of the process of eq.(7). The main difficulty is the mathematical structure which includes both the linear process φ_i^L and it's integral in the expression of φ_i. There is also a physical difficulty due to the absorbing barrier at the origin.. This absorbing state will be reached asymptotically by any realization of the process. As a consequence the steady state distribution has an (integrable) singular behaviour in the origin, close to the critical point $((k+j) < \varepsilon/2)..$

In the present work we propose to study the relaxation from an unstable state in terms of the average withespect to a finite time of the process φ_i :

$$\Phi_i = \frac{1}{t} \int_0^t \varphi_i(t') \, dt' \tag{8}$$

This new process aims to fulfil two requirements: the first is related to a measure process for a physical quantity, which implies always a time average; the second is related to the smoothing on the fluctuations due to the time average.

As a consequence we expect also a non-ergodic behaviour of the process φ_i in a short time-scale.

Moreover the mathematical structure of $\Phi_i(t)$ appears simplified because $\Phi_i(t)$ is given in terms of the time integral of the linearized process. By using eq.(4) we have:

$$\Phi_i(t) = \frac{1}{gt}\ln\left(1 + g\int_0^t \varphi_i(t')\,dt'\right) \tag{9}$$

The statistical properties of the process $\Phi_i(t)$ will be conveniently expressed by the generating function:

$$G_{\Phi_i}(\lambda) = \left\langle e^{-\Phi_i(t)\lambda}\right\rangle = \sum_{k=0}^{\infty}\binom{q}{k}\left\langle\left[g\int_0^t \varphi_i(t')\,dt'\right]^k\right\rangle \tag{10}$$

with $q = -(\lambda/gt)$. Therefore our problem is reduced to the calculation of the moments of the process $z(t)$, so defined:

$$z(t) = g\int_0^t m(0)e^{\delta t' + \sqrt{\varepsilon}W(t')}\,dt' \tag{11}$$

and $m(k,t) = \langle z^k(t)\rangle$. These moments are known from ref.[5]. We find however interesting to derive their expression with a different method. Because of the properties of the Wiener process, we can write:

$$z(t) = \lim_{\Delta t \to 0}(g\,m(0)\,\Delta t(1+xy)) \tag{12}$$

where:

$$x(t) = e^{\delta\Delta t + \sqrt{\varepsilon}\Delta W_1(t)}$$

$$y(t) = (1 + e^{\delta\Delta t + \sqrt{\varepsilon}\Delta W_2(t)} + e^{2\delta\Delta t + \sqrt{\varepsilon}\Delta W_3(t)} + \ldots)$$

$x(t)$ and $y(t)$ are independent processes, and $y(t)$ has the same probability distribution of the process $z(t)$. We can obtain a recurrence relation from eq.(12) by using the identity:

$$z_N = 1 + \sum_{n=1}^{N-1}e^{n\delta\Delta t}\prod_{\ell=1}^{n}e^{\sqrt{\varepsilon}\Delta W_\ell}$$

and the positions: $\langle x^k\rangle = f(k)$, $\langle z_N^k\rangle = g_N(k)$, and because of the statistical independences of the x and z processes. We obtain therefore:

$$g_N(k) + \sum_{n=1}^{\infty}\binom{k}{n}(-1)^n g_N(k-n) = f(k)\,g_{N-1}(k) \tag{13}$$

Because of the identity:

$$m(k,t) = <z(t)^k> = \lim_{\substack{\Delta t \to 0 \\ N \to \infty}} [g\, m(0)]^k \Delta t^k\, g_N(k)$$

we obtain, after simple calculations a differential equation for the moments of the process $z(t)$:

$$\frac{\partial m(k,t)}{\partial t} = (\delta k + \frac{\varepsilon}{2}k^2)\, m(k,t) + k\, m(k-1,t) \tag{14}$$

From this eq. we easily derive a differential equation for the generating function of the process $z(t)$:

$$\frac{\partial G_z(x,t)}{\partial t} = -x G_z + (\delta + \frac{\varepsilon}{2})x\frac{\partial G_z}{\partial x} + \lambda\frac{^2\varepsilon\partial^2 G_z}{2\ \partial x^2} \tag{15}$$

The associated probability distribution is eventually obtained by Laplace antitrasformation of the generating function G_z, or alternatively solving the Fokker Planck equation which can be obtained by direct transformation of the eq.(15):

$$\frac{\partial P}{\partial t} = -\frac{\partial P}{\partial z} + (3\frac{\varepsilon}{2} - \delta)z\frac{\partial P}{\partial z} + \frac{\varepsilon}{2}z\frac{_2\partial^2 P}{\partial z^2} + (\frac{\varepsilon}{2} - \delta)\, P \tag{16}$$

EXACT ANALYTICAL RESULTS

We can obtain easily a mapping between the two generating functions $G_z(x,t)$ and $G_{\Phi_i}(\lambda,t)$ and then after solving the eq.(15) we have an exact expression for $G_{\Phi_i}(\lambda,t)$. From the properties of the gamma function we have:

$$G_{\Phi_i}(\lambda,t) = <[1+z(t)]^{-q}> = \sum_{k=0}^{\infty}(-1)^k\frac{\Gamma(q+k)}{\Gamma(q)}<z^k> \qquad , q = \frac{\lambda}{gt} \tag{17}$$

then using the integral representation of the function $\Gamma(z)$ and resumming we obtain:

$$(1+z)^{-q} = \frac{\displaystyle\int_0^{\infty} e^{-zx} e^{-x} x^{q-1} dx}{\displaystyle\int_0^{\infty} e^{-x} x^{q-1} dx} \tag{18}$$

and after averaging:

$$G_{\Phi_i}(\lambda) = \frac{\int_0^\infty G_z(x,t)e^{-x}x^{q-1}dx}{\int_0^\infty e^{-x}x^{q-1}dx} \tag{19}$$

It is possible to show, by using the methods of group analysis[7], that a particular solution of the eq.(15) is :

$$G_z(x,t) = e^{\alpha t}f(x)$$

With the product transformation: $f(x) = u(x)\, x^{-\frac{a}{2b}}$ we can obtain the Bessel differential equation, and we have:

$$G_z(x,t) = e^{\alpha t}x^{-\frac{\delta}{\varepsilon}}J_{+2\sqrt{\delta^2/\varepsilon^2+2\alpha/\varepsilon}}\left(2i\sqrt{\frac{2x}{\varepsilon}}\right) \tag{20}$$

Because of the linearity of the eq.(15) and rejecting the second linearly independent solution of the diff.eq. to avoid divergences when x approaches to zero, we have:

$$G_z(x,t) = \sum_{n=0}^{\infty} e^{\alpha_n t}x^{-\frac{\delta}{\varepsilon}}c_n J_{v_n}\left(2i\sqrt{\frac{2x}{\varepsilon}}\right) \tag{21}$$

To satisfy the constraint of analiticity for the generating function when x approaches to zero, we obtain a discrete set for the parameter v_n :

$$v_n = 2n + 2\frac{\delta}{\varepsilon} \tag{22}$$

To derive the expansion coefficients C_n we need an exact expression of the moments of the process z(t). This can be done solving by Laplace transform the eq.(14), obtaining:

$$<z^n(t)> = \sum_{k=0}^{n}(-1)^{n-k}\binom{n}{k}\frac{\Gamma\left(\frac{2\delta}{\varepsilon}+k\right)\left(\frac{2}{\varepsilon}\right)^{n+1}(\delta+k\varepsilon)\,e^{\left(\delta k+k^2\frac{\varepsilon}{2}\right)t}}{\Gamma\left(\frac{2\delta}{\varepsilon}+k+n+1\right)} \tag{23}$$

Then using eq.(23) we get finally:

$$G_z(x,t) = \sum_{n=0}^{\infty}(-1)^n\left(\frac{2}{\varepsilon}\right)^{n+1}\sum_{k=0}^{n}\frac{(-1)^k}{(n-k)!}\frac{\Gamma\left(\frac{2\delta}{\varepsilon}+k\right)(\delta + k\varepsilon)\,e^{\alpha_k t}x^n}{\Gamma\left(\frac{2\delta}{e}+k+n+1\right)n!} \tag{24}$$

where $\alpha_k = k\delta + (\varepsilon/2)k^2$. Now putting eq.(24) into (19) it is easy to show that:

$$G_{\Phi_i}(\lambda,t) = \sum_{k=0}^{\infty}\left(\frac{2}{\varepsilon}\right)^k\frac{\Gamma(q+k)}{k!}\frac{\Gamma\left(\frac{2\delta}{\varepsilon}+k\right)(\delta + k\varepsilon)\,e^{\alpha_k t}}{\Gamma\left(\frac{2\delta}{\varepsilon}+2k+1\right)\Gamma(q)}\,{}_1F_1(q+k;\frac{2\delta}{\varepsilon}+1+2k;\frac{2}{\varepsilon}) \tag{25}$$

where ${}_1F_1$ is the confluent hypergeometric function. Inserting in series (25) the following integral representations:

$$\exp\left[k^2\frac{\varepsilon}{2}t\right] = \frac{1}{\sqrt{\pi}}\int_{-\infty}^{+\infty}dv\,\exp\left[-v^2 + kv\sqrt{2\varepsilon t}\right]$$

$${}_1F_1(a,b,z) = \frac{\Gamma(b)}{\Gamma(a)\Gamma(b-a)}\int_0^1 du\,e^{zu}u^{a-1}(1-u)^{b-a-1}$$

we obtain finally a compact analytical expression for G_{Φ_i} , which is very useful for numerical evaluation:

$$G_{\Phi_i}(\lambda,t) = \frac{\varepsilon\,\Gamma\left(\frac{2\delta}{\varepsilon}\right)\Gamma\left(\frac{\delta}{\varepsilon}+1\right)}{\sqrt{\pi}\,\Gamma\left(\frac{\delta}{\varepsilon}\right)\Gamma\left(\frac{2\delta}{\varepsilon}-q+1\right)\Gamma(q)}\int_{-\infty}^{+\infty}dv\int_0^1 du\,e^{-v^2}\frac{2u}{\varepsilon}u^{q-1}(1-u)^{\frac{2\delta}{\varepsilon}-q}\,{}_2F_2\left(\frac{2\delta}{\varepsilon},\frac{\delta}{\varepsilon}+1;\frac{2\delta}{\varepsilon}+1-q,\frac{\delta}{\varepsilon};y\right) \tag{26}$$

where:

$$y = \frac{2}{\varepsilon}u\,(1-u)e^{\delta t + v\sqrt{2\varepsilon t}} \tag{27}$$

and ${}_2F_2$ is a generalized hypergeometric function.

ACKNOWLEDGEMENTS

The authors wish to thank Prof. F.de Pasquale for enlightening suggestions and interesting discussions. This work is supported in part by the Italian Ministry of Education and the National Group of Structure of Matter.

REFERENCES

1. F.de Pasquale, P. Tartaglia and P. Tombesi, Il Nuovo Cimento 69B, 228 (1982).
2. A. Schenzle and H. Brand, Phys. Rev. A20, 1628 (1979).
3. S. Kalashina, S. Kogure, T. Kawakubo, and T. Okada, J. Appl. Phys. 50, 6296 (1979).
4. A.P. Siebesma and L. Pietronero, Eur. Phys. Lett. 4, 597 (1987).
5. M. Suzuki, S. Takesue and F. Sasagawa, Proc. of Theor. Phys. 68, 98 (1982).
6. L. Brenig and N. Banai, Physica 5D, 208 (1982).
7. P. Barrera and T. Brugarino, Il Nuovo Cimento 92B, 142 (1986).

THE EFFECTS OF STATIC DISORDER

ON POLARON TRANSPORT

Dora Izzo*

Physics Department
Massachusetts Institute of Technology
Cambridge, Massachusetts, 02139

INTRODUCTION

The motion of charge carriers in chalcogenide glasses presents very
particular properties. Emin[1] suggested that the transport properties of
these low mobility solids can be described by a model in which a self-
trapped carrier, the small polaron, hops between lattice sites. The steady
state transport properties of the hopping system are described by the
mobility μ and by the diffusion coefficient D. For an ordered lattice, D
can be written as $D = W\underline{a}^2$, where W is the jump rate between neighboring
sites and $\underline{a}$ is the lattice constant. In a disordered lattice, W is not
identical for all pairs of adjacent sites, therefore we have to find a
suitable averaging procedure. We solve the problem by calculating coherent
hopping rates which result from an effective medium average.

THE HAMILTONIAN AND THE EQUATION OF MOTION

A model Hamiltonian for carrier hopping must include the carrier subsystem, the phonon subsystem, the carrier-phonon interaction and the
interaction with external fields. The carrier Hamiltonian can be written as

$$H_e = \sum_m \varepsilon_m a_m^\dagger a_m + \frac{1}{2} J \sum_{m,m'} a_m^\dagger a_m , \qquad (II-1)$$

where a_m ($a_m^\dagger$) is the annihilation (creation) operator for a Fermi system,
ε_m is the carrier energy at the mth site and J expresses the overlap of
wavefunctions at neighboring sites. In hopping transport, J is so small

*Present address:
Instituto de Física da Universidade de Sao Paulo
Departamento de Física Experimental
Caixa Postal 20516 - CEP 01498
Sao Paulo - SP - BRAZIL

Large-Scale Molecular Systems, Edited by W. Gans *et al.*
Plenum Press, New York, 1991

that the second term in equation (II-1) can be considered as the perturbing term of the full Hamiltonian.

We choose one of the simplest kinetic models to describe the hopping motion of a charge carrier. It is the Pauli Master Equation (PME),

$$\frac{\partial}{\partial t} P_m(t) = \sum_{m'=m\pm 1} [W_{mm'}P_{m'}(t) - W_{m'm}P_m(t)] \ , \tag{II-2}$$

where $P_m(t)$ is the probability of finding the polaron on site m at time t and the quantity $W_{m'm}$ is the hopping rate between sites m and m'. The latter quantity is obtained by a thermal average of the transition probability of an assisted hop between sites (Fermi Golden rule). We can express $W_{m'm}$ as $W_{m'm} = [\exp(-\beta/2(\varepsilon_{m'}-\varepsilon_m))]w_{m'm}$. In this equation, $w_{m'm}$ is proportional to J^2. It is also a function of the phonon density of states and of the carrier-phonon coupling. Its dependence on the indices is of the type $w_{m'm} = w(|\varepsilon_{m'}-\varepsilon_m|)$.

EFFECTIVE MEDIUM APPROXIMATION

Our goal is to calculate the drift velocity V (V = μ.E , where E is the external electric field), and the diffusion coefficient D. In the one-dimensional lattice,

$$V(t) = \frac{d}{dt} \langle n(t) \rangle$$

and $\tag{III-1}$

$$D(t) = \frac{d}{dt} \frac{1}{2} [\langle n^2(t) \rangle - \langle n(t) \rangle^2] \ ,$$

where $\langle n(t) \rangle$ and $\langle n^2(t) \rangle$ are the mean displacement and the mean-square displacement respectively. It is expedient to rewite the PME (equation II-2) in terms of fluxes of probabilites $J_n(t)$ and $Q_n(t)$,

$$\frac{\partial}{\partial t} P_n = - J_n + Q_n + J_{n-1} - Q_{n-1} \ , \tag{III-2}$$

where

$$J_n \equiv (P_n - P_{n+1})F_{n,n+1} \qquad \text{and} \qquad Q_n \equiv (P_n + P_{n+1})H_{n,n+1} \ . \tag{III-3}$$

The quantities $F_{n'n}$ and $H_{n'n}$ are the redefined hopping rates:

$$F_{n'n} \equiv 1/2 \ (W_{nn'} + W_{n'n}) \qquad \text{and} \qquad H_{n'n} \equiv 1/2 \ (W_{nn'} - W_{n'n}).$$

The definition of the generalized fluxes as in (III-3) is interesting because we can show that

$$\langle n(t) \rangle = \lim_{k \to 0} L^{-1} [(\tilde{J}_k(\varepsilon) - \tilde{Q}_k(\varepsilon))] \tag{III-4a}$$

and

$$\langle n^2(t)\rangle = \lim_{k\to 0} L^{-1} [(2i\frac{\partial}{\partial k} + 1)(\tilde{J}_k(\varepsilon) - \tilde{Q}_k(\varepsilon))] , \qquad \text{(III-4b)}$$

where $\tilde{J}_k(\varepsilon)$ $(\tilde{Q}_k(\varepsilon))$ is the Fourier-Laplace transform of $J_n(t)$ $(Q_n(t))$. Following Kundu et al.[2], we obtain a set of two equations in the unknowns $\tilde{J}_k$ and $\tilde{Q}_k$, proceeding as follows. First we write two equations of motion, one for $(P_n - P_{n+1})$ and another for $(P_n + P_{n+1})$. Next we multiply and divide the first one by $F_{n,n+1}$ and the second one by $H_{n,n+1}$, which yields a set of two differntial equations in the variables J_n and Q_n. Finally we Fourier-Laplace transform these equations, obtaining

$$\begin{vmatrix} \tilde{J} \\ \tilde{Q} \end{vmatrix} = G \begin{vmatrix} 1 - e^{iq} \\ 1 + e^{iq} \end{vmatrix} . \qquad \text{(III-5)}$$

A formal expression for G can always be obtained. In practice, though, equation (III-5) cannot be solved because G describes all the hopping rates. We solve this problem using an effective medium approximation. Rather than working with G, we obtain an ensemble averaged $\langle G\rangle$. A convenient choice is to write G as $G = (1/\varepsilon)[g^{-1} - t]^{-1}$, where g^{-1} is known and t contains all the disorder. We write t in terms of fluctuations about the mean inverse rates $(1/F_{mm'} - 1/c_1)$ and $(1/H_{mm'} - 1/c_2)$, where $1/c_1$ and $1/c_2$ are coherent inverse rates to be determined. The mean field approximation claims that, in average, the fluctuations cancel each other, so that $\langle t\rangle = 0$. Given a distribution of site energies $\{\varepsilon_m\}$, the condition $\langle t\rangle = 0$ determines c_1 and c_2. In an ordered chain, $\langle t\rangle \equiv 0$, and we can show that $D = c_1$ and $V = 2c_2$, where c_1 and c_2 are uniform rates. Here we extend these relations as a criterion for determining D and V for the disordered chain, except that in the present case, c_1 and c_2 are coherent inverse rates.

RESULTS AND DISCUSSION

We consider that a band of optical phonons assists the hopping motion. We choose a bivalued distribution of site energies: at $-\sigma/2$ and $\sigma/2$ with probabilities P and (1-P) respectively. Figures 1 and 2 show the overall feature of decreasing V with increasirg disorder σ. In figure 1, the plot of V as a function of the dimensionless electric field Δ, shows that in the case of vanishing disorder, V vanishes linearly with Δ, whereas in the finite disorder case, the approach to zero is more abrupt. In figure 2 we plot V as a function of P, where we obtain the expected behavior: the smallest V occurs at P=1/2; the highest values of V, corresponding to P=0 and P=1, coincide with V in the non-disordered case.

The mean field solution that we derive here gives the expected behavior. Some of the results agree well with values obtained from other methods[3]. The advantage of this solution is that it can be extended to higher

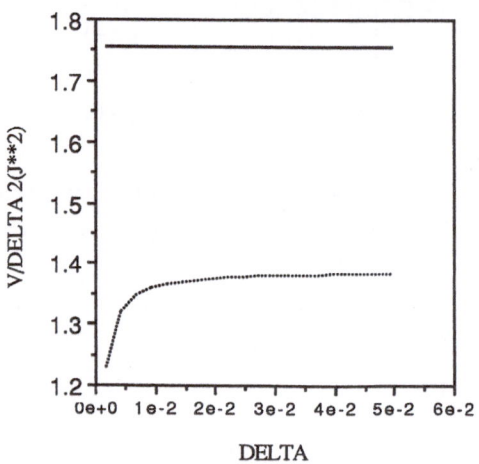

Figure 1. dimensionless coupling constant g=1.; phonon bandwidth= 6.10^{-2}; $\sigma=4.10^{-4}$ continuous line, $\sigma= 4.10^{-2}$ dotted line.

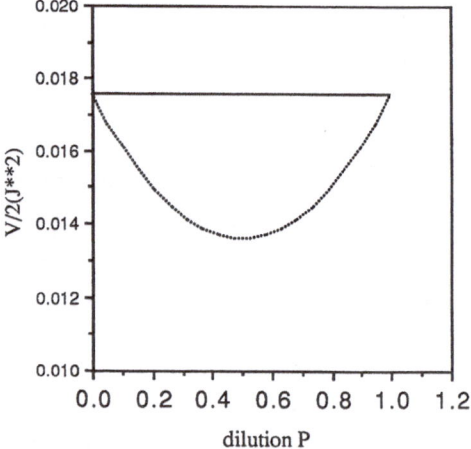

Figure 2. dimensionless coupling constant g=1.; phonon bandwidth= 6.10^{-2}; dimensionless temperature T=1.; Δ=.01; $\sigma=4.10^{-4}$ continuous line, $\sigma= 4.10^{-2}$ dotted line.

dimensions. It can be used to obtain the frequency dependent quantities: we can explore the similarity between the PME and the Generalized Master Equation in the Laplace domain.

REFERENCES

(1) D.Emin, Comments Solid State Phys. 11, 35, 59 (1983).
(2) K.Kundu, P.Parris and P.Phillips, Phys. Rev. B 35, 3468 (1987).
(3) D.Izzo, Phd. Thesis, Cambridge, 1990.

DIFFUSION IN A DISORDERED LINE

A. Valle, M.A. Rodríguez and L. Pesquera

Departamento de Física Moderna, Facultad de Ciencias
Universidad de Cantabria, 39005 SANTANDER (SPAIN)

1. INTRODUCTION

There are many physical phenomena in which the diffusion process takes place. In most cases these phenomena are modelled by using particles moving in an homogeneous medium. The intrinsic heterogeneity of the matter is then taken as a second order effect. This assumption is frequently valid under laboratory conditions but in general fails in real world materials.

Consideration of heterogeneity can be essential in certain situations in which the medium has a strong disorder or the physical phenomenum under consideration is very sensitive to weak disorder. Examples of current interest are the problem of diffusion in fractal structures, diffusion through porous media and in general phenomena related with anomalous diffusion[1].

The study of diffusion in random media has been performed mainly with discrete models. A tipical one can be a random walk with random transition probabilities, generally taken as independent between sites[1]. As we show elsewhere[2] the continuum limit of this model does not reproduce a proper case of diffusion in a standard random field. On the other hand models of disorder in continuous media are scarce and sometimes present unphysical situations. In fact some authors have used a Gaussian noise as random diffusion coefficient[3] and we show in this paper that such a choice presents physical and mathematical inconsistencies. Specifically our results indicate that the positivity of the random coefficient is essential to obtain results with a clear physical meaning. This requirement adds complexity to our analysis because implies the use of bounded random quantities and does not permit the advantage of working with well known processes as for example Gaussian ones.

In this paper we consider an one dimensional model of diffusion in a disordered medium. Mathematically this model consists of a diffusion equation with a random positive coefficient. Our analysis is valid for a general random coefficient but in this paper we restrict ourselves to the case of weak disorder, that is random coefficients with finite inverse moments.

To average the random diffusion equation we use a method based on projection operators and ordered cumulants. We have used succesfully this method in discrete models[4] and here we adapt it to work on the continuum model. As a consequence of this adaptation we are able to define an effective medium approximation in a continuum medium which is the equivalent of the E.M.A. in discrete models. Finally we obtain exact results for the long time behaviour of the time dependent diffusion coefficient in the cases in which the random diffusion coefficient is modelled by a dichotomic process and by a generalised Poisson process. In both situations it is possible to analyse the effect of the correlation length. As an interesting result we show that in the Poisson case a change of the correlation length can produce changes from weak to strong disorder or viceversa. This result seems to be similar to the one recently obtained with a Sinai model with a two level correlated process[5].

2. THE AVERAGED DIFFUSION EQUATION

Let us consider our model of diffusion in a disordered one dimensional medium by means of the following equation:

$$\frac{\partial P(x,\,t)}{\partial t} = D\frac{\partial^2 P(x,\,t)}{\partial x^2} + \frac{\partial}{\partial x}\xi(x)\frac{\partial}{\partial x}P(x,\,t) \tag{1}$$

where $P(x,t)$ is a probability density and $D+\xi(x)$ is the random diffusion coefficient. We take $\xi(x)$ with zero mean value, $<\xi(x)>=0$, and we assume the positivity of the diffusion coefficient so that $\xi(x)$ must be a bounded process $\xi(x)\geq-D$.

The average of equation (1) can be obtained following the method of reference 4 here applied to a continuum equation. We give a brief sketch of the method. For the sake of simplicity we take two operators defined by

$$O_\xi = \frac{\partial}{\partial x}\xi(x)\frac{\partial}{\partial x} \qquad \text{and} \qquad Mf(x,\,t) = \int_{-\infty}^{\infty}dx'\int_0^t dt'\,G(x,\,t|\,x',\,t')\,f(x',\,t')$$

being $G(x,t|x',t')=G(x-x',t-t')$ the Green function associated to the deterministic diffusion equation.

In terms of these operators the random diffusion equation (1) can be written as

$$P(x,\,t) = M\delta(t-t_0)P(x,\,t_0) + MO_\xi P(x,\,t) \tag{2}$$

Then we define the projection operator $\mathcal{P}$ as $\mathcal{P}F(\xi(x)...\xi(z))=<F(\xi(x)...\xi(z))>$, that is, this operator acts averaging over all random functions placed to its right.

Now we apply to eq. (1) first the operator $\mathcal{P}$ and further $(1-\mathcal{P})$ so obtaining two coupled equation for $\bar{P}=\mathcal{P}P$ and $(1-\mathcal{P})P$. Solving formally the second equation and substituting into the first one, we obtain

$$\frac{\partial\bar{P}}{\partial t} = D\frac{\partial^2\bar{P}}{\partial x^2} + \mathcal{P}O_\xi\bar{P} + \mathcal{P}O_\xi M(1-\mathcal{P})O_\xi\bar{P} + \mathcal{P}O_\xi M(1-\mathcal{P})O_\xi M(1-\mathcal{P})O_\xi\bar{P} + \dots \tag{3}$$

and after integrating by parts and rearranging it reads:

$$\frac{\partial\bar{P}}{\partial t} = D\frac{\partial^2\bar{P}}{\partial x^2} + \frac{\partial}{\partial x}\sum_{n=0}^{\infty}\int_{-\infty}^{\infty}dx_1\dots\int_{-\infty}^{\infty}dx_n\int_0^t dt_1\dots\int_0^{t_{n-1}}dt_n <\xi(x)\xi(x_1)\dots\xi(x_n)>_T \times$$

$$\times\frac{\partial^2 G(x,\,t|\,x_1,\,t_1)}{\partial x_1^2}\dots\frac{\partial^2 G(x_{n-1},\,t_{n-1}|\,x_n,\,t_n)}{\partial x_n^2}\frac{\partial P(x_n,\,t_n)}{\partial x_n} \tag{4}$$

where $<\xi(x)\xi(x_1)\dots\xi(x_n)>_T$ denotes Terwiel's cumulants defined as $<\xi(x)\xi(x_1)\dots\xi(x_n)>_T = \mathcal{P}\xi(x)(1-\mathcal{P})\xi(x_1)\dots(1-\mathcal{P})\xi(x_n)$[6].

This equation is our starting point. Note that despite its apparent complexity it can lead to simple analytical solutions in the usual cases in which Terwiel's cumulants only depend on the difference of arguments $<\xi(x)\xi(x_1)\dots\xi(x_n)> =\vartheta(x-x_1,\dots,\,x_{n-1}-x_n)$. Effectively in this case the second term is a sum of convolutions in space and time. An expression for the generalised diffusion coefficient $D(k,s)$ defined as

$$\bar{P}(k,\,s) = \frac{1}{s+k^2 D(k,\,s)} \tag{5}$$

being $\bar{P}(k,s)$ the Fourier Laplace transform of $\bar{P}(x,t)$, can be easily calculated as

426

$$D(k, s) = D + \sum_{n=1}^{\infty} \int_{-\infty}^{\infty} dy_1 \cdots \int_{-\infty}^{\infty} dy_n e^{ik(y_1 + \cdots + y_n)} \vartheta(y_1 \cdots, y_n) \frac{\partial^2 G(y_1, s)}{\partial y_1^2} \cdots \frac{\partial^2 G(y_n, s)}{\partial y_n^2} \quad (6)$$

3. EFFECTIVE MEDIUM APPROXIMATION

From eq. (4) one can find expressions like (6) for other transport coefficients.[4] In this paper we restrict ourselves to the analysis of the frequency dependent diffusion coefficient $D(k=0,s)$ for long times (small s). A direct calculation of this coefficient from eq. (6) involves an infinite number of terms, even for the low order in s (s^0). In order to avoid this difficulty we introduce a perturbative expansion around an effective medium.

The Laplace transformed Green function is easily obtained, giving:

$$G(x, s) = \frac{1}{2 (Ds)^{1/2}} e^{-(s/D)^{1/2} |x|} \quad (7)$$

and also it is easy to check the following relation

$$\frac{\partial^2 G(x, s)}{\partial x^2} = \frac{1}{D}(sG(x, s) - \delta(x)) \quad (8)$$

Substituting (8) in (6), integrating the δ terms and rearranging we obtain:

$$D(k, s) = D + <\psi> + \sum_{n=1}^{\infty} \left(\frac{s}{D}\right)^n \int_{-\infty}^{\infty} dx_1 \cdots \int_{-\infty}^{\infty} dx_n e^{-ikx_n} < \psi(0)\psi(x_1) .. \psi(x_n) >_T \times$$
$$\times G(-x_1, s) G(x_1 - x_2, s) \ldots G(x_{n-1} - x_n, s) \quad (9)$$

with $\psi(x) = \dfrac{\xi(x)}{1+(1-P)\xi(x)/D}$ and $D(0,s) = D + <\psi> + \sum D_i(s)$.

The order of the n term in this sum is zero (s^0) because $<\psi> \neq 0$ and as a consequence the cumulant does not vanish in all the interval of integration. Things would be very different if $<\psi> = 0$ because a cumulant with only one argument x far enough the other points, $\{x_1, \ldots x_{i-1}, x_{i+1}, \ldots, x_n\}$, vanishes. Then it is easy to see that the order of the i th term would be:

$$D_i(s) \approx s^{\frac{i+2}{4}} \quad (\text{i even}) \quad \text{and} \quad D_i(s) \approx s^{\frac{i+1}{4}} \quad (\text{i odd})$$

This is our desired expansion. Hence, our following task is to get a transformation $\psi \to \tilde{\psi}$ in which $<\tilde{\psi}> = 0$. This transformation can be obtained by introducing from the beginning an effective medium, adding and substracting a quantity Γ in (1)

$$\frac{\partial P}{\partial t} = (D + \Gamma)\frac{\partial^2 P}{\partial x^2} + \frac{\partial}{\partial x}(\xi - \Gamma)\frac{\partial}{\partial x}P \quad (10)$$

and following, step by step, exactly the same calculation. For the generalised diffusion coefficient we obtain

$$D(k, s) = D + \Gamma + \sum_{n=1}^{\infty} \left(\frac{s}{D+\Gamma}\right)^n \int_{-\infty}^{\infty} dx_1 \ldots \int_{-\infty}^{\infty} dx_n e^{-ikx_n} < \tilde{\psi}(0) \tilde{\psi}(x_1) .. \tilde{\psi}(x_n) >_T \times$$
$$\times \tilde{G}(-x_1, s) \tilde{G}(x_1 - x_2, s) .. \tilde{G}(x_{n-1} - x_n, s) \quad (11)$$

where $\tilde{G}$ is the Green function G with the change $D \to D+\Gamma$ and

427

$$\tilde{\psi} = \frac{\xi - \Gamma}{1 + (1 - \mathcal{P})(\xi - \Gamma)/(D + \Gamma)}$$

The precise value of Γ will be fixed by the condition $\langle\tilde{\psi}\rangle = 0$, that is:

$$\Gamma = \frac{\left\langle \dfrac{\xi}{D + \xi} \right\rangle}{\left\langle \dfrac{1}{D + \xi} \right\rangle} \qquad (12)$$

This condition is equivalent to the selfconsistency condition of the E.M.A. found in discrete models.

Concluding with this section we remark our main results. We have obtained a perturbative expansion for short frequencies (11) in the generalised diffusion coefficient $D(k,s)$. The perturbation appears to be around an effective homogeneous medium with a diffusion coefficient that can be calculated by means of a condition similar to the autoconsistency condition of discrete models. Application of (11) to particular cases is the subject of the following sections.

4. EXACT RESULTS FOR A TWO LEVEL PROCESS

Our first example is a disordered medium such that the diffusion coefficient $D+\xi(x)$ has two possible states $D+\Delta$ and $D-\Delta$ with the same probability, that is $\langle\xi(x)\rangle = 0$. For the sake of simplicity we assume $\xi(x)$ to be an exponential correlated process, so that its correlation function has the form

$$\langle \xi(x)\xi(x') \rangle = \Delta^2 e^{-\frac{|x - x'|}{l}} \qquad (13)$$

being l the correlation length. In order to keep the positivity of the diffusion coefficient the amplitude of the process ξ must be smaller than D, $\Delta < D$. The limiting case $D=\Delta$ leads to strong disorder and it will not be treated in this paper.

The application of eq.(12) to this example inmediatly gives an exact effective diffusion coefficient $\tilde{D}$ as:

$$\tilde{D} = D + \Gamma = \frac{D^2 - \Delta^2}{D} \qquad (14)$$

From this formula we can see that the disorder acts reducing the diffusion of the particle. An interesting thing is that the effective diffusion coefficient $\tilde{D}$ is only dependent of the intensity Δ. The independence of the correlation length is a rather surprising fact because it means, for example, that the limit of zero correlation $l \to 0$, which leads to a delta Kronecker correlation noise $\langle\xi(x)\xi(x')\rangle = \Delta^2 \delta_{x,x'}$, has the same effect over the effective diffusion coefficient as other cases with nonzero correlation. We note that a delta correlated noise also appears in the continuum limit of discrete random walks with uncorrelated disorder.[2] Finally we remark the necesity of taking positive random diffusion coefficients $\tilde{D} > 0$ in order to obtain correct results. A violation of this rule $\Delta > D$ would lead in this case to negative diffusion coefficient in (14).

The frequency dependent terms of $D(s)$ are obtained from expression (11). In the case of a process with fixed levels it is not difficult to obtain directly an expression for $\tilde{\psi}$ in terms of ξ. In our case the expression is

$$\tilde{\psi} = \left(1 - \frac{\Delta^2}{D^2}\right) + \left(\frac{1}{D} - \frac{\Delta^2}{D^3}\right)\xi P\xi \qquad (15)$$

As we remarked in a previous section we can obtain low orders in s by calculating only a few terms in (11). Hence, the lowest order $s^{1/2}$ is easily calculated from the first term in (11), giving:

$$D_1 = \left(\frac{D^2 - \Delta^2}{D^3}\right)\left(\frac{sD}{D^2 - \Delta^2}\right)^{1/2} \frac{\Delta^2}{\left(\frac{1}{l} + \left(\frac{sD}{D^2 - \Delta^2}\right)^{1/2}\right)} \qquad (16)$$

$D_1(s)$ can be considered as the first dynamical correction to the static coefficient $\tilde{D}$. We see that as in contrast with the static coefficient now $D_1(s)$ depends strongly on the correlation length. When l is small also the coefficient D_1 is small being zero in the limiting case $l \to 0$. The static limit $l \to \infty$ obviously must reproduce a static diffusion coefficient $D(0,s) = D$ but it is reproduced in a singular form, with the contribution of all frequency dependent coefficients $D_t(s)$ that in this limit become static coefficients.

Finally and after straightforward but lengthly calculations it is possible to obtain higher order terms in s. As an example we show the term of order s which is given by two contributions, one coming from the

$$D_2(s) = \frac{\Delta^4}{D^4} s \frac{1}{\left(\frac{1}{l} + \left(\frac{sD}{D^2 - \Delta^2}\right)^{1/2}\right)^2} \qquad (17 - a)$$

and the other given by

$$D_3(s) \approx \frac{5}{6} \frac{\Delta^4}{D^4} s l^2 \qquad (17 - b)$$

which comes from the third term.
The assymptotic limits of $D_2(s)$ are similar to that of $D_1(s)$.

5. EFFECTIVE DIFFUSION COEFFICIENT FOR A GENERALISED POISSON PROCESS

Our second example is a disordered medium with a generalised Poisson process as diffusion coefficient. This process can be thought of as composed by elementary functions located at points x_i. The x_i points are distributed according to a Poissonian distribution of parameter λ. The elementary functions also follow a stochastic distribution affecting some magnitude of its shape. In this example we have chosen exponential functions with a height ω and decay parameter $1/l$, being ω a random variable exponentially distributed and with mean value ω_0. With this choice we obtain a process $\vartheta(x)$ exponentially correlated with $\lambda \omega_0$ as mean value.
It is more usual to consider process with zero mean value which is easily done taking $\xi(x) = \vartheta(x) - \lambda\omega_0$. Now the correlation function for the process $\xi(x)$ is given by

$$<\xi(x)\xi(x')> = \frac{\lambda\omega_0^2}{l} e^{-|x-x'|/l} \qquad (18)$$

and its probability distribution is:

$$P(\xi) = \frac{e^{-\lambda l}}{\left(\frac{\omega_0}{l}\right) \Gamma(\lambda l)} e^{-\xi/w_0} (\xi + \lambda\omega_0)^{\lambda l - 1} \qquad (19)$$

The positivity of the stochastic diffusion coefficient $D + \xi(x)$ is guaranteed if the condition $D > \lambda\omega_0$ holds.
The effective diffusion coefficient $D + \Gamma$ can be calculated from (12) and (19) giving:

$$\tilde{D} = D + \Gamma = (D - \lambda\omega_0)\left(\frac{(al)^{-\lambda l} e^{-al}}{\Gamma(1 - \lambda l, al)}\right) \qquad (20)$$

429

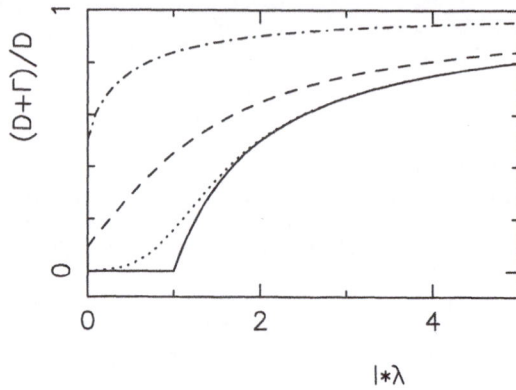

Figure 1.- Plot of the effective diffusion coefficient $(D+\Gamma)/D$ vs correlation length in units of λ^{-1} for different values of D:solid line, D=1;dotted line, D=1.001;dashed line, D=1.1;dotdashed line, D=2. The parameters are taken: $\lambda=1$ and $\omega_0=1$.

where $\quad a = \left(\dfrac{D-\lambda\omega_o}{\omega_o}\right)$ and $\Gamma(\alpha, x) = \displaystyle\int_x^\infty e^{-t}\, t^{\alpha-1} dt.$

In this case the coefficient $\tilde{D}$ depends strongly of the correlation length. A plot of the relative change of the diffusion coefficient $(D+\Gamma)/D$ versus the correlation length in units of λ^{-1} is given in figure 1 for different values of $\lambda\omega_o/D$. In all cases the disorder reduces the diffusion coefficient like in our preceeding example. Now the correlation length plays an important role in the determination of the diffusive properties. The effective diffusion coefficient grows when increasing the correlation length. It can be shown that $(D+\Gamma)/D \to 1$ when $l \to \infty$ (static limit). In the limiting case $\lambda\omega_o/D=1$ (solid line) we can see a transition from weak disorder ($l\lambda>1$) to strong disorder ($l\lambda<1$). This transition is also observable at the level of the probability density $P(\xi)$ in (16) as a change from finite to infinite inverse moments.

REFERENCES

1.- S. Alexander, J. Bernasconi, W.R. Schneider and R. Orbach, Rev. Mod. Phys. 53, 175 (1981), J.W.Haus and K.W.Kher, Phys. Rep. 150, 263 (1987).
2.-E. Hernández García, L. Pesquera, M.A. Rodríguez and M. San Miguel, J. Stat. Phys. 55, 1027 (1989).
3.-J.Heinrichs, Phys. Rev. Lett. 52, 1261 (1984), G. Nicolis, V.J. Altares, J. Phys. Chem., 93, 2861 (1989).
4.-M.A. Rodríguez, E. Hernández García, L. Pesquera and M. San Miguel, Phys. Rev. B40, 4212 , (1989), E. Hernández García, M.A. Rodríguez, L. Pesquera and M. San Miguel, submitted to Phys. Rev. B.
5.-S.Havlin,M.Schwartz, R.Blumberg, A. Bunde, H.E.Stanley. Phys Rev. A40,1717 (1989).
6.-R.H.Terwiel, Physica 74, 248 (1974).
7.-E.Hernández García, L.Pesquera, M.A. Rodríguez, M.San Miguel, Phys Rev. A36, 5774 (1987).

BIMOLECULAR DIFFUSION-LIMITED REACTION KINETICS AT STEADY-STATE

Eric CLEMENT[†], Leonard SANDER[*], Raoul Kopelman[**]

[†]Laboratoire d'Optique de la Matière Condensée, Univ. Pierre
et Marie Curie, 4 place Jussieu 75005 PARIS, FRANCE
[*]Department of Physics & [**]Deparment of Chemistry, University
of Michigan, 48109 Ann Arbor, USA

INTRODUCTION

Physics of condensed matter offers numerous examples of reactive systems where the transport of reactants such as atoms, molecules or any localized excitation, is of the diffusive type. In general these systems exhibit two distinct time scales. One is a typical time of reaction between reactants and the other is a characteristic time associated with the microscopic erratic movements of the particles, often referred to as a time of jump The limiting process is called *diffusion limited* when at the time scale characterizing a microscopic jump, particles seem to react "instantaneously" when they are in contact at a microscopic distance (reaction radius). Bimolecular diffusion-limited reactions of the type $A_i + A_j -> Product$, where A_i and A_j are distinct or similar rectants, are extensively investigated since they represent many important physical situations. The product formed by the reaction can be an aggregate, a third particle A_k, a particle A_i or A_j, or both particles may annihilate in pairs and leave the system. The range of applications spanned by these models covers particle aggregation, gelation of sols, coalescence of aerosols, electron-hole recombination in semi-conductors, soliton-antisoliton annihilation in polymers, exciton fusion in molecular crystals, etc.... Classically, the point of view associated with these different physical situations is essentially the same. The reaction rate Q_{ij} is:

$$Q_{ij} = K_{ij}\, \rho_i \rho_j \qquad (1)$$

where K_{ij} is the a constant, dependent only on the diffusion constant and the reaction radii of the particles, and ρ_i, ρ_j are respectively the concentrations in reactants i and j. This is the classical rate equation or mean field reaction law. The classical order of reaction X, is the sum of the exponents for each concentration: $X = 1+1 = 2$. In this picture, the local distribution of reactants is regarded as homogeneous. In fact, the real phenomenology has sometimes very little resemblance to the predictions of the classical theory. In general, each situation has an associated critical dimension below which spatial fluctuations of density drive the system to a new behavior (the critical dimension may even be infinite). In this paper we are only interested in three elementary reactions of the type A+A->0, A+B->0 and A+T->T where A and B are different mobile reactants and T is a fixed trap. We propose here to investigate the steady state properties of these reactions taking place on regular euclidean spaces and on fractal structures, in the diffusion limited regime. This work is a natural extension of the activity dealing with relaxation phenomena and transient reaction kinetics problems in disordered media[1]. Its domain of application spans various areas of physics and chemistry of condensed matter. For example, reactions of the type A+A->0 or A+T->T are models describing exciton fusion kinetics in disordered molecular crystals or polymer blends[2]. Reactions of the type A+B->0 are found in solid state physics in the case of electron-hole annihilation or defect fusion. Furthermore, all the concepts emphasized here may also find applications in the field of surface catalysis. We show here the departure of these systems from the classical picture and we propose a new unified vision solely based on intrinsic properties of the diffusion process.

COMPACT AND NON COMPACT RANDOM WALK

In previous work[3-6] we have found that the relevant parameter describing the steady state of the reaction kinetics is the *spectral dimension* d_s[7]. The spectral dimension is an intrinsic parameter of a medium characterizing energy transfer properties, and in particular, diffusion. For euclidean structures, d_s is the euclidean dimension d and the case of euclidean spaces is viewed as an extension of the fractal case when we take $d = d_s$. The reason for the influence of the spectral dimension on reaction kinetics is contained in the fact that d_s controls the time dependence of the number of *distinct* sites visited by a random walker S_N. For a medium with a spectral dimension $d_s > 2$ (a 3D euclidean space for instance), the number of distinct sites visited by a walker during a N step random walk is:

$$S_N \approx N$$

(2a)

Thus S_N grows linearly with respect to time and the walk is called non-compact[8], i.e. there is only few revisitations of the sites explored and the walker escapes easilly for its original starting point. For $d_s < 2$, S_N grows sublinearly with time, we have:

$$S_N \approx N^{d_s/2}$$

(2b)

The walk is called compact[8], i.e. there is a lot of revisitations of the sites and the escape probability of a walker from a given site goes to zero a long time. The case $d_s = 2$ is marginal and we have the logarithmic behavior:

$$S_N \approx N/\ln N$$

(2c)

The following will show the fundamental aspect of the quantity S_N in the reaction problem.

UNIFIED VISION AT STEADY STATE

In bimolecular diffusion limited processes the overal balance between reaction rates Q and steady state densities is accounted for by the Smoluchowski boundary condition[9]:

$$Q \approx \frac{\rho_1 \rho_2}{\Lambda}$$

(3)

where ρ_1 and ρ_2 are respectively the steady state densities of reactants 1 and 2 (1 and 2 can be identical species).and Λ is a typical scale of separation between reactants that we defined to be the *self-organization scale.*. In the classical picture, Λ is on the order of radius of reaction *a*, which is the microscopic size of the system. We have studied theoretically each reaction process on euclidean spaces and fractal structures using various approaches like equation hierarchy decoupling and langevin equation solving etc... We came to the conclusion that the scale Λ could be mesoscopic or even macroscopic depending on the spectral dimension of the medium. This phenomenon was called a self-organization of reactants and is splitting the unity of the classical description into a zoology of particular cases depending on the system considered. However, we have found[3-6] that an unified description could be recovered if we claim for each system, the existence of a characteristic time τ which can be interpreted as an average lifetime for the particles in the medium, then self-organization scale Λ can be cast into the form:

$$\frac{\Lambda}{a} \approx \frac{V_\tau}{S_\tau}$$

(4)

where V_τ ithe total or cumulative volume swept out by a particle during τ and S_τ is the effective volume explored by a particle (number of distinct sites visited) during τ. In figure I we sketch these volumes. Therefore, for spectral dimension $d_s < 2$ we have a self organization of reactants up to a scale Λ such that :

$$\Lambda \approx \tau^{1-d_s/2}$$

(5)

For $d_s > 2$, Λ is microscopic and independent of τ, therefore no large scale structure exists and the reaction kinetics is classical. The case $d_s = 2$ is found to be the critical dimension of the problem. Then we find a marginal logarithmic dependence of Λ with τ. Below the critical dimension, large scale density fluctuations become relevant and each situation has its own phenomenology. In particular, we may find macroscopic reaction laws with anomalous reaction orders (bigger than 2) or anomalous rate constants . In the following we will consider for each reaction case the specific implications of equation (4).

HOMOMOLECULAR ANNIHILATION A+A->0

In the case of homomolecular annihilation[4], A+A->0, Λ is a typical scale of *depletion* around each reactant and τ is the typical reactant life-time with:

$$\tau \approx \frac{\rho}{Q}$$

(6)

where ρ is the steady state density of A. Combining equations (3), (5) and (6) we obtain an anomalous effective reaction order in the low density limit:

$$X = 1 + 2/d_s \quad \text{for } d_s < 2$$

On figure II we show a snap-shot of a distribution of reactants on a percolation cluster.

TRAPPING PROBLEM

For the trapping problem[5], A + T -> T, the fluctuation of the trap distribution is found to be irrelevant to the leading scaling behavior of the self organization length Λ at low densities. The relevant fact is that we have, for $d_s < 2$, a *depletion* of particles A around the traps. The typical lifetime at steady state is

$$\tau \approx \frac{\rho}{Q}$$

with ρ the density of A and Q the reaction rate. The scale of the trap-particle organization is:

$$\Lambda \approx c^{1-2/d_s} \quad \text{for } d_s < 2$$

where c is the trap concentration. We have the anomalous rate law:

$$Q \approx \rho \, c^{2/d_s}$$

with an anomalous order relatively to the trap concentration:

$$X = 2/d_s$$

On figure II we show a snap-shot of a distribution of reactants and traps on a percolation cluster. We note that the overall reaction order is $1 + 2/d_s$, the same as for the A+A->0 case.

HETEROMOLECULAR ANNIHILATION A+B->0

In the case of heteromolecular annihilation[3-6], A + B ->0, Λ is the scale of a self-organization phenomenon called *segregation*. At steady state, domains of identical species with sizes comparable to Λ, build up in the medium. The situation is more complex than in the homomolecular reaction case and τ is found to be dependent either on source conditions or on some intrinsic particle lifetime. We separated the source terms into two main categories. In a first category we consider sources for which at any time an identical number of As and Bs is conserved in the medium. If reactants are created at random, we find:

$$\tau \approx L^2 \qquad (7a)$$

where Λ is the system size. We observe a size dependent segregation. With the same conservation constraint, if the particles are created as A-B pairs with A and B separated by a distance δ, we have[1]:

$$\tau \approx \delta^2 \qquad (7b)$$

The segregation scale dependson δ. It is important to notice that for a geminate creation (reactants created at a distance δ on the order of a), we obtain a microscopic segregation scale and this situation becomes analogous to classical kinetics. In a second category, we consider sources where the conservation constraint is removed. If no other decay mechanism is present, fluctuations in particle difference grow untill we have a complete *saturation* of the medium with one of the species. There is no reactive steady state. If an extra decay mechanism is considered, fluctuations grow up to a size defined by τ which turns to be the intrinsic lifetime of the decay mechanism. If we consider vertical annihilation[3] (different particles react when landing on the top of each others) with an external rate of particles R, we have:

$$\tau \approx R^{-1}$$

In this case we obtain at low density an effective reaction order:

$$X = 4/d_s$$

On figure IV we show a snap-shot of a distribution of reactants A and B on a percolation cluster in the vertical annihilation case.

If the decay is controlled by an intrinsic mechanism A->0 and B->0, with the same rate constant K, then we have

$$\tau \approx K^{-1}$$

We induce a K dependent segregation but no anomalous order of reaction. These last three cases are important for practical applications because, besides geminate particle creation, it is difficult to find a source satisfying the exact conservation constraint. However, though the conservation is not exact, these cases lead to a mesoscopic segregation.

It is important to realize that all the results presented here on a scaling form are mostly valid in the low reactant density limit. We have also shown that larger densities sometimes imply significant deviations from the scaling laws.

CONCLUSION

In this paper we present a self-organization mechanism triggered by the spectral dimension characterizing the diffusion of reactants on a medium. More specifically, we identify the compactness of the walk (number of distinct sites visited) to be at the origin of the phenomenon. The upper critical dimension is found to be 2. It is important to realize that the type of self-organization presented here is by nature different from complex chemical reaction processes like the Belousov-Zhabotinsky reaction. For such processes, the reactant organization originates from the complex interplay between diffusive transport and non linear local reaction laws, and though even a large scale ordering may be found, a complete mixing of reactants is assumed at the *local* level (classical limit). In constrast, in our case, a mesoscopic organization exists at the local level and may significantly transform the local laws of reaction (anomalous order of reaction).

This work was supported by NSF grants 88-42001, DMR 88-01120 (RK) and DMR 88-15908 (LMS). One of us (EC) thanks CNRS UA-800 its financial support.

---

[1] The generalization of conditions (7a) and (7b) to fractal structures, is a conjecture based on consistency with the general picture presented in this paper. It is proven to be correct for euclidean spaces only.

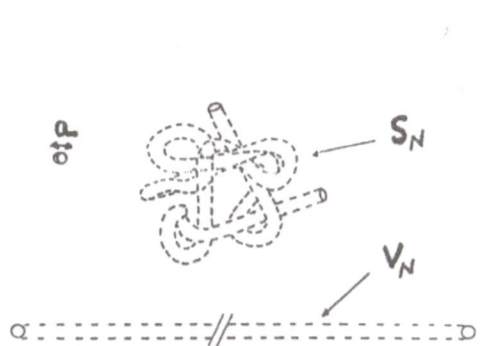

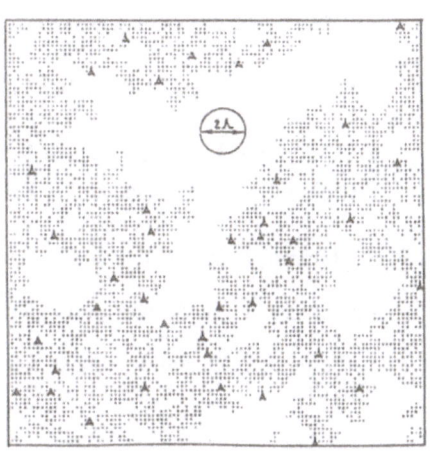

Fig. 1. Sketch of volumes S_N and V_N swept out by a particle of size a during a N steps random-walk. S_N does not account for multiple revisitations, V_N does.

Fig. 2. Distribution of reactants A for the A+ A annihilation at steady state, on a percolation cluster. The radius of the circle has the size of the depletion length Λ

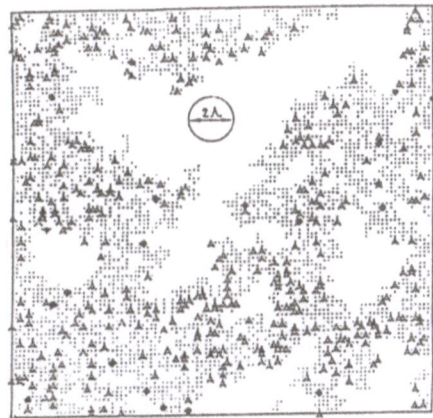

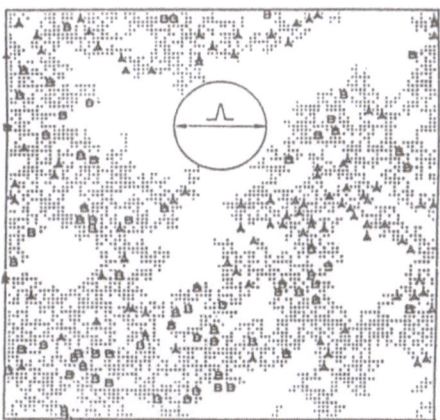

Fig. 3. Distribution of reactants A for the A+ T ->T reaction at steady state, on a percolation cluster. The dot symbols are the fixed traps. The radius of the circle has the size of the depletion length Λ

Fig. 4. Distribution of reactants A and B for the A+ B annihilation at steady state, on a percolation cluster. The source is with vertical annihilation. The radius of the circle has the size of the segregation length Λ.

REFERENCES

1. A.Blumen, J.Klafter and G.Zumofen, in *Optical Spectrocopy of glasses*, ed. I. Zschokke (Reidel Publ. Co., Dordrecht, Holland(1986)).

2. R.Kopelman, Science **241**, 1620 (1988).

3. a) E.Clément, L.M. Sander and R.Kopelman, Phys.Rev.A **39**, 6455 (1989); b) E.Clément, L.M. Sander and R.Kopelman, Phys.Rev.A **39**, 6466 (1989).

4. E.Clément, L.M. Sander and R.Kopelman, Phys.Rev.A **39**, 6472 (1989).

5. E.Clément, L.M. Sander and R.Kopelman, Euro .Phys.Lett.**11** (8), 707 (1990).

6. E.Clément, L.M. Sander and R.Kopelman, *to be published in* Chemù Phys (1990).

7. a) S. Alexander and R. Orbach, J. Physique Lett., **43**, 625 (1982). b) R. Rammal and G. Toulouse, J. Physique. Lett., **54**, 44, L13 (1983).

8. P.G. de Gennes, J. Chem. Phys. **76**, 3316 (1982).

9. Von Smoluckowski, Z.Phys. Chem. **29**, 129 (1917).

DYNAMIC PERCOLATION THEORY FOR DIFFUSION OF INTERACTING PARTICLES:
TRACER DIFFUSION IN A MULTI-COMPONENT LATTICE-GAS

Rony Granek

School of Chemistry, Sackler Faculty of Exact Sciences
Tel-Aviv University
Tel-Aviv 69978, Israel

ABSTRACT

Dynamic percolation theory is used to obtain the tracer diffusion coefficient in multicomponent mixtures of "non interacting" lattice-gas (with only blocking interactions, i.e. double occupancy of a lattice site is forbidden) within the effective medium approximation (EMA). Our approach is based on regarding the background particles as a changing random environment for the tracer. The result is expressed in terms of local fluctuation time parameters, which we attempt to determine from the lattice-gas dynamics. Special attention is given to the single component and the binary mixture cases, were we compare two possible choices for these parameters. The resulting tracer diffusion coefficient for both choices compares well with numerical simulations whenever single bond dynamics and single bond EMA are expected to be reliable.

I. MOTIVATION AND APPROACH

Diffusion of independent particles in static percolating networks has been thoroughly investigated in the past two decades.[1,2] However, in reality the diffusing particles interact among themselves, thus limiting the validity of these studies to extremely low concentrations. A more general situation is that of a binary mixture of diffusing particles. The latter case reduces to the former when particles of one kind are infinitely slow relative to the other kind.

In this work we examine the applicability of the recently developed dynamic bond percolation (DBP) theory to these problems. The DBP theory was originally designed to calculate the effective diffusion coefficient of a single random walker in a dynamically changing neighborhood. In the model developed by Druger, Ratner and Nitzan[3-8] the whole network is "renewed" with a given waiting time distribution, while in the model of Harrison and Zwanzig[9] (HZ) the flucuations within single bonds are considered. The simplest versions of both theories give identical results for the diffusion rate. The HZ theory was later extended by Granek and Nitzan[10] to include many bond transition rates and bond-bond correlations.

The HZ formalism and it's extensions, which are cast in the framework of effective medium theory (EMA), are more easily adapted to the present applications and are therefore used here. Our approach is based on the observation that the background particles can be viewed as a changing random environment for the tracer. We limit ourselves here to the so called "non interacting" lattice-gas (LG) where only

Large-Scale Molecular Systems, Edited by W. Gans *et al.*
Plenum Press, New York, 1991

blocking interactions are taken into account (namely double occupancy of a site is forbidden). The situation becomes more complicated when longer range interactions are included and first steps to deal with this problem were already described.[12]

II. THE MULTICOMPONENT MIXTURE

Consider an n component (not counting vacancies) "non interacting" lattice-gas (blocking interactions only) whose constituents are $A_1, A_2, A_3, ..., A_n$ with corresponding elementary jump rates $\Gamma_1, \Gamma_2, \Gamma_3, ..., \Gamma_n$. We consider the diffusion of a tracer O particle with an elementary jump rate Γ_0 embedded in this LG. Focussing on a nearest-neighbor (NN) bond to the tracer O particle we can define a stochastic state variable ξ for this bond which is related to the kind of particle that occupies the corresponding NN site to the tracer, either $A_1, A_2, A_3, ..., A_n$ or neither of them - V (for vacancy). Thus, ξ can take the symbolic "values" $\xi = A_1, A_2, A_3, ..., A_n, V$. The stochastic (dimensionless) jump rate of the tracer particle to that NN site is given by the function $\sigma(\xi)$ defined as

$$\sigma(\xi) = \begin{cases} 1 & \text{if } \xi = V \\ 0 & \text{if } \xi = A_1, ..., A_n \end{cases} \qquad (\text{II.1})$$

Thus each bond is associated with only two basic rates, 0 and 1, however $\sigma(\xi)=0$ corresponds to many states of a bond. The transitions between the bond states is *assumed* to be described by a characteristic *Markovian* rate equation

$$\frac{\partial}{\partial t} f_\alpha(\xi,t) = \sum_{\xi'} \Omega_\alpha(\xi,\xi') \, f_\alpha(\xi',t) \qquad (\text{II.2})$$

where $f_\alpha(\xi,t)$ is the probability that bond α is in state ξ at time t and $\Omega_\alpha(\xi,\xi')$ is the characteristic rate matrix. Its elements have to be found from the lattice-gas dynamics itself.

The elements $\Omega(\xi,\xi')$ are determined in the following way: First, we do not allow a direct exchange between the particles. This is impossible because of the blocking interactions. Secondly, we choose them to obey detailed balance conditions. Thus, given the concentrations $\{c_i\}$ of the $\{A_i\}$ (i=1,...,n) components, the matrix elements $\Omega(\xi,\xi')$ are is given by

$$\Omega(A_i, V) = c_i/\tau_i \ ,$$
$$\Omega(V, A_i) = c_V/\tau_i \ ,$$
$$\Omega(A_i, A_j) = -(c_V/\tau_i) \, \delta_{ij}$$

$$\Omega(V, V) = -\sum_i c_i/\tau_i \qquad (\text{II.3})$$

where $c = \Sigma_i c_i$ is the total concentration of particles and $c_V = 1-c$ is the vacancy concentration. The fluctuation time parameters $\{\tau_i\}$ (i=1,...,n) are in principle functions of the jump rates Γ_i and the concentrations $\{c_i\}$. The simplest choice for these parameters is obtained from the following mean-field approach. First we find the mean-field jump rate of an i type particle from a site which is NN to our tracer particle. This rate is $c_V(Z-1)\Gamma_i$, where Z-1 instead of Z appears because the tracer site is excluded. Comparing these to the corresponding rate implied by Eqs. (II.2) and (II.3), c_V/τ_i, we find that

$$\tau_i^{-1} = (Z-1)\Gamma_i \qquad ; i=1, ..., A_n . \qquad (\text{II.4})$$

The tracer random walk is approximately described by a stochastic master equation for the walker probability $P_i(t)$ to be at site i at time t

$$\frac{d}{dt} P_i(t) = \Gamma_0 \sum_{j \in \{i\}} \sigma[\xi_{ij}(t)] \left[P_j(t) - P_i(t) \right] \tag{II.5}$$

where $\{i\}$ denotes the group of sites nearest-neighbors to i.

Eqs. (II.2) and (II.5) involve three approximations: (a). we use a bond fluctuation model for what is actually a site fluctuation dynamics, (b). we neglect correlations between occupying-deoccupying events on neighbouring sites, and (c). we disregard the difference between the dynamics of the NN sites which is dominated by $Z-1$ neighbours (the tracer site is not counted) and the dynamics of the other sites which is dominated by the availability of Z neighbours.

Our aim is to average over Eq. (II.5) in conjunction with Eq. (II.2). This generally leads to the effective-medium equation

$$\frac{d}{dt} \langle P_i(t) \rangle = \Gamma_0 \sum_{j \in \{i\}} \int dt' \, \tilde{\psi}(t-t') \left[\langle P_j(t') \rangle - \langle P_i(t') \rangle \right] \tag{II.6}$$

and the problem is to determine the effective rate (memory kernel) $\tilde{\psi}(t)$. An EMA solution for this problem has been obtained by Granek and Nitzan as an extension of the HZ treatment[10,11]. The result is that the frequency dependent (dimensionless) effective jump rate $\psi(\omega)$, the Fourier-Laplace transform of $\tilde{\psi}(t)$, is determined from a secular equation that involves the eigenvalues and the left eigenvectors of the transition matrix Ω, and the lattice Green's functions at the origin.[2] The dependence of $\psi(\omega)$ on ω and on these eigenvalues λ_ℓ enter *only through combinations of the form* $i\omega + \lambda_\ell$. The tracer diffusion coefficient is then given by

$$D_{to}(\omega) = \Gamma_0 \, \psi(\omega) \, a^2 \tag{II.7}$$

where a is the lattice constant (henceforth we use $a=1$).

III. THE SINGLE COMPONENT CASE

Assume a tracer O particle with jump rate Γ_0 to diffuse among otherwise identical A particles whose jump rate Γ ($\equiv \Gamma_A$) and concentration c ($\equiv c_A$). Thus Eq. (II.2) reads

$$\frac{\partial}{\partial t} \begin{pmatrix} f(A,t) \\ f(V,t) \end{pmatrix} = \frac{1}{\tau} \begin{pmatrix} -(1-c) & c \\ 1-c & -c \end{pmatrix} \begin{pmatrix} f(A,t) \\ f(V,t) \end{pmatrix} \tag{III.1}$$

For this case the formalism leads to the following self-consistent equation for the effective hopping rate ψ[9-12]

$$\psi(\omega) = 1 - \frac{c}{1 - p_c + p_c \, \epsilon g(\epsilon)} \tag{III.2}$$

where

$$\epsilon = \frac{i\omega + \tau^{-1}}{\Gamma_0 \, \psi(\omega)} \tag{III.2a}$$

and

$$p_c = 2/Z \tag{III.2b}$$

is the EMA percolation threshold, and where $g(\epsilon)$ is the lattice Green's function of the origin.[2]

In order to use Eq. (III.2) to find the effective hopping rate ψ and therefore the associated tracer diffusion coefficient, we need an explicit expression for τ. Here we consider two choices:

(a). Make a mean-field approximation for the motion of a background particle which is NN to the tracer. This leads to[12] (c.f., Eq. (II.4))

$$\tau^{-1} = (Z-1)\Gamma \tag{III.3}$$

(b). Make a mean-field approximation to the equation describing the site occupation dynamics, which is governed by the chemical diffusion coefficiet, to obtain[12] an equation of the type (III.1). Thus τ can be identified with the relaxation time for *density fluctuations* in sites NN to the the tracer particle and is therefore related to the chemical diffusion coefficient. In the present case however the chemical diffusion coefficient is independent of particle concentration and this procedure yields again[12] Eq. (III.3).

In the DC ($\omega=0$) and low vacancy concentration ($c\rightarrow 1$) limits, Eqs. (III.2) and (III.3) yield (with $\gamma=\Gamma_0/\Gamma$)

$$\psi = \frac{Z-1}{Z-1+2\gamma}(1-c) \tag{III.4}$$

It is interesting to note that Kikuchi[13] has obtained the same result using the path probability method in the pair approximation.

Numerical results of our approximation on an FCC lattice ($Z=12$) were compared[12] to numerical simulations of Kehr, Kutner and Binder.[14] Fig. 1 shows the DC ($\omega=0$) correlation factor $f=(1-c)^{-1}\psi$ for self-diffusion ($\Gamma_0=\Gamma$) as a function of particle density c. The full line is the theoretical result of Sankey and Fedders[15] (based on diagramatic methods) and the dashed line is our result based on Eqs. (III.2) and (III.3) with $\Gamma_0=\Gamma$. The simulation results are taken from Ref. 14a.

We have also compared[12] the correlation factor f for a tracer particle O in a background of A particles, for a few orders of magnitude change in $\gamma=\Gamma_0/\Gamma_A$ and for different concentrations c_A. It was found that there is a good agreement between our simple theoretical approximation and the simulation results on the NILG both for the self-diffusion problem and for the mixed tracer diffusion problem provided that the background particle density c_A is not close to the percolation thershold. In other cases, a substantial improvement of the results was obtained when we replaced in Eq. (III.2) the EMA bond percolation threshold $p_c=1/6$ (c.f. Eq. (III.2b)) by the exact site percolation threshold $p_c=0.199$.[1]

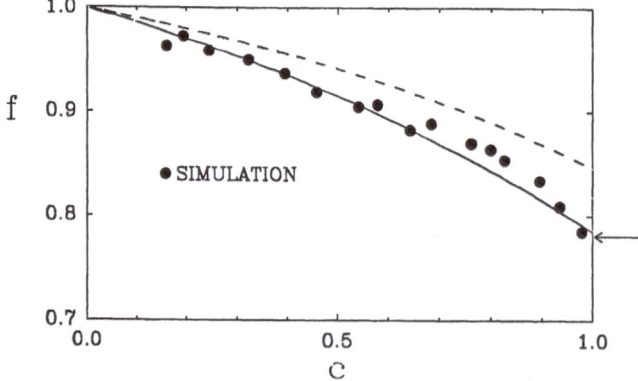

Fig. 1.   Correlation factor f plotted against concentration c for tracer self diffusion in a NILG system on FCC lattice ($Z=12$). The dashed line is a numerical solution to Eq. (III.2) (with $\omega=0$) using Eq. (III.3) for τ. The simulation results are from ref. (14a). The full line is the theory of Sankey and Fedders[15]. The arrow denotes the exact result for $c\rightarrow 1$.

IV. THE BINARY MIXTURE

Here we consider a mixture of two types of particles, A and B, with jump rates Γ_A and Γ_B respectively. Thus, the bond state variable ξ can take the values A, B or V. The dynamical matrix Ω is therefore 3x3. The elements of Ω are given by Eq. (II.3). Thus, with the concentrations of the two components denoted by c_A and c_B and the (yet undetermined) fluctuation times denoted by τ_A and τ_B, the bond state dynamics is described by[11]

$$\frac{\partial}{\partial t} \begin{bmatrix} f(A,t) \\ f(V,t) \\ f(B,t) \end{bmatrix} = \begin{bmatrix} -c_V/\tau_A & c_A/\tau_A & 0 \\ c_V/\tau_A & -(c_A/\tau_A + c_B/\tau_B) & c_V/\tau_B \\ 0 & c_B/\tau_B & -c_V/\tau_B \end{bmatrix} \begin{bmatrix} f(A,t) \\ f(V,t) \\ f(B,t) \end{bmatrix} \tag{IV.1}$$

where $c = c_A + c_B$ is the total concentration of particles and $c_V = 1-c$ is the vacancy concentration.

A solution to this model has been is obtained using our general formalism. In the remaining we shall discuss the fluctuation times τ_A and τ_B. Consider again the two choices for these parameters:

(a). Use the mean–field jump rates of the A and B particles from a site which is NN to the tracer particle. This leads to[11] (c.f., Eq. (II.4))

$$\tau_A^{-1} = (Z-1)\Gamma_A \tag{IV.2a}$$

and

$$\tau_B^{-1} = (Z-1)\Gamma_B \tag{IV.2b}$$

This is a reasonable approximation in many situations. We note however that this choice disregards any possible dependence of τ_A and τ_B on the composition of the mixture. In particular, in the static A limit $\tau_A \to \infty$, one expects some effect of the percolation properties of the static random network made by the A particles on the fluctuation time τ_B. Such effects are absent in Eqs. (IV.2).

(b). Relate the fluctuations times to the chemical diffusion coefficients of the system because they govern concentration fluctuation relaxations. In our binary mixture there are four such chemical diffusion coefficients: D_{AA}, D_{AB}, D_{BA} and D_{BB}. They are defined as the coefficients of the two coupled phenomenological diffusion equations[16,17]

$$\frac{\partial}{\partial t} \begin{bmatrix} c_A(x,t) \\ c_B(x,t) \end{bmatrix} = \begin{bmatrix} D_{AA} & D_{AB} \\ D_{BA} & D_{BB} \end{bmatrix} \begin{bmatrix} \nabla^2 c_A(x,t) \\ \nabla^2 c_B(x,t) \end{bmatrix} \tag{IV.3}$$

Here the diffusion coefficients may depend in a complicated way on the concentrations $c_A = \langle c_A(x,t) \rangle$ and $c_B = \langle c_B(x,t) \rangle$, on the jump rates Γ_A and Γ_B, and on interaction parameters (in the interacting LG case). Using the discrete form of Eq. (IV.3), applying to it a mean–field approximation and comparing with Eq. (IV.1) we arrive at[11]

$$\tau_A^{-1} = (Z-1)\left[D_{AA} + \frac{c_B}{c_A} D_{AB} \right] \tag{IV.4a}$$

$$\tau_B^{-1} = (Z-1)\left[D_{BB} + \frac{c_A}{c_B} D_{BA} \right] \tag{IV.4b}$$

Like Eqs. (IV.2), Eqs. (IV.4) reduce to $\tau_A^{-1} = \tau_B^{-1} = (Z-1)\Gamma$ when $\Gamma_A = \Gamma_B \equiv \Gamma$ and are therefore consistent with the single component treatment.

In Figs. 2 and 3 we present some numerical solutions to the EMA equation[11] obtained for this case and for both choices for τ_A and τ_B. We show the theoretical tracer diffusion coefficient of B particles ($\Gamma_0 = \Gamma_B$) on a square lattice for the static case $\gamma = \Gamma_B / \Gamma_A = \infty$ (Fig. 2) and for $\gamma = 10$ (Fig. 3), and numerical simulation results for this case. The simulations were performed on a 100x100 square lattice with periodic boundary conditions using standard Monte-Carlo technique. Since in both cases ($\gamma = \infty$ and $\gamma = 10$) the motion of the tracer particle is expected to be sensitive to the percolation threshold, we use (in our analytical results that produce Figs. 2 and 3) the exact site percolation threshold $p_c = 0.59275$ rather then the (EMA) bond percolation threshold $p_c = 0.5$.

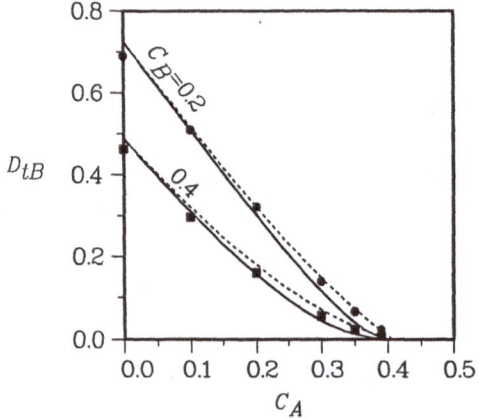

Fig. 2.   Tracer diffusion coefficient D_{tB} (in units of $\Gamma_B a^2$) in a static A background on a square lattice, plotted against c_A for different concentrations c_B. The lines are theoretical results using Eqs. (IV.2) (full lines) and (IV.4) (dashed lines) for τ_B. The symbols represent our simulation results for $c_B = 0.2$ (circles) and $c_B = 0.4$ (squares). The estimated error of these results is within 5% for $c_A \leq 0.35$ and about 25% for $c_A = 0.39$.

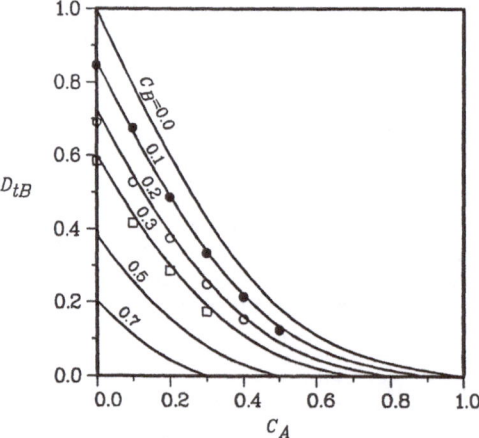

Fig. 3.   Tracer diffusion coefficient D_{tB} for $\gamma = \Gamma_B / \Gamma_A = 10$ in a square lattice. The lines are theoretical results using Eqs. (IV.4) for τ_A and τ_B. The results of using Eqs. (IV.2) for these times are less then 1% different from those plotted. The symbols are our simulation results: filled circles - $c_B = 0.1$, open circles - $c_B = 0.2$, open squares - $c_B = 0.3$.

442

V. CONCLUSIONS

In this work we applied dynamic percolation theory to develop an effective medium approximation for the diffusion of mixtures of particles with hard core interactions. The resulting EMA expression for the tracer diffusion coefficient is expressed in terms of relaxation times for local concentrations fluctuations in the vicinity of the tracer. The latter are expressed in terms of the chemical diffusion coefficients of the system; however, simpler mean-field expressions also work quite well. A treatment in the DBP framework of a non-interacting lattice-gas in a static percolating *bond* network[18] leads to similar results.

The method advanced here can be improved consistently by using many bond EMAs (cluster EMAs) to include correlations that exist in the bond network. 'Static' correlations are a result of the site randomness property while 'dynamic' correlations exist between occupation-deoccupation events in nearest-neighbour sites. An example of partially incorporating the dynamical correlations is already described in Ref. 12. Taking into account the static correlation can help to avoid the nessecity for an artificial replacement (in the EMA equations) of the EMA percolation threshold by the actual site percolation threshold.

ACKNOWLEDGEMENT

I am greatful to Prof. Abraham Nitzan for close colaboration. I also thank Marvin Silverberg for the use of his computer programs for lattice diffusion of interacting particles and for useful discussions. Discussions with Mark A. Ratner, S.D. Druger, J. Klafter, M. Bixon, K.W. Kehr and B. Whaley are also greatfully acknowledged, and special thank is given to Prof. Mark Ratner for hospitality when part of this work was done.

REFERENCES

(1). D. Stauffer, *Introduction to Percolation Theory* (Taylor & Francis, London, 1985), chap. 5.

(2). M. Sahimi, B.D. Hughes, L.E. Scriven and H.T. Davis, J. Chem. Phys. **78**, 6849 (1983).

(3). S.D. Druger, A. Nitzan and M.A. Ratner, J. Chem Phys. **79**, 3133 (1983).

(4). S.D. Druger, M.A. Ratner and A. Nitzan, Phys. Rev. *B* **31**, 3939 (1985).

(5). S.D. Druger in *Transport and Relaxation Processes in Random Materials*, edited by J. Klafter, R. J. Rubin and M.F. Shlesinger (World Scientific, Singapore, 1986).

(6). R. Granek, A. Nitzan, S.D. Druger and M.A. Ratner, Solid State Ionics **28-30**, 120 (1988).

(7). S.D. Druger and M.A. Ratner, Chem. Phys. Lett. **151**, 434 (1988)

(8). S.D. Druger and M.A. Ratner, Phys. Rev. *B* **38**, 12,589 (1988)

(9). A.K. Harrison and R. Zwanzig, Phys. Rev. *A* **32**, 1072 (1985).

(10). R. Granek and A. Nitzan, J. Chem. Phys. **90**, 3784 (1989)

(11). R. Granek and A. Nitzan, to be published.

(12). R. Granek and A. Nitzan, J. Chem. Phys. **92**, 1329 (1990).

(13). R. Kikuchi, Prog. Theor. Phys. Suppl. **35**, 1 (1966).

(14). (a). K. W. Kehr, R. Kutner and K. Binder, Phys. Rev. *B* **23**, 4931 (1981).
    (b). R. Kutner and K. W. Kehr, Phil. Mag. *A* **48**, 199 (1983).

(15). O. F. Sankey and P. A. Fedders, Phys. Rev. *B* **15**, 3586 (1977).

(16). S. R. De Groot and P. Mazur, *Non-Equilibrium Thermodynamics* (North-Holland, Amsterdam, 1969).

(17). K. W. Kehr, K. Binder and S. M. Reulein, Phys. Rev. *B* **39**, 4891 (1989).

(18). M. Silverberg, M.A. Ratner, R. Granek and A. Nitzan, J. Chem. Phys. (in press).

CROSSOVER FROM DISPERSIVE TO DIFFUSIVE ENERGY TRANSPORT

T. Kirski and C. von Borczyskowski

Freie Universität Berlin
Fachbereich Physik
Arnimallee 14
D-1000 Berlin 33

INTRODUCTION

Excitation energy transport in disordered solids is one of the relaxation phenomena studied intensively during the past ten years. Especially transport in glasses has attracted considerable interest[1], but also some porous systems[2], polymers[3] and disordered crystals[4] have been investigated. Disordered crystals offer the possibility to model complex relaxation behavior in a distinct and well-defined manner. For this reason we have chosen a chemically mixed crystal (CMC) of p-dichloro-(DCB) in p-dibromobenzene (DBB) as a representative example of a substitutionally disordered crystal. In recent publications we have shown that triplet excitation energy transport exhibits dispersive character[5] as a function of doping concentration. Instead of using isotopically mixed crystals[6] CMC show due to the formation of induced energy funnels[7,8] a richer energy landscape which gives rise to a broad waiting time distribution for jumps of excitation energy described by random walk processes[9]. Phosphorescence measurements indicated[10] that at temperatures above about 6 K transport becomes strongly temperature dependent. This gave rise to a comparison of transport in disordered crystals with relaxation processes in spin glasses[11]. Within this concept energy transport exhibits as a function of temperature a kind of glass transition which is manifested in a cross-over from ergodic to non-ergodic behavior[11]. Above the glass transition, the phosphorescence decay of the excitation energy can be described by a stretched exponential Kohlrausch-Williams-Watts (KWW) behavior. In this communication we like to report on the temperature dependence of the parameters β and τ, which enter the KWW description via the phosphorescence intensity $I_{Phos} \sim \exp[-(t/\tau)^{\beta}]$.

CHARACTERIZATION OF THE SYSTEM

DCB replaces DBB molecules in a substitutional way[12] and forms statistically mixed crystals at any concentration[5]. Strong guest-host interactions give rise to the formation of energy funnels including distorted DBB host molecules which have triplet energy levels approximately in the middle of the energy separation of 53 cm^{-1} between DCB guest molecules and the DBB exciton band[7]. Increasing concentration results in the build-up of guest aggregates such as dimers and trimers, which serve as energy sinks when excitation energy initially deposed into DCB monomers is migrating over the crystal.

Fig. 1 shows schematically that at high concentrations energetically lowered clusters are formed which consist of overlapping energy funnels having statistically distributed energy sinks incorporated in form of DCB aggregates. Donor states within the cluster are separated by energy barriers of 26 cm^{-1}. Phosphorescence and ODMR experiments show that below 6 K energy transport is not thermally activated and may be described by a tunneling process[13]. Within one cluster or even among those clusters which may be reached within the DCB triplet liftetime of about 16 ms energy transport is described by a KWW behavior. Increasing temperature will give rise to thermally activated transport between more distant clusters (valeys), which will result in a dense network of connected valeys at sufficiently high temperatures. The activation energy for this process has been found to correspond with the energy separation of 53 cm^{-1} between DCB monomers and the DBB host exciton band[13].

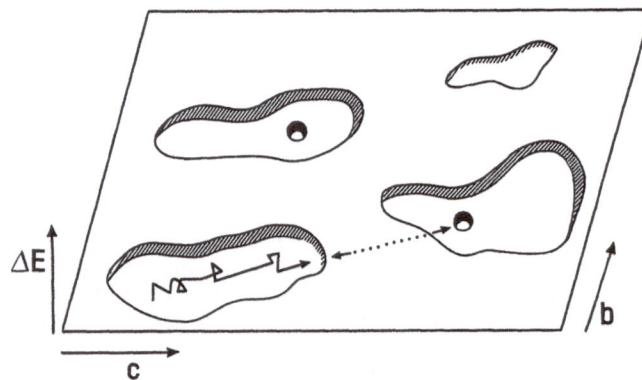

Fig. 1 Schematic view of the energy landscape of DCB energy
funnels in DBB. The plateau represents the DBB exciton
band and the valleys correspond with overlapping DCB
energy funnels. Within the valeys DCB aggregates serve
as energy sinks. Tunneling among the valeys is indicated.

TEMPERATURE DEPENDENT PHOSPHORESCENCE EXPERIMENTS

We have performed time resolved phosphorescence experiment as a function of temperature at various concentrations of DCB. DCB monomer triplet states have been selectively monitored by a monochromator. At concentrations above 1 % and low temperatures the decay is non-exponential. We have fitted the decay numerically by assuming a KWW decay superimposed by a monoexponential decay with the intrinsic triplet lifetime τ_0 of DCB of 16 ms resulting in

$$I_{phos}(t) = A \exp\left[-(t/\tau)^{\beta} - t/\tau_0\right] + B \exp(-t/\tau_0) \tag{1}.$$

A more recent analysis revealed[14] that Eq. 1 is only an approximation and the exact description especially at low concentrations and low temperatures is still an open question. However, in the sense of Fig. 1 we believe that the exponential part B represents isolated monomers or clusters which cannot reach a trap within the triplet excited state lifetime.

Fig. 2 shows the fit parameters τ and β of Eq. 1 as a function of temperature for a concentration of 8 %. The hatched area indicates that the decay cannot be fitted with B = 0. Above about 10 K β becomes temperature dependent which corresponds to a crossover to a diffusive behavior.

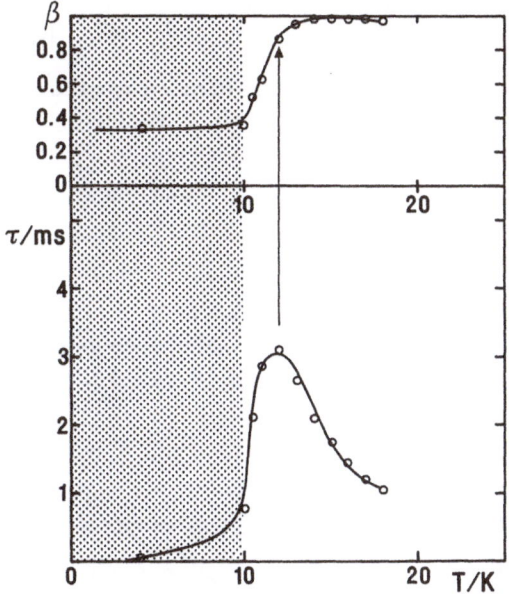

Fig. 2 KWW fit parameters β and τ for 8% DCB in DBB as a function of temperature.

Qualitatively the same behavior is observed for a wide concentration range. Fig. 3 contains the result for β at various concentrations. Introducing a reduced temperature scale T/T_m, where T_m corresponds to the temperature when $\tau(T)$ reaches a maximum results in a mastercurve also shown in Fig. 3. Only for concentrations (1.7%) far below the percolation threshold deviations are observed below 0.9 T_m.

To unravel the temperature dependence of τ for $T>T_m$ we used an Arrhenius type plot of log (τ) versus T_m/T which results in $\tau = \tau_\infty e^{\Delta E/kT}$. Straight lines in Fig. 4 indicate that τ is temperature activated with activation energies decreasing with increasing concentration. From these plots we have evaluated ΔE and τ as a function of concentration. The results are presented in Fig. 5.

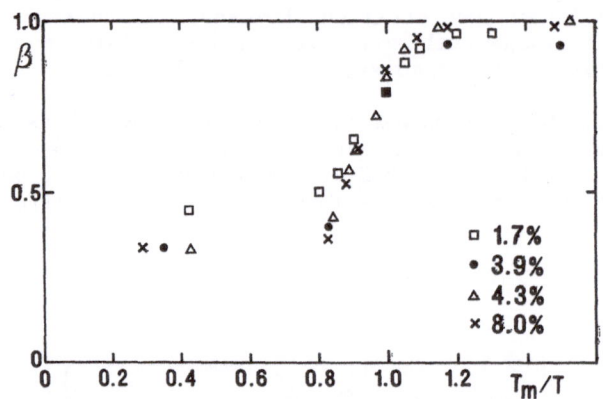

Fig. 3 β as a function of temperature on a reduced temperature scale T/T_m

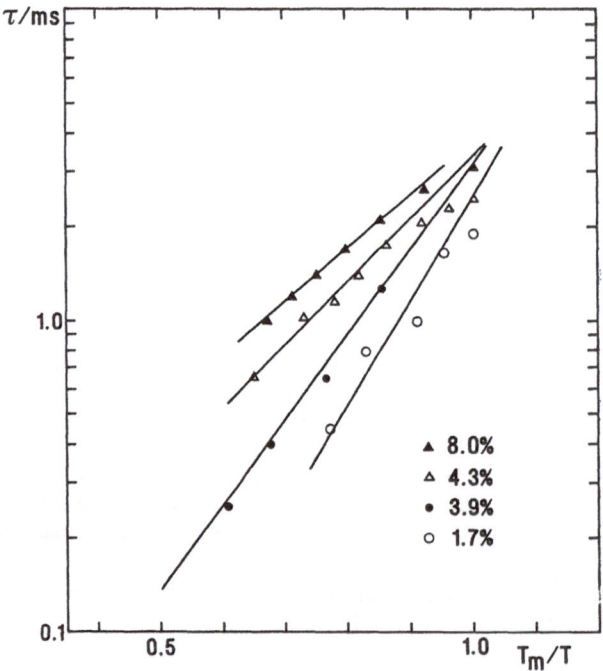

Fig. 4 τ as a function of T_m/T

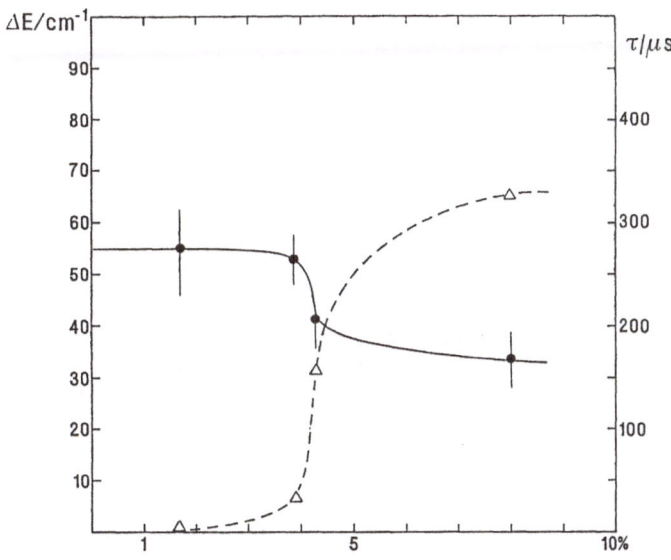

Fig. 5 ΔE and τ_∞ revealed from an Arrhenius type behavior in
Fig. 4 as a function of DCB doping concentration.

DISCUSSION

As recently shown[5,9] the observed KWW behavior of the phosphorescence
decay stems from a random walk with trapping of the excitation energy,
which can be described by a stretched exponential with $\beta = d_s/2$. d_s corresponds to the spectral dimension in case of a fractal structure of the
underlying lattice. We could show[9] that the random walk indeed occurs on
a fractal because we observe (dynamic) percolation at a doping concentration of 4.5 %. In addition to this substitutional disorder manifested in
a percolating lattice the temperature dependence indicates the presence
of additional energy disorder due to energy barriers which are hindering
the transport process.

The maximum of $\tau(T)$ in Fig. 2 and the breakdown of a pure KWW behavior below T_m indicates the presence of a transition from ergodic to
non-ergodic relaxation, where non-ergodicity means that the excited triplet manifold does not reach thermal equilibrium within the intrinsic
lifetime. This concept has been outlined recently[11]. We will restrict the
following discussion to the ergodic region above T_m.

In this region there is still dispersive energy transport. As can be
seen from Table 1 the temperature T_m depends on the concentration of DCB.
According to Fig. 5 the activation energy ΔE for the temperature dependence of τ goes from ΔE close to the DCB trap depth at low concentrations
(53 cm^{-1}) to the energy separation between DCB traps and induced DBB
funnels (26 cm^{-1}). It reveals a pronounced crossover at the critical
concentration for percolation found at 4.5 % DCB from other experiments[9].

Table 1. T_m as a function of doping concentration

c/%	T_m/K
1.7	10
3.9	12
4.3	13
8.0	12

$\tau(T)$ is related to the detrapping from DCB monomers resulting in a final trapping at DCB aggregates. As we have shown recently[13] at low temperatures tunneling between monomers separated only by distorted DBB molecules is faster than thermally activated jumps across the barrier height of 26 cm^{-1}. Increasing the temperature results in depopulating DCB monomers (energy funnels) and monomer clusters (overlapping energy funnels) via the undistorted DBB exciton band. At low DCB concentrations the number of exciton states N_{ex} per energy funnel N_F is very high resulting in a low preexponential factor $\tau_\infty \sim N_F/N_{ex}$ according to the description of detrapping processes derived for very low concentrations[16]. Within the DBB exciton band transport will be very fast.

Increasing DCB concentration results in an increase of N_F/N_{ex} and thus of τ_∞ as can be seen from Fig. 5. On the other hand the connectivity of energy funnels will be increased finally reaching a network of energy funnels only separated by 26 cm^{-1} barriers at a concentration of about 8 % DCB. Both effects will have opposite influence on the absolute value of T_m.

CONCLUSION

We have shown that energy transport above a critical temperature T_m is characterized by thermally activated dispersive transport among DCB induced energy funnels. The activation energy depends on the DCB concentration and can be identified with the spectroscopically observed energy landscape.

ACKNOWLEDGMENT

Financial support from the Deutsche Forschungsgemeinschaft (Sfb 337) is gratefully acknowledged.

REFERENCES

1. A. Blumen, J. Klafter and G. Zumofen, in: "Optical Spectroscopy of Glasses", I. Zschokke, ed., Reidel, Dordrecht (1986)
2. J. Prasad and R. Kopelman, Phys. Rev. Lett. 59:2103 (1987)
3. G. Peter, H. Bässler, W. Schrof and H. Port, Chem. Phys. 94:445 (1985)
4. G. B. Talapatra, D. N. Rao and P. N. Prasad, Chem. Phys. 101:147 (1986)
5. T. Kirski, J. Grimm and C. von Borczyskowski, J. Chem. Phys. 87:2062 (1987)

6. R. Kopelmann, in: "Laser Spectroscopy of Solids", W. M. Yen and P. Selzer, eds., Springer Verlag, Berlin (1981)
7. J. Grimm, T. Kirski and C. von Borczyskowski, Chem. Phys. Lett. 128:569 (1986)
8. J. Kolenda and C. von Borczyskowski, J. Luminesc. 42:217 (1988)
9. C. von Borczyskowski and T. Kirski, Phys. Rev. Lett. 60:1578 (1988)
10. C. von Borczyskowski and T. Kirski, J. Luminesc. 38:295 (1987)
11. C. von Borczyskowski and T. Kirski, Phys. Rev. B 40:11335 (1989)
12. C. von Borczyskowski, M. Plato, P. Dinse and K. Möbius, Chem. Phys. 35:355 (1978)
13. S. A. Gilbert, T. Kirski, H. Brenner and C. von Borczyskowski, Chem. Phys.Lett., subm.
14. C. von Borczyskowski and T. Kirski, Ber. Buns. Gesell. Phys. Chem. 93:1373 (1989)
15. C. von Borczyskowski, in: "Relaxation in Complex Systems and Related Topics", I. A. Campbell and C. Giovanella, eds., Plenum Press, New York (1989)
16. M. D. Fayer and C. R. Gochanour, J. Chem. Phys. 65:2472 (1976)

THE COUPLING SCHEME FOR RELAXATIONS IN COMPLEX CORRELATED

SYSTEMS

K.L. Ngai

Naval Research Laboratory
Washington, D.C. 20375 USA

INTRODUCTION

In this NATO Advanced Study Institute many different problems of large-scale molecular systems were discussed. The range of topics covered in this ASI is immensely broad. In view of the very nature of this ASI, what I addressed in lecture and elaborated further here is only a subset of all the large-scale molecular systems discussed in the Proceedings. I am primarily concerned with irreversible processes (relaxation) in correlated systems in which some identical constituents, molecules, ions, or their analogues, are interacting in either the quantum or classical mechanical sense, whichever is appropriate. Additional randomness caused by possible factors such as the presence of not identical constituents and fluctuations of local environments makes the problem even more complex. Correlated systems with additional complications such as distribution and randomness will be referred to as complex correlated systems. I am interested in the dynamics of irreversible processes in these systems which require solutions to these many body problems that give the time developments of either macroscopic (e.g. stress, strain, and electric polarization) or microscopic (e.g. orientation of tagged molecules, center-of-mass vector of a probe polymer chain in a polymer matrix) dynamical variables. When interactions between the constituents are strong, the system becomes highly correlated and solution is extremely difficult. In this paper I shall focus on three examples of such highly correlated systems. These are: (1) a glass forming viscous liquid that is made up of molecular units that are densely packed together and hence interacting strongly with each other (e.g. O-terphenyl, 1,3,5 trinaphthalbenzene, and toulouene); (2) a vitreous ionic conductor (e.g. the alkali oxide trisilicate and triborate glasses, Na_2O-$3SiO_2$ and Li_2O-$3B_2O_3$ respectively and also defect crystalline ionic conductors (e.g. Na β-alumina) that contain a propensity of interacting ions; and (3) polymer melts of long linear or star branched macromolecules that are fully entangled with each other and noncrossability of these densely packed macromolecules implies strong interaction. In these highly correlated systems motion of each relaxation species has to be correlated or cooperative with those of the others. There is no established method of solution to this kind of problems. A popular approach is by way of generalized Langevin equation (GLE) in which the mutual interactions between the relaxing species contribute to dynamic memory terms that involve the total force exerted on one species by the others. The exact GLE obtained is formal and not useful until approximations are made. It is not clear that the approximations usually made is adequate. In recent works in the area of polymer melts, after an approximate GLE has been obtained, the dynamic memory function is computed by the mode-mode coupling

(MMC) techniques. Finally, in order to arrive at the results some delicate approximations have to be made. The results obtained are still at variance with some key experimental data (to be discussed elsewhere). It is too early to tell whether this approach can be built into a useful and reliable method for solving relaxations in correlated systems. Thus we may conclude that at this time there is no firmly established theoretical method to address these problems. On the other hand, results of experimental investigations carried out over the period of the last several decades have accumulated to such an extent that patterns of behavior have emerged. From these patterns, phenomenologies of relaxations in correlated systems (CS's) and complex correlated systems (CCS's) have been established although theoretical explanations are lacking. For rescue from this dire situation a less ambitious and even unconventional theoretical approach may be necessary as a substitute for the illusive first principle theory. In the following I shall describe such a substitute which my co-workers and I have been advocating for years and is now generally referred to as the coupling scheme.[1-8] The rest of the paper is organized as follows. First, I point out that the interesting models proposed by Prof. Klafter and Prof. Blumen, which have applications in many large-scale molecular systems,[9-10] are not appropriate to use for the description of relaxations in CS's and CCS's. After this, the coupling scheme is brought out by presenting a new approach that is only sketched out here. This approach captures the basic and essential features of the coupling scheme and the discussion brings out the difference with all other models. It also exposes the limitation of the coupling scheme. However, I shall go on to show that this limitation is is not serious and similar situations occur in other branches of physics. The coupling scheme is justified as an acceptable procedure in physics. In this manner we have constructed a pragmatic theoretical approach to the problem that gives results or predictions in good agreement with experimental data. Monte Carlo computer simulations[11-15] performed recently in some of the model CCS's provide beautiful confirmations of the coupling scheme.

2. Models Inappropriate For Relaxations in CS's and CCS's

In my 1979 papers[1] on nonexponential relaxations and developments thereafter I pointed out among others the frequent occurrence of the fractional exponential relaxation function

$$\exp-(t/\tau^*)^\beta \tag{1}$$

now more widely known as the Kohlrausch-Williams-Watts (KWW) function. These results stimulated a number of workers to propose models designed primarily to explain the KWW function. It is a pleasure to acknowledge the contributions by Profs. Blumen, Klafter, and coworkers.[9,10] Their works include the Förster direct energy transfer model, the defect diffusion model and the defect diffusion-reaction model. KWW functions are derived in the contexts of these models. Nonexponential relaxations that are well approximated by the KWW functions have wide occurrence in physics, chemistry and materials science. It is not difficult to understand because after all the KWW is a two parameters function that often provides adequate fit to a nonexponential relaxation function whatever the origin of the nonexponentialty. Therefore KWW relaxation in two different systems may arise from entirely different physical mechanism. For example both the complex correlated systems (CCS's) of my present interest and the Förster model and the defect diffusion models have relaxation functions of the KWW form.

It is useful to clarify that among many models of KWW relaxation which one is appropriate and which ones are inappropriate for a certain class of problems. The preference is to put the emphasis on obtaining first conceptual understanding before looking for the mathematics, but not vice versa. For relaxations in CCS's a conceptual understanding is not clear at this time and the mathematics to go with it is even more remote. On the other hand for the Föster problem or the defect-diffusion problem, they are conceptually clear and the mathematical techniques needed to solve them are at hand, although they are nontrivial and extremely challenging. In the Förster model, a donor is initially excited, and the excitation is transferred to the acceptors with rates $W(r)$ which depend on the donor-acceptor distance r. A donor -acceptor pair that is characterized by dipol-dipole interactions has an energy transfer rate

$$W(r) = (R_0/r)^6 / \tau \qquad (2)$$

where τ is the fluorescence lifetime of the isolated donor and R_0 is the initial Föster radius, the donor-acceptor distance at which the energy transfer rate equals $1/\tau$. The fluorescence decay function $\phi(t)$ of the donor is given by

$$\phi(t) = \exp [-(t/\tau) -p\int dr\rho(\mathbf{r})\{1-\exp[-tW(\mathbf{r})]\}] \qquad (3)$$

where p is the density of acceptors and $\rho(\mathbf{r})$ is the site-density distribution function. On substituting Förster's expression for $W(r)$ of Eq.2 into the above equation (3) a KWW form for $\phi(t)$ was obtained. This is a very transparent model for KWW function but it has nothing to do with relaxation of correlated systems. Although a correlated system is characterized by mutual interactions between the relaxing species and at first sight is anologous to the dipole-dipole interactions of a donor-acceptor pair, the anology is only apparent. In Förster model, the dipole-dipole interaction of a donor-acceptor pair gives immediately an energy transfer rate $W(r)$ of the initially excited donor. However we do not enjoy such luxury in a correlated system. For two interacting relaxing species, like two interacting ions in a vitreous ionic conductor, two entangled chains in a polymer melt or two structural units in a glass-forming viscous liquid, there is no such pair wise relaxation rate. The physics of relaxations in CCS's is entirely different from the donor fluorescence decay problem. The former is still intractable as a many-body problem while the latter is a soluble problem by superpositions (e.g. Eq. (3)) of two-body problems (e.g. Eq. (2)).

Currently there is another proposal[16,17] that KWW relaxation may be obtained from a model based on Glarum's defect diffusion by calculating the flux of defects arriving at a target (i.e. a relaxing species). This model assumes that the random walks of defects follow a waiting time distribution function $\psi(t) \sim t^{-1-\alpha}$, and the relaxation of a target is directly proportional to the flux of defects arriving at the location of the target. The mathematics of continuous time random walks (CTRW)[18] leads to KWW functions for relaxation of the target. This model may have applications in other problems but probably not for relaxations in CCS's. The important element of correlations between the relaxation species (e.g. targets) are lost. In some CCS's it is difficult to see what the defects are and why the stochastics of one body (defect) can solve a many body problem.

I believe that these KWW models have wide applicability to relaxations in some large-scale molecular systems but not CS's and CCS's.. I wish to express my appreciation to Prof. Blumen and Prof. Klafter for clarifying discussions on this and other subjects during this NATO ASI.

The most interesting properties of relaxations in CCS's do not consist of the KWW form of the relaxation functions alone. Actually there are many interesting properties that comes with the relaxation time τ^* that usually are difficult to explain. For entangled monodisperse polymer melts[19], the interesting properties of τ^* include the $M^{3.4}$ and the $M^{3.0}$ dependence on molecular weight M respectively for the terminal shear viscoelastic relaxation time and the self diffusion relaxation time defined by $\langle R_g^2 \rangle / D$, where $\langle R_g^2 \rangle$ is mean square radius of gyration and D is the self-diffusion coefficient. For structural relaxations in viscous liquids such as O-terphenyl, the properties of τ^* include its fragile behavior as a function of T_g/T as defined by Angell[20] and the correlation[21] of fragile behavior with strong temperature dependence of the KWW exponent β. For vitreous ionic conductors, the most intriguing and fundamental property of τ^* is its anomalous isotope mass dependence that deviates strongly from the classical $\sqrt{m}$ dependence[3,22]. All these three CS's have KWW relaxation functions. A viable theory for relaxation of CS's and CCS's must not only arrive at the KWW form but also correctly explain these interesting properties of τ^* and others. Such a theory is not easy to come by because there is no established method of solution to this many body problem. There is no foundation to build on and not even an indication of how to start. Until the day someone comes up with a rigorous solution to the entire problem, critics should bear in mind that this is an unsolved problem. Progress could be slow and painful.

3.   A Classical Phase Space Diffusion Model of Relaxations in CCS's

As discussed, relaxations in CCS's are intractable many body problems that have defied an exact solution by established theoretical methods. In 1979 I made an attempt to gain some insight into the problem by following the relaxation of any one of the identical species with time. At sufficiently short time the relaxation proceeds as if the species were unaffected by its interaction with others except possibly for a renormalization of the independent species relaxation rate due to the "mean field" effect of the presence of the others. The renormalized independent species relaxation rate has been denoted by W_0. From the first kind of models based on level spacings distribution from random matrix theory, a time scale $t_c \equiv \omega_c^{-1}$ appears naturally. At times t after t_c the species independent relaxation rate W_0 will be slowed down by the mutual interactions between the species. In fact $\omega_c = E_c/h$ where E_c is of the order of the maximum level spacing and hence is characteristic of the nature of the interactions between the species. It is easy to see that stronger interactions correspond to larger E_c, and hence a shorter crossover time t_c from independent relaxation to slowed-down correlated relaxation. Intuitively it is clear interactions among the species commonly encountered in CCS's (as in the three examples given in the Introduction) impose correlations, constraints or cooperatively conditions on the relaxation of each individual species. As a consequence the independent rate W_0 will be slowed down for $t > t_c$. The early approaches allowed us to calculate the time-dependence of the rate slowing down and we found a time-dependent relaxation rate $W(t)$ of each species of the form

$$W(t) = \begin{cases} W_0 & \omega_c t < 1 & (4) \\ \\ W_0(\omega_c t)^{-n} & \omega_c t > 1 & (5) \\ \\ 0 < n < 1 & & (6) \end{cases}$$

where n is called the coupling parameter. The size of n is proportional to the coupling or interaction of one species with the others. Subsequent to the publications of these initial models[1,2,6] and in the course of time, several theoretical attempts have been made to approach the problem by different ways[2-8], all leading to the same conclusions (1)-(3). No claim is made that the models have been described with the clarity and rigor necessary to satisfy the theoretical community. I view these efforts as less ambitious attempts to obtain some insight into the difficult problem of relaxations in CCS's. In my opinion there is no rigorous and/or viable theory.

The essential results obtained as embodied by Eqs. (4)-(6) are intuitively reasonable and actually supported by recent results of Monte Carlo simulations[11-15]. In Monte Carlo simulations of (a) center-of-mass diffusion of entangled polymer melts and (b) of diffusion of ions that are interacting Coulombically with each other, when the number of Monte Carlo steps taken (or time) is not large the dynamics of diffusion of a chain in (a) and an ion in (b) are of the same nature as of an independent chain and ion respectively. That is, in the center-of-mass mean square displacement correlation function

$$g_{cm}(t) = <[\mathbf{R}_{cm}(t) - \mathbf{R}_{c.m.}(o)]^2> \tag{7}$$

obtained by Monte Carlo simulations has the time dependence of an independent Rouse chain, e.g. $6D_0t$, for short times. Similarly in (b), the mean square displacement $<r^2(t)>$ if an ion for small times increases linearly with t like that for an independent ion. In either case (a) or (b), there is a characteristic time-scale t_c when the independent relaxation ceases because an anomalous time dependence

$$g_{cm}(t), <r^2(t)> \quad \sim \quad t^a \quad t_c<t<t^*, \quad o<a<1 \tag{8}$$

takes over after t_c. Moreover the t^a dependence in Eq. (8) is consistent with the rate-slowing down in the form given by Eq. (5). In fact it can be shown that

$$a = 1-n \tag{9}$$

During the period when I was working on this problem using the level spacings distribution from random matrices, it was also the beginning of the study by others of semiclassical, but non-classical, behavior characteristics of systems whose classical motion exhibits chaos[23]. Berry and coworkers found the statistics of the energy level (given by the eigen values of the energy operator) for systems whose classical counterparts are chaotic are those of the random matrices. This connection suggests the classical counterpart of the quantum-mechanical model of relaxations in CCS's may be found from the properties of classical chaos. In the following I shall sketch the physical ideas behind an approach based on classical chaos that R.W. Rendell and I have proposed. Whatever the advance made so far, it is limited to a conceptual understanding only and a full fletched theory has not yet been developed.

It is expedient to present the ideas in a context analogous to the procedures usually followed in Monte Carlo simulations (MCS). To start with we consider the collection of a large number K identical relaxation species (e.g. polymer chains, ions or structural units) are at equilibrium with each other. In MCS this is equivalent to the procedure of equilibration of the initial state to ensure that it has correct static properties before dynamics ar simulated by MC steps. In the description of relaxation of the i-th species, a number of macroscopic quantities $A^{(r)}$ are considered. Each A^r is a function $A^r(q^i,p^i)$ of the f coordinates $q_1^i q_2^i,.... q_f^i$ and f canonically conjugate momenta $p_1^i,p_2^i,........,p_f^i$. We use van Kampen's phase space model[24] diffusion in of irreversible processes to describe the relaxation of a species. Moreover we shall adopt notations and nomenclature close to his. van Kampen (vK) partition the phase space Γ^i into regions determined by

$$a_v^{(r)} < A^r (q^i p^i) < a_v^{(r)} + \Delta a \tag{10}$$

which he called phase cells. These phase cells are labelled by J_i. VK defined a dot in Γ^i space as the set of values of the 2f coordinates $q_1^i,...,p_f^i$ that change with time according to the classical Hamiltonian's equation

$$\dot{q}^i_k = \frac{\partial H(q^i, p^i)}{\partial p^i_k}, \quad \dot{p}^i_k = -\frac{\partial H(q^i, p^i)}{\partial q^i_k} \tag{11}$$

where $H(q^i, p^i) \equiv H_i$ is the Hamiltonian of the i-the species that include the coordinates of the "matrix" in which the relaxation species are embedded. For polymer chains, the matrix are the solvent molecules if polymer solutions are considered. For alkali ions in silicate or borate glasses, the matrix is the glass-forming network modified by the alkali oxide. vK argued that the variation of the position of the dot in Γ^i space is so turbulent and irregular that on the coarse space of the macroscopic observer the motion of the dot looks like a succession of random jumps, similar to Brownian motion. This feature has led vK to apply the methods of diffusion theory to the motion in Γ^i space. This theory is formulated in terms on the probability $T_t(J_i|J_i')$ to be in any cell J_i at time t if at $t = o$ the system is at cell J_i'. vK further argued that T_t satisfies the Chapman-Kolmogorov equation

$$T_{t_1 + t_2}(J_i|J_i') = \sum_{J''} T_{t_2}(J_i|J''_i)T_{t_1}(J''_i|J'_i) \tag{12}$$

In other words, the stochastic process of diffusion in phase space used to describe the system macroscopically is a Markov process. Equation (12) leads immediately to the master equation

$$dP_J/dt = \sum_{J_i} \{W_{J_iJ_i'}P_{J_i'} - W_{J'J_i}P_{J_i}\} \tag{13}$$

for the probability $P_{J_i}(t)$ that at time t the system is in cell J_i. The coefficients $W_{JJ'}$ in Eq. (13) can be interpreted as the transition probabilities per unit time from J_i' to J_i. For simplicity let us consider the case that there are only two cells J_i and J'_i and assume $W_{J'_iJ_i} = W_o/2$. Then $P_{J_i'} - P_{J_i} = P_i$ satisfies the relaxation rate equation

$$dP_i/dt = -W_oP_i \tag{14}$$

If there is no interaction between the species, then each species relaxes independently of each other and exponentially according to Eq. (14). However, if the species are interacting say pairwise with interaction Hamiltonian

$$H_{int} = \sum_{i \neq j} H_{ij}(q^i, p^i; q^j, p^j) \tag{15}$$

we expect the relaxation of all species to be correlated. The diffusion processes also become correlated. The transition rate $W_{J_i,J_i'}$ or W_{J_i',J_i} of the i-th process now depends on its own progress as well as the progress of the other processes because of interaction Hamiltonian. The problem becomes correlated diffusion in $(2f)^K$ dimensional phase space which is another intractable problem. Nevertheless some qualitative results can be obtained from the properties of chaotic behavior in nonlinear classical mechanics. We shall extend vK's phase space diffusion picture of irreversible processes to include mutual interaction between the species. The nature of the mutual interactions between species in correlated systems are in general nonintegrable in the classical mechanics sense. The introduction of such terms will destroy the remaining regular trajectories of the H_i's and give rise to additional chaos. Coulombic interaction between ions, in vitreous ionic conductors, van der Waal's interaction between structural units of a viscous liquid, as well as the entanglement interactions for dense packed chains are likely to be nonintegrable. The phrases "remaining regular trajectories" and "additional chaos" are used above because the molecular chaos hypothesis has been made by vK for H_i by itself. We shall examine what that additional chaos introduced by the interactions H_{ij}'s will have on phase space diffusion in vK's formulation of irreversible processes. The nature of the additional chaos in phase space caused by the interactions H_{ij}'s is different from that of the molecular chaos in H_i. The latter, from a macroscopic (coarse-grained) point of view as implied by Eq. (12), makes the flow in phase space like a diffusion process starting at times before the former has to be considered sequentially in modifying the diffusion. To find out how the diffusion is modified, we ask first the question of what are the changes of the trajectories associated with H_i caused by the perturbations of the H_{ij}'s? Consider the natures of the interactions H_{ij}'s in the examples of correlated systems cited here. For entangled polymer chains, the entanglement interactions can be modelled by the excluded volume terms which cause a hindrance in the motion of the chain segments when they collide. For a dense collection of ions in an ionic conductor, interactions between the ions with Coulomb forces or nearest neighbor forces will retard the motion of the ions when they come closer to each other. A similar situation holds for rotations of structural units in a viscous liquid. All these situations suggest that the diffusive motion in Γ_i-space of the i-th species will be impeded because the trajectories of H_i, which was viewed by vK on the coarse scale as a succession of random jumps, are scattered by the interactions H_{ij}'s, biased the random jumps and as a result reduce vK's transition rates $W_{J'J}$. We can now see the cause of the slowing down of the transition rate of the individual species by the mutual interactions between them based on vK's model. To make further progress in finding out how this slowing down of $W_{J'J}$ proceeds with time, we consider the trajectories corresponding to the non interacting Hamiltonian $\sum_i H_i$ side by side with the trajectories corresponding to the interacting Hamiltonian $\sum_i H_i + H_{int} \equiv H_{cs}$ of the correlated system. Based on the noninteracting Hamiltonian vK constructed an ensemble of dots over a cell J' with constant density in considering the quantity $T_{\Delta t}(J \mid J')$ which is the fraction of these dots that will have moved into cell J after Δt. The ensemble of dots in cell J' represent the ensemble of trajectories corresponding to the H_i's emanating from cell J'. At t = o these trajectories of H_i's are the analogues of the prepared initial equilibrated state in computer simulations before Monte Carlo steps are taken to make moves allowed by the mutual constraints acting between the units. Although we prepare the initial state as the ensemble of trajectories of H_i's, each of these trajectories will start to be switched back to a true trajectory of the actual Hamiltonian $\sum_i H_i + H_{int}$ at some later time. The average of this later time over the ensemble of trajectories will be called t_c. It is clear that the magnitude of t_c depends on the two Hamiltonians and their difference (i.e. H_{int}), and not on $T_{\Delta t}(J \mid J')$ or $W_{JJ'}$. It is natural to expect that stronger interaction H_{int} has shorter t_c. At comparable densities of the relaxation species in the three examples of correlated systems discussed, intuitively it seems reasonable that the Coulomb interactions between the ions is stronger than the van der Waal interactions between liquid structural units, which in turn is stronger than the excluded volume interactions between the segments of a macromolecule. If this idea is true then t_c (ions) $< t_c$ (liquid structural units) $< t_c$(chains). After the characteristic time t_c, trajectories of the total Hamiltonian H_{cs} enter into the consideration of diffusion in phase space. As discussed earlier, these trajectories of H_{cs} contribute reduced transition rates $W_{JJ'}$ between cells J and J'. Hence we obtain the first result that the relaxation rate of macroscopic coarse grained quantities will be slowed down starting at a time t_c characteristic of H_{int}. In the simplified version, Eq. (11), of the master equation, these results can be restated as follows. Initially, at short times, each species relaxes independently with rate W_0 as if the mutual interactions between them were not there. This continues until at a certain time-scale t_c when the independent relaxation rate W_0 starts to be reduced. Next we shall find out how the slowing down of the relaxation rate will proceed with time after t_c. Writing out the rate reduction as a product $W_0 f(t)$ where $f((t_c) = 1$ and $f(t) < 1$ for $t > t_c$, our task is to understand the origin as well as the nature of the time-dependence of $f(t)$.

In the previous discussions we have attributed the rate slowing down to the additional chaos caused by H_{int}. Chaotic trajectories attributable to the presence of H_{int} are responsible for impeding the diffusive motion in Γ_i-space. Starting at $t = 0^+$ with the prepared states of the $i = 1,..., K$ species located initially in some phase cells J'_i where the effect of H_{int} is minimal. As diffusion in the hyperphase space $\Gamma_1 \otimes \Gamma_2 \otimes \ldots \otimes \Gamma_K$ proceeds the nature of H_{int} in correlated systems generally implies increasing interaction energy with time. This is because diffusion from J'_i to other phase cells generally leads to an enhancement of the interaction energy. For coarse grained coordinates, there is an analogy that is often easy to visualize. For example, a collection of entangled Rouse chains initially equilibrated as done in Monte Carlo simulation and subsequently set in diffusive motion will eventually have segments of different chains "dragging" each other along and increasing the interaction energy. As time increases, the diffusive motions of all chains become more correlated with each other and the interaction energy increases monotonically. Increase in interaction energy is accompanied by a corresponding increase in total energy E. It is a general property of nonintegrable Hamiltonians that more trajectories become chaotic at higher energies E. This property has been demonstrated for a number of standard nonintegrable Hamiltonians including that of Henon-Heiles on coupled oscillator systems, and is expected to have generally validity. Since chaotic trajectories associated with H_{ij} are responsible for the slowing down of diffusion rate in phase space, we are led to the conclusion that the rate reduction factor $f(t)$ defined earlier is a monotonic decreasing function of time. How exactly $f(t)$ decrease as a function of time cannot be determined by the qualitative arguments offered here without a rigorous treatment. The latter has to be rather sophisticated in order to track all the processes (diffusion and chaotic transitions) going on at the same time. It is also likely that vK's crude model may not be sufficient to serve as the basis of a rigorous theory. Another approach may be needed.

Even though we do not know exactly what function $f(t)$ is except that it is a monotonic decreasing function of time, the results obtained, i.e. a time-dependent relaxation rate

$$W(t) = \begin{cases} W_0 & t/t_c < 1 \quad (16) \\ W_0 f(t) & t/t_c > 1 \quad (17) \end{cases}$$

are in accord with what have been obtained from the quantum-mechanical model calculation based on the Gaussian Orthogonal Ensemble distribution of energy levels. In that model we found

$$f(t) = (\omega_c t)^{-n}, \quad 0 < n < 1 \qquad (18)$$

where $\omega_c \equiv t_c^{-1}$. In other attempts including the present one whenever the explicit time-dependence of $f(t)$ cannot be derived, we deduced $f(t)$ by imposing various requirements that must be satisfied by the relaxation function $\phi(t)$ obtained as solutions to the rate equation

$$d\phi/dt = -W(t)\phi \qquad (19)$$

One such example is the use of the requirement of "thermorheological simplicity"[7] of $\phi(t)$. Another example is the stability of the frequency spectrum to superposition. It would be nice if a step-by-step rigorous approach is available. Since it eludes everybody, I do not think one needs to apologize for doing whatever one can even with means that may be considered meager. In the next section I shall argue for the value of such an approach that is a combination of conceptual theoretical results and experimental data.

4. A Viable Approach: Conceptual Theory Augmented by Experimental or Computer Simulation Data, the Coupling Scheme

Even to this date of writing this manuscript a first principle, parameterless and rigorous theory of relaxations of correlated systems that is problem solving has not been proposed by anyone. In the past few decades more and more experimental data have accumulated that have remained unexplained. Ten years ago, I recognized that meaningful progress can be made by an approach based on the results obtained[1], merging them with some basic experimental data to determine the coupling parameter n. The resultant "theory" is able to predict additional data and phenomenological patterns. The conceptual results obtained recaptured partly here in Section 3 can be summarized by Eqs. (16) - (19). The approach deviced goes as follows. Take an available and reliable theory for the independent relaxation of a species without having to consider the dynamic correlation with others. If such a theory is not available, then construct one. The task is possible because this is a soluble problem. With a theory of independent relaxations at hand, the mode of relaxation as well as its rate W_0 are known along with its dependences on thermodynamic variables such as temperature, intrinsic material variables such as molecular weight M, isotope mass m, concentration c, and extrinsic physical variables such as the wave vector Q of the mode. If there are more than one mode of relaxation, we also know the relative contribution of the various modes to a relaxation process as well as an intuitive feel of which relaxation mode is most susceptible to be slowed down when dynamic correlations are in force. Note in some, more complicated relaxation process, W_0 may not be a constant and can have a time-dependence of its own.

With a reliable theory of uncorrelated relaxation and its rate W_0 at hand, we attack the problem of relaxation in a CS by way of Eqs. (16) - (19). From results of conceptual theoretical developments including that given in Section 3, I know t_c is thermodynamical variable independent and f(t) is a monotonic decreasing function of time. Let me take f(t) = $(\omega_c t)^{-n}$ to be the general form although it has not been proven with the clarity and rigor to satisfy everybody.

I am fully aware that two parameters ω_c and n have been introduced No one prefers a theory with parameters. Nevertheless there are nontrivial predictions that will guide us through to establish a viable approach. The first prediction from Eqs. (16) - (19) is the KWW form for the normalized relaxation function

$$\phi(t) = \exp -(t/\tau^*)^{1-n} \tag{20}$$

This prediction that all CS have the KWW form, or other close approximations to it, is not trivial. In some CS's including the terminal relaxation of monodisperse polymer chains, this prediction was made before it was verified subsequently by comparisons with experimental data done first by us and subsequently by others. It is also supported by Monte Carlo computer simulations. Also Eq. (20) is a strong prediction because it says that if a system is genuinely a CS as defined in Section 1, then the relaxation function must have this form. This is remarkably the case for many CS's. We should be careful to distinguish CS from CCS's. The latter are complex CS's which, in addition to correlated relaxation, have additional complications such as site randomness, percolation, distribution of W_0's and distribution in n's that need to be considered. These additional considerations on top of Eq. (20) may modify the latter to be non-KWW for any n or an apparent KWW for an unphysical n. The procedure to extract the true coupling parameter or parameters may not be straightforward in CCS's.

After passing the first test on the KWW shape of the relaxation function with the experimental data or computer experimental data of a CS, the relaxation time τ^* that comes with the KWW function has usually a few interesting and often intriguing properties that cannot be understood easily[5]. For example in entangled monodisperse polymer melt, τ^* for shear viscoelastic response and self-diffusion have the $M^{3.4}$ and M^3 molecular weight dependence respectively. For an archetypal ionic conductor $LiO_2 \bullet 3B_2O_3$, the ionic conductivity relaxation time τ^* has the anomalous $m^{1/2(1-n)}$ Li ion isotope mass dependence[22] rather then the normal $m^{1/2}$ dependence. A critical test of the approach is whether it can explain these anomalous dependences based on the known dependences of the uncorrelated relaxation rate W_0 and the value of n already determined for that particular correlated relaxation processes. The second relation

$$\tau^* = \{(1-n)\omega_c^n/W_0\}^{1/(1-n)} \tag{21}$$

which is a consequence of Eqs. (17) - (19) and referred to as the "second" relation offers a critical test of the whole approach because it makes a prediction on the dependence of τ^* on a variable U from the corresponding known dependence of W_0 and the already determined value of n, ie.

$$\tau^*(U) \sim \{W_0(U)\}^{-1/(1-n)} \tag{22}$$

Ther are cases in which we can explain the experimental observed dependences of τ^* on more than one variable, say, U,V,..., T, from the corresponding known dependences of W_0 on these variables and the second relation (19) with the same n, i.e.

$$\tau^*(U,V,..., T) \sim \{W_0(U,V,...,T)\}^{-1/(1-n)} \tag{23}$$

Note that from the conceptual basis of the theory ω_c is dependent on the interactions between the relaxing species but not on variables U that do not alter the interactions. The predicted dependences of τ^* on U according to Eq. (22) for a number of CS's are in good quantitative agreement with experimental data. It is important in making a test of the second prediction (22) not to involve ambiguities, additional assumptions, or, complications. Otherwise it will not be a critical test.

In several cases the W_0's are known even quantitatively. With W_0 and n known and τ^* taken from experimental data, the second relation enables ω_c to be deduced[5]. The values of ω_c's obtained are physically reasonable and depend on the nature of the mutual interactions. We found the order of magnitude of ω_c is roughly $10^{12}s^{-1}$ for Coulomb interactions, $10^{11}s^{-1}$ for van der Waal interactions and $`10^{10}s^{-1}$ for excluded volume interactions. The orders of magnitude of ω_c are physically reasonable and they decrease with weaker interactions as expected. There are situations in which the experimental measurement does not provide the relaxation function either in the time domain or in the frequency domain. An example is the measurement of probe or self diffusion of polymer chains which furnish only the diffusion constant D. To determine n we use the second relation (23) for one variable say U to determine n. That is, if the product forms

$$1/W_0(U,V,..., T) \equiv g(U)h(V)...z(T) \tag{24}$$

$$\tau^*(U,V,..., T) \equiv g^*(U)h^*(V)...z^*(T) \tag{25}$$

hold, n can be determined from

$$g^*(U) \sim \{g(U)\}^{1/(1-n)} \tag{26}$$

After n has been determined, it is nontrivial to verify if the other predictions like

$$h^*(V) \sim \{h(V)\}^{1/(1-n)} \tag{27}$$

are valid for the same n.

These procedures of the approach may be foreign to some researchers who are accustomed to problems in which all quantities are calculable from first principles like electronic band structure. Before too harsh a criticism is being leveled at the approach, the critic is reminded first that at the present time there is no better theory of relaxation in CS's which contains no parameter. Second, not all quantities in a theory are calculable or need to be calculable in order to qualify the theory to be viable. This has precedents before in physics. For example in the venerable field of quantum electrodynamics (QED)[25] the beautiful theory of S. Tomonaga, J. Schwinger, R.P. Feynmann and F. Dyson cannot calculate the mass of the electron (including the self-energy) and the charge of the electron. Even Tomonaga proposed the "principle of renunciation "to explain that one has to give up the hope that the theory is perfect and that everything can be calculated from it. Instead, one should make a definite distinction between things that can be calculated and those that cannot. But as a result of the theory, an answer can be found to any electromagnetic process. The theory made perfect predictions of the magnetic moment of the electron, the famous Lamb shift and others. We are fully aware that QED is a more difficult fundamental problem when compared with relaxation in CS's. But these features of the theory of QED bear some resemblence to our mode of operations in the coupling scheme of relaxations in correlated systems.

Recent advances in Monte Carlo and molecular dynamics computer simulations[11-15] of two kinds of CS's have yielded data that provide strong support for the conceptual foundations as well as the predictions of the approach which is now often referred to as the coupling scheme. The first kind of computer simulations are on dynamics of densely packed polymer chains. Jeff Skolnick and I[26] have reviewed the results of three major simulations, by him and his collaborators[11], by Pakula et al[12] and by Kremer and Grest[13], on the mean-square displacement of the center-of-mass $g_{cm}(t)$ defined by Eq. (7). If the dynamics of polymer chains were the same as described by the Rouse model[19] (independent relaxations without cooperativity with other chains), then at all times t

$$g_{cm}(t) = 6D_0 t \tag{28}$$

On the other hand if the dynamics is governed by reptation, the tube model[27] predicts

$$g_{cm}(t) = \begin{cases} t & \tau < \tau_e \tag{29} \\ t^{1/2} & \tau_e < t < \tau_d \tag{30} \\ t & t > \tau_d \tag{31} \end{cases}$$

where $\tau_e = \tau_b N_e^2$ and $\tau_d = \tau_b N^3/N_e$ where τ_b is the time required to diffuse a distance equal to the effective bond length b and N_e is the average number of monomers between entanglements. The computer simulation results of the three groups on $g_{cm}(t)$ are consistent and in quantitative agreement with each other, but deviate strongly from the predictions of both the Rouse dynamics and the reptation dynamics. Remarkably the results are in accord with the coupling scheme. Here I summarize the results that are given in detail in a recent work by Skolnick and I that is to be submitted shortly to a refereed journal.

In the Monte Carlo studies of Kolinski, Skolnick and Yaris, they found three time regimes in a plot of log $g_{cm}(t)$ vs log t. At short times and distances such that $g_{cm}(t)<b^2$, it was found (though results are unpublished) that $g_{cm}(t)$ is well described by Rouse dynamics, eq. (28). For distances such that $b^2<g_{cm}(t)<2<R_g^2>$ where $<R_g^2>$ is the mean-radius of gyration, it was found that

$$g_{cm}(t) \sim t^a \qquad (32)$$

where the exponent a is a monotonic decreasing function of chain length N, decreasing from the value of 0.91 when N = 64 to 0.69 ± 0.02 when N = 800. Finally, in the long time limit when $g_{cm}(t) > 2<R_g^2>$, $g_{cm}(t)$ had the time-dependence of

$$g_{cm}(t) = 6Dt + c \qquad (33)$$

where c is a small positive constant that reflects the faster motion of the center of mass at shorter times. The diffusion coefficient D decreases with increasing N. If D for two nearest neighbor values of N are represented locally in the power-law form of

$$D \sim N^{-\alpha} \qquad (34)$$

then for N = 64 to 100, α = 1.4; for N = 100-216, α = 1.6; and for N = 216-800, α = 2.05. The exponents a and α correlate with each other. As N increases, a decreases while α increases. These results are almost quantitatively reproduced in the other major computer simulations. In the molecular dynamics simulation by Kremer and Grest, as N increases from 25 to 200, the exponent a decreases down to a value of about 0.70 while the exponent α increases from 1.0 to 2.0. The results obtained by three groups are almost the same, the magnitudes of the exponents a and α and their dependences on N are similar. The minimal value that a reaches at the highest N studied in each simulation when α approaches 2 is about the same for all simulations, lying in the neighborhood of $0.70 > \alpha > 0.68$. All computer simulations have obtained $g_{cm}(t)$'s that are irreconcilable with the prediction (30) of the tube-reptation model. The exponent a is still far from the predicted value of 1/2 even at sufficiently large N such that $D \sim N^{-2}$, the signature of reptation, becomes valid. On the other hand, these results are in agreement with the coupling scheme. According to the coupling scheme, initially at short times any chain diffuses independently. Since the Rouse model describes well independent chain motion, we use it to calculate the initial time dependence of $g_{cm}(t)$ which is given by Eq. (28). Diffusion of chains can be formulated in terms of continuous time random walks of Montroll and Weiss.[16] Let Q(t) be the probability that the center-of-mass, initially at a certain site, will remain at the same site at a later time. We consider random walks of distances $c_0 = <R_g^2>^{1/2}$. For Rousean diffusion, Q(t) satisfies the rate equation

$$dQ/dt = -\lambda_0 Q \qquad (35)$$

where

$$\lambda_0 = 3D_0/R_g^2 \tag{36}$$

The coupling scheme further suggests that the Rousean diffusion rate λ_0 is slowed down in the manner according to Eqs. (17) and (18) as

$$W_D(t) = \lambda_0(\omega_c t)^{-n_D} \tag{37}$$

The transition time distribution function $\psi(t)$ is defined as $\psi(t) \equiv -dQ/dt$ where Q satisfies now the slowed-down rate equation

$$dQ/dt = -W_D(t)Q \tag{38}$$

The result for $\psi(t)$ is given by

$$\psi(t) = \lambda_0(\omega_c t)^{-n_D} \exp{-(t/\tau_D^*)^{1-n_D}} \tag{39}$$

where

$$\tau_D^* = \{(1-n_D)\,\omega_c^{n_D}\lambda_0^{-1}\}^{1/(1-n_D)} \tag{40}$$

With $\psi(t)$ determined by Eqs. (39) & (40), $g_{cm}(t)$ can be obtained by solving the CTRW equation with a memory function related in a complicated way to $\psi(t)$. The techniques for solving $g_{cm}(t)$ actually has been given in a paper by Ngai and Liu[28], and details can be found in Ref. (28). The results obtained are

$$g_{cm}(t)/R_g^2 = \begin{cases} [(1-n_D)\Gamma(1-n_D)/\Gamma(2-n_D)\tau_D^{*\,1-n_D}]t^{1-n_D}, & t<\tau_D^* \tag{41} \\ \\ [\Gamma(2-n_D)/(1-n_D))\,\tau_D^*]t, & t>\tau_D^* \tag{42} \end{cases}$$

The coupling scheme indeed has reproduced the a t^a intermediate time dependence for $g_{cm}(t)$ where a can be identified as

$$a = 1-n_D \tag{43}$$

The self-diffusion constant D obtained by comparing Eq. (42) with $g_{cm}(t) = 6Dt$ is

$$D = (<R_g^2>/6\tau_D^*)\,[1/\Gamma(2-n_D)/(1-n_D)] \tag{44}$$

If we determine n_D by Eq. (43) and the computer experimental value of a, then Eqs. (44) and (40) will determine D. These Equations predict

$$D \propto N^{1-2/(1-n_D)} \tag{45}$$

With the minimum value of 0.68 for n_D, we found

$$D \propto N^{-1.94} \tag{46}$$

which is in good agreement with computer simulations.

A few years ago in collaboration with McKenna and Plazek, I[29,5] have used the real experimental data of self-diffusion in polyethylenes and hydrogenated polybutadienes to determine n_D by relating the temperature dependence of $\lambda_o^{-1} \sim \exp(E_a/RT)$ where E_a is the known internal rotation isomerism energy barrier to the measured temperature dependence $\exp(E_a^*/RT)$ of τ_D^* or $1/D$.

$$E_a^* = E_a/(1-n_D) \tag{47}$$

Actually

$$D \propto N^{1-2/(1-n_D)}\exp(-E_a/(1-n_D)RT) \tag{48}$$

which is an example of Eq. (23). In this manner McKenna, Plazek and I determined n_D to be 0.32. It is remarkable that this value of n_D obtained from actual experimental data from the coupling scheme is reproduced exactly by computer simulations performed by three different groups.

Recently Monte Carlo simulations of the correlated diffusion of ions interacting with Coulomb forces in ionic conductors were performed by Dieterich, Bunde[14,15] and coworkers.(DB). They considered the lattice-gas Hamiltonian

$$H = (1/2) \, \Sigma e^2 n_l n_{l'} / |l-l'| + \Sigma \in_l n_l \tag{49}$$

where l are the lattice sites and n_l is the occupation number. The Hamiltonian contains the Coulomb interactions between the ions and a local contribution allowing for different local site energies $\in_l$. This system is similar to the ions in a crystalline or nitreous ionic conductor which are examples of complex correlated systems. They calculated the mean-square displacement $<r^2(t)>$ of a tracer ion. The results are remarkably similar to $g_{cm}(t)$ of a tracer polymer chain in an entangled polymer melt just discussed. They found, in their notations,

$$<r^2(t)> \sim \left\{ \begin{array}{lll} 6D_{st}t & \text{, short } t & (50) \\ t^{k'}, & k'<1, \text{ intermediate } t & (51) \\ 6Dt & \text{, long } t & (52) \end{array} \right.$$

These and especially the existence of an intermediate time-regime in which $r^2(t) \sim t^{k'}$ with exponent $k'<1$ are the same as the behavior of $g_{cm}(t)$ given by Eqs. (28), (32) and (33). Moreover they found the tendency of k' to decrease as the Coulomb interaction is enhanced. The dispersion (e.g. $k'<1$) is observed for ordered lattices ($\in_l \equiv$ o) and thus the anomalous diffusion at intermediate times is caused solely by interparticle interactions. The dispersion is enhanced (smaller k') by the influence of Coulomb interactions and disorder. However little dispersion was observed if the interaction was switched off but the disorder was kept. All these properties indicate the importance of the mutual interactions in the dynamics of the ions. These results for a correlated ionic system are isomorphic to the entangled polymer melt, supporting the contention that a universal description of relaxation of CS's and CCS's may exist as the coupling scheme has advocated. The expressions for

$\langle r^2(t) \rangle$ in the three time regimes, can be derived by the coupling scheme (Eqs. (16)-(19)) by using CTRW just as done in Ref. 28 and in the above for diffusion of polymer chain. We can identify

$$k' = 1 - n_D \qquad \qquad (53)$$

where n_D is now the coupling parameter for motion of an ion. In contrast to the simulations of polymer chains diffusion which are athermal, the Monte Carlo simulations of ionic conductors have introduced temperature into the procedure. DB found that the time t_c of crossover from independent diffusion as short times to anomalous diffusion at intermediate times is independent of temperature. This temperature independence of t_c is in agreement with the coupling scheme. Also DB found that on varying the temperature, the upper crossover time (τ in DB and τ^* in our notation) correlates with the long-time diffusion constant D such that $D \sim \tau^{-1}$ or τ^{*-1}. This result follows immediately from Eq. (42). The other equation (41) has the scaling form $D(t) \sim \tau^* f(t/\tau^*)$ with $f(t/\tau^*) \sim (t/\tau^*)^{k'-1}$ which is as obtained by DB.

Thus Monte Carlo simulations in two CCS's have reproduced all the physics proposed by the coupling scheme. These recent developments provide the most positive signs that we are in the right direction. Thus the coupling scheme is not only a viable approach to understand the experimental data but also embodie the correct physics.

5. Conclusions

Since there is no first principles and parameterless theory of relaxation in correlated systems and complex correlated systems, as far as we know, it is counter productive to sit and wait for such a theory to appear while rejecting any approach that produces some results. We must not insist that everything has to be calculated from first principles. As discussed in Section 4, in the venerable field of quantum electrodynamics (QED) even Tomanaga[25] had to invoke the principle of renunciation to avoid possible unreasonable demand that every quantity in QED must be calculated before the theory can be accepted. I am not promoting an approach which involves undeterminable parameters with ambiguous physical meaning.

Guided by conceptual theoretical developments, we have constructed the coupling scheme which has proven to be a viable approach to relaxations in CS's and CSS's. All parameters can be determined systematically by experimental measurements or computer simulations. After the key parameters have been determined, the scheme can explain the properties of the relaxation which are often anomalous. I sincerely believe that this coupling scheme has brought us out of the wilderness in the frontier research field of relaxations in complex correlated systems. The method and approach employed perhaps not generally be appreciated because in this area of research a wide variety of scientists and engineers are involved whose natural priorities, ultimate goals and scientific trainings are quite distinct. Some, unaware of how difficult the problem is, may reject outright any approach that is short of being a perfect theory. In spite of these road blocks, the time is right that a knowledgeable person in this field will find the coupling scheme an acceptable development and a stepping stone in the attempt of a perfect theory. In the context of the article on scientific attitudes by Chandrasekhar[30] I think I am trying my best to paint the face of the Madonna. But if my painting is interpreted instead as Mrs. Pelham feeding chickens, then I agree obligingly. However, in closing, I must quote the passage from Chandrasekhar: "If one recognizes that one can never come to painting a Madonna, what, then, are the satisfactions and the rewards? I suppose that one must count them in those brief moments of sudden insight which occur to one on rare occasions. One may never come to painting a Madonna. But, perhaps, in capturing on canvas, the rugged lines in the face of Mrs Pelham, etched by the toils of her life, the painter may have experienced a sudden insight into the sadness of the human condition which he may cherish all his life. And so it is in all other walks of creative effort".

Acknowledgement

I would like to thank the program committee and Prof. A. Blumen for inviting me to lecture at this NATO ASI, and Profs. Blumen and J. Klafter for discussions. This work is supported in part by ONR contract N0001489WX24074. I would like to thank A.K. Rajagopal for a critical reading of the manuscript and for valuable suggestions.

References

1. K.L. Ngai, Comm. Solid State Phys. 9, (1979) 127 and ibid 9, (1980) 141.

2. A.K. Rajagopal and K.L. Ngai in Relaxations in Complex Systems, (Eds. K.L. Ngai and G.B. Wright) National Technical Information Service, Springfield, V(1984) 275.

3. A.K. Rajagopal, R.W. Rendell, K.L. Ngai, and S. Teitler, Ann. N.Y. Acad. Sci. 484, (1986) 321.

4. A.K. Rajagopal, K.L. Ngai, R.W. Rendell and S. Teitler, Physica 149, (1988) 358.

5. K.L. Ngai, R.W. Rendell, A.K. Rajagopal, and S. Teitler, N.Y. Acad. Sci. 484, (1986) 150.

6. A.K. Rajagopal, S. Teitler, and K.L. Ngai, J. Phys. C.: Solid State Phys. 17, (1984) 6611.

7. K.L. Ngai, A.K. Rajagopal, and S. Teitler, J. Chem. Phys. 88, (1988) 5086.

8. A.K. Rajagopal, K.L. Ngai and S. Teitler, Nucl. Phys. (B) 5A, (1988) 97 and also K.L. Ngai, A.K. Rajagopal and S. Teitler, Ibid, 103.

9. A. Blumen, J. Klafter and G. Zumofen, in Optical Spectroscopy of Glasses (I. Zschakke ed.) D. Reidel Publ. Co. (1986), p. 199.

10. See lecture notes by A. Blumen and J. Klafter in this Proceeding.

11. A. Kolinski, J. Skolnick and R. Yaris, J. Chem. Phys. 86, 1567, (1987); ibid 7164.

12. T. Pakula and S. Geyler, Macromolecules 20, 2909 (1987).

13. K. Kremer and G. Grest, J. Chem. Phys.(1990).

14. W. Dieterich, J. Peterson, A. Bunde and H.E. Roman, to appear in Solid State Ionics (1990).

15. A. Bunde, in Proc. 3rd Bar-Ilan Conference on Frontiers in Condensed Matter Physics, Physica A (1990).

16. E.W. Montroll and M.F. Shlesinger, Proc. Natl. Acad. Sci. USA 81, 1280 (1984).

17. J.T. Bendler and M.F. Shlesinger, Macromolecules,

18. E.W. Montroll and G. Weiss, J. Math. Phys. $\underline{6}$, 167 (1965).

19. See J.D. Ferry, "Viscoelastic Properties of Polymers", John Wiley & Sons, NY (1980).

20. C.A. Angell in "Relaxations in Complex Systems", (Eds. K.L. Ngai and G.B. Wright) National Technical Information Service, Port Royal Road, Springfield, VA 1984) 3.

21. K.L. Ngai, J. Non-Cryst. Solids $\underline{95}$, & $\underline{96}$, 969 (1987); K.L. Ngai, R.W. Rendell and D.J. Plazek, J. Chem. Phys. (in press).

22. K.L. Ngai, R.W. Rendell and H. Jain, Phys. Rev. B $\underline{30}$, 2133 (1984).

23. For a review see M.V. Berry, Proc. R. Soc. Lond. A413, 183 (1987).

24. N.G. van Kampen in "Fundamental Problems in Statistical Mechanics" (Ed. E.G.D. Cohen) North-Holland (1962) 173.

25. Y. Nambu, "Quarks", World Scientific 1985.

26. K.L. Ngai and J. Skolnick to be published.

27. P.G. DeGennes, "Scaling Concepts in Polymer Physics" Cornell University, Ithaca, (1979).

28. K.L. Ngai and F.S. Liu, Phys. Rev. B24, 1049 (1981).

29. G.B. McKenna, K.L. Ngai and D.J. Plazek Polymer 26 1651 (1985).

30. S. Chandrasekhar, Nature $\underline{344}$, 285 (1990).

SIMULATION OF EXCITATION TRANSPORT IN DISORDERED MEDIA

R.Richert, L.Pautmeier and H.Bässler

Fachbereich Physikalische Chemie and Zentrum für
Materialwissenschaften, Philipps - Universität
D 3550 Marburg, FRG

INTRODUCTION

Diffusivity and mobility are characteristic quantities of transport phenomena easily correlated to the transition rates and the biasing force in regular systems. In the absence of regularity any quantification in terms of a rate is obscured by the disorder induced distribution of rates or waiting times. It is this characteristic property of random media together with the stochastic and often hierarchical nature of relaxational processes which prevents a straightforward analytical approach to rate controlled quantities in these systems. If a model for the underlying microscopic individual transition is at hand computer simulations yield a powerful tool for quantitatively accessing a wide spectrum of observables.

At present we consider the problems of transport in organic glasses and polymers as examples for disordered condensed media. As transporting species we regard charge carriers and triplet excitations, both subject to short range transitions, i.e. mainly nearest neighbor jumps, via exchange interaction. The relevant experimental problems covered are thus spectral and spatial diffusion of triplet excitons and current patterns of time-of-flight (TOF) measurements where the diffusion is biased by an external field. For each case the importance of considering energetic (diagonal) as well as spatial (off-diagonal) disorder will be indicated assuming normal distributions for site specific energies and wavefunction overlap parameters for modelling the statistics in organic systems. Since it is our intent to emphasize the role disorder plays in transport phenomena the simulations do not involve traps.

SIMULATION TECHNIQUE

Simulations are carried out employing the Monte Carlo method for realizing jumps within a 3-dimensional cubic array of 70x70x8000 hopping sites of lattice spacing a with periodic boundary conditions for the directions perpendicular to the field. A 7x7x7 cube is considered as manifold of target sites

Large-Scale Molecular Systems, Edited by W. Gans *et al.*
Plenum Press, New York, 1991

with transition rates from site i to site j according to

$$v_{ij} = v_o \cdot \exp[-\Gamma_{ij} \cdot \delta r_{ij}/a] \cdot \exp[-(\epsilon_j - \epsilon_i)/kT] \qquad (1)$$

$$\Gamma_{ij} = \Gamma' = \Gamma_i + \Gamma_j \ , \quad \delta\epsilon = \epsilon_j - \epsilon_i$$

δr is the jump distance. For non-activated jumps, i.e. $\delta\epsilon < 0$, the last exponential term is set to 1, anticipating that downward jump are not impeded by an energy matching condition. Diagonal disorder enters this model by selecting site energies ϵ at random from a Gaussian density of width σ. Off-diagonal disorder is accounted for via variations of the electronic exchange interaction quantified in terms of Γ also subject to Gaussian statistics but with standard deviation Σ and centered at $\Gamma_o = 5$ to obtain a mean overlap parameter $\langle\Gamma'\rangle = 10$. The electrostatic energy of an external electric field E is cast into site energies. Typically, the runs of 100-200 excitations, each started at a random position, serve for achieving averaged quantities as the mean energy, number N(NSV) of new sites visited and diffusivity. In the case of an external field further results are the mean displacement, current and arrival time at the exit contact.

RESULTS AND DISCUSSION

Focussing on a Gaussian density of states (DOS) for the site energies implies that excitations can attain thermal equilibrium reached at the mean energy σ^2/kT below the center of the DOS. In course of this relaxation the number of sites which still contribute to the transport is increasingly deminished, provided that $\sigma \gg kT$. In this dispersive transport regime the mean jump rate decreases similar to the KWW pattern $v \sim \exp(-t^\alpha)$ [1]. Finally, steady state conditions and, concomitantly, a time independent diffusivity is established obeying a non-Arrhenius type temperature dependence of the form $D = D_o \cdot \exp[-(T_o/T)^2]$ as confirmed by effective medium approach (EMA) theory [2].

Triplet Excitons (E=0)

Triplet states in a neat aromatic condensed system are subject to diffusive migration during their lifetime. Two techniques without spatial resolution can be employed for monitoring the motion of excitons, i) the depopulation due to capture in traps of chemical or structural origin [3] and ii) spectral diffusion mapping the energetic relaxation within the DOS after a random population of site energies [4,5]. Experimentally, benzophenone served as model glass former exhibiting a preexponential frequency of only $10^8 s^{-1}$ which shifts the time domain of energetic relaxation to an accessable range. Employing time resolved spectroscopy the spectral diffusion has been recorded at 4.2 and 80 K revealing quantitative accord with the simulation results, also for the regime of frustrated relaxation at 4.2 K due to trapping in the low energy wing of the DOS [5]. A similar agreement with the calculation is found for the temporal decrease in diffusivity observed via trapping in the same system indicative of the KWW decay pattern as expected. For the above results variations Σ of the exchange interactions have been disregarded.

In the case of pure off-diagonal disorder ($\sigma=0, \Sigma>0$) relaxational processes are inhibited. Allowing for symetrical varations in the parameter Γ_i leaves the mean value $\langle \Gamma_i + \Gamma_j \rangle$ unaffected yet raises the mean transition rate $\langle v_{ij} \rangle = v_o \cdot \exp[-\langle \Gamma' \rangle] \cdot \exp[\Sigma^2/2]$. At large Σ a second effect sets in because the migrating species may find spatial detours accidentally having high transition rates, similar to finding the most favorable path in a network of random resistors. The latter notion emphasizes the percolative aspect of this type of motion. The experimental counterpart to this simulation result is the observation of an increase in the diffusion constant of excitons upon melting a crystalline structure which obviously increases fluctuations of overlap parameters. Quantitatively, the simulation predicts that a variation of Γ of only 30% is needed to account for a rise of the diffusivity by more that two orders of magnitude [6].

Charge Carriers (E>0)

The typical experiment involving motion of charge carriers is the TOF technique where a sheet of carriers is generated instantaneously followed by drift towards the exit contact as directed by an external electric field. In regular systems the detected current, sampling the product of mean velocity and carrier concentration, can be expected to show an initial plateau followed by a sharp decay at times where the majority of carriers arrive at the counter electrode as observed [7]. For disordered media it critically depends on the actual situation, namely electric field, extent of disorder, mobility and sample length, whether a current plateau is observed or pure dispersive transport prevails during the transit time. If a plateau is in fact observed Gaussian transport characteristics may be expected contrasting the large span of arrival times generally observed [8]. The above problem together with the field and temperature dependence of mobility will be addressed in the following.

Dispersive Regime. On the basis of the simulations a transition from dispersive to non-dispersive transport arises readily via the condition that transport becomes time independent after equilibration of the mean energy within the DOS [9]. The computation yields an energetic relaxation time τ which follows $\tau = 10 \cdot t_o \cdot \exp[(\sigma/kT)^2]$, $t_o = 1/[6 \cdot v_o \cdot \exp(-\Gamma_o)]$ being the dwell time in the ordered counterpart system. Realizing that the current equilibrates at $0.1 \cdot \tau$ already implies that TOF signals must become non-dispersive for transit times $t \geq 0.1 \cdot \tau$. Under typical conditions, i.e. for $d = 10 \mu m$, $E = 10^5 V/cm$ and $T = 300K$, this case is realized for disorder parameters $\sigma \leq 4.4 kT$.

Non-dispersive Regime. We focus now on the understanding of TOF signals exhibiting a plateau in the current trace well before the first carriers have arrived. Neglecting the initial dispersive current spike the mobility is given by $\mu = (d/\langle t_{tr} \rangle)/E$ where $\langle t_{tr} \rangle$ is the mean transit time, d is the sample length and E the electric field. The diffusion constant D is monitored indirectly within TOF signals by the distribution of arrival times governing the shape of the trailing edge of the current profile. For Gaussian transport the idealized relation between current $I(t)$ and diffusion constant is

$$I(t) = 1 - \tfrac{1}{2} \, \mathrm{erfc}[(d - \mu Et)/(4Dt)^{\frac{1}{2}}] \qquad (2)$$

Using Einstein's relation $eD = \mu kT$ the profile $I(t)$ is defined by d, μ and E which usually are well defined quantities but commonly fail to match the experimental data if inserted in eqn.(2) in the sense that the trailing decay exceeds the

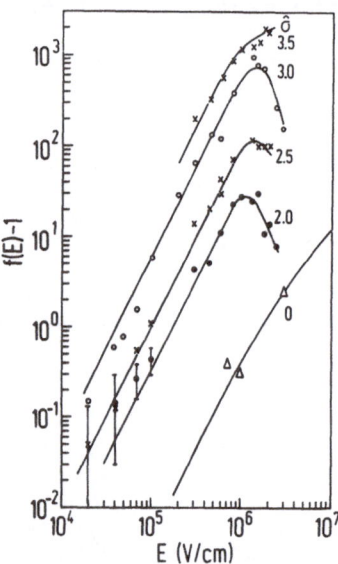

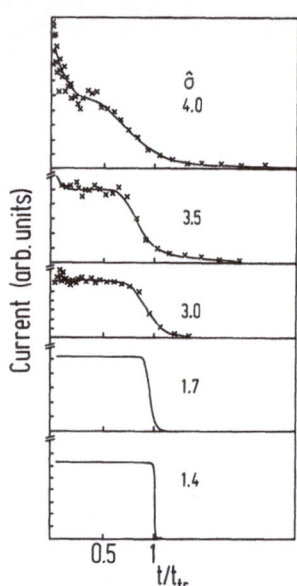

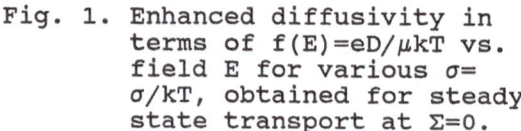

Fig. 1. Enhanced diffusivity in terms of f(E)=eD/μkT vs. field E for various σ̂= σ/kT, obtained for steady state transport at Σ=0.

Fig. 2. Simulated TOF signals normalized to the transit time t_tr. Disorder is indicated by σ/kT (Σ=0).

calculated time range of carrier arrivals significantly. For a system devoid of disorder a ratio eD/μkT in the range 1 to 3 is plausible for realistic high fields beyond the ohmic regime which is still 2-3 orders of magnitude below diffusion constants needed to rationalize experimental TOF signals in glasses. For equilibrium hopping transport in a Gaussian DOS the simulations indicate marked deviations from an eD=μkT behaviour [10] represented e.g. by eD/μkT≈1000 for a DOS of width σ=3.5kT and in experimentally relevant field ranges as depicted in fig.1 . Additionally invoking variations of the overlap parameter Γ, i.e. Σ>0, further enhances the discrepancy relative to Einstein's relation. The effect of eD≫μkT on the normalized TOF signal upon increasing the disorder as well as the transition to the pure dispersive transport becomes obvious from fig.2 .

In an isoenergetic system the mobility drops with increasing field because velocity becomes field saturated. Increasing the disorder parameter σ reverses this effect due to partial elimination of activation barriers in the direction of the field which is paralleled by a field induced raise of the mean carrier energy by ≈σ/2 at E=2MV/cm. The sign of dμ/dE thus changes upon varying σ. Qualitatively, the same effect is observed at fixed σ if Σ is lowered. At Σ=0 mobility increases with field as above but turning on fluctuations of Γ (Σ>0) leads to an overcompensation of the previous effect again changing the sign of dμ/dE. Recalling that the mobility is enhanced by Σ in the E=0 limit via favorable detour routes, the above effect becomes obvious noting that such paths are eliminated by a sufficiently large biasing field. For a fixed σ/kT=3 and various Σ the field dependence of mobility is shown in fig.3 . Within the indicated field range Gill's [11] empirical E and T

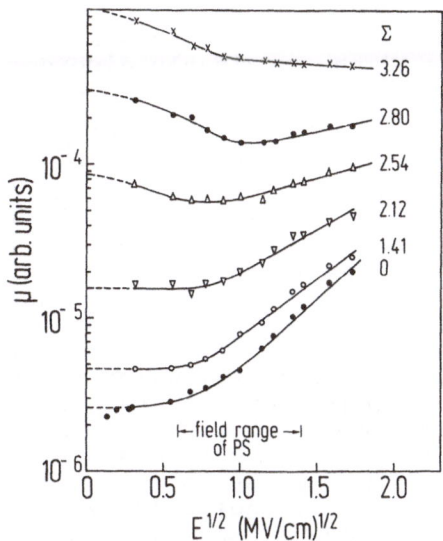

Fig. 3. Simulation results for mobility as a function of
electric field E for various degrees Σ of spatial
disorder and for a constant energy disorder of
σ/kT=3 (see text).

dependence of mobility which is experimentally confirmed by
Peled and Schein [12] can be mapped on the basis of disorder
alone if the thermal approach to T_g is translated empirically to
an increase of spatial fluctuations [13]. A quantitative
analysis of the general case σ>0 and Σ>0 indicates that the
field dependence of mobility can be resembled within a factor of
2 by regarding the contributions related to σ>0,Σ=0 and Σ>0,σ=0
independently.

It should be noted that the features of disorder enhanced
diffusivity and the negative slope of μ versus E for the σ>0 and
Σ>0 case are not recovered on the basis of the distribution of
transition rates alone which can be formulated analytically for
the simulated process. We therefore conclude that these
transport qualities must be related to the topology of the
system considered, i.e. on the spatial structure of paths and
detours and the memory of a diffusing particle to its history of
migration.

Under the realistic assumtion that the first two moments of
the distribution of waiting times for the random walk exist the
spatial profile of carriers must attain a Gaussian shape in the
long time limit assuring the validity of eqn.(2). Even for the
electrode distance of 8000 sites which corresponds to a
macroscopic sample thickness of several μm the Gaussian shape is
not attained resulting in longer current tails relative to the
profile of eqn.(2).

CONCLUDING REMARKS

The simulation results for transport in disordered media
employing experimentally relevant ranges for all parameters

indicates drastic effects relative to systems devoid of disorder. As we have shown previously disorder alone does recover current profiles in polymer samples, the transition from dispersive to non-dispersive transport upon varying the temperature and the change in sign of the $d\mu/dE$ slope. Stressing the evident importance of disorder for rationalizing transport phenomena in organic glasses we wish to emphasize that it is not the intent of this work to claim trapping or polaronic effects being negligible. For a detailed understanding of experimental findings it is the combination of the effects which is to be disentangled which is beyond the scope of the simulation results and beyond resolution and achieved parameter ranges of present data.

Apart from the EMA theory [2] which recovers many features found in the present simulation, several aproaches for delineating transport in disordered systems exist which map energetic disorder in terms of multiple trapping or a CTRW formalism. The crucial parameter entering these latter pictures is the trap distribution or, equivalently, the distribution of waiting times. A direct comparison between multiple trapping and the present simulations is impossible since the trapping pictures disregard the topology of the system with the obvious advantage of results appearing in a closed analytical form [14]. With an appropriate waiting time distribution $\Phi(t)$ the CTRW formalism can yield a similar representation of experimental data as achieved with the simulations but leaving the physical contributions to $\Phi(t)$ obscured.

REFERENCES

1. H. Bässler, phys. stat. sol. (b) 107:9 (1984)
2. B. Movaghar, J. Mol. Electronics 3:183 (1987)
3. R. Richert and H. Bässler, J. Chem. Phys. 84:3567 (1986)
4. R. Richert and H. Bässler, Chem. Phys. Lett. 118:235 (1985)
5. R. Richert, H. Bässler, B. Ries, B. Movaghar, and M. Grünewald, Phil. Mag. Lett. 59:95 (1989)
6. L. Pautmeier, B. Ries, R. Richert, and H. Bässler, Chem. Phys. Lett. 143:459 (1988)
7. N. Karl, E. Schmid, and M. Seeger, Z. Naturforsch. 25a:382 (1970)
8. H.-J. Yuh and M. Stolka, Phil. Mag. B 58:539 (1988)
9. L. Pautmeier, R. Richert, and H. Bässler, Phil. Mag. Lett. 59:325 (1989)
10. R. Richert, L. Pautmeier, and H. Bässler, Phys. Rev. Lett. 63:547 (1989)
11. W. G. Gill, J. Appl. Phys. 43:5033 (1972)
12. A. Peled and L. B. Schein, Chem. Phys. Lett. 153:422 (1988)
13. L. Pautmeier, R. Richert, and H. Bässler, Synthetic Metals, in press
14. H. Schnörer, D. Haarer, and A. Blumen, Phys. Rev. B 38:8097 (1988)

DYNAMICAL EXPONENTS FOR 1-D RANDOM-RANDOM DIRECTED WALKS

Claude ASLANGUL[a], Marc BARTHELEMY[a],
Noëlle POTTIER[a], Daniel SAINT-JAMES[b,c]

(a) GPS(*), Tour 23, Université Paris VII, 2, Place Jussieu, 75251 PARIS
Cedex 05 (FRANCE)
(b) Laboratoire de Physique Statistique, Collège de France, 3, rue d'Ulm
75231 PARIS Cedex 05 (FRANCE)
(c) Also at Université Paris VII
(*) Laboratoire associé au C.N.R.S. (U.A. n°17) et aux Universités Paris
VII et VI

INTRODUCTION

We consider the one-dimensional random directed walk on a disordered lattice described by the following master equation :

$$\frac{dp_n}{dt} = -W_n p_n + W_{n-1} p_{n-1} \tag{1}$$

This equation is the directed version of the usual master equation which describes a random process in which the variable of interest can only increase (see for instance ref. 1 for the symmetric case and refs. 2 and 3 for recent reviews on asymmetric models). In eq. (1) $p_n(t)$ denotes the probability to be at site labelled n at time t, the W's are non-negative quantities chosen independently at random in a given probability distribution $\rho(W)$. The W's are assumed to be time-independent (quenched disorder). The general both-way asymmetric walk is believed to be asymptotically similar to a directed walk on a renormalized lattice [4,5]; the directed walk is thus expected to generate the basic features of the general problem in a simplified framework.

We first find the asymptotic behavior of the average over disorder of the quantities :

$$\overline{x(t)} = \sum_{n=0}^{+\infty} n\, p_n(t) \tag{2}$$

$$\overline{x^2(t)} = \sum_{n=0}^{+\infty} n^2\, p_n(t) \tag{3}$$

In all the following, overbarring means an average computed with the p's, still a priori depending on the particular sampling of the W's, and <...> stands for a disorder average taken through the use of $\rho(W)$. In a second step we calculate $\overline{x(t)}$ for a given sample.

Large-Scale Molecular Systems, Edited by W. Gans *et al.*
Plenum Press, New York, 1991

The nature of $\rho(W)$ basically characterizes the final dynamics and the existence of dynamical phases (see ref. 6). We set :

$$\rho(W) = C_\mu W^{\mu-1} f_c\left(\frac{W}{W_m}\right) \qquad (W_m > 0, \mu > 0) \qquad (4)$$

C_μ is the normalisation constant and f_c a cut-off function basically specified by the fixed frequency W_m. In a thermally activated hopping model (W_n proportional to $\exp(-\varepsilon_n/kT)$), μ is the ratio of the thermal energy over the typical height of the energy barriers. Note that the smaller μ, the higher the probability to find a quasi-broken link.

DYNAMICAL EXPONENTS FOR $< \overline{x(t)} >$ AND $< \overline{\Delta x^2(t)} >$

To calculate the disorder-averaged quantities at large time we solve eq. (1) by a Laplace transformation. A generating function method allows to find the disorder-averages :

$$< x_1(z) > = \frac{1}{z^2 R(z)} - \frac{1}{z} \qquad (5)$$

$$< x_2(z) > = \frac{2}{z^3 R^2(z)} - \frac{3}{z^2 R(z)} + \frac{1}{z} \qquad (6)$$

where $R(z)$ is the Stieltjes transform of $\rho(W)$ and where $x_1(z)$ and $x_2(z)$ are the Laplace transforms of $\overline{x(t)}$ and $\overline{x^2(t)}$.

The disorder-averaged mean square dispersion $< \overline{\Delta x^2(t)} >$ is obtained by a Laplace inversion of the quantity $< \Delta_2(z) >$ defined as :

$$< \overset{.}{\Delta_2}(z) > = < x_2(z) - (x_1 * x_1) > \qquad (7)$$

(In the latter equation, * denotes the convolution in the Laplace variable z).

In order to get a closed expression for the averaged convolution $<x_1*x_1>$, we note the functional relation valid for any given sampling of the lattice :

$$z\, x_1(z; W_0, W_1, W_2, ...) = \frac{W_0}{z + W_0}\, [1 + z\, x_1(z; W_1, W_2, W_3, ...)] \qquad (8)$$

By applying twice this relation, using the fact that the W's are dummy uncorrelated variables when disorder average is taken, we can find an expression for $< x_1(z)x_1(z') >$; combined with $<x_2(z)>$, as shown in eq. 7, it eventually leads to the disorder-averaged mean square displacement $< \overline{\Delta x^2(t)} >$ (see ref. 7).

The disorder-averaged coordinate turns out to have the following asymptotic behaviours ($T = W_m t$ and Γ is the Euler function) :

$$0 < \mu < 1 \qquad < \overline{x(t)} > \approx \frac{\sin \pi\mu}{\pi\mu\, \Gamma(\mu+1)} T^\mu \qquad (9)$$

$$1 < \mu \qquad < \overline{x(t)} > \approx \frac{\mu - 1}{\mu} T \qquad (10)$$

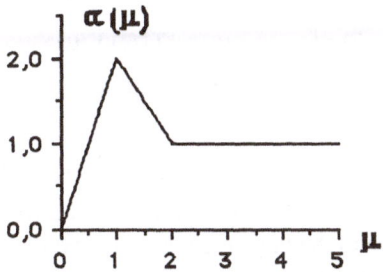

Fig.1. Variations as a function of μ of the dynamical exponent $\alpha(\mu)$ characterizing the dominant term in $< \overline{\Delta x^2(t)} >$.

The results expressed by eqs. (9) and (10) are in agreement with ref. 3, obtained independently by another method. For $\mu > 1$, a finite velocity exists; on the contrary, for $\mu < 1$, the coordinate increases slower than t due to the greater weight of quasi-broken links.

The analysis of the mean square dispersion is, as usual, much more involved. After some algebra and using contour integration, we find the following results :

$\underline{0 < \mu < 1}$

$$< \overline{\Delta x^2(t)} > \approx \frac{1}{\Gamma(2\mu)} \left[\frac{\sin \pi\mu}{\pi\mu} \right]^3 I(\mu) \, (W_m t)^{2\mu} \tag{11}$$

where $I(\mu)$ is a known function of μ.

$\underline{1 < \mu < 2}$

One finds (F is the hypergeometric function) :

$$< \overline{\Delta x^2(t)} > \approx \frac{(\mu-1)^3 \, \Gamma(\mu)}{\mu(3-\mu) \, (2-\mu)} \, F(1, \mu+1; 3, -1) \, (W_m t)^{3-\mu} \tag{12}$$

$\underline{\mu \geq 2}$

Now, the two first inverse moments of $\rho(W)$ exist and, as expected, the regime is a standard drift-diffusion one. One finds :

$$< \overline{\Delta x^2(t)} > \approx 2 \frac{(\mu-1)^3}{2 \, \mu^2 \, (\mu-2)} \, W_m t \tag{13}$$

For $\mu < 1$, the numerous broken links slow down the motion of the center of the packet. For $\mu < 1/2$, they are even so numerous as to hinder the spreading. For $1/2 < \mu < 1$, they become less efficient: the center speeds up while the spreading becomes superdiffusive. For $1 < \mu < 2$, the spreading is still superdiffusive while the number of quasi-broken links is not large enough to forbid ordinary drift. Finally, for $\mu > 2$, these links are so rare that the standard drift-diffusion regime is restored (see fig. 1).

NON SELF-AVERAGING PROPERTY OF $\overline{x(t)}$ FOR $\mu < 1$.

The sample-to-sample fluctuations of $\overline{x(t)}$ can be displayed by analyzing the relative dispersion linked to disorder, δ :

$$\delta \equiv \frac{< \left[\overline{x(t)}\right]^2 > - \left[< \overline{x(t)} >\right]^2}{\left[< \overline{x(t)} >\right]^2} = \frac{\mu}{2^{2\mu-1}} \frac{\Gamma(1/2)\,\Gamma(\mu)}{\Gamma(\mu+1/2)} - 1 \tag{14}$$

δ is a monotonically decreasing function for μ between 0 and 1, assuming its maximum value (equal to unity) for $\mu \to 0_+$ and vanishing at $\mu = 1$ (see fig. 3 in ref. 7). This latter result establishes the fact that, due to the large disorder, the particle coordinate still fluctuates from one sample to another, even in the final dynamics. The self-averaging property is recovered for $\mu > 1$.

CALCULATION OF $\overline{x(t)}$ FOR A GIVEN SAMPLE

We now show that the dynamical exponent for the coordinate, (calculated above for an average over disorder), is already present in any realization of a given sample.

Setting $\Gamma(z) = z^2 x_1(z)$, one deduces from eq. 8 that the probability distribution for the random variable Γ, $P(\Gamma,z)$, obeys the following integral equation :

$$P(\Gamma, z) = \int_0^{+\infty} dW\, \rho(W)\, \frac{z+W}{W}\, P(\frac{z+W}{W} \Gamma - z, z) \tag{15}$$

In order to analyze eq. (15), we first Laplace transform $P(\Gamma, z)$ with respect to Γ by defining $\Pi(\tau, z)$ which satisfies another integral equation; the latter is solved by assuming an entire series expansion :

$$\Pi(\tau, z) = 1 + \sum_{n=1}^{+\infty} \alpha_n(z)\, \tau^n \tag{16}$$

where the dominant contribution to α_n can be found [8].

The differences between the two cases $\mu < 1$ and $\mu > 1$ arise from the behaviour of $<(z+W)^{-1}>$ at small z.

$\underline{\mu < 1}$

For small z, $\Pi(\tau,z)$ can be written as :

$$\Pi(\tau,z) = F_\mu \left[Z(\tau,\mu)\right] \qquad Z(\tau,\mu) = \frac{\sin\pi\mu}{\pi\mu}\, W_m^\mu z^{1-\mu}\, \tau \tag{17}$$

where the function $F_\mu(Z)$ has a known [8] expansion in powers of Z. Using the scaling law provided by (17) and denoting now by $f_\mu(X)$ the Laplace inverse of the function $F_\mu(Z)$, we find for any sample :

$$\overline{x(t)} \approx x_0\, (W_m t)^\mu \tag{18}$$

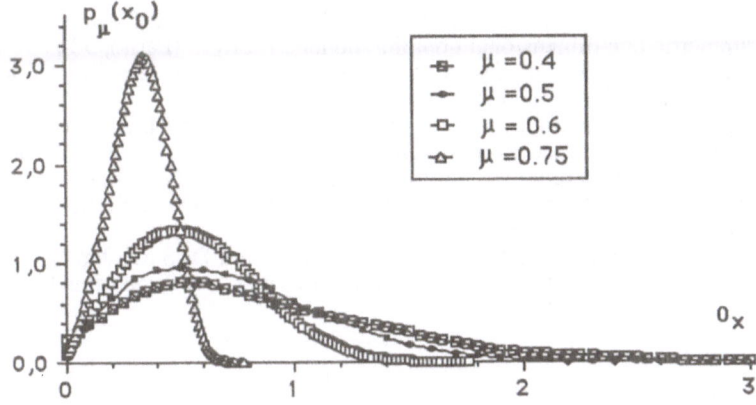

Fig.2 .Variations of the distribution function p_μ as a function
of x_0 for several values of μ.

where x_0 is a random number with a distribution $p_\mu(x_0)$ given by :

$$p_\mu(x_0) = \frac{\pi\mu}{\sin\pi\mu} \; \Gamma(\mu+1) \; f_\mu\left[\frac{\pi\mu}{\sin \pi\mu}\Gamma(\mu+1) \, x_0\right] \qquad (19)$$

$p_\mu(x_0)$ is explicitly known by all its moments [8] and has been numerically computed; the results are reported on fig. 2 for several values of μ. As expected, when μ increases, a peak occurs which is more and more pronounced and moves towards the origin. In the limit $\mu \rightarrow 1$, this fact yields the $\delta(x_0\text{-}0+)$ distribution. This phenomenon may be viewed as the precursor of the settling of the self-averaging property at $\mu = 1$.

$\mu > 1$

Now, due to eq. (15) and to the behavior of α_n for small z, one finds :

$$P(\Gamma, z) = \delta \left(\Gamma - \frac{1}{<W^{-1}>} \right) \qquad (20)$$

This shows that the limit of the derivative of $\overline{x(t)}$ with respect to time, i.e. the velocity, tends to $<W^{-1}>^{-1}$ with probability 1 at large times. This thus quickly establishes the existence of a finite ordinary drift characterized by a self-averaging velocity, a result already obtained by other methods [9].

References

1. S. Alexander, J. Bernasconi, W.R. Schneider and R. Orbach, "Excitation dynamics in random one-dimensional systems", Rev. Mod. Phys., 53 : 175 (1981).
2. J. P. Bouchaud, A. Comtet, A. Georges, and P. Le Doussal, "Classical diffusion of a particle in a one-dimensional random field of force", submitted to Annals of Physics.
3. J. P. Bouchaud and A. Georges, "Anomalous diffusion in disordered media : statistical mechanisms, models and physical applications", Phys. Rep., to be published.
4. J. Bernasconi and W.R. Schneider, "Self-similar temporal behavior of random walks in one-dimensional random media", in "Fractals in Physics", L. Pietronero and E. Tosatti, Eds., (Elsevier Science Publishers, 1986).

5. C. Aslangul, N. Pottier and D. Saint-James, "Velocity and diffusion coefficient of a random asymmetric one-dimensional hopping model", J. Phys. (Paris), 50 : 899 (1989).

6. B. Derrida, "Velocity and diffusion coefficient of a periodic one-dimensional hopping model", J. Stat. Phys., 31 : 433 (1983).

7. C. Aslangul, M. Barthélémy, N. Pottier and D. Saint-James, "Dynamical exponents for one-dimensional random-random directed walks", J. Stat. Phys., 59 : 11 (1990).

8. C. Aslangul, M. Barthélémy, N. Pottier and D. Saint-James, "Microscopic dynamical exponents for random-random directed walks on a one-dimensional lattice with quenched disorder", to appear in J. Stat. Phys..

9. C. Aslangul, J. P. Bouchaud, A. Georges, N. Pottier and D. Saint-James, "Exact results and self-averaging properties for random-random walks on a one-dimensional infinite lattice", J. Stat. Phys., 55 : 461 (1989).

INTERACTIONS BETWEEN POLY(STYRENE-CO-STYRENE SULFONIC ACID) AND POLY(METHYLMETHACRYLATE-CO-4-VINYL PYRIDINE) IN DIMETHYL SULFOXIDE SOLUTION BY PHOTON CORRELATION SPECTROSCOPY

A. Rizos, G. Fytas

Research Center of Crete and
Department of Chemistry, University of Crete
P.O. Box 1527, 71110 Iraklion, Crete, Greece

and

A. Eisenberg

Department of Chemistry, McGill University
Montreal, Quebec, Canada H3A 2K6

ABSTRACT

Photon correlation spectra of the polarized scattered light from poly(methyl methacrylate-co-4-vinylpyridine) poly(MMA-co-4VP), poly(styrene-co-styrene sulfonic acid) poly(S-co-SSA) and mixtures of both have been studied in the temperature range from 25 to 150°C. From the observed correlation functions the distribution of relaxation times is computed by means of a direct inverse Laplace transformation. The extracted $L(\ln\tau)$ distributions reveal a two peak structure for poly(S-co-SSA) which exhibit a q dependence characteristic of diffusional dynamics. The mixture of the two components above 85°C shows $L(\ln\tau)$ distributions very similar to that of poly(MMA-co-4VP). We have attributed this similarity to interpolymer complex formation via proton transfer from the SSA group to the 4VP ring and subsequent coil overlap as suggested by the NMR experiment. The observed single peak structure is dominated by the larger poly(MMA-co-4VP) and is probably the result of the screened electrostatic interactions of the SSA groups in the blend.

INTRODUCTION

A series of recent studies[1] concentrated on polymer blends are receiving considerable attention, partly because of their extensive industrial applications. Unfortunately, due to the small entropy of mixing the majority of polymer blends form incompatible mixtures, therefore miscibility enhancement[2] has been a very active field of current research. Several enhancement techniques have been utilised, such as copolymerization of the two materials to be mixed, crosslinking, hydrogen bonding, formation of charge transfer complexes and many others. In a broad range of polymer pairs miscibility can be induced by introducing appropriate ionic groups into each polymer chain. Interpolymer complexes may be formed by mixing polyelectrolytes of opposite sign. It is obvious that the ratio of interacting ionic groups has to be equimolar in order to obtain maximum interaction[2].

In a recent study[3], NMR spectroscopy was used as a tool for the determination of miscibility in local scale, as this technique can offer information about differences as small as a few A. In dimethyl sulfoxide at 85°C mixtures of partially sulfonated (SSA) polystyrene with copolymers of poly(methylmethacrylate) (PMMA) with 4-vinylpyridine (4VP) become soluble. The presence of aromatic groups of poly(S-co-SSA) influences the chemical shift of the methoxy protons of PMMA. The resulting ionic interaction of the two chains is followed by NMR by monitoring the evolution of the methoxy signal. One and two-dimensional NOE-correlated spectra prove the presence of coupling between the two polymers leading to an interpolymer complex at 85°C with an interchain distance of about 4A. As the temperature is further increased the upfield NMR shift - induced by the aromatic shielding of the poly(S-co-SSA) chain - is hardly observed. This finding however may not necessarily suggest polymer-polymer dissociation.

On the other hand, the specific polymer blend interactions can exert a measurable effect on the translational diffusion coefficient of the blend with regard to the individual polymers. Detailed information on diffusional dynamics of polymers in solution can be obtained from dynamic light scattering measurements. The purpose of the present photon correlation ˎspectroscopic (PCS) study is to provide strong evidence on specific interactions between poly(S-co-SSA) and poly(MMA-co-4VP) in dimethyl sulfoxide solutions under the same conditions of the recent NMR investigation. An additional objective to examine the applicability of PCS was brought about after we noticed that only a few measurements in reactive systems have been published[4-6].

EXPERIMENTAL METHOD

The time correlation functions $G(q,t)$ of the polarized scattering intensity arising from concentration fluctuations were measured at a wave vector q and temperatures between 25 and 150°C. The light source was an Ar^+ laser (Spectrum Physics 2020) operating at a single mode at 488 nm with a stabilized power of 100 mW. Both the incident beam and the scattered light were polarized vertically (V) with respect to the scattering plane. The VV intensity correlation functions over 4.3 decades in time were measured with a 28-channel logarithmic-linear single clipped correlator (Malvern K7027) in one run. In the homodyne limit the desired normalized correlation function $g(q,t)$ of the scattered electic field is related to $G(q,t)$ by

$$G(q,t) = A[1+f \cdot \frac{1+k}{1+<n>} \mid ag(q,t) \mid^2] \tag{1}$$

where A is the base line, $<n>$ is the average number of photocounts per sampling interval, k is the clipping level, and f is the instrumental factor. The factor a is the fraction of the polarized light scattering intensity arising from concentration fluctuations with correlation times longer than about 10^{-6}s. For all solutions studied, the value of a was close to one.

MATERIALS

The two partially sulfonated polystyrenes (poly(S-co-SSA)) and one copolymer of poly(methylmethacrylate) with 4-vinyl pyridene (poly(MMA-co-4VP)) were the same with those used in Ref.3. The two low charge PSSA samples contain 11.06 mol% (molecular weight, MW = 30000) and 8.92 mol% styrene sulfonic acid (SSA) (MW = 250000) whereas the 4-VP content in copolymer PMMA-4VP (MW = 100,000) amounts to 9.56 mol%

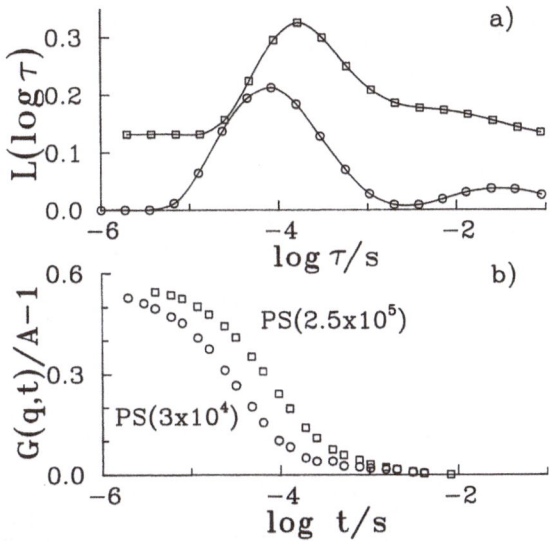

Fig.1.   a) Net correlation functions (G(q,t)/A-1 for the polarized light scattering intensity at a scattering angle of 90°C for the two parent polystyrenes solution in dioxane ($c_{PS} = 10^{-3}$ g/cc) at 25°C.

b) Distribution of relaxation times obtained from the inverse Laplace transform analysis of the time correlation functions for the two parent polystyrene samples in dioxane (C = g/cc). The distribution L(logτ) is a measure of the molecular weight distribution.

4-VP.   Owing to the complicated translational diffusion dynamics of polyelectrolyte solutions[7], it is useful to examine the behavior of the neutral polymers in dilute solution. Only the parent polystyrene (PS) samples are available, as the partial sulfonation follows the radical polymerization of PS. Figure 1a shows semilog plots of the normalized G(q,t)/A-1 (eq.1) for the two parent PS samples dissolved in p-dioxane ($c_{PS} = 4.7 \times 10^{-3} g/cm^3$, 2.7×10^{-3} g/cm^3 respectively for the low and high MW) at 25°C and at a scattering angle of 90° ($q = 2.6 \times 10^{-3} A^{-1}$). The nonexponential shape of $G(q,t)$ for dilute polymer solutions is caused by a distribution of molecular sizes and hence of molecular weights[8]. In this case, $C(q,t) = G(q,t)/A-1$ can be represented by

$$C(q,t) = \int_{-\infty}^{\infty} d\ln\tau \ L(\ln\tau)e^{-t/\tau} \qquad (2)$$

where $L(\ln\tau)$ denotes a spectrum of relaxation times which can be transformed to a distribution of molecular weights. An algorithm developed by Provencher[9], that has become a standard method of analysis, is used to extract $L(\ln\tau)$ from the experimental $C(q,t)$. Figure 1b shows the inverse Laplace transform of $C(q,t)$ displayed in Fig.1a. As can be inferred from this Figure, these samples exhibit considerable polydispersity usually found in thermally polymerized polymers. From the position of the dominant peak in Fig.1b one can obtain the translational diffusion coefficient $D = 1/\tau q^2$ and hence the equivalent hydrodynamic radius R_h using the Stokes-Einstein formula

$$D = \frac{k_B T}{6\pi\eta R_h} \qquad (3)$$

where h is the solvent shear viscosity and k_B the Boltzmann constant. As this equation is valid in the limit $c \to 0$, the values of 96A and 203A, respectively for the low and high MW parent PS samples, are apparent hydrodynamic radii. We found, however, that the measured diffusion coefficient D of both PS chains in dioxane for two concentrations in the range $(2.6-10) \times 10^{-3}$ g/cm^3 at 25°C was virtually independent of c.

The $L(\log\tau)$ distribution for the corresponding low charge poly(S-co-SSA) samples in dimethyl sulfoxide (DMSO) at 25° reveal also a two peak structure shown in Fig.2. However, the amplitude ratio between the long and short time peak is very different from that of Fig.1b; here, the slow mode dominates. The relaxation times associated with the two peaks exhibit a q^2 dependence characteristic of diffusional dynamics. The physical origin of the two modes is not yet clear[7,10-12]. The slow relaxation process is more pronounced at low scattering angle (and hence low q) and at lower polymer concentration where electrostatic interactions are not screened. The "slow" diffusion coefficient increases with increasing concentration and decreasing MW whereas the fast mode seems to be insensitive to MW (Fig.2) and c.

For salt-free polyelectrolyte solutions in the dilute regime, the polyelectrolyte chains are predicted[13] to be fully stretched with average dimensions proportional to the contour length L. This extreme dilute regime is hardly accessible, and the most experimental studies have been carried out at $c > c^*$, the critical concentration for which the persistence length approaches L. The concentrations used were all above c^* defined as[14]

$$c^* = M_0/(16\pi QALN_A) \qquad (4)$$

where M_0 is the molar mass of the monomeric unit, Q is the Bjerrum length, N_A the Avogado number and A is the average distance between two successive charges along the chain. For $M = 3.10^4$ g/mol and $M = 3.10^5$ g/mol with 10 mol% ion content the values of c^* in DMSO is about 4×10^{-5} g/cm^3 and 4×10^{-6} g/cm^3 respectively. Thus the Poly(S-co-SSA)/DMSO solution of Fig.2 are in semidilute regime for which D should increase with c[13,14].

Dynamic and static light scattering investigations have been reported for highly charged polystyrenesulfonate (PSS)[10-12] and quaternized poly(2-vinyl-pyridine)[7] in semidilute and concentrated water solutions with rather inconsistent results as far as the concentration dependence of D is concerned. As to the two modes structure of the concentration correlation function C(q,t) a reliable analysis requires broad dynamic time range and hence knowledge of the baseline A in Eq.1. The more recent study[7] has properly met suitable conditions and clearly shown the existence of two relaxation processes in semidilute solutions of 40% quaternized poly(2-vinyl-pyridine). The analysis of C(q,t) was however model dependent. The long and short time decay of C(q,t) were represented by single exponential functions and a stretched exponential was fitted to the broad decay in the middle. The slow diffusion coefficient was found to decrease with increasing concentration in contradiction to the present results, whereas the fast diffusion constant shows similar behavior in both cases. In semidilute and concentrated regimes the correlation length ξ is predicted[14] to decrease and hence $D(\sim\xi^{-1})$ to increase with c.

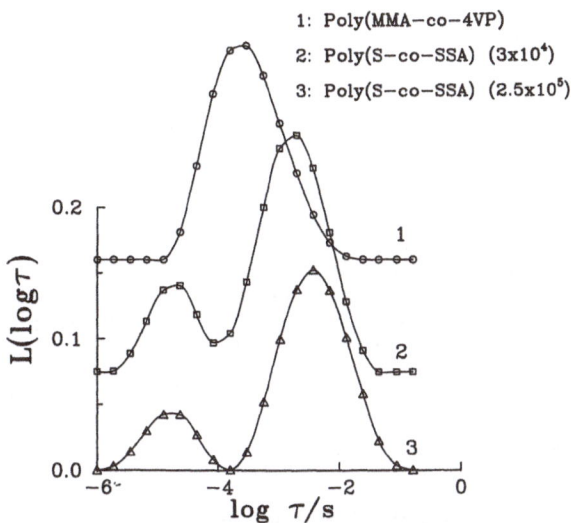

Fig. 2. Distribution of relaxation times L(logτ) for poly(MMA-co-4VP) (c = 7×10^{-3} g/cm^3) and two different molecular weight (number in parentheses) PSSA (C = 7×10^{-3} g/cc) in dimethoxy sulfoxide at 25°C.

Nevertheless it is, likely, in view also of the data of Fig.1b, that the double feature of L(logτ) is a characteristic feature of the polyelectrolyte solutions without added salt. However for a thorough study, the diffusional dynamics of polyelectrolytes in solution using monodisperse samples with different molecular weight and ion content should be investigated. It is interesting to note, that L(logτ) for Poly(MMA)-co-4VP) displays (Fig.2) a single-peak structure, due probably to screened intermolecular interactions in the polar DMSO. From the diffusion coefficient and Eq.3 an effective R_h = 200 A was evaluated.

Mixing of the poly(S-co-SSA)/DMSO and poly(MMA-co-4VP)/DMSO solutions at 25°C, with equivalent numbers of interacting SSA and 4VP groups, produces immediately precipitation. The mixture of the two solutions became clear above about 85°C for the concentrations and samples used in NMR[3] and present investigation. Distributions L(logτ) (Eq.2) of the two polymer solutions and of the blend are shown in Fig.3.

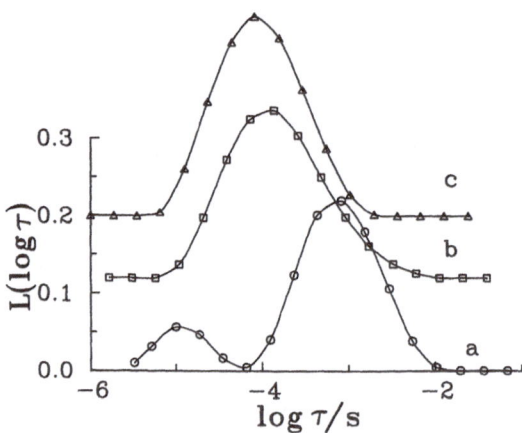

Fig. 3. Distributions of relaxation times L(logτ) at 85°C for a) poly(S-co-SSA) (MW = 3×10^4, c = 4.3×10^{-3} g/cm^3). b) poly(MMA-co-4VP) (c = 5.36×10^{-3} g/cm^3) and c) their blend solution in DMSO 65 min. after the sample was set at 85°C.

It is evident that the spectrum L(logτ) of the blend is very similar to that of the poly(MMA-co-4VP). We have attributed this similarity to interpolymer complex via proton transfer from SSA group to the 4VP ring and subsequent coil overlap as suggested by the NMR experiment. Apparently, the observed single peak structure is dominated by the larger poly(MMA-co-4VP) and is probably the result of the screened electrostatic interactions of the SSA groups in the blend. For total concentration c = 9.7×10^{-3} g/g, we obtained constant light scattering intensity about one hour after the blend solution was set at 85°C in agreement with the time evolution of NMR spectrum[3]. The rate determining step is probably the coil overlap which among other parameters might depend on concentration. In fact, the equilibration time at 85°C is much shorter for a total concentration c = 2.23×10^{-3} g/g. In principle, light scattering intensity could be used to study the kinetics of the process of coil overlap.

Based on the plots of L(logτ) in Fig.4, interpolymer complex formation occurs at 150°C as well. At about this temperature, the NMR shielded due to the proximity of S at a distance of about 4A - OCH$_3$ signal was practically gone[3]. Alternatively, the only difference between Figs.3 and 4 is limited to the shift of L(logτ) to shorter times due simply to the raising of temperature from 85° to 150°C. Thus, ion pair formation between SSA groups and 4VP ring is possible even at 150°C in DMSO, an aprotic solvent with high dipole moment and dielectric constant which decreases the Bjerrum length.

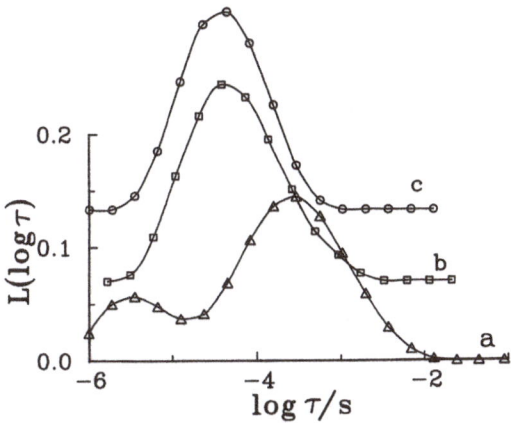

Fig. 4. Distributions of relaxation times L(logτ) at 150°C for the samples of Fig.3.

The changes that are witnessed upon mixing the two solutions can be contrasted to those observed when blend solution is formed from neutral PS and PMMA solutions in p-dioxane. The static intensity at the polarized light scattered from binary solutions of PS(MW=16.700) and PMMA (MW = 7.5x10^4) in dioxane and the corresponding blend solution was measured for three different concentrations at a scattering angle of 90° and 25°C. The concentration dependence of the relative intensity $I^* = (I_{solu}-I_{solv})/I_B$, where the subscripts solu, solv and B denote respectively the samples solution, solvent and the standard neat benzene, is depicted in Fig.5a. As expected, in the absence of specific interactions, the intensity of the ternary solutions fall between the intensities of the corresponding binary solutions. In contrast, the intensity I^* (Fig.5b) of the poly(S-co-SSA)/poly (MMA-co-4VP) blend in DMSO at 85°C is more than twice as large as the average I^* for the solutions of the blend components. This increase is a composite effect, owing to the higher MW and larger radius of gyration of the interpolymer complex. Similar changes in the static light scattering intensity were observed in an earlier study of poly(methacrylic acid)/poly(N-vinyl-2pyrolidone) system[2]. Depending on the solvent used, the MW of the interacting blend was four to eight times larger than the MW's of the individual polymers. The picture that emerged from Fig.3 to 5 was that the particular interacting blend reflects more or less the solution properties of the larger poly(MMA-co-4VP) chain which bears about one hundred 4VP rings. For comparison, the low MW poly(S-co-SSA) contains about thirty SSA ionic groups per chain. We should therefore expect an entirely different behavior if the high MW poly(S-co-SSA) chain, with about two hundred SSA groups, is used for the formation of a blend where the concentrations of interacting groups (4VP, SSA) are the same. This situation is

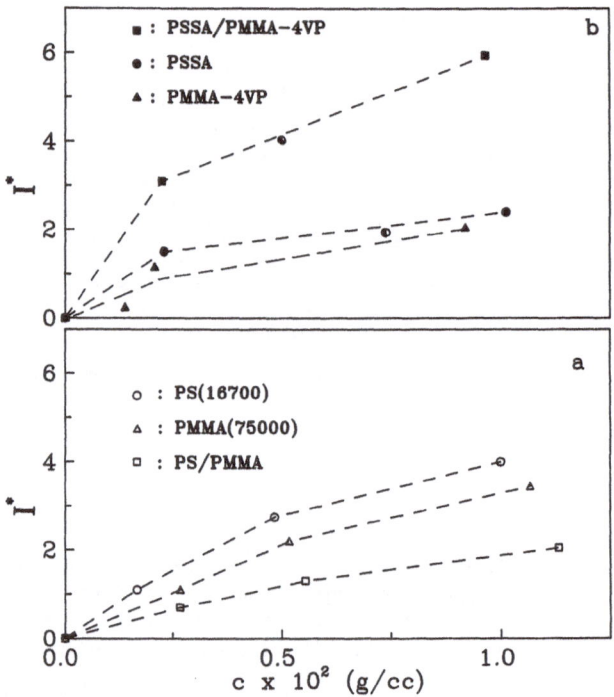

Fig. 5. a) Relative polarized light scattering intensities I^* of binary solutions of PS(O), PMMA(Δ) and their blend solution (□) in p-dioxane at 25°C.

b) I^* versus polymer concentration in binary solutions of poly (MMA-co-4VP) (▲), poly(S-co-SSA) (●) and their blend solution (■) in DMSO at 85°C. The solid symbols ◐, are referred to the high MW poly(S-co-SSA) and the corresponding blend respectively.

demonstrated in Fig.6, where the typical double peak structure in the L(logτ) spectrum of the poly(S-co-SSA) persists in the blend as well. In this case, the size of the interpolymer complex is determined by the longer poly(S-co-SSA) chain. However, the more efficient screening of the interactions between the SSA groups in the blend may decrease the average dimension of

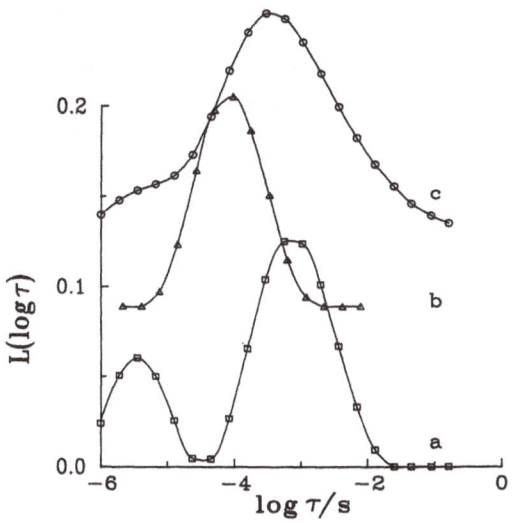

Fig. 6. Distributions of relaxation times L(logτ) at 120°C for
a) poly(S-co-SSA) (MW = 2.5×10^5, c = 2.8×10^{-3} g/cm^3)
b) poly(MMA-co-4VP) (2.2×10^{-3} g/cm^3) and c) their blend solution in DMSO.

poly(S-co-SSA) and hence increase its effective diffusion D in the blend. In fact, the main peak in L(logτ) of the blend is shifted to shorter times when compared to the poly(S-co-SSA) solution. Finally, it is worth mentioning that the short time peak of L(logτ) in Fig.6a is also present in the blend (Fig.6c). If this mode is related to "a cooperative" motion of the entangled network[11,15] formed by the poly(S-co-SSA) chains, then the position of the fast mode should be less affected by the formation of the interpolymer complex. Clearly, this explanation is quite speculative and further studies of the double feature of the L(logτ) spectrum and its dependence on MW, ionic group content and polymer concentration are definitely needed.

REFERENCES

1. Natasohn, A., Murali R., Eisenberg A., Macromol. Chem., Macromol. Symp. 16, 175 (1988) and references herein.
2. Ohno, H., Abe, K., Tsuchida, E., Macromol. Chem. 179, 755 (1978).
3. Natasohn, A., Eisenberg, A., Macromolecules 20, 323 (1987).
4. Patkowski, A., Chu, B., J. Chem. Phys. 73, 3082 (1980).
5. a) Chu, B., Fytas, G., Macromolecules 15, 561 (1982).
   b) Lee, D.C., Ford, J., Fytas, G., Chu, B., Hagnaner, G.L. Macromolecules 19, 1586 (1986).
6. Willmonth, F.M., Rance, D.G., Henman, K.M., Polymer 25, 1185 (1984).
7. Schmidt, M., Macromol. Chem., Rapid Commun. 10, 89 (1989).
8. Gulari, E., Gulari, E., Tsunashina, Y., Chu, B., Macromolecules 12, 599 (1979).
9. Provencer, S.W., Comput. Phys. Commun. 27, 213, 229 (1982).
10. Grüner, F., Lehman, W.P., Fahlbush, Weber, R., J. Phys. A: Math. Gen. 14 L307 (1981).
11. Koene, R.S., Mandel, M., Macromolecules 16, 973 (1983).
12. Drifford, M., Dalbiez, J.P., J. Chem. Phys. 88, 5368 (1984).
13. de Gennes, P.G., Pincus P., Velasco, R., Brochard, F., J. Phys. (Fr.) 37, 1461 (1976).
14. Odijk, T., Macromolecules, 12, 688 (1979).
15. Brown, W., Johnson, R., Macromolecules 19, 2002 (1986).

INTERACTIONS OF SIGMA CONJUGATED POLYMERS

WITH STRONG OPTICAL FIELDS

Jonathan R. G. Thorne, John M. Zeigler[+]
and Robin M. Hochstrasser

Department of Chemistry
University of Pennsylvania
Philadelphia, PA 19104, USA

+ Silchemy
2208 Lester Dr NE Albuquerque
NM 87112, USA

INTRODUCTION

The polysilanes $(RR'Si)_n$ are a new class of sigma-conjugated linear-chain polymers that have strong absorption and fluorescence in the near ultraviolet and potential for non-linear optical applications.(See ref.[1] for a review.) A description of the optical excitations as delocalized excitons whose coherent spatial extent and whose incoherent energy transfer properties are controlled by the static disorder of the medium, is presented elsewhere in this volume.[2] In this work we extend these conclusions by examining the behaviour of the polymer under intense laser irradiation. The experiments reviewed here are of two types, those that use non-resonant light (conventional non-linear responses) and those that use resonant light(excited state absorption). The polymers are studied at room temperature in thin film and solution.

RESULTS AND DISCUSSION

Non-resonant Excitation

Poly(di-n-hexylsilane) is the most widely studied by optical techniques of the sigma-conjugated polysilane polymers: the spectra below are discussed at length in references[3,4]. The absorption spectrum of $[(C_6H_{13})_2Si]_n$ thin film on quartz substrate at room temperature is shown in figure 1, and has a resonance at $\sim 27000 cm^{-1}$. (The peak in the spectrum at $\sim 32000 cm^{-1}$ results from a disordered phase of the polymer.[1]) The 3-photon resonance was revealed by third-harmonic generation from the polymer film of second Stokes Raman shifted dye-laser light at $\sim 9000 cm^{-1}$(1-photon energy of 27000 cm^{-1}). The quartz substrate provides a reference for the absolute magnitude of the signal. The 2-photon excitation spectrum in the figure was recorded by detection of the total fluorescence as a function of dye laser excitation frequency in the range 18000 - $12000 cm^{-1}$. The resonance in this experiment occurs at $17000 cm^{-1}$ (one photon energy $\sim 34000 cm^{-1}$), $7000 cm^{-1}$ above the coincident 1- and 3-photon values.

The most general conclusion we can draw is that the third order polarizabilities that give rise to the second and third spectra in the figure are large and indicative of a system having a considerable degree of electron delocalization.

Large-Scale Molecular Systems, Edited by W. Gans *et al.*
Plenum Press, New York, 1991

Our results are similar to those found for the pi-conjugated polydiacetylenes where the 2-photon accessible state is ~ 10000 cm^{-1} above the 1-photon accessible state at energies of 16000-20000cm^{-1}.[5] By contrast the excitations in polyacetylene have been found to be degenerate at the optical band edge at an energy of ~ 14000cm^{-1} for 1-, 2- and 3-photon resonances.[6] The results for polysilanes are interpretable in terms of strong electron correlations leading to splitting of the excited charge density B and spin wave A states responsible both for 1- and 3- on the one hand and for 2-photon resonance on the other. In polyacetylene these correlations are much weaker.

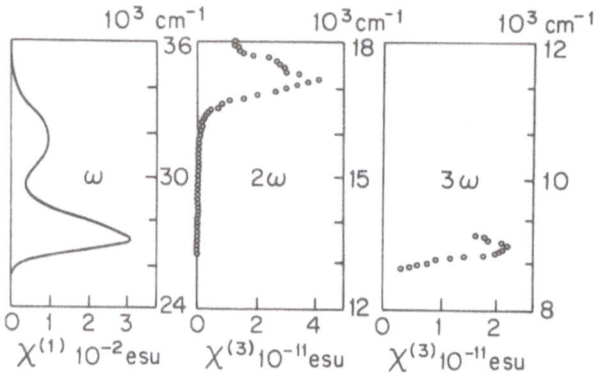

Fig. 1. Multiphoton resonances for poly(di-n-hexylsilane)
Absorption:two-photon fluorescence:third-harmonic generation.

Resonant Excitation

We turn now to the behaviour of the polymer under strong excitation with light in the absorption region of the spectrum. The absorption spectrum of poly(di-n-hexylsilane) in room temperature solution has a peak at 317nm(31500cm^{-1}) having a linewidth of ~ 3500 cm^{-1}. The fluorescence is red-shifted by ~ 2000 cm^{-1} and peaks at ~ 29500cm^{-1}, with a width of ~ 2000cm^{-1}. The position of this latter maximum depends slightly upon excitation wavelength.[1] Evidence from hole-burning studies and the much smaller (<10cm^{-1}) shift found in low-temperature glasses shows that an individual chromophore upon excitation does not strongly relax due to coupling to the lattice.[2] Negligible phonon structure is seen in absorption and emission spectra, and the excitation does not have the character of a strongly bound polaron. Electron-phonon coupling in the polymers is weak. The process of spatial energy transfer in the polymers has been followed by the technique of time-resolved fluorescence depolarization[7] which reveals highly dispersive kinetics characteristic of relaxations occurring on many different timescales from tens of picoseconds to nanoseconds and longer. We believe these energy transfer processes are responsible for the large apparent absorption-emission spectral shifts.

We have recorded transient absorption spectra excited by a 70fs amplified CPM laser system at 312nm. A full experimental description and a discussion of the nature of the high-lying electronic states responsible is in the course of publication.[8] Figure 2 shows the transient absorption spectrum recorded at pump-probe delay times of 100fs and 4ps as it evolves from that due to absorbing to that due to emitting states.

494

Two excited state absorptions are seen at ~27000cm⁻¹ in the ultraviolet and ~10000cm⁻¹ in the infra-red. The latter may well correspond to excitation to the two-photon accessible state already discussed.[3] The temporal evolution of the transient absorption provides a means of following the ultrafast relaxation dynamics of the excitation: the kinetics indicate a relaxation time of 700fs. The pump-probe technique allows much improved time resolution compared to that we have been able to achieve with time correlated single photon fluorescence studies.[7]

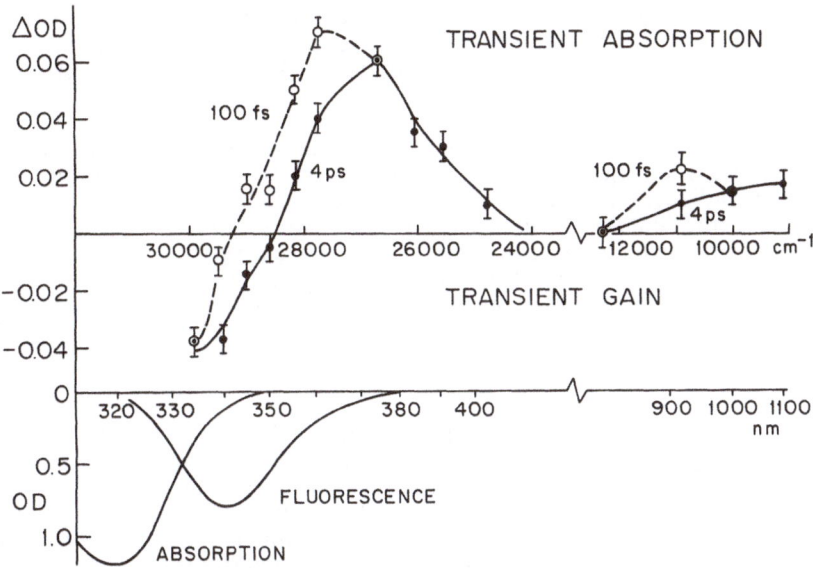

Fig. 2. Transient absorption/gain spectrum of poly(di-n-hexylsilane) excited at 312nm (Absorption and fluorescence spectra are also shown)

States of the polymer are excited in our experiment on the high energy side of the absorption peak at 312nm(32000cm⁻¹). We interpret the ultrafast relaxation observed in terms of relaxation of the initially created excitations in a disordered band. We can draw parallels with the situation in semiconductors pumped above the band-edge where one cooling mechanism is thought to be carrier LO-phonon scattering with sequential (cascade) emission of optical phonons.   The resonance integral for silicon-silicon site coupling along the polysilane chain is large and generally accepted to have a value ~10,000cm⁻¹. In a disordered chain model, the pump laser generates a distribution of eigenstates of each disordered chain in the ensemble.[2] Phonon assisted(incoherent) energy transfer rates between these Born-Oppenheimer states are then determined by the nuclear momentum operators. The result is spatial energy transfer and we consider this the likely relaxation occurring in 700fs. Ultrafast relaxation processes have been observed in other polymeric systems[9,10], but generally ascribed to structural deformation processes, leading to excitation localization. In polyacetylene lattice deformation is believed to occur on a 0.1 ps time scale. Soliton pair recombination is a subsequent spatial energy transfer process occurring at the speed of sound in ~0.5ps.[9] In polydiacetylene early time processes are again thought to be dominated by structural relaxations to "kink" states taking place in ~1ps.[10] Significant local distortion is unlikely in the polysilane case since fluorescence and absorption are not much shifted and changes in structure can be delocalized over the extent of the excitation.   It will be worthwhile to develop a model for the non-Born-Oppenheimer coupling of the nearby exciton states of disordered linear chains. Calculation of the rate of such a process will depend upon the coupling to the relevant promoting phonon mode and the magnitude of the energy defect. We can, however, address the question of whether 700fs is a reasonable value for the phonon-assisted energy transfer time in the polymer at room temperature, based on other experimental measurements. Spectral holes burnt in the polymer in glasses at low temperature have widths of about $4cm^{-1}$,[2] suggesting a dephasing time T_2 of ~5ps. This places an upper limit on the energy relaxation rate at low temperature corresponding to an energy loss process having a T_1 time of 2.5ps. Single phonon emission processes, having energy defects of several hundred wavenumbers would be expected to have weak temperature dependences, varying as $(1+n)$ where n is the phonon occupation number: a room temperature value of the transfer rate ~4 times larger than this would not be unlikely. Our time resolved measurements are thus consistent with the frequency domain studies described in this volume[2] and suggest that holewidths are determined by T_1 relaxation processes of the initially created excitation.

We turn lastly to a discussion of the coherent exciton length in polysilane. We have two estimates of this quantity. The first comes from the shortening of the radiative lifetime[7] (super-radiance) associated with exciton delocalization. The second comes from the saturation of transient absorption at high pump powers[8] (exciton phase-space filling). Both these estimates give an upper limit to the coherence length of ~25 silicon atoms in the chain. Whether the full extent of this conjugation length is reflected in the magnitude of the third order non-linear optical responses is a question that is as yet unanswered.

CONCLUSIONS

Observation of the 1-,2- and 3-photon resonances in polysilanes has led to the conclusion that the optical excitations are strongly correlated electron-hole pairs having a small degree of charge-transfer character. The third order polarizabilities for the molecule are large. Fast energy transfer in the polymer, poly(di-n-hexylsilane) in solution is observed to occur on a time scale of 700fs after excitation at short wavelength. We believe this process populates a distribution of lower energy states through phonon assisted relaxations. Both our time domain measurements and our estimates of the exciton coherence length are in good agreement with those derived from hole burning studies, discussed elsewhere in this volume.

This work was supported by the Sandia National Laboratories and the US Department of Energy under grant DEAC04-76-DP00789.

REFERENCES

1. R. D. Miller and J. Michl, Chem. Rev., 89:1359 (1989)
2. A. Tilgner, H. P. Trommsdorff, J. M. Zeigler and R. M. Hochstrasser, this volume and J. Luminescence, 45:373 (1990)
3. J.R.G.Thorne, Y.Ohsako, J.M.Zeigler and R.M.Hochstrasser, Chem. Phys. Lett., 162:455 (1989)
4. J.R.G.Thorne, Y.Ohsako, R.M.Hochstrasser and J.M.Zeigler, J. Luminescence, submitted
5. R.R.Chance, M.L.Shand, C.Hogg and R.Silbey, Phys. Rev. B, 22:3540 (1980)
6. W.S.Fann, S. Benson, J.M.J.Madey, S.Etemad, G.L.Baker and F.Kajzar, Phys. Rev. Lett., 62:1492 (1989)
7. Y. R. Kim, M. Lee, J. R. G. Thorne, R. M. Hochstrasser and J. M. Zeigler, Chem. Phys. Lett., 145:75 (1988)
8. J. R. G. Thorne, S. T. Repinec, S. A. Abrash, R.M. Hochstrasser and J. M. Zeigler, Chemical Physics, in press
9. L. Rothberg, T. M. Jedju, S. Etemad and G. L. Baker, Phys. Rev. B, 36:7529 (1987)
10. B. I. Greene, J. Orenstein, R. R. Millard and L. R. Williams, Chem. Phys. Lett., 139:381 (1987)

RANDOMLY BRANCHED POLYMERS

M. DAOUD

Laboratoire Léon Brillouin[*]
C.E.N.Saclay – 91191 Gif/Yvette cedex, France

INTRODUCTION

Polymers may be linear when the monomers are bifunctional, and branched when they are multifunctional. In the latter case, polymerisation is usually accompanied by gelation, i.e. the formation of an infinite, elastic network.In the following, we will by interested in the finite, eventually very large polymers that constitute the sol phase. This is characterized by a very broad distribution in the molecular weights, or polydispersity. The latter is similar to the cluster distribution in the percolation problem. Because of this, effective fractal dimensions are measured, related to both the fractal dimension of every polymer,and to the distribution of molecular weights. after recalling the main results for polydispersity, we review the recent ideas concerning both static and dynamic properties of branched polymer solutions.

POLYDISPERSITY

Consider a vessel filled with multifunctional monomers at initial time. At later times, reaction progresses,and randomly branched monomers are synthesized. This was modelled by percolation[1,2], following early mean field work by Flory[3], Stockmayer[4] and others[5]. As in percolation, there is a threshold where an infinite network appears. This is called the gel. The finite clusters constitute the sol. The latter is very polydisperse: the number distribution function, that is the number $P(N,\varepsilon)$ per unit volume of polymers made of N monomers at a distance ε from the threshold,is

$$P(N,\varepsilon) \sim N^{-\tau} f(\varepsilon N^{\sigma}) \qquad (1)$$

where τ and σ are the percolation exponents. This implies that two diverging characteristic molecular weights may be defined, namely the weight-average and Z-average molecular weights N_w and N_z respectively.

$$N_w = \frac{\int N^2 \, P(N,\varepsilon) \, dN}{\int N \, P(N,\varepsilon) \, dN} \sim N^{-\gamma} \tag{2}$$

and

$$N_z = \frac{\int N^3 \, P(N,\varepsilon) \, dN}{\int N^2 \, P(N,\varepsilon) \, dN} \sim \varepsilon^{-1/\sigma} \tag{3}$$

Note that N_z corresponds to the cut-off mass in the distribution. Note also that N_z and N_w are related through the fractal dimension D_p of percolation

$$N_z \sim N_w^{D_p/(2D_p-d)} \sim N_w^{5/4} \tag{4}$$

where we used Flory's approximation for the fractal dimension of percolation clusters[6], and d=3. Thus in the reaction bath, and in the absence of any solvent, we have, for the radius R of a polymer

$$R \sim N^{D_p} \tag{5}$$

and, within Flory approximation, for d =3

$$D_p = 5/2 \tag{6}$$

This form of the distribution function was checked by Schosseler and Leibler on Polystyrene cross-linked by irradiation[7], Adam et al. on polyurethane[8], Patton et al.[9] on branched polyesters, and Lapp et al.[10] on chemically end-linked polystyrene, with similar results, i.e. $\tau = 2.3 \pm 0.1$, showing the universality of the result, which is independent of the local Chemistry.
We turn now to the influence of this polydispersity on static measurements. In all the following, we assume that the synthesis is finished and that the distribution is quenched, whatever happens to the sol.

STATIC PROPERTIES

Dilute solutions

We consider first dilute solutions where the sol has been diluted in an excess of a good solvent. In such solvent, excluded volume interactions are present and swell every polymer. Thus the fractal dimension changes, and goes from D_p to the dimension D_0 of an animal. For d=3, we have

$$D_0 = 2 \tag{7}$$

However, because of polydispersity, what is measured is an average over the whole distribution. Thus, the average radius in a scattering experiment is

$$\langle R_z^2 \rangle = \frac{1}{N_W} \int R^2(N) \, N^2 \, P(N,\varepsilon) \, dN \tag{8}$$

Using relations (2),(4) and (7), we get[11]

$$\langle R_z^2 \rangle \sim N_W^{5/8} \tag{9}$$

Similarly, the scattered intensity[12] in a light or neutron scattering experiment is, for one mass and for the distribution respectively

$$S_1(q) \sim q^{-2} \tag{10a}$$
$$S_t(q) \sim q^{-8/5} \tag{10b}$$

relation (9) was checked by quasi-elastic light scattering by Candau et al.[13], by static light scattering by Schosseler and Leibler[14], and Patton et al.[9], whereas relations (10) were checked by small angle neutron scattering by Bouchaud et al.[15], who exhibited the difference between a fractionated solution and a native,polydisperse, one.Their result is shown on figure 1 and allows to check the respective fractal dimensions of a single polymer and a polydisperse solution.
The above results are in very good agreement with the predictions. These are valid in dilute solutions, as long as the various polymers do not overlap each other. Above a cross-over concentration C*, because of overlap, a different behavior is observed.

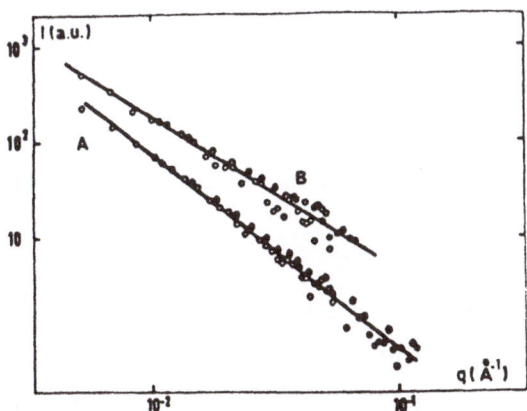

Figure 1. the scattered intensity in a small angle neutron scattering experiment by a fractionated (a), and a native(polydisperse)(b) samples of polyurethane in deuterated tetrahydrofurane. Slopes are 1.98±0.03 and 1.59±0.05. From reference (15).

Semi dilute solutions

The overlap concentration C* corresponds[16] to space filling by the polydisperse swollen polymers:

$$C*^{-1} \sim \int R^3(N) \ P(N,\varepsilon) \ dN \sim N_w^{-3/8} \tag{11}$$

where we used relations (1) and (4). Above C*, in addition to the average radius, one may introduce a screening length ξ corresponding to the local swollen behavior. This may be found by a scaling argument. The characteristic length may be written in the following form

$$L(C) = N_w^{5/8} \ f(C/C*) \tag{12}$$

Assuming that f(x) behaves as a power law, this is determined by the condition that it is independent of the average molecular weight, because it is a local property. We find

$$\xi \sim C^{-5/3} \tag{13}$$

The average radius is determined either by a scaling argument following the same lines as above(the condition is that the N dependence is the same as in the reaction bath, because the polymers interpenetrate each other, as they do in the latter) or by saying, equivalently, that if one defines blobs made of g monomers such that $\xi \sim g^{5/8}$, and takes g and ξ as unit length and distance, then the solution is similar to a melt. Then we have

$$R_z \sim \left(\frac{N_w}{g}\right)^2 \xi \tag{14}$$

where the exponent $2 = D_p(3-\tau)$ takes into account both the fractal dimension of a cluster and the polydispersity of the distribution, exactly as for dilute solutions.
Relation (11) was checked by light scattering by Adam et al.[17] and Patton et al.[9] who also checked it indirectly by intrinsic viscosity measurements. Relation (14) was tested by Delsanti et al.[18], with reasonable agreement though not as good as previous results.

DYNAMICS

The dynamical properties were considered in the reaction bath first by Durand et al.[19] who measured the complex modulus $\bar{G}(\omega)$ as a function of frequency. For High frequencies, this has a power law behavior

$$\bar{G}(\omega) \sim (i\omega)^{-0.72 \pm 0.05} \tag{15}$$

which may be understood following a scaling argument by Efrös and Schklovskii[20] for the conductivity of a mixture of good and poor conductors close to the percolation threshold. With our notations, this reads

$$\bar{G}(\omega) = \varepsilon^{\mu} f_{\pm}(i\omega\varepsilon^{-s-\mu})$$ (16)

where the exponents s and μ are percolation exponents. Then, in the high frequency regime, $\omega\varepsilon^{-s-\mu} \gg 1$, one finds

$$\bar{G}(\omega) \sim (i\omega)^{-\mu/(s+\mu)}$$ (17)

The experimental results are in good agreement with the percolation results. But there is still controversy[21,22] whether s and μ should have the percolation values or not, related to the existence of hydrodynamic effects in the reaction bath. In all cases, the complex viscosity, $\bar{\eta}(\omega) \equiv \bar{G}(\omega)/i\omega$, is directly related[23] to the distribution of relaxation times $H(\tau)$:

$$\bar{\eta}(\omega) = \int \frac{H(\tau)}{1+i\omega\tau} d\tau$$ (18)

Using relation (16), we get

$$H(\tau) \sim \tau^{-\mu/(s+\mu)} h(\tau\varepsilon^{s+\mu})$$ (19)

This distribution may be characterized by two diverging times, for instance the moments of order zero and one. Thus one finds

$$T \sim \eta \sim \varepsilon^{-s}$$ (20)

and

$$T_z \sim \varepsilon^{-s-\mu}$$ (21)

This in turn implies power law behaviors for relaxation phenomena at intermediate time scales, for $\tau \ll T_z \sim \varepsilon^{-s-\mu}$.
Such broad distribution of relaxation times is also present in dilute solutions. Work is still under way[24] to determine all the rheological properties of these solutions as well as those of a gel close to the gelation threshold.

REFERENCES

* Laboratoire commun C.N.R.S.- C.E.A.
1. P.G. de Gennes Scaling concepts in polymer physics, Cornell University Press, 1979.

2. D. Stauffer, Introduction to percolation theory, Taylor and Francis, 1985.
3. P.J. Flory, Principles of Polymer Chemistry, Cornell University press, (1953).
4. W.H. Stockmayer, J. Chem. Phys.11, 45 (1943).
5. M. Gordon,, S. B. Ross-Murphy, Pure Appl. Chem. 43: 1 (1975).
6. J. Isaacson, T. C. Lubensky, J. Phys. 42: 175 (1981).
7. F. Schosseler , L. Leibler , Macromolecules , 18: 398, (1985) .
8. M. Adam, M. Delsanti, D. Durand, Macromolecules 18: 2285, (1985).
9. E. Patton, J.A. Wesson, M. Rubinstein, J.C. Wilson, L.E. Oppenheimer, Macromolecules, 22:1946, (1989).
10. A. Lapp, L. Leibler, F. Schosseler, C. Strazielle, Macromolecules, 22: 2871, (1989).
11. M. Daoud, F. Family, G. Jannink, J. Physique Lett., 45:199, (1984).
12. J.E. Martin, B.J. Ackerson, Phys. Rev. A31:1180, (1985).
13. S. J. Candau, M. Ankrim, J. P. Munch, P. Rempp, G. Hild, R. Osaka, in Physical Optics of Dynamical Phenomena in Macromolecular Systems, W. De Gruyter,Berlin,145,(1985).
14. L. Leibler, F. Schosseler, Phys. Rev. Lett., 55: 1110, (1985). See also in Physics of Finely Divided Matter, Springer Proc. Phys. 5, Springer, 135,(1985).
15. E. Bouchaud, M. Delsanti, M. Adam, M. Daoud, D. Durand, J. Physique Lett., 47: 1273, (1986).
16. M. Daoud, L. Leibler, Macromolecules, 21: 1497, (1988).
17. M. Adam, M. Delsanti, J.P. Munch, D. Durand, J. Physique, 48: 1809, (1987).
18. M. Delsanti, J.P. Munch, D. Durand, J.P. Busnel, M. Adam, to be published.
19. D. Durand, M. Delsanti, M. Adam, J.M. Luck, Europhys. Lett., 3: 297, (1987).
20. A.L. Efrös, B.I. Schklovskii, Physica Status Solidi, B76:475,(1976).
21. M. Rubinstein, R.H. Colby, J.R. Gillmor, Polymer preprint 30: 1, (1989). See also in Space-Time Organization in Macromolecular Fluids, F. Tanaka, T. Ohta and M. Doi Eds, Springer Verlag, (1989).
22. J.E. Martin, D. Adolf, J.P. Wilcoxon, Phys. Rev. Lett., 61:2620, (1988).
23. M. Daoud, J. Phys. A21: L237, (1988).
24. M. Daoud, J.E. Martin, in The fractal approach to heterogeneous Chemistry, D. Avnir Ed., J. Wiley, 105,(1989).

MULTI–PARTICLE RELAXATION IN ELECTRONICALLY EXCITED POLYMERS: DISTRIBUTION OF TRANSITION RATES FROM FLUORESCENCE DATA – A NUMERICAL APPROACH

H. F. Kauffmann*, G. Landl*, and H. W. Engl°

* Institut für Physikalische Chemie,
Universität Wien, Währingerstraße 42,
A–1090 Wien, Austria
° Institut für Mathematik
Johannes-Kepler-Universität Linz
A–4040 Linz, Austria

I. Introduction

Recently, there has been increased activity devoted to understanding the dynamics of electronic excitation in disordered polymeric materials by optical techniques.[1,2] Time resolved fluorescence spectroscopy has become a powerful kinetic tool for probing polymer relaxation events in the range of a few picoseconds to some tens of nanoseconds. Typical physical processes in polymer-bound chromophores that have been investigated by transient fluorescence are excitation energy transport, rotational sampling and trap-controlled interconversion.[3] Among the various transient configurations measuring such short time-profiles, the statistical single-photon time correlation (SPT)[4] is the most sensitive and now routinely applied in polymer photophysics.

However, despite the considerable advance in the measurement of high-quality fluorescence patterns, the correct analysis of fluorescence data is faced with a couple of severe problems. One of these stems from an inherent mathematical difficulty, namely the "ill-posedness" of the underlying integral equation.[5] If $F(t)$ represents the δ-pulse fluorescence response of the polymeric system, then the impulse fluorescence response $h(t)$ to an experimental optical pulse $l(t)$ (termed the instrument response function) is given by the convolution integral

$$h(t) = \int_0^t l(t')F(t - t')dt' \qquad (1)$$

Since the finite temporal width of the optical pulse $l(t)$ in most (experimental) cases interferes with the sharp profile of the system's δ-pulse response $F(t)$, some of the characteristics of $F(t)$ is smoothed out and a deconvolution is necessary to recover $F(t)$

Large-Scale Molecular Systems, Edited by W. Gans *et al.*
Plenum Press, New York, 1991

from smooth data $h(t)$.[6] However, the inversion has the property that small noise may be amplified arbitrarily much, which, of course, results in a severe ill-conditioning of any numerical algorithm that does not take this effect into account; algorithms tailored to handling ill-posed problems are called "regularization methods".[5] Apart from the smoothing property of the integral operator in (1) accounting for the ill-posedness, the serious numerical difficulty of performing the inverse procedure in (1) is the superposition of *non-random* noise in the collected SPT data, $h(t)$. Because of the systematic (and non-reproducible) contamination of the Poissonian noise level in experimental fluorescence data caused by non-linearities in the time-to-amplitude converter, scatter, PMT-color effects etc., both $l(t)$ and $h(t)$ are incomplete representations and thus, the reconstruction of $F(t)$ by inversion, even in the case of simple decay laws, is hard to accomplish.

To overcome this difficulty, typical forward convolution techniques have been worked out that are fitting raw data $h(t)$ to the δ-pulse solutions of preconceived models in iterative fit-&-compare cycles using chi-squares statistics. While *multiexponential* trial functions as typical δ-pulse solutions to linear kinetic interconversion schemes may be quite appropriate for modelling processes in small molecules — at least at a low level of data precision[7] — they must fail, in general, when interpreting excitation dynamics [8,9] and rotational sampling[10] in polymer morphologies. Therefore, for many photophysical situations of fluorescent polymers, the results of multiexponential reconvolution (best-fit amplitudes A_i and apparent lifetimes T_i) have no real physical meaning, unless one interprets the number of exponentials to represent a lower limit for the number actually involved in the profile and thus, as the minimal set in a pure curve parameterization of a typically *non-exponential* fluorescence pattern.

II.   Nonexponential Fluorescence – Distribution of Lifetimes

Nonexponential relaxation behavior in polymer fluorescence has received considerable attention in recent years. Nonexponentiality generally arises from a *distribution* of lifetimes which is caused by the pronounced heterogeneity and the multi-particle properties of chromophoric sites in polymer systems.[3] Therefore, a correct evaluation of polymer fluorescence data h(t), i.e., the translation of static and dynamic disorder to specific parameters of polymer structure and dynamics must imply, necessarily, the use of a *distributed* fluorescence function

$$F(t) = \int\limits_0^\infty \Phi(\tau) \exp[-t/\tau] \, d\tau \qquad (2)$$

which is the ensemble-and configurational average of the donor chromophore δ-pulse response expressed in terms of a continuous summation of single exponential relaxations. $\Phi(\tau)$ is the distribution function of fluorescence lifetimes which contain the corresponding event-times $1/k$ by the relation $(1/\tau - 1/\tau_0)^{-1} = 1/k$ (τ_0, natural lifetime of an isolated chromophore). Since $F(t)$ represents the quasi-Laplace transform of the underlying distribution, things get even more complicated in data analysis. Substituting (2) in (1) we have

$$h(t) = l(t) * \int\limits_0^\infty \Phi(\tau) exp[-t/\tau] \, d\tau \qquad (3)$$

504

where the unknown distribution function $\Phi(\tau)$ is subject to the smoothing properties of both the convolution and the Laplace transform operator. Clearly, the evaluation of lifetime-distributions (event-times) from fluorescence data of polymers is therefore a non-trivial problem in numerical analysis, as, in particular, the inverse Laplace transform is severely ill-posed (it exceeds the ill-posedness of the deconvolution operator, by far), so that its numerical implementation is notoriously ill-conditioned. Thus, analyzing fluorescence data, we are actually trying to solve *two* ill-posed problems, simultaneously. To face the problem, methods using definite trial-distributions in the minimization of chi-square might be applied, in principle (reconvolution, forward Laplace transform). However, aside from the preconceived bias inherent in the method, the drawback is that the actual distribution of physical event-times is rather unknown in polymers, so the parameters of so called "reasonable" target distributions hold only little physical relevance, if, in fact, the data arise not from this "plausible" model, but rather from a quite different distribution. Similar arguments are also valid when using a sum-of-exponentials trial function as a "multi-peak" distribution function in reconvolution. Since multiexponential fits pass the chi-square criterion, quite satisfactorily, and thus, belong to the (large) set of feasible solutions, at least at a precision level not too high, the parameter obtained from such multicomponent fits have no physical meaning, but rather represent appropriate averages over a true distribution.

From this point of view, the reconstruction of distributions as well as the differentiation of SPT fluorescence data made up of a set of discrete components from those that have a continuous distribution, have become an even more pressing problem in polymer photophysics. More recently, numerical techniques that avoid specific trial distributions, but rather reconvolute the raw data by a coarse discretization of the Laplace transform (Eq. (3)) are arousing ever increasing interest. For pure decay curves of fluorescence, both the Exponential Series Method (ESM)[11] and the Maximum Entropy Method (MEM)[12,13] are very promising. While in the MEM technique an entropy-like function (a regularization term) is maximized subject to the additional boundary that chi-square equals approximately 1, in the ESM method chi-square is minimized. A common feature of both methods is the lack of bias and the capacity for recovering the amplitudes of an exponential series with fixed lifetimes.[11] Very recently, we worked out a modified ESM which has shown a remarkable stability in data simulations.[14] Inherent in our ESM algorithm is the possibility of combining it with a Tikhonov regularization[5] for adequately handling the ill-posedness as well as the potential of resolving *negative* preexponentials which allows, as a novel result, to recover even *rise-time* distributions from growing fluorescence profiles. While in our recent work, first of all, the structure of the algorithm and the mathematical aspects have been treated,[14] in this paper, we will demonstrate some results obtained by this new ESM version with special emphasis given to two typical photophysical situations in polymer photophysics. The one is concerned with distributed fluorescence quenching of donor chromophores by traps or excimer-forming-sites (EFS) located in *two* different domains of polymer disorder, and modelled in terms of two (Gaussian) input distributions. The second situation deals with the simulation of distributed trap-fluorescence, the evolution of which being prepaired by (distributed) donor deactivation and generated by one *negative* and one positive (Gaussian) distribution function. We want to show that ESM combined with regularization is able to recover smooth input distributions, quite satisfactorily. The problem one is encountered in analyzing negative amplitude functions, namely, the presence of an interfering oscillation between rise and decay, is discussed.

III. Method of Computation

In the Exponential Series Method[11] the distributed fluorescence $F(t)$ in Eq. (2) is approximated by a coarse discretization of the Laplace transform.

$$F(t) \approx \sum_{n=1}^{N} \phi_n \exp[-t/\tau_n] \tag{4}$$

In the analysis of fluorescence convolution data $h(t)$ (Eq. (3)), typically, 75 exponentials have been set in our ESM trial function, with τ_n fixed and, in general, logarithmically distributed along the relaxation scale of $h(t)$. Our ESM minimizes the function

$$T(\phi) = \chi^2 + \gamma \|\varphi''\|^2 \tag{5}$$

where chi-square, in case of Poissonian counting statistics, equals

$$\chi^2 = \sum_{i=1}^{M} \frac{(h_i - h(t_i))^2}{h_i} \tag{6}$$

with h_i representing the number of synthetic (experimental) photons in the i-th channel and $h(t_i)$ standing for the calculated value for time t_i, according to the discrete form of Eq. (3)

$$h(t_i) = \sum_{n=1}^{N} \phi_n \left(\frac{1}{2} l_i + \sum_{j=1}^{i-1} l_j \exp[-\frac{t_i - t_j}{\tau_n}] \right) \tag{7}$$

The second term in Eq. (5) is the regularization function. It comprises the mathematical norm of the second derivative which as a discretized approximation takes the form

$$\|\varphi''\| \approx \sum_{n=2}^{N-1} \frac{\frac{\varphi_{n+1} - \varphi_n}{\tau_{n+1} - \tau_n} - \frac{\varphi_n - \varphi_{n-1}}{\tau_n - \tau_{n-1}}}{\frac{\tau_{n+1} - \tau_{n-1}}{2}} \tag{8}$$

where φ_n is such, that

$$\phi_n = \varphi_0 + \varphi_n^2 \tag{9}$$

for an arbitrary (non positive) lower bound φ_0 and a regularization parameter γ appropriately chosen. The procedure allows the set of amplitudes $\{\phi_n\}$ corresponding to $\{\tau_n\}$ to be evaluated from Eq. (7) in a non-linear, iterative free-parameter optimization. Our algorithm follows a hybrid-method similar to that of Fletcher and Xu[15] which combines a Newton and a quasi-Newton technique. Details of the method are given elsewhere.[14]

Only synthetic fluorescence data have been used in this study. According to Eq. (3), profiles have been generated by convolving sufficiently smooth input distributions with an experimental (noisy) pulse. In addition, the patterns have been superimposed by Gaussian noise and data quality was adjusted to $3 \cdot 10^5$ counts in the peak channel (CPC). Typical Gaussian input distributions with mean values sufficiently separated from each other and with equal statistical weights have been used in the simulations spanning up relaxation profiles over at least three decades. In all cases, a rectangular distribution with logarithmically spaced lifetimes between 1 ns and 110 ns has been used as the initial guess in the ESM probe function.

IV. Results and Discussion

Fig. 1 deals with the recovery of two (positive) Gaussian distribution functions from noisy fluorescence data and is intended to show the positive effect of regularization. The data used for our reconstruction were generated by convolving an experimental pulse by a decay generated by two Gaussian lifetime distributions (central tendency at $\mu_1 = 10$ ns and $\mu_2 = 80$ ns, respectively, with standard deviations $\sigma_1 = 3$ ns and $\sigma_2 = 7$ ns, respectively). This decay profile was synthezised with $3 \cdot 10^5$ CPC (counts/peak channel) and then superimposed by Gaussian counting noise with a variance equal to the data (Fig. 1a). Figures 1b), 1c), and 1d) show the lifetime-distributions reconstructed with our algorithm (dots: recovered values, smooth solid lines: input distributions), with values $\gamma = 0$ (Fig.1b)), 100 (Fig.1c), and 1000 (Fig.1d) for the regularization parameter. The unregularized, i.e., "pure ESM" algorithm is able to recover the second Gaussian only, while the first Gaussian is reconstructed with considerable error (100 %, Fig.1b)). In addition, a severe artefact appears at very short lifetimes. However, with increasing level of regularization this artefact disappears, and the recovery of the Gaussian short-lifetime group improves considerably. Note that a further increase of the regularization parameter will eventually lead to a "solution" which is very insensitive to noise, but whose resolution becomes worse. Both signal and noise are "smoothed away". For references concerning optimal choices of the level of regularization, see Ref.:[14]

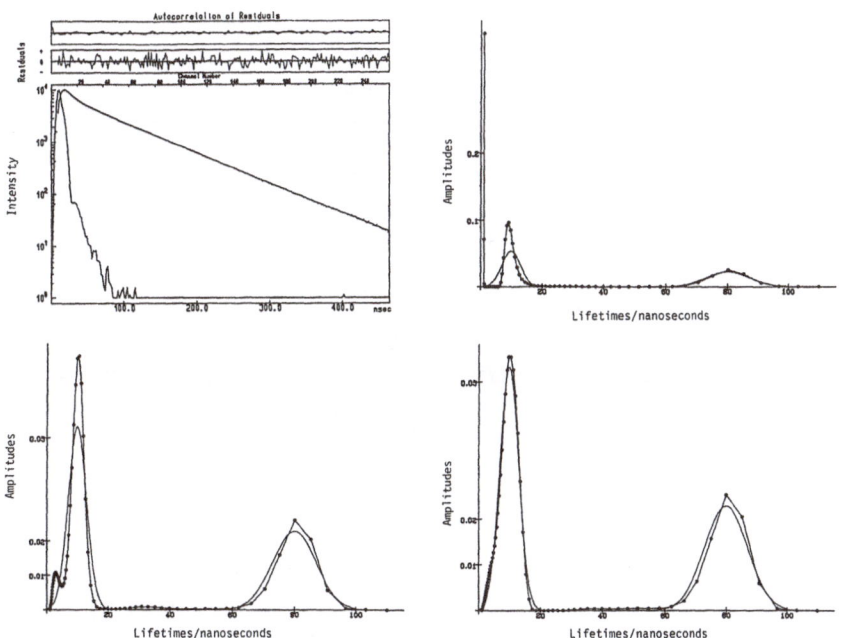

Figure 1. Analysis of lifetime-distributions from synthetic fluorescence data by ESM. a) experimental pulse – decay profile generated according to Eq. (3) with two Gaussian input functions (see the text for details), CPC: $3 \cdot 10^5$. Recovered histograms (dots) in b), c) and d) (solid curves: input distributions): b) without regularization ($\gamma = 0$), $\chi^2 = 1.000$; c) regularization parameter $\gamma = 100$, $\chi^2 = 1.009$; d) $\gamma = 1000$, $\chi^2 = 1.015$.

In Fig. 2a the data were generated as in example 1, namely, from two Gaussians, but, now, the first one with negative amplitudes. This situation may correspond to trap– (or excimer) fluorescence in a distributed pair of monomeric and trap-like states, where after monomer-excitation the traps (excimer-forming-sites) will be prepaired by migrational and/or rotational sampling in a sequential process. While the unregularized version in Fig. 2b (besides having a low resolution) produces an artefact at short lifetimes , regularization with $\gamma = 100$ and $\gamma=1000$ gradually removes this artefact. However, the resolution does not improve substantially, whereas the typical low-frequency oscillation between the negative and the positive parts remains present. The latter is due to the fact that we use the norm of φ'' as the regularization term, thus penalizing large total curvature. This forces the recovered function to be rather smooth. The same effect is known from spline interpolation termed "overshooting". It can probably be avoided by using a different regularization function; work on this subject is in progress. However, care has to be taken in doing this. When solving ill-posed problems, one can basically reconstruct (nearly) whatever one wants by using the "right" a-priori information (which translates into the corresponding regularization function). Therefore, in the analysis of experimental data, one has to make sure to use only a-priori information about the unknown lifetime–distribution that is justified by theoretical treatments or supported by additional experimental facts.

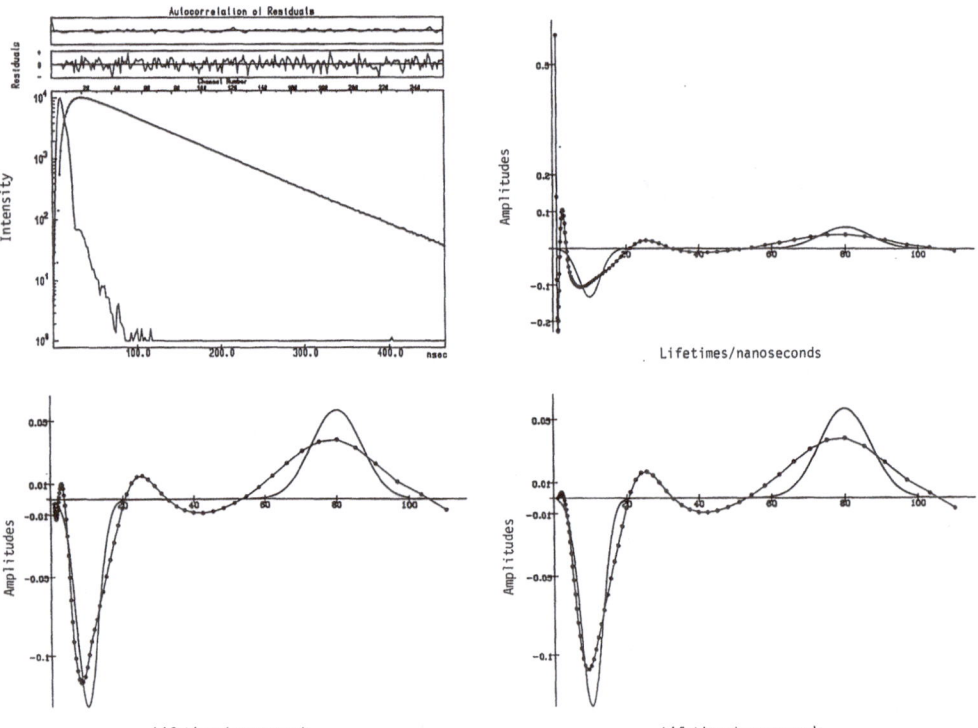

Figure 2. Analysis of lifetime-distributions from synthetic fluorescence data by ESM. a) experimental pulse – rise-and-decay pattern generated , according to Eq. (3) with one negative and one positive Gaussian input function (see the text for details), CPC: $3 \cdot 10^5$. Recovered histograms (dots) in b), c) and d) (solid curves: input distributions): b) without regularization ($\gamma = 0$), $\chi^2 = 0.964$;   c) regularization parameter $\gamma = 100$, $\chi^2 = 0.967$;   d) $\gamma = 1000$, $\chi^2 = 0.969$.

In addition, we have systematically modelled a couple of hypothetical situations in polymers by considering smooth distributions (and/or discrete exponentials) of various shapes of the amplitude-functions. However, the scope is too great to deal with in this format. In general, the recovery of input the functions strongly depends on the data quality, the length of relaxation scale and the temporal width of the optical pulse (additional smoothing!). Nevertheless, these numerical results obtained by synthetic data should stimulate further analysis on high–resolution, high–precision polymer fluorescence data and thus, promote experimental verification of hopping– and rotational frequency distributions from experimental profiles in disordered polymer morphologies.

Acknowledgements

The authors like to thank the Fonds zur Förderung der Wissenschaftlichen Forschung (FWF), Wien, Austria, for financial support (Projekt P6101, P7182, S32/03 and P7869-PHY). Furthermore, acknowledgement is made to Österreichische Forschungsgemeinschaft, Wien, Austria, for providing funds from the program "Internationale Kommunikation" (H.F.K.).

References

1. H. BÄSSLER, Site-selective fluorescence spectroscopy of polymers, in: *Optical Techniques to Characterize Polymer Systems*, H. BÄSSLER, ed., Elsevier, Amsterdam (1990).

2. *Molecular Dynamics in Restricted Geometries*, J. KLAFTER, J. M. DRAKE, eds., J. Wiley, New York (1989).

3. H. F. KAUFFMANN, Electronic coupling and relaxation patterns in polymers, in: *Photochemistry and Photophysics*, Vol.2, J. RABEK, ed., CRC, Boca Raton (1990), p.57

4. D. V. O'CONNOR and D. PHILLIPS, *Time-Correlated Single Photon Counting*, Academic, London (1984)

5. *Inverse and Ill-Posed Problems*, H. W. ENGL, C. W. GROETSCH, eds., Academic, Orlando (1987)

6. *Deconvolution and Reconvolution of Analytical Signals, Applications to Fluorescence Spectroscopy*, M. BOUCHY, ed., ENSIC-INPL, Nancy (1983)

7. E. M. BUCHBERGER, B. MOLLAY, W. D. WEIXELBAUMER, H. F. KAUFFMANN, and W. KLÖPFFER, Excited-state relaxation in bichromophoric rotors: time-resolved fluorescence of 1,3-di(N-carbazolyl)propane — a three-state analysis, *J.Chem.Phys.* **89**:635 (1988)

8. G. H. FREDRICKSON, and C. W. FRANK, Nonexponential transient behavior in fluorescent polymeric systems, *Macromolecules* **16**:572 (1983)

9. a) W.-D. WEIXELBAUMER, J. BÜRBAUMER, and H. F. KAUFFMANN, Excitation energy transport and reversible trapping in aromatic vinylpolymers: transient long-time behavior of a dissociative monomer-excimer system. A deterministic treatment, *J.Chem.Phys.* **83**:1980 (1985); b) H. F. KAUFFMANN, B. MOLLAY, W.-D. WEIXELBAUMER, J. BÜRBAUMER, M. RIEGLER, E. MEISTERHOFER, and F. R. AUSSENEGG, Electronic energy transport in aromatic vinyl-polymers: Nonexponential picosecond trapping in poly-(N-vinylcarbazole), *J.Chem.Phys.* **85**:3566 (1986)

10. B. MOLLAY, G. LANDL, and H. F. KAUFFMANN, Distributed electronic relaxation and nonexponential fluorescence in polymers: Reversibility in donor-excimer pairs — A perturbation theory treatment, *J.Chem.Phys.* **91**:3744 (1989)

11. a) D. R. JAMES and W. R. WARE, Recovery of underlying distributions of lifetimes from fluorescence decay data, *Chem.Phys.Lett.* **126**:7 (1986); b) A. SIEMARCZUK and W. R. WARE, Complex excited-state relaxation in (9-anthryl)-N,N-dimethylaniline derivatives evidenced by fluorescence lifetime distributions, *J.Phys. Chem.* **91**:3677 (1987);

12. a) A. D. LIVESEY and J. C. BROCHON, Analyzing the distribution of decay constants in pulse-fluorometry using the maximum entropy method, *Biophys.J.* **52**:693 (1987); b) F. MEROLA, R. RIGLER, A. HOLMGREN, and J. C. BROCHON, Picosecond Tryptophan fluorescence of thioredoxin: evidence for discrete species in slow exchange, *Biochemistry,* **28**:3383 (1989).

13. a) A. SIEMARCZUK and W. R. WARE, Temperature dependence of fluorescence lifetime distributions in 1,3-di(1-pyrenyl)propane with the maximum entropy method, *J.Phys.Chem.* **93**:7609 (1989); b) A. SIEMARCZUK and W. R. WARE, A novel approach to analysis of pyrene fluorescence decays in sodium dodecylsulfate micelles in the presence of Cu^{2+} ions based on the maximum entropy method, *Chem.Phys.Lett.* **160**:285 (1989).

14. G. LANDL, T. LANGTHALER, H. W. ENGL, and H. F. KAUFFMANN, Distribution of event times in time-resolved fluorescence: The exponential series approach — algorithm, regularization, analysis, *J.Comp.Phys.,* in press .

15. R. FLETCHER and C. XU, Hybrid methods for nonlinear least squares, *IMA J. Num.Anal.* **7**:371 (1987).

UNDERSTANDING HEAT CONDUCTION IN ORIENTED POLYMERS

Martin Pietralla

Abteilung für Experimentelle Physik
Univertität Ulm, Einstein-Allee 11
D-7900 Ulm GERMANY

ABSTRACT

It is shown why heat conduction generally becomes anisotropic in oriented polymers. In semicrystalline polymers the smallness of the crystals introduces additional size effects, e.g. the existence of non-propagating acoustical modes. In the superstructure phonon focussing appears. The influence of these effects on the conduction mechanism is discussed.

RUBBERS AND GLASSES

An amorphous solid can be regarded as a snapshot of a liquid. The structures are identical but the convective motions are absent in the solid. Heat conduction at high temperatures (above the Dulong Petit limit) considered here is successfully described by binary collisions [1]. Convective motions contribute less than 2%. Thus heat conduction below and above the glass transition temperature T_g mainly differs by the slightly different coefficient of thermal expansion [2]. Considering a glass from chain molecules two different attempt frequencies of collisions of atoms or stiff motifs can be assumed: one to next neighbours within the same chain and one to the next neighbours within a neighbouring chain due to the local anisotropic oscillators leading to the collisions. The overall orientation of the segments if not random introduce an anisotropic heat transport. Formally we can define a "segmental heat conduction" using

$$k = G\rho c_p lv \qquad (1)$$

(k: heat conduction, ρ: density, c_p: specific heat at constant pressure, l mean free path, v: velocity of energy transport, G geometric factor) for intrachain k" (parallel to segments) and interchain k^+ (perpendicular to segments) heat transport. The overall orientation parameter of the segments is given by

$$P_2 = (3<\cos^2\theta> - 1)/2 \qquad (2)$$

With the definition of the anisotropy ratio $A = k"/k^+$ we find if averaging over conductivities [3]

$$A=(1+2qP_2)/(1-qP_2) ; \quad q=(A_o-1)/(A_o+2) \quad (3)$$

where A_o is this ratio taken for the segments. The averaging over resistivities is quite similar. In amorphous polymers the orientation is well characterized by birefringence via $\Delta n = \Delta n_o P_2$ where Δn_o is the "intrinsic birefringence of the segment. Various orientations are usually obtained by stretching a sample to different stretching ratios $\lambda = l_{final}/l_{initial}$. In a rubber where permanent networks exist, the orientation varies reversibly with the draw ratio. In glassy polymers, depending on the stretching conditions, flow may contribute to the deformation, and it is to be preferred to compare results on the birefringence scale which in any case is an orientational scale. In fig.1a the anisotropy of a stretched rubber is plotted versus draw ratio. The development of anisotropy is obvious. We should have in mind that a rubber still is a liquid. The full drawn curve is calculated with the inverse Langevin approach for the segmental orientation of a single chain being comprised of 135 segments. The orientation achieved is small and so is the anisotropy. Nevertheless it is comparabel with a quartz crystal (A=1.7 at room temperature). The segmental anisotropy has been set to A_o=7.7 which is an estimate by comparing van der Waals fluids with covalently bonded amorphous solids. Setting $A_o=\infty$ one in turn gets a lower limit of the orientation achieved. In fig.1b the results on polycarbonate (PC), a glassy polymer, is depicted. The orientation developed is much higher ($P_2 \leq 0.5$) and in turn is the anisotropy [4]. The simple model outlined works very well at high temperatures but is not applicable at low temperatures. A macroscopic picture can be developed using equ,(1). The measured velocity of hypersound (low frequency phonons) is used for v, c_p is set equal for all directions and $l \propto \Lambda^u$ (Λ phonon wave length $\propto$ v at fixed frequency) is assumed being due to structure scattering of a slowly varying local density. The anisotropy is then given by

$$A=(c_{33}/c_{11})^{(2+u)/2} \quad (4)$$

From fig.1b we see that this simple formula describes the result very well. (u=3 (PC), u=2 (PMMA), u=0.208 (PET)). Its usefulness for lower temperatures cannot be judged yet on account of the lack of data both thermal and elastic ones.

SEMICRYSTALLINE POLYMERS

Many polymers have semicrystalline structures being comprised of crystals and amorphous phases. Crystals are mostly of lamellar shape with a thickness distribution at some ten nanometer range and may be arranged in typical superstructures. Upon deformation the superstructure is destroyed in a certain range of deformation ($1.3 < \lambda < 2.7$ for polyethylene (PE)) and a new fibrillar one develops. There the crystals are small (~10nm) and of nearly cubic shape arranged on a statistical (paracrystalline) linear lattice [5]. On further stretching -if possible at all- the amorphous regions become more ordered and eventually (on superdrawing) they merge with the crystals to a common continous structure [6]. The orientation as measured by X-ray scattering or approximately by birefringence (average over phases with different effective monomer birefringence [7]) is quickly saturating (fig.2a) whereas heat conduction [8] as well as elastic moduli [9] into the stretching direction linearly

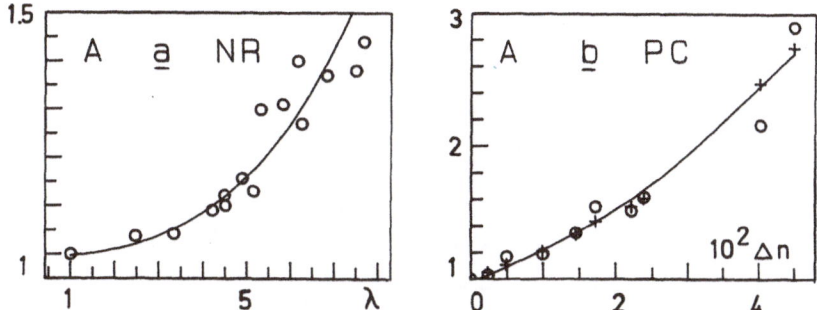

Fig.1 Heat conduction anisotropy in oriented amorphous polymers
a) Natural rubber at T=50°C. Solid curve calculated for a chain
length of 135 segments with segmental anisotropy A_o=7.7.
b) Polycarbonate at T=30°C. o measured directly, + calculated
from equ.2 with u=3.

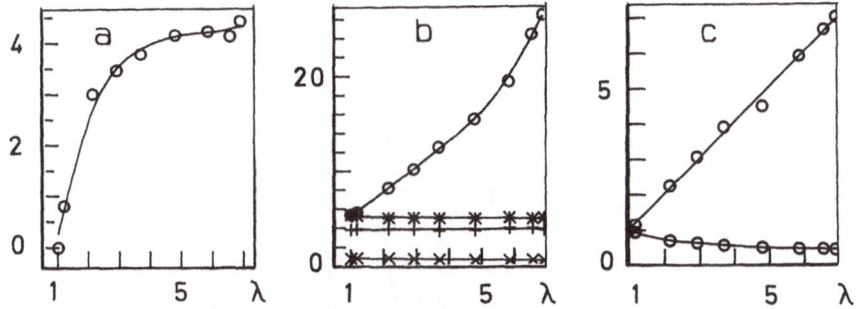

Fig.2 Properties of oriented low density polyethylene
a) birefringence (ordinate $10^2 \Delta n$)
b) elastic constants in GPa (from top to bottom $c_{33}, c_{11}, c_{13}, c_{44}$)
c) heat conduction (upper curve k''/k_o , lower curve k^+/k_o)

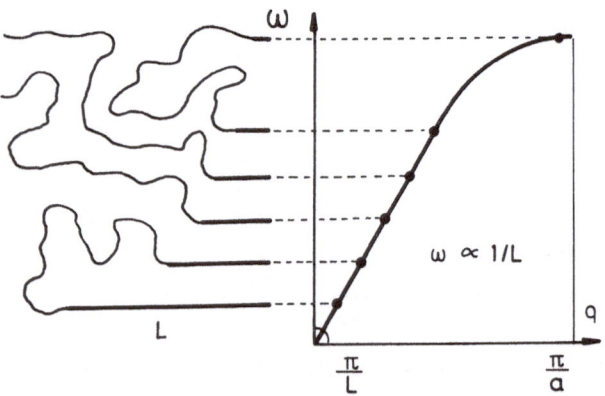

Fig.3 Acoustic phonon branch and the LAM-modes of a linear chain. The chain length increases in multiples of the uppermost (1,2,3,4,5,10). Propagating modes exist only for q<π/L.

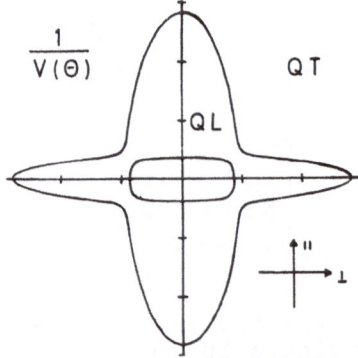

Fig.4 The slowness surface of PE stretched to λ=6.9. The outer curve is the QT-mode, the inner curve the QL-mode. Energy flux by a mode is directed normal to this surface (Phonon focussing).

increase (fig.2b,c). With elasticity unfolding of chains enhances the average unperturbed length of fully stretched ones whose limiting modulus is about 300 GPa. Heat conduction however, cannot be directly explained as well.

- the conception of binary collisions cannot be used since the crystals have their own dynamics of periodic structures in a confined space changing its size upon stretching.

- we cannot use mixture rules as for macroscopic phases. It is not possible to define different equilibrium temperatures at either end of a crystal or amorphous region and thus one cannot use Fouriers law of heat conduction. The point is that the finite size of the crystals establishes a localized dynamics in the material. Depending on chain length we can observe on the average the whole acoustical branch of an infinitely long crystal by RAMAN-scattering (fig.3). Hence the scattering is due to local modes and not to acoustical phonons. The momentum transfer of light is bound to $\Delta p < 2\hbar k$ (backscattering). The phonon wave lenght would be much greater than the crystals. These so called LAM-modes [10] (acronym for: longitudinal acoustical modes, but better: localized acoustical modes) are very sensitive to chain distortions. A kink within a crystal divides the chain into two different parts. Thus coupling to other modes must be weak. The situation within the amorphous parts is not so clear cut. However, local modes must exist but they are not bound to a straight one dimensional object. In both phases transverse modes must exist as well as torsional ones [11] but the only propagating modes are the acoustical modes of the superstructure. The direction dependence of their velocity can be measured by BRILLOUIN-scattering. The plot of $v^{-1}(\theta)$ (slowness surface) gives the shape of the constant energy surface. The normals to it indicate the direction of energy transport by the group velocity v_g. From fig.4 it is obvious that the shape of this surface (for the QL-mode at least) gives rise to a considerable focussing of the energy transport (phonon focussing). Let us judge the importance of this effect. Equ.(1) with constant mean free path results in an anisotropy $A=2.5$ at $\lambda=6.9$ for PE compared to the measured value of $A=16$! Correction of the mean free path however is still in need after taking care of the focussing effect. This we have done in [8] where the contributions to the heat currents by all branches in parallel and perpendicular direction are summed up numerically. The basic idea $l \propto \Lambda^u$ is still retained but now $\Lambda(\theta)$ is taken explicitly into account. Physically meaningful results only evolve if the components of v_g are used. This finding can be interpreted as the result of the particle motions in the non purely polarized phonons (QL:quasi-longitudinal, QT:quasi-transverse). The scattering exponents found group at $u=2$ and tend to decrease with increasing draw ratio. Having fitted the anisotropy we can compare the direction dependent values in terms of the isotropic ones. The systematic difference found is due to neglecting the scattering strength. $l \propto v^2/<\delta v^2>L(\Lambda/L)^u$ would better fit to the mean free path [8]. Upon stretching the structure becomes increasingly more homogeneous, hence the prefactor rises. (A similar effect is observed in light scattering. The opaque samples become clear and transparent.)

TEMPERATURE DEPENDENCE

One problem remains. The absolute magnitude of thermal conductivity cannot be explained by superstructure phonons. Their contribution to the specific heat is rather small. Using the parameters of the sample with $\lambda=6.9$ we can compare with literature values of the temperature dependence of thermal conductivity. We use $k(T) = K_o > c_p(\Theta_D, d) l v_g$ where $c_p(\Theta_D, d)$ are contributions to the specific heat described by non-integer (d) Debye-continua [12]. The mean free path is regarded as being constant since structure scattering is still assumed in the temperature range accessible. The temperature dependence of sound velocity has been neglected. The result is shown in fig.5. The parameters of the Debye-functions clearly demonstrate that other mechanisms than phonons must contribute to the heat transport. These mechanisms lead to lower dimensional Debye-continua. We may conjecture that nearly one -dimensional modes couple to the superstructure phonons via anharmonic processes. But this question remains open until more data are available. One further consequence of our treatment is that at very low temperatures the anisotropy should become very small. The faster the modes i.e. the higher Θ_D(mode) the higher the temperature where it ceases to contribute significantly to the specific heat. At the end only transverse phonons are left having a nearly isotropic contribution. The other extrapolation to the polymer with totally extended chains is even more important. The latest measurements on "ultraorientd" PE show that k" increases linearly at least up to λ=50 arriving there at the value of stainless steel. In the ultradrawing process the molecules become unentangled and the numbers of folds and kinks are decreasing upon stretching at about constant average orientation. The differences between crystals and amorphous regions gradually vanish. Indeed the idea of a continuous crystal-like structure has been put forward [6]. For heat conduction this is equivalent to merging propagating and non propagating modes on the acoustic branches. The heat transport will become more and more effective eventually becoming that of a macrocrystal with a lot of distortions (point-, line- and other scattering) where the usual picture of heat conduction within a dielectric crystal holds. The anisotropy may become very

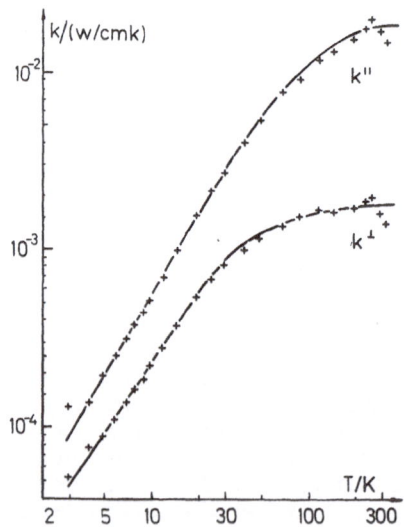

Fig.5 The temperature dependence of the directional thermal conductivities of PE (data from [13]). The curves are calculated with direction dependent Debye-temperatures and dimensionalities [8]. The low temperature part is mainly due to the (quasi)- transverse phonons. Changes at the glass transition have been neglected.

516

large. This is supposed from the ratio of the elastic constants $c_{33}/c_{11}=64$.

For the ideal crystal with Umklapp-processes anisotropy ratios as high as $A=3000$ have been calculated [14]. The effects discussed here with the example of PE will be encountered in any micro-heterogeneous systems like layered structures. Whether the low dimensional modes in systems of chain molecules [15] deliver a bridge to the fracton concept [16] is open to discussion.

ACKNOWLEDGEMENT

The author likes to thank Prof.H-G Kilian for his encouraging discussions and steady interest in problems of thermal conductivity.
This work was supported by the DFG Sonderforschungsbereich 239.

REFERENCES

[1] McLaughlin E   in "Thermal conductivity", Tye RP (ed)
                    Academic Press, London, (1969)
[2] Eiermann K,   Kolloid ZuZ Polym 198:5 (1964)
[3] Pietralla M,   Colloid & Polym Sci 259:111 (1981)
[4] Pietralla M, Schubach HR, Dettenmaier M, Heise B
                    Prog Colloid & Polym Sci 71:125 (1985)
[5] Heise B, Kilian H-G, Pietralla M
                    Prog Colloid & Polym Sci 62:16 (1977)
[6] Porter RS   Amer Chem Soc Polymer Prepr 12:39 (1971)
[7] Pietralla M, Kilian H-G J Polym Sci Polym Phys Ed 18:285
                    (1980)
[8] Pietralla M, Weeger RM, Mergenthaler DB
                    Z Phys B-Condensed Matter 77:219 (1989)
[9] Weeger RM   Dissertation, Ulm (1986)
[10] Rabolt JF   CRC Crit Rev Solid State & Mater Sci 12:165
                    (1985)
[11] Baur H   Kolloid ZuZ Polym 250:1000 (1972)
[12] Baur H   Kolloid ZuZ Polym 241:1057 (1970)
[13] Engeln I, Meissner M in "Non-Metallic Materials and Com-
                    posites at Low Temperatures 2". Hartwig G,
                    Evans D (eds). New York, Plenum (1982)
[14] Choy CL, Wong SP, Young K
                    J Polym Sci Polym Phys Ed 23:1495 (1985)
[15] Cheban YuV   Sov Phys J 21:1053 (1978)
[16] Alexander S, Laermans C, Orbach R, Rosenberg HM
                    Phys Rev B 28:4516 (1983)

STATIC CORRELATIONS OF POLYMER CHAINS IN NETWORKS

R. Oeser

Institut von Laue Langevin
BP 156x
F-38042 Grenoble, France

INTRODUCTION

The conference held in Maratea, Italy was about large scale
systems and it was implicitly assumed that the scales regarded
were greater than the size of one molecule. This paper is about
systems were the important length scales are greater than those
in a normal liquid yet cannot be on a supramolecular scale:

Which ever piece of rubber one takes is usually one
molecule: for example in each car tyre (if one forgets about
the filler) most likely one can go from each atom A to whatever
atom B using chemical bonds as bridges. Now this is exactly the
definition of a molecule. Nevertheless this does not tell us a
lot about the properties of this network since it is obviously
important how many possibilities there are to go from A to B
i.e. topology should play an important role for the
understanding of 3-dimensional networks.

METHOD

The only method to study correlations in the bulk for these
systems is small angle neutron scattering combined with prepar-
ative chemistry. The different scattering lengths of hydrogen
and deuterium for neutrons allow to create a scattering contrast by isotopic substitution. Small angle neutron scattering
is sensitive to the length scales which are of interest in
these systems, i.e. from several bond lengths to several times
the size of an elementary mesh (i.e. a chain connecting two
crosslinks). The coherent scattering cross section is in principle proportional to the Fourier transform of the pair correlation function. Because of the rather small range of scattering vectors this transformation is difficult to do and in this
paper only deals with some qualitative aspects of the scattering data.

Large-Scale Molecular Systems, Edited by W. Gans *et al.*
Plenum Press, New York, 1991

SYSTEMS

The networks which were used in this study were made by
"endlinking" of polydimethylsiloxan chains terminated by two
reactive endgroups. The crosslinking agent was a tetravinyl
siloxan compound. Complete chemical reaction with ideal stoi-
chiometry is assumed[1]. Therefore all of the junctions in the
networks were crosslinks i.e. four chains start at each junction point. The unattached labeled chains are introduced into
these matrices by simple swelling or mixed as unreactive components with the reaction moiety.

EXPERIMENTS

All experiments were done at the ILL, Grenoble (France)
using the small angle cameras D11 and D17. The spectrum in
fig.1. was recorded with a stripe of dimensions $1 \cdot 15 \cdot 60$ mm^3
which assured a uniaxial extension. The sample was stretched in
the horizontal direction. The matrix had an average meshsize of
10000 (M_n, polydispersity ≈ 0.8) filled with 10% of its volume
by a melt of deuterated PDMS with a molecular weight of 6000.

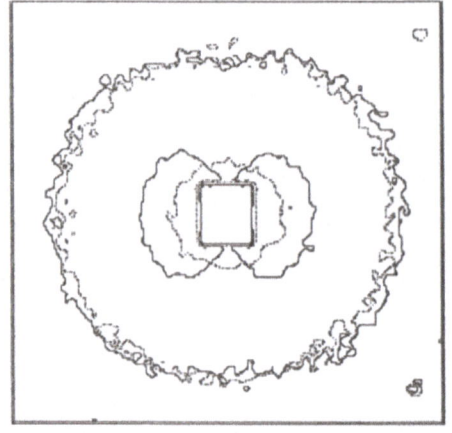

stretching direction: ⟷

Figure 1. superposition of 2-dimensional spectra of the same sample
at rest (dotted line) and an elongation rate of 1.62 (continuous
line). The spectra are normalized and the contourlines correspond to 6
and 16 cm^{-1} for the outer and inner lines, respectively. The rectangle in
the middle is masked by the trap of the direct beam and limits the lowest
accessible scattering value to ≈ 0.008Å^{-1}. The borders of the spectrum
correspond to 0.06 Å^{-1} .

The two dimensional spectrum of the sample at rest (dotted
lines in fig. 1) consists of concentric circles comparable to
those of a melt of labeled chains and unlabeled chains of the

same molecular weight as the mesh in the networks. Thus the samples are microscopically isotropic as expected. However the absolute intensities deviate from that of the melt towards higher values starting from about 0.06Å^{-1} when approaching zero scattering angle.

What happens to the spectrum when the sample is uniaxially stretched?

At <u>large scattering angles</u> $> 0.06\text{Å}^{-1}$ the contourlines of the macroscopically isotropic and anisotropic sample coincide, i.e. neutrons cannot distinguish a stretched from an isotropic sample at larger scattering values. (Fig. 1).

On the contrary there is a large difference in systems where all chains are attached at both ends. This is illustrated in Figure 2 where a large mesh size of 16000 (H and D) was used. The anisotropy of each chain is even more pronounced with smaller mesh sizes.

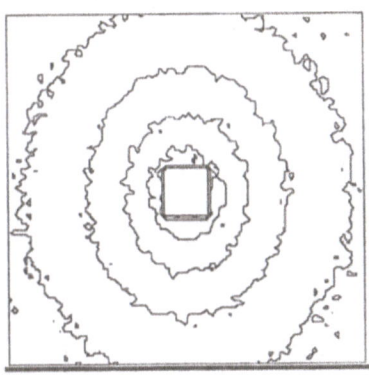

Figure 2. For comparison the spectrum of a completely crosslinked system made of a mixture of hydrogenated and deuterated meshes of mass 16000 is given at an extension of 1.5. Stretching direction is also in the horizontal plane. Note the orientation of the ellipses and the absolute intensities of 3.5, 6, 9 and 12 cm^{-1} increasing toward the centre. $0.01\ \text{Å}^{-1} < q < 0.07\text{Å}^{-1}$

stretching direction: ⟷

At <u>small angles</u> the spectrum is dominated by a strong increase in intensity in stretching direction while in the perpendicular direction the intensity is decreasing relativ to the isotropic state. This gives rise to the unusual spectrum shown with continuous lines in Figure 1 which we called "butterfly isointensity pattern"[2]. The "wings" of this spectrum are oriented perpendicularly to the ellipses of completely crosslinked systems.

To illustrate the strong dependance of this increase on the wave vector we show the normalized intensity $I(q)_{||}$ and $I(q)_\perp$ for three different absolute values of q. It is typical for these systems that at qR_g-values[3] greater than 2 the scattering function does no longer depend on $\mathbf{q}$ but only on the absolute value of the scattering vector q.

PHYSICAL INTERPRETATION

If one considers that the sample used for the anisotropic spectrum of Fig.1 can be obtained by a simple "Gedankenexperiment"[4] from that used for Fig.2 the lack of similarity between the two spectra is astonishing:

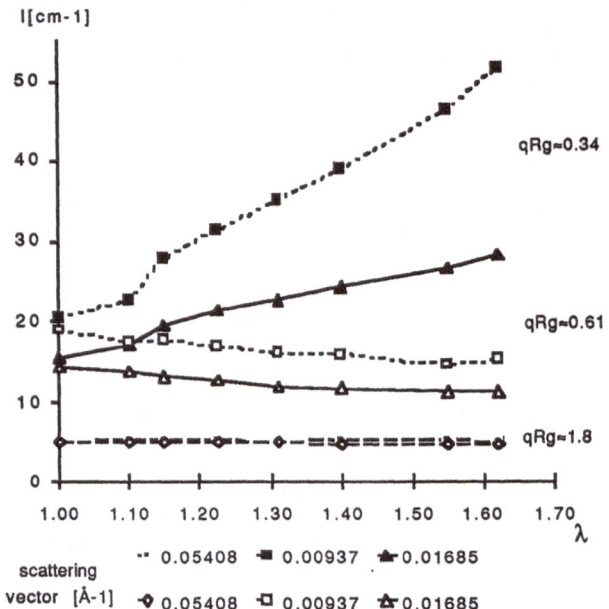

Figure 3 coherent scattering intensities as a function of stretching degree λ (in cm^{-1} for 3 different scattering vectors 0.054, 0.0094 and 0.017 Å^{-1} corresponding to $qR_G \approx 0.34$, 0.61 and 1.8, respectively.
Open symbols q-values perpendicular to stretching direction
Filled symbols q-values parallel to stretching direction.

The results at length scales where the correlations within the elementary chains are seen are understood by the most naive picture one can imagine about this system namely that the unattached chains no longer feel the constraint of the stretched network They are no longer stretched (i.e. their end

to end vector is on the average as long as in a melt) nor
oriented (distribution of the end to end vectors is isotopic in
space). So entropy driven relaxation seems to determine the
system.

The strong increase in scattering at small scattering vectors seen parallel to the stretching direction must be attributed to the interchain structure factor. That means the
distances between the centres of mass of the chains are no
longer distributed randomly as in the melt or in the equivalent
completely crosslinked system. The deviation from randomness
becomes even more pronounced when the system is stretched.
Again entropy might help to explain the observed scattering
increase. It has been observed from light scattering experiments in gels (i.e. networks swollen by a solvent) that polymer
density fluctuations exist at rather large spatial scales. This
was attributed[5] to natural fluctuations in crosslink density.
The swelling process is due to the dilution of the polymer
chain in the solvent. It is limited by the elastic contribution
of the network which depends on the length of the meshes and
therefore on crosslink density. Both effects are of entropic
nature but since the elasticity is fluctuating the result is
the existence of "hard" and "soft" regions. The "soft" regions
are supposed to be swollen to a much higher extend than the
"hard" regions thus creating some contrast because of different
polymer concentration.

Free polymer chains in a network act like a solvent except
that their mixing entropy is much smaller which means that they
very much prefer the "soft" regions with small crosslink density. This would explain the excess intensity in the isotropic
state.

The regions with high crosslink density have been proposed
to emerge[6] from clusters which are being formed during the
crosslinking process. This point of view can be adopted for the
model networks which are prepared without any solvent. These
clusters are linked together at the gelation threshold and give
regions with less crosslinks between them after completion of
crosslinking.

This may be compared to the the models of DLA at the beginning of the gelation process. Applied to the systems used in
this study it might even be possible to use the concept of a
competition between chemical reactivity and diffusion to get an
idea about a possible fractal dimension of these clusters.[7]
Finally gelation should take into account the cluster-cluster
aggregation. The simulations on these models show that there
are fractal structures which are eventually connected between
themselves leaving less dense spaces between them.

Uniaxial extension: Generally swelling can be regarded as
3-dimensional elongation. This idea is applied and the "hard"
regions are believed to deform much less than the average. Then
the fluctuations in the parallel direction will increase if the
system is allowed to rearrange the hard regions on a large
scale as shown in Fig. 3. Since the freely moving chains will
preferably be situated in these regions this will label these
large fluctuations.

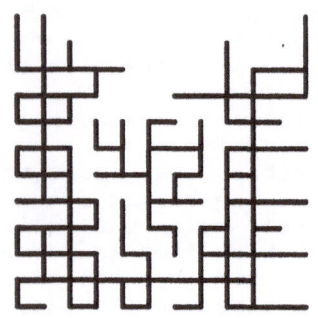 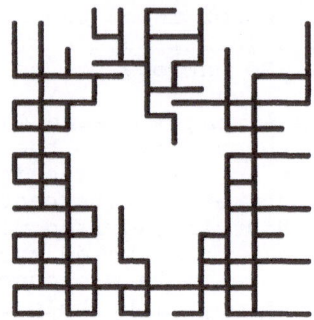

Figure 3. clusters which may "disentangle" upon elongation giving rise to large spatial fluctuations.

Perpendicular to the stretching direction the fractal structures are thought to interpenetrate because of their tree like structure. This diminishes the fluctuations in that direction leading to a decrease in scattering.

These "geometrical" constraints act in the same way as an attractive force between the chains thus allowing to describe the interchain structure factor with an Ornstein Zernicke approximation[8] which gives the "excess intensity" at small scattering vectors.

REFERENCES

[1]This is of course the subject of numerous research but the effects this paper deals with seem to be related to a very basic property of networks

[2]R. Oeser, C. Picot, J. Herz in "Molecular motion in dense polymer systems" Proceedings in physics 29, 104 Springer Verlag(1988)

[3]R_G is the Radius of gyration of the unattached chains.

[4]"thought-experiment" Starting from a completely crosslinked system where each chain is attached at both ends to a crosslink and some of these chains are labeled: It is only necessary to cut the ends of the labelled chains at the crosslinks to obtain the system yielding "butterfly patterns"

[5]J. Bastide, L. Leibler Macromolecules 21, (1988)

[6]J. Bastide, L. Leibler and J. Prost, Macromolecules 23, 1821 (1990)

[7]experiments on the pre gel have been proposed by several groups

[8]R. Oeser submitted to Phys. Rev. Letters

THEORY OF POLYMERS ON FRACTAL LATTICES

Apurba Kumar Roy* and Alexander Blumen

Physikalisches Institut and BIMF, University of Bayreuth
D-8580 Bayreuth, Federal Republic of Germany

I. INTRODUCTION

Nowadays, There is much interest in the statistics of polymers (both of linear chain type and also branched) in good solvents[1]. The most salient feature of real polymers is the "excluded-volume" effect : the physical fact that no two different monomers composing the polymer can occupy the same spatial position at the same time. In the case of a linear chain polymer without the excluded-volume constraint the chain is Markovian (Gaussian) and can be modelled as a random walk (RW); whereas with inclusion of this constraint the chain becomes non-Markovian and corresponds to a self-avoiding random walk (SAW). In the case of branched polymers we call the analogous case excluded-volume-branched-polymers (EVB) and distinguish them from simple randomly branched polymers (RB), where no such restrictions apply. The statistical properties of SAW and EVB on several kinds of fractal lattices have been extensively discussed in recent years[2-7]. The motivation for analysing the statistical features of polymers on fractals comes from the wish to understand the controversial (and much studied) problem of establishing the critical behaviour of SAW on random media[8-10].

---

* On leave from Santipur College, Nadia - 741404, India.

Large-Scale Molecular Systems, Edited by W. Gans *et al.*
Plenum Press, New York, 1991

In statistical studies the principal quantity of interest is the mean-square radius of gyration $<R^2>$ of a polymer consisting of N monomers; $<R^2>$ for large N scales asymptotically as :

$$<R^2> \approx N^{2\nu}, \qquad (1.1)$$

where we denote the SAW and the EVB exponents by ν_{SAW} and ν_{EVB}. In Euclidean spaces a very good mean-field estimate for these quantities is provided by the Flory formula; one has[1,11]: $\nu_{SAW}=3/(2+d)$ and $\nu_{EVB}=5/2(2+d)$. Analytical[2,6] and computer simulation calculations[10] for SAW and EVB on fractals showed that the corresponding $\hat{\nu}_{SAW}$ and $\hat{\nu}_{EVB}$ values (defined in accordance to Eq.(1.1)) differ from ν_{SAW} and ν_{EVB} values which hold for on the Euclidean space which embeds the fractal. In this paper, we derive approximate formulas for $\hat{\nu}_{SAW}$ and $\hat{\nu}_{EVB}$; for this we make use of the properties of linear chain and branched polymers and of the topology of fractal lattices. As a support to our heuristic method, we show that mean-field estimates obtained from our formulas reproduce nicely the values of $\hat{\nu}_{SAW}$ and $\hat{\nu}_{EVB}$ calculated by various methods on different fractal lattices.

II. PHENOMENOLOGICAL THEORY OF POLYMERS ON FRACTALS

We start from $G_N(r)$, the number of configurations of polymers consisting of N monomers, whose *radius of gyration* is r (*not* the end-to-end distance). We focus on $G_N(r)$ for fractal lattices. The form of $G_N(r)$ is constrained through the basic geometrical and topological properties of the underlying fractal structure and through the polymer statistics. Here we follow the spirit of the arguments given by Lhuillier[12] in the case of pure lattice case. Thus, $G_N(r)$ must be exceedingly low in two extreme situations : a) when $r<N^{1/d_F}$, where d_F is the fractal dimension of the lattice[13]. This is so because the minimum radius of gyration corresponds to a collapsed polymer. b) when $r>N^{x/d_{min}}$, where d_{min} is the fractal dimension of the shortest (chemical) path on the lattice[14] and x is some exponent characterising the topology of the polymer. We thus assume that the probability distribution function $P_N(r)$, defined by $P_N(r)=G_N(r)/G_N$, where G_N is the total number of configurations, has the form (cf. ref. 5-7, 12) :

$$P_N(r) \approx \exp\left[-N\left\{ C_1\left(\frac{N}{r^{d_F}}\right)^\alpha + C_2\left(\frac{r}{N^{x/d_{min}}}\right)^\delta \right\} \right]. \qquad (2.1)$$

Thus for small r the free energy of the polymer is dominated by a term $(N/r^{d_F})^\alpha$, which may be thought of as being the two-body repulsive energy between distant basic units. For large r the free energy is dominated by the term $(r/N^{x/d_{min}})^\delta$, which represents the contribution from the configurational entropy. In Eq. (2.1) α and δ are unknown exponents, to be determined at a later stage. Now, the most probable radius of gyration of a polymer is given by the maximum of $P_N(r)$. Thus :

$$\hat{\nu} = \frac{1}{d_F} \frac{(d_F + x\, K\, d_L)}{(d_F + K)} \qquad (2.2)$$

where K $(=\delta/\alpha)$ is a (positive) exponent ratio and d_L $(=d_F/d_{min})$ is the spreading (or connectivity) dimension of the fractal[14]. Now, for SAW, $x = 1$ (as it can be stretched to its maximal length N in the pure lattice case); and for EVB, $x = y$ (to be determined later), so that we obtain :

$$\hat{\nu}_{SAW} = \frac{1}{d_F} \frac{(d_F + K\, d_L)}{(d_F + K)} \qquad (2.3a)$$

$$\hat{\nu}_{EVB} = \frac{1}{d_F} \frac{(d_F + y\, K\, d_L)}{(d_F + K)} \qquad . \qquad (2.3b)$$

It is to be noted that, apart from the fractal lattice properties, $\hat{\nu}_{SAW}$ and $\hat{\nu}_{EVB}$ depend on the ratio K, a typical property of the radius of gyration distribution function, but not on the individual values of δ and α .

III. MEAN-FIELD ESTIMATE FOR $\hat{\nu}_{SAW}$

To the best of our knowledge, there does not exist any precise estimate for the distribution of the radius of gyration Eq. (2.1) for SAW on fractals. This prevents us from having any precise values for the exponents α and δ. We are thus restricted to calculate $\hat{\nu}_{SAW}$ by using mean-field estimates for α and δ. The term $N(r/N^{1/d_{min}})^\delta$ in Eq. (2.1) with $x = 1$ for SAW, can be thought of as being the elastic energy part, which arises from the configurational entropy. In the mean-field approximation one can write this elastic energy term as :

$$N \left(\frac{r}{N^{1/d_{min}}} \right)^\delta \approx - \ln \Phi_N(r) \qquad (3.1)$$

where $\Phi_N(r)$ is the probability distribution of random walks on fractals. $\Phi_N(r)$ is not Gaussian [5], but follows rather :

$$\Phi_N(r) \approx \exp \left[- C \left(\frac{r}{N^{1/d_W}} \right)^u \right] , \qquad (3.2)$$

where u ($\neq 2$) is an unknown exponent and d_W is the dimension of RW on fractals. In terms of d_S, the spectral dimension of the fractal[14], $d_W = 2d_F/d_S$ [13]. Inserting Eq. (3.2) into Eq. (3.1) and comparing r- and N- terms we get $\delta = u = 2d_F/(2d_L - d_S)$. With this value of δ and the mean-field value of $\alpha = 1$ we get for $K = \delta/\alpha$ the mean-field result :

$$K = 2 d_F / (2 d_L - d_S). \qquad (3.3)$$

Putting Eq.(3.3) in Eq. (2.3a) we obtain the following mean-field expression for $\hat{v}_{SAW}$:

$$\hat{v}_{SAW} = \frac{1}{d_F} \left[\frac{4 d_L - d_S}{2 d_L - d_S + 2} \right] \qquad (3.4)$$

The same expression was found recently by Aharony and Harris[3], who used a more standard way to derive the Flory formula. Table 1 summarises the numerical results on diferent fractals and compares them to Eq. (3.4).

Table 1 Exponents for $\hat{v}_{SAW}$ and $\hat{v}_{EVB}$ on different fractals and comparison between numerical (or exact) values and formulas in the text.

d	d_F	d_L	d_S	$\hat{v}_{SAW}$		$\hat{v}_{EVB}$	
				Eq.(3.4)	other	Eq.(4.1)	other
Percolation fractal (only backbone):							
2	1.61	1.43	1.25	0.77	0.767[*]		
3	1.75	1.26	1.23	0.66	0.67 $\pm$0.04[*]		
4	1.90	1.18	1.18	0.62	0.63 $\pm$0.02[*]		
5	1.93	1.14	1.18	0.56	0.54 $\pm$0.02[*]		
6	2	1	1	1/2	1/2		
Sierpinski-gasket:							
2	ln3/ln2	ln3/ln2	2ln3/ln5	0.825	0.798[#]	0.694	0.716[§]
3	2	2	2ln4/ln6	0.724	0.718[#]	0.612	0.605[§]
Koch-curve:							
2	ln5/ln3	ln5/ln3	1.2427	0.855	0.891[#]		
Given-Mandelbrot fractal:							
2	ln6/ln3	ln6/ln3	2ln6/ln(90/7)			0.684	0.706[§]

* Ref.10; # Ref.2; § Ref.4.

IV. MEAN-FIELD ESTIMATE FOR $\hat{v}_{EVB}$

To obtain $\hat{v}_{EVB}$, we have to know the mean-field values for y and K. Recently, Vilgis[15] has found from a mean-field type argument that $y = 1/\hat{d}$ where $\hat{d}$ ($=4/3$) is the spectral dimension of the EVB itself, i.e., $y = 3/4$. In a similar way as in section III we can identify in the mean-field limit the elastic energy term for EVB (with $y = 3/4$) $N (r / N^{y/d}_{min})^{\delta}$ with $- \ln \Phi_N(r)$; here $\Phi_N(r)$ is the probability distribution of randomly branched polymers (RB) on fractals. For Euclidean spaces RB show the same distribution as simple RW. We assume here that analogously to the RW case for RB $\delta = 2d_F/(2d_L - d_S)$ also holds. With this δ value and the mean-field value of $\alpha = 1$, we get $K = 2d_F/(2d_L - d_S)$. Thus putting this value of K and $y=3/4$ in Eq. (2.3b) we get the following mean-field expression for $\hat{v}_{EVB}$:

$$\hat{v}_{EVB} = \frac{1}{2d_F} \left[\frac{7d_L - 2d_S}{2d_L - d_S + 2} \right]. \tag{4.1}$$

Comparison of Eq. (4.1) to exact results found for Sierpinski gaskets and for Given-Mandelbrot fractals is displayed in Table 1. The agreement is quite good : we find our expression to be within 3% accurate when compared to the exact data.

ACKNOWLEDGEMENT

We acknowledge helpful comments from Dr. B. K. Chakrabarti. We thank the Deutscher Akademischer Austauschdienst (DAAD) for financial support through the award of a research fellowship to AKR. Grants from the Deutsche Forschungsgemeinschaft (SFB 213) and from the Fonds der Chemischen Industrie are gratefully acknowledged.

REFERENCES

1. P. G. de Gennes, in **Scaling Concepts in Polymer Physics** (Cornell University, Ithaca, New York, 1979).

2. R. Rammal, G. Toulouse and J. Vannimenus, J. Phys. (Paris) **45**, 389 (1984).

3. A. Aharony and A. B. Harris, J. Stat. Phys. **54**, 1091 (1989).

4. M. Knežević and J. Vannimenus, Phys. Rev. B **35**, 4988 (1987).

5. A. K. Roy and A. Blumen, J. Stat. Phys. **59**, 1581 (1990).

6. A. K. Roy and A. Blumen, J. Chem. Phys., submitted (1990).

7. A. K. Roy, B. K. Chakrabarti and A. Blumen, J. Stat. Phys., submitted (1990).

8. A. K. Roy and B. K. Chakrabarti, J. Phys. A **20**, 215 (1987).

9. S. B. Lee and H. Nakanishi, Phys. Rev. Lett. **61**, 2022 (1988).

10. Y. Meir and A. B. Harris, Phys. Rev. Lett. **63**, 2819 (1989).

11. J. Isaacson and T. C. Lubensky, J. Phys. (Paris) **41**, L 469 (1980).

12. D. Lhuillier, J. Phys. (Paris) **49**, 705 (1988).

13. D. Stauffer, in **Introduction to Percolation Theory**, (Taylor and Francis, London, 1985).

14. H. E. Stanley, in **On Growth and Form**, H. E. Stanley and N. Ostrowsky, eds. (NATO ASI Series E, No. 100, 1986).

15. T. Vilgis, J. Phys. (Paris) **49**, 1481 (1988).

SELF CONSISTENT INTERPRETATION OF PERCOLATION IN A MICROEMULSION

Rolf Hilfiker[1], Hans-Friedrich Eicke[1] and Harry Thomas[2]

[1]Institut für Physikalische Chemie, Universität Basel
   Klingelbergstrasse 80, CH-4056 Basel
[2]Institut für Theoretische Physik, Universität Basel
   Klingelbergstrasse 82, CH-4056 Basel

INTRODUCTION

Microemulsions are thermodynamically stable multicomponent systems with well defined nano-heterogenities in the colloidal size range. Apart from many other interesting properties they have attracted attention as model systems for supermolecular fluids.

A particularly well-suited system for physical investigations is the H_2O/AOT(sodium di-2-ethylhexylsulfosuccinate)/iso-octane system [1], being one of the very few ionic three-component microemulsions (i.e. no cosurfactant is needed). It forms in a certain range of temperature and composition nanometer-sized water-droplets covered by a monomolecular layer of surfactant dispersed in oil. The nanodroplet radius is controlled by the ratio $w_0=[H_2O]/[AOT]$ and has a typical value of 10 nm. This system exhibits a lower consolute point and, even more interesting, shows an increase in electrical conductivity by several orders of magnitude if the temperature or nanophase concentration is increased (see Figure 1). This increase has been ascribed to a <u>percolation</u> transition, where an infinite cluster of nanodroplets is formed.

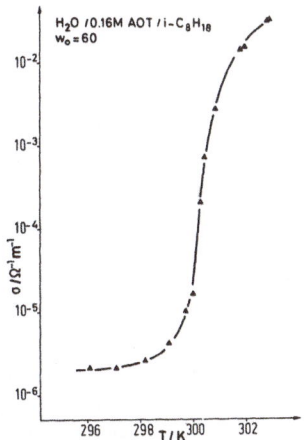

Fig. 1. T-dependence of conductivity System: H_2O/AOT/iso-octane, $w_0=60$. $c_{AOT} = 0.16$ mol/l.

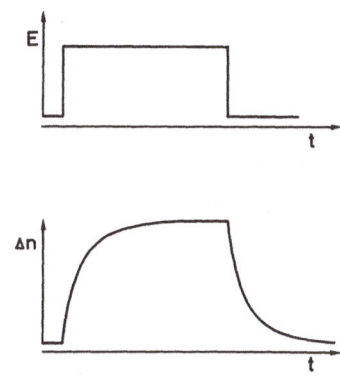

Fig. 2. Transient of electric bi-refringence and corresponding field-pulse

Nanodroplet cluster formation implies a considerable polydispersity of cluster sizes. The concentration distribution function (c_s) of cluster sizes (s), as well as the relation between cluster mass and cluster radius (R_s) [2] is predicted by percolation theory, i.e.

$$c_s = C_0 \, s^{-\Theta} \exp(-C_2 \, s^{\zeta}),$$

(1) and

$$R_s = C' \, s^{\rho} \ .$$

(2)

The parameters C_0, C_2 and C' have to be determined by experiment and depend only on the values $T-T_c$ or $P-P_c$, respectively. (T_c is the critical percolation temperature and P_c the critical percolation density.) The exponents are predicted by theory [2].

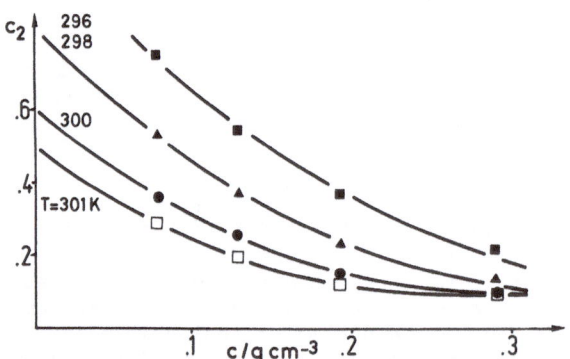

Fig. 3. Value of C_2 as a function of concentration of solute. System: H_2O/AOT/iso-octane, w_0=60. Parameter: temperature.

KERR EFFECT

Kerr effect measurements offer the possibility to determine the parameter C_2, which essentially characterizes the distribution function, in the T, c - plane. The decay of the birefringence (Δn) (see Figure 2) can be described as a sum of Debye-relaxation processes corresponding to the various cluster sizes [3], i.e.

$$\Delta n(t) = \sum_{s=1}^{\infty} \exp\{-t/(\tau_0 \, s^{3\rho})\} \, C_1 \, s^{2\chi-\theta} \exp(-C_2 \cdot s^{\zeta}) \ ,$$

(3)

where t is the time after the pulse is switched off, C_1 is a constant, τ_0 the Debye relaxation time of a monomeric subunit (aqueous nanodroplet), and χ the exponent relating the electrical and optical polarizability with the cluster mass.

From a fit of Eq. (3) to the transients of the birefringence, the parameter C_2 is obtained (Figure 3).

VISCOSITY

A polynomial expansion of the Einstein formula, like e.g. Guth's approach (Eq. (4)), can be used to describe the relation between the relative viscosity (η_{rel}) of a mixture of spheres and the volume fraction of spheres (ϕ) at moderately high ϕ.

$$\eta_{rel} = 1 + 2.5\,\phi + 14.1\,\phi^2 \quad . \tag{4}$$

Note that in a percolating system ϕ is no longer the (weighed-in) volume fraction of surfactant plus water (ϕ_0), but instead the _apparent_ volume fraction of the fractal nanodroplet clusters [4]. From the ratio of ϕ and ϕ_0 a "ramification factor" (α_{visc}) is obtained. For $c \to 0$, where no aggregates are present, α_{visc} extrapolates to 1.3 and not to 1 due to solvation of the nanodroplets. α_{Kerr} can be calculated up to a constant factor ($C_R \sim C'$) from the C_2 values determined from the Kerr-effect measurements, i.e.

$$\phi = C_R\,\phi_0\,\frac{\displaystyle\sum_{s=1}^{\infty} s^{3\rho-\Theta}\exp(-C_2 s^{\zeta})}{\displaystyle\sum_{s=1}^{\infty} s^{1-\Theta}\exp(-C_2 s^{\zeta})} \quad . \tag{5}$$

The ϕ_0 – dependence of α_{visc} and α_{Kerr} is displayed in Figure 4 at three different temperatures.

DIELECTRIC PROPERTIES

Another proof for the fractal nature of the clusters is obtained from dielectric measurements. The observed increase of the static dielectric constant ε_s with temperature cannot be explained by simply assuming a mixing formula (e.g. Clausius-Mossotti), because the volume fraction of the nanodroplets, as well as ε and σ of the nanodroplets and the oil, are only weakly temperature dependent.

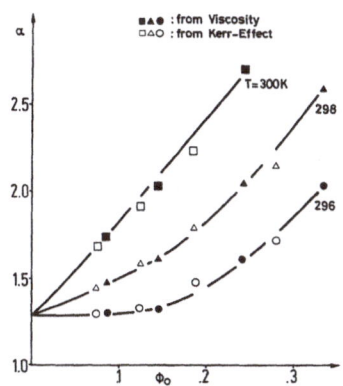

Fig.4. α_{visc} (closed symbols) and α_{Kerr} (open symbols) versus ϕ_0. System: H_2O/AOT/ i-octane, $w_0=60$.

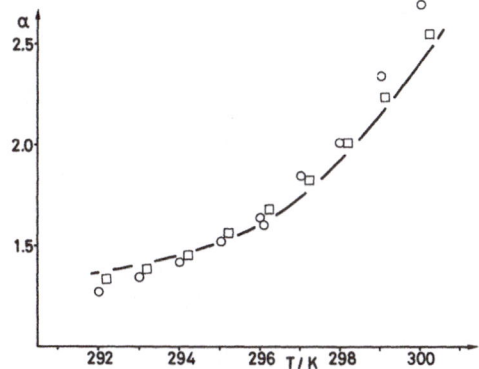

Fig. 5. T-dependence of α-values from Kerr-effect (line), viscosity (circles) and dielectric (squares) measurements. System:H_2O/AOT/i-octane, $w_0=60$, $c_{AOT}=0.16$ mol/l ($\Phi_0=0.23$).

The observed behaviour can, however, be understood by taking into account the fractal nature of the clusters. As the temperature approaches the percolation threshold, the clusters grow in size and consequently more and more oil is included in the clusters, i.e. the effective volume fraction of the volume occupied by the clusters grows, which leads in turn to an increase of the static dielectric constant. The ramification of the clusters is characterized by the ramification factor α_{diel}.

Quantitatively, ε_s is calculated by applying Bruggeman's formula (6) (as an approximation) first to the clusters which are thought to represent an effective medium of oil inclusions in a highly conducting (aqueous) matrix. Then the clusters with the just calculated properties, dispersed in a continuous oil medium, are treated again by Bruggeman's formula. α_{diel} is treated as an adjustable parameter (see Figure 5).

$$\frac{(\varepsilon-\varepsilon_1)^3}{\varepsilon} = \frac{(\varepsilon_2-\varepsilon_1)^3}{\varepsilon_2} (1-\Phi)^3 \qquad . \qquad (6)$$

ε_1, ε and ε_2 are the complex dielectric constants of the solute, the solution, and the solvent, respectively. Φ is the volume fraction of the solute.

With the same assumptions, the dispersion of the dielectric constants (ε' and ε'') are correctly predicted [5].

LIGHT SCATTERING

The excess Rayleigh ratio (R'_θ) of a system of polydisperse hard spheres is given by Eq. (7), where M_s, $(\partial n/\partial c)_s$, and c_s are the molar mass, the refractive index increment, and the concentration in $g cm^{-3}$ of an aggregate of s nanodroplets, respectively. K ($=4\pi^2 n^2/(\lambda_0^4 N_A)$) is an optical constant, c^0 ($\sim c'$) an adjustable parameter, and C_{PD} a factor which is a simple function of ϕ and the third to sixth moments of the radius distribution function [6].

$$R'_\theta = c^0 \cdot K \cdot C_{PD} \cdot (1-\phi)^4 (1+2\phi)^{-2} \sum_{s=1}^{\infty} (\partial n/\partial c)_s^2 M_s c_s \qquad . \qquad (7)$$

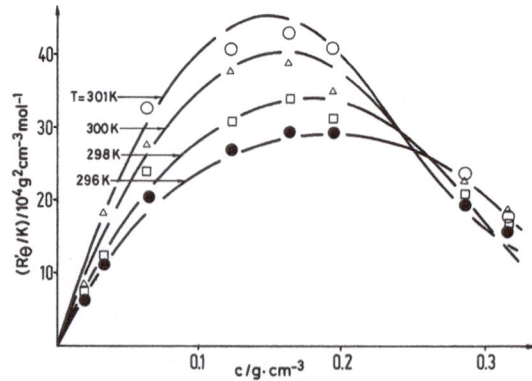

Fig. 6. Experimentally determined Rayleigh ratios at various temperatures versus concentration of solute. System: H_2O/AOT/iso-octane, $w_0=60$. The solid lines represent the calculated values. Parameter: temperature.

From the C_2 values which were determined from Kerr-effect results, R'_θ can be calculated [4] with one adjustable parameter (C^0) which must be the same for all temperatures. Calculated and measured values of R'_θ are displayed in Figure 6.

CONCLUSIONS

Results from dielectric, viscosity, light-scattering and Kerr-effect measurements could all be explained with one set of parameters obtained from Kerr-effect measurements, taking the percolation model as a basis. Physically most relevant is the fact that fractal clusters are formed. Due to the loose fractal structure of these clusters, the effective volume fraction of the solute increases as the percolation threshold is approached by either increasing the temperature or the concentration of nanodroplets. This in turn leads to the increase of the viscosity and the dielectric constant as observed. On the other hand, the high apparent volume fraction is responsible for the comparatively low Rayleigh ratio as the percolation threshold is approached, since a high volume fraction of hard spheres decreases the scattered intensity (Percus-Yevick Equation).

Acknowledgements

The authors are grateful to the Swiss National Science Foundation for financial support. One of us (R.H.) acknowledges generous support by the Treubel Foundation and the Deutsche Bunsengesellschaft für Physikalische Chemie.

REFERENCES

[1] H.-F.Eicke, in "Interfacial Phenomena in Apolar Media", H.-F.Eicke and G.D.Parfitt, Eds., Marcel Dekker, New York 1987.
[2] D.Stauffer, Phys.Rep. 54 (1979) 1.
[3] H.-F.Eicke, R.Hilfiker and H.Thomas, Chem.Phys.Lett. 120 (1985) 272.
[4] R.Hilfiker and H.-F.Eicke, J.C.S.Faraday Trans I 83 (1987) 1621.
[5] H.-F.Eicke, S.Geiger, F.A.Sauer and H.Thomas, Ber.Bunsenges.Phys.Chem. 90 (1986) 872.
[6] P.N.Pusey, H.M.Fijnaut and A.J.Vrij, J.Chem.Phys. 77 (1982) 4270.

EXPERIMENTAL EVIDENCE OF FRACTAL AGGREGATES IN DENSE MICROEMULSIONS

Salvatore Magazu'*, Domenico Majolino, Francesco Mallamace,
Norberto Micali*, Cirino Vasi*

Dipartimento di Fisica Universita' di Messina, 98166 vill. S.Agata Messina
* Istituto di Tecniche Spettroscopiche del CNR, 98166 vill. S.Agata Messina

INTRODUCTION

The phase in which a microemulsion is of water in oil type, as shown by S.A.N.S. data, can be considered as a colloidal suspension. The pair potential presents a repulsive hard–core plus a Yukawa tail representing the attractive interaction[1] $V(r) = V_A(r) + V_R(r)$. This potential form, similar to that used in the DLVO (Derjaguin, Landau, Verwey and Overbeek)[2] theory for colloids, shows two minima with a barrier and gives origin to interesting phenomena such as a phase transition with an upper cloud point temperature and a percolation–like transition[1,3] that suggest aggregation processes can be present in our system. Furthermore, the packing fraction of the droplets, keeping constant their sizes, can be easily changed[1], giving rise to a very dense liquid. From a microscopic point of view, the motion of the individual droplet is constrained by the interaction among its neighbours. At normal densities, the probability of an entrapment of the particle, in a cage formed by its nearest neighbours, is low and the particle can diffuse over large distances. For high concentrations (very high packing) the diffusional motion of the particle is dominated by a continuous and, for long time trapping into structural cages, translational motion is possible only if a hole is opened in these cages (for high dense systems the probability of a hole to be opened is very small). This latter process, which corresponds to a slowing–down in the density correlation function, is the configurational or structural arrest, well described by mode–mode coupling theories[4] on glassy state. The glass–transition can be studied by dynamic light scattering as a function of the microemulsion concentration; in particular we measure the dynamic structure factor $S(k,\tau)$ proportional to the autocorrelation function of the scattered field $g^1(k,\tau)$. Its initial slope is the mean linewidth $\langle \Gamma \rangle$ of spatial fluctuations of wavevector k. Care measurements[5] of this latter quantity have shown in the system AOT–water–decane the slowing–down of the density–density correlation function supporting the idea of large clustering effects among the spherical droplets in agreement with structural models of simple glasses generated by the assembly of hard spheres. Different theories[6], in particular the well–known free volume theory, indicate that glass transition is a cooperative phenomenon where all particles are involved; in particular several models invoke the presence of clusters. Molecular Dynamics[7] experiments in densely packed hard spheres give evidence that large clusters can arise spontaneously. Therefore, we have several suggestions that at high concentrations large clusters can originate from the droplets aggregation and light scattering (elastic and quasi elastic) experiments can give a direct way of studying the system, particularly in the region where it presents a glass transition verifying if ordered structures are present, and of measuring their dimensions and the kinetics of the aggregation. Also in an indirect way viscosity measurements can give information about such a process; in particular as shown in the final part of this work the data of this quantity as function of temperature and of volume fraction ϕ present a well pronounced peak that in the frame of the current theories can be ascribed to an aggregation process.

Large-Scale Molecular Systems, Edited by W. Gans *et al.*
Plenum Press, New York, 1991

For a system in which aggregation processes are present, kinetic models for random aggregation give us in terms of scaling arguments the k dependence of the measured mean linewidth $\langle\Gamma\rangle$ which may depend on the center–mass motion or on the cluster internal modes. In general it is shown that[8]: $\langle\Gamma\rangle = k^2 D_M F(kR_M)$, where D_M is the average diffusion coefficient and R_M is the average radius of gyration of the clusters. For $kR_M \ll 1$, it is shown that $F(kR_M)=1$ and the linewidth shows the well known k^2 dependence: $\langle\Gamma\rangle = D_M k^2$. For $kR_M \gg 1$ scaling arguments for self similar objects[8,9] give $F(kR_M) \sim kR_M$ and, in this large kR_M limit, the mean linewidth becomes independent from the correlation range and we have $\langle\Gamma\rangle \sim k^3$. If we obtain such a k dependence in a real experiment is possible to obtain by scattered intensity measurements detailed information about the properties of the structures and about the growth mechanism determining them. In other words if in an experiment we obtain such a k^3 dependence in the linewidth $\langle\Gamma\rangle$ we have the indication that the studied system has a fractal structure.

In fig.1a we show the relaxation rate $\langle\Gamma\rangle/k^2$ versus k for our microemulsion at different ϕ. It is evident that in the k range for $\phi = 0.1$, and 0.2 used, the condition $\langle\Gamma\rangle \sim k^2$ holds, while for $\phi = 0.56$ and 0.58 we are in the $\langle\Gamma\rangle \sim k^3$ case; for $\phi = 0.7$ we reobtain the k^2 dependence. Therefore, we can conclude that for low and very high ($\phi = 0.7$) concentrations we are in the Guinier regime where the scattered intensity is insensitive to the structure of the scatterers. For $\phi = 0.56$, 0.58 we are in the Porod regime where the angular dependence of the scattering intensity can give information about the cluster structure and the cluster–size distribution. Since these are concentrations where the slowing down is observed, this k^3 dependence in the mean linewidth shows that interesting structural processes are present.

The measured intensity can be written as: $I(K) \propto F(k)S(k)$ where $F(k)$ is the form factor of the scatterers (water droplets) and $S(k)$ is the interparticle structure factor which is the Fourier transform of the pair correlation function $g(r)$. Since the true dimensions of the droplets[1], all over the concentration range ϕ, are very small (60 Å), the form factor can be considered, in our experimental k range, nearly constant ($kR_0 \ll 1$, where R_0 is the droplet radius). Therefore the k dependence of the scattered intensity is entirely in $S(k)$. Because dynamical scattering data give evidence of large aggregates at the concentrations where is present the structural arrest, it is possible to analyze the intensity data in a way that directly connects the experimental results to the particular aggregation process. In particular, dispersed system can originate large structural aggregates whose mass (M) scales as their radius (R) following the power law[10]: $M \propto R^D$ where D is the fractal dimension. Consequently, it can be shown[10] that, on a length scale smaller than the clusters dimension, the scattered intensity is related to the exachanged wavevector k, through the simple relation:

$$I(k) \propto k^{-D} \text{ (for } kR \gg 1) \qquad (1)$$

We can use this simple relationship to measure D. However, we expect for our system that the constitutive droplets build up, for high concentration ϕ, aggregates having â finite range of correlation ξ. It is shown [11] that this effect can be taken into account by introducing an exponential cut–off factor $\exp(-r/\xi)$, analogous to the one used in critical phenomena, for the pair correlation function $g(r)$; in such a way the structure factor $S(k)$ assumes the following expression:

$$S(k)= 1+ \frac{1}{(kR_0)^D} \frac{D\Gamma(D-1)}{\left(1+\frac{1}{k^2\xi^2}\right)^{(D-1)/2}} \sin[(D-1)\text{tg}^{-1}(k\xi)] \qquad (2)$$

where $\Gamma(x)$ is the gamma function. This expression for $S(k)$ has been developed and successfully applied to fit S.A.N.S. spectra in protein complexes in D_2O solutions [11] and furnishes a simple and accurate method of determining the fractal dimension D and the value of ξ by the intensity profiles. For its complete application a very extended k range ($k < 1/\xi$ up to $k > 1/R_0$) is necessary compared with the one allowed by visible light measuments. In any case it has been shown that in the range $1/\xi \ll k \ll 1/R_0$, i.e. on a length scale smaller than the cluster dimension and greater than the constitutive monomer (single droplet in our case), we again have $S(k) \sim (kR_0)^{-D}$ and we obtain that the scattered intensity is related to the exchanged wavevector k through the results $I(k) \sim k^{-D}$. This latter result outside the k limits, $k\xi \gg 1$ and $kR_0 \ll 1$, cannot be applied for the D determination. In any case, the k region, in which holds, is strongly dependent[11] on the ratio ξ/R_0.

and a good determination of D is obtainable for $\xi/R_0 \sim 50$. In a such case the obtained values of D can be directly connected to the peculiar aggregation kinetics; for example, values of D=1.75 and D=2.1 respectively indicate a cluster–cluster diffusion limited aggregation and the reaction limited aggregation[12]. In fig. 1b we show our intensity scattered profiles ln I(k) versus lnk for the concentrations $\phi = 0.1, 0.2, 0.56, 0.58$ and 0.7; for $\phi = 0.1, 0.2$ and 0.7, we are in the Guinier regime and the intensity is constant in the explored k range, eventual aggregates, if present, have dimensions smaller than k^{-1}. For 0.56 our data are well fitted by $I(k) \sim k^{-D}$ (continuous line), the obtained slope is the fractal dimension D that in our case is $D \sim 2.1 \pm 0.05$ indicating that the kinetic of growth of the clusters is a reaction limited. Our experimental angular range is $30 \leq \theta \leq 120$ degree corresponding to the k range $7.19\mu m^{-1} \leq k \leq 24.1\mu m^{-1}$. Such a k interval, available with the use of the light scattering, is too narrow to accurately define a fractal behaviour; usually the region of self–similarity for fractal systems is observed on a larger scale of lengths. In fact, such a k interval, is narrow for the application of eq. 1 that requires k ranges of several orders of magnitude; nevertheless our results are sufficient, to point out the existence of a scaling behaviour. The problem, however, can be overcome by the obtained k dependence of the mean linewidth $\langle \Gamma \rangle$.

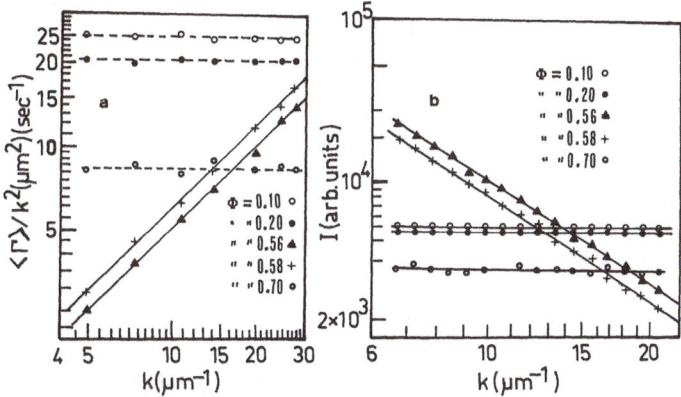

Fig. 1 — a) Plot of the relaxation rate of the mean linewidth $\langle \Gamma \rangle/k^2$ versus k at different volume fraction ϕ. — b) Intensity profiles ln I(k) vs ln k at different ϕ.

VISCOSITY DATA

We have also studied the shear viscosity as a function of the concentration at different temperatures in the same microemulsion, in order to have different insights on the aggregations phenomena already evidenciated by light scattering data. The viscosity is measured using a standard Ubbelohde viscometer and in fig. 2, the ratio (relative viscosity) of the measured microemulsion' s viscosity to decane' viscosity η_0 ($\eta_S = \eta/\eta_0$) is shown. With the exception of the data taken at $\phi = 0.1$, all plots end at the phase separation temperature. As can be seen the data show an increase in the viscosity, starting at about 25 °C for lower concentrations and about 12 °C for the higher, with a maximum, whose position is well defined for volume fractions $\phi > 0.25$. The temperature value of the maximum decreases when the concentration increases. For all the concentrations we have a lower temperature region where the relative viscosity is temperature independent; such a T interval is more extended for the lower concentrated microemulsion.

The entire behaviour of the data indicates some kind of thermally activated process, in particular as can be shown in the following to the well defined aggregation process among the droplets.

Therefore, the data are analyzed in terms of a recent two fluid model[13] developed by M.J. Grimson and G.C. Barker for colloidal systems, that explains how the interparticle interaction plays the fundamental role in determining the shear viscosity of a concentrated suspension like ours, giving a direct account of the contribution of the repulsive potential among the particles, while the contribution of the attractive potential takes into account in a qualitative way the effect on the viscosity by the formation of aggregates.

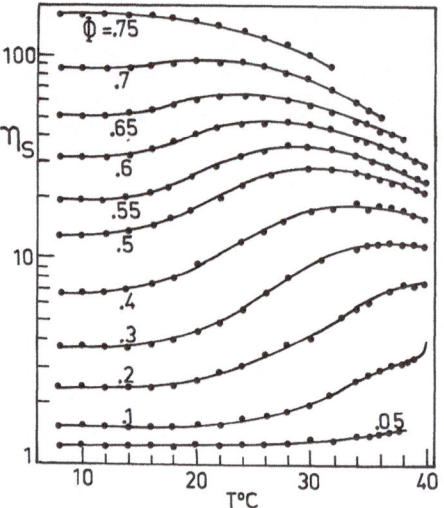

Fig. 2 . Plot of the relative viscosity $\eta_S = \eta/\eta_0$ as a function of temperature for different ϕ values. Continuous lines are guide for eyes.

Being the interacting colloidal system in the dispersion coupled to the solvent through the hydrodynamic forces between the particles, the viscosity of the dispersion can be taken as the sum of the solvent viscosity η_0 and the viscosity of the interacting system η_1 together with a coupling contribution η_2 which contains all the hydrodynamic contributions, namely;

$$\eta = \eta_0 + \eta_1 + \eta_2 . \qquad (3)$$

As far as the contribution η_1 is concerned, it has been shown[13] that for a concentrated colloidal suspension in which holds the Maxwell model of viscoelasticity ($\eta_1 \sim G_\infty \tau$), under assumptions of local ordering and nearest neighbour interactions, η_1 can be easely calculated, when the repulsive interaction dominantes, from an hard core potential form: $V(r) \sim (\sigma/(r - \sigma))^n$ (a "hard" core of diameter $\sigma = 2R_0$, R_0 radius of the particle, together with an inverse power). Considering that at the lower temperatures used, where as shown by R. F. Berg et al.[14] and verified by us aggregation (temperature) effects are not present, taking into account also hydrodinamics effects to the first order (Einstein contribution) and using the cited calculation for η_1 we have fitted our data (η_S measured at T=10 °C) using the following expression:

$$\eta_S = 1 + \frac{1}{\eta_0}(\eta_1 + \eta_2) = (1 + 2.5\ \phi) + A \left(\frac{\phi}{\phi_m}\right)^{(n+1)/3} \left(1 - \left(\frac{\phi}{\phi_m}\right)^{1/3}\right)^{-n} \quad (4)$$

where ϕ_m is the random close packing volume fraction. Using as free parameters the multiplicative constant A, the random close packing volume fraction ϕ_m, and the exponent of the potential form n. The result of such a fit is shown in fig. 3 and the obtained values for the parameters that are : A = 0.56, ϕ_m = 0. 81 and n = 1.97. Such a result well agrees with the chemical and physical properties of water in oil microemulsion[1].

540

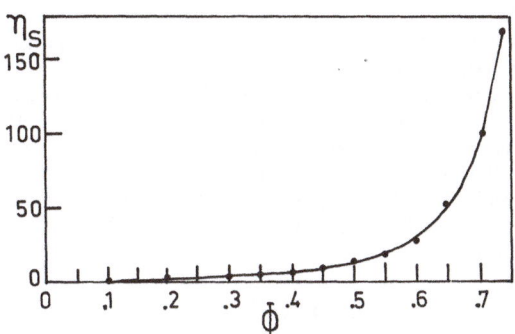

Fig. 3 Best fit of the η_S for T = 10 ˚C with the Eq. 4.

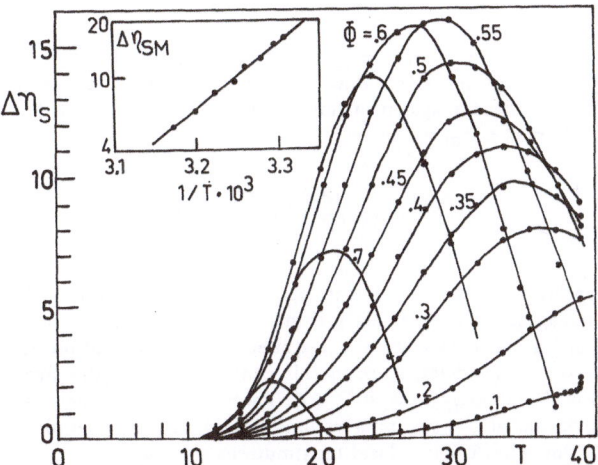

Fig 4 Plot, as a function of T and for different ϕ, of the excess of the viscosity $\Delta\eta_S$. In the inset is shown the logarithm plot vs 1/T, for $\phi \leq \phi_S$, of the maximum value of the excess of viscosity $\Delta\eta_{SM}$.

Finally, we discuss the effects of the aggregation process in the viscosity as due to the attractive potential. Since, the measured viscosity at T=10 °C can be entirely attributed to the effect of the repulsion among the particles, in order to obtain some insight on the effects of the attractive potential we consider the excess of viscosity $\Delta\eta_S$ (the difference between the η_S measured at higher T and their constant value for low T (10 °C)) at different ϕ and against T; the result of such a procedure is shown in fig. 4. As can be seen for each volume fraction we have curves with a well defined maximum, whose position is lowered in T, increasing ϕ. The $\Delta\eta_{SM}$ (the value of $\Delta\eta_S$ in correspondence to each maximum) increases its value, increasing ϕ up to $\phi \simeq 0.55$, then decreases sharply. In particular the behaviour of $\Delta\eta_S$ as a function of concentration and temperature can be connected to the observed glass transition and to the "structural arrest". In any case the larger values of $\Delta\eta_{SM}$ correspond to the $\phi < \phi_m$ values in which light scattering data show the presence of a fractal structure. Our goal is to show that such maxima are directly connected with the aggregation processes evidenciated by light scattering data and in particular with their fractal structure. In particular, the presence of these maxima in the excess of viscosity and their dependence with T and ϕ agrees very well with the experimentally observed behaviour of the concentrated suspensions of colloidal particles and is consistent with the used two fluid model for a colloidal system with particles of diameter σ that have a short–ranged attractive interaction that leads to an aggregation. In particular, it is shown that, according to the model of cluster–cluster aggregation[15], the growth mechanism produces an M–component system of clusters of particles with an effective volume fraction ϕ_{eff} that is related with the true ϕ in terms of the R_M (radius of the Mth cluster) and of the fractal dimension D by:

$$\frac{\phi_{eff}}{\phi} \sim \left(\frac{R_M}{\sigma}\right)^{(d-D)} \qquad (5)$$

where d is the Euclidean dimension ($D \le d$). Calling ϕ_s the volume fraction that correspond the largest possible cluster that, increasing ϕ, just spans the entire system we can write:

$$\frac{\phi_{eff}}{\phi_S} \sim \left(\frac{R_M}{\sigma}\right)^{(d-D)} \qquad (6)$$

if $D < d$, then $\phi_S < \phi_m$. When $\phi_S < \phi < \phi_m$ the dispersion is able to aggregate with a self–similar structure up to an upper cut–off length scale λ with $\lambda < \Lambda$, where Λ^3 is the volume for the system. Packing constraints impose for the system a homogeneity with a uniform density for length scales above λ, therefore for $\phi_S < \phi < \phi_m$ we have:

$$\frac{\phi_{eff}}{\phi} \sim \left(\frac{\lambda}{\sigma}\right)^{(d-D)} \sim \frac{\phi_{eff}}{\phi_S} \qquad (7)$$

For $\phi \to \phi_m$, the packing causes $\lambda \to \sigma$ and $\phi_{eff} \to \phi$. In such a way, it is shown that in a colloidal system where aggregation effects are present, $\phi_{eff} > \phi$ and ϕ_{eff} will display a maximum at ϕ_S. These results affect the shear viscosity of the system in the contribution η_2 of the hydrodynamic interaction of the clusters of particles, since in this contribution the effective volume fraction ϕ_{eff} must be used instead of ϕ and $\phi_{eff} \to \phi$, if $D < d$. Therefore, for aggregates with fractal structures, ϕ_{eff}/ϕ displays a maximum at the spanning volume fraction ϕ_S and the plot of the shear viscosity versus the volume fraction will display a local maximum at $\phi = \phi_S$.

The temperature (fig. 4) dependence of $\Delta\eta_{SM}$ shows that the maximum value for $\Delta\eta_{SM}$ is obtained for the volume fraction value $\phi \simeq 0.54$ as predicted by the kinetic theory for a hard spheres glass–forming liquid of Bengtzelius, Götze and Sjölander[4] and by computer simulation[7]. In any case, light scattering measurements confirm the entire physical picture with which are analysed viscosity data, in particular the spanning volume fraction ϕ_S value corresponds to the largest structure built up in the system.

Finally, we give the experimental evidence that the aggregation phenomenon is thermally activated using the concepts of the DLVO theory for the colloid stability. The potential V_T of our system shows the well known form of the DLVO theory : two minima with a barrier. When the particles are able to overcome the barrier, the primary minimum is reached giving rise to the aggregation process. Assuming that the number of particles in the secondary minimum N(t) at a given time obeys the rate equation it results:

$$\frac{d}{dt} N(t) = - K N(t) \qquad (8)$$

As shown in the above model for the aggregation effects to the viscosity, the effective volume fraction ϕ_{eff}, and therefore the measured excess of viscosity $\Delta\eta_{SM}$, at the spanning volume fraction ϕ_S, can be considered as connected with the number N_i of particles aggregated in a given cluster and more directly to the probability K that two single particles in the colloidal suspension stick after the escape of the potential barrier. In such terms utilising for K the well known Chandrasekhar's expression[16] we can write:

$$\Delta\eta_{SM} \; \alpha \; N_i \; \alpha \; k \; = \frac{(V"_{min} \; V"_{max})^{1/2}}{2\pi\beta} \; e^{- \Delta V/k_B T} \qquad (9)$$

Such a behaviour obviously holds only for $\phi \leq \phi_S$, while in the case in which $\phi_S < \phi < \phi_m$, packing constrains impose the gradual fragmentation of the fractal structure up to the close packing structure with an Euclidean dimension $d = 3$. The corresponding sharp decrease, in this concentration region, in $\Delta\eta_{SM}$ can be connected with such packing effects.

In the inset of fig. 4 we show, for $\phi \leq \phi_S$ and for each volume fraction studied, the logarithm plot of the excess viscosity maximum value $\Delta\eta_{SM}$ versus $1/T$. As can be seen, the data well satisfies Eq. (9) and the least square best fit gives $\Delta V \sim 8 \; k_B T$. This value, for the potential barrier height, agrees with the ones estimated for the interaction of two spherical particles[17] in a colloidal solution.

In conclusion, we observe that such a behaviour in the viscosisity is, in some way, also connected with the percolation effects that are observed in the system. In the model presented here, the viscosity behaviour has some similarity to the percolation, but differs in the treatement of the spanning clusters. However, recent analysis of viscosity–percolation effects are based on a narrow link between the polymeric growth processes[18] or on an aggregation process[19] with percolation.

REFERENCES

1. M. Kotlarchyk,S.H. Chen,J.S. Huang and M.W. Kim, Phys. Rev. Lett. 53 941 (1984); Phys. Rev. A 29, 2054 (1984).
2. E.J. Verwey and J. Th. Overbeek, "Theory of the stability of lyofobic colloids" (Elsevier, Amsterdam, 1948).
3. M.A. van Dijk, Phys. Rev. Lett. 55 , 1003 (1985); M.W. Kim and J.S. Huang , Phys. Rev. A 34 , 719. (1986)
4. U. Bengtzelius, W. Götze and A. Sjölander, J. Phys. C, 17, 5915 (1984)., U. Bengtzelius, Phys. Rev. A, 34, 5059 (1986).
5. S.H. Chen and J.S. Huang, Phys. Rev. Lett. 55, 1888 (1965); E. Sheu, S.H.Chen, J.S. Huang and J.C. Sung, Phys. Rev. A. 39, 5867 (1989).
6. see J. Jäckle , Rep. Prog. Phys. , 49 , 171 (1986).
7. H. Jónsson and H.C. Andersen, Phys. Rev. Lett.,60, 2295 (1988).
8. J. E. Martin and D.W. Shaefer, Phys. Rev. Lett. 53, 2457 (1984).
9. D. W. Shaefer and C.C. Han, in Dynamic light Scattering edited by R. Pecora (Plenum, New York, 1985).
10. J. E. Martin and J. Ackerson Phys. Rev. A 31, 1180 (1985)
11. S.H. Chen and J. Teixeira , Phys. Rev. Lett. 57, 2583 (1986).
12. D.A.Weitz, J.S. Huang, M.Y. Lin and J. Sung, Phys. Rev. Lett. , 54 , 141 (1985).
13. M.J. Grimson and G.C. Barker, Europhys . Lett., 3, 511 (1987).
14. R.F. Berg , M.R. Moldover. and J.S. Huang, J. Chem. Phys., 87, 3687 (1987).
15. See for example "Scaling phenomena in disordered System" edited by R. Pynn and A. Skjeltorp (Plenum, New York 1985).
16. S. Chandrasekhar, Rev. Mod. Phys., 1943, 15 , 1 (1943).
17. D. Eagland in "Water a comprensive treatise" edited by Franks F., Vol. 5 (Plenum, New York, 1975).
18. R. Botet, R. Jullien and M. Kolb J. of Phys. A, 17, L75 (1984).
19. P. Mills, J. Phys. (Paris) Lett. 46, L301 (1985).

Fractal Dynamics of the Catalytic CO - Oxidation -

Application of fractal cellular automata models

Peter Jörg Plath

Institut für Angewandte und Physikalische Chemie, Universität Bremen, Bibliothekstraße NW 2, D - 2800 Bremen 33

Experimental Introduction

Let us regard the apparently simple looking oxidation reaction of CO to yield CO_2. As a catalyst we use palladium single crystals distributed in amorphous Al_2O_3- carriers or located inside zeolite crystals [1 -3].

The catalyst powder is placed on a sieve forming an almost two-dimensional layer. We observe the time dependence of the CO-conversion by measuring the CO_2 content in the outlet of the reactor by means of an ir-signal [4].

There are constraints on running this reaction in an oscillatory regime. One characteristic feature of the reaction that could be observed was the fractal structure of the product oscillations. Fig. 1 depicts a representative self-affine time series, which was obtained by using a zeolite catalyst (22 mg) containing 14.7 wt% Pd at the reactor temperature T_R = 207 °C. The experimental conditions can be changed over a sufficiently wide range, resulting always in different, but piecemeal self-affine time series. Varying the experimental constraints does not significantly alter the complexity of the temporal patterns obtained.

The time series patterns of CO-conversion are characterized by:
a) very short breaks in conversion:
    - even breaks succeeding one another are of very different sizes;
b) the sequence of the breaks:
    - it is self-affine in relation to the sizes of the breaks;
c) the periodic behaviour of the piecemeal self-affine pattern:
    - after a while the complex self-affine pattern as a whole becomes periodical.
These are the three essential features for modelling the underlying chemical systems.

Dimension analysis of the experimental data

So far, the qualitative statement of the self-affinity of the time-series has been based solely on the visual impression gained by the observer. This statement has now to be confirmed by numerical analysis.

For this purpose the amplitude x of any break is measured, providing a sequence of numbers. This can be mapped onto the ordered set of the natural

Large-Scale Molecular Systems, Edited by W. Gans *et al.*
Plenum Press, New York, 1991

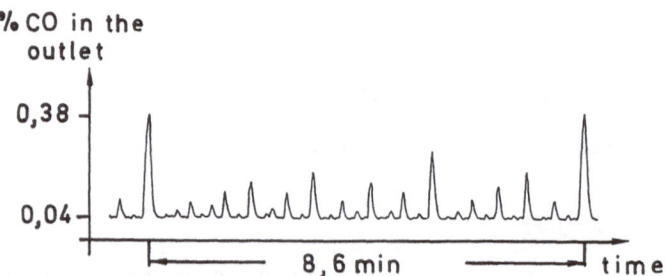

Fig. 1   A self-affine interval of the time series during the catalytic CO-oxi-
          dation using synthetic air and a Pd-crystal loaded zeolite catalyst (22
          mg catalyst, 14.7 % Pd, 0.565 vol% CO in the feed, temperature of the
          reactor T_R = 207°C , flow rate 11,1 l/h, volume of the reactor 14 ml).

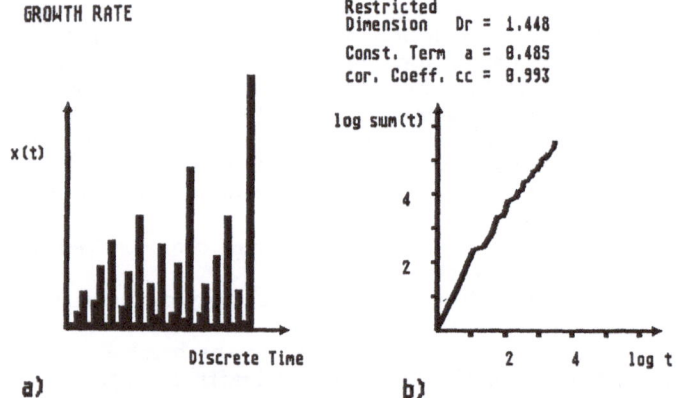

Fig. 2   Analysis of the growth rate dimension of the histogram of the breaks
          in the CO-conversion: a) This histogram belongs to the time series
          shown in Fig. 1 ; b) Log-Log-plot of the function (4) : log (sum(t)) ≡
          log S(t) = D_R log t; the restricted dimension is given by the slope of
          the "normalized" regression line : D_r = 1.448.

numbers ($t_i \in \mathbb{N}$, $i \in I$), which represent the discrete time evolution. Since only the minima of the time series (which are equivalent to the maxima of the breaks of conversion or the maxima of the unconverted CO) are mapped onto the discrete set of the natural numbers, the continuous time series becomes a temporal discrete sequence. The values x_i can now be understood as the weights of the discrete times i. Plotting the logarithm of the sum S(t) of the amplitudes x_i as a function of the logarithm of t: log S(t) = f(log(t)),

$$S`(t) = \sum_{i=1}^{t} x_i \tag{1}$$

a straight line representing a graph of the linear function is obtained:

$$\log S`(t) = G \log t + a \tag{2}$$

The slope G of this line is the scaling exponent of the function

$$S`(t) \sim t^G$$
$$S`(t) = a \, t^G \tag{3}$$

In order to interpret this scaling as a dimension, one has to "normalize" the sequence S'(t) yielding S(t) by setting S(1) = 1. The graph of this new function

$$\log S(t) = D_r \log t \tag{4}$$

becomes a straight line through the origin, where the slope D_r is now the growth rate of the sequence of breaks in the time interval under discussion. By this "normalization" the self-affine pattern of the time series becomes self-similar, because the normalized weights can now be regarded as similar to columns in a histogram covering a discrete net plane. If this histogram is fractal one can estimate its growth rate dimension [5] (see Fig. 2).

It is clear that the sequence of breaks can always be considered within only one period in order to correlate the pattern of the time series with a scaling factor which can be interpreted as a fractal dimension. If one were to extend the analysis to a large number of periodic intervals, the growth rate D_r would tend to one.

On the other hand, the value of D_r is related to the number of breaks within any one period. Only in the case of an infinite number of breaks - this means in an infinite non-periodic time interval - the value D_r could be identified with the fractal growth rate dimension D of the underlying self-similar pattern. The value $D_r = D$ for an infinite number of steps is merely the limit of D_r if the pattern is self-similar on all scales.

An artificial time series

For example, the growth rate $D_r(n+1)$ of the numbers of corners in the triangular Sierpinsky gasket is given by:

$$z(0) = 2$$
$$z(n+1) = 3 \, z(n) - 3 \qquad\qquad n = 0,1,2,3,...$$

$$t(n) = 2^n + 1 \qquad\qquad t(n) \in T = \mathbb{N}$$

$$D_r(n+1) = \frac{\log z(n+1)}{\log t(n)} \tag{5}$$

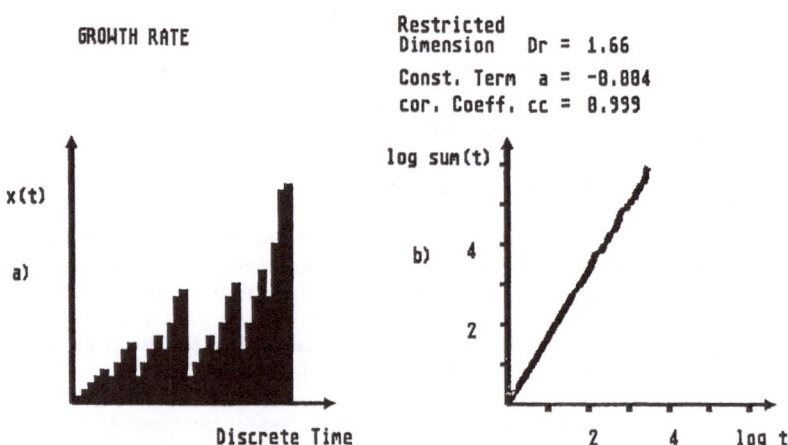

Fig. 3   Estimation of the growth rate of the sequence of vertices x(t) in the graph of the Sierpinsky gasket, where time t is the kathetus of this triangle. The restricted dimension D_r = 1.660 is estimated from the slope of the corresponding regression line. a) Histogram; b) Graph of the function log sum(t) ≡ log S(t) = D_r log t.

548

This growth rate $D_r(n+1)$ equals the fractal dimension D of the gasket as measured for instance by the total edge length only for n = 0 and in the limit if n equals infinity: n = ∞. So, $D_r(n+1)$ is not scale invariant for all values of n.

Furthermore, for small variations of t ϵ T = $\mathbb{N}$ one can observe small fluctuations in D_r. So, for a given range of t the value of D_r represents the growth behaviour of the pattern, which is invariant under scaling in just this limited interval up to t (see Fig. 3). Let us call D_r the restricted dimension of the pattern, since its scale-invariance is restricted only to the interval under consideration. In addition, I should mention that $D_r(n+1)$ is not necessarily the same as the D_r , because D_r is estimated by means of the regression line on all possible sums.

$$S(t) = \sum_{k=1}^{t} \Sigma(k) \tag{6}$$

$$\Sigma(k) = \sum_{i} z(i,k) \tag{7}$$

whereas $D_r(n+1)$ takes into consideration only those sums where k is equal to t(n).

We can now return to the experimental observations. The length of the non-periodic "self-affine" intervals in the periodic time series can be taken as the range of t for which D_r is the restricted dimension of the sequence of breaks. The length of t can be measured by counting the breaks of interest. Carrying out a large number of measurements, it turns out that the restricted dimension D_r of the temporal sequences of breaks of the CO conversion is bound to the range between about $1.2 \leq D_r \leq 1.6$.

Modelling

From the time series one can simply deduce that there should exist parts of the catalyst which are either productive or non-productive [4]. Therefore one may assume that the large breaks of the conversion are correlated to a large number of parts of the catalyst, all of which are non- productive at the same time step, whereas in the case of high conversion nearly all parts of the catalyst seem to be productive.

Translating this assumption into convenient mathematics let us divide the whole catalyst into a finite number of elementary reactors of macroscopic scale which can be identified with the catalyst grains or pellets. The most simple mathematical description of such an elemenary reactor is a cell in an automaton [6 - 9]. For simplicity these cells may have only two states z(t):

$$z(t) = 0 \text{ productive state}$$
$$z(t) = 1 \text{ non-productive state.} \tag{8}$$

Since the restricted dimension D_r of the time series is always essentially less then D_T = 2, for modelling purposes there are good reasons for taking the most simple arrangement of the cells to form a linear cellular automaton.

In addition, each cell i may have only one nearest neighbour - for instance its lower one (i-1). The dynamics of the automaton is now given by the very simple translation rule of the states z(i,t) of the i-th cell at time t [10]:

$$z(i,t+1) = \Big(z(i,t) + z(i-1,t) \Big) \bmod 2 \qquad (9)$$

In this way an irreversible dynamics has been introduced which gives rise to structure formation. The temporal development of this automaton results in a fractal pattern of the spreading excitations in space-time (see. Fig. 4). Projecting this pattern onto the time axis by summing up all states $z(i,t)$ of the cells i of the automaton at time t:

$$\Sigma(t) = \sum_i z(i,t) \qquad (10)$$

one can construct a sequence of state sums $\Sigma(t)$, which approximates qualitatively to the experimental time series.

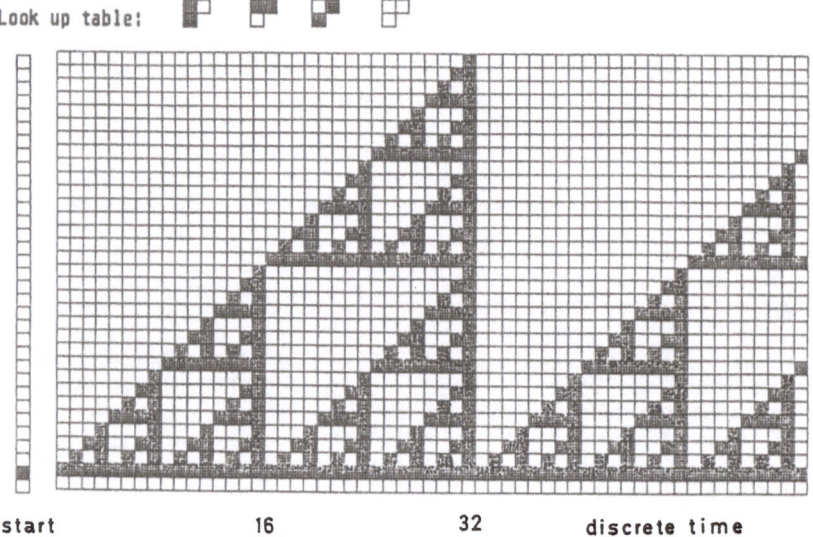

Look up table:

Fig. 4   Spatial and temporal development of the one-dimensional cellular automaton described above by the ex-or rule. The length of the automaton is L = 32, so that the length of the periodic interval is also t =32.

Again, one can estimate the growth rate dimension D = 1.5849.. of the unlimited sequence of sums $\Sigma(t)$, as well as the restricted dimensions D_r of the temporarily restricted time intervals (see Fig. 5). For t $\rightarrow$ ∞ the restricted dimension D_r tends to D = 1.5849... The value of D_r is always less than or equal to D, whereas the value $D_r(n)$ equals D for all values of the doubling number n.

Two-dimensional automaton -- a generalized transformation rule

In the case of this cellular automaton, both the growth rate dimensions— the fractal dimension D and the restricted dimensions D_r and $D_r(n)$ of the sequences of the state sums — equal the corresponding dimensions of the underlying two-dimensional pattern of excitation of the one-dimesional automaton in its space time.

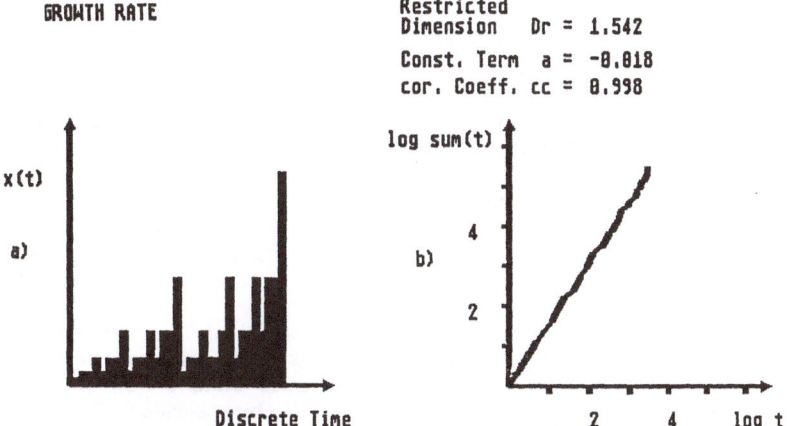

Fig. 5   a) Time series of the sums $\sum(t)$ of the states for the ex-or rule automaton defined above. b) This histogram possess the restricted dimension D_r = 1.542 which is estimated by the slope of the regression line of the function log sum(t) $\equiv$ log S(t) = D_r log t, where S(t) is defined by the functions (6) and (7).

However, this powerful but simple one-dimensional model does not reflect the more or less two-dimensional arrangement of the catalyst. For this reason the transformation rules are now extended to the dynamics of an *"exposed adsorption and cooperative annihilation"* [11]:

exposed adsorption
$$z(i,t) = 0 \rightarrow z(i,t+1) = 1 \text{ if } *\left(z(i/j, t)=1 \right) = 1$$

(11)

cooperative annihilation
$$z(i,t) = 1 \rightarrow z(i,t+1) = 0 \text{ if } *\left(z(i/j, t)=1 \right) \geq 2$$

where $\#\left(z(i/j,t) = k \right)$ is the number of cells j which neighbour the cell i at time t and which are in the state k. The i-th cell neighbours itself.

Observing this process, one can discern the incessant birth of temporarily growing clusters (see Fig. 6). After a very short time the clusters vanish by cooperative annihilation, whereas adsorption takes place at the borderlines of theses clusters only at those cluster cells which are exposed.

This dynamics is a generalization of the simple ex-or rule which has been used in the one-dimensional automaton. On the two-dimensional square lattice this generalized dynamics also reproduce a fractal pattern of excitations in space-time. But there are now two space and one time variables, so patterns occur inside a quadratic pyramid which open with time. The top of the pyramid is the initial point of the excitation.

If one forms the sequence of state sums Σ :

$$\Sigma (t) = \sum_{i,j} z(i,j,t) \tag{12}$$

this sequence is fractal too, but the limit of its restricted growth dimension is $D_r(n) = D = 2.323..$ if n tends to infinity..

$$D_r(n) = \frac{\log 5^n + \log 2}{\log 2^n} = \frac{\log 5}{\log 2} + \frac{\log 2}{n \log 2} \tag{13}$$

For any finite time t = n the restricted dimension $D_r(n)$ is greater than D. $D_r(n)$ can never be smaller than D, if one commenses with one cell at time t=1, and if the unlimited automaton is restricted to a finite time t . In this simple version the model cannot reflect the experimental time series, the restricted dimensions of which are less than two.

But if the grid is limited, the sequence $\Sigma(t)$ of the states necessarily becomes periodic. The period length depends upon the size of the quadratic grid and on the position of the starting point (see. Fig. 7). The sequence $\Sigma(t)$ within such a period differs from the sequence at the beginning, which is not influenced by the border of the limited automaton. To evaluate the model, one has to compare the periodic intervals of the model with the corresponding intervals of the experimental time series which are almost periodic. It is obvious that the periodicity of the model is too strong and the dimension is too high, although for small grids it can be less than two. Nevertheless, the model behaves correctly in qualitative terms.

An amorphous two—dimensional cellular automaton model

To model the catalyst in a more realistic way, one has to take into account the fact that the catalyst forms an almost two-dimensional layer but that the geometrical arrangement of the neighbouring elementary reactors may vary from one elementary reactor to another, and that they may not be densly packed or arranged quadratically. The distribution of the elementary reactors forms an amorphous catalytic layer, so to speak. Therefore, one should diminish the number of neighbouring cells in the model and should arbitrarily arrange them on a square lattice but in such a way that they touch each other.

Rather than random distributing the cells within the original square lattice, the regularity of the neighbourhood of each cell should be preserved. The regularity is the number of cells which are direct neighbours of the cell under discussion. The resulting grid of the cellular automaton should be almost regular and fully connecting, but without any translation symmetry with respect to any possible unit cell.

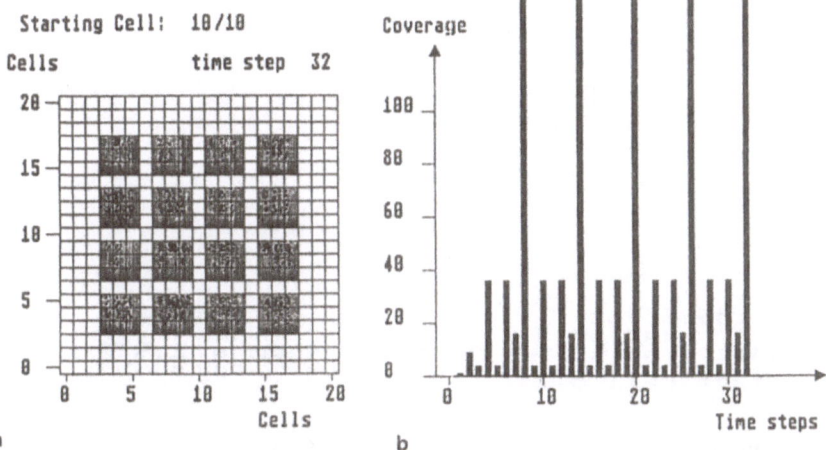

Fig. 6　a) State of the two-dimensional cellular automaton at time t = 32 based on the *"exposed adsorption and cooperative annihilation"* rule. b) The function (7) : $\Sigma(k) = \sum_i z(i,k)$ belonging to the temporal development of this automaton. The restricted dimension of the first interval (k = 1 to 8) is D_r = 2.444 .

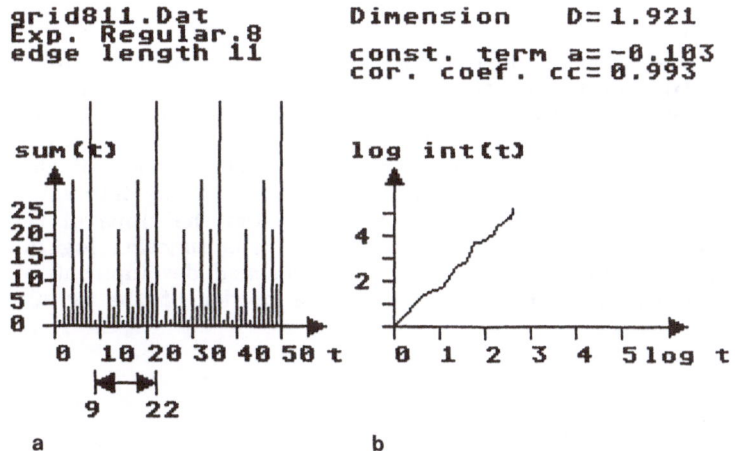

Fig. 7　a) Time series: sum(t) $\equiv \Sigma$(t) = f(t) of the sequence of the states of the cellular automaton. The inner cells are regular to degree eight. The edge length of the automaton is 11 cells. A periodic interval is formed, the length of which is 14 time steps. The starting cell was (-1,-1) with respect to the central cell (0,0). b) Estimation of the restricted dimension D_r = 1.921, which is correlated to the time interval depicted in a).

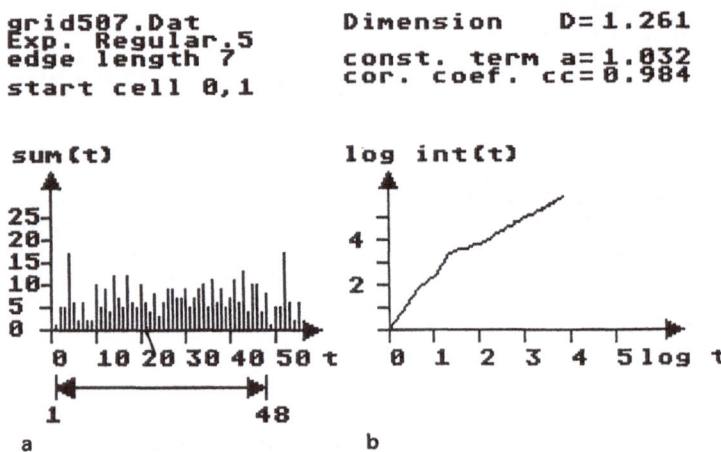

Fig. 8   a) Time series: sum(t) ≡ ∑(t) = f(t) of the sequence of states of a
cellular automaton. The inner cells are regular to degree five. The
edge length of the automaton amounts to 7 cells. A periodic interval
is formed in the range of t = 1 to t = 48. The starting cell was (0,1)
with respect to the central cell (0,0). b) Estimation of the restricted
dimension D_r = 1.261, which is correlated to the interval depicted in
a). The restricted dimension is taken from the slope of the regression line. The slope of the first part of the function log (int(t)) ≡ log
S(t) = f (log(t)) is related to the unchecked development of the excitation which is not restricted to the border of the cellular automaton.

Such a grid can be formed using a special growth mechanism working on the original square lattice. Starting with any cell, new cells can be "adsorbed" randomly, taking into account the desired regularity of the cells under discussion, and of those neighbouring cells the regularity of which could be influenced by the intended adsorption. By this means, an almost regular lattice can be set up which may serve as a grid for a cellular automaton. It should be mentioned that the inner part of such a grid is regular.

Now, using the transformation rule of the *"exposed adsorption and cooperative annihilation"*, fractal patterns of excitation arise which are embedded in three dimensional space-time. These patterns exhibit restricted dimensions D_r less two. Depending upon the position of the initial excited cell, the automaton can become periodic (see Fig. 8). But because of the amorphous arrangement of the cells these periods are very seldom. Even for small grids the periodic intervals are very long in most cases, and the fractal character of their pattern is often not as well developed as in the experimental time series.

For these reasons a proper model should be limited to a small grid. Furthermore, it should be based on an almost eight-regular grid with only very few lattice defects, which might be located preferentially close to the borderline of the grid. Such restrictions will assure a fractal pattern of the sequences of the state sums $\Sigma(t)$ within almost periodic intervals, the restricted dimensions of which are less than two.

Conclusion

In this article several cellular automata models are presented for describing the catalytic CO-oxidation, the dynamic of which is of high complexity. All these models are based on a discrete macroscopic ansatz dividing the catalyst into a finite number of elementary reactors. The numerical analysis of the experimental time series of the CO_2-production showed that they could be characterized by the restricted dimension D_r which is essentially less than two. This fact encouraged us to model the catalytic reaction by a one-dimensional and finite cellular automaton.

A more realistic model has also been proposed which takes into account the finite and almost flat arrangement of the catalyst and the amorphous distribution of its elementary reactors. This cellular automaton model works on a finite and almost eight-regular grid which is embedded in a two-dimensional square lattice. Using the transformation rule of *"exposed adsorption and cooperative annihilation"* the resulting restricted dimension D_r will also be less than two, as in the case of the one-dimensional automata.

There are a very great number of such stochastic but regular and finite grids which can be used to create a variety of cellular automata. Such cellular automata could be excellent models for the dynamic behaviour of heterogeneous catalytic reactions, because the amorphous arrangement of the elementary reactors in the catalyst can be well modelled in this way.

References

[1] N.I. Jaeger, K. Möller, P.J. Plath, J. Chem. Soc., Faraday Trans. I, 86 (1986) 3315-3330.
[7] P.J. Plath, K. Möller, N.I. Jaeger, J.Chem. Soc., Faraday Trans. I, 84,6 (1988) 1751-1771.
[3] N.I. Jaeger, K. Möller, P.J. Plath, Ber. Bunsenges. Phys. Chem. 89 (1985) 633-637.

[4] K. Möller, Thesis: "Untersuchung des dynamischen Verhaltens der CO-Oxidation an Pd-Trägerkatalysatoren", Universität Bremen, Fachbereich Biologie/Chemie, 1984.

[5] St.J. Wilson, Physica 10D (1984) 69-74.

[6] M. Gerhardt, Thesis: "Mathematische Modellierung der Dynamik der heterogen katalysierten Oxidation von Kohlenmonoxid: Numerische Behandlung eines diskreten mathematischen Modells von über Diffusion miteinander gekoppelter chemischer Speicher", Universität Bremen, Fachbereich Biologie/Chemie, 1987.

[7] P.J. Plath, in: "Irreversible Prozesse und Selbstorganisation" (editors; W. Ebeling, H. Ulbricht), Teubner Texte zur Physik Bd. 23, Teubner Verlag Leipzig (1989) pp. 162-170.

[8] P.J. Plath, "Spatial and Temporal Fractals in Heterogeneous Catalysis", in "Synergetics, Order and Chaos", edit. M.G. Velarde, World Scientific, Singapore, New Jersey, London, Hong Kong, (1988) p. 331-348.

[9] T. Toffoli, N. Margolus, Cellular Automata Machines, MIT Press, Cambridge (Mass.), London (1987) pp. 146.

[10] P.J. Plath, "Optimal Structures in Heterogeneous Reaction Systems" Springer Series in Synergetics, Vol. 44, Springer-Verlag, Berlin, Heidelberg, New York, London, Paris, Tokyo, Hong Kong (1989).

[11] P.J. Plath, in: "Nonlinear Wave Processing in Excitable Media" (editor: M. Markus), Proceedings of the Nato Advanced Research Workshop, Leeds, Sept. 1989.

ELECTRODEPOSITION: FRACTAL AND MULTIFRACTAL MEASURES

Francesc Sagués, Josep M. Costa, Francesc Mas,
Marta Vilarrasa and Laura López-Tomàs

Departament de Química Física. Universitat de Barcelona
Facultat de Química. C/ Martí i Franquès, 1
E-08028-Barcelona, Spain

I. INTRODUCTION

Growth processes taking place in systems operated far away from equilibrium constitute a subject of considerable current interest[1]. Particularly interesting are those phenomena which give rise to irregular and highly ramified structures commonly analyzed in terms of fractal measure[2]. We are here specifically concerned with an example of such aggregation processes occurring when a metallic deposit is electrochemically grown on an electrode from an aqueous solution of an appropriate electrolyte. This electrodeposition phenomenon has been particularly studied during these past years since it is easily amenable to experimentation, and additionally it displays a wide variety of different morphologies ranging from regularly dendritic to desorderly fractals[3-12].

Actually, electrodeposition must be analyzed from two complementary points of view. There is first the question of a precise characterization of the fractal structures it originates under appropriate experimental conditions concerning the applied potential and electrolyte concentration. Certainly more complicated and far less pursued is the analysis of the electrodeposition process from a dynamical point of view. In this respect, even if it is generally accepted that the phenomenon belongs to the general class of growth processes governed by a diffusion field with boundary values specified at an interface driven by the normal velocity of the Laplacian field, we are still needing a full understanding of the appropriate specialization of this general framework to the specific conditions under which electrodeposition takes place.

In our presentation here we address both structural and dynamical questions. On what respects to the quantitative characterization of the deposits, we propose the use of the concept of generalized fractal dimensions[13] in order to discuss their degree of self-similarity. Finally, the particular dynamical aspect here reported concerns the experimental determination of the growth velocities of the electrodeposited phases and their eventual dependence on the pair of experimentally controlled parameters, namely the applied potential and concentration.

Large-Scale Molecular Systems, Edited by W. Gans *et al.*
Plenum Press, New York, 1991

II. FRACTAL MEASURES

Essentially, three different methods of calculating fractal dimensions of aggregates have been utilized in the past. They are respectively associated to the cluster or comulative mass fractal dimension D_c[14], the density-autocorrelation dimension $D_{corr.}$[14] and the radius of gyration fractal dimension D_g[15], All of them are defined in terms of appropriate scaling laws: $M(r) \alpha r^{D_c}$, $C(r)_{r \to 0} \alpha r^{-(d-D_{corr.})}$, $M \alpha R_g^{D_g}$, where respectively $M(r)$ represents the cumulative mass within areas of linear dimensions r, $C(r)$ denotes an averaged autocorrelation for densities of the aggregate evaluated at points separated a distance r, and finally M stands for the mass of the aggregate of radius of gyration R_g.

An extended discussion of the fractal measures of a typical aggregate is better presented in terms of what are called the generalized fractal dimensions D_q[13]. Several numerical routines have been developed to compute D_q[16]. The one that will be here illustrated, previously employed for electrodeposition patterns in Ref. 9 and 12 is based on a box-counting algorithm especially appropriate to the evaluation of D_q with $q \geq 0$. It consists in covering the pattern with a mobile grid composed of square boxes of variable size ε. The set of D_q ($q \neq 1$) are obtained from the scaling relation

$$Z_q \underset{\varepsilon \to 0}{\alpha} \varepsilon^{(q-1)D_q} \tag{1}$$

for the partition function defined as

$$Z_q \equiv \sum_{i=1}^{N(\varepsilon)} p_i^q(\varepsilon) \tag{2}$$

Where p_i is the relative portion of the cluster contained in the i-th cell.

$$\sum_{i=1}^{N(\varepsilon)} p_i(\varepsilon) = 1 \tag{3}$$

and $N(\varepsilon)$ is the total number of boxes covering the aggregate. The previously introduced dimensions D_c and $D_{corr.}$ correspond respectively to D_0 and D_2. The equality of the set of D_q dimensions is a direct signature of the self-similarity of the aggregate.

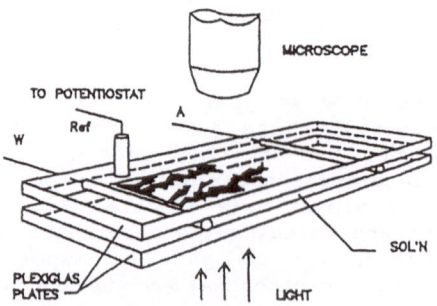

Fig. 1. Schema of the electrochemical cell for experiments conducted at a constant potential.

III. EXPERIMENTAL GROWTH OF ELECTRODEPOSITS

Experiments on electrodeposition are carried out in a cell, filled with the solution of the ion to be deposited, which contains a cathode (working electrode) where the deposition takes place, an anode (auxiliary electrode) and, for specific operation conditions (see below) a reference electrode. A suitable device, usually a potentiostat, is set to drive the electrodeposition process. Electrochemical measurements can be made according to three different techniques: constant voltage, constant potential and constant current. The constant voltage technique consists in applying a constant potential across the working and auxiliary electrodes. In experiments conducted at constant potential, the potential of the working electrode, which is monitored relative to the reference electrode, is maintained at a fixed value. In the third technique, current flowing through the cell is kept constant.

To develop quasi two dimensional aggregates using any of such techniques, the electrochemical cell schematically shown in Fig.1 is used.

Pure metal wires act as both working (*W*) and auxiliary (*A*) electrodes; They are placed between two parallel plates and separated by a short distance. The reference electrode (*Ref*), a saturated calomel electrode (*SCE*), is positioned near the working electrode, as indica-ted in Fig.1. This arrangement can be modified in a number of ways, depending on the scope of the experiment. Particular conditions are given in Sect.IV.

The electrolytic solutions fill the space limited by the working and auxiliary electrodes and the plates. Care was taken in order to prevent solution loses during the experiment. The experimental system is illuminated from below and the growning structures are photographed from above. The photographs are processed using a standard image analysis system, and digitized with a resolution of 512 X 512 pixels. The digitized patterns were then transferred onto a computer for further analysis.

IV. STRUCTURE AND DYNAMICS OF ELECTRODEPOSITS

Electrodeposits of copper at room temperature were obtained at constant electrode potential, whose values range from -3.0 V to -5.0 V vs. *SCE* and copper sulphate concentrations are fixed between 10^{-2}M and 10^{-1}M.

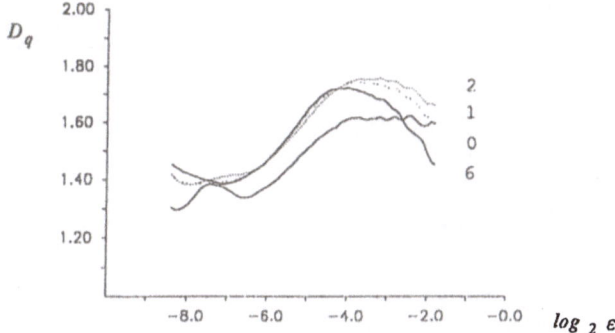

Fig. 2.   Values of D_q as expressed by the local slopes corresponding to $\log_2 Z_q/(q-1)$ vs. $\log_2 \varepsilon$ plots, for the electrodeposition pattern of fig. 5b ($c= 0.05$M and $E = -4.0$V vs. SCE). The selected indices q are: $q=0$ (——); $q=1$ (·······); $q=2$ (·····) and $q=6$ (–··–·) .

Supporting electrolyte is absent, and the anode and cathode electrodes are 25 mm copper wires, 0.2 mm diameter, and separated 50 mm. Experiments are stopped when the metallic trees develop on the cathode up to a size of several milimeters, depending on experimental conditions. The particular experimental conditions here chosen actually give rise to specific morphologies of the deposits ranging between homogeneous and open fractal. In analogy with what is known from electrodeposition experiments with Zn[6,7] one would reasonably expect that a whole set of very different textures could be obtained when varying the experimental controled values of concentration and applied potential.

For a particular set of potential and concentration values the fractal measures of a digitized pattern ((b) in Fig.5 below) were calculated according to the method outlined in the previous section. Graphs for the local slopes corresponding to double logarithmic plots of $log_2 Z_q/(q-1)$ vs. $log_2 \varepsilon$ are presented in Fig.2 for different values of q. An appreciable degree of self-similarity is apparent from these representations when limited to an intermediate range of ε values far from the

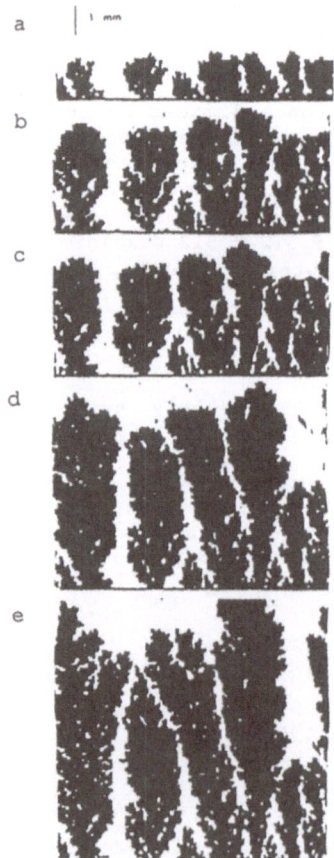

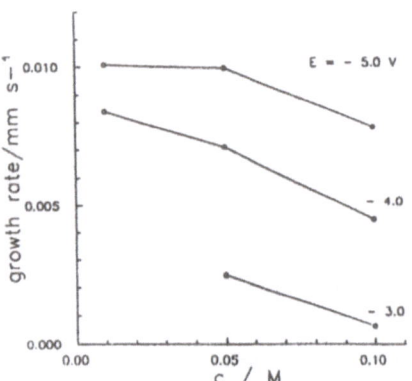

Fig.4. Plot of growth velocity vs. concentration ($c=[Cu^{2+}]$) for different electrode potentials (E).

Fig. 3. Patterns of electrodeposited Cu (c = 0.05 M, E = -4.0V vs. SCE) taken at different times: (a) 180 s; (b) 430 s; (c) 480 s; (d) 720 s and (e) 900 s.

560

extreme ε regions where intrinsic limitations arise either in finite-size effects or in the projection and digitization procedures employed. The averaged value of D_q are 1.71 ± 0.01 for the range of length scales $2^{-5.5} \le \varepsilon \le 2^{-2.5}$.

In Fig.3 we show different patterns of electrodeposited Cu taken at different times. From such representations we measure the growth velocity from the dependence of the average thickness, here expressed in terms of the average hight of the so-called "upper surface", on time. The interface velocity is apparently constant over much of the cell. Different values at various electrode potentials and concentrations are summarized in Fig.4. In going to lower potentials and larger concentrations the growth velocity decreases. The potential dependence is easily explained since higher potentials represent faster electrode kinetics according to the well-known Butler-Volmer phenomenological rate equation[17]. The effect of concentration is more complex involving probably several effects either originated in a variation on the free separation distance or in their width. In any case, in going to very high concentrations the deposits are very compact, and they correspondingly grow very slowly.

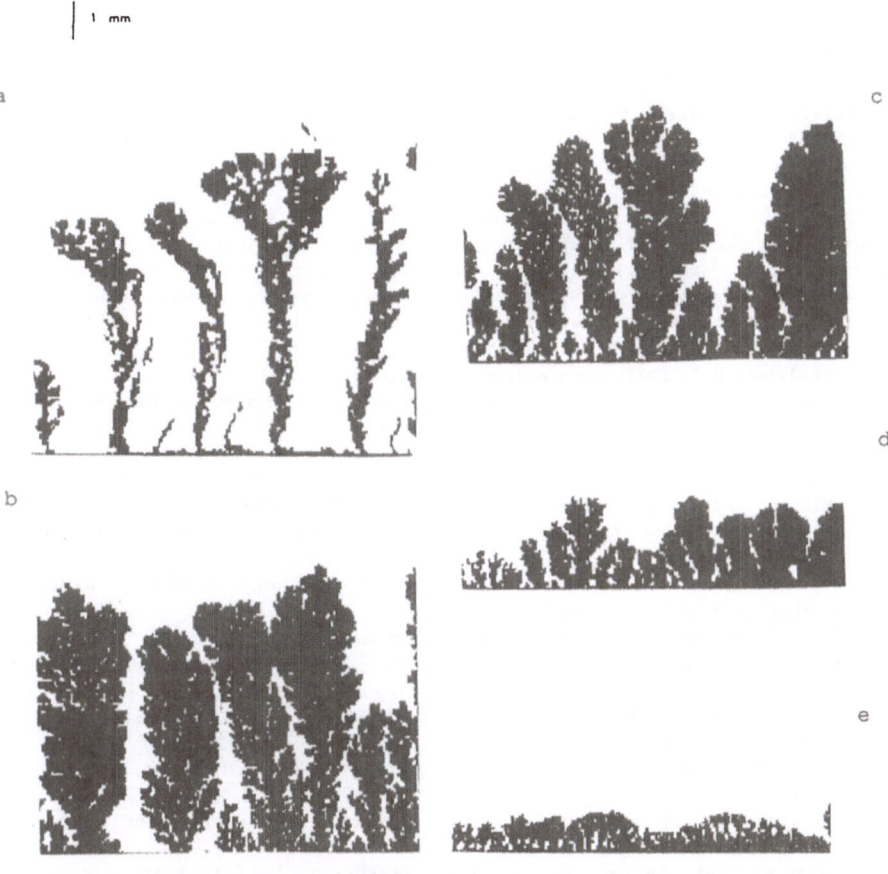

Fig. 5. Patterns of electrodeposited Cu: (a) c=0.01 M, E=-4. V vs. SCE, t = 12 min., $\langle D_q \rangle$ = 1.53 ± 0.03; (b) c = 0.05 M, E = -4.0 V vs. SCE, t = 12 min., $\langle D_q \rangle$ = 1.71 ± 0.01; (c) c = 0.1 M, E = -5.0 V vs. SCE, t = 9 min., $\langle D_q \rangle$ = 1.71 ± 0.05; (d) c = 0.1 M, E = -4.0 V vs. SCE, t = 12.5 min., $\langle D_q \rangle$ = 1.57 ± 0.04 and (e) c = 0.1 M, E = -3.0 V vs. SCE, t = 15 min., $\langle D_q \rangle$ = 1.54 ± 0.02.

We stress on the expected, although not totally understood, correlation between structure (morphology) and dynamics (growth velocities) by composing Fig.5 from different patterns corresponding to the range of electrode potential and concentrations in Fig.4. An averaged fractal dimension is also given for each pattern. In the region of low concentrations (0.01 M) (fig. 5a) patterns with a well-defined planar front growth rapidly, filling space at a typical length scale of a few trees much little densely (low fractal dimension) that those much more ramified patterns (higher fractal dimension) growing more slowly at higher concentrations (0.05 M) (fig. 5b). In the limit of the highest concentrations studied (0.1 M) and for intermediate potentials (-4 V vs. SCE) (fig. 5d), the aggregate tends to be more compact and enclosed within a more regular boundary what determines again a low fractal dimension. On the other hand, in going to higher potentials (from fig. 5e to fig. 5c), larger growth velocities result in more ramified textures whose fractal dimensions tend to increase.

Acknowledgements: The authors are gratefully indebted to L. Solé, S.Martinez and R. Pascual, from *J. Almera Institute (CSIC)*, Barcelona, for their kind help with the digitization procedures.

V.- REFRENCES

1. "Fractals in Physics", L. Pietronero and E. Tosatti, eds. North Holland, Amsterdam, 1986, and references quoted therein; "Random Fluctuations and Pattern Growth", H.E. Stanley and N. Ostrowsky, eds. Kluwer Acad. Pub., Dordrecht, 1988, and references quoted therein.
2. B. Mandelbrot, "Fractals: Form, Chance and Dimension", Freeman, San Francisco, 1977; "The Fractal Geometry of Nature", Freeman, San Francisco, 1982; J. Feder, "Fractals", Plenum Press, New York, 1988; T. Wicsek, "Fractal Growth Phenomena", World Scientific, Singapore, 1989; "The Fractal Approach to Heterogeneous Chemistry: Surfaces, Colloids and Polymers", D. Avnir (ed.), J. Wiley, New York, 1989.
3. R. Tamamushi and H. Kaneko, Electrochim. Acta, 25 (1980) 391.
4. R.M. Brady and R.C. Ball, Nature (London), 309 (1984) 225.
5. M. Matsushita, M. Sano, Y. Hayakawa, H. Honjo and Y. Sawada, Phys. Rev. Lett., 53 (1984) 286.
6. Y. Sawada, A. Dougherty and J.P. Gollub, Phys. Rev. Lett., 56 (1986) 1260.
7. D. Grier, E. Ben-Jacob, R. Clarke and L.M. Sander, Phys. Rev. Lett., 56 (1986) 1264.
8. D.B. Hibbert and J.R. Melrose, Phys. Rev. A, 38 (1988) 1036.
9. F. Argoul, A. Arneodo, G. Grasseau and H.L. Swinney, Phys. Rev. Lett., 61 (1988) 2558.
10. P. Garik, D. Barkey, E. Ben-Jacob, E. Bochner, N. Broxholm, B. Miller, B. Orr and R. Zamir, Phys. Rev. Lett., 62 (1989) 2703.
11. G.L.M.S. Kahanda and M. Tomkiewicz, J. Electrochem. Soc., 136 (1989) 1497.
12. F. Sagués, F. Mas, M. Vilarrasa and J.M. Costa, J. Electroanal. Chem., 278 (1990) 351.
13. P. Grassberger and I. Procaccia, Physica D, 13 (1984) 34; T.C. Halsey, M.H. Jensen, L.P. Kadanoff, I. Procaccia and B. Shraiman, Phys. Rev. A, 33 (1986) 1141.
14. P. Meakin, Phys. Rev. B., 30 (1984) 4207; F. Sagués and J.M. Costa, J. Chem. Educ., 66 (1989) 502.
15. S.R. Forrest and T.A. Witten, J. Phys. A, 12 (1979) L109; T. Witten and L. Sander, Phys. Rev. Lett., 47 (1981) 1400.
16. G. Grasseau, Thèse de Doctorat, Bordeaux (1989); T. Tél, A. Fülöp and T. Vicsek, Physica A, 159 (1989) 155.
17. A.J. Bard and L.R. Faulkner, "Electrochemical Methods. Fundamentals and Applications", Wiley, New York, 1980; J.M. Costa, "Fundamentos de Electródica", Alhambra, Madrid, 1981; Southampton Electrochemistry Group, "Instrumental Methods in Electrochemistry", Ellis Horwood, Chichester, 1985; D.A. Kessler, J. Koplik and H. Levine, Adv. Phys., 37 (1988) 255, Section 8.3.

HIGH RESOLUTION SPECTROSCOPY OF LANGMUIR-BLODGETT FILMS

Michel Orrit, Jacky Bernard

Centre de Physique Moléculaire Optique et Hertzienne, CNRS et
Université de Bordeaux I, 351 Cours de la Libération
F - 33405 Talence (France)

and

Dietmar Möbius

Max Planck-Institut für biophysikalische Chemie
Postfach 2841   D-3400 Göttingen (F.R.G.)

INTRODUCTION

Persistent spectral hole burning (HB) is now in use for about fifteen years in the investigation the low-temperature dynamics of condensed organic materials [1-3]. It is well established that the hole width is very sensitive to matrix structure. In particular, a striking difference is found between crystals, where the holes are broadened according to the activation of a libration mode, and glasses or polymers where the temperature dependence is much slower (of the type T^{α}, $1 \leqslant \alpha \leqslant 2$) [4]. This behavior has been attributed to the broad spectrum of the specific low-energy excitations of glasses known as two-level-systems (TLS). The study of new 'exotic' matrices for dopant dye molecules may on the one hand bring valuable information on the dynamics and structure of these systems as compared to more conventional phases, on the other hand help us shed light on the mechanisms behind dephasing and spectral diffusion.

Langmuir-Blodgett films (LBF) are fascinating systems obtained by piling up monomolecular layers of amphiphilic molecules. These layers are prepared at the air-water interface [5]. As such, LBF present an opportunity to manipulate molecules at the microscopic level through macroscopic techniques. They are now extensively studied in many laboratories around the world [6,7] for their importance in fundamental science (models for surfaces or biological structures) as well as their potential applications (molecular engineering and electronics, construction of non-centrosymmetric solids, specific sensors for molecules, etc...). During the past few years, almost all characterization methods of materials science (infra-red, Raman spectroscopies, X-ray, neutron and electron scattering and diffraction, scanning tunneling microscopy, etc...) have been successfully applied to LBF, thereby improving our knowledge of their properties and structure. However, the conventional optical spectroscopy of LBF containing absorbing dye molecules is limited by broad bands that persist at low temperatures. In our laboratory, we have recently begun a HB study of dye molecules imbedded in LBF. As matrices for HB, LBF present specific advantages :

i) Flexibility at the molecular and at the layer scale, by changing the matrix molecule and the adjacent layer in a multilayer assembly.

ii) Possibility of interface and surface probing using a single dye doped monolayer at the interface.

iii) Possibility of studying molecules with well-defined orientation through the Stark shift under a normal electric field.

iv) Dimensionality effects : Since LBF are ill-ordered 2D systems, the study of dephasing and spectral diffusion can show specific effects which can be helpful to test the current theories.

SURFACE PROBING USING A MONOLAYER

When a molecule goes from a gaseous to a condensed phase, its transition frequency is usually red-shifted due to the polarization of neighboring solvent molecules. Using a resorufin [8] or cyanine monolayer with dye molecules close to the sample surface (separated from the vacuum by the saturated chains about 25 A thick), we could see upon immersion in helium a similar red shift due to the polarization of superfluid helium. The shift (about 0.5 GHz) was smaller than the hole width but reproducible and reversible when reverting to the gas atmosphere. The order of magnitude of the shift can be explained using a simple model of electrostatic images.

However, the broadening of the hole upon immersion shows that this picture is still too simple. The hole broadening could thus inform us about the charge distribution in the disordered microscopic surrounding of the dye molecule. These charges give rise to random static image fields when the helium half-space is added.

HOLE-BURNING KINETICS AND HOLE SHAPE

When burning holes in a free-base porphyrin-doped multilayer, we found a nearly linear dependence of the hole depth on the logarithm of the burning fluence. This very simple dependence has been explained in the recent literature [9,10] by assuming a very broad distribution of the HB rates, due to microscopic disorder in a glassy matrix.

Using a similar model [11], we reproduced the logarithmic burning kinetics and obtained a hole width at half-depth varying as the power 0.25 of the burning fluence. This dependence was observed in different 3D samples for which the model should apply too. We also studied the hole shape and found it to change from Lorentzian at low fluences to logarithm of Lorentzian for strong fluence. Moreover, the very broad saturated holes should be rather insensitive to spectral diffusion. Therefore the deep holes could allow us to determine the 'true' homogeneous width as measured by the burning photon during the absorption process. When applied to the free-base porphyrin under study, the model indicates a high sensitivity of the double proton transfer to the microscopic structure of the matrix. We are now extending our model to account for dispersive power broadening as observed in some polymers [12]. This effect can lead to logarithmic burning kinetics, but apparently not to the power law with exponent 0.25 for the width-fluence dependence.

The hole width in LBF multilayer and its temperature dependence resemble very closely those in usual 3D glasses or polymers, with perhaps somewhat larger temperature exponents. In contrast to these results the holes burnt and measured in monolayer samples, or assemblies consisting of only a few layers, present very broad holes at low temperature. At higher temperatures, the hole width becomes comparable to that of the multilayers. This 'residual' width effect was found in all matrix-dye couples studied so far (resorufin/dialkylammonium [8], porphyrin/polymeric cyanoacrylate [13], amphiphilic cyanine/cadmium arachidate [14]). The 'residual' width at the lowest temperature studied (1.6 K) amounts to a few GHz, which is much larger than all possible population relaxation T_1 times for the dye molecules under study. Thus we attributed this width to the matrix dynamics at low temperature, resulting in either dephasing if the fluctuations are faster than the dephasing time T_2, or in spectral diffusion for slower fluctuations. Our set-up does not allow us to probe holes at times shorter than 100 s, but we checked that at longer times we observe no significant additional broadening by spectral diffusion. The time scale of the fluctuations responsible for the residual width is thus shorter than 100 s.

In order to clear up this surprising difference between monolayer and multilayer, we prepared multilayer assemblies where the dye occupies different positions (see Fig. 1) : Directly against the glass backing (position 1), in the middle of the stack (positon 5) or in the surface monolayer (position 11). The results for the width as a function of temperature are presented on Fig. 1. The hole width of sample 5 closely follows that of the multilayer sample. Sample 1 shows a similar behavior but with a larger width. This is easily explained if the interface between the glass and the first monolayer is strongly disordered, or if the glass itself contributes to dephasing through its own TLS. The holes of the last surface position (n°11) on the other hand, show the characteristic residual width of the monolayer. We may thus conclude that the low temperature dynamics underlying the residual width take place in the surface monolayer.

We also tried to measure pressure-induced shifts of the holes by passing from superfluid helium at nearly zero pressure to liquid helium at atmospheric pressure. We assume that the hole changes due to the temperature change are negligible, as observed at higher temperatures. While the multilayer holes just shifted to the red at higher pressure, at a rate similar to that of other 3D matrices (about 0.3 GHz for 1 bar) [15], the holes burnt in the surface monolayer broadened dramatically (of a few GHz) under pressure changes. The pressure broadening was much larger than the shift. This sensitivity to pressure changes is also a characteristic of the surface layer.

We feel that the residual width is difficult to explain by a 'soft' (low energy $\hbar\Omega$) vibration because this should give rise to a strong T^2 dependence at $k_B T > \hbar\Omega$, which is not observed. Therefore we assume that the broadening is caused by two-level-like excitations, and hope that a surface monolayer could represent a good and reasonably simple model to study these still mysterious excitations. We are now trying to devise new experiments to solve the problem of the origin of these dynamics and would appreciate any suggestion of new experiments or explanations.

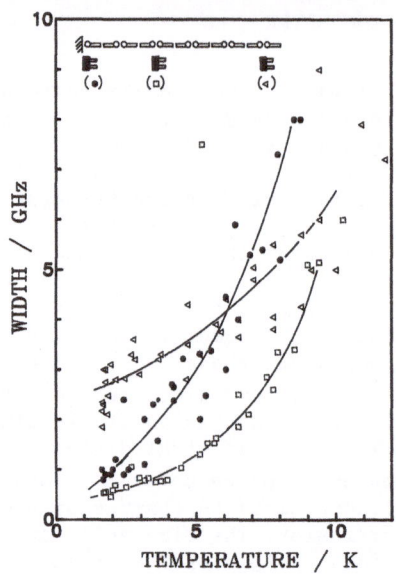

Figure 1.   *Temperature dependence of the hole width for a cyanine doped LBF.
The widths were obtained by extrapolation to zero hole depth of
series of holes burnt at increasing fluences [8]. In the three
samples studied here, the dye-doped monolayer was placed at different positions in a multilayer assembly (see inset). Only
when the dye is close to the surface do we recover the residual
width of the monolayer [14] at low temperature. The solid lines
are just guides for the eye.*

REFERENCES

[1] I. Zschokke, ed., Optical Spectroscopy of Glasses (Reidel, Dordrecht, 1986).

[2] Optical Linewidths in Glasses, J. Luminescence 36, n° 4,5 (1987).

[3] E. W. Moerner, ed., Topics in Current Physics, vol. 44, Persistent Spectral Hole-Burning, Science and Applications (Springer, Berlin, 1988).

[4] R. M. Macfarlane and R. M. Shelby, J. Luminescence 36 (1987) 179.

[5] H. Kuhn, D. Möbius and H. Bücher, in : Physical Methods of Chemistry Vol. 1, part III B, eds. A. Weissberger and B. Rossiter (Wiley, New York 1972).

[6] D. Möbius, ed., Langmuir-Blodgett 3, Thin Solid Films Vol. 159,160 (1988).

[7] Proceedings of the 4th International Conference on Langmuir-Blodgett Films, Tsukuba (Japan), Thin Solid Films Vol. 178,179 (1989).

[8] M. Orrit, J. Bernard and D. Möbius, Chem. Phys. Letters 156 (1989) 233.

[9] R. Jankowiak, R. Richert and H. Bässler, J. Phys. Chem. 89 (1985) 4569.

[10] Y. Kanematsu, R. Shiraishi, A. Imaoka, S. Saikan and T. Kushida, J. Chem. Phys. 91 (1989) 6579.

[11] H. Talon, M. Orrit and J. Bernard, Chem. Phys. 140 (1990) 177.

[12] B. L. Fearey, T. P. Carter and G. J. Small, Chem. Phys. 101 (1986) 279.

[13] J. Bernard, M. Orrit, R. I. Personov and A. D. Samoilenko, Chem. Phys. Letters, 164 (1989) 377.

[14] J. Bernard and M. Orrit, Proceedings of the DPC'89 Conference in Athens Ga. (U.S.A.), August 1989, to appear in J. Luminescence (1990).

[15] Th. Sesselmann, W. Richter, D. Haarer and H. Morawitz, Phys. Rev. B 36 (1987) 7601.

MOLECULAR DYNAMICS SIMULATION OF THE TRANSPORT OF SMALL MOLECULES

ACROSS A POLYMER MEMBRANE

R.M. Sok, H.J.C. Berendsen and W.F. van Gunsteren

Laboratory of Physical Chemistry, University of Groningen
Nyenborgh 16, 9747 AG Groningen, The Netherlands

1. INTRODUCTION

The diffusion of small molecules (such as H_2, N_2, O_2, CH_4, CO_2, H_2O etc.) through polymer solid membranes has been studied extensively in the past decades, mainly because of the large number of industrial applications in which this process plays a major role. These include protective coatings, electronic and biomedical devices, packing materials and selective separation of gaseous and liquid mixtures.

Even though many theories on this subject have been developed[1,2,3] a complete understanding of this complex process, especially at the molecular level, is still missing. Molecular dynamics (MD) simulations[4,5,6] can give insight into the fundamental processes which play a role in the diffusion and transport, because MD is in principle able to predict macroscopic transport properties based on atomic interactions. The direct simulation of molecular transport driven by concentration, pressure or osmotic gradient is possible using NEMD (Non–equilibrium molecular dynamics), but use of NEMD for transport through polymer membranes is likely to be limited because of the long time scale on which these processes reach steady state conditions.

We have therefore chosen a different approach that will allow the prediction of transport properties on a much longer time scale than the time scale on which MD simulations can be carried out. This approach uses a stochastic description of the transport process: the particle (small molecule) is supposed to diffuse through a *thermodynamic potential profile* which is determined by the driving force due to a gradient across the membrane and by the solvation properties of the molecule in the polymer matrix. The latter can be described as an excess thermodynamic potential which can be determined by MD simulations.

In section 2 the relation between transport coefficients and excess thermodynamic potential will be derived. In section 3 the MD simulation and the preparation of the starting configuration of the polymer is described and sections 4 and 5 describe how diffusion coefficient and free energy profile can be derived from the simulations. In the present preliminary state of our simulations, results can not yet be presented.

2. THEORY

The flux J of particles through a stationary membrane is determined by the average particle velocity v and their concentration c:

$$J(r) = c(r)\ v(r) \qquad (1)$$

Large-Scale Molecular Systems, Edited by W. Gans *et al.*
Plenum Press, New York, 1991

In the limit of linear irreversible processes, the average velocity can be written as

$$v(r) = \frac{1}{\zeta} \; F_{th}(r) \tag{2}$$

where F_{th} is the thermodynamic force and ζ a friction coefficient which we assume to be independent of the position r in the homogeneous polymer. The thermodynamic force on a particle is the negative gradient of its chemical potential. If the chemical potential of a particle in the membrane can be written as

$$\mu(r) = \mu^0 + RT \ln c(r) + \mu_{ex}(r) \tag{3}$$

in which μ^0 is the standard chemical potential and $\mu_{ex}(r)$ is the excess chemical potential of the system compared to the ideal gas phase, then the flux velocity can be given by

$$v(r) = \frac{-1}{\zeta} \; \nabla \left(\mu^0 + RT \ln c(r) + \mu_{ex}(r) \right) \tag{4}$$

and the flux becomes

$$J(r) = \frac{-RT}{\zeta} \; \nabla c(r) - \frac{c(r)}{\zeta} \; \nabla \mu_{ex}(r) \tag{5}$$

This can be rewritten as

$$J(r) = \frac{-RT}{\zeta} \; e^{-\mu_{ex}(r)/RT} \; \nabla \left(c(r) \; e^{\mu_{ex}(r)/RT} \right) \tag{6}$$

If we solve this equation for a gradient in the x–direction and for the stationary state ($J(x)$ is constant), apply this to a membrane of thickness d, separating two ideal gas phases with concentrations c_2 and c_2 respectively, assume that $\mu_{ex}(x)$ is constant throughout the membrane and take $\mu_{ex}(0) = \mu_{ex}(d) = 0$, then we find

$$J = -D \frac{(c_2 - c_1)}{d} \; e^{-\mu_{ex}/RT} \tag{7}$$

where we identified RT/ζ as the diffusion constant D.
So all the information needed from MD is the diffusion constant D and the excess thermo-dynamic potential μ_{ex} of the membrane compared to the gas phase.

3. MOLECULAR DYNAMICS SIMULATION OF A POLYMER

The specific properties of (amorphous) polymers make the normal way of preparing a starting configuration, that is melting a crystal–like structure, useless. Because of the long correlation lengths and times of the equilibration processes in polymers, which are some orders of magnitude larger than feasible within MD, we need some other way of preparing a starting sample.

A commonly used method is based on a self–avoiding random walk[7] In this method a chain is built up site by site. The position of the next site is chosen on the basis of the position of the previous site, using information about bond length and bond angle and choosing a random dihedral angle. This position is then either accepted or rejected with a Monte Carlo like criterion. The disadvantage of this method is that at higher densities the probability to add another site is practically zero.

One possible way to avoid this problem is based on MD rather than Monte Carlo. One starts with a very dilute system of several chains subject to periodic boundary conditions which interact through the normal bonded forces. The non–bonded forces are not van der Waals like, but instead represent a soft–core repulsive potential

$$V(r) = V_{max} \left(1 - \left(\frac{r}{r_0} \right)^2 \right)^2 \qquad\qquad 0 \leq r \leq r_0 \qquad\qquad (8)$$

In the preparation phase V_{max} is set such that the molecules are able to move through each other. Though unphysical, it prevents the chains from getting trapped. Every MD step the size of the periodic box is decreased by a small amount until the correct density has been reached. (One might even consider going to a higher density to allow for pressure relaxation in the first normal MD steps). At this time an energy minimization is performed with the normal non–bonded interactions to release the excess stress in the system after which a normal MD simulation can be performed.

We have followed this procedure and made a few samples of dimethyl–siloxane (without cross links) starting of at a density of 0.01 g/cm³ and going to 1.0 g/cm³. If we then performed constant pressure and temperature MD we found that in the first 20 ps the system expanded almost 10% (going from a density of 1.0 g/cm³ to 0.92 g/cm³), but after 20 ps density and energies remained constant. Thus this method of preparation proved to be quite practical.

4. DIFFUSION FROM MD

The diffusion constant can be calculated from either the velocity autocorrelation function

$$D = \frac{1}{3} \int_0^\infty < \boldsymbol{v}(0) \cdot \boldsymbol{v}(t) > dt \qquad\qquad (9)$$

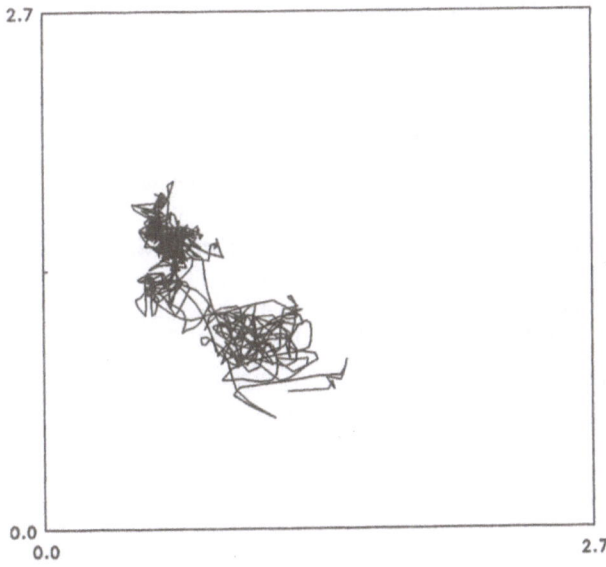

Fig. 1.   80 ps trajectory of a guest molecule (CH$_4$) in a dimethyl–siloxane polymer membrane (not shown). Dimensions in nm.

or the root mean square displacement of the particle.[5]

$$D = \lim_{t \to \infty} \frac{1}{6t} < [\ \boldsymbol{r}(0) - \boldsymbol{r}(t)\ ]^2 > \qquad (10)$$

Because the long time tail of the velocity autocorrelation in molecular dynamics simulations is statistically inaccurate, the diffusion coefficient is normally calculated from equation (10), or more precisely from the slope of the root mean square displacement in time. But as clearly can be seen in figure 1, the diffusion process is a hopping between cavities. In fact in these 80 ps the particle visits four or five cavities. Therefore very long simulations are required to obtain the global diffusion constant.

5. EXCESS CHEMICAL POTENTIAL FROM MD

The calculation of the excess thermodynamic potential is complicated, because this quantity cannot be determined directly from any ensemble average. Many methods to find the free energy indirectly have been suggested[5] but none of them is generally applicable. We propose a combination of two commonly used methods: the *particle insertion method* and the *thermodynamic integration method*.

The particle insertion method itself is suitable for systems of low and moderate density, but breaks down for dense systems. The reason is that attempts to insert a particle yields high interaction energies and hence low Boltzmann factors.

Thermodynamic integration can be used to "grow" particles slowly or through several intermediate equilibrium stages from zero interactions to their full interaction with the matrix. In practice this process turns out to be difficult in the first steps of the growth process, because the arbitrary choice of starting conditions provides a large statistical fluctuation.

Therefore we propose to combine the best of the two methods: first insert a "reduced" particle (with reduced interactions) using the particle insertion method and then slowly "grow" the particle to its full size. During the latter part of the process the polymer matrix is allowed to relax around the growing particle.

1. A.E. Chalykh and V.B. Zlobin, *Russ. Chem. Rev.* <u>57</u>(6) (1988) 504

2. S.A. Stern and H.L. Frish, *Ann. Rev. Mater. Sci.* <u>11</u> (1981) 523

3. M. Doi and S.F. Edwards, *The Theory of Polymer Dynamics*, Clarendon, Oxford 1986

4. M.P. Allen and D.J. Tildesley, *Computer Simulation of liquids*, Clarendon, Oxford 1987

5. C.R.A. Catlow, S.C. Parker and M.P. Allen ed., *Computer Modeling of Fluid Polymers and Solids*, Kluwer Academic Publishers, Dordrecht, The Netherlands 1989

6. W.F. van Gunsteren and H.J.C. Berendsen, *Angew. Chemie.*, to be published

7. J.H.R. Clarke and D. Brown, *Molec. Sim.* <u>3</u> (1989) 27

REACTIONS IN MICROEMULSIONS: FRACTAL MODELING

P. Lianos

University of Patras
School of Engineering   and
26000 Patras Greece

P. Argyrakis

Department of Physics 313-1
University of Thessaloniki
54006 Thessaloniki Greece

ABSTRACT

We use a fractal picture to understand the kinetics of the A+B reaction in microemulsion droplets, micelles, lipid vesicles, and similar systems. We examine the case where [A]<[B], using equations previously utilized for the intepretation of luminescence quenching in such systems. We find that a percolation model of clusters below the critical threshold with a specified range in cluster size provides a good example of an inhomogeneous space, such is the case for these systems. The [A]/[B] density ratio was either constant or varying as a function of p (p=the probability for an open site). Our conclusions show that the resulting spectral dimension of the reaction process is always smaller numerically than the customary value $d_s=4/3$, and that it is strongly dependent on the size and shape of the reaction space.

INTRODUCTION

Fractal modeling has been successfully applied in recent years not only to describe transport and interactions in solid state physics but also to explain the experimental luminescence decays in the excited states in such organized assemblies such as micelles, microemulsions, and vesicles in the presence of quenchers. As it has been shown recently [1-5] the decay profile of an excited probe in the presence of quenchers obeys the following equation:

$$I=I_0\exp(-k_0t)\exp(-k'Qt^f+k''Qt^{2f}) \qquad (1)$$

where k_0 is the luminescence decay rate constant in the absence of quenching, k' and k'' are constants, Q is the quencher concentration, and f is a positive number smaller than unity. This equation is amenable to the cases where the reacting species are restricted by the geometry or the viscosity of the reaction medium so that the displacement of the reactants during the lifetime of the excited state is very limited. The same equation should apply for any reaction A+B ==> products, where the concentration [A] is substantially smaller than the concentration [B]. If [A] is monitored by some other means than luminescence then Eq. (1) should be used without the factor $\exp(-k_0t)$. The term $k'Qt^f$ is exact, but the term $k''Qt^{2f}$ is an approximation of a series with an infinite number of terms [3,4]. A similar equation has been successfully applied to describe multistep energy transfer, where f equals one half of the spectral dimension

[6]. Notice that in this case, since Q can not be directly defined in the usual molarity units, we simply calculate here the product k'Q and k''Q. The verification of Eq. (1) has been done by fitting it to experimental decay profiles of luminescent molecules solubilized in water-in-oil microemulsions or phospholipid vesicles [3,4]. We do the same here but utilize the decay profiles as derived from computer simulations of the A+B reaction. Our main argument is that if such a fitting of equation (1) to both experimental data and computer simulation data is possible at all, then its functional form must be correct, at the very least.

The major conclusion from recent *kinetic studies in fractal and other* inhomogeneous spaces is that the kinetics in these spaces is highly non-classical and the rate laws are dominated by the so-called spectral dimension. This is because equation (1) can be re-written in the following form, after Blumen, Klafter [6]:

$$I=I_0 \exp(-k_0 t)\exp(-\lambda a t^{d_s/2} + \lambda^2 b t^{d_s}) \qquad (2)$$

where $\lambda=\ln(1-p)$, p being the probability that a given site is an open (allowed) site. Equations (1) and (2)$_2$ are essentially the same, with the substitutions that k'Q=λa, and k''Q=$\lambda^2 b$. The exponent d_s is the spectral dimension of the reaction medium, and as is well known [14], it is:

$$d_s = 2d_f/d_W \qquad (3)$$

where d_f is the fractal dimension, and d_W is the random-walk dimension, thus incorporating both the static (fractal, geometric) and dynamic (random-walk, diffusion) aspects of the process. For example, for the 2-dim percolation cluster [14] exactly at the critical threshold d_f=1.89, and d_W=2.87.

In the past for the A+B reaction mostly the [A]=[B] simpler case has been studied, and practically all effort has been placed in understanding the segregation of reactants that takes place in this type of reaction [7,8,10]. In the present work [A] is much smaller than [B]. The impetus for such a situation is found in numerous works of luminescence probing of organized modecular assemblies [9]. In the present computer simulation procedures the reaction medium has been chosen to be the clusters formed by a 2-dim binary lattice which has a concentration p of open (allowed) sites less than p_c, the critical percolation threshold. Even though most of the work on reactions up to now is done exactly at the critical threshold we choose here to utilize clusters of small sizes and avoid the infinite cluster at p_c. This is because real organized assemblies, such as micelles, microemulsion droplets, vesicles, etc., have a lower size limit of the order of a few tens of Angstroms. Thus we use as a lower limit of cluster size s=25 sites. As far as the upper limit is concerned we arbitrarily choose s=50, 250 and ∞ , since the above systems are also found in a great variety of sizes. Thus we restrict all reactions within these bounds in our effort towards understanding the luminescence mechanisms of several experimental systems. In the next section we describe the techniques that we used for the computer simulations. Following this we present our results, and finally our conclusions.

COMPUTATIONAL METHODS

The computer simulation of the chemical reactions was done using the techniques previously reported [10]. Briefly, the lattice clusters are generated on a 300x300 square lattice at a given p, the probability for an open site. As stated above a minimum and maximum cluster size are also specified, and only the clusters that fall within this prescribed s-range are kept, while all the rest become non-accessible sites. A certain number of A and B particles are placed at random on the available clusters at time t=0. Reaction proceeds in the known way, i.e. all particles perform random walks with the stipulation that when an A and B particle occupy the same site they annihilate and are removed from the system, while nothing happens between two A or two B particles. We use excluded volume principles, i.e. we do not allow more than one particle to occupy a given site at any one time. We monitor the density of A particles, [A], as a function of time. The basis for our time unit is one Monte Carlo Step (MCS), which is defined as the time it takes for all present particles to move one step to one of their nearest neighbors. In these calculations time goes from 0 to 200 MCS. This may correspond to a realistic situation when 1 MCS=1 nsec. For example, in pyrene excimer probing of micelles formed with tetradecyltrimethylammonium halides [11] the monomer decay time ranges between 160 and 300 nsec. We monitored several interactions with several [A]/[B] ratios. We noticed that when this ratio is less than 0.20 the reaction proceeds too fast to allow appropriate analysis, as the number of A reactants is too small. Subsequently, we carried out all calculations by using a ratio [A]/[B]=0.20. In real luminescence quenching experiments the [L*]/[Q] ratios are substantially smaller, but this does not change the conclusions of this work. Our choice of the [A]/[B] ratio is dictated here by the available number of lattice sites, and using a larger lattice could easily accomodate for different ratios, albeit more time consuming.

We distinguish two cases: (I) The A and B concentartions with respect to the total volume, [A] and [B], vary as a function of p. (II) The total number of A (and B particles) remains constant, independently of p, so that [A] and [B] are constant.

VALUES OF THE EXPONENT f AND THE CONSTANT k'Q

Equation (2) was fitted to all annihilation decay curves by non-linear regression, least-square analysis. In case I, where the total number of A and B species increased with p, we monitored reactions at clusters generated at several different p values in the range p=0.34 - 0.58. The fit was very good, as seen in the example of Fig. 1. It was equally good for both this case and the second case presented below. This is a sound verification of the validity of this equation. Fitting gave the values of the corresponding parameters f, k'Q, and k''Q. The time dependence of [A] always has a decaying form, as it is always true that $k'Qt^f > k''Qt^{2f}$. This situation has been shown experimentally, but it is also the consequence of the fact that $k'Qt^f$ is the leading term of an infinite series, as discussed above.

Various values for this case are shown in Table I. There is a problem with defining appropriate units for the product k'Q, since it is multiplied by time raised to a non-integer power. Furthermore, this value does not remain constant, thus preventing comparison of the data at different p values. Fot this reason and in order to get some representrative reaction rates we have adopted the time-dependent expressions used previously:

$$k'(t)Q = fk'Q/t^{1-f} \tag{4}$$

k(t)Q is not defined for t=0, but this presents no problem since our data refer to discrete time values (channels). In Figure 2 we show experimental results of the quenching rate k(t) variation as a function of time [4] for the system dipalmitoylphosphatidylcholine-L-α-phosphatidylglycerol (DPPC-LαPG). We observe that k(t) decreases with time for a given Q value and tends to a limited k_∞ which is shown in Table I as $k_\infty Q$.

k'(t)Q and $k'_\infty Q$ are always given in s^{-1} and they correspond to second-order reactioin rate constants multiplied by concentration (expressed in M). We have chosen the given k'Q as representative values proportional to the real reaction rates.

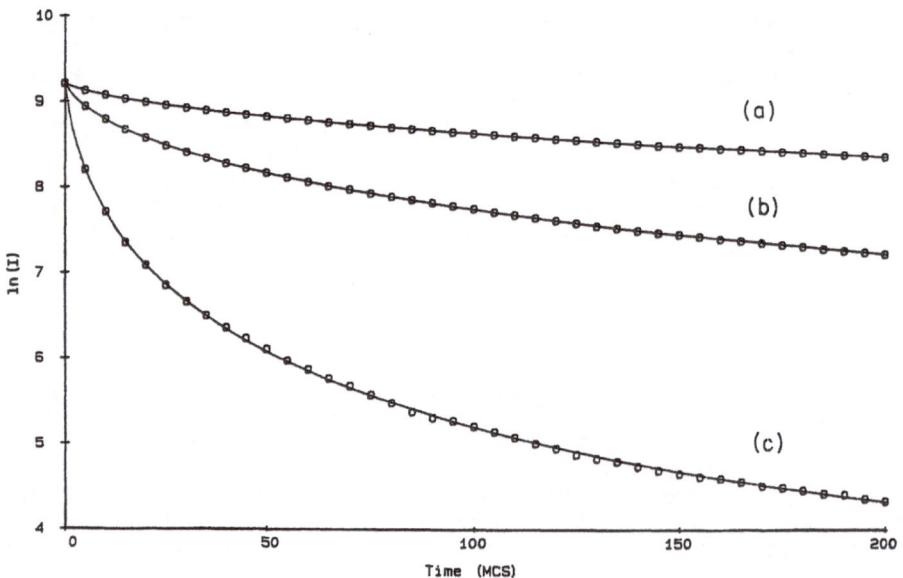

Figure 1. Decay of the [A] density for the A+B==>0 reaction. Solid lines are solutions of eqn. 2. Circles are results from the computer simulated reactions at p=0.58. Three different cluster size ranges are given: (A) $25<s<\infty$, (B) $25<s<250$, and (C) $25<s<50$.

However, we still need to make an estimation of the concentration Q which, of course, in our case is proportional to the concentration of the majority species [B]. We proceed to calculate the average size I_{av} for the available clusters and the number of sites N on which reaction can occur. I_{av} is calculated by using the standard method as the second moment of the cluster distribution [12]. The proper equation that was used is the following:

$$I_{av} = \sum_{s_{min}}^{s_{max}} i_s s^2 / \sum_{s_{min}}^{s_{max}} i_s s \qquad (5)$$

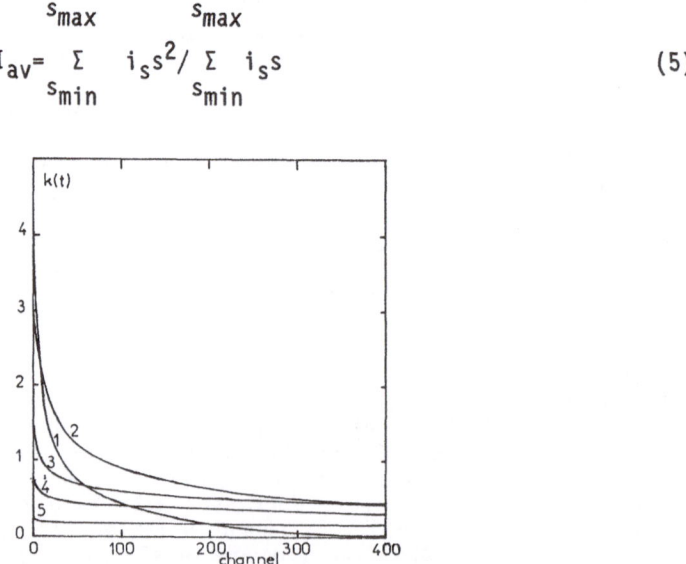

Figure 2. Variation of k(t) with time for various concentration mixtures of the system DPPC-LaPG [3].

Table I. Values of f, $k_\infty Q$, $k_\infty Q/N$ for case I ([B]/N=constant)

p	f	$k_\infty Q$	$10^4 k'Q/N$
Cluster size 25<s<∞			
0.34	0.54	9.30	110.8
0.37	0.50	8.23	37.1
0.40	0.44	6.72	13.8
0.45	0.48	8.00	6.2
0.47	0.48	7.90	4.5
0.48	0.45	6.98	3.5
0.49	0.47	7.80	3.5
0.50	0.47	7.77	3.1
0.55	0.45	7.55	2.0
0.57	0.50	9.29	2.2
0.58	0.49	8.74	2.0
Cluster size 25<s<250			
0.34	0.57	10.29	122.6
0.37	0.46	7.13	32.1
0.40	0.45	6.68	13.7
0.45	0.48	7.76	6.0
0.47	0.48	7.91	4.6
0.48	0.47	7.68	3.9
0.49	0.47	7.62	3.5
0.50	0.48	8.15	3.5
0.55	0.48	8.33	4.2
0.57	0.49	8.97	6.7
0.58	0.49	8.94	9.0
Cluster size 25<s<50			
0.34	0.57	10.29	125.8
0.37	0.46	7.13	34.7
0.40	0.45	6.68	16.2
0.45	0.48	7.76	9.7
0.47	0.45	7.18	8.2
0.48	0.48	7.94	9.0
0.49	0.45	7.18	8.2
0.50	0.48	8.03	9.6
0.55	0.44	8.80	18.1
0.57	0.50	9.35	27.9
0.58	0.52	9.90	36.9

Units of $k_\infty Q$ are: s^{-1} on an arbitrary scale. The units of the quantity $10^4 k'Q/N$ are $M^{-1}s^{-1}$ also on an arbitrary scale.

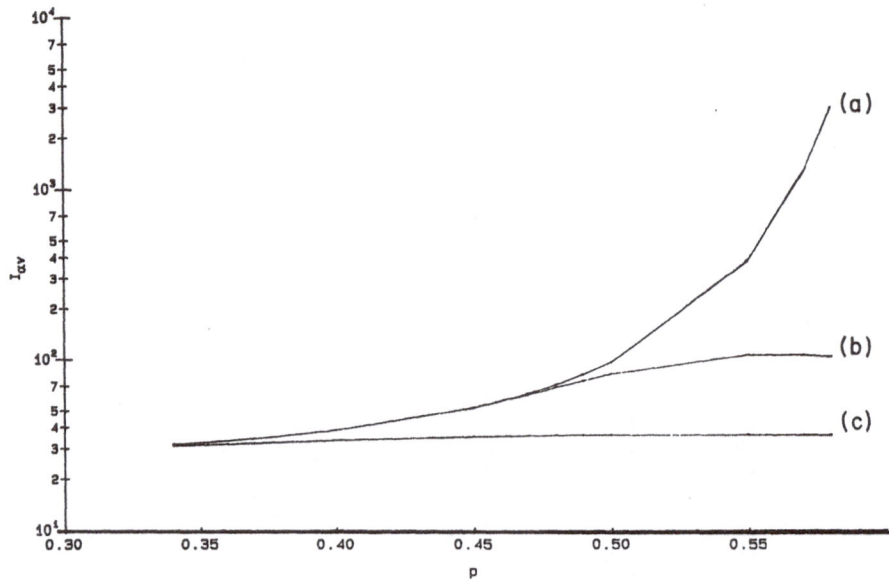

Figure 3. I_{av} vs. p for the three cluster size ranges studied. The function I_{av} is calculated using eqn. 3. The designations are the same as in Figure 1.

where s is the cluster size and i_s is the frequency of its appearance. The lower and upper limits in these sums are the limits of our size ranges. In Figure 3 we plot the average cluster sizes I_{av} vs. p for the three different ranges that we used. We see that the average cluster size I_{av} increases as p increases, as expected. At large p the clusters become larger and less numerous. At p=0.58, i.e. close to the percolation threshold, practically only one cluster of very large size is about to be formed. As we expect s ==>∞ when p approaches the percolation threshold. The variation, however, of the average number of sites N on which reaction can occur is not monotonous. In Figure 4 we plot this number N vs. p. We observe here that the case 25<s<∞ behaves as expected for cluster sizes without an upper limit. The other two cases show a maximum since the cut-off of large size clusters at high p excludes an increasing number of reaction sites. Thus, by properly choosing the limits in the values of s we have a model in which we directly control the size and shape of the reaction medium and later can apply it to specific experimental systems. Having obtained the N values we proceed with the following reasoning: In case I we keep the [B]/N ratio constant. Therefore, Q∼[B]∼N. It is then obvious that k'_∞∼k'Q/N. Therefore, the values of k'_∞Q/N also shown in Table I represent an equivalent second-order reaction rate constant calculated for infinite time. We consider this to be our long time limit (last few MCS). In fact, the average k'(t) values calculated over 200 MCS show exactly the same trend as k'.

Table II shows corresponding values calculated for the second case where [A], [B] are constant. Because Q∼[B], Q is also constant. Therefore, k'_∞∼k'$_\infty$Q and we did not need to calculate the column k'_∞Q/N in this case. In other words, the k'_∞Q values represent the equivalent second-order reaction rates in this case. Finally, the calculated values of k''Q (not given) in both cases were always two orders of magnitude smaller that k'Q, but they remained in this range and they never reached or tended to zero.

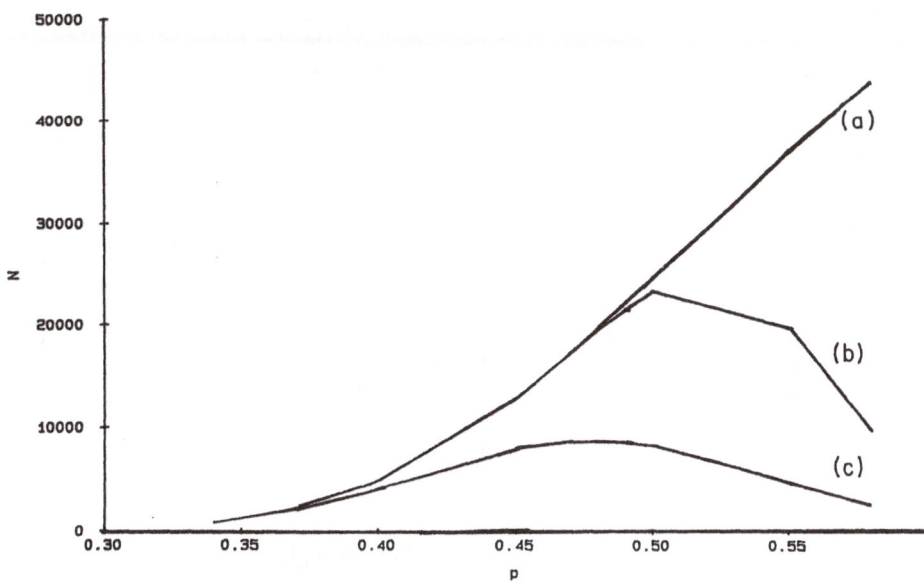

Figure 4. N vs. p for the three cluster size ranges studied. N gives the absolute number of sites on which the reaction can occur. The designations are the same as in Figure 1.

DISCUSSION

As we have already pointed out the annihilation reaction between a minority species A and a majority species B proceeds in such a manner that the survival probability of [A] is proportional to an exponential with the argument being a sum of an infinite series of terms. This series in real situations can be adequately represented by the zero order approximation term only, i.e. only $k'Qt^f$, and this has been shown in various experimental situations. This decay has been shown in the present work to be also in agreement with a percolation cluster picture. In isolated clusters, such as micelles or lipid vesicles, we have experimentally found that a first order approximation is necessary for the interpretation of our data. The new results here verify the second assumption, i.e. the $k''Q$ values were in all cases different from zero, and furthermore, their ratio to the $k'Q$ terms were constant. The good fit of equation (2) to the computer simulation data, as well as to the experimental data showes that the first order picture (and not higher order) is adequate.

We have previously suggested that difusion-controlled luminescence quenching reactions are equivalent to quenching by multistep energy transfer, when diffusion is treated via random walks of the reaction species. We have a similar situation here, and therefore, equation (2) becomes equivalent to a similar equation suggested by Blumen and Klafter, where $f=d_s/2$, and d_s is the spectal dimension of the reaction process. Our calculated values of f are then proportional to the spectral dimension, which is in turn proportional to the fractal dimension of the reaction domain. In our case I, where [B]/N=constant, the f values show some variation with p. The value of f is larger for small p, it decreases as p increases, and finally increases again as p approaches the percolation threshold. This is true for all cluster sizes with lower limit, and with or without upper limit. Such behavior indicates a complex way with which d_s depends on the reaction and its environment. In fact d_s is inversely proportional to the

Table II. Values of f and k'Q for case II ([A], [B] =constant)

p	f	$k_\infty Q$			
Cluster size $25<s<\infty$			Cluster size $25<s<50$		
0.37	0.24	11.8	0.37	0.21	10.6
0.40	0.39	10.9	0.40	0.45	16.4
0.45	0.50	6.5	0.45	0.45	9.0
0.47	0.49	4.9	0.47	0.46	8.4
0.48	0.53	4.8	0.48	0.47	8.6
0.49	0.51	4.1	0.49	0.47	8.8
0.50	0.53	3.8	0.50	0.46	9.0
0.55	0.53	2.8	0.55	0.46	15.3
0.57	0.54	2.6	0.57	0.38	16.8
0.58	0.55	2.6	0.58	0.34	17.3
Cluster size $25<s<250$					
0.37	0.23	10.9			
0.40	0.39	10.8	Units of $k_\infty Q$ are: s^{-1} on an arbitrary scale.		
0.45	0.50	6.5			
0.47	0.49	4.9			
0.48	0.52	4.7			
0.49	0.49	3.9			
0.50	0.50	3.7			
0.55	0.52	5.0			
0.57	0.48	6.4			
0.58	0.48	8.5			

random walk exponent (dimension) d_w. A small increase of d_f (the fractal dimension) and a small decrease of d_w can account for the variation of d_s. Nevertheless, f remains below the value f=2/3, the original conjecture for fractals.

The situation was different for our case II, where both [A] and [B] are constant. As seen in Table II, the variation of f is different for the clusters with upper size limit from those without upper size limit. Thus for $25<s<\infty$, f increases monotonously with p, and it is twice as large for p=0.58 than for p=0.37. However, for $25<s<250$ and $25<s<50$ f achieves a maximum for intermediate p values and it is smaller both at small or high p values. This behavior is in the same line as the variation of the number N of sites on which reaction can occur, with p. Apparently, the environment is more restricted when the number N is smaller, as long as the number or reacting A and B species remains unchanged. This model corresponds to real experimental situations where the global reactant concentration is fixed but the number of solubilization sites changes. For example, we encounter a similar situation described by the above case II when in a solution containing micelles or microemulsion droplets we change the concentration of micelles keeping constant the global concentration of intramicellar reactants.

The equivalent second-order reaction rate constants given by k'Q/N of Table I and by k'Q of Table II show a uniform behavior in all cases and size ranges examined. Reaction rates seem to be dominated only by N. Thus for $25<s<\infty$ they decrease monotonously, while for $25<s<250$ or $25<s<50$ they go through a minimum. This is simple to explain: when the number of available sites is small, the reactants are closer to each other and the reaction is facilitated. This is leading to larger reaction rates when N is smaller, and vice-versa.

Summarizing, the fractal behavior of the reaction between a minority species A and a majority species B follows the model of equation (2) applicable to a reaction domain consisting of isolated (non-percolating) clusters. When the global concentration varies proportionally to the number N of sites on which reaction can occur the fractal behavior is complex and does not show an important variation. When, however, the global concentration is constant but N varies, we observe a dramatic variation of the fractal behavior. This models several experimental situations. The second order reaction rates are dominated only by the value of N, and they are independent of the concentartion of the reactants.

References

(1)   P. Lianos and P. Argyrakis, Phys. Rev. A, 39,4170 (1989).
(2)   P. Lianos and S. Modes, J. Phys. Chem., 91,6088(1987).
(3)   P. Lianos, J. Chem. Phys., 89,5237(1988).
(4)   G. Duportail and P. Lianos, Chem. Phys. Lett., 165,35(1990).
(5)   S. Modes, P. Lianos, and A.Xenakis, J. Phys. Chem., in press.
(6)   J. Klafter and A. Blumen, J. Chem. Phys., 80,875(1984).
(7)   R. Kopelman, J. Stat. Phys. 42,185(1986)
(8)   D. Toussaint and F. Wilczek, J. Chem. Phys., 78,2642(1983);
(9)   J. Lang, A. Jada and A. Malliaris, J. Phys. Chem., 92,1946
        (1988); See also references there in.
(10)   P. Argyrakis and R. Kopelman, J. Phys. Chem., 91,2699(1987); J.
        Phys. Chem., 93,225(1989).
(11)   P.Lianos, M.L.Viriot, and R.Zana, J. Phys. Chem., 88,1098(1984).
(12)   J. Hoshen, R. Kopelman, and E. M. Monberg, J. Stat. Phys.,
        19,219(1978).
(13)   S. Alexander and R. Orbach, J. Phys. Lett., 43,L625(1982).
(14)   S.Havlin and D.ben-Avraham, Adv. Phys., 36,695(1987).
(15)   P.Meakin and H.E.Stanley, J. Phys. A, 17,L173(1984).

ENHANCED MEMBRANE RIGIDITY IN CHARGED LAMELLAR PHASES

Paul G. Higgs and Jean-François Joanny
Institut Charles Sadron
6 rue Boussingault, 67083 Strasbourg, France

1. Introduction

Lamellar phases formed by amphiphilic molecules consist of a stack of alternating solvent layers and surfactant membranes. The stability of the structure is governed by the interactions between neighboring membranes mediated by the solvent layers, these interactions include the Van der Waals attractive forces, the hydration forces and the electrostatic forces. It was argued by Helfrich [1] that due to the very low surface tensions, the out-of-plane undulations of the membrane are important and induce a repulsion of entropic origin between membranes. The strength of the undulation repulsion is controlled by the bending elasticity of the membranes that is usually described, by the bending constants k_C and k_C for the mean and Gaussian curvatures.

For neutral surfactants the bending energy arises from the rearrangement of the molecular packing when the membrane is curved. The bending constants are typically of order the thermal energy $k_B T$. In organic solvents, if k_C is small enough, the repulsive undulation repulsion may dominate the Van der Waals attraction and stabilize the lamellar structure up to very large intermembrane spacings. For charged surfactant membranes in water, the repulsive electrostatic interactions contribute to the stability of the structure. They also reduce the out-of-plane fluctuations and thus give a positive contribution to the bending constant which in some cases may be much larger than the intrinsic contribution. This reduces considerably the undulation repulsion.

In order to study the role of the electrostatic interactions, Pincus et al [2] have identified three important length scales in the problem:

    i.   the mean membrane separation in the lamellar phase $2d$;

    ii. the Debye screening length $\kappa^{-1} = (8\pi n_\infty L)^{-1/2}$;

    iii. the Gouy-Chapman length $\lambda = e/(2\pi L\sigma)$;

Large-Scale Molecular Systems, Edited by W. Gans *et al.*
Plenum Press, New York, 1991

where n_∞ is the bulk electrolyte concentration, σ is the surface charge density, and $L = e^2/(4\pi\varepsilon T)$ is the Bjerrum length.

Winterhalter and Helfrich [3] have studied the limit of high salt concentrations. The linearised Debye-Huckel equation is valid for the potential. The electrostatic contributions to the bending constants are evaluated by comparison of the energy of cylindrically and spherically shaped membranes with flat planes. It is found that

$$k_C = \frac{3\sigma^2}{4\varepsilon\kappa^3} \sim \frac{T}{\kappa^3 L \lambda^2} \qquad , \quad (\kappa d > 1, \kappa\lambda > 1) \tag{1}$$

Duplantier [4] has shown that the same result applies if the potential is calculated about an undulating surface. If $\kappa d > 1$ but $\kappa\lambda < 1$ we are in the 'intermediate' regime of reference [2] where the Debye-Huckel theory is no longer valid and the non-linear Poisson-Boltzmann equation must be solved. This has been done by Mitchell and Ninham [5] and gives

$$k_C = \frac{T}{\pi\kappa L} , \qquad (\kappa d > 1, \kappa\lambda < 1) \tag{2}$$

Here we consider the limit of high surface charge and zero salt concentration: $\lambda < d < \kappa^{-1}$, referred to in [2] as the Gouy-Chapman region.

2. The Gouy-Chapman Region

We consider limit of high surface charge where the Gouy-Chapman length λ is smaller than the spacing d and the added salt concentration is negligible. We note that d is still important in this case since it is the finite value of d which imposes the finite concentration of counterions. The corresponding Debye-Hückel case with $\kappa^{-1} \ll d$ becomes a one-membrane problem since there is a finite salt concentration independent of d.

We wish to find the potential between two membranes which we take to be curved with a displacement from their mean position $u(x) = u \cos qx$, with a small amplitude $u \ll d$. For simplicity we consider the two membranes to have opposite displacements so that there is a plane of symmetry between the two. The electrostatic potential potential satisfies the Poisson-Boltzmann equation

$$\nabla^2\phi = -\frac{n_0 e}{\varepsilon} \exp(-\frac{e\phi}{T}) \tag{3}$$

We assume that there is a fixed surface charge and hence the normal field is known at the surface. Due to symmetry the z component of the field is zero at the midplane. In order to calculate the electrostatic correction to the bending energy k_C, we need to solve the potential up to second order in u since the curvature energy is of the form $F_{bend} = (k_C u^2 q^4)/4$. Expanding the potential in in Fourier series, we look for a solution

$$\phi(x,z) = \phi_0(z) + u^2\phi_{02}(z) + u\phi_1(z)\cos qx + u^2\phi_2(z)\cos 2qx + \dots \qquad (4)$$

The bending constant obtained by this method [10] is $k_c \sim \frac{d}{L}T$ as conjectured in reference [2]. We have not however calculated the constant of proportionality due to mathematical complexity. It is desirable to have an estimate of the numerical constant to enable comparison with experiment. We therefore consider the simpler geometry of two membranes curved into concentric cylinders with the radius of curvature R >> d. The solution of the Poisson-Boltzmann equation between two cylinders has been given by Fuoss et al.[6]. We apply this [10] in the highly charged limit where $\lambda \ll d$, giving the result

$$k_c = (\frac{1}{\pi} - \frac{\pi}{12})\frac{2d}{L}T \sim 0.06\frac{2d}{L}T, \qquad (\lambda < d < \kappa^{-1}) \qquad (5)$$

Thus k_c does not depend on the charge density in the limit of high charge density.

3. Discussion

The bending constant has now been calculated in all the cases of the problem except in the ideal gas region where the distance between membranes is small $(d \ll \lambda)$ and the electrostatic interactions are not screened. The bending constant is however known in both regions which border the ideal gas region. The only scaling of the lengths which has a smooth crossover to the known results at the boundaries $\lambda \approx d$ and $\kappa^2\lambda d \approx 1$ is that proposed in [2]:

$$k_c = \frac{\lambda}{L}T, \qquad (d < \lambda < \kappa^{-1}) \qquad (6)$$

Lamellar phases have been studied experimentally by dynamic light scattering (Nallet et al.[7,8]) and by synchrotron X ray scattering (Roux & Safinya [9]). Dilute lamellar phases exist with a repeat distance of several hundred Ångstroms. If such systems are highly charged in the absence of added salt the result (5) implies a $k_c \gg T$, and hence extremely flat, rigid membranes.

The above calculations suggest several possible experiments. Firstly, it should be possible to see the linear dependence of k_c on d in the Gouy-Chapman region by gradual dilution of the lamellar phase. Lamellar phases have been found which are stable over a wide range of repeat distances eg. 50Å - 800Å for the sodium dodecyl sulphate systems discussed by Nallet et al.[7]. Secondly, we may look for the salt dependence of k_c by gradual addition of salt at a fixed d. We expect k_c independent of n_∞ until $\kappa^{-1} \sim d$, and then $k_c \sim n_\infty^{-1/2}$ (from eqn. 2). It may also be possible to investigate the dependence on the charge density σ by forming membranes of a mixture of charged and uncharged molecules. For small enough σ we will be in the ideal gas region $(\lambda > d)$ even for relatively large d.

The calculation of the bending energy has also been done at constant charge density. If the membrane is made by a mixture of charged and uncharged surfactants, the local composition of the membrane fluctuates and these fluctuations are coupled to the undulations of the membrane. We expect this to reduce the electrostatic contribution of the bending energy.

References

[1] Helfrich, W. (1973) *Z. Naturforsch.* **28C** 693.

[2] Pincus, P; Joanny, J-F; Andelman, D. (1990) to appear *Europhys. Lett.*

[3] Winterhalter, M; Helfrich, W. (1988) *J. Phys. Chem.* **92** 6865.

[4] Duplantier, B. To appear in proceedings of Bar-Ilan conference on Frontiers in Condensed Matter Physics (Tel Aviv Jan. 1990).

[5] Mitchell, D.J; Ninham, B.W. (1989) *Langmuir* **5** 1121.

[6] Fuoss, R.M; Katchalsky, A; Lifson, S. (1951) *Proc. Nat. Acad. Sci.* **37** 579.

[7] Nallet, F; Roux, D; Prost, J. (1989) *J. Phys. (France)* **50** 3147.

[8] Nallet, F; Roux, D; Prost, J. (1989) *Phys. Rev. Lett.* **62** 276.

[9] Roux, D; Safinya, C.R. (1988) *J. Phys. (France)* **49** 307.

[10] Complete version of this article submitted to *J. Phys. (France)*.

Words of Thanks

Ladies and Gentlemen,

The conference has not yet come to its end, but we just heard a talk of dangerous content dealing with enhanced diffusion. In the afternoon I expect an exponentially enhanced diffusion of the participants away from Acquafredda. So I would like to take the opportunity now to say a few words.

On behalf of the audience I would like to thank A. Blumen, and the organizing committee A. Amann, W. Gans and R. Silbey but also Mrs. I. Noss very much for the excellent work they did in the organization of the conference. They selected good speakers and interesting topics and they also took great care of our well-being. I would like to thank the organizers too for selecting such a patient and interested audience.

In some respects the conference was a difficult one because of its interdisciplinary character on a high level. This asks considerably more both from the speakers and the audience. Sometimes there was the danger of getting drowned in an infinite sea of boson operators, but we survived. Sometimes the fractals or multifractals dominated and set traps to catch the audience in their infinite network, which could only escape by localizing in the nonexistent exponents of Anderson's space. There the audience was menaced by reptating polymers. There was no chance to fall asleep because of the noise which was generated when it was employed to replace the phonon couplings. Finally one last word. Let us hope that experiment is going to survive in view of so much theory the experimentalists cannot explain!

But let me leave the ironical remarks. The conference was a great one, not only because of the enhanced knowledge and not only because all of us go back home with additional ideas for research. It was a great one because we could live together, think together and enjoy life together. It was a good time!

Thank you.

W. von Niessen

Lecturers

Anton Amann
Laboratorium für physikalische Chemie
ETH-Zentrum
CH-8092 Zürich
Switzerland

Alexander Blumen
Theorie molekularer Festkörper
Universität Bayreuth
Universitätsstr. 30
D-8580 Bayreuth
Germany

Werner Gans
Institut für physikalische
und theoretische Chemie
Freie Universität Berlin
Takustr. 3
D-1000 Berlin 33
Germany

Gérard Jannink
Laboratoire Léon Brillouin
C.E.N. Saclay
F-91191 Gif-sur-Yvette Cedex
France

J. Klafter
Dept. of Chemistry
Tel Aviv University
Ramat Aviv
69978
Israel

Max Kolb
École Normale Supérieure
46, allée d'Italie
F-69364 Lyon Cedex 07
France

Jan Noolandi
Xerox Research Center of Canada
2660 Speakman Drive
Mississauga
Ontario
L 5 K 2 L 1
Canada

Philip Pechukas
Dept. of Chemistry
Columbia University
New York
NY 10027
USA

Peter Pfeifer
Physics Department
University of Missouri
Columbia
MO 65211
USA

Peter Reineker
Universität Ulm
Abteilung für theoretische Physik
Oberer Eselsberg
D-7900 Ulm
Germany

Alfred Rieckers
Institut für theoretische Physik
Universität Tübingen
Auf der Morgenstelle 14
D-7400 Tübingen
Germany

Geoffrey L. Sewell
Queen Mary and Westfield College
Department of Physics
University of London
Mile End Road
London
E1 4NS
Great Britain

Robert Silbey
Department of Chemistry
MIT
Cambridge
MA 02139
USA

James L. Skinner
Department of Chemistry
1101 University Ave
University of Wisconsin
Madison
WI 53706
USA

Participants

Jean-Pierre Aimé
Groupe de Physique des
Solides de l'ENS
Université de Paris VII
tour 23
2, place Jussieu
F–75251 Paris
France

Panos Argyrakis
Department of Physics 313-1
University of Thessaloniki
GR-54006 Thessaloniki
Greece

Marc Barthélémy
Université de Paris VII
GPS-ENS
Tour 23
2 place Jussieu
F-75251 Paris Cedex 05
France

Ivan Barvík
Institute of Physics
Charles University
Ke Karlovu 5
12116 Prague 2
Czechoslovakia

Franco Battaglia
Istituto di Chimica
Università della Basilicata
Potenza
Italy

George L. Bleris
Dept. of Physics 313-1
University of Thessaloniki
GR-54006 Thessaloniki
Greece

Edgar M. Blokhuis
Dept. of Physical and
Macromolecular Chemistry
Gorlaeus Lab.
University of Leiden
P.O.B. 9502
Leiden
The Netherlands

Jan Boeyens
Dept. of Chemistry
University of the Witwatersrand
Johannesburg
PO Wits 2050
South Africa

Christian von Borczyskowski
Freie Universität Berlin
Fachbereich Physik
Arnimallee 14
D-1000 Berlin 33
Germany

Mustafa Cebe
Uludag Universitesi
FEN-ED-Fakültesi
Kimya Bölümü ÖG.Üyesi
Görükle-Bursa
Turkey

Tsun-Mei Chang
Dept. of Chemistry
1101 University Ave.
University of Wisconsin
Madison
WI 53706
USA

Uwe Claussen
Bayer AG
ZF-FGF, Geb. Q 18
D-5090 Leverkusen
Germany

Eric Clément
Université Pierre et Marie Curie
Laboratoire d'Optique
de la Matière Condensée
tour 13
4, place Jussieu
F-75252 Paris Cedex 05
France

Rob Coalson
Dept. of Chemistry
Univ. of Pittsburgh
Pittsburgh
PA 15260
USA

Josep M. Costa
Dept. of Physical Chemistry
University of Barcelona
Av. Diagonal 647
E-08028 Barcelona
Spain

Mohamed Daoud
Lab. Léon Brillouin
C.E.N. Saclay
F-91191 Gif-sur-Yvette
France

Spiros Evangelou
FO.R.T.H.
Institute of Electronic
Structure & Laser
P.O.Box 1527
Heraklion 711 10
Crete
Greece

Deborah Glynis Evans
Chemistry Department
University of the Witwatersrand
Johannesburg
PO Wits 2050
South Africa

Jose Fernando Ferreira Mendes
Fac. Ciencias
Universidade do Porto
Praca Gomes Teixeira
P-4000 Porto
Portugal

Idalino Franco
Univ. Nova di Lisboa
SGAAF
Quinta da Torre
P-2825 Monte de Caparica
Portugal

Thomas Gerisch
Institut für theoretische Physik
Universität Tübingen
Auf der Morgenstelle 14
D-7400 Tübingen
Germany

Antonino Giacalone
Via Isidor La Lumia 8
I-90139 Palermo
Italy

Gerhard Glatting
Theoretische Physik
Universität Ulm
D-7900 Ulm
Germany

Rony Granek
School of Chemistry
The Sackler Faculty of Exact Sciences
Tel-Aviv University
Tel-Aviv
69978
Israel

Elvira Graziano
Dipartimento di Fisica
dell' Università di Salerno
I-84100 Salerno
Italy

Malte Gross
Freie Universität Berlin
Fachbereich Physik
Arnimallee 14
D-1000 Berlin 33
Germany

Jochen Hertle
Institut für theoretische Physik
Universität Tübingen
Auf der Morgenstelle 14
D-7400 Tübingen
Germany

Paul G. Higgs
Institut Charles Sadron
6 rue Boussingault
F-67000 Strasbourg
France

Rolf Hilfiker
Inst. für physikalische Chemie
Universität Basel
Klingelbergstr. 80
CH-4056 Basel
Switzerland

Reinhard Honegger
Institut für theoretische Physik
Universität Tübingen
Auf der Morgenstelle 14
D-7400 Tübingen
Germany

Dora Izzo
Dept. de Física Experimental
Instituto de Física da
Universidade de São Paulo
Caixa Postal 20516
CEP 01498
São Paulo - SP
Brazil

Naeem Jan
Cavendish Laboratory
Madingley Road
Cambridge
CB3 0HE
Great Britain

Robert Johnson
IBM Almaden Research
650 Harry Road
San Jose
CA 95120-6099
USA

Harald Kauffmann
Institut für physikalische Chemie
der Universität Wien
Währingerstr. 42
A-1090 Wien
Austria

Günther Köhler
Universität Bayreuth
Experimentalphysik IV
Postfach 101 251
D-8580 Bayreuth
Germany

Joachim Komorowski
I.N.Stranski-Institut für Physikalische
und Theoretische Chemie
TU Berlin
Ernst-Reuter-Platz 7
D-1000 Berlin 10
Germany

Thorsten Koslowski
Technische Universität Braunschweig
Inst. für physikalische
und theoretische Chemie
Hans-Sommer-Str. 10
D-3300 Braunschweig
Germany

Harald Krbecek
Universität Ulm
Abt. Experimentelle Physik
Albert-Einstein-Allee 11
D-7900 Ulm
Germany

David E. Logan
Physical Chemistry Laboratory
University of Oxford
South Parks Road
Oxford
OX1 3QZ
England

Ann Rachel Leheny
Dept. of Chemistry
Columbia University
Box 864
New York
NY 10027
USA

Stefan Luding
Universität Bayreuth
Experimentalphysik IV
D-8580 Bayreuth
Germany

Salvatore Magazù
Istituto Tecniche Spettroscopiche
del CNR Messina
Contrada Papardo
Salita Sperone
Vill.S.Agata
I-98166 Messina
Italy

Domenico Majolino
Dip. di Fisica dell' Università
degli Studi
Contrada Papardo
Salita Sperone
Vill S.Agata
I-98166 Messina
Italy

Graziano Marano
Università di Catania
Dip. di Fisica
Corso Italia 57
I-95100 Catania
Italy

Detlev Mergenthaler
Universität Ulm
Abt. Experimentelle Physik
Albert-Einstein-Allee 11
D-7900 Ulm
Germany

Michael Messina
Dept. of Chemistry
Univ. of Pittsburgh
Pittsburgh
PA 15260
USA

Norberto Micali
Istituto Tecniche Spettroscopiche
del CNR Messina
Contrada Papardo
Salita Sperone
Vill.S.Agata
I-98166 Messina
Italy

K.L. Ngai
Electronic Science and
Technology Division
Naval Research Laboratory
Washington DC
20375-5000
USA

Wolfgang von Niessen
Inst. für physikalische
und theoretische Chemie
Technische Universität Braunschweig
Hans-Sommer-Str. 10
D-3300 Braunschweig
Germany

Ralf Oeser
Institut Laue-Langevin
156 X centre de tri
F-38042 Grenoble Cedex
France

Michel Oostwal
Dept. of Physical and
Macromolecular Chemistry
Gorlaeus Laboratories
State Univ. of Leiden
P.O.Box 9502
Leiden
The Netherlands

Michel Orrit
CPMOH 351
Cours de la Libération
F-33405 Talence
France

Philip Phillips
Dept. of Chemistry
Room 6-223
Massachusetts Institute of Technology
Cambridge
MA 02139
USA

Martin Pietralla
Abteilung experimentelle Physik
Universität Ulm
Einstein-Allee 11
D-7900 Ulm
Germany

Peter Jörg Plath
Inst. für Angewandte und
Physikalische Chemie
Universität Bremen
Bibliothekstr. NW2
D-2800 Bremen 33
Germany

Ralf Quadt
Universität zu Köln
Institut für Theoretische Physik
Zülpicher Str. 77
D-5000 Köln 41
Germany

Anton K. Rauscher
Institut für physikalische Chemie I
Universität Bayreuth
Universitätsstr. 30
D-8580 Bayreuth
Germany

Ranko Richert
Fachbereich physikalische Chemie
Universität Marburg
Auf den Lahnbergen
D-3550 Marburg
Germany

Apostolos Rizos
University of Crete
Chemistry Department
Iraklion
Crete
Greece

Apurba Kumar Roy
Experimentalphysik IV
Universität Bayreuth
Postfach 101 251
D-8580 Bayreuth
Germany

Barbara Rüdiger
Via Nizza 53
I-00198 Roma
Italy

Francesc Sagués
Dept. of Physical Chemistry
University of Barcelona
Av. Diagonal 647
E-08028 Barcelona
Spain

Jeffery G. Saven
Dept. of Chemistry
1101 University Ave.
University of Wisconsin
Madison
WI 53706
USA

Horst Schnörer
Universität Bayreuth
Experimentalphysik IV
Postfach 101 251
D-8580 Bayreuth
Germany

Michael Schreiber
Institut für physikalische Chemie
Johannes-Gutenberg-Universität
Jakob-Welder-Weg 11
D-6500 Mainz
Germany

Gerhart Schroff
Institut für theoretische Physik
Universität Tübingen
Auf der Morgenstelle 14
D-7400 Tübingen
Germany

Hannah Sevian
Dept. of Chemistry
1101 University Ave.
University of Wisconsin
Madison
WI 53706
USA

Patrizio Severino
Via San Domenico N.15
I-80127 Napoli
Italy

Rob Sok
Physical Chemistry
Dept. of Chemistry
University of Groningen
Nyenborg 16
NL-9747 AG Groningen
The Netherlands

Bernardo Spagnolo
Dip. di Energetica ed
Applicazioni di Fisica
Viale delle Scienze
I-90128 Palermo
Italy

Alberto Suárez
Dept. of Chemistry
6-234 A
MIT
77, Massachusetts Av.
Cambridge
MA 02139
USA

Jonathan Thorne
Dept of Chemistry
Univ. of Pennsysvania
Philadelphia
PA 19104
USA

Andreas Tilgner
Laboratoire de Spectrométrie Physique
BP 87
F-38402 St. Martin d'Hères
France

Pasquale Tomasello
Dip. di Fisica
Università di Catania
Corso Italia 57
I-95129 Catania
Italy

Angel Valle Gutierrez
Departamento de Fisica Moderna
Universidad de Cantabria
Avenida los Castros S/N
E-39005 Santander
Spain

Cirino Vasi
Istituto Tecniche Spettroscopiche
del CNR Messina
Contrada Papardo
Salita Sperone
Vill.S.Agata
I-98166 Messina
Italy

Valentina Vassilenko
Centro de Fisica Molecular
dos Universidades de Lisboa
Av. Rovisco Pais
Complexo I - I.S.T.
P-1000 Lisboa
Portugal

Giuseppe Vitiello
Physics Dept.
Università di Salerno
I-84100 Salerno
Italy

Siynet Voit
Universität Bayreuth
Experimentalphysik IV
D-8580 Bayreuth
Germany

Stephan Zanzinger
Institut für theoretische Physik
Universität Tübingen
Auf der Morgenstelle 14
D-7400 Tübingen
Germany

Gert Zumofen
Laboratorium für physikalische Chemie
ETH-Zentrum
CH-8092 Zürich
Switzerland

STUDY INSTITUTE ON LARGE-SCALE MOLECULAR SYSTEMS
March 25 - April 7, 1990
Acquafredda di Maratea, Italy

Index